16th Personal Armour Systems Symposium (PASS 2023)

Personal Armour

Dresden, Germany
11-15 September 2023

ISBN: 979-8-3313-1852-9
DOI: https://doi.org/10.52202/080042

Printed from e-media with permission by:

Curran Associates, Inc.
57 Morehouse Lane
Red Hook, NY 12571

Some format issues inherent in the e-media version may also appear in this print version.

Royal Military Academy (Belgium)
Department of Weapon Systems
Renaissance Avenue 30
1000 Brussels
Belgium

Additional copies of this publication are available from:

Curran Associates, Inc.
57 Morehouse Lane
Red Hook, NY 12571 USA
Phone: 845-758-0400
Fax: 845-758-2633
Email: curran@proceedings.com
Web: www.proceedings.com

TABLE OF CONTENTS

ARMOUR MATERIALS

BEHIND ARMOUR BLUNT TRAUMA

BLAST INJURY AND MIGITATION

CASUALTY REDUCTION AND OPERATIONAL ANALYSIS

HUMAN FACTORS

PERSONAL ARMOUR END ITEMS

TEST STANDARDS AND METHODS OF ASSESSMENT

THREATS (INCLUDING LASER)

What is the effect of composite covers on ceramics in hard armour plates?

B.B. Johnsen[1] and D.B. Rahbek[1]

[1]Norwegian Defence Research Establishment (FFI), Instituttveien 20, NO-2007 Kjeller, Norway
bernt.johnsen@ffi.no

Abstract. Hard armour plates in body armours are often constructed of a single ceramic tile, which is covered by a high-strength sheet material and backed with several layers of aramid and/or ultra-high molecular weight polyethylene (UHMWPE) fibres. The main role of the hard ceramic strike face is to erode, blunt and shatter an impacting projectile, such as hard-cored 7.62 mm armour piercing projectiles. However, ceramics are inherently brittle materials and they will fracture in a brittle manner when subjected to high stress loads. To mitigate this brittle fracture, the ceramics may be covered with one or more layers of high-strength fibres in polymer matrices in order to maintain the integrity of the ceramic. This can improve the ballistic performance of the first strike, will definitely improve the multi-hit properties and will ultimately increase the robustness of the plate. Several studies have looked into how the failure of the ceramic tiles is affected by composite covers and how different designs affect the penetration performance. In all the studies, however, the target design (e.g. the ceramic, sheet material, number of layers, etc.) and the experimental conditions (e.g. type of projectile, impact velocity, ballistic test procedure, etc.) were different. Different effects have been found from wrapping the ceramic in a glass fibre fabric, or from adding front or back covers. Hence, the studies do not fully agree on how the ballistic performance changes with addition of a composite cover to a ceramic tile. A meaningful comparison of the studies can therefore be difficult. This paper explores the findings in the literature of the effect of composite covers on ceramics in hard armour plates, and also suggests potential future developments in hard armour plate designs.

1. INTRODUCTION

The two main components in a typical body armour system for torso protection are a soft ballistic panel and a hard armour plate [1, 2]. In a military context, the soft panel is designed to give protection against various types of relatively small primary and secondary fragments from indirect fire munitions, and in many cases handgun rounds. The hard armour plate, on the other hand, provides protection against projectiles with higher kinetic energy, of which high-velocity armour piercing (AP) rifle bullets is the most serious threat. The hard armour plate is typically designed to be used in conjunction with the soft panel to give the required level of protection. This is effective in reducing both the lethality and injuries. Body armour systems for soldiers are made of combinations of materials, with the aim of making them as light-weight as possible given the specific threats they are designed to protect against. This is to reduce burden and thus improve the mobility, endurance and comfort of the soldier, at the same time as keeping the cost at an acceptable level. Hence, there is always a drive amongst armour manufacturers to implement new, more efficient materials, and to improve designs and manufacturing methods to be able to provide systems with improved performance. The design of hard armour plates with ceramics as the strike face is one area that has received some attention over the last few decades. One way to improve penetration resistance of the hard armour plates is to cover the ceramic with a high-strength composite sheet material.

This paper discuss the findings in the literature of composite covers on ceramics in hard armour plates. The literature gives some indication of what designs of composite-covered ceramics that give the most efficient solutions. Recent work that has been conducted at the Norwegian Defence Research Establishment (FFI) and collaborating partners on this topic is also summarised. Some recent results seem to indicate an unexplored potential of new designs and manufacturing methods in the development of composite sheets in hard armour plates.

2. CONSTRUCTION OF HARD ARMOUR PLATES

2.1 Typical design

The hard armour plate typically consists of a monolithic, double-curved ceramic tile which is wrapped in a thin composite material, Figure 1. The tile thickness is around 10 mm, but will vary with the defined threat level and type of ceramic. Additionally, at the back of the wrapped ceramic tile, there is a backing material consisting of several layers of high-performance ballistic fibres, such as aramid or ultra-high molecular weight polyethylene (UHMWPE). Although some hard armour plates are designed to be used

https://doi.org/10.52202/080042-0001

as a stand-alone plate, it is more common that it must be used in conjunction with (icw) the soft panel in order to give the required protective performance.

Many hard armour plates are designed to protect against hard-cored AP projectiles up to caliber 7.62 mm, with the typical impact velocities being around 900 m/s. The ceramic tile acts as the strike face and is usually composed of alumina (Al_2O_3), silicon carbide (SiC) or boron carbide (B_4C). The main reason for using ceramics as the strike face is that they are very hard materials with a relatively low density.

The soft panel is also made of aramid or UHMWPE. Used on its own, the soft panel typically provides protection against various types of relatively small primary and secondary fragments from indirect fire munitions, and in many cases handgun rounds. This is achieved from the fibres' ability to absorb and disperse the kinetic energy of the fragments or the bullets. The same mechanisms apply when the soft panel absorbs the kinetic energy of fragments formed during impact on the hard armour plate.

2.2 Failure mechanisms

When a projectile impacts on the hard armour plate at a given velocity it is desirable, from a protection point of view, that the core material of the projectile is eroded at the tip and fragmented as much as possible, Figure 2, and that the core fragments are considerably slowed down. This reduces the kinetic energy and makes it easier to prevent the projectile from perforating the armour system. The high hardness of ceramics makes them beneficial for eroding and fragmenting the projectile.

However, ceramics are brittle materials and they will fracture during impact due to the high stress loads that are involved, hence failure will be observed. Figure 3(a-c) shows some of the failure mechanisms that occur in the ceramic tile. These particular tests were conducted at relatively low velocities, but still serve as a good illustration of the mechanisms. The impact causes the formation of radial cracks, cone cracks and fragmentation of the ceramic [3, 4]. At the point of impact, the formation of a comminuted ceramic takes place in the area in front of the projectile. Here, the ceramic becomes less confined, resulting in lower resistance to penetration. Fragmentation of the steel-core of an AP projectile is also shown in the figure.

For ceramic tiles that are not well supported or confined, the ceramic will shatter into relatively large fragments, which will reduce its ability to stop the projectile. To improve the ballistic performance of the ceramic tile, it can be covered by, or wrapped in, a sheet of high-tenacity fibre composite material [5-12]. When the ceramic is covered by a sheet material, the ceramic will be partly held together both during and after impact from the projectile. The fractured ceramic is then prevented from moving and is held in place in the path of the projectile. This may give more time for erosion and fragmentation of the core of the projectile. The sheet material may also contribute to improved multi-hit performance, since it introduces radial constraint on the ceramic surface. This prevents through-thickness cracks from opening and confines larger ceramic fragments, hence improving the integrity of the tile after impact. This is illustrated in Figure 3(b), where it is obvious that longer interaction times with the projectile also results in the formation of smaller ceramic fragments. Another possible effect of the sheet cover is to reduce the angle of spall that is ejected from the strike face.

2.3 Importance of composite covers in hard armour plates

Some insight into the importance on composite covers can be found from studies on hard armour plates [11, 13, 14]. In these studies, the V50 ballistic limit for hard armour plates with cracks in the ceramic tile, and the results were compared with intact, un-damaged plates. A reduction in V50 from 3 to 10 % was observed, also when the shots were placed directly on or in the vicinity of the cracks. This suggests that the protective ability was not catastrophically affected by a limited number of cracks in the ceramic. The tested plates in these studies all had composite covers on the ceramic. The cover probably contributed to maintaining some integrity of the ceramic during impact by holding the ceramic fragments in place.

However, a more recent study showed that the presence of cracks in the ceramic significantly affected the protective ability [15]. A reduction in V50 of around 30 % was observed. An important difference from the other studies was that hard armour plates designed for single-hit impacts were employed, as opposed to plates designed for multi-hit in the other studies. The lower performing single-hit plates did not have composite covers on the front and back of the ceramic, which may have influenced the performance. A similar effect was observed in an older study [16]. Cracks with a relatively large crack opening, i.e. cracks where the fracture surfaces were not in contact, resulted in reduced protective ability. This might have been avoided if the ceramic was surrounded by a composite material.

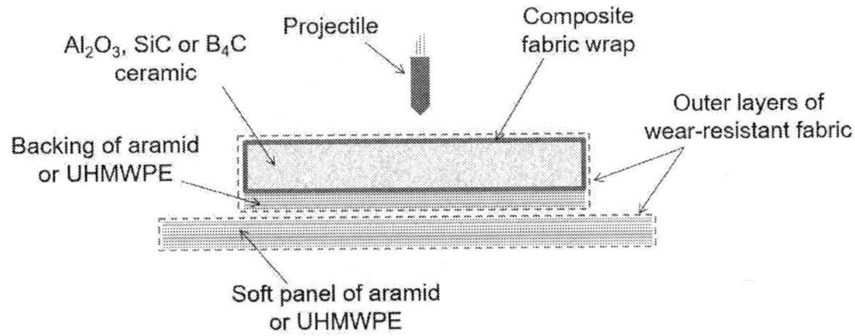

Figure 1. Cross-sectional view of what a typical construction of a hard armour plate can look like. Hard armour plates are often designed to be used in conjunction with a soft panel; also shown.

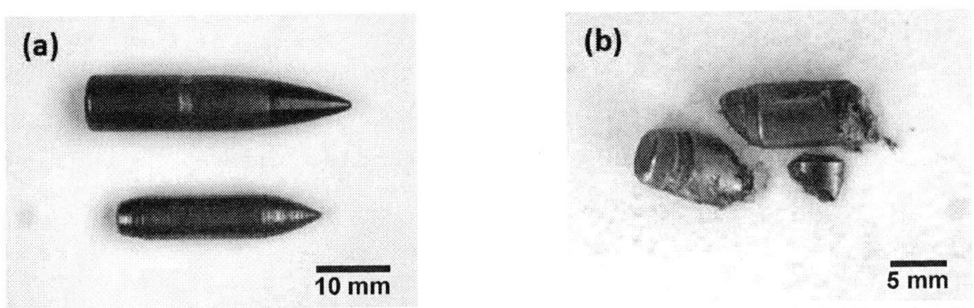

Figure 2. Example of fragmentation of the hard steel core of the 7.62×63 mm M2 AP projectile after impact on a ceramic tile: (a) intact core prior to impact (mass 5.19 g), and (b) fragmented core.

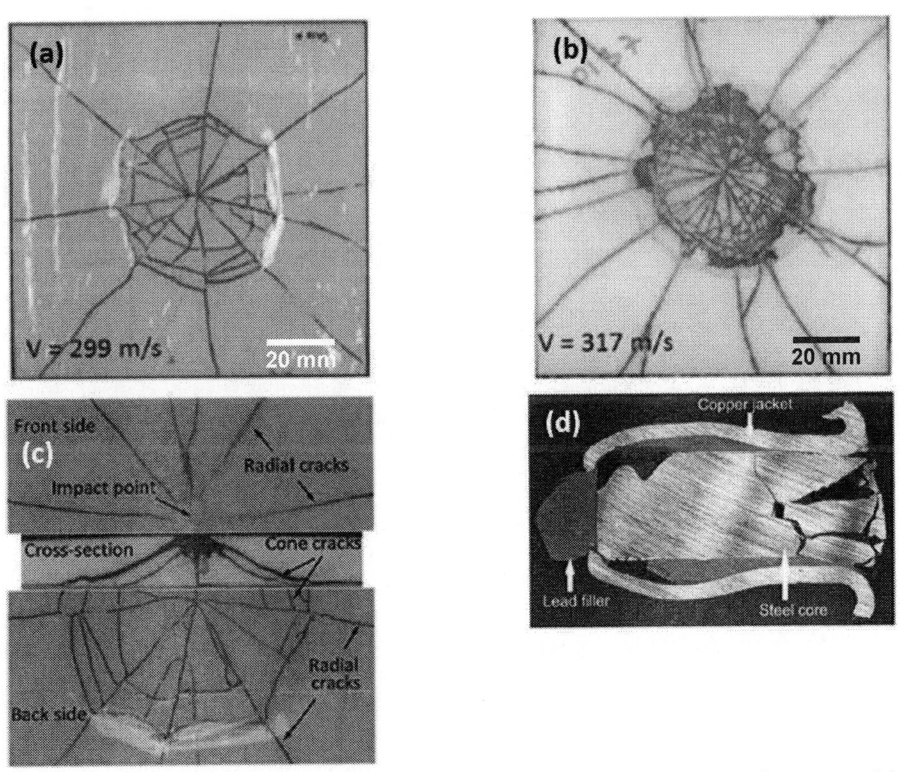

Figure 3. Relatively low-velocity impacts at around 300 m/s on alumina/composite targets: (a) Back side of alumina in (a) un-covered and (b) covered targets. (c) Ceramic damage; front, cross-sectional and rear view after impact on an alumina tile. (d) Cross-section of a 7.62 mm AP M61 projectile after impact. This work was a collaboration between FFI and SRI International. Adapted with permission from Rahbek et al. [8].

https://doi.org/10.52202/080042-0001

3. STUDIES ON COMPOSITE COVER DESIGNS

Several studies have performed experiments in efforts to better understand how ceramic tiles fail, and how failure is affected by a composite cover. An overview of this literature is given in Table 1. Other sheet materials have been investigated, but the focus here will be on composite materials.
One complicating matter is that the target design (e.g. the ceramic, sheet material, number of layers) and the experimental conditions (e.g. type of projectile, impact velocity, ballistic test procedure) were different. Some studies were performed using regular AP projectiles at typical muzzle velocity of around 800-900 m/s, while other studies have employed other projectile geometries at lower velocities. The studies therefore do not fully agree on how the ballistic performance changes with addition of a composite cover to a ceramic tile.

3.1 Role of projectile

Ceramic hard armour plates are often required to protect against AP projectiles with hard core materials. Although important lessons can be learned by using other types of projectiles, the most interesting studies from a 'real-life' scenario are therefore the ones that use AP, or similar, projectiles. It has been shown that the perforation ability of AP projectiles can be directly correlated with the hardness of the core material [17]. Harder core materials will generally have a higher penetrative ability, while more core erosion and fragmentation will occur particularly when the ceramic has a higher hardness than the core material. On the other hand, projectiles with lower hardness, for example Ball projectiles, will result in other deformation modes, such as e.g. mushrooming or petalling.
In addition to the penetrator material, the penetrator shape will also have an effect, meaning that a different shape of the nose of the projectile may lead to different penetration mechanisms and perforation velocities [18]. Conical or ogival tipped projectiles may penetrate at lower velocities than cylindrical/blunt projectiles. To complicate things further, the same projectile core with and without the jacket and lead cap can also have different penetration ability [10, 19].

3.2 Role of cover

3.2.1 Ceramic failure

The application of a composite cover has a significant effect on the failure of the ceramic during impact. Crouch has shown that the addition of an aramid fibre-reinforced composite cover to the ceramic tile in a hard armour plate may lower the back-face deformation upon multi-hit [11]. In the same study, Crouch observed that the addition of the composite cover affected the failure mechanisms of the ceramic. An increase in the number of radial cracks from an average of 10.8 without cover to 16 for a covered ceramic was observed. Similar effects have also been observed in other studies, including Rahbek et al. [8], which observed an increase in radial cracks from 9 to 14 for tests conducted at ~300 m/s. The density of cracks in the cone region was also much higher in the covered tile and, as a result, a much higher number of incipient fragments were formed in the covered target. In another study, Reddy et al. [7] found that the size distribution of the ceramic debris created during impact changed toward smaller fragment sizes when a ballistic fibre front cover was added (the ceramic was backed by a glass fibre composite).

3.2.2 Front constraint

Several studies have shown that covering of ceramic tiles with fibre composites (glass or carbon fibres) or ballistic fibres (aramid or UHMWPE) may lead to improved ballistic performance in terms of increased core fragmentation, reduced residual velocity of the core fragments, and increased kinetic energy-loss. The studies do not fully agree on how the performance changes with addition of a composite cover, but positive effects on an areal density basis has been observed.
One proposed mechanism which contributes to the increased performance is the constraining effect on the ceramic debris at the front of the ceramic (strike face). This constraint increases the flow of ceramic debris towards the penetrator, gives more time for penetrator-ceramic interaction since the ceramic is kept in front of the penetrator for a longer period of time, and slows down propagation of cracks in the ceramic [5-7]. This effect was observed in a study conducted by Sarva et al. [5], which found a significant improvement in the ballistic performance from front covers made of different materials. Significantly higher kinetic energy-loss and more projectile erosion were observed. However, Sarva et al. employed a

cylindrical flat-ended tungsten heavy alloy penetrator in their impact experiments. This penetrator was continuously consumed by erosion by the penetration, and did not shatter or break.

3.2.3 Pointed projectiles

Pointed AP projectiles made of hard steel, with one example being the 7.62×63 mm M2 AP, have been observed to fragment during impact due to their high hardness [7, 19]. Still, noteworthy effects of a composite front cover or wrapping on ceramic damage, core erosion, fragmentation, and back-face deformation, have been observed also for pointed AP 7.62 mm projectiles of hardened steel [6, 7, 11]. Nunn et al. [6] found a >40% increase in the V50 of a boron carbide tile by adding a composite cover that led to an increase in areal density by 9%. The suggested mechanisms that contribute to the improved performance were quite similar to those proposed by Sarva et al. [5], i.e. increased flow of ceramic debris against the projectile.

Contrary to the observations by Sarva et al., some studies have found that adding a composite cover layer to the ceramic tile does not have a significant effect on the projectile core [8, 10]. This is interesting, since erosion of the projectile is an important mechanism for defeating such threats. Hence, it is not obvious that ceramic flow against pointed AP projectiles will always be a noteworthy effect.

3.3 Role of target design

The effect of target design was investigated in a study at FFI where the main aim was to isolate the possible differences of adding front or back composite covers on ceramic tiles [12]. Very few studies have tried to isolate the effects of front and back covers, since the composite layer was usually present on both sides, or the composite was wrapped around the tile. In this study, two or four layers of a glass fibre-reinforced composite material were applied to the front and/or backside of an alumina tile, which resulted in a maximum increase in areal density of 9.5% compared to a bare tile. The composite-covered targets were tested with a 7.62 mm AP projectile at 800 m/s, which always resulted in perforation and fragmentation of the hard projectile core (as illustrated in Figure 2).

The results showed that the core fragmentation and the kinetic energy-loss of the projectile were most significant for the targets with the composite-cover on the back of the alumina; the core mass was reduced by up to 61%, while the kinetic energy was reduced by up to 84% (mainly as an effect of reduced mass, and to a less extent reduced velocity), Figure 4. It was obvious that the target configuration had a significant influence on the fragmentation of the projectile core, and that front covers did not give increased fragmentation.

The observed effects were somewhat different from several studies in the literature, as discussed above, where positive effects were mainly attributed to front covers or to wrapping of the ceramic. For example, Sarva et al. [5] found minor additional improvements when also adding a back cover, although increased energy absorption from a back cover has been reported for very thin (2 mm) alumina tiles [9]. The most likely mechanism is that the support and restraint of the back cover contributes to a time-delay in the opening of tensile cracks on the back of the ceramic, perhaps in combination with reduced reflection of stress waves into the ceramic. This time-delay gives more time for interaction with the penetrator, hence improving the ballistic performance of the target. If this hypothesis is true, then the composite cover on the rear may be more important than the cover on the front. In hard armour plates, the aramid or UHMWPE backing may provide the sufficient support.

3.4 Numerical modelling

Numerical modelling has been used to describe the effect of a composite wrapping around alumina tiles impacted by 7.62 mm AP projectiles. Rahbek et al. [8] found significantly more damage in covered tiles at sub-muzzle velocities (180 and 310 m/s), compared with bare tiles. The modelling successfully reproduced the experimentally observed failure mechanisms; radial cracks were first initiated at the back of the tile beneath the impact zone, followed by the formation of cone cracks that connected the radial cracks. Although covering resulted in more ceramic damage, an influence on projectile erosion was not observed.

Projectile impact at muzzle velocities and higher velocities was modelled by Guo et al. [20] for bare and covered tiles. The modelling showed that the composite cover resulted in increased resistance to perforation and that it had an effect on the fracture process of the ceramic tile. The fracture cone was slightly larger in the covered tile, which helped to distribute the load to a larger area on the backing plate that was employed. The fracture cone was also formed faster and, similar to the observations by Rahbek et al. [8], more ceramic damage was observed inside the cone.

https://doi.org/10.52202/080042-0001

Table 1. Experimental work in the literature that discuss the effect of adding composite covers to ceramic tiles. Adapted from Rahbek and Johnsen [12].

Reference	Projectile; impact conditions; ceramic	Cover material; target design	Main conclusions on effect of cover
Nunn et al., 2005 [6]	7.62 mm AP M61; V_{50} (625-878 m/s); 6.2 mm PAD B_4C	UD carbon fibre/epoxy, PBO/vinylester; Front and back cover, 2-8 layers on each side, 1 mm spall cover and 7.25 mm backing of UHMWPE	Penetration resistance increases with increasing number of plies
Sarva et al., 2007 [5]	6 mm tungsten heavy alloy (WHA), cylindrical flat-ended, mass 10.68 g; ~900 m/s; 12.7 mm 99.5% Al_2O_3, 12.7 mm SiC	UD glass fibre tape, E-glass/epoxy weave, carbon fibre/epoxy weave; Mainly front cover, 1-9 composite layers	Front-face restraint gives reduction in projectile residual kinetic energy, mass and velocity
Reddy et al., 2008 [5]	7.62 mm AP projectile, mass 10.44 g, 5.2 g hardened steel core; 820 ± 10 m/s; 7 mm 99.5% Al_2O_3	UD UHMWPE, aramid fabric; Wrapping in 2-4 layers, i.e. 2-4 layers on each side, backed by 10 mm E-glass laminate	Higher energy absorption, lower projectile residual mass and velocity with increasing number of wrapping layers
Crouch, 2014 [11]	7.62 mm M2 AP; 868-887 m/s, multi-hit; 9 mm SiC	Elastomeric film, polyester or polypropylene fibre/epoxy, aramid fibre/epoxy; Hard armour plates with clad ceramic, UHMWPE backing and nylon fabric wrap	Increase in number of radial ceramic cracks, and decrease in back-face deformation with fibre-reinforced cover
Crouch et al., 2015 [10]	7.62 mm AK-47, mass 7.91 g, 3.59 g mild steel core; 697-739 m/s, reverse ballistics; 3, 4 and 5 mm B_4C	Aramid fibre/epoxy; Front and back cover, with or without 30 ply UHMWPE backing, some targets with air gap between backing and sabot	Little effect on projectile deformation and erosion
Öberg et al., 2015 [9]	8 mm spherical projectile, 52100 type steel, mass ~2 g; 220-230 m/s; 2 mm Al_2O_3	Carbon fibre/PET weave; 1 layer back cover	Higher energy absorption from back cover
Rahbek et al., 2017 [8]	7.62 mm AP M61, 3.7 g hardened steel core; 176-351 m/s; 10 mm 98% Al_2O_3	Glass fibre/PET fabric; Front and back cover, 2 layers on each side	Increased ceramic damage, no difference in projectile core damage
Rahbek and Johnsen, 2019 [12]	7.62 mm M2 AP; ~800 m/s; 8.2 mm 98% Al_2O_3	Glass fibre/PET fabric; Front and/or back cover, 2 or 4 layers on each side	Back cover gave higher projectile core fragmentation
Rahbek et al., 2023 [21]	7.62 mm M2 AP; V_{50} (376-802 m/s); 7.0 mm 98% Al_2O_3	Fabric wrap or filament-wound yarn of glass fibre/PET; One 0°/90° layer on each side, with 8 mm PC backing	Filament winding gave significantly higher V_{50}

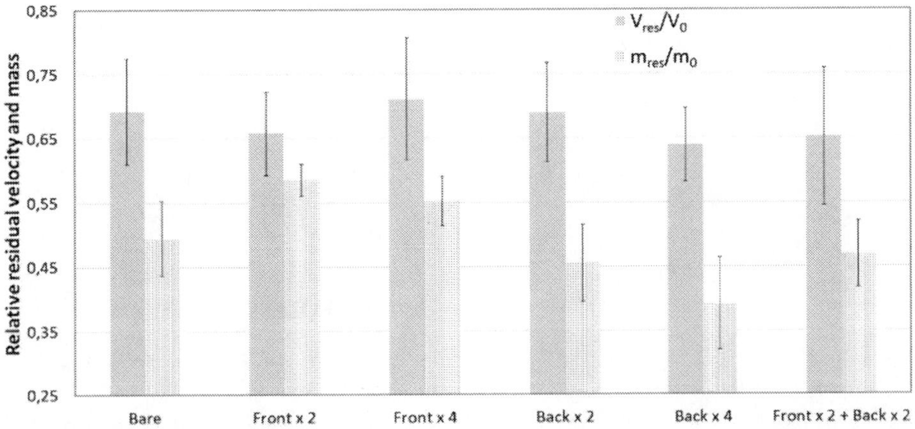

Figure 4. Relative residual mass, mres/m0, and relative residual velocity, Vres/V0, for the largest core fragments. Average values ± 95% confidence intervals. Adapted from Rahbek and Johnsen [12].

4. POTENTIAL FUTURE DEVELOPMENTS

In the studies discussed above, the targets were all produced by manufacturing technologies where the method of application of the composite covers or the fabric wraps are well known. It was more the effects of using different sheet materials, designs and lay-ups that were investigated. One area that has not received a lot of attention, at least in the open literature, is different manufacturing technologies. This is, however, an area that should have the potential to significantly improve the protective ability of hard armour plates, and there are a few examples of this in the recent literature.

4.1 Pre-tensioning of fabrics

The number of studies that discuss the effect of pre-tensioning of composite materials layers on ceramics is very low or non-existent. This is not to be confused with radial pre-tensioning which has been shown to improve the resistance to penetration [22]. However, Jaitlee [23] investigated the effect of a composite cover on the rear side of 4 mm silicon carbide tiles. In this work, the cover was an aramid fabric that was pre-tensioned and then held in place by a cured epoxy. Ballistic testing with a 7.62 mm mild steel core projectile indicated that the pre-tensioning in the fabric gave a reduction in back face signature. (No V50 data were presented.) No explanation for this observation was given.

A few studies have investigated the effect of pre-tensioning in composite laminates and woven fabrics (not as covers on ceramics) [24, 25]. For example, decreased V50 has been observed for a glass fibre-reinforced plastic (GFRP) that was loaded in uniaxial tension, partly explained by less energy absorption in the preloaded composite, while the V50 of a high-strength fabric that was pre-tensioned in the warp direction was increased up to a critical level of pre-tension, after which the ballistic limit was reduced. These studies are not directly relevant for the understanding of the pre-tensioned composite covers on ceramics but they may, as discussed below, provide some supporting insight.

4.2 Filament winding with yarns

In a recent study, a new method of applying the fibre-composite cover around the ceramic tile was investigated in a collaboration between FFI and NFM Group [21]. The method consisted of filament winding of glass fibre/polyester yarns under some pre-tension, Figure 5, around an alumina tile. The filament winding was conducted in two directions around the tile to produce a composite sheet layer consisting of two plies with a 0o/90o lay-up on each side of the tile. These targets were compared with targets where a similar glass fibre/polyester fabric was wrapped around the alumina, which is a more traditional way of producing hard armour plates, and also with bare alumina targets. All targets were backed by polycarbonate and had the same overall areal density. Ballistic testing was conducted using the 7.62×63 mm M2 AP projectile, and the projectile fragments were collected and analysed.

The target with the filament-wound composite cover gave a much higher V50 than the target with the fabric wrap; values of 622 m/s and 536 m/s, respectively, were measured, Table 2. The 16% difference between the targets was remarkable considering the identical areal density. The bare alumina target had a V50 of 586 m/s, which was also considerably lower than the filament-wound target.

 https://doi.org/10.52202/080042-0001

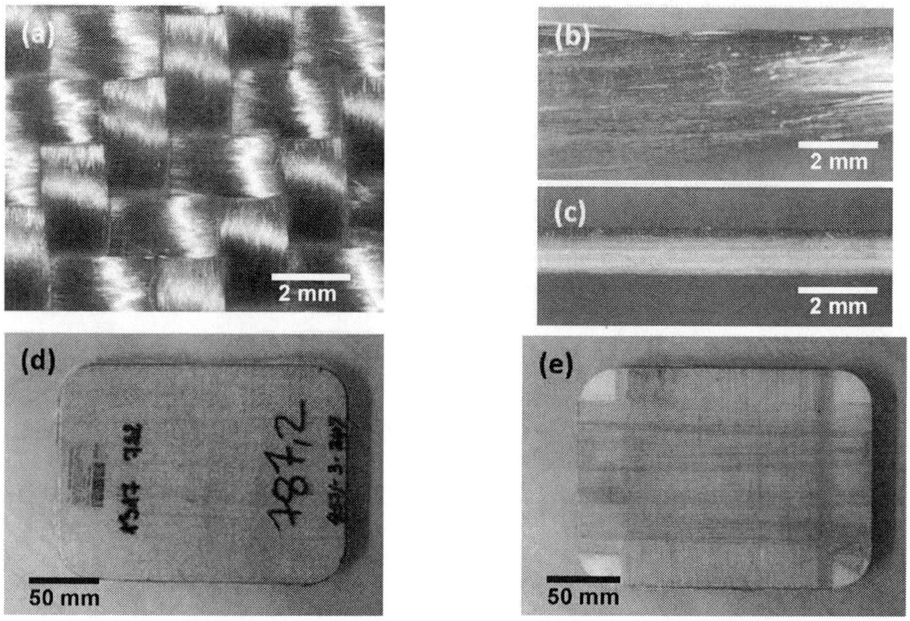

Figure 5. Microscopy images of (a) the commingled glass/PET woven fabric, and (b) un-tensioned and (c) pre-tensioned commingled glass/PET yarn. Targets made of (d) fabric and (e) pre-tensioned yarn.

Table 2. V50 ballistic limit velocity with 90% confidence interval (CI) for targets where the composite cover was applied by different manufacturing methods. Depending on the type of target, the alumina thickness was 7.4 or 7.0 mm. A PC backing with an areal density of 9.6 kg/m3 was bonded at the back of all targets, giving an overall areal density of 38.1 kg/m3. The average residual core mass, mres, of the largest fragment of perforating (CP) shots is also provided. Adapted from Rahbek et al. [21].

Type of composite cover	Areal density of alumina (kg/m3)	Areal density of composite (kg/m3)	V50 (m/s)	90% CI at V50 (m/s)	mres of CP shots just above V50 (g)
Bare alumina	28.5	n/a	586	[574-596]	~3.4
Fabric wrap	27.0	1.5	536	[506-552]	~3.7
Filament-wound yarn	27.0	1.5	622	[608-637]	~2.5

The two production methods gave a very different residual projectile core mass for shots at velocities just above the V50. For the fabric-covered target, the mass was ~3.7 g, while the mass was much lower at ~2.5 g for the filament-wound target. It could be argued that the lower residual mass of the latter target was due to higher impact velocities, giving more tip erosion and higher fragmentation, but the difference existed even at similar impact velocities. Hence, it seemed clear that composite covers produced by filament winding gave higher core fragmentation of perforating bullets.

The only difference between the two composite-covered targets was the application method of the composite resulting in different lay-ups. The filament-wound composite had plies with unidirectional fibres in a 0o/90o lay-up, while the fabric-covered-targets had woven fibres with a certain degree of waviness, Figure 5. The most likely explanation for the higher performance of the filament-wound target was related to the higher degree of fibre orientation and, to some extent, the fibre pre-tension during manufacture. It is known that fibre waviness can reduce the mechanical properties of fibre-reinforced composites [26]. It has also been observed that the energy absorption and V50 of glass fibre-reinforced laminates can be decreased with increasing preload [24]. This suggests that straining of the fibres may be negative for the capability of absorbing projectile energy. On the other hand, due to the high fibre alignment, the unidirectional fibres in the composite cover may be 'activated' and able to take up load more quickly during projectile impact. One hypothesis is that the initiation of radial cracking from tensile forces may then be delayed due to the support on the backside of a ceramic. In addition to the fibre orientation, the pre-stress in the fibres may be also be beneficial since the fibres have already partly been strained during the manufacturing process. Overall, these results, together with the study by Jaitlee [23], seem to indicate a potential effect of pre-tensioning of composite sheets in hard armour plates.

5. CONCLUSION

The resistance to perforation of hard armour plates can clearly be increased when the ceramic tile is covered with a sheet layer of a fibre-composite material, at the expense of only small increases in areal density and little additional weight added. Typical failure of the ceramic is the formation of radial cracks and cone cracks, and a composite cover increases the amount of radial cracking and results in smaller fragments inside the cone area. Through-thickness cracks are prevented from opening and the ceramic fragments are confined, hence the integrity of the ceramic after impact is improved.

Front covers are usually added to manage spall ejection. However, one proposed mechanism for the increased ballistic performance is related to the front constraint of the ceramic strike face. The constraint increases the flow of ceramic debris towards the penetrator, and gives more time for penetrator-ceramic interaction since the ceramic is kept in front of the penetrator for a longer period of time. This effect has also been observed for pointed AP projectiles. On the contrary, front covers have also been observed to give no significant effect on projectile erosion or core fragmentation.

There are indications that the target design and manufacturing method may have a significant effect on the ballistic performance. More core fragmentation and higher kinetic energy-loss have been observed when the composite cover is placed on the back of the ceramic. This was explained by increased support and restraint introduced by the back cover, which contributes to a time-delay in the opening of tensile cracks on the back of the ceramic.

Recently, positive effects were observed from composite covers with some added pre-tension or with a high degree of fibre orientation (less waviness). These effects are currently unexplored, but suggest that new designs and manufacturing methods may improve penetration resistance. If these effects can be utilized, hard armour plates with a lower areal density can be produced. It should also be acknowledged that the double-curvature of the ceramic tile makes manufacturing more challenging.

One relevant topic that has not been discussed in this paper is the effect of adhesion between the ceramic and the composite cover. This may also influence on the occurrence of ceramic/composite delamination and the ballistic performance.

Acknowledgments

The authors would like to acknowledge a number of people who have contributed to this work in different ways, either through valuable discussions and support or assistance with the experimental testing; Gwyn E. Roberson and Toivo Horvei at NFM Group, Jeffrey W. Simons, Takao Kobayashi and Donald A. Shockey at SRI International, as well as Lasse Sundem-Eriksen, Ole Andreas Haugland and Torbjørn Olsen at FFI.

References

[1] P. J. Hazell, *CERAMIC ARMOUR: Design and Defeat Mechanisms.* Argos Press: Canberra, Australia, 2006.

[2] I. G. Crouch, "Body armour - New materials, new systems", *Defence Technology,* vol. 15, pp. 241-253, 2019.

[3] D. A. Shockey, J. W. Simons, D. R. Curran, "The Damage Mechanism Route to Better Armor Materials", *International Journal of Applied Ceramic Technology,* vol. 7, no. 5, pp. 566-573, 2010.

[4] B. G. Compton, E. A. Gamble, F. W. Zok, "Failure initiation during impact of metal spheres onto ceramic targets", *International Journal of Impact Engineering,* vol. 55, pp. 11-23, 2013.

[5] S. Sarva, S. Nemat-Nasser, J. Mcgee, J. Isaacs, "The effect of thin membrane restraint on the ballistic performance of armor grade ceramic tiles", *International Journal of Impact Engineering,* vol. 34, no. 2, pp. 277-302, 2007.

[6] S. D. Nunn, J. G. R. Hansen, B. J. Frame, R. A. Lowden, "Improved ballistic performance by using a polymer matrix composite facing on boron carbide armor tiles", *Advances in Ceramic Armor*. The American Ceramic Society: J. J. Swab Ed., pp. 287-292, 2005.

https://doi.org/10.52202/080042-0001

[7] P. Reddy, V. Madhu, K. Ramanjaneyulu, T. Bhat, K. Jayaraman, N. Gupta, "Influence of polymer restraint on ballistic performance of alumina ceramic tiles", *Defence Science Journal,* vol. 58, no. 2, pp. 264-274, 2008.

[8] D. B. Rahbek, J. W. Simons, B. B. Johnsen, T. Kobayashi, D. A. Shockey, "Effect of composite covering on ballistic fracture damage development in ceramic plates", *International Journal of Impact Engineering,* vol. 99, pp. 58-68, 2017.

[9] E. K. Oberg, J. Dean, T. W. Clyne, "Effect of inter-layer toughness in ballistic protection systems on absorption of projectile energy", *International Journal of Impact Engineering,* vol. 76, pp. 75-82, 2015.

[10] I. G. Crouch, G. Appleby-Thomas, P. J. Hazell, "A study of the penetration behaviour of mild-steel-cored ammunition against boron carbide ceramic armours", *International Journal of Impact Engineering,* vol. 80, pp. 203-211, 2015.

[11] I. G. Crouch, "Effects of Cladding Ceramic and Its Influence on Ballistic", in *28th International Symposium on Ballistics,* Atlanta, GA, USA, 22-26 September 2014, pp. 1084-1094.

[12] D. B. Rahbek, B. B. Johnsen, "Fragmentation of an armour piercing projectile after impact on composite covered alumina tiles", *International Journal of Impact Engineering,* vol. 133, 103332, 2019.

[13] I. Horsfall, D. Buckley, "The effect of through-thickness cracks on the ballistic performance of ceramic armour systems", *International Journal of Impact Engineering,* vol. 18, no. 3, pp. 309-318, 1996.

[14] C. Watson, T. Ringrose, S. Champion, I. Horsfall, B. Mallon, "The effect of cracks on the ballistic performance of contoured protective body armour plates", in *23rd International Symposium on Ballistics*, Tarragona, Spain, 16-20 April 2007, pp. 1357-1354.

[15] A. Azevedo, M. Vervoort, F. Coghe, "Experimental assessment of the effect of damage on the ballistic resistance of ceramic inserts", in *Personal Armour Systems Symposium 2018 (PASS 2018)*, Washington, DC, USA, 1-5 October 2018, pp. 351-360.

[16] J. P. F. Broos, R. Gunters, "Study on the ballistic performance of monolithic ceramic plates", in *23rd International Symposium on Ballistics*, Tarragona, Spain, 16-20 April 2007, pp. 1561-1567.

[17] G. Di Benedetto, P. Matteis, G. Scavino, "Impact behavior and ballistic efficiency of armor-piercing projectiles with tool steel cores", *International Journal of Impact Engineering,* vol. 115, pp. 10-18, 2018.

[18] S. Dey, T. Børvik, O. S. Hopperstad, J. R. Leinum, M. Langseth, "The effect of target strength on the perforation of steel plates using three different projectile nose shapes", *International Journal of Impact Engineering,* vol. 30, no. 8, pp. 1005-1038, 2004.

[19] M. S. Burkins, D. J. Little, M. S. Love, "A potential methodology for evaluating cermic quality", in *30th International Symposium on Ballistics*, Long Beach, CA, USA, 11-15 September 2017, pp. 2491-2501.

[20] G. Guo, S. Alam, L. D. Peel, "An investigation of the effect of a Kevlar-29 composite cover layer on the penetration behavior of a ceramic armor system against 7.62 mm APM2 projectiles", *International Journal of Impact Engineering,* vol. 157, 104000, 2021.

[21] D. B. Rahbek, G. E. Roberson, B. B. Johnsen, "Improved ballistic limit velocity from filament-wound fibres as composite cover on ceramic tiles", *submitted for publication,* 2023.

[22] P. J. Hazell, H. Wang, T. Lin, M. Saleh, V. Luzin, M. A. Kader, M. Saadatfar, A. Ameri, C. Ellis, "On the improvement of the ballistic performance of a silicon carbide tile through pre-stress: Experiments and simulations", *International Journal of Impact Engineering,* vol. 151, 103836, 2021.

[23] R. Jaitlee, "Physical protection: Interdependence between hard armour and soft armour", Doctor of Philosophy (PhD), RMIT University, Melbourne, Australia, 2013.

[24] M. V. Zhikharev, S. B. Sapozhnikov, O. A. Kudryavtsev, V. M. Zhikharev, "Effect of tensile preloading on the ballistic properties of GFRP", *Composites Part B: Engineering*, vol. 168, pp. 524-531, 2019.

[25] E. Tapie, E. S. L. Tan, Y. B. Guo, V. P. W. Shim, "Effects of pre-tension and impact angle on penetration resistance of woven fabric", *International Journal of Impact Engineering,* vol. 106, pp. 171-190, 2017.

[26] P. Kulkarni, K. D. Mali, S. Singh, "An overview of the formation of fibre waviness and its effect on the mechanical performance of fibre reinforced polymer composites", *Composites Part A: Applied Science and Manufacturing,* vol. 137, 106013, 2020.

https://doi.org/10.52202/080042-0001

Towards a Fully Circular Aramid Yarn

Stan Maassen[1], B. Gebben[1], V. Srinivas[1] and R. Hartert[2]

[1]*Teijin Aramid Tivolilaan 50 6802 ED, Arnhem The Netherlands*
stan.maassen@teijinaramid.com
[2]*Teijin Aramid Kasinostraße 19-21, Wuppertal Nordrhein-Westfalen 42103 Germany*
ruediger.hartert@teijinaramid.com
+49 173 9448751

Abstract. When it comes to armour systems, ballistic and protective performance is the key parameter. However, in today's world, also the topic of sustainability increasingly gains attention. Not only to reduce the waste of end-of-life armour materials, but also by lowering the eco-footprint of these materials at the production side. Key future developments will therefore be in the combined effects of circularity and reduction of greenhouse gas emissions in scope 1, 2 and 3. Aramids are one of the main materials used in personal protection, for example in vests, shields, and helmets. As one of the main aramid producers, Teijin Aramid wants to accommodate both topics by producing yarns with excellent performance in a sustainable way. To reach this goal, we are developing various new recycling routes: 1. Mechanical recycling: a route that we are already operating in the production of pulp and want to extend to other products. 2. Physical recycling: a completely new recycling route that allows production of aramid yarn using aramid from a recycle source as input. 3. Chemical recycling: by using a depolymerization process, we aim to convert end-of-life aramids to resources that can be used in our polymer factory for the production of new polymer that will be used to produce aramid yarn.
Besides these recycling routes, Teijin Aramid puts significant effort on gaining access to sustainable resources for the production of our yarn. By joining research consortia and by collaboration with business partners, we study the possibilities of sourcing our raw materials from a renewable source, such as bio-based or plastic waste-based sources, to move away from fossil resources. With these and other developments, Teijin Aramid aims to lower its carbon footprint to net zero, thus making a significant impact on the sustainability of aramid-based products such as armour systems. And while doing this, Teijin Aramid ensures thatv the high quality and performance of the aramid materials stay intact. Here, the latest developments with regards to sustainability and circular economy in the personal protection market are discussed, covering the steps Teijin Aramid has made, and is making, to lower our eco footprint and invite all value chain partners to join the discussion on how we can collaborate in becoming fully circular over the whole market.

1. INTRODUCTION

Ballistic and protective performance is the key parameter in the design and production of armour systems such as vests, shields, helmets, and panels. A lot of effort goes into improving the design of these products and the optimization of the material(s) that are used aiming to further increase their performance, and comfort in use.
However, in contrast to only a few years ago, sustainability and recycling of armour systems is increasingly gaining attention. First and foremost, solutions are sought for reusing or recycling End-of-Life (EoL) products. At the moment, most are still landfilled or incinerated at the end of their lifetime. Only a small fraction of the EoL products, primarily 100% aramid and clean, such as aramid ballistic vests, are currently recycled.
But also in the production phase changes need to be made to lower the eco-footprint of the various armour systems. This can be achieved by moving from non-renewable, fossil resources to renewable raw materials. These can, for example, be bio-based sources but they may also derive from plastic waste recycling or include recycling of EoL aramid yarn back into new aramid yarn. Furthermore, by increasing the efficiency of the production processes involved the footprint is reduced further.
Aramids are one of the main materials used in personal protection, for example in vests, shields, and helmets. As one of the main aramid producers, Teijin Aramid wants to accommodate both topics by producing yarns with excellent performance in a sustainable way. Figure 1 presents the fully sustainable and circular value chain for aramid products that Teijin Aramid envisions for the future.

https://doi.org/10.52202/080042-0002

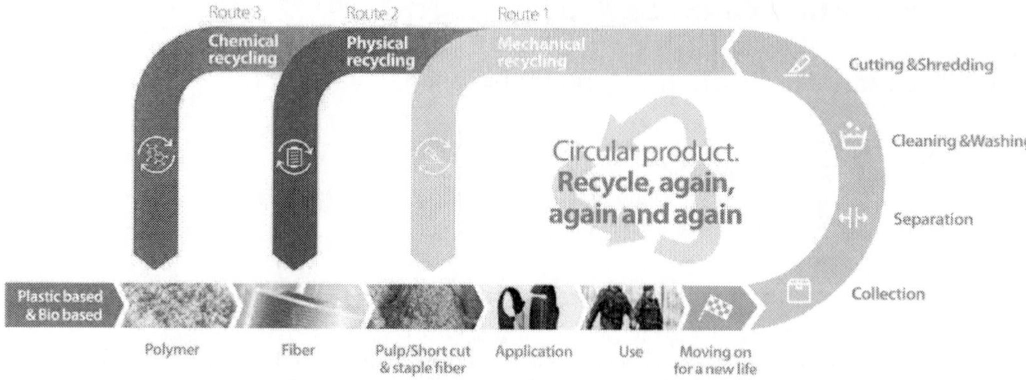

Figure 1. A fully sustainable and circular value chain for aramid products [1]

As described in Figure 1, the chain starts with renewable carbon resources, for example from biobased or EoL plastic sources. From these, aramid polymer (Poly-(para-phenylene-terephthalamide), PPTA) and yarn are produced which are used in applications, *e.g.* in armour systems. To lower the eco-footprint in the production phase, all processes involved need to be operated as efficient as possible and using 100% renewable and clean energy. After the use phase, all applications are given a new life. For this, EoL products need to be collected and different materials need to be separated, think for example of separating the inner aramid layers from the outer fabric of a protective ballistic vest. After separation, various pretreatment steps, *e.g.* washing and cutting, are required to prepare the materials for the three recycling routes that Teijin Aramid has or is developing:

1. Mechanical recycling: a route that we have already been operating for over 20 years in the production of pulp and which we want to extend to other products.
2. Physical recycling: a completely new recycling route that allows production of aramid yarn using aramid from a recycle source as input.
3. Chemical recycling: by using a depolymerization process, we aim to convert EoL aramids to resources that can be used in our polymer factory for the production of new polymer that will be used to produce aramid yarn.

With these three recycling routes, aramid from different production left-over and EoL sources can be reintroduced in the production chain to produce new aramid products.

2. RECYCLING ROUTES

2.1. Mechanical recycling

The first route to recycle aramid materials is to reprocess into new products using mechanical processes. In general, such processes involve cutting, milling, or refining of the fiber to yield a short fiber product. This can be a cut or milled fiber of a specific length, or aramid pulp. At the moment, ballistic fabrics are the main source for recycling through this route.

At Teijin Aramid, this mechanical recycling route has already been in place for over 20 years, yielding Twaron pulp (Figure 2). Due to its chemical and physical properties, *e.g.* high abrasion and temperature resistance, aramid pulp can be used as a replacement for asbestos in applications such as brake pads, gaskets, and clutch plates. Although this is a very relevant recycling route for various applications, due to the fact that it does not yield new aramid fiber it can be considered a downcycling route.

https://doi.org/10.52202/080042-0002

Figure 2. Twaron pulp [2]

2.2. Physical recycling

To become fully circular, there is a need for recycling of aramid fiber not only to a different product, *i.e.* pulp as described in section 2.1, but also back into new aramid fiber. One way to achieve this is through the physical recycling route that is currently being developed at Teijin Aramid.

The physical recycling process involves obtaining aramid yarn, *e.g.* from EoL or production left-over sources, and processing it such that it can be dissolved and reintroduced into the aramid yarn spinning process. This technology has already been proven at a pilot stage at Teijin Aramid's research centre in Arnhem, The Netherlands and a patent on this technology is pending [3]. Here, a process has been developed to recycle aramid yarn by dissolving the yarn in sulphuric acid and using the obtained spinning solution in Teijin Aramid's wet-spinning process to produce new aramid yarn. As input, aramid yarn from various sources and having different morphologies have been recycled. Furthermore, different recycle contents have been obtained producing yarn from a combination of the recycle feedstock and virgin PPTA, mixed at different ratios. Interestingly, the dissolution behaviour of aramid yarn differs from the dissolution of PPTA polymer powder. This means that the dissolution conditions, such as time and temperature, need to be tuned to the recycle content. Figure 3 shows mechanical properties of yarn produced under similar conditions with varying recycle content.

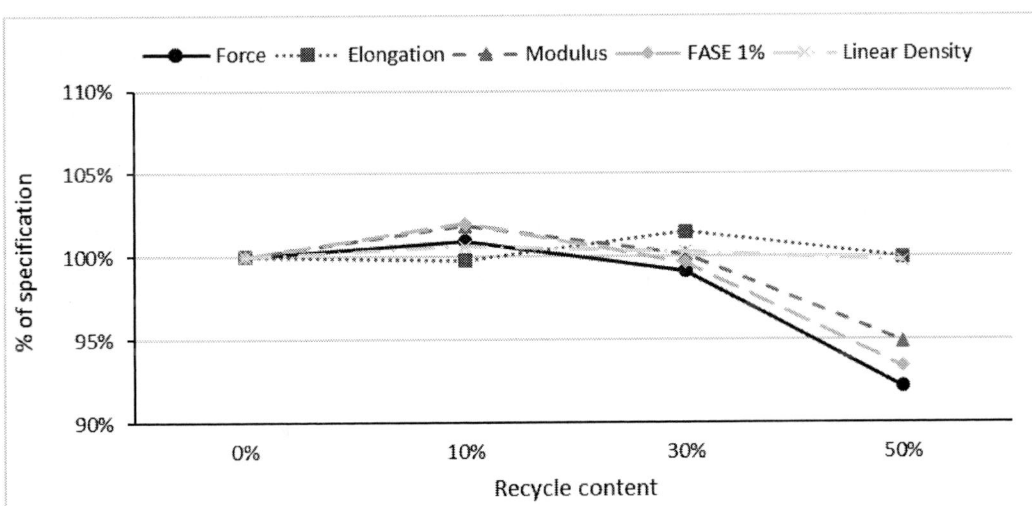

Figure 3. Mechanical properties for Twaron yarn with 0-50% recycle content, introduced using physical recycling and produced under the same conditions. Dots show the force at break (Force), squares show the elongation at break (Elongation), Triangles show the modulus of the yarns (Modulus), Diamonds show the force at 1% elongation (FASE 1%), and Crosses show the linear density (Linear Density). All data is normalized relative to the virgin (0% recycle content) yarn.

As observed from Figure 3, at recycle contents $\leq 30\%$ no significant changes in mechanical properties are measured. At 50% recycle content, the linear density and elongation at break of the yarn remain unaffected, while the Force, Modulus, and FASE 1% show a significant reduction. The cause of this, and

https://doi.org/10.52202/080042-0002

how to avoid it, is still under investigation. Further development and optimization of this recycling route, especially for high recycle contents, is currently under investigation.
Although the mechanical properties presented in Figure 3 already give a good impression on the quality of the yarn, further evaluation was performed to understand the yarn's performance in applications. As an example of this, Figure 4 shows "Time-to-Failure" (TTF) results for yarns with varying recycle contents.

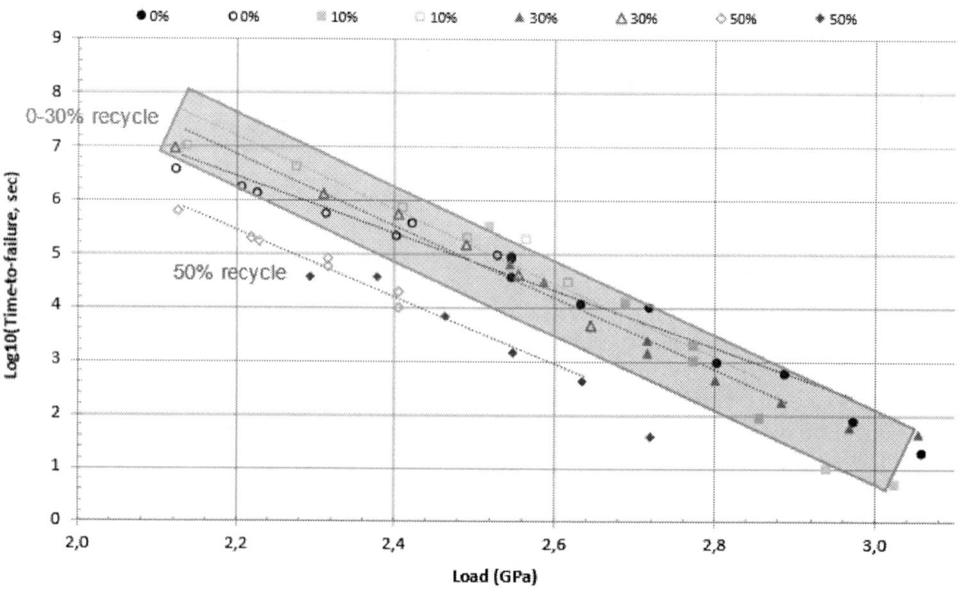

Figure 4. Time-to-Failure data of aramid yarn containing 0% (circles), 10% (squares), 30% (triangles), and 50% (diamonds) recycle content, introduced using physical recycling under the same process conditions.

TTF data gives information on the performance over longer times and the lifetime of synthetic fibers [4]. Similar to the data presented in Figure 3, from Figure 4 can be seen that no significant changes in performance are observed between aramid yarns with a recycle content of 0%, 10%, and 30% that were produced under the same process conditions. Similar yarns with a recycle content of 50% clearly show poorer TTF performance, as can be observed by the shift down of the data set for this yarn (blue diamonds) indicating shorter time to failure at a given load as compared to the other samples.
To evaluate the performance of these yarns with recycle content in high-demanding, real-life applications, Teijin Aramid collaborated with the companies FibreMax, located in The Netherlands, and Hampidjan, based in Iceland. First results have shown that also in these applications, no difference in performance is observed between virgin yarn and yarn with 30% recycle content [5-8].
Also for ballistic applications further research towards the performance of these aramid yarns with recycle content has been performed.
A theoretical estimate of the ballistic potential of a yarn can be provided by the characteristic velocity reported by Cunniff [9]. This parameter is a function of the primary yarn properties in the longitudinal direction. A physical interpretation can be best thought of as the product of the yarn specific toughness and the longitudinal wave speed or simply the energy absorption rate of a yarn. Based on the mechanical properties measured, Twaron containing 30% recycled content shows a comparable characteristic velocity. Furthermore, ballistic evaluations reveal the energy absorption due to addition of recycled content also does not vary, so does the speed of sound though the yarn/fabric. These results, shown in Figure 5, indicate that a yarn with 30% recycle content shows the same ballistic performance as a virgin aramid yarn.

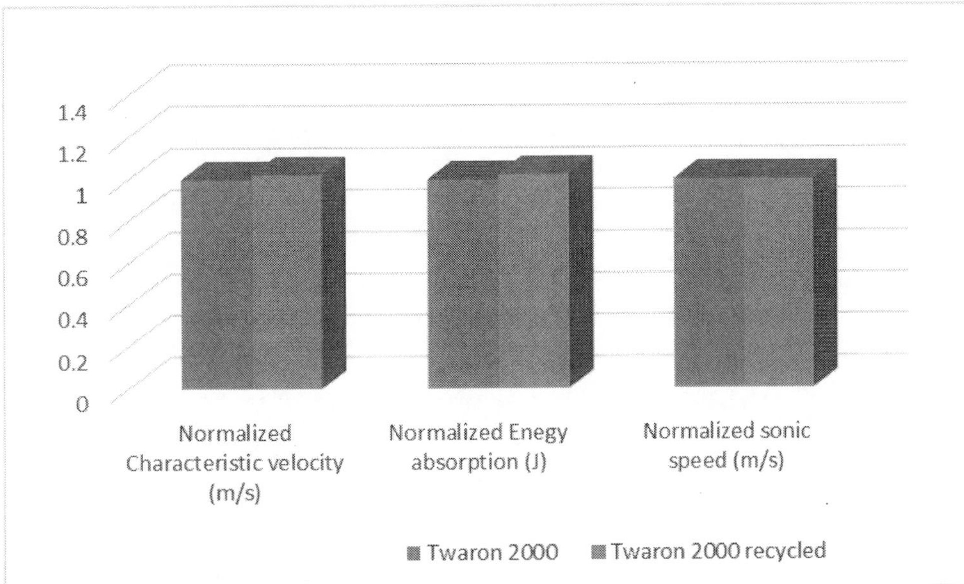

Figure 5. Normalized ballistic performance characteristics comparing standard Twaron 2000 yarn with a similar yarn containing 30% recycle content.

Currently, Teijin Aramid is working on moving this development from a R&D to a production scale. If successful, a Twaron yarn with recycle content can be produced commercially in the near future.

2.3. Chemical recycling

The third recycling route that is currently being studied and developed by Teijin Aramid involves chemical recycling of aramid products. This process comprises depolymerization of the PPTA polymer into its monomers that can, after isolation, be reused to produce new, virgin-grade PPTA that can be spun into aramid fiber.

Similar to the physical recycling route described in section 2.2, this route also allows for the production of a yarn with a recycle content. However, in contrast to introducing the recycle material in polymer form during the fiber spinning process as is done in the physical recycling route, in the chemical recycling process the recycle content is introduced even further back in the value chain, replacing part of the raw materials required for the production of PPTA polymer.

Going further back in the production chain is inevitably associated with a higher footprint since more subsequent process steps are required. However, an advantage of this recycling approach is that it allows for the production of PPTA polymer that is of the same quality as virgin-based polymer while replacing fossil-based feedstock with recycle-based raw materials, and thus does not affect the aramid yarn that is produced from this recycle-based PPTA polymer.

3. PRETREATMENT

To facilitate the recycling routes discussed in sections 2.1, 2.2, and 2.3 various pretreatments of the production left-over and EoL products are required. In general, recycling of any material is easier the purer a material is. Separating the various different components of a product, *e.g.* metals, fibers, coatings, resins etc., before recycling the materials improves the quality of products that can be obtained through recycling. For this reason, for recycling of aramid fiber Teijin Aramids includes separation and cleaning steps to improve the quality of the input for recycling. Further research towards these separation technologies is done to enable recycling of heavily contaminated aramid products to high quality aramid products. An example of such a development is the mechanical process to remove resin from helmets, to recover aramid fiber with a lower contamination level [9].

To reduce the required amount of pretreatment required before recycling, designing our products beforehand with recycling in mind is crucial. An example of such a development is our proprietary fluor-free coating that can replace persistent and potentially harmful PFAS (per- and polyfluoralkyl substances) being current industrial standard as a water-repellent treatment for ballistic fabrics. We were able to maintain fragment and bullet resistance in wet and dry testing on similar level at reduced overall shoot-pack water take-up.

https://doi.org/10.52202/080042-0002

4. RENEWABLE RESOURCES

The last aspect discussed here is the switch that Teijin Aramid aims to make from non-renewable, fossil-based resources to sustainable alternatives. For this, Teijin Aramid is looking into alternative sources of raw materials obtained from bio-based resources or produced in the recycling of plastic waste. Already in 2019, Teijin Aramid partnered with the Dutch company BioBTX to produce a completely bio-based aramid fiber. In this project, an 92% bio-based aramid fiber was produced at a lab scale which had similar performance as conventional yarns.

Furthermore, Teijin Aramid is actively involved in research consortia, for example the "InREP"-project, to convert plastic waste into resources for the production of aramid yarn [1]. These, however, are long term developments and will require more time before being applicable at a large scale.

5. CONCLUSIONS

Besides ballistic and protective performance, also sustainability is gaining importance for amour systems. This paper discusses the opportunities for aramid-based products to contribute to producing more sustainable armour systems without compromising on performance. With the developments discussed here, Teijin Aramid works to lower the eco-footprint of aramid products and reaches out to all value chain partners to collaborate on activities to become fully sustainable over the entire market.

References

[1] Teijin Aramid, Scaling for Impact, Sustainability Report 2021, 2021

[2] Teijin Aramid, Twaron Product Brochure

[3] Boerstoel, H., Nijenhuis, W., La Faille, R., Günes, R., Boudebah, S., Gebben, B., Maassen, S., Komiya, N., Continuous process for recycling, WO 2021/123210 A1, 2021

[4] Knoester, H., Den Decker, P., Van den Heuvel, J., Tops, N., Elkink, F., Creep and Failure Time of Aramid Yarns Subjected to Constant Load, Macromol. Mater. Eng., 2012, 297: 559-575

[5] Mary Bailey, Teijin Aramid demonstrates fiber-recycling breakthrough, Chemical Engineering, 2021

[6] FibreMax, FibreMax and Teijin A Sustainable Long-term Relation, https://fibremax.nl/projects/fibremax-teijin/

[7] Hampidjan, Terra Slings The Environmental Solution, 2021

[8] Teijin Aramid, Teijin Aramid demonstrates potential of circular high-performance fibers, 2021

[9] Cunniff, P. Dimensionless Parameters for Optimization of Textile-Based Body Armor Systems. In Proceedings of the 18th International Symposium on Ballistics, San Antonio, TX, USA, 15–19 November 1999; Volume 2, pp. 1303–1310.

[10] Korte, H., Process for recycling material containing high-tenacity fibers impregnated with a resin matrix, US 20010023903 A1, 2000

Development and simulation of Protective Systems for Energy Absorption under Ballistic Loading Conditions

Lennart Alkemade[1], Jan Bohlen

1TSS International BV Zuideinde 2991 LK Barendrecht Nederland
lennart.alkemade@tssh.com
+31 (0) 180618922

Abstract.
This work presents an experimental and simulative investigation of the protective capabilities of SKYDEX® material solutions subjected to ballistic loading. In this work, the application of a new DYNEEMA® / SKYDEX® material approach to reduce BABT (behind armour blunt trauma) when used in bulletproof vests is simulated, tested and evaluated. For direct comparison, a reference set-up is used, which equally fulfils the VPAM 6 protection level (7.62 x 39 FeC projectile). The assessment criteria for the evaluation of the corresponding human injuries, the 44mm indentation criterion in ballistic clay, is part of different current test standards like the VPAM BSW 2006. For the simulation-based development and iterative evolution of this novel approach to ballistic protective vests, IMPETUS Afea® is used for the numerical simulation of the interactions of the projectile with the protective structure and to visualise causal relationships for further structural optimisations with the objective to further minimize BABT effects. For this purpose, a numerical representation study of fibre composites or fabric materials is presented to determine a suitable modelling type for this kind of materials. Additionally, the ballistic clay material model, which is adapted to the requirements of the VPAM for determining the BFS (Back-Face-Signature) is also presented. This is followed by the comparison and evaluation of the two protective vest configurations regarding the existing impact mechanism, penetration depth, impulse transfer and trauma severity by simulation and ballistic testing. This proves that the new DYNEEMA® / SKYDEX® approach offers a significant improvement in terms of penetration depth, general BFS and BABT and therefore lower injury severity risk and higher probability of unharmed survivability compared to the reference protective vest.

1. MOTIVATION FOR SOPHISTICATED DEVELOPMENTS IN BALLISTIC PROTECTION

As with all certifications and test specifications, these are limited to a specific range of load cases, which are intended to represent as many real scenarios as possible but are far from being able to capture all potential threats, which means that the protective effect for certain areas of application or attack situations is already reduced. Even in the case of scenarios that are covered by the certification and are therefore averted by the vest, this does not mean that human injury is ruled out. The certifications allow certain puncture or trauma depths during the test, which certainly cause injuries to people.
The "BABT", which stands for "Behind Armor Blunt Trauma", is decisive in terms of ballistics. This describes the trauma that the body experiences behind the body armour. If the case required by the certification occurs that the projectile does not penetrate the protective layer, its energy is partially absorbed by the protective layer in the form of failure modes and deformation while deaccelerating the projectile. However, according to [1] the rest is conducted to the body and occurs there as BABT. It thus describes, according to [2], a "non-penetrating injury due to the rapid deformation of the body armour", caused by the impact of a bullet or other projectile. According to [3], if the residual energy is sufficiently high, this can cause not inconsiderable injuries to the body such as lesions in the lungs, bone and tissue damage in the chest or even ruptures of the heart. In extreme circumstances, the BABT can even result in death. Therefore, it is critical to reduce the Back Face Signature (BFS) of the trauma plate, as reducing the BFS will reduce BABT and thus increase the chance of survivability.

1.1 Development of a New Type of Ballistic Trauma Plate to Reduce the BABT

Due to the already existing cooperation between EDAG Engineering GmbH & TSS International BV in the field of shock mitigation for armoured vehicles, it has been examined whether a variant of the SKYDEX® panels can make a positive contribution to reducing the BABT, since it has outstanding energy absorbing properties. Since it does not provide ballistic protection DYNEEMA® plates can be used to stops the projectile threats. Combined these materials complement each in terms of ballistic protection and energy absorption. Therefore, the composite of DYNEEMA® & SKYDEX® is the most optimum protection composite for both soldier personal protection systems and armoured vehicles with mine protection flooring. The aim of this simulation-driven development, using IMPETUS AFEA

https://doi.org/10.52202/080042-0003

Solver® Simulation Software, is to demonstrate how the different materials react when combined and thus further reduce the risk of BABT-related injuries.

1.2 Approaches to cover the 7,62 x 39 FeC Threat

Two main designs of protective vests have prevailed, the soft ballistic protective vest and the hard ballistic plate carrier:

- The soft ballistic protective vest: Mainly composed of many layers of ballistic textiles and fibre fabrics. These can have different interweaving structures at the layer level and different connection techniques between the layers. This type of body armour is used for both ballistic and stab protection applications, as modern fabrics, usually made of aramid or polyethylene, offer very high mobility and low weight on the one hand, but also have a high potential for energy absorption on the other.
- Hard ballistic armour: Consists of one or more rigid plates attached to a flexible fabric panel called plate carrier. These plates usually consist of steel, ceramic, or pressed fibre composites such as "DYNEEMA® HB212" UHMWPE plates from AVIENT. They are typically used for higher ballistic threats, such as rifle attacks, and offer good protection in terms of both penetration and trauma depth minimization. In combination with cushioning layers, however, these are often heavy and limit mobility.

The project partners have set themselves the goal of developing effective protection to cover the VPAM 6 threat (7,62 × 39mm FeC) and have pursued three approaches as part of the simulative design:

- Monolithic Steel approach (Domex Protect 500): To be able to classify the development progress better and to make the differences more visible, the classic monolithic steel plate is taken into account as a slide-in solution for a plate carrier in the context of the simulative design and reference. With this approach, the kinetic energy of the projectile is converted into the plastic deformation of the protection plate, which destroys/deforms the projectile core with its hardness.
- UHMWPE-Plate approach (DYNEEMA® HB212): The second approach pursued as part of the simulative design aims to stop the projectile by deformation and catching it inside the different layers of the DYNEEMA® HB212 trauma plate UHMWPE material. The advantage is a lighter and therefore more comfortable to wear trauma plate, but it comes with the disadvantage of a larger BABT than heavier monolithic steel solutions.
- Stacked approach (DYNEEMA® HB212 / SKYDEX® Trauma Reduction Layer): In order to compensate for the described disadvantages of the UHMWPE-Plate compared to the monolithic steel plate, the third pursued approach of the simulative design is to combine strong stopping power and the enhanced weight benefit of the bare UHMWPE-DYNEEMA-Plate with a highly reduced BABT by using SKYDEX® as trauma reduction layer on the inner side of the trauma plate. As shown in Figure 1.

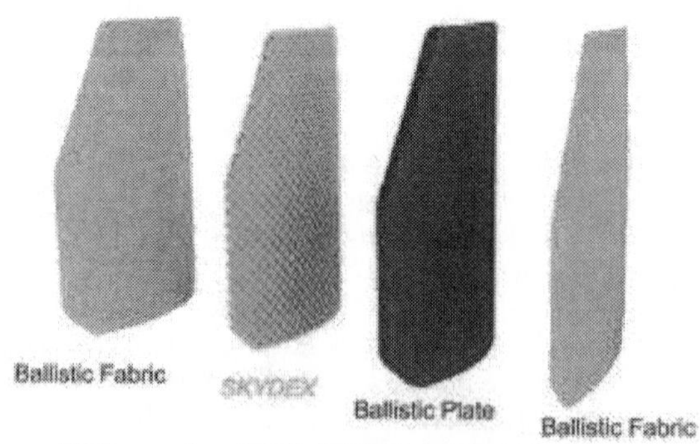

Figure 93 Example of DYNEEMA® HB212 & SKYDEX®

The IMPETUS Afea Solver®, specially developed by IMPETUS Afea AS for highly non-linear and highly dynamic tasks, is used to develop an effective layering of the selected material combination. This explicit FE solver, with Lagrangian discretization, is particularly suitable for the simulative depiction of stresses and ballistic effects on structures and thus for the numerical depiction of large structural dynamic deformations under extreme load conditions. Only volume elements are used since shell elements would lead to inaccurate results under these loading conditions according to [4].

2. MODELLING APPROACH OF THE PROTECTION AND EVALUATION MATERIALS

In this paragraph we like to present the different modelling approaches of the protective and evaluation materials and their corresponding constitutive material formulations when necessary. For the reference setup Domex Protect 500® of „SSAB AB" is used in the MAT-METAL material formulation of IMPETUS [5]. Like the chosen projectile materials (mild steel core, brass jacket and lead filler) this materials were calibrated internally prior to this project and are therefore not part of this technical report.

2.1 DYNEEMA® HB212 UHMWPE of „AVIENT Corporation"

First of all, there are different ways of mapping a fibre composite on a geometric level, independently of the material model. The volume elements specified by IMPETUS as an element type are very well suited for modelling the fabric materials. The distinction to be considered much more closely is the choice between "macro", "meso" and "micro" level according to [6]. These three types of modelling describe how exactly the fibre strands and the matrix are discretized by elements.

- Macro Level: Modelling of the entire network as a single continuum. Failure of fibre strands, the matrix or delamination of individual layers can or must be calculated and mapped solely by the material model. An intralaminar failure cannot be represented.
- Meso Level: Modelling of single or multiple layers as a continuum layer. Failure of fibre strands and matrix must be represented by the material model. Delamination of individual layers or stacks of layers can be realized through contacts and MPC conditions.
- Micro Level: Discretization of the individual fibres and the matrix with elements. Both failure of fibre strands and matrix as well as delamination of individual layers and fibre strands can be imaged.

Figure 2 schematically shows the difference between the three types of modelling in IMPETUS.

Figure 94 Differences between the three types of modelling

Modelling at the micro level is obviously the most accurate variant for mapping a fibre composite or fabric material. However, it must be noted that a very high number of elements is required for the networking of individual filament strands and the possibly complex, perforated matrix structure around the fibres, which increases the computing time and resource consumption. Nevertheless, the meso level modelling of multi-layered fabric material is also perfectly suited for the evaluation of ballistic impact scenarios and therefore used for the development of the new DYNEEMA® / SKYDEX® approach. In figure 3, the direct comparison between meso and micro level modelling is shown under ballistic impact loading of a soft ballistic layered aramid structure penetrated with a 9mm projectile.

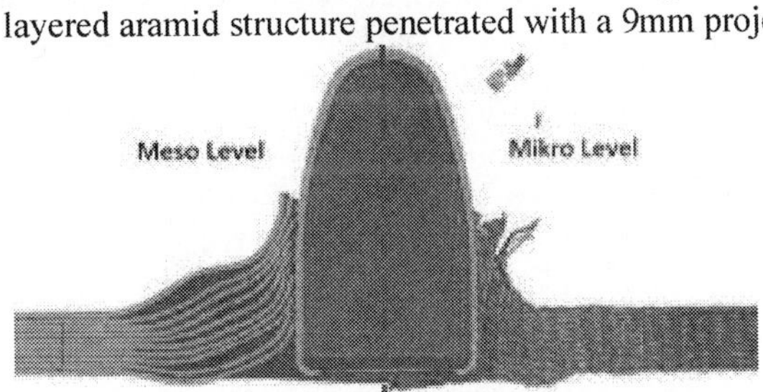

Figure 95 Meso and Micro level modelling of multi-layered fabric

The MAT_FABRIC material model implemented in IMPETUS according to [7] is used to simulate the HB212. This model is designed to simulate fabric materials as a continuum. It enables the input of matrix and fibre parameters, which are then converted into a common constitutive law. Figure 4 shows the characteristic of the stress-strain curve of this material model, whereby it should be noted that the fibre stress σ_i is dependent on the fibre strain ε_i in direction i.

https://doi.org/10.52202/080042-0003

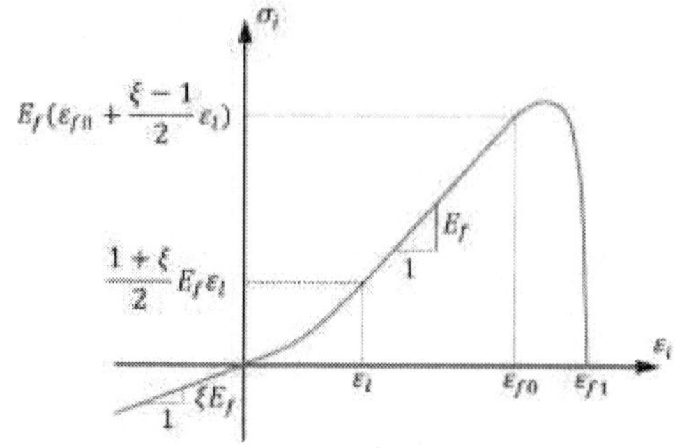

Figure 96 Characteristics of the stress-strain curve

The linear fibre modulus of elasticity E_f can be clearly seen here, although this does not start at the origin, but from the locking strain (fibre locking strain) ε_l. This locking strain describes the alignment of the fibres within the yarn or the strand and the alignment of the strands within the composite or fabric until the material is fully resilient. In addition to specifying the locking strain, it is also possible to enter an initial stiffness component ξ, which influences the reduced slope below the blocking strain, as well as the compression stiffness and the plasticization at the end of the linear law. The two parameters ε_{f0} and ε_{f1}, which represent the elongations at failure of the fibres, are also related to this. These can also be made dependent on the strain rate via the reference strain rate $\dot{\varepsilon}_0$ and a strain rate exponent c_ε. Two of these strains can thus be used to define the onset of failure and a point at which all fibres are torn, which significantly improves the characteristic of element failure. In addition, a matrix failure parameter and an erosion strain ε_e, can be configured. The total stress within the element is calculated according to Equation 1 by separating the hydrostatic, deviatoric and a damping part, which is set via the dynamic viscosity μ_{dyn}. In addition, the equation takes the previously calculated fibre stresses σ_i in the four principal directions into account.

$$\sigma = 2 \cdot G \cdot \varepsilon^{e}_{dev} - p \cdot I + \left(\sum_{i=1}^{4} f_{F,i} \cdot \sigma_i \cdot v_i \otimes v_i\right) + 2 \cdot \mu_{dyn} \cdot \dot{\varepsilon}_{dev} \quad (1)$$

This material model is very well suited for modelling fabrics since it can numerically represent the volume fraction and the locking strain of a real fabric material. The HB212 used in this development project was initially derived from an existing material model for a comparable material and then validated based on ballistic tests.

2.2 SKYDEX® Ballistic Trauma Reduction Layer (BTRL) of „SKYDEX Technologies"

In order to avoid unnecessary complexity in the simulation of the SKYDEX® ballistic trauma reduction layer (BTRL), a geometric representation of the specific, characteristic cell structure of the SKYDEX® -material was avoided. Instead, the BTRL was mapped to a closed continuum reflecting the mechanical stress-strain properties of the material itself. For this purpose, the *MAT_VISCOUS_FOAM material model [8] is used as part of this project, since it enables the characteristic properties by specifying the stress-strain curve under compression loads as shown schematically in figure 5.

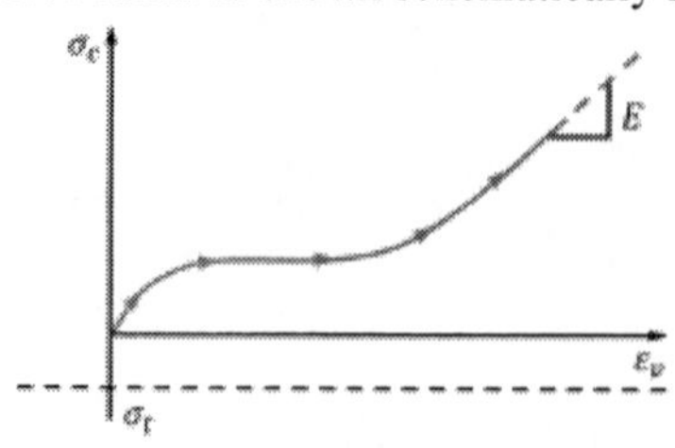

Figure 97 Characteristic stress-strain curve under compression loads

SKYDEX Technologies was able to provide the results of compression tests with speeds between 0.01m/s and 1m/s, which were extrapolated to the expected speed range of the back face of the strike face in the event of a projectile impact. The test procedure and evaluation were simulated and iteratively adjusted to the real results. Figure 6 shows the material test simulation set-up together with the extrapolated test curve for a punch speed of 200m/s and the associated stress-strain curve of the material.

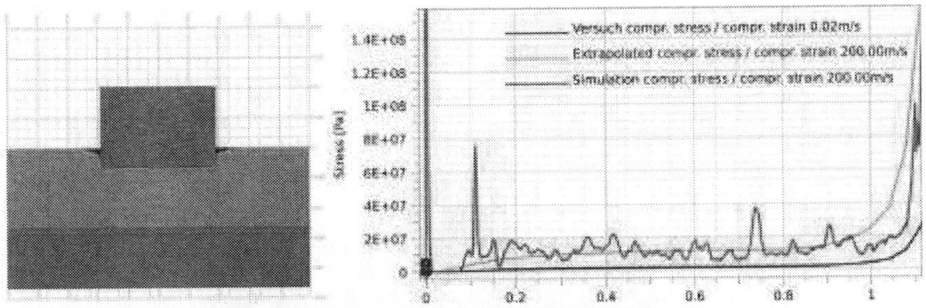

Figure 98 Punch-test simulation set-up & extrapolated test curve

2.3 Ballistic plasticine of „Carl Weible KG"

It should be noted in advance that the terms plasticine and ballistic clay are used synonymously in this report. The MAT-Metal is also used to represent the plasticine in this project, since, as described, it is a material model that can be used very flexibly for ductile materials. Due to the advantage of specifying the plastic flow directly as a stress-strain-curve and not having it calculated using predefined model parameters, this model makes it possible to numerically map the elastic-perfectly plastic behaviour of the ballistic clay directly. For this material model, it is also possible to convert the continuum discretized by finite elements into SPH particles with the same material properties, to receive reliable and numerically accurate results even with the expected very large deformations of the ballistic clay.
The plasticity test is carried out for all regulations by means of a ball drop test, whereby essentially only the drop heights, the associated expected intrusions and the number of tests differ. It can be seen that the plasticine from "Carl Weible KG" used in the VPAM test guidelines [9] and the " Roma Plastilina® No. 1" used in HOSDB, CAST and NIJ test standards behaved very similarly with regard to the required intrusions, which means that these measurements appear to be well comparable. As specified at the beginning, the VPAM test guidelines were used to validate the model of the ballistic clay.
In order to test the plasticity of the plasticine a sphere drop system with a steel sphere (diameter 63,50 ± 0,05 mm, mass 1.039 ± 5,00 g) is to be used [9]. The distance between the lower edge of the sphere and the surface of the plasticine is to be 2.000 ± 5,00 mm, which gives an impact velocity of 6,26m/s. The plasticine with its applied conditioning temperature is acceptable when the depth of each depression is 20,00 ± 2,00 mm [9]. In accordance with VPAM BSW 2006 a simulation model for the sphere drop test onto the plasticine was set up and the material parameter of the ballistic clay where iterative adjusted to provide the prescribed indentation depth, which is shown together with test results in figure 7.

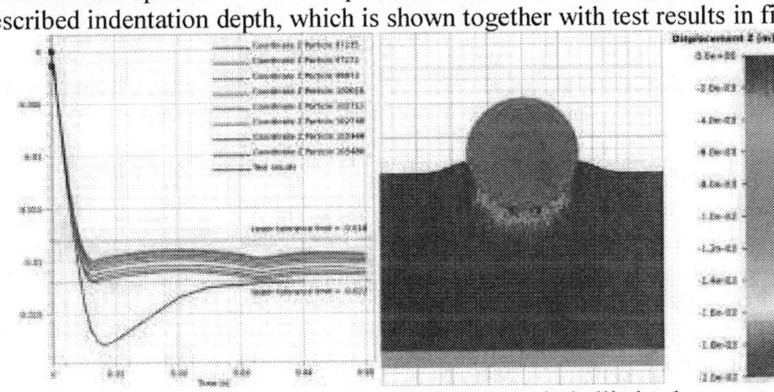
Figure 99 Impact test results of a sphere in ballistic clay

Since the behaviour of the ballistic clay is a critical factor for the subsequent evaluation of the BABT we additionally focussed on a correct behaviour of the displaced clay material, which forms a small circular hill on the edge of the impacting sphere which is also shown in figure 11. Hence the good correlation the chosen parameter set for the plasticine is suited for indicating the plastic deformation and the BABT with its corresponding maximum admissible indentation depth of 44,00mm [9].

3. CLASSICAL APPROACHES VS. THE DYNEEMA® / SKYDEX® STACKED APPROACH

The protective layer (steel or UHMWPE) and the ballistic trauma reduction layer (if used) form together the protective system. A sample size 300mm x 300mm sample is used to eliminate all edge and size effects. The protective system is located on top of the ballistic clay, which has a total thickness of

https://doi.org/10.52202/080042-0003

80,00mm supported by support surface, which is fixed in all translatory and rotatory directions. The protective system is refined in the impact area of the Projectile to achieve sophisticated material and failure behaviour. In total the simulation model, shown in figure 8 consists of ~ 1.000.000 SPH particles representing the ballistic clay and up to 79.000 quadratic elements representing the protective system.

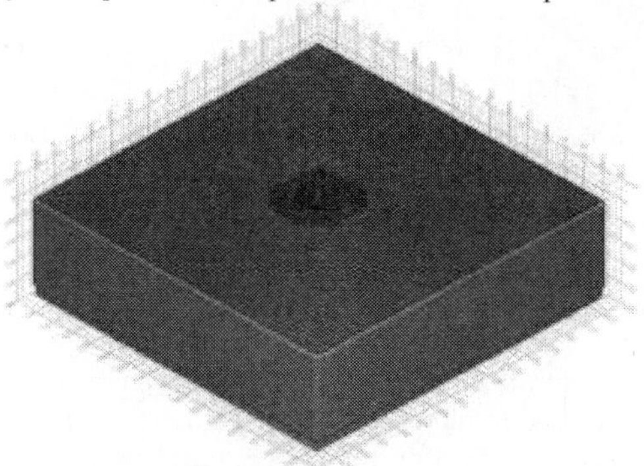

Figure 100 Simulation model with SPH particles and quadratic elements

The initial velocity of the 7,62 × 39mm FeC projectile is set to 720,00m/s according VPAM APR [10], which results in a kinetic muzzle energy of 2.074J, since it has a mass of 8,00g. The projectile itself is represented by around 2.000 cubic elements. The cross section of the real projectile in accordance with [11] and the discretized simulation model is shown in figure 9.

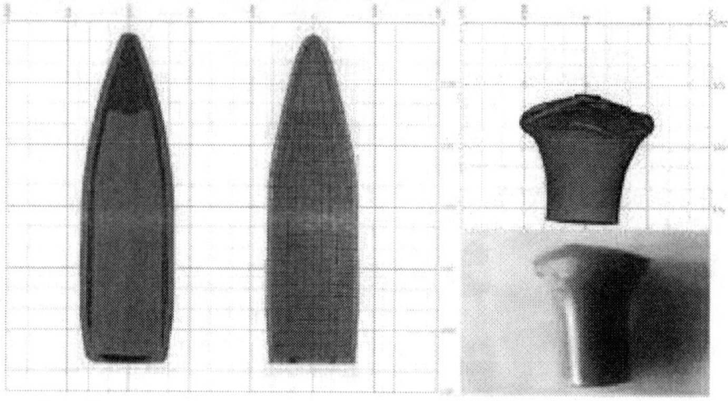

Figure 101 7,62 × 39mm FeC cross section acc. to [11] and core deformation in UHMWPE acc. to [12]

The AK47, one of the most widely used firearms in the world, is chambered for the 7,62 × 39mm cartridge. Therefore, it was the goal to recreate one of the most common threat scenarios below the entrance level for armour piercing projectiles with hardened steel cores. The chosen projectile consists of a mild steel core, a brass jacket and lead filler. The deformation behaviour of the mild steel core in UHMWPE layered material is also shown in figure 9 in comparison for reality according to [12] and the simulation model. As clearly visible the mild steel core is partly eroded and shows a small mushrooming effect by enlarging its diameter on the impacting side.

To evaluate the results and the possibly occurring benefits several different evaluation criteria are chosen. With the areal density and thickness on the one hand side two values are chosen to assess the wearing comfort of the protective system based on design and manufacturing parameters. For the comparative evaluation of the protective capabilities on the other hand three ballistic performance parameters are chosen. First the maximum Force value between the protective system and the ballistic clay material, representing the human body. Secondly the BFS (back face signature) of the protective system is chosen to evaluate the plastic intrusion of the protective layer. It is estimated from the displacement plot of the backside of the protective layer without trauma reduction layer. Lastly the BABT (behind armour blunt trauma) of the protective system is chosen to evaluate the dynamic intrusion of the protective layer. It is determined from the displacement plot of the backside of the protective layer with trauma reduction layer thickness included if installed. The main goal by setting the evaluation criteria is to identify possibly occurring injury reduction potentials in the different approaches and to generate measurable values for the comparison.

For all validation tests of the DYNEEMA® HB 212 we rely on the worst-case scenario results regarding the number of penetrated layers. The results were obtained with the 7,62 × 39mm surrogate from "Sellier & Bellot" described within the CAST 2017 [13] to achieve the most aggressive test configuration. In the simulation the 92 plies are represented by eight layers. Each of the layers consists of two of finite elements in thickness direction, which means each element represents 5,75 plies of DYNEEMA HB 212. The simulation results with number of penetrated layers for validation purposes is shown in figure 10.

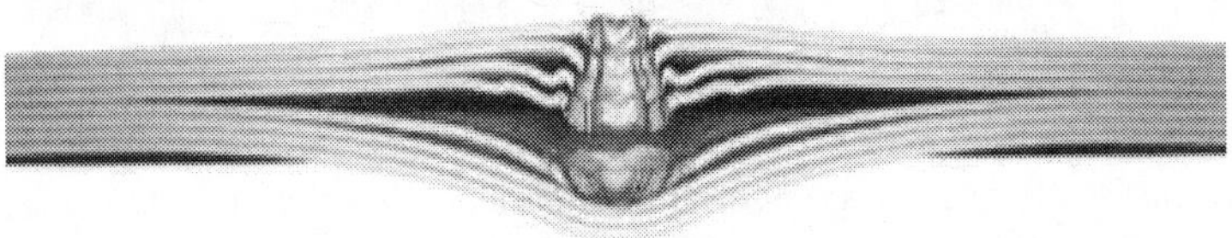

Figure 102 Simulation result of 92 plies DYNEEMA with number of penetrated layers

In the test setup a mean value of 54,28 penetrated layers occurred, which is equal to 59,0% of the pressed UHMWPE fibre structure. In the simulation the material model was therefore iteratively adapted to provide a mean value of 57,50 penetrated layers, which is equal to 62,5% for extra safety margin. The overall surface behaviour of the DYNEEMA® HB 212 panel is shown in figure 17 together with the damage behaviour of the SKYDEX® BTRL behind 92 plies of DYNEEMA® HB 212 after the impact of the 7,62 × 39mm projectile.

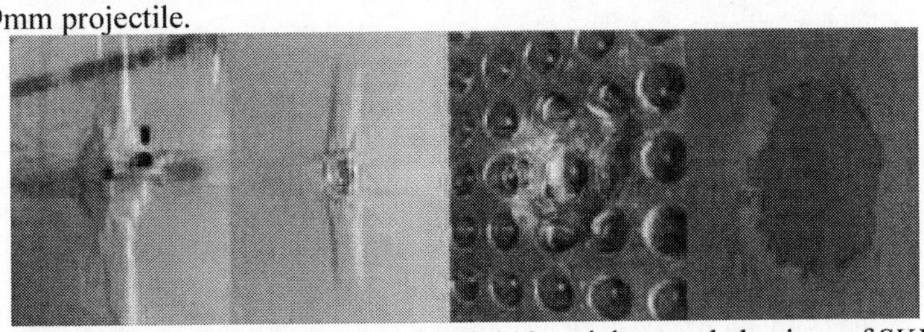

Figure 103 Surface behaviour of DYNEEMA® HB 212 and damage behaviour of SKYDEX® BTRL

On the strike-face of the DYNEEMA® HB 212 panel (left picture in figure 11) a change in the structure in the 0°/90° fibre orientation is noticeable. This elongation effect of the most heavily loaded fibres was also determined in the simulation regarding shape and size (second to left picture in figure 11). The SKYDEX® ballistic trauma reduction layer shows maximum plastic compression of the structural deformation chambers behind 92 plies of DYNEEMA® HB 212 after the impact of the 7,62 × 39mm projectile (second to right picture in figure 11). This amount of deformation and the affected elements size is not suited for explicit dynamic simulation, due to the inversely proportional behaviour of the element edge length to the possible time step size. Therefore, heavily compressed elements of the SKYDEX® BTRL will be eroded and deleted from the simulation (right picture in figure 11).

3.1 Monolithic Steel approach (Domex Protect 500)

The first classic approach numerically investigated is the monolithic steel approach, represented by Domex Protect 500 material in a thickness of 4,00mm, resulting in an areal density of 31,40kg/m². In figure 12 the general simulation setup is shown together with the mesh density in the impact region.

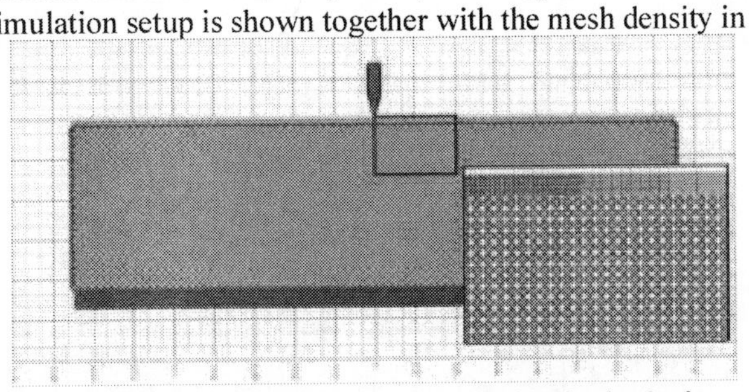

Figure 104 General simulation setup and mesh density in the impact region

In the impact area a mesh density of 0,67mm in thickness direction and 0,83mm in plane was chosen to achieve sophisticated material and failure behaviour of the armouring material. In figure 13 a row of six

https://doi.org/10.52202/080042-0003

pictures, representing six characteristic time frames within the simulation are shown, to give a better feeling of the course of the events.

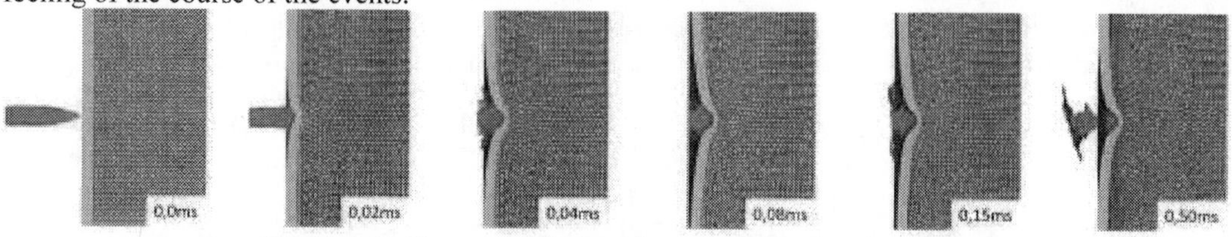

Figure 105 Characteristic time frames within the simulation

To cover all characteristic time frames within the simulation a description of the associated events was prepared and is shown in table 1.

Table 1. Associated events within the simulation of Domex Protect 500

Time	Investigation
0,00ms	Start of the simulation, the initial projectile velocity is 720m/s.
0,02ms	The tip of the projectile is already eroded and the deformation of the mild steel core has begun. The enlargement of the cross section of the projectile (mushrooming) is visible, as the local deformation of the monolithic steel plate. The projectile velocity is ~ 690m/s.
0,04ms	The deformation of the mild steel core is completed, resulting in a noticeable cross section enlargement of the projectile (mushrooming). The jacket flows off on the outside of the projectile and tears open multiple times. The projectile velocity is ~ 250m/s.
0,08ms	The deformation of projectile and armour plate are completed. Starting of the rebound of the monolithic steel plate and the projectile which has therefore a velocity of ~ 0m/s.
0,15ms	The projectile detaches from the decelerating plate.
0,50ms	End of the simulation

With an areal density 31,40kg/m² this approach leads to a SAPI-plate weight (Small Arms Protective Insert) in medium size (241 x 318 mm) [14] of 2,41kg for the wearer. Covering his chest and back a system like this would lead to a total weight of 6,32kg, assuming the plate carrier weights 1,50kg.
The force maximum which is transferred from the protective system to the ballistic clay was 77,86kN. Together with the BABT as dynamic deformation of the protective system and the BFS as plastic deformation this value is taken as reference value for the evaluation of the further investigated approaches. Considering the 44,00mm BABT threshold value for the permitted dynamic deformation of a body armour the achieved 6,43mm and the estimated ~ 3,00mm for the BFS seems to be very reasonable results and be taken as reference for the other approaches. Since steel is an extremely hard material compared to the ballistic clay and to the human body in general it is believed that one main goal of this study has to be the lowering of the transferred force maximum to further reduce the BABT injury probabilities and severities.

3.2 UHMWPE-Plate approach (DYNEEMA® HB212)

The next classic approach investigated is the UHMWPE-plate made of 92 hot-pressed plies of DYNEEMA® HB212. In figure 14 the general simulation setup is shown together with the mesh density in the impact region.

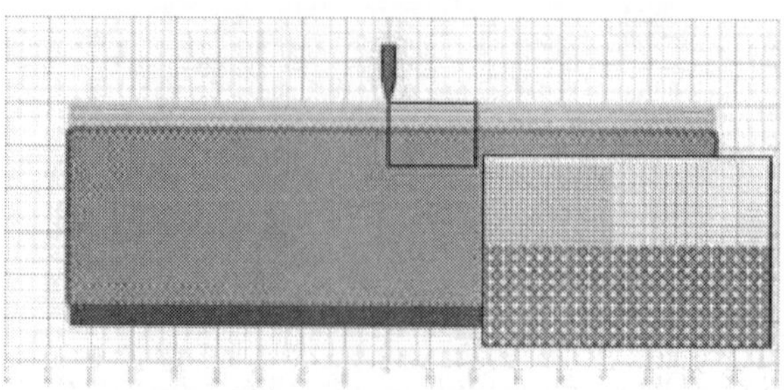

Figure 106 General simulation setup and mesh density in the impact region

In the impact area a mesh density of 0,82mm in thickness direction and 0,83mm in plane was chosen to achieve sophisticated material and failure behaviour of the hot-pressed fibre material. In figure 15 a row of six pictures, representing six characteristic time frames within the simulation are shown, to give a better feeling of the course of the events.

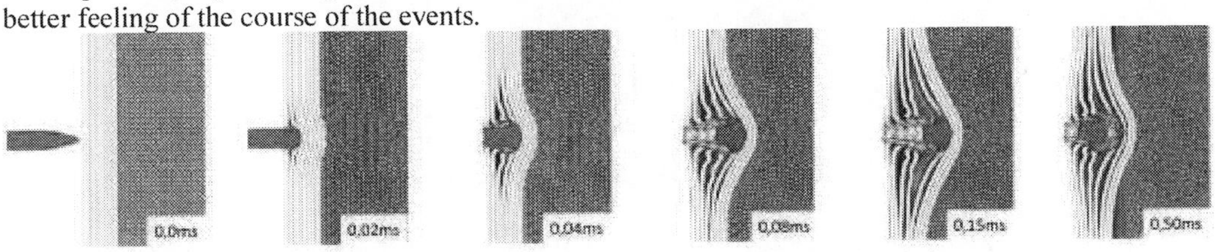

Figure 107 Characteristic time frames within the simulation

To cover all characteristic time frames within the simulation a description of the associated events was prepared and is shown in table 2.

Table 2. Associated events within the simulation of DYNEEMA® HB 212

Time	Investigation
0,00ms	Start of the simulation, the initial projectile velocity is 720m/s.
0,02ms	Penetration of the first layers of DYNEEMA® with visible tip erosion of the lead filler and beginning deformation of the mild steel core. The projectile velocity is ~ 650m/s.
0,04ms	The mushrooming and therefore the enlargement of the cross section of the projectile is visible while more and more layers of DYNEEMA® HB 212 are penetrated. This comes in combination with an increased loss in projectile velocity which is now ~ 450m/s.
0,08ms	The deformation of the mild steel core is almost finished, resulting in an enlargement of the cross section of the projectile (mushrooming). This effect leads to a higher penetration resistance, completing the penetration of DYNEEMA® plies at this point. The projectile itself (and therefore also the pressed UHMWPE-plate) has a velocity of ~ 70m/s, leading to further energy consumption by delamination and interlaminar failure.
0,15ms	The deformation of the projectile and the armour plate are completed. Starting of the rebound of the steel plate together with the projectile, which has a velocity of ~ 0m/s.
0,50ms	End of the simulation

With an areal density 12,84kg/m² this approach leads to a SAPI-plate of 0,98kg for the wearer. Covering his chest and back a system like this would lead to a total weight of 3,46kg, assuming the plate carrier weights 1,50kg. The weight saving is possible due to a significantly increased thickness of the 92 Plies of DYNEEMA® HB212 of 13,10mm compared to the 4,0mm thickness of the monolithic steel approach. The transferred force maximum from the protective system to the ballistic clay is 72,29kN in this approach which is just a little less than within the monolithic steel plate. Originally a significant reduction was expected here and is needed to lower the injury probability and severity of BABT. In addition to this the dynamic deformation as direct indicator for BABT doubles from 6,43mm to 13,89mm in comparison with the monolithic steel plate. This applies also for the estimated plastic deformation as BFS that is now ~ 7,00mm.

3.3 Stacked approach (DYNEEMA® HB212 / SKYDEX® Ballistic Trauma Reduction Layer)

The final approach investigated in this technical study is the combination of the UHMWPE-plate made out of 92 hot-pressed plies of DYNEEMA® HB212 and the SKYDEX® Ballistic Trauma Reduction Layer (BTRL). This approach is intended to combine the positive properties of the two classic approaches investigated before and to lower the probabilities and severities of BABT injuries even further than these. In figure 16 the general simulation setup is shown together with the mesh density in the impact region.

https://doi.org/10.52202/080042-0003

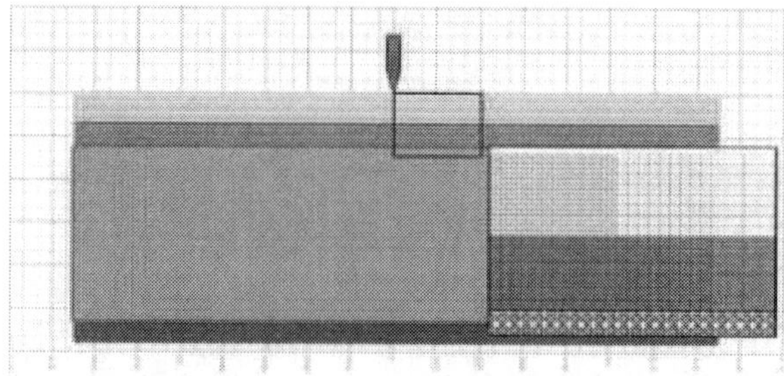

Figure 108 General simulation setup and mesh density in the impact region

In the impact area a mesh density of 0,82mm in thickness direction and 0,83mm in plane was chosen to achieve sophisticated material and failure behaviour of the hot-pressed fibre material and the BTRL. In figure 17 a row of six pictures, representing six characteristic time frames within the simulation are shown, to give a better feeling of the course of the events.

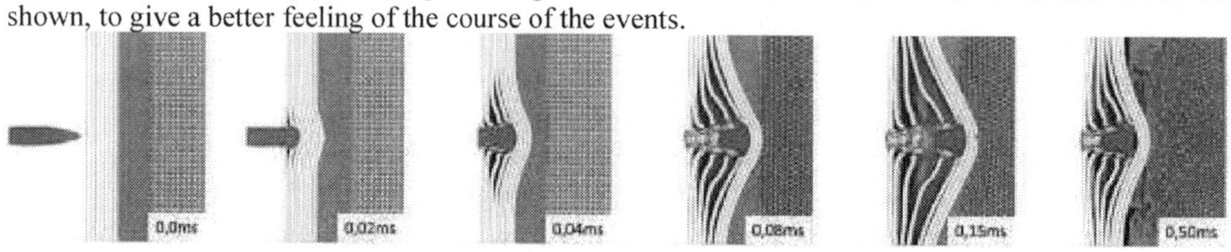

Figure 109 Characteristic time frames within the simulation

To cover all characteristic time frames within the simulation a description of the associated events was prepared and is shown in table 3.

Table 3. Associated events within the simulation of DYNEEMA® HB 212 / SKYDEX® BTRL

Time	Investigation
0,00ms	Start of the simulation, the initial projectile velocity is 720m/s.
0,02ms	Penetration of the first layers of DYNEEMA® with visible tip erosion of the lead filler and beginning deformation of the mild steel core. Local deformation of the SKYDEX® BTRL. The projectile velocity is ~ 660m/s.
0,04ms	The enlargement of the cross section of the projectile (mushrooming) is visible while more and more layers of DYNEEMA® HB 212 are penetrated. The projectile velocity is ~ 480m/s. Further deformation of the SKYDEX® BTRL with first erosion of elements.
0,08ms	The deformation of the mild steel core is almost finished, resulting in a noticeable enlargement of the cross section of the projectile. This effect leads to a higher penetration resistance, which is the reason why the penetration of DYNEEMA® plies is completed at this point. The projectile (and therefore also the pressed UHMWPE-plate) has a velocity of ~ 130m/s at this point, leading to further energy consumption by delamination and interlaminar failure. Apparently a slower deacceleration of the projectile leads to smaller forces on the body of the wearer of the protective system.
0,15ms	The deformation of the projectile and the armour plate are completed. Starting of the rebound of the steel plate together with the projectile which has therefore a velocity of ~ 0m/s. The remaining layers hold the mushroomed projectile, thus further enlargement of the impact area. The erosion of elements from the SKYDEX® BTRL is now completed.
0,50ms	End of the simulation

With an areal density 18,74kg/m² this approach leads to a SAPI-plate of 1,44kg for the wearer. Covering his chest and back a system like this would lead to a total weight of 4,38kg, assuming the plate carrier weights 1,50kg. This is 1,85kg less than the monolithic steel approach, which will have positive effects for the wearer regarding moveability, agility and endurance. Otherwise, this amount of saved weight could be substituted by gear, ammunition or other equipment.

The BABT is with its 6,73mm of dynamic deformation on the level of the monolithic steel plate like the BFS with the estimated plastic deformation of ~ 3,00mm as well which were ambitious targets, but it is possible to achieve them with this configuration of the protective system. In addition, it has been possible to reduce the maximum force by more than 50% to 35.91kN. For this purpose, the force-time curves transferred from the protective system to the ballistic clay are shown in Fig. 18 (left chart).

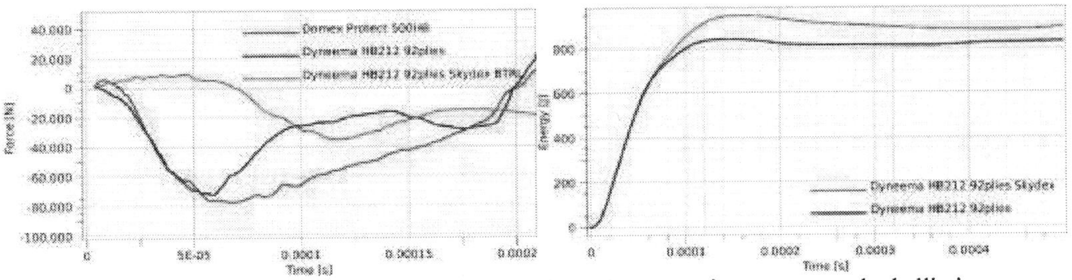

Figure 110 Force-time curves transferred from the protective system to the ballistic

By using the SKYDEX® BTRL as an additional layer between the 92 layers of DYNEEMA® HB212 and the ballistic clay, the force curve increases much later and the slope is generally less steep, which results from the slower deceleration of the projectile, since it is now decelerated over a longer distance and time. Furthermore, the locally impacting projectile energy is now distributed over a larger area using SKYDEX® BTRL. Due to the increased distance to the body, the protective capacities of the UHMWPE fibres can be used more effectively, and more energy can be absorbed through the inter- and intralaminar interactions within the UHMWPE plate. This is also illustrated in Fig. 18 (right chart) in which the combined elastic energy, plastic work and delamination energies for the UHMWPE DYNEEMA® HB212 layers of the two approaches are shown in comparison.

4. RESULTS AND CONCLUSION

In general, very positive results were achieved, all of which are within the permissible limit values of the test guidelines used. Following the results achieved in the simulations for the three different approaches examined are listed in table 4 below, with the monolithic steel approach serving as reference.

Table 4. Results of the comparative simulations

	Monolithic Steel approach (Domex Protect 500)		**UHMWPE-Plate approach** (DYNEEMA® HB212)		**Stacked approach** (DYNEEMA® HB212 / SKYDEX® BTRL)	
Areal density	31,40 kg/m²	100,00%	12,84 kg/m²	40,90%	18,74 kg/m²	59,70%
Thickness	4,00 mm	100,00%	13,10 mm	327,50%	26,10 mm	652,50%
Force maximum	77,86 kN	100,00%	72,29 kN	92,80%	35,91 kN	46,10%
BFS	~ 3,00 mm	100,00%	~ 7,00 mm	233,30%	~ 3,00 mm	100,00%
BABT	6,43 mm	100,00%	13,89 mm	216,02%	6,73 mm	104,67%

By combining DYNEEMA® HB212 and SKYDEX®, it was possible to develop an approach with a significantly reduced areal density in contrast to the monolithic steel plate, which enables despite of it the same low deformation values. Compared to the pure UHMWPE plate approach (DYNEEMA® HB212), the areal density and the overall thickness of the protective structure increase by 5,90 kg/m² and 13,00mm due to the additional ballistic trauma reduction layer (BRTL). However, this reduces the maximum contact force between the protective structure and the body (represented by the ballistic clay) by ~ 50% (36,38kN) compared to the UHMWPE plate approach, which further minimizes the risk of injury for the wearer of such vests. Further ballistic tests with the stacked approach (DYNEEMA® HB212 / SKYDEX® BTRL) are currently being carried out to confirm the initial positive results and establish this combined DYNEEMA® HB212 / SKYDEX® BTRL approach as a sophisticated solution for further development in personal armour systems.

References

[1] Prat, Nicolas, Frederic Rongieras, JeanClaude Sarron, Alain Miras und Eric Voiglio (2021). Contemporary body armor: technical data, injuries, and limits. BrétignysurOrge.
[2] Carr, Debra, Ian Horsfall und Chris Malbon (2016). Is behind armour blunt trauma a real threat to users of body armour? A systematic review. Cranfield: Cranfield University.
[3] Cannon, Leslie (2001). Behind Armour Blunt Trauma an emerging problem. Cosham.
[4] Olovsson, Lars. 2018. RE: General information about IMPETUS. [E-Mail] Huddinge, Schweden : s.n., 4. Juni 2018. 09:58.

 https://doi.org/10.52202/080042-0003

[5] IMPETUS Afea AS (2021). MAT_METAL. URL:
https://www.impetus.no/support/manual/?command=MAT_METAL (Retrieved: 16-11-2021).
[6] CADFEM GmbH (2016). Composite. URL: http://www.cae-wiki.info/wikiplus/index.php/Composite (Retrieved: 24-09-2021).
[7] IMPETUS Afea AS (2021). MAT_FABRIC. URL:
https://www.impetus.no/support/manual/?command=MAT_FABRIC (Retrieved: 27-10-2021).
[8] IMPETUS Afea AS (2021). MAT_VISCOUS_FOAM. URL:
https://www.impetus.no/support/manual/ ?command=MAT_VISCOUS_FOAM (Retrieved: 17-12-2021).
[9] Vereinigung der Prüfstellen für angriffshemmende Materialien und Konstruktionen (2009). Prüfrichtlinie „Ballistische Schutzwesten ". Münster.
[10] Vereinigung der Prüfstellen für angriffshemmende Materialien und Konstruktionen (2010). Allgemeine Prüfgrundlagen für ballistische Material, Konstruktions und Produktprüfungen. Münster.
[11] Vereinigung der Prüfstellen für angriffshemmende Materialien und Konstruktionen (2021). ANSCHLUSSDOKUMENT zur VPAM–APR „Prüfmittel ". Münster.
[12] Precision Ballistics Laboratories UK (2022). 'Mild Steel' cores from 7.62x39mm PS Ball (MSC) bullets. URL: https://www.linkedin.com/posts/precision-ballistics-laboratories-uk_pbl-precision-ballistics-activity-7037429715405533184-ZMjl/?utm_source=share&utm_medium=member_desktop (Retrieved: 21-03-2022).
[13] Payne, Thomas, Sarah O'Rourke und Chris Malbon (2017). Body Armour Standard (2017). Sandridge: Centre for Applied Science und Technology (CAST).
[14] BAE Systems (2007). Hard Body Armor Insert Plates, Phoenix, AZ USA (Retrieved: 2008-11-11)

https://doi.org/10.52202/080042-0003

Effect of Backing Material Stiffness on Ballistic Performance of Ceramic/UHMWPE Personal Body Armour

Naresh Bhatnagar[1], Kartikeya Kartikeya[1], Dhruv Narayan[1], Hemant Chouhan[1], Khushi Ram[1] and Aisha Ahmed[1]

[1]*Department of Mechanical Engineering, Indian Institute of Technology Delhi, New Delhi, India*
nareshb@mech.iitd.ac.in

Abstract. Ceramics erode hardened steel core (HSC) projectile of a bullet and ultrahigh molecular weight polyethylene (UHMWPE) composite catches the eroded projectile in a Ceramic/UHMWPE composite armour plate. Another important role of the backing UHMWPE composite is to provide sufficient stiffness to the ceramic to allow comminution. In this study, the stiffness of the backing UHMWPE composite is varied by changing the consolidation pressure while keeping the prepreg material grade and number of plies the same. High-velocity impact tests of all the configurations were conducted on a single-stage gas gun using AK 47 7.62 × 39 mm HSC projectiles (Kirkee bullets) at a velocity of 700 ± 15 m s^{-1} using helium gas. The only difference in the impacting projectile was the lack of a rifling effect in this gas gun. The effect of stiffness on the ballistic performance of ceramic/UHMWPE composite is characterized using a novel contraption by mounting an impact force sensor at the back of armour panel fixation inside the gas gun vacuumized target chamber. The peak force vs stiffness relation helps understand the future personal body armour design requirements and manufacturing insights.

1. INTRODUCTION

A bullet fired from small arms generally contains a core. The core can be manufactured from, lead, mild steel, hardened steel or tungsten carbide [1–3]. A 7.62 × 39 mm hardened steel core (HSC) projectile is classified at threat level 5 out of six threat levels in the Indian personal body armour standard i.e., IS 17051:2018 [4]. This projectile contains a hardened steel core with a hardness of up to 45 HRC. Composite armour is required to defeat the 7.62 × 39 mm HSC threat where ceramic erodes the projectile's core and backing of ultrahigh molecular weight polyethylene (UHMWPE) composite catches the eroded projectile [5]. Boron carbide (B_4C), silicon carbide and aluminium oxide are popular choices to fabricate ceramic/UHMWPE composite armours which can defeat similar threats [6–10].

Wilkins tested an alumina/aluminium composite armour in ballistic impact [11]. It was noticed that as alumina's fracture conoid extended to the interface between it and aluminium, the aluminium experienced maximum compressive force in the line of impact and deformed. The deformed aluminium then separated from the alumina resulting in a tensile stress state in the ceramic. The tensile stresses failed the ceramic. Thus, the failure of a ceramic depends upon the stiffness of the backing material. The stiffer material resulted in a greater time delay in the breaking up of ceramic [1,11]. Wang et al. [12] designed a functionally graded armour to defeat 12.7 mm AP projectiles realizing the importance of stiffness of the backing material [12]. The armour had four layers i.e., a ceramic layer to blunt the projectile's nose, an intermediate metallic layer to provide stiffness to the ceramic, another intermediate layer of lightweight UHMWPE composite to arrest deflection of the metallic layer and a final metallic layer to reduce overall back face deformation of the armour plate.

Savio et al. [13] explained the failure mechanisms of hard steel core projectiles after their ballistic impact with B_4C tiles. Two different mechanisms were noted. First, failure originated from the target-projectile interface due to very high contact stresses that was responsible for the erosion of the projectile. This mechanism remains active till the ceramic cracks and is thus dependent on the stiffness of the ceramic's backing material. Second, failure also originated from the back of the projectile which was responsible for the breaking of the projectile. Although, the reason for the second failure mechanism could not be determined.

Puente et al. [14] tried to determine the optimum thickness of alumina/aluminium composite armour designed to defeat 7.62 mm tungsten carbide core projectiles. They observed that the ceramic's damage and failure were less with thinner adhesive layers. However, ceramic detached from aluminium when a very thin adhesive layer was used, reducing the multi-hit capability of the armour. The authors suggested an optimum thickness of 0.3 mm for their armour. Seifert et al.[15] tested the effect of adhesives, inter-tile gap width and impact location on the ballistic performance of a ceramic/metal armour impacted with a tungsten carbide projectile [13]. The ballistic limit velocity was higher in the case of an epoxy-based adhesive than in the case of a modified polyurethane-based adhesive. In the case of both adhesives, an increase in inter tile gap resulted in higher residual velocity and lower damage to

 https://doi.org/10.52202/080042-0004

the projectile. It was also observed that shots on the tile's edges resulted in higher residual velocity than shots in the middle of the inter-tile gap.

Liu et al. [16] pointed out that pressure and time of application of pressure are crucial for the manufacturing of composite laminates. They suggested minimal changes in the temperature schedule suggested by manufacturers of polymers during manufacturing as that can drastically alter the structural properties of polymers as is also suggested by Zeng et al. [14]. They fabricated carbon/epoxy composites at five levels of pressure using autoclave moulding. The void contents decreased as pressure increased enhancing the mechanical properties of the composites [15]. Greenhalgh et al. [18] also observed that fabrication temperature and pressure played a key role in the impact performance of UHMWPE composite materials. In their study UHMWPE composites were at two different pressures i.e., 165 bar and 300 bar. They noticed that UHMWPE composite fabricated at higher pressure had better ballistic properties. Chocron et al. [19] noted an increase in mechanical properties of UHMWPE composites with the increase in applied confinement pressure during testing. Lassig et al.[20] explained that increased consolidation pressure reduces void density, cracking in the matrix, fibre-volume fraction of composite and fibre-fibre bonded joints. They observed UHMWPE composites fabricated at higher pressure had significantly improved ballistic limit, but the limit of higher pressures is still unknown.

Zulkifli et al.[21] strategically placed carbon fibre fabric into UHMWPE fibre-reinforced composites at different locations. The flexural modulus of composites increased for configurations where carbon fibre fabric was loaded in compression as it is much stiffer than UHMWPE fibre in compression. The configuration with carbon fibre fabric layers in front of the UHMWPE composite performed best in ballistic tests. Zhang et al. [22] also noticed a similar effect in B_4C/carbon-epoxy/UHMWPE composite armour.

It is evident from the relevant literature survey that a study on the understanding of the effect of consolidation pressure on the ballistic performance of ceramic/UHMWPE armour is still absent. Thus, in this preliminary study, the consolidation pressure of backing UHMWPE composite is changed leaving all other parameters at the same values. The fabricated armours were then subjected to a high-velocity impact test using a 7.62 × 39 mm HSC projectile in a single-stage gas gun. The cores of projectiles were collected after each test and their residual mass was recorded. The force of the impact was also recorded using a high impact force sensor. The following sections present the results and discussions of the study.

2. MATERIALS AND METHODS

This section explains armour fabrication methodology and ballistic test setup.

2.1 B_4C/UHMWPE Composite Armor

The B_4C/UHMWPE composite armours were fabricated for this study. The B_4C tiles were procured locally in the form of regular hexagons of 6.5 mm thickness (areal density 16.5 kg m^{-2}) and 17 mm edge length and 30 mm edge-to-edge distance as suggested by lead ceramic tiles suppliers. These hot-pressed B_4C tiles were known to have better ballistic efficiency than reaction bonded B_4C tiles [16].

The UHMWPE cross-ply fabric was procured from Honeywell International Inc, USA. First, several plies of UHMWPE cross-ply fabric of 200 mm × 200 mm dimensions were cut. These plies were then placed in a preheated mould to fabricate a laminated composite (areal density 8 kg m^{-2}) under pressure according to the manufacturer's recommended cycle. The fabrication pressure was varied as 250 bars, 500 bars and 750 bars which are designated further in the text as low pressure (LP), medium pressure (MP) and high pressure (HP), respectively. Three identical UHMWPE composites were fabricated at each pressure making a total of 9 armour plates.

The B_4C tiles were placed on prefabricated UHMWPE composite as shown in Figure 1. The joining was done using polyurethane-based adhesive at a maximum temperature of 80 °C. The setup was placed in a vacuum bag and a pressure of 13 bars was applied during autoclave joining.

Figure 1. B_4C tiles adhesively bonded to prefabricated UHMWPE composite in an autoclave

2.2 Ballistic Test Setup

The ballistic tests were conducted on a single-stage gas gun installed at COE-Personal Body Armour Lab at Indian Institute of Technology Delhi (IITD) as shown in Figure 2. Only one impact test was performed on one armour plate in this preliminary study, however there are three armours fabricated at a pressure value. Helium gas was used to propel the projectile in a sabot. The projectile was put in a sabot which was screwed to a high-speed valve. This sabot-valve assembly was put into the reservoir end of the gas gun. Subsequently, the armour plate was placed in the impact chamber. First, the impact chamber is evacuated to 500 mbar of pressure, then a precalculated amount of Helium is filled in the reservoir. In this study, 78 bar of Helium pressure was required to achieve the required velocity of 700 m/s. The high-speed valve was then actuated by pneumatic action which suddenly shears its plastic screw joint with the sabot and the sabot gets accelerated in the barrel by expanding of Helium gas. The sabot gets broken by a sabot trapper in the impact chamber and only the projectile is impacted on the armour plate. The yaw of the projectile was not measured. Further details of the test setup are disclosed in the following reference [5]. After each test, both armour and projectile were studied to understand their deformation and failure mechanisms.

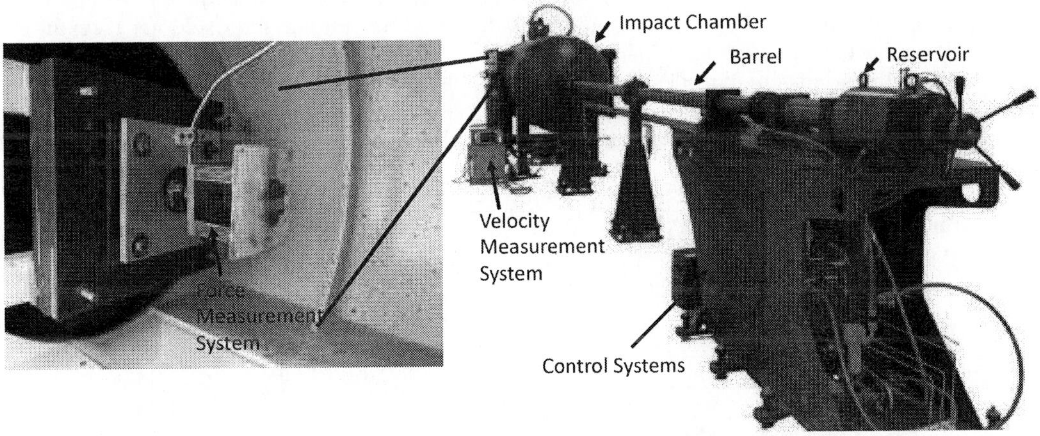

Figure 2. The single-stage gas gun used to conduct high-velocity impact tests

The schematic of the force measurement system (make: Kistler) used in this study is shown in Figure 3. There were two challenges in designing such a system. The first was to protect the force sensor in case of armour perforation and the second was to extract force-time data from a sealed impact chamber using an optical fibre cable. Figure 3 shows a schematic of the force measurement system where it is apparent that the impacted force is transmitted to the force sensor using columns. These columns allow proper transmission of force and allow the back of the armour to deform freely. The second challenge was solved by designing and fabricating a feed-through system that allowed the optical fibre cable to pass through without breaking the vacuum seals.

https://doi.org/10.52202/080042-0004

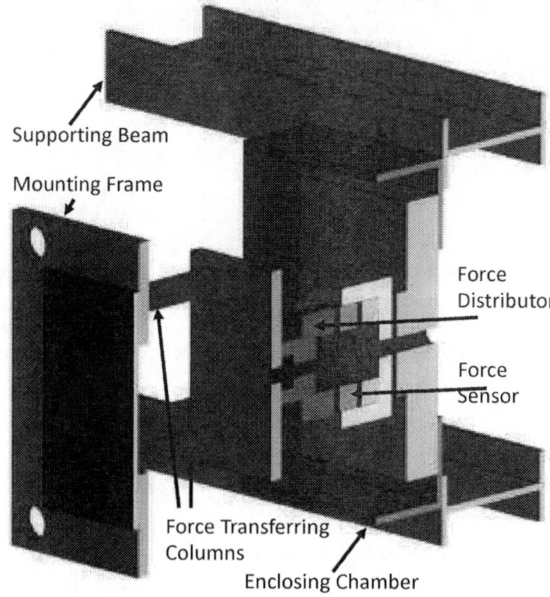

Figure 3. Schematic of force measurement system

A 7.62 × 39 mm HSC projectile was impacted on the fabricated B_4C/UHMWPE composite armours. This projectile weigh 7.5 g and its core weighs 3.5 g. The length of the core is 17.8 mm. A total of 9 tests were performed and analysed.

3. RESULTS AND DISCUSSIONS

The results of high-velocity impact tests are summarized in Table 1. The tested velocities were in the expected range i.e., 700 ± 15 m s^{-1}. All the bullet impact forces were measured as shown in Table 1. All the cores were recovered except one as it got stuck deep into the armour and could not be extracted without damaging the armour. Figure 4 shows the front and back views of the armours after testing. The impact resulted in a separation of tiles from the UHMWPE composites. The impacted tiles along with adjacent tiles were damaged. The UHMWPE composite behind the tiles was deformed. The middle of each edge of the UHMWPE composite was also drawn in slightly as apparent in Figure 4.

Table 1. Summary of high-velocity impact tests

Armour Type	Velocity (m s^{-1})	Impact Force (Max in kN)	Result	Residual Core Weight (g)	Backface Deformation (mm)
HP A	717	25.6	Not perforated	*	29.98
HP B	706	33.6	Not perforated	2.29	24.28
HP C	715	34.3	Not perforated	2.56	23.88
MP A	709	32.6	Not perforated	2.57	23.38
MP B	714	30.9	Not perforated	2.48	23.18
MP C	706	36.9	Not perforated	2.59	23.68
LP A	712	16.8	Perforated	2.64	22.28
LP B	711	32.8	Not perforated	2.37	24.18
LP C	708	32.3	Not perforated	2.50	24.38

*Core could not be recovered

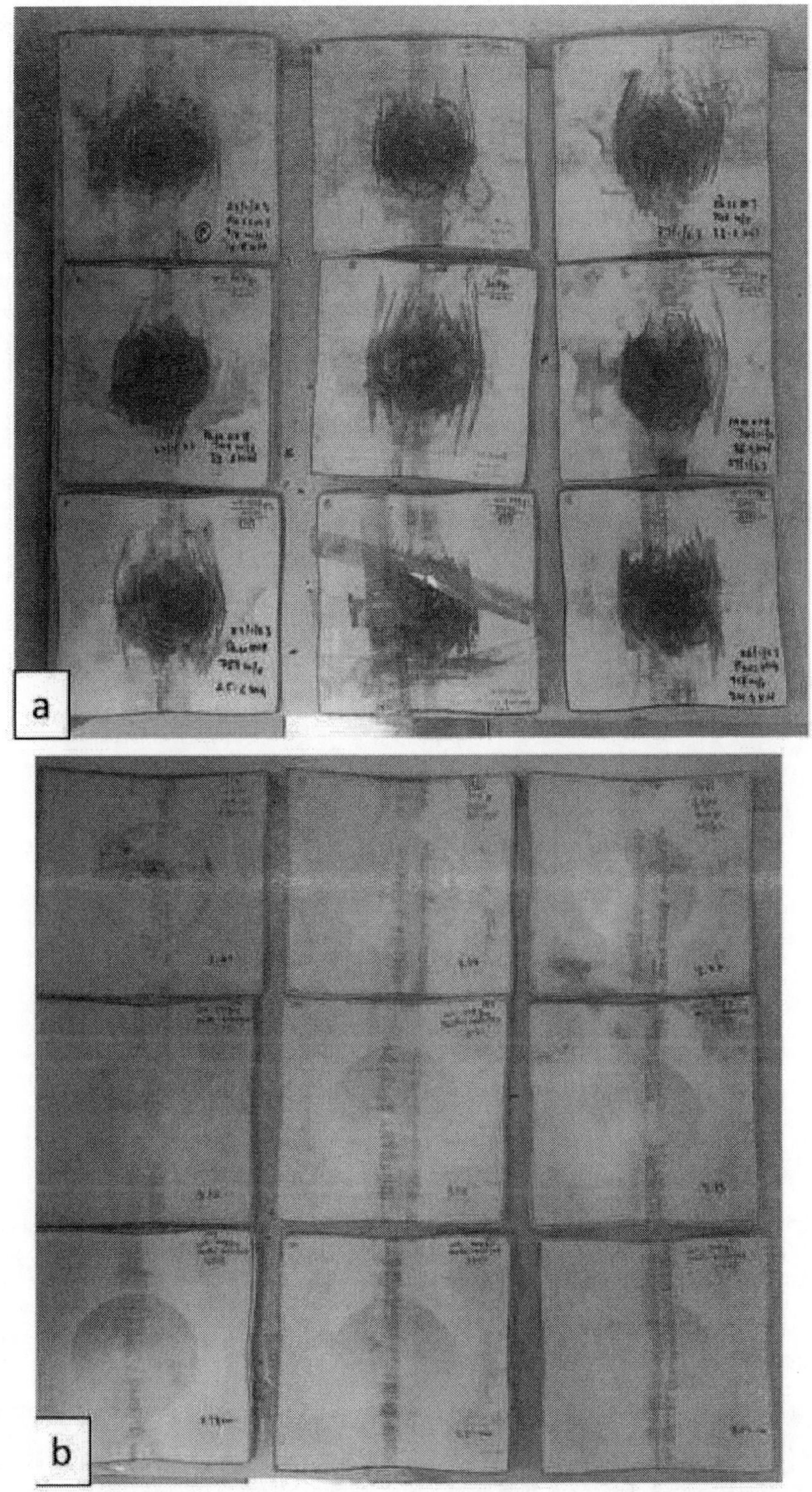

Figure 4. Images of armour after testing (a) Front view and (b) Back view

Only one perforation observed in the study. An armour with LP backing was perforated there were no perforations in MP and HP backings. Although only one perforation was observed out of three tested LP-backed armours, it can be suggested that B_4C/UHMWPE composite armour with LP backings cannot reliably always defeat the threat. The perforated armour had a back face deformation (BFD) of 22.28 mm. The UHMWPE composite defeats a projectile by membrane resistance. Generally, BFD is

https://doi.org/10.52202/080042-0004

lower when the projectile perforates the armour. The low BFD in case of perforation can be attributed to lower engagement time of the projectile with the armour.

Figure 5 shows the bullet impact force time history as recorded from the force dynamometer mounted in the impact chamber. It shows the data from eight tests as for one test i.e., HP B armour, only peak force was recorded due to initial technical issues. The force–time curves are similar for all the cases except two i.e., one perforation of LP-backed armour and one HP-backed armour. Thus, there isn't a significant difference in the recorded forces when a projectile is defeated by the armour. In case the projectile perforates the armour, it engages less with the armour and thus the recorded force amplitude is comparatively less. Also, the force recorded in one of HP-backed armour is less due to turning of the projectile away from the impact direction. In this case, the projectile perforated the B_4C tiles but turned away from the impact direction and got embedded deeply into the armour. The higher BFD was attributed to a turning of the projectile after penetrating the armour not off-axis impact. As the projectile turns it loses its momentum and its perforating capability. However, it engages for higher time with the armour which may result in higher BFD.

The average force recorded in the current study for cases where armour plates defeated the HSC projectile is 32.4 kN. It can be appreciated that this force is high enough to cause significant behind-armour blunt trauma to the wearer of this armour [17]. The time for reaching peak force is just about 100 microseconds (μs) for almost all the backing stiffness.

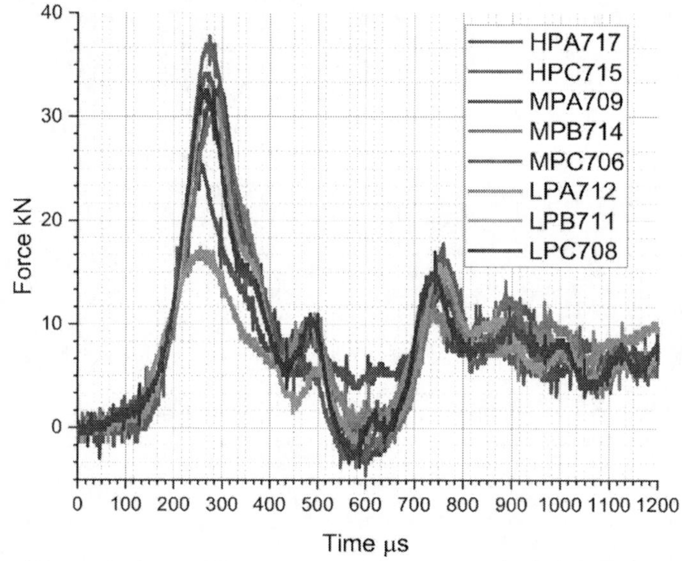

Figure 5. Force (kN) – time (μs) history from the ballistic tests

Figure 6 shows the residual HSC cores after ballistic tests. The erosion of cores is apparent in Figure 6. The first core in Figure 6 (a) is the one which perforated the armour. It can be noticed from Table 1 that this core has the highest weight i.e., 2.64 g. However, the erosion is random and approximately similar for all cases. Since there is no appreciable difference between the residual weights of cores for all defeated cases, it can be concluded that an armour fabricated with either MP or HP backing is just suitable to defeat a 7.62 × 39 mm HSC projectile.

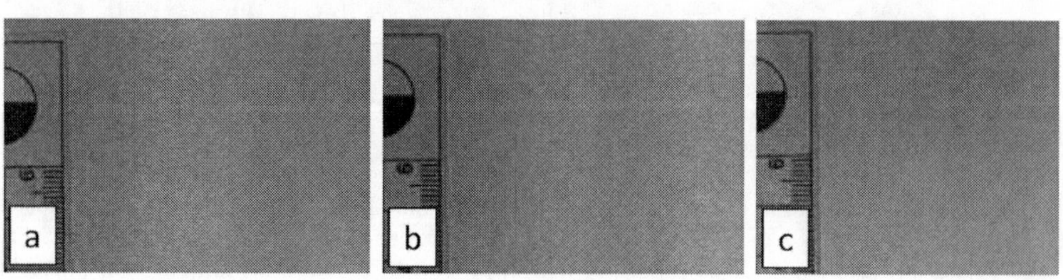

Figure 6. Residual HSC cores after ballistic test on composite armours with (a) LP backing, (b) MP backing and (c) HP backing

4. CONCLUSIONS

In this preliminary study, the effect of consolidation pressure of UHMWPE composite on the ballistic performance of B_4C/UHMWPE composite armour was studied. A total of nine ballistic tests were conducted with three replicates of three consolidation pressures. The following can be suggested from this study:

- Only one perforation was observed. The armour fabricated with LP backing had perforation. Thus, a UHMWPE composite fabricated at LP pressure may not be suitable for manufacturing B_4C/UHMWPE composite armour plates. However, further tests may need to be conducted at different velocities to confirm this observation.
- The force–time history is always similar for the case where armour plates defeat the HSC projectile.
- On analysis of residual cores, no significant difference between MP and HP pressure consolidated UHMWPE composite-backed armours was observed, which indicates stiffness of a backing is vital for the success of an armour plate. Thus, either MP or HP pressure consolidated UHMWPE composite backing should be used in mass manufacturing, even if it's not going to be cost effective to the industry.
- A higher tonnage press is required for fabricating either HP or MP pressure consolidated UHMWPE composite plates as backing material stiffness do plays an important role in the success of personal body armour.
- Also, the variation in results can be due to the variation in performance of the projectile and further studies may be conducted.

Acknowledgments

The authors acknowledge the use of manufacturing and testing facilities at the Centre of Excellence in Personal Body Armour at the Indian Institute of Technology Delhi.

References

[1] Crouch IG. The Science of Armour Materials. 1st ed. Woodhead Publishing Limited; 2017.

[2] Hazell PJ, Appleby-Thomas GJ, Philbey D, Tolman W. The effect of gilding jacket material on the penetration mechanics of a 7.62 mm armour-piercing projectile. Int J Impact Eng 2013;54:11–8. https://doi.org/10.1016/j.ijimpeng.2012.10.013.

[3] Peroni L, Scapin M, Fichera C, Manes A, Giglio M. Mechanical properties at high strain-rate of lead core and brass jacket of a NATO 7.62 mm ball bullet. EPJ Web of Conferences, vol. 26, 2012. https://doi.org/10.1051/epjconf/20122601060.

[4] Bureau of Indian Standards. Textiles — Bullet Resistant Jackets — Performance Requirements. IS 17051 : 2018 2018.

[5] Kartikeya K, Chouhan H, Ram K, Prasad S, Bhatnagar N. Ballistic evaluation of steel/UHMWPE composite armor system against hardened steel core projectiles. International Journal of Impact Engineering 2022;164:104211. https://doi.org/10.1016/j.ijimpeng.2022.104211.

[6] Medvedovski E. Ballistic performance of armour ceramics: Influence of design and structure. Part 1. Ceramics International 2010;36:2103–15. https://doi.org/10.1016/j.ceramint.2010.05.021.

[7] Moynihan, Thomas J.; Chou, Shun-Chin; Mihalcin AL. Application of depth-of-penetration test methodology to characterize ceramics for personnel protection. 2000.

[8] Krishnan K, Sockalingam S, Bansal S, Rajan SD. Numerical simulation of ceramic composite armor subjected to ballistic impact. Compos B Eng 2010;41:583–93. https://doi.org/10.1016/j.compositesb.2010.10.001.

[9] Crouch IG, Appleby-Thomas G, Hazell PJ. A study of the penetration behaviour of mild-steel-cored ammunition against boron carbide ceramic armours. Int J Impact Eng 2015;80:203–11. https://doi.org/10.1016/j.ijimpeng.2015.03.002.

[10] Savio SG, Madhu V. Ballistic performance evaluation of ceramic tiles with respect to projectile velocity against hard steel projectile using DOP test. Int J Impact Eng 2018;113:161–7. https://doi.org/10.1016/j.ijimpeng.2017.11.020.

[11] Wilkins ML. Third Progress Report of Light Armor Program. Livermore, California: 1968.

[12] Wang Q, Chen Z, Chen Z. Design and characteristics of hybrid composite armor subjected to projectile impact. Mater Des 2013;46:634–9. https://doi.org/10.1016/j.matdes.2012.10.052.

[13] Seifert W, Strassburger E, Dolak M, Schaare S. Experimental study on the dependency of the ballistic performance of tiled ceramic/metal targets on inter tile gap width and projectile impact position. Int J Impact Eng 2018;122:50–9. https://doi.org/10.1016/j.ijimpeng.2018.08.006.

[14] Zeng Z, Guo B, Li M, Li J, Zhou XD. Experimental and simulated investigation of temperature distribution of UHMWPE laminated composites during hot pressing process. J Appl Polym Sci 2018;135. https://doi.org/10.1002/app.45874.

[15] Liu L, Zhang BM, Wang DF, Wu ZJ. Effects of cure cycles on void content and mechanical properties of composite laminates. Compos Struct 2006;73:303–9. https://doi.org/10.1016/j.compstruct.2005.02.001.

[16] Savio SG, Rao AS, Reddy PRS, Madhu V. Microstructure and ballistic performance of hot pressed & reaction bonded boron carbides against an armour piercing projectile. Advances in Applied Ceramics 2019;118:264–73. https://doi.org/10.1080/17436753.2018.1564416.

[17] Allsop DL, Perl TR, Warner CY. Force/Deflection and Fracture Characteristics of the Temporo-parietal Region of the Human Head. JOURNAL OF PASSENGER CARS 1991;100:2009–18.

Impact of mechanical stress on ballistic performance of body armour materials

C. Boettger[1], M. Dombrowski [2], M. Fietkau[1], H. Knoester[3]

[1]*Teijin Aramid GmbH, Kasinostrasse 19-21, 42103 Wuppertal, Germany*
[2] *Pro-Systems S.p.A., Via al Corbé 63/65, 22076 Mozzate, Italy*
[3] *Teijin Aramid BV, Tivolilaan 50, 6824 Arnhem, The Netherlands*

Abstract. Ballistic resistant vests usually are used for a period of 5 – 10 years and in some cases even longer. It is of the utmost importance to the user that the vest, over the course of its wearable lifetime, offers reliable protection in accordance with the performance standard it had originally been designed and certified to.
With the introduction of the NIJ 0101.06 standard in the year 2008 a tumble test was instituted with the intention to provide some indication of the armor's ability to maintain ballistic performance after being exposed to conditions of heat, moisture and mechanical wear. This test requires the tumbling of bullet resistant vests for 10 days at a climate of 65°C and 80% RH, simulating mechanical stress potentially introduced to such vests during real use. While the protocol does not predict the service life of the vest nor does it simulate an exact period of time in the field, the belief is that if the sample armor can still stop a bullet after the tumble test, then the production armor *should* withstand normal use wear and tear and still be strong enough to protect the wearer.
Teijin Aramid utilizes the tumble test to investigate and compare the impact of mechanical stress at elevated temperature and humidity on various armor materials and panel constructions.
During a first series of baseline trials, monolithic ballistic panels were constructed from both woven and Uni-Directional (UD) fabrics made from Teijin Aramid's Twaron® para-aramid as well as several ballistic UDs made from Ultra-High Molecular Weight Polyethylene (UHMWPE). In a second series of tests, several hybrid panel constructions were made from a combination of woven Twaron® fabrics in conjunction with UHMWPE UD. All armor panels were subject to the tumble test in accordance with the NIJ 0101.06 standard. Ballistic testing of panels was then conducted with the 9mm DM41 in both "new" (un-tumbled) and post-tumbled conditions and analyzed utilizing logistic regression.
Test results observed by Teijin Aramid reveal statistically significant differences in tumbler (aging) resistance of the individual ballistic materials. The same holds true for the different hybrid constructions, even though the ratio of woven Twaron® fabric to UHMWPE-UD material content was held constant between them.
Details about the test method will be provided and all ballistic results generated are compared in graphical form using S-curves, followed by final conclusions.

1. BALLISTIC PERFORMANCE MEASUREMENT METHOD

During ballistic limit testing, test articles are repetitively subjected to projectiles in a range of impact velocities. This range is balanced in the sense that both complete and partial perforations are required. Generally, the objective is to measure a specific impact velocity, called V_{50}, for which the probability of observing a complete perforation is 50%. There are methods providing procedures for such testing including methodologies to calculate V_{50}. V_{50} is a good measure to enable calculation of the specific energy absorption, which allows comparison of the ballistic efficiency of different materials. V_{05} is a better indicator to assess the safety margin of body armour. It represents the impact velocity for which the probability of a complete perforation is 5%. We use the method of logistic regression to determine V_{50} and V_{05}, which boils down to applying linear regression to the logarithm of the odds of a complete perforation. The quality of the logistic model can be checked afterwards by a goodness-of-fit test (chapter 1.2), testing how well model predictions mirror observed data.

1.1 Logistic regression

The method of logistic regression allows estimation of the probability of complete perforation $p(v)$ for any impact velocity v. Another advantage is that logistic regression comes with instruments to determine confidence boundaries. By definition, $0 \leq p(v) \leq 1$, and $p(v)$ must have a sigmoidal, non-linear character. Therefore, applying logistic regression, the probability is linearized by transformation to the logit of the probability $\ln(p/1-p)$, which is unbounded to the positive and negative side. The logit is approximated by a straight line where impact velocity is the independent variable:

$$\ln\left(\frac{p}{1-p}\right) = \beta_0 + \beta_1 v, \quad \text{or: } p(v) = \frac{e^{\beta_0+\beta_1 v}}{1+e^{\beta_0+\beta_1 v}} \tag{1}$$

https://doi.org/10.52202/080042-0005

The above, explicit solution for $p(v)$ is a sigmoid function which we henceforth denote as S-curve. Standard linear regression is not possible because, for individual observations, probability p takes on two values ($p = 0$ and $p = 1$) only. Estimation of the parameters β_0 and β_1 can be done by, for instance, maximum likelihood estimation. This method also renders the variances of the estimates (σ_0^2, σ_1^2) and their covariance (σ_{01}). The standard deviation σ of the estimate for the logit $\beta_0 + \beta_1 v$ itself can be expressed in terms of these variances and covariance:

$$\sigma = \sqrt{\sigma_0^2 + 2v\sigma_{01} + \sigma_1^2} \quad (2)$$

The true logit corresponding to velocity v is contained in the interval $\langle \beta_0 + \beta_1 v \pm z_{\alpha/2} \cdot \sigma \rangle$ with probability $1 - \alpha$ and $z_{\alpha/2}$ is the z-score defining how many standard deviations one has to be away from the center of the interval to have a probability of only $\alpha/2$ for the logit to be even further away from the mean. Since the sigmoid function is strictly increasing, the predicted $p(v)$ will be contained (again with probability $1 - \alpha$) within the boundaries:

$$p_{LB}(v) = \frac{e^{(\beta_0+\beta_1 v - z_{\alpha/2} \cdot \sigma)}}{1 + e^{(\beta_0+\beta_1 v - z_{\alpha/2} \cdot \sigma)}} < p(v) < \frac{e^{\left(\beta_0+\beta_1 v + z_{\frac{\alpha}{2}} \cdot \sigma\right)}}{1 + e^{\left(\beta_0+\beta_1 v + z_{\frac{\alpha}{2}} \cdot \sigma\right)}} = p_{UB}(v) \quad (3)$$

Functions $p_{LB}(v)$ and $p_{UB}(v)$ are chosen as confidence bounds of the S-curve. Figure 1 visualizes the S-curve of a ballistic armor, including its confidence bounds. Clearly, the uncertainty in V_{05} is larger compared to V_{50}. Meaningful estimation of V_{05} requires a large amount of measurements, particularly for impact velocities with low probability of complete perforation. Figure 1 also shows other characteristics of the ballistic test results:

VLCP the lowest velocity for which a complete perforation was observed
VHPP the highest impact velocity for a partial perforation was observed
ZMR Zone of Mixed Results ranging from VLCP to VHPP
velocity range where partial and complete perforations alternate

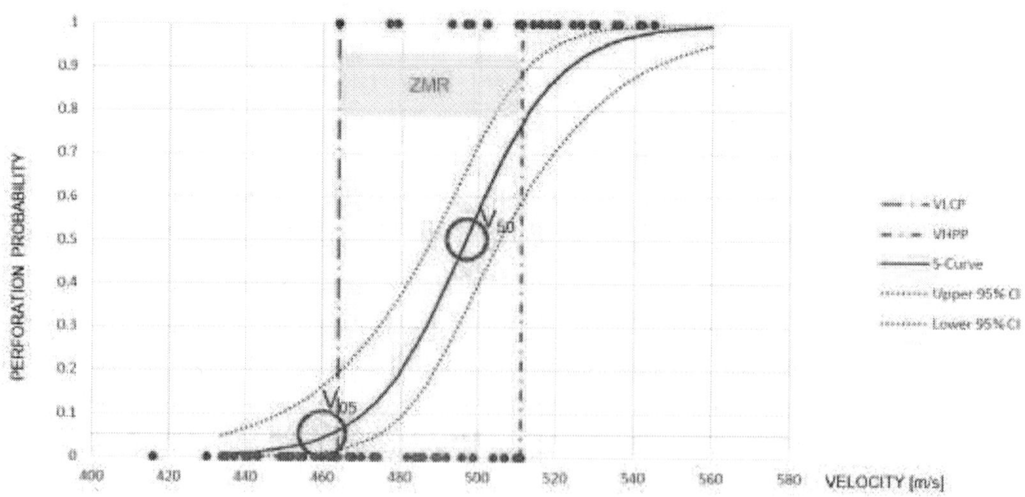

Figure 1. S-curve with 95% confidence intervals and ZMR

Ballistic articles with a steep S-curve, or narrow ZMR, are more desirable, as the confidence bounds for V_{50} and V_{05} will be relatively narrow and therefore the ballistic performance of the armor is well predictable. As can be seen from Figure 2, not just V_{50}, but also the slope of the S-curve determines the safety margin of an armour. While the V_{50} of the blue and yellow armour are the same, the slope of the yellow armour, due to a smaller Zone of Mixed Results (ZMR), is steeper, resulting in a substantially higher V_{05}. The figure also contains V_{ref} and V_{refmax} which illustrate the required and maximum test velocities specified in the official test standard (e.g. VPAM). The figure below illustrates the test velocities typically used for 9 mm DM41. While the yellow armour has a V_{05} substantially higher than

$V_{refmax,}$ the blue one has a V_{05} below V_{refmax}. Even so the V_{50} is similar, safety margin of the yellow armour is much higher compared to the blue armour.

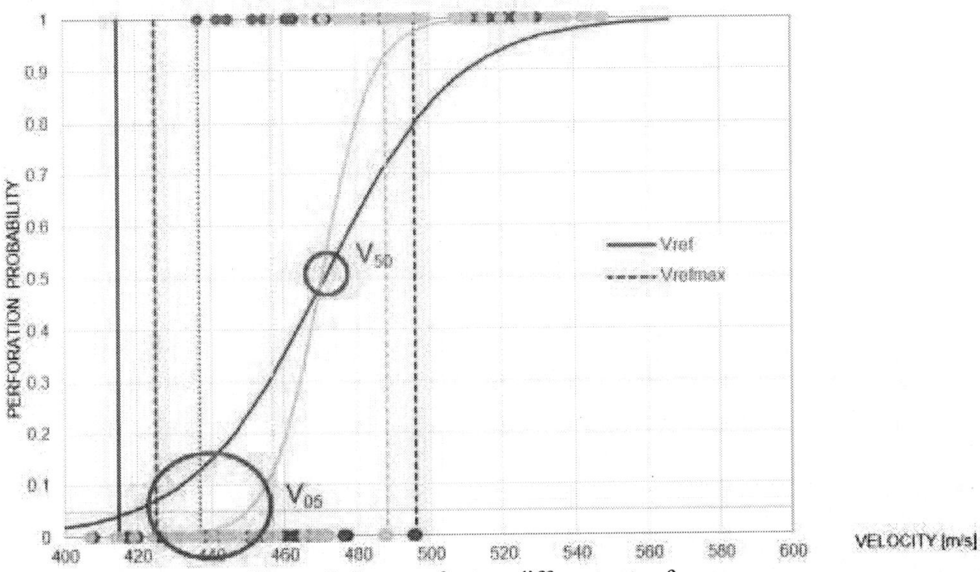

Figure 2. S-curve for two different sets of armour

1.2 Goodness-of-fit test

Standard linear regression minimizes the distance between model predictions and observations and further has the benefit that fit quality is easily visualized. Testing the quality of a logistic regression model is less trivial. Application of logistic regression anyway assumes that the probability of perforation as function of impact velocity is described by a point symmetric S-curve. If the physics of stopping bullets changes for high impact velocity, for instance due to bullet deformation, symmetry of the S-curve may be lost and hence the quality of logistic regression is impaired. In such case one could for example focus on experiments with low to moderate impact velocities only and still use logistic regression. Obviously, such model is then unreliable in predicting probability of perforation for high bullet velocity. Still, the V05 / V50 prediction capability of such a model can be fine. In a general sense, with no prior knowledge on the fit quality of the logistic regression model, the goodness-of-fit of the model can be tested. We apply the Hosmer-Lemeshow test, which was specially developed for this purpose. The expected probability of perforation for each observation is recorded. All observations are grouped in in a finite number of groups with respect to their expected probability of observation. Next, the expected number of observations per group is compared with the actual number of observations. A sum-of-squares test statistic then determines if the distance between 'actual' and 'expected' is small enough in order to accept the fit quality of the model. The logistic regression models in this paper were tested in this way. In all cases the fit quality of the models were found to be acceptable.

2. INFLUENCE OF MECHANICAL STRESS ON BALLISTIC PERFORMANCE OF TWARON® PARA- ARAMID AND UHMWPE

2.1 Tumbler test

Ballistic resistant vests usually are used for a period of 5 – 10 years and in some cases even longer. It is of the utmost importance to the user that the vest, over the course of its wearable lifetime, offers reliable protection in accordance with the performance standard it originally had been designed for and certified to. With the introduction of the NIJ 0101.06 standard [1] in 2008 a tumble test was instituted, with the intention to provide some indication of the armor's ability to maintain ballistic performance after being exposed to conditions of heat, moisture, and mechanical wear. According to this test standard, panels have to be conditioned for 10 days at 65 °C and 80 %RH while being tumbled. The tumbler simulates mechanical stress potentially introduced to ballistic vests during real use. At a tumbling frequency of 0.083 Hz, there are in total 72,000 revolutions during the conditioning period of 10 days. During this

https://doi.org/10.52202/080042-0005

process the ballistic panel is protected by a heat-sealed pouch and an additional garment stitched around the edges. Figure 3 (right) shows how a panel, removed from its garment, looks like after being tumbled for 10 days. Tumbling typically introduces wrinkles and creases. Those can cause lower velocity perforations during ballistic testing, resulting in a flattening of the S-curve. As a consequence, V_{50} and V_{05} may decrease.

While the NIJ 0101.06 protocol does not predict the service life of the vest nor does it simulate an exact period of time in the field, it is expected that if the sample armor can still meet the requirements after the tumble test, then the production armor *should* withstand normal use wear and tear and still be strong enough to protect the wearer. Forster et al. [2] observed that the retained mechanical properties of yarn material after the NIJ 0101.06 conditioning protocol coincides with the retained mechanical properties of the worst performing fraction of yarn materials collected from field-worn armor. The incentive of this study is to better understand how conditioning affects the ballistic performance of woven fabric and UD made from Twaron® Para Aramid, as well as UDs made from UHMWPE.

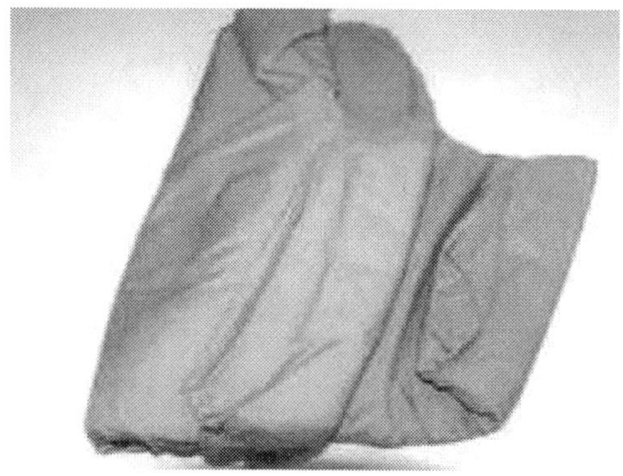

Figure 3. Tumble drum IAW NIJ 0101.06 (left) and tumbled armour panel (right) (reproduced from [3])

2.2 Test program

We tested 4 materials as depicted in the table below. This choice of materials allows a comparison between Twaron and UHMW-PE based UDs as well as a comparison between (Twaron) UD and woven fabrics.

Table 1. Constructions tested

Name	**AD [g/sqm]**	**Material construction**	**Panel construction**	**Test pack AD [kg/sqm]**
Twaron® CT612LS	125	Woven, plain	Quilted	4.5
Twaron® UD	112	2ply UD, no film	Corner tacked	4.2
UHMW-PE UD1*	-	2ply UD, with film	Corner tacked	3.9
UHMW-PE UD2*	-	2ply UD, with film	Corner tacked	3.9

*3rd generation high performance UHMW-PE UDs soured from 2 different manufacturers

2.3 Test procedure

Testing was done using new (as-manufactured) and conditioned ballistic panels. Conditioning was performed in accordance with the NIJ 0101.06 standard, at the Application Competence Center (ACC)

of Teijin Aramid (Wuppertal, Germany). Ballistic testing was conducted at the ballistic shooting range of ACC, using 9 mm DM41 (steel jacketed round). The shooting pattern and sequence were in accordance with a Teijin Aramid method. At least 10 new and conditioned panels were tested per construction. Each panel was shot 8 times, resulting in at least 80 shots per set of samples. The panel size has been 40 x 40cm and minimum shot-to-shot distance was 90 mm. Analysis was done using logistic regression.

2.4. Test results

The logistic regression analysis provides us with estimates for V_{50} and V_{05} before and after conditioning for each material. See Table 2 and Table 3. As can be seen from Table 2, the V_{50} of Twaron® CT612LS woven fabric is not affected by conditioning. For Twaron UD, a small drop of about 5% in V_{50} was observed, whereas the effect on both UHMW-PE UDs is substantially greater (11 – 14% drop in V_{50}). The effect of tumbling on V_{05} is similar to what was found for V_{50}. While Twaron® CT612LS woven fabric does not show any change, V_{05} is substantially reduced for both UHMW-PE UDs.

Table 2. Effect of tumbling on the V_{50} of different materials

Name	V_{50new} **[m/s]**	V_{50cond} **[m/s]**	**Δ [m/s]**	**Δ [%]**
Twaron® CT612LS *woven fabric*	470.3	469.2	-1.1	-0.2%
Twaron® UD	514.2	490.8	-23.4	-4.6%
UHMW-PE UD1	566.5	484.8	-81.7	-14.4%
UHMW-PE UD2	567.5	505.6	-61.9	-10.9%

Table 3. Effect of tumbling on the V_{05} of different materials

Name	V_{05new} **[m/s]**	V_{05cond} **[m/s]**	**Δ [m/s]**	**Δ [%]**
Twaron® CT612LS woven fabric	449.8	450.0	0.2	0.0%
Twaron® UD	480.1	438.0	-42.1	-8.8%
UHMW-PE UD1	516.8	327.1	-189.7	-36.7%
UHMW-PE UD2	507.5	371.8	-135.7	-26.7%

Figure 4 illustrates the S-curves for the new and conditioned materials. Here solid lines are for the new materials and the dashed lines for the conditioned materials.

https://doi.org/10.52202/080042-0005

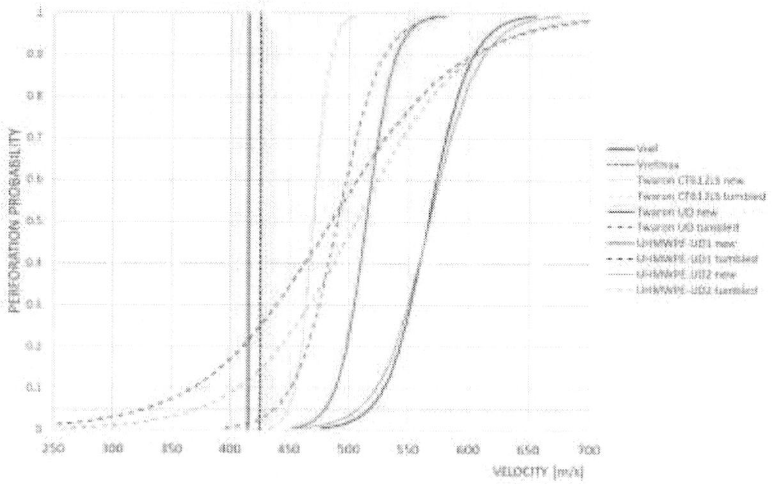

Figure 4. S-curves for Twaron® CT612LS woven fabric, Twaron® UD and UHMW-PE UDs before and after tumbling

The transformation of the S-curves of the tested UHMW-PE UDs after conditioning is eye-catching. Not only do they shift to the left, the flattening is also substantial. This flattening explains that V_{05} is stronger reduced than V_{50}. Twaron® UD also shifts and flattens after conditioning, but these effects are rather mild. Reduction of V_{05} and V_{50} remain below 10 %. It is striking that the S-curve of the Twaron woven material after conditioning is indistinguishable from the S-curve of the new material.

The observed performance change of the UHMW-PE UDs after the NIJ 0101.06 conditioning is statistically significant. Although this inference is already clear from Figure 4, addition of confidence bounds in Figure 5 provides hard evidence. To avoid confusion, we only visualized Twaron® CT612LS and UHMW-PE UD1 with their confidence bounds in Figure 5.

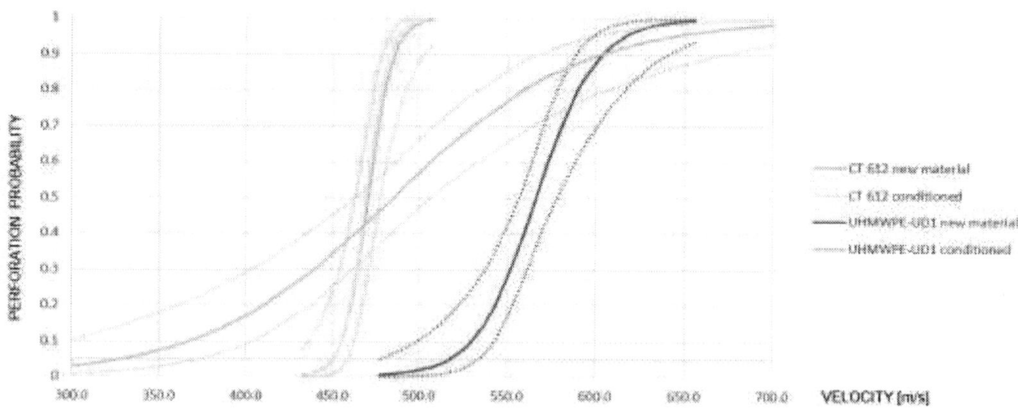

Figure 5. S-curves for Twaron® CT612LS *woven fabric,* and UHMW-PE UD1 before and after tumbling also including confidence boundaries.

3. INFLUENCE OF MECHANICAL STRESS ON BALLISTIC PERFORMANCE OF HYBRIDS MADE FROM TWARON® WOVEN FABRIC AND UHMWPE UD

Both para-aramid woven fabric and UD, as well as UHMWPE UDs, have their pros and cons. While UHMW-PE UD often exhibits higher V_{50} against bullets compared to woven Aramids, the S-curve of woven Aramid is steeper and less affected by mechanical stress/tumbling. The intention of the hybrid testing was to learn whether hybridizing woven Twaron® with UHMW-PE UD could result in constructions having high ballistic performance, combined with good resistance against mechanical stress.

3.1 Test program

We constructed 3 different hybrids (Table 5) from 2 materials (Table 4). All three hybrids have the same weight per meter squared.

Table 4. Materials used for hybrid testing

Name	**AD [g/sqm]**	**Material construction**	**Panel construction**	**Proportion [kg/sqm]**
Twaron® CT612LS	125	Woven, plain	Quilted	2.25
UHMW-PE UD2	-	2ply UD, with film	Corner tacked	2.02

Table 5. Constructions used for hybrid testing
Conditioning and testing in accordance with paragraph 2.3.

Component 1	**Component 2**	**Component 3**	**Test pack AD [kg/sqm]**
strike face	***back face (body side)***		
UHMW-PE-UD2	Twaron® CT612LS		4.27
Twaron® CT612LS	UHMW-PE-UD2	Twaron® CT612LS	4.27
Twaron® CT612LS	UHMW-PE-UD2		4.27

3.2 Test results for hybrid constructions

The results in Figure 6 show that the order of Twaron® and UHMW-PE UD in hybrids seems to have a substantial impact on their resistance against mechanical stress. The construction using Twaron® CT612LS woven fabric at the strike face and UHMW-PE UD at the back face resulted in the highest V_{50}, steepest S-curve and highest V_{05}, both before and after tumbling. The results suggest that by a smart combination of Aramid and UHMW-PE UD materials, ballistic constructions can be well optimized.

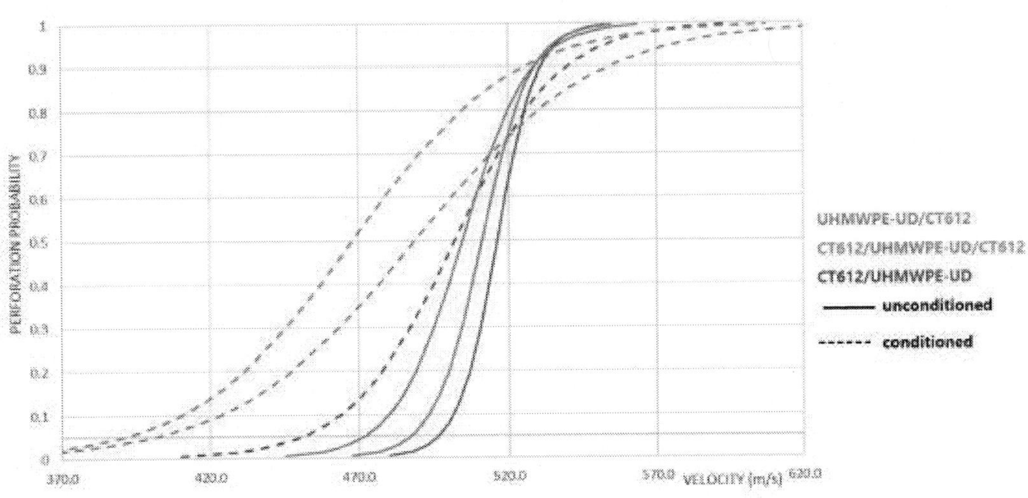

Figure 6. S-curves for various Twaron® CT612LS/UHMW-PE UD hybrids before and after tumbling

4. CONCLUSIONS

https://doi.org/10.52202/080042-0005

We showed that the tested woven fabric and UD made from Twaron® offer high resistance against tumbling/mechanical stress. On the contrary, the tested UHMWPE UDs are significantly affected after tumbling. Smart hybridizing of woven Twaron® fabric with UHMWPE UD enhances ballistic performance and improves resistance against tumbling. The improvement heavily depends on the construction and the way the materials are combined with each other.

References

[1] Ballistic Resistance of Body Armor, NIJ Standard-0101.06, NCJ 223054, July 2008

[2] Forster A.L., Leber D.D., Engelbrecht-Wiggans A., Landais V.,Chang A., Guigues E., Messin G., Riley M.A., J Res Natl Inst Stan, 2020, 125:125026

[3] Guidelines for Completion of Armor Conditioning by NIJ Standard-1010.06 Section 5, National Institute of Justice Compliance Testing Program (NIJ CTP), U.S. Department of Justice

Disclaimer

The Fundamental Limitations of Clay for Assessing Human Response for Behind Armour Blunt Trauma

J. Op 't Eynde[1,2], D. Y. Pang[2], C. F. Morino[2], M. Z. Abrams[2], J. R. Kait[2], R. S. Salzar[3], T. B. Bentley[4], B. S. Shender[5], C. R. Bass[2]

[1]*joost.opteynde@duke.edu*
[2]*Injury Biomechanics Laboratory, Duke University,*
101 Science Drive 1427 CIEMAS BME Box 90281, Durham, NC 27708, USA
[3]*Center for Applied Biomechanics, University of Virginia,*
4040 Lewis and Clark Drive, Charlottesville, VA 22911, USA
[4]*Office of Naval Research, 875 N Randolph St, Arlington, VA 22217, USA*
[5]*Naval Air Warfare Center Aircraft Division,*
48110 Shaw Road, Building 2187 Suite 2203, Patuxent River, MD 20670, USA

Abstract. Current assessment of both military and civilian body armour uses Roma Plastilina #1 (RP-1), an oil/wax-based modelling clay, as the surrogate for Behind Armour Blunt Trauma (BABT). A depth of 44 mm is the threshold for unacceptable armour backface deformation. In this study, high speed x-ray backface deformation data from hard plate body armour tests in human cadavers and clay are compared to data from simulated rifle BABT indenter tests utilizing porcine models and clay, to evaluate the current clay assessment method. Fifty-two clay indenter impacts were performed with impact energy ranging from 175 J to 508 J, resulting in plastic clay deformations depths from 30.7 mm to 65.3 mm. The indenter velocities during these tests ranged from 31.7 m/s to 55.1 m/s, equivalent to 7.62 x 51 mm rifle round velocities of 651 m/s to 1106 m/s. In contrast to the cadaver and animal models, clay exhibited a strong rebound effect. This effect reduces the final deformation by 5-25% depending on the velocity of the impact, obscuring the actual peak dynamic deformation in the clay by a significant unknown fraction of the residual plastic deformation. When comparing the indenter clay results to experiments with similar indenters on live pigs, the indenter in clay requires over 30% higher energy to achieve deformations similar to those seen in the pig torso, demonstrating that deformation might exacerbated substantially in living tissue. A scaling relationship is developed to relate indenter impact velocity to rifle round velocity based on the body armour as a kinetic energy dissipater. Results imply significant differences between clay and tissue, reiterating that RP-1 is not a suitable surrogate and that the current composition and testing procedures involving RP-1 possess neither the complete plasticity nor a comparative or equal deformation depth to that of living tissue.

1. BACKGROUND

Development of body armour capable of protecting the user from ballistic threats has resulted in the creation and adoption of armour mechanisms that rely on mitigating acute damage through spreading out the impact across a wider area through deformation of armour materials. Body armour has been found to greatly increase survivability of law enforcement officers shot in the torso by reducing the risk of penetrating trauma [1, 2]. However, this distribution of force results in deformation that occurs on the rear surface of the armour plate. Extrusion of armour material due to this deformation is recognized as backface deformation (BFD). As this surface impinges with a high rate upon the physiological body of the wearer, BFD has been shown to cause injuries to the ribcage and internal organs of the thorax, which is recognized as Behind-Armour Blunt Trauma (BABT) [3]. Initially ballistic gelatine was utilized as the tissue surrogate for BABT characterization, but analysis utilizing the gelatine model requires costly high-speed video/camera equipment to capture the maximum deformation, and raised concerns about discerning the correct displacement due to the refractive index of the air-gelatine interface. Examination of the current tissue surrogate utilized to characterize BABT mechanism of injury began in the 1970s [4], with the aim of developing an expedient, low-cost alternative to ballistic gelatine for assessing VIP soft body armours against handgun rounds. The contexts of these earlier studies all address relatively low-velocity impacts from traditional pistol cartridges and thus fail to address characteristics of increasingly relevant high-velocity impacts on armour specifically designed to combat rifle threats.

These early tests performed in the 70s resulted in the adoption of Roma Plastilina #1 (RP-1), an oil/wax-based modelling clay, as the standard tissue surrogate for deformation testing, with a maximum deformation depth of 44 mm being the threshold for unacceptable BFD. In Prather 1977 [4], RP-1 is recognized as a relatively plastic, affordable alternative to ballistic gelatine as it possesses comparable

 https://doi.org/10.52202/080042-0006

maximum deformations. This deformation data is based on a previous BABT test utilizing 45 kg caprine models which were assumed equivalent to an adult human, in which no fatalities were recorded. Recognizing these limitations, the study states the "data is limited and hence no solid conclusions can be drawn as yet regarding the effect of deformation depth". An underlying assumption in these tests is that maximum deformation of the clay is equal to the residual deformation, eliminating the requirement for expensive equipment needed in the gelatine tests. Considering the changes in RP-1 material composition from the 1970s to present day, the limitations of the assumptions made, and the use of low-velocity .38 Special as the sole test metric, the fidelity of RP-1 as a tissue simulant in contemporary BABT testing should be investigated.

As an art modelling clay, the composition of RP-1 is relatively inconsistent and the material has become progressively stiffer over time due to changes in both clay and wax composition [5]. Current testing methodologies developed by the NIJ [6] attempt to address this through substantially varying temperature to ~38^0 C from the original ~20^0 C, partially melting the wax constituents, and utilizing pass/fail calibration drop tests, but this compensatory method is not representative of the original RP-1 composition and potentially changes other material characteristics of the clay. Inhomogeneous working of the clay is an additional limitation of RP-1 as a tissue simulant. Unlike gelatine, it is difficult to visually recognize or otherwise confirm uniformity of the material between tests. Further, maintaining a consistent clay temperature is difficult: the temperature and thus material properties of the clay in one region may be substantially different from another once exposed to an environment of different temperature. While testing in a room-temperature environment, the surface of the clay is often noticeably cooler than the interior and will possess different mechanical properties.

Advances in arms and body armour have also rendered the .38 Special inadequate as a sole metric for the evaluation of all body armour. Rifle cartridges have traditionally defined high-velocity applications, but handgun or pistol cartridges have also been developed since development of the standard that can meet traditionally rifle cartridge velocities. Further, this same standard of 44 mm BFD in RP-1 is utilized for all body armour evaluation with no distinction between hard or soft body armours. While aramid fibres and other soft armour materials defeat the projectile through direct deformation of the material [7], hard plates commonly composed of ceramics such as boron or silicon carbide dissipate energy through fracture [8], and hard Ultra-High Molecular Weight Polyethylene (UHMWPE) plates also dissipate energy through fracture and delamination [9]. This difference in mechanism of action may lead to further inconsistencies when evaluating both with the same standard.

Despite these limitations and change in scope, the RP-1 standards developed for soft aramid body armour in the 70s is still the preeminent technique for body armour assessment. The difference between high-velocity and low-velocity applications has been demonstrated in numerous relatively recent animal model studies, including a study done by Gryth et al., 2007 [10] in which 22 porcine models were used as a tissue surrogate in testing high-velocity impact BABT characteristics with the 7.62 x 51 mm NATO rifle cartridge on hard ceramic armour plates with soft armour backers, with 50% and 25% mortality rates from BABT for 40 mm and 34 mm respective maximum deformation depths. Through scaling, the porcine model is more representative of adult human mass than the original caprine model, further emphasizing questions on the fidelity of RP-1. The current study incorporates data from recent porcine tests with an indenter representative of the deforming backface of armour, serving to compare porcine and clay models [11].

Methods for evaluating the fidelity of RP-1 clay outside of direct comparison to animal or cadaver models have been assessed in past studies through employing measures described in current standards developed by the NIJ. Studies employing such controls evaluate material characteristics by either directly inducing deformation through using a shoot-pack under body armour systems with known characteristics or employing an indenter to simulate the impact of the armour backface against tissue. Utilizing a test box with the same dimensions as described in the standard and creating indenters based on the deformation profiles of armour BFD, Graham and Zhang clarify the dynamic behaviour of RP-1 under impact, observing unrepresentative material characteristics in the results such as clay extrusion post-impact, a rebound causing differences between maximum and residual deformation depth, and impact or penetration characteristics of the indenter [12]. More recently, in 2022 Zhang et al. go on to develop a model of the rebound effect in RP-1 [13].

2. METHODS

2.1 Clay impact tests

In this study, a high-pressure launch tube and indenter were used to provide the impacts to clay representative of armour backface deformation. 3D-printed polycarbonate indenters of masses between

341 g and 350 g with densities and profiles representative of armour deformation profiles found in high-speed x-ray images in rifle-context hard plate (Ultra-High-Molecular-Weight Polyethylene) body armour tests [3, 11] were utilized to evaluate the RP1 clay model under high-velocity impact. Two indenter designs were used utilizing an identical profile head but with different bodies, a cylinder design with straight walls and a wasp-waisted design attempting to remove mantle surface that would impede a rebound effect in the clay.

Figure 1. A) High pressure launch tube, B) Wasp-waisted type Indenter on clay box with deformation profiles, note the extrusion of the clay from the indenter impacts and leftover clay residue along the indenter head from the rebound effect of the clay 'grabbing' the indenter.

The impacts were recorded utilizing high-speed video (Phantom v711, Vision Research) perpendicular to the path of the indenter at the muzzle. This high-speed video was then utilized to determine the velocity and rebound of the clay. Post impact, a depth camera (Intel RealSense) and 3D-scanning software (Dot3D, DotProduct) were used to capture the residual deformations in clay to determine volume, area, and profile of the deformations. Residual deformation depth was also measured using a depth micrometer. Clay temperature was measured at 2 cm depth following each impact utilizing a temperature probe.

The clay target 56 cm x 56 cm x 14 cm aluminium box with plywood base followed the NIJ 0101.06 standard for body armour testing. Clay was calibrated with spherical calibrators dropped from a 200 cm height with a resultant acceptable deformation depth of 17 mm to 21 mm. Each clay box sustained 4 indenter impacts during the testing phase before being refilled and planed for reheating, and each reheating phase was held at 40-42 °C for at least an overnight period. Should the box not pass calibration, the box would be planed and reheated or allowed to cool depending on whether calibration deformations were under or over acceptable limits respectively. Heating was performed to soften the clay according to the testing standard such that desired calibration depths were achieved.

After reworking the clay to a uniform flat surface, four equally spaced impact tests were performed in succession. The indenter was propelled utilizing high pressure helium gas. Velocity of the indenter was determined using high speed video. Clay deformation profiles were measured relative to the edge of the box. Once testing was concluded, and the clay was flattened and placed in the oven overnight to achieve equilibrium temperature.

The indenter-clay results were compared with data from simulated rifle BABT indenter tests utilizing porcine models [11] to evaluate the RP1 clay model by comparing clay and living tissue under a high-velocity BABT impact context.

https://doi.org/10.52202/080042-0006

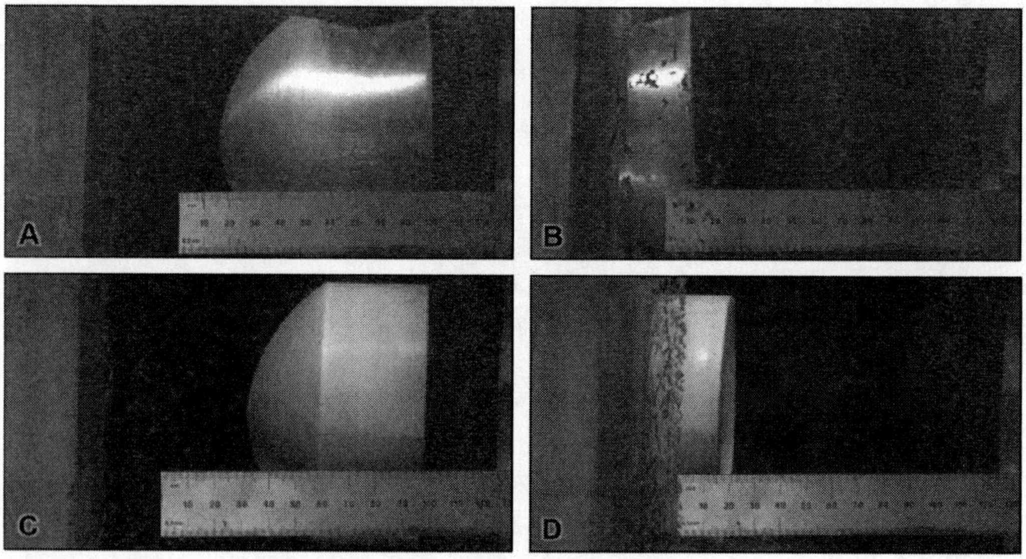

Figure 2. High-Speed Video Frames. A) Wasp-waist type in flight. B) Wasp-waist type impact, note extrusion of material. C) Cylinder type in flight. D) Cylinder type impact.

2.2 Equivalent Rifle Velocity

The fidelity of the clay indenter model for rifle rounds into hard armour (developed from hard armour x-ray profiles into pigs and human cadavers was evaluated using similar human cadaver [3] and porcine models [11]. An estimate of equivalent rifle round velocity into hard body armour for an indenter impact was determined based on an assumption that the armour dissipates a fraction of the incoming kinetic energy through the fracture or delamination of UHMWPE material. As the body armour material's capacity for energy dissipation is compromised, the remainder of the energy is transferred into the thorax as kinetic energy of the armour backface. The indenter serves to model this residual kinetic energy of the armour and the attacking projectile into the thorax as the thorax deforms. We model this residual energy as an energy fraction (EF) of the incoming energy to match projectile-armour impacts with indenter impacts. With similar displaced volumes and depths in clay between the indenter and projectile-armour models, the equivalent projectile velocity of a 9.5 g 7.62x51 mm NATO (M80 Ball) projectile on UHMWPE can be estimated assuming the indenter kinetic energy is equal to the residual kinetic energy of the armour backface upon impact.

$$\frac{1}{2} m_{indenter}\, v_{indenter}^2 = EF * \frac{1}{2} m_{round}\, v_{round}^2 \tag{1}$$

$$v_{indenter} = \left(EF * \frac{m_{round}}{m_{indenter}}\right)^{1/2} v_{round} \tag{2}$$

To find this energy fraction, regression fits for both the deformation depth in clay and volume of displaced clay were compared between the indenter-clay and projectile-armour-clay and the EF was optimized for closest match between the regression fits.

3. RESULTS

3.1 Clay impact tests

Fifty-two clay indenter impacts were performed with impact energy ranging from 175 J to 508 J, resulting in plastic clay deformations depths from 30.7 mm to 65.3 mm. The indenter velocity upon impact during these tests ranged from 31.7 m/s to 55.1 m/s, equivalent to 7.62 x 51 mm rifle round velocities of 651 m/s to 1106 m/s. In contrast to the cadaver and animal models, clay exhibited a strong rebound effect. This effect reduces the final deformation by 5-25% depending on the velocity of the impact, obscuring the actual peak dynamic deformation in the clay by a significant unknown fraction of the assumed clay residual plastic deformation. The measurements from these impacts can be found in table 1. Blank values

https://doi.org/10.52202/080042-0006

for the rebound are present for impacts where the indenter fully entered the clay and visual contact was lost.

Table 1. Indenter impact test measurements.

Velocity (m/s)	Plastic Deformation (mm)	Rebound (mm)	Momentum (Kg*m/s)	Energy (J)	Area (cm^2)	Volume (cc)
42.9	54.3	3.4	15.0	321.0	98.8	272.4
42.9	53.0	3.5	15.0	321.0	95.3	269.6
44.8	56.1	4.7	15.6	350.0	97.9	312.6
51.5	65.3	7.3	18.0	463.0	104.8	407.7
35.1	41.0	7.4	12.2	215.0	91.8	187.1
39.2	45.8	6.0	13.7	268.0	99.9	232.0
33.2	40.0	8.5	11.6	192.0	98.2	192.4
33.6	39.3	6.4	11.7	197.0	90.3	175.9
31.7	30.7	6.0	11.1	175.0	100.3	119.7
40.1	43.4	8.6	14.0	281.0	93.2	215.3
38.4	37.6	6.5	13.4	257.0	95.7	171.4
40.6	40.1	9.5	14.2	288.0	98.6	197.1
40.6	45.0	5.6	14.0	285.0	94.4	238.1
44.8	44.0	9.9	15.5	348.0	96.6	303.0
43.0	37.0	10.3	14.9	320.0	99.7	280.3
41.2	45.0	9.9	14.3	294.0	93.7	244.9
41.9	48.6	7.3	14.5	303.0	104.4	273.4
46.9	51.9	7.7	16.2	381.0	105.4	316.8
40.1	43.4	7.6	13.9	279.0	102.5	239.9
41.0	47.0	9.5	14.2	291.0	98.4	263.0
42.4	47.0	7.7	14.7	312.0	100.1	289.0
43.2	48.2	9.9	14.9	323.0	93.8	263.2
38.5	41.8	6.0	13.3	256.0	126.9	190.0
44.4	45.9	7.3	15.4	342.0	105.9	242.3
44.8	51.2	7.3	15.5	348.0	112.7	272.9
45.4	55.2	8.1	15.7	357.0	107.2	316.7
46.6	53.3	9.4	16.1	375.0	104.3	288.8
48.0	57.0	9.4	16.6	399.0	113.4	342.5
53.5	51.5	9.6	18.5	496.0	115.6	301.0
38.3	43.5	6.0	13.4	257.0	91.4	214.8
38.6	42.7	8.7	13.5	260.0	90.4	214.0
40.5	44.6	6.9	14.2	287.0	87.0	225.0
42.1	45.7	9.8	14.7	310.0	97.8	256.7
51.4	62.8		17.7	455.0	115.4	386.4
45.6	51.3		15.7	359.0	128.9	296.2
43.0	47.9	6.9	14.8	318.0	115.9	260.8
39.0	40.9	8.0	13.4	262.0	102.8	211.5
40.3	44.4	7.3	13.9	280.0	118.7	228.4
43.9	48.3		15.2	333.0	100.0	261.1
46.6	53.0		16.1	375.0	106.3	302.8
49.0	55.7		16.9	413.0	116.2	347.3
41.0	46.4	8.1	14.1	289.0	103.1	234.1
39.9	43.6	8.5	13.8	275.0	122.9	222.4
49.8	56.4		17.2	428.0	101.2	327.3
49.7	60.2		17.2	427.0	130.1	385.6
52.7	60.4	10.3	17.5	462.0	111.2	399.2
52.6	57.3	5.1	17.5	460.0	106.1	354.2
52.9	59.0	11.1	17.6	466.0	109.3	404.6
54.2	61.4	11.1	18.1	491.0	109.1	422.8
54.0	61.1	7.3	18.0	487.0	111.5	420.4
55.1	64.7	12.1	18.4	508.0	117.3	483.2
48.8	57.4	9.6	17.2	419.0	109.2	353.3

When comparing the effect of the wasp-waisted design with the cylindrical body, both body designs had equivalent results that could not be separated. For further analysis the results are grouped together with consistent results.

https://doi.org/10.52202/080042-0006

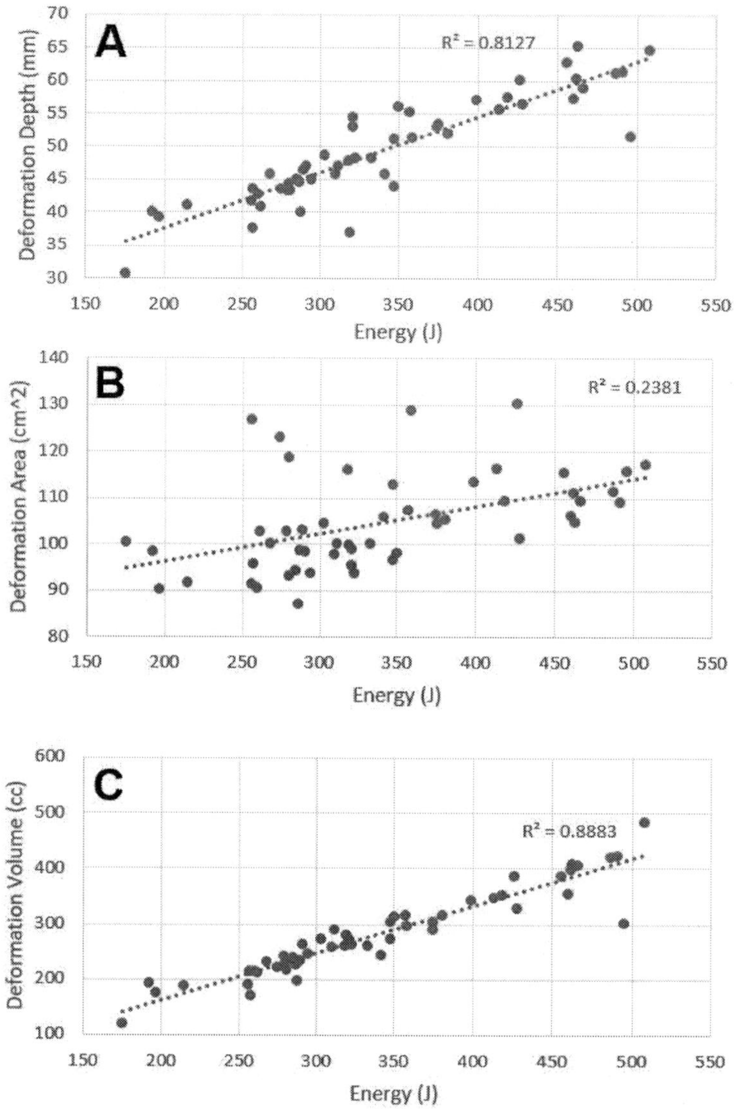

Figure 3. Residual deformation depth (A), area (B), and volume (C) of displaced clay for all indenter tests as a function of indenter kinetic energy. Volume of displaced clay correlates best with impact energy.

When comparing the residual depth, area and volume of the deformation in the clay in Figure 3, both the depth and volume show a good correlation with the impact energy, with the volume slightly outperforming the depth. Deformation area is more variable and reflective of the limited surface area of the indenter, but still correlates with impact energy.

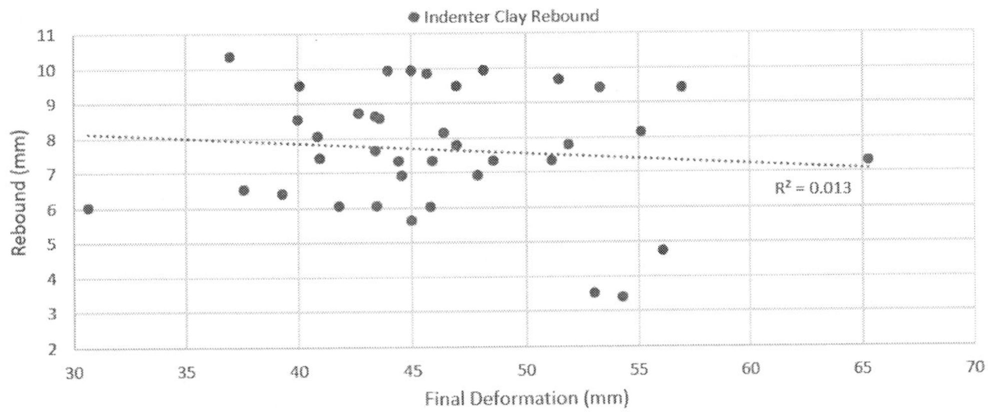

Figure 4. A strong rebound effect showed no correlation to the total residual deformation depth.

A strong rebound effect was observed for all impacts, resulting the in the final deformation differing substantially from the maximum achieved dynamic deformation during impact. These rebounds can be seen in Figure 4. The rebound distance was quite variable between impacts and showed no correlation to impact energy or to residual deformation. As a percentage of the final deformation, the rebound comprised anywhere from 5% to 25% of the total deformation depth. This shows that clay is a dynamic material with transient material properties.

3.2 Equivalent Rifle Velocity

After scaling the energy fraction of the rifle round impacts on clay performed in a previous study [3], energy fractions were obtained for both a residual depth regression and a residual volume regression. The energy fraction of the simulated UHMWPE armour is best fit as 12% on a depth basis and 8.7% on a volume basis, and the regression fits for these can be seen in Figure 5. Using the obtained energy fractions we can then apply them to Equation (1) and (2) to obtain scaling relationships:

$$v_{indenter} = 0.049 * v_{round} \text{ (Volume basis)} \quad (3)$$

$$v_{indenter} = 0.058 * v_{round} \text{ (Depth basis)} \quad (4)$$

This model was verified with data from high velocity rifle round impacts on armour plate equipped human cadavers [3] and live indenter impact porcine test fracture data [11], comparing the 50% risk of fracture round velocity (700 m/s) in cadaver relative to the 50% risk of fracture indenter velocity in porcine (32.75 m/s) and its equivalent round velocity (668 m/s) on a volume basis. This suggests hard body armour functions as a dissipater of 88-91.3% of the kinetic energy of the incoming round.

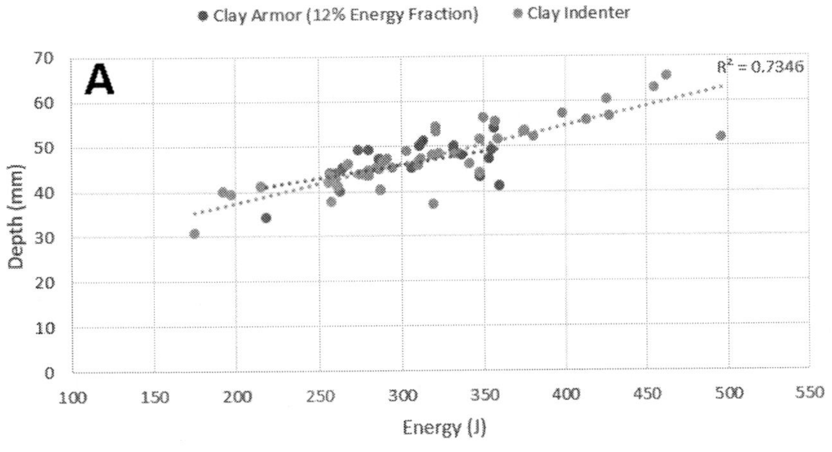

https://doi.org/10.52202/080042-0006

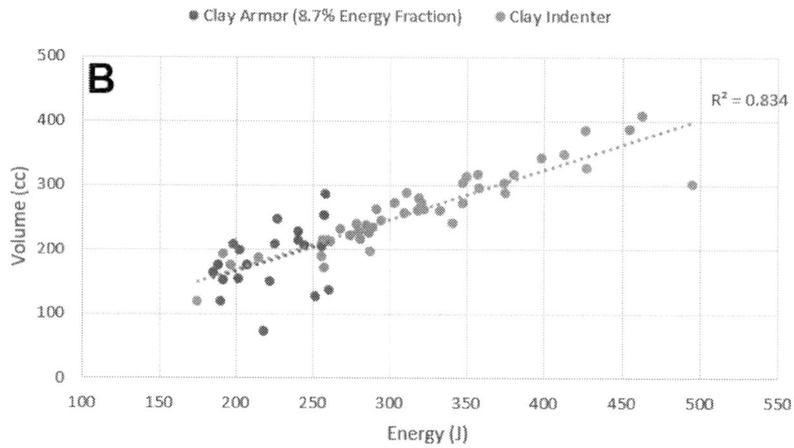

Figure 5. Regression fits of kinetic energy of the indenter and scaled rifle round with respect to A) residual clay depth, and B) residual clay volume. Rifle round kinetic energy was scaled down to 12% (A) and 8.7% (B).

3.2 Comparison to live porcine impacts

When comparing the deformation depth of the impacts to those seen live porcine impacts in a previous study [11], the indenter in clay requires over 30% higher energy to achieve deformations of similar magnitude to those seen in the pig torso. These results can be seen in Figure 6. Impacts that resulted in clay deformation depths below 44 mm resulted in significant injuries in the animals that would require immediate medical attention in the field.

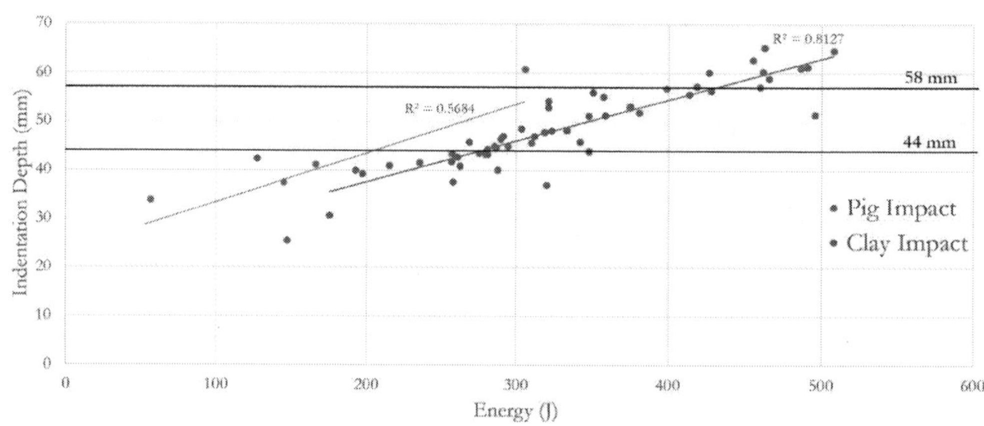

Figure 6. The relationship between the Porcine and Clay models using indenter impacts. This highlights the difference between clay and live tissue, with clay being stiffer and requiring more kinetic energy to achieve a similar deformation depth as the porcine impacts.

4. DISCUSSION

This study elucidates several parameters that are used to evaluate the fidelity of the RP-1 clay model of BABT. First, elastic volumetric rebound of the clay varies by incoming velocity and invalidates the assumption that the clay responds fully plastically under dynamic deformation. Indeed, the elastic rebound for this indenter is 5-25% of the total depth, implying that direct clay measurements under predict actual clay deformation by a variable amount sufficient to substantially increase injury risk (cf. [5]). This increase in risk is presumably dependent on area characteristics of the backface and cannot be assessed using clay response alone.

Second, this study finds different residual clay depths in this study compared with previous porcine and human cadaver depth measurements in indenter and armour tests at similar impact energies. This finding invalidates the second assumption of clay testing, the concordance between human deformation depths and clay deformation depths under similar conditions. Using energy as a metric accounts for the variable indenter mass between the different tests and provides a dynamically appropriate comparison between porcine and clay models. For live pig indenter tests under similar conditions [11], the indenter was found to cause a greater peak deformation in living tissue when compared to the residual plastic deformation of the clay model. This is especially concerning as it demonstrates that the extent of deformation in clay may be exacerbated in living tissue. Other contexts find indentation in clay may be reduced compared with porcine models [11]. This also highlights the need for a more accurate assessment method for high-velocity high-energy contexts such as those for hard armour.

Third, assessing the relationship between the impact energy of the 7.62 x 51 mm NATO M80 Ball round used in the armour-clay tests and the indenter-clay test performed in this study demonstrates the applicability of the indenter model for the evaluation of RP-1. A series of energy fractions were examined with respect to residual deformation depth and the UHMWPE armour utilized in the armour tests was found to permit 8.7-12% of the M80 Ball projectile's kinetic energy to be transferred to the wearer to create the back face deformation depth, or residual deformation depth in clay.

Fourth, the use of the two different indenter body profiles to examine the rebound effect of RP-1 was also investigated. During testing, it was noted that the plastic clay extrudes outwards to 'grab' the indenter upon impact, and no corresponding difference was seen in the percentage of indenter rebound between the wasp-waisted and cylindrical indenters at the same kinetic energies. No difference is noted between residual deformation volumes between the two indenter types; it can therefore be assumed that indenter behaviour is relatively independent of indenter mantle shape. This study also shows that increasing energy generates larger deformation volumes and depths, but deformation area is loosely correlated with impact energy. This may be a limitation of the indenter model given the constant cross-sectional area and frontal profile.

5. CONCLUSION

In conclusion, this study demonstrates the strong limitations of the fidelity of RP-1 as a tissue surrogate in the context of BABT. By following the universal test procedures utilized in evaluating body armour, this study elucidates differences in BABT from high-velocity and low-velocity impacts on the RP-1 clay model under these standards. Results from the clay and porcine tests imply significant differences between the two models, reiterating that RP-1 is not a suitable tissue surrogate especially in the context of high-velocity impact and that the current composition and testing procedures involving RP-1 possess neither the complete plasticity nor a comparative or equal deformation depth to that of living tissue. Major points of difference include the different mechanical structure between clay and physiological ribs or internal organs, response at different impact energies with respect to deformation depth, volume, and area.

These implications may also be applied to current and future body armour development noting that the current 44 mm deformation depth in RP-1 standard may not prevent BABT injury, the current use of RP-1 as a tissue surrogate being based on low-velocity impact is inadequate for evaluating high-velocity impact rated hard armour plating, and that RP-1 possesses characteristics not reflective of living tissue such as the lack of rib response, more centred deformation, and greater extrusion of material. Further, this study contributes new methodologies to the field of injury biomechanics, including a new methodology in comparing BABT tissue surrogates to other BABT models through utilization of 3D-printed indenters based on cadaver deformation profiles for rapid expedient characterization of a BABT tissue surrogate.

References

[1] LaTourrette, T., J Occup Environ Hyg, 2010; 7(10); pp. 557–562.
[2] Liu, W. and Taylor, B., J Occup Environ Hyg, 2017; 14(2); pp. 73-80 .
[3] Bass, C.R., et al., Int J Occup Saf Ergo, 2006; 12(4); pp. 429–442.
[4] Prather, R.N., et al., Backface Signatures of Soft Body Armors and the Associated Trauma Effects. 1977, Chemical Systems Laboratory: Aberdeen Proving Ground, MD, USA.
[5] National Research Council, Testing of body armor materials: Phase III. (National Academies Press, Washinton, DC, USA, 2012).
[6] NIJ Standards - 0101.06. 2008, National Institute of Justice.
[7] Merkle, A.C., et al., J Trauma Acute Care Surg, 2008; 64(6); pp. 1555-1561.

https://doi.org/10.52202/080042-0006

[8] Wen, Y., et al., J Mech Behav Biomed Mater, 2015; 45; pp. 11-21.
[9] Gilson, L., et al., Int J Impact Eng, 2020; 136; p. 103417.
[10] Gryth, D., et al., Mil Med, 2007; 172(10); pp. 1110-1116.
[11] Op 't Eynde, J., et al., Behind armour blunt trauma (BABT) indenter simulating high-velocity impacts from rifle rounds on hard body armour., PASS 2020, Copenhagen, Denmark.
[12] Graham, M.J. and T.G. Zhang, Experimental Investigation of Geometric Effects on Ballistic Clay Backing Material. 2019, CCDC Army Research Laboratory Aberdeen Proving Ground.
[13] Zhang, T.G., et al., Modeling the Rebound of Ballistic Roma Plastilina No. 1 Clay. 2022, CCDC

https://doi.org/10.52202/080042-0006

Scaling animal to human injury response for use in improved behind armor blunt trauma injury criteria

Op' t Eynde, J.[1], Shah, A.S.[2], McMahon, J.A.[3], Pang, D.Y.[1], Stemper, B.D.[2], Yoganandan, N.[2], Salzar, R.S.[3], McEntire, B.J.[4], & Bass, C.R[1].

[1]*Duke University, 2138 Campus Drive, Durham, NC*
(joost.opteynde@duke.edu)
[2]*Medical College of Wisconsin, 8701 Watertown Plank Road, Milwaukee, WI, USA*
[3]*University of Virginia, 1827 University Avenue, Charlottesville, VA*
[4]*U.S. Army Aeromedical Research Laboratory, 6901 Farrel Rd, Fort Rucker, AL*

Abstract. Developed more than four decades ago, the current behind armor blunt trauma (BABT) evaluation standard based on plastilina clay has limited biofidelity for assessing thoracic injury from backface deformation (BFD) from high-velocity rifle threats. Further, the standard relies on complex and uncertain analogies between an animal model and human thoracic and injury response. To improve the biomechanical basis for future BABT injury assessments, we have performed representative BABT impacts on swine and human cadavers for BFD velocities representing high-velocity rifle rounds on hard armor plates. Impactor dynamics were determined using an onboard accelerometer and high-speed video, and rib fractures were assessed using post-test micro-CT imaging and necropsy. The kinetic energy of the impact was scaled according to body mass based on equal velocity scaling, widely used in injury biomechanics. This scaling was used in logistic survival analysis to determine rib fracture injury risk for cadaveric swine and humans. Scaled impact energy to produce a 50% risk of rib fracture was 113.9 Joules (J) (Confidence Interval [CI]: 90.3, 137.6) for the human cadavers and 143.9 J (CI: 103.8, 184.1) for the porcine cadavers. Confidence intervals of injury risk curves substantially overlap for the human and swine cadavers, suggesting that this scaling is appropriate for transferring risk across these species. Residual energy differences of 20 to 30% for similar injury risk between the human and swine cadavers suggest an additional bone quality scaling is desirable since the swine cadavers are generally at an earlier developmental age than the available human cadavers. This is the first comprehensive study to provide scaling to humans from a porcine model of hard armor BABT. The structural scaling relationships between the human and swine cadavers are valuable in developing transfer functions for injury risk curves from planned live swine BABT impact experiments assessing the pathophysiology.

1. INTRODUCTION

Body armor provides effective protection from penetrating trauma for military and law enforcement personnel [1,2] using a 'passive defeat' mechanism that transfers localized momentum from the incoming round into the regional momentum of the deforming body armor. This mechanism slows and often fragments or deforms the projectile, greatly reducing the potential for penetration of the armor. However, a defeated round can still cause the armor backface to deform into the thorax or other body regions and cause damage to the underlying anatomy, known as behind armor blunt trauma (BABT). This BABT to the skeletal anatomy and internal organs, such as lungs, heart, and liver, can cause severe morbidity or death [3-6].

Initially developed more than four decades ago, the current BABT evaluation standard based on a maximum of 44 millimeter (mm) deformation measured in plastilina clay has limited biofidelity for its current uses, such as assessing thoracic injury from backface deformation (BFD) resulting from high-velocity rifle threats. Further, the standard relies on complex and uncertain analogies between the animal model and human thoracic and injury response based on limited testing with goat thoraces. Goats with mass 40-50 kg (assumed equal to a human) were exposed to a handgun round impacting on a soft body armor. Roma Plastilina #1 (RP1) clay was found to deform at different (stiffer) rate, but the final deformation was similar to the final deformation in goats as well as in gelatin based on high speed video. The soft armor was then shot with a gelatin backing in a configuration that was estimated to produce a 10% lethality based on a general blunt impact model, but no actual lethality in animals was tested [5, 7-9]. The average maximum deformation in the gelatin was 44 mm which was then assumed to be equal to the clay, and representative of the human torso response [9]. No correlation between BFD and injuries was established. The original researchers and many later researchers have called for additional animal and surrogate tests to improve these initial assessments [10], with a particular focus on the potential to improve and optimize hard armor characteristics based on a systematic assessment of BABT injury biomechanics. The standard was developed to be conservative, and as such modern hard body armor provides high levels of protection, while potentially being excessively heavy. A critical target for

https://doi.org/10.52202/080042-0007

optimization is reducing soldier-borne mass and bulk while maintaining effective ballistic protection. The weight, bulk, and thermal load of armor reduce the mobility of the Warfighter, which negatively impacts operational performance [11-13]. Region-based thoracoabdominal injury criteria are necessary to provide tools to avoid under-designing (i.e., increased risk of penetration or BABT injury for a given threat) or over-designing armor (i.e., carrying unnecessary mass).

Currently, unavailable injury criteria for BABT include injuries to the human skeletal components under direct backface loading and injuries to underlying organ systems from stress transmitted through the wall of the thorax. Injuries may include acute damage, such as bony fracture, and injuries that require physiological processes to develop, such as lung contusion or commotio cordis [14]. Experiments using human cadaveric specimens can provide accurate kinematic and dynamic responses of the body and bony fracture tolerances; however, live animal model testing is needed to provide systemic injuries and injuries that require physiology to develop. Since there are differences between living humans, human cadavers, and live animal models, scaling principles are needed to relate dynamic responses, injury structures, and physiological results between human and swine models.

Adult swine are the most analogous animal model to humans when studying BABT due to the similarities in size and thorax anatomy [15-18]. The swine model is also widely used in automotive standards to closely represent human thoracoabdominal organs in a car crash scenario [19]. These scaling relationships will be essential for translating results from future in vivo animal research. For the current study, physiological scaling based on allometry (power law scaling relative to body mass) and the concept of physiological time (species specific time scale based on heart rate, respiratory rate, immune response, development rate) suggests that physiological scaling between adult swine and humans is unity [20]. Structural mechanical scaling is used between swine and humans of different sizes. The current study develops structural mechanical scaling for rib fractures resulting from BABT between porcine and human cadavers based on the equal velocity approach outlined by Eppinger et al. (1984) [21].

2. METHODS

2.1 Cadaveric Impactor Tests

A BABT impactor was designed to match mass and shape from BFD profiles collected from high-speed flash x-rays of human cadaveric surrogate response during hard armor defeats of a realistic battlefield threat [3, 22]. This indenter was 3D printed in polycarbonate, with a diameter of 100 mm, dome height of 25 mm, and mass of 0.22 to 0.36 kilograms (kg). A tri-axial linear accelerometer (Endevco 7284A-60k) was mounted on the back surface of the indenter. The impactor was launched at a range of velocities (13 to 52 meters per second [m/s]) using pressurized helium gas to simulate BABT impacts.

Whole-body unembalmed human cadavers and recently sacrificed adult swine cadavers were exposed to impacts to the ribcage at representative BFD velocities. Impacted areas included anterior and posterior lungs, posterior kidneys, and lateral covered liver. In the current analysis, no distinction is made between specific impact sites on the ribcage. Both surrogates were tested with increasing velocities, and x-rays were obtained in some specimens between tests; however, palpations and clinical assessments were done in all specimens between tests to ensure structural integrity before additional testing in other body regions. Linear strain gages (Micro-measurements C4A-09-060SL-350-39P) and acoustic sensors (Physical Acoustics S9225) were mounted to the ribcage, and pressure transducers (Millar Mikro-Tip SPR-524) were inserted into the lungs through the trachea. Tri-axial linear accelerometers (Endevco 7270A) were mounted on the spine. Data was gathered at a sampling rate of 100 kilohertz (kHz) or more, and high-speed video cameras (Phantom V711, Vision Research) were positioned at different planes to capture the impactor and surrogate kinematics. Following the final tests, a high-resolution computed tomography studies (i.e., CT scan) and a detailed necropsy were performed focusing on assessments of skeletal and soft tissue injuries. A bony fracture was classified as an injury, whereas the absence of bony fractures was classified as a non-injury.

2.2 Data Analysis

Impactor velocity was obtained from high-speed video imaging. The impact kinetic energy was calculated using the impactor mass and velocity, and this was used as an input variable for the injury risk calculation. In total, 73 rib impacts on 18 male cadaveric human specimens and 44 rib impacts on 16 cadaveric swine specimens were included. The body mass (± SD) for these specimens was 80.1 ± 10.9 kg for the human cadavers and 44.0 ± 10.0 kg for the swine cadavers. To normalize for specimen size, the impact energy was scaled using an equal velocity approach [20] according to the body mass of the tested specimen in both swine and human cadavers with unity scaling for allometry.

$$E_{scaled} = \frac{E_{specimen}}{\lambda} \text{ with } \lambda = \left(\frac{M_{specimen}}{M_{reference}}\right)^{a} \quad (1)$$

The reference mass, $M_{reference} = 80$ kg, represents the average male Warfighter body weight. For scaling between swine and humans, the allometric scaling exponent $a = 1$. These scaled kinetic impact energy results were then used to perform a survival analysis for rib fracture injury risk. Non-injury points were considered right censored, and injury points were considered interval censored between 0 Joules (J) and the scaled impact energy. Anderson-Darling coefficients were used to determine the optimal fit among logistic, log-logistic, Gaussian, and Weibull distributions.

2.3 Hard Armor Rifle Round Equivalence

To compare the impacts performed by the impactor to BFD of hard body armor, impacts were performed with the impactor onto RP1 clay contained in an aluminum sided box (56 cm x 56 cm x 14 cm) with plywood base according to the NIJ 0101.06 Ballistic Resistance of Body Armor Standard. These impacts were compared to impacts from a 7.62-by-51 mm-class threat round on hard polyethylene body armor on the same clay standard. Volume of deformed clay was found to have the best correlation to kinetic energy for the impactor as well as for the 7.62 round. To estimate a rifle round equivalence to the blunt impactor, the body armor was assumed to dissipate a certain amount of energy from the incoming round, with the remaining energy fraction (EF) being translated into kinetic energy of the armor backface. This percentage of energy not dissipated by the body armor was calculated by calculating the least squares fit for a residual kinetic energy to volume of displaced clay regression. This energy fraction was found to be $EF = 0.087$, meaning 92.3% of the kinetic energy of the bullet was dissipated by the armor. The impactor velocities can then be related to equivalent rifle round velocities as follows:

$$\frac{1}{2} m_{impactor}\, v_{impactor}^2 = EF * \frac{1}{2} m_{round}\, v_{round}^2 \quad (2)$$

$$v_{impactor} = \left(EF * \frac{m_{round}}{m_{impactor}}\right)^{1/2} v_{round} \quad (3)$$

The velocities of the impactor in the current study simulate BFD into hard personal protective armor from a 7.62-by-51 mm-class threat round at 206 to 890 m/s.

3. RESULTS

Of the 73 impacts (impact energy 100.4 ± 64.0 J) on the human cadavers, 58% (n = 42) had no skeletal injuries, and 42% (n = 31) had rib fractures. Of the 44 swine impacts (impact energy 89.0 ± 80.1 J), 52% (n = 23) had no injury, and 48% (n = 21) had rib fractures. The energy of the impacts by injury type before and after scaling is shown for the human and swine cadavers in Figure 1. Additional data from individual tests is available in Appendix 1 and 2. A separate survival analysis risk function was calculated for the scaled human cadavers and scaled porcine models. The resulting injury risk curves with 95% confidence intervals are shown in Figure 2. A parametric logistic distribution was the best fit among the distributions tested. While not identical, confidence intervals of injury risk curves substantially overlap for the human and swine cadavers, suggesting that this scaling may be appropriate for comparing risk across species. A 50% risk of rib fracture is obtained at a scaled impact energy of 113.9 J (CI: 90.3, 137.6) for the human cadavers and at 143.9 J (CI: 103.8, 184.1) for the porcine cadavers. Depending on the risk level, a porcine specimen required, on average, a 20 to 30% higher scaled energy to achieve the same injury risk. For example, at 20% risk the porcine energy is 21% higher, at 50% risk it is 26% higher, and at 80% risk it is 28% higher. Residual energy differences of 20 to 30% for similar injury risk between the human and swine cadavers suggest an additional bone quality scaling is desirable since the swine cadavers are generally at an earlier developmental age than the available human cadavers.

https://doi.org/10.52202/080042-0007

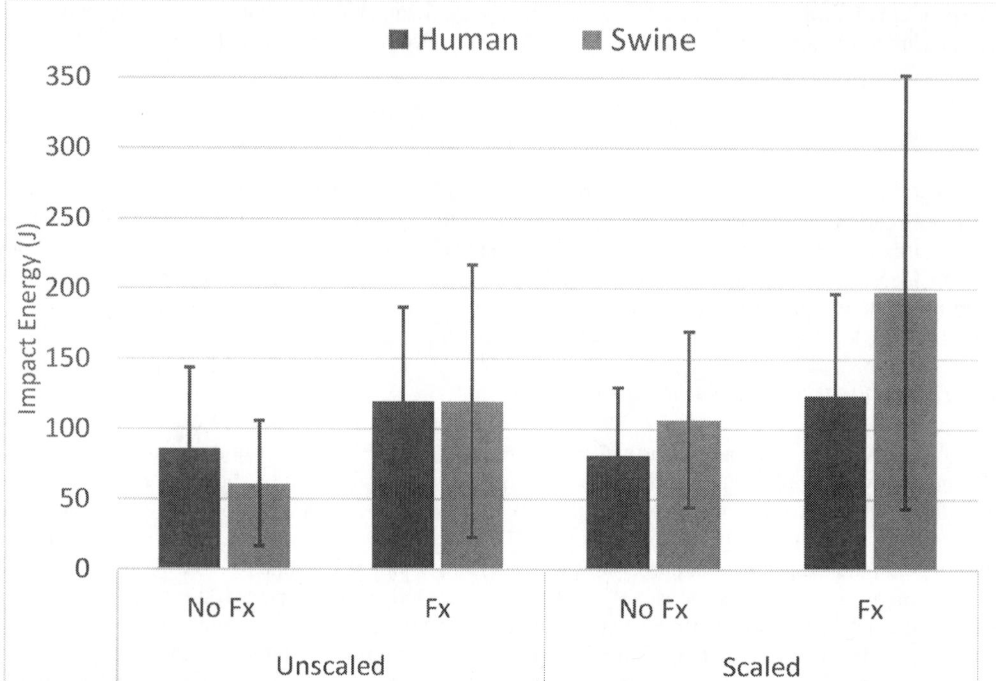

Figure 1. Kinetic energy of the impacts for which rib fractures (Fx) or no rib fractures (No Fx) occurred, for both human and swine cadavers, before and after body mass scaling. The bars indicate standard deviations. Scaling takes into account the body mass of the specimens, resulting primarily in higher scaled impact energy for the porcine cadavers.

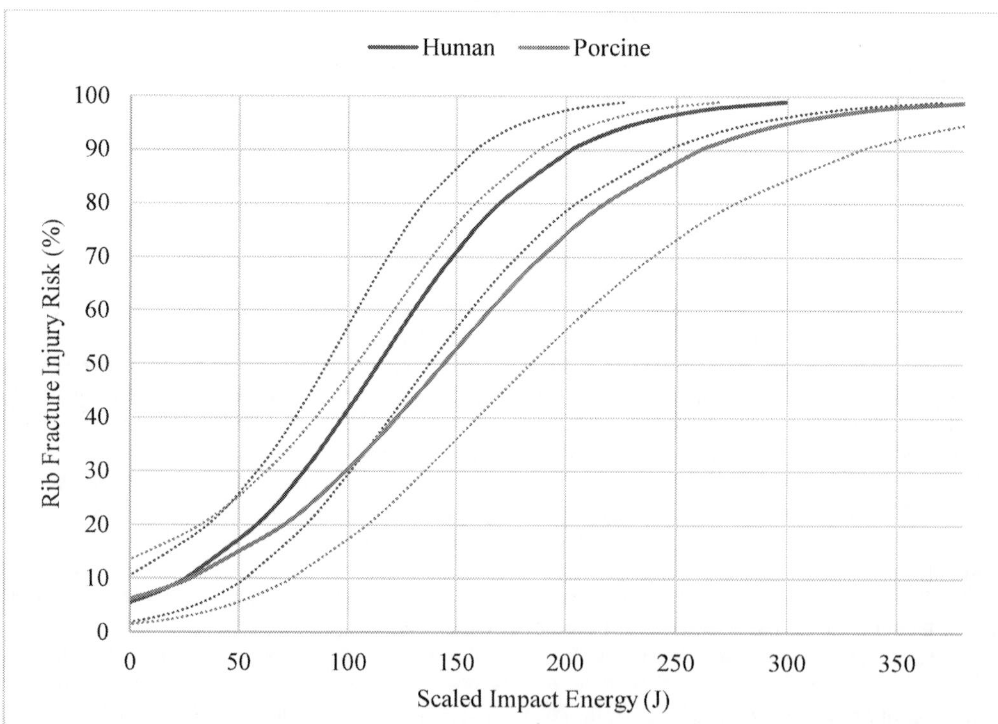

Figure 2. Rib fracture injury risk curve for BABT impacts to the ribcage in human and porcine cadavers. The dotted lines indicate the 95% confidence interval for the injury risk curves of the corresponding model.

4. DISCUSSION

Body mass scaling accounts for most of the differences in rib fracture injury risk in porcine cadavers compared to human cadavers, suggesting that a swine of approximately equal mass is a good skeletal surrogate for a human thorax, based on the overlapping injury risk curve confidence intervals. Since swine of equal or lesser weight than adult male humans are much younger in their developmental cycle than the human cadavers (58 ± 10.9 years old) tested, the ribcage structure contains more cartilaginous structures and is less brittle. This difference in brittleness may have contributed to the increased fracture resilience in the porcine specimens. The same is not necessarily true for internal organ injuries, as a rib fracture might dissipate more of the impact energy that is not transferred further into the thorax. Further analysis of internal organ injuries, especially for *in vivo* animal experiments, is needed.

These investigations including human and porcine cadavers, as well as *in vivo* porcine experiments, are currently underway. Thoracoabdominal regions investigated included the right and left lungs, heart, protected and unprotected liver, abdomen, kidney, and thoracic spine. The goal of these studies is to develop region-specific risk curves for BABT injuries.

In the current study, only male cadavers were included. To ensure applicability to the female Warfighter, subjecting female cadavers to an equivalent BABT regime should be a goal of future studies, especially considering the body mass scaling predicts increased risk of rib fractures for lower body mass.

One limitation of using a logistic distribution to predict injury is the asymptotic behavior near the lower end of the risk scale, combined with the interval censoring of injury data points. This results in elevated injury risk for impacts with 0 J energy, which does not represent reality. However, the logistic distribution still provided the best overall fit for the rib fracture injury risk across the range of impacts performed. Narrowing down injury censoring intervals using injury timing data from strain gauges and acoustic sensors attached to the ribs might improve these censoring intervals and the confidence intervals for the injury risk curves with further analysis.

5. CONCLUSIONS

This study is the first comprehensive approach to provide scaling for a porcine model between matched experiments simulating hard armor BABT events that incorporated BFD from the armor to whole-body human cadavers. The swine experimental model is widely used and accepted in automotive standards. The structural scaling relationships between the human and swine cadavers will be valuable in developing transfer functions from incapacitation-based injury risk curves from planned live swine BABT impact experiments exploring the physiology of BABT. Injury risk curves presented in this study may guide armor design, ensuring safety in the pursuit of lighter weight armor alternatives even when bullet penetration is prevented.

Acknowledgments

The authors gratefully acknowledge the U. S. Army Medical Research and Development Command (USAMRDC) and the U. S. Army Aeromedical Research Laboratory for funding and collaboration through MTEC-21-06-MPAI-031, Behind Armor Blunt Trauma Consortium: Injuries, Mechanisms, and Biomedical Injury Criteria.

Disclaimer

The views, opinions, and/or findings contained in this manuscript are those of the author(s) and should not be construed as an official Department of the Army position, policy, or decision, unless so designated by other official documentation. Citation of trade names in this report does not constitute an official Department of the Army endorsement or approval of the use of such commercial items.

References

[1] Peleg, K., et al., Does body armor protect from firearm injuries? Journal of the American College of Surgeons, 2006. 202(4): p. 643–648.

[2] LaTourrette, T., The life-saving effectiveness of body armor for police officers. Journal of occupational and environmental hygiene, 2010. 7(10): p. 557–562.

 https://doi.org/10.52202/080042-0007

[3] Bass, C.R., et al., Injury risk in behind armor blunt thoracic trauma. International journal of occupational safety and ergonomics, 2006. 12(4): p. 429–442.
[4] Cannon, L., Behind armour blunt trauma-an emerging problem. Journal of the Royal Army Medical Corps, 2001. 147(1): p. 87-96.
[5] Carroll, A.W. and C.A. Soderstrom, A new nonpenetrating ballistic injury. Annals of surgery, 1978. 188(6): p. 753.
[6] Liden, E., et al., Some observations relating to behind-body armour blunt trauma effects caused by ballistic impact. The Journal of trauma, 1988. 28(1 Suppl): p. S145-8.
[7] Montanarelli, N., et al., Protective garments for public officials. 1973, ARMY LAND WARFARE LAB ABERDEEN PROVING GROUND MD.
[8] Goldfarb, M.A., et al., A method for soft body armor evaluation: medical assessment. 1975, National Technical Information Service: Springfield, VA, USA.
[9] Prather, R.N., C.L. Swann, and C.E. Hawkins, Backface Signatures of Soft Body Armors and the Associated Trauma Effects. 1977, Chemical Systems Laboratory: Aberdeen Proving Ground, MD, USA.
[10] National Research Council, Testing of body armor materials: Phase III. 2012, National Academies Press: Washinton, DC, USA.
[11] Larsen, B., et al., Body armor, performance, and physiology during repeated high-intensity work tasks. Military medicine, 2012. 177(11): p. 1308-1315.
[12] Shanley, L.A., B.L. Slaten, and P.S. Shanley, Military protective clothing: Implications for clothing and textiles curriculum and research. Clothing and Textiles Research Journal, 1993. 11(3): p. 55-59.
[13] Ricciardi, R., P.A. Deuster, and L.A. Talbot, Metabolic demands of body armor on physical performance in simulated conditions. Military medicine, 2008. 173(9): p. 817-824.
[14] Council, N.R., Testing of body armor materials: Phase III. 2012, National Academies Press: Washinton, DC, USA.
[15] Gryth, D., et al., Severe lung contusion and death after high-velocity behind-armor blunt trauma: relation to protection level. Military medicine, 2007. 172(10): p. 1110-1116.
[16] Mirzeabasov, T., et al. Further investigation of modelling system for bullet-proof vests. in Personal armour systems symposium. 2000.
[17] Bowen, I.G., E.R. Fletcher, and D.R. Richmond, Estimate of man's tolerance to the direct effects of air blast. 1968, Lovelace Foundation for Medical Education and Research: Albuquerque, NM.
[18] Prat, N., et al., Intrathoracic pressure impulse predicts pulmonary contusion volume in ballistic blunt thoracic trauma. Journal of Trauma and Acute Care Surgery, 2010. 69(4): p. 749-755.
[19] Klinich, K.D., et al. Factors associated with abdominal injury in frontal, farside, and nearside crashes. Stapp car crash journal 54 (2010): 73.
[20] Boxenbaum, H., Interspecies scaling, allometry, physiological time, and the ground plan of pharmacokinetics. Journal of pharmacokinetics and biopharmaceutics, 1982. 10: p. 201-227.
[21] Eppinger, R.H., J.H. Marcus, and R.M. Morgan, Development of dummy and injury index for NHTSA's thoracic side impact protection research program. SAE transactions, 1984: p. 359-387.
[22] Sarron, J., et al. Physiological results of NATO BABT experiments. Personal Armour Systems Symposium. 2000. Colchester, UK.

https://doi.org/10.52202/080042-0007

Appendices

Appendix 1: Cadaveric Human BABT Test Data

Specimen	Test	Specimen Mass (kg)	Impactor mass (kg)	Impact Velocity (m/s)	Impact Energy (J)	Reference Scaled Energy (J)	Rib Fracture (0/1)	Scaled Rifle Velocity (m/s)
Human_01	1	81.2	0.36	21.4	82.7	81.4	0	447
Human_01	2	81.2	0.36	21.6	83.8	82.6	1	450
Human_01	3	81.2	0.36	40.9	301.4	296.9	1	853
Human_02	1	79.4	0.36	21.3	81.9	82.5	1	445
Human_02	2	79.4	0.36	20.8	77.7	78.2	1	433
Human_03	1	90.7	0.36	19.4	67.7	59.8	0	404
Human_03	2	90.7	0.36	39.6	282.4	249.1	0	826
Human_04	1	81.6	0.36	19.1	65.5	64.2	1	397
Human_04	2	81.6	0.36	18.8	63.3	62.0	1	391
Human_05	1	89.8	0.22	26.0	74.2	66.1	0	423
Human_05	2	89.8	0.22	33.3	121.7	108.4	1	542
Human_06	1	73.9	0.22	23.5	60.8	65.8	1	383
Human_06	2	73.9	0.22	20.3	45.2	48.9	1	330
Human_07	1	86.6	0.22	17.4	33.2	30.7	0	283
Human_07	2	86.6	0.22	23.3	59.7	55.1	1	379
Human_08	1	77.1	0.22	22.3	54.6	56.7	0	363
Human_08	2	77.1	0.22	21.0	48.6	50.4	0	342
Human_09	1	82.5	0.22	14.8	24.0	23.3	0	241
Human_09	2	82.5	0.22	15.4	26.2	25.4	0	251
Human_10	1	77.1	0.22	18.7	38.5	40.0	0	305
Human_10	2	77.1	0.22	18.8	38.9	40.3	0	306
Human_11	1	52.3	0.24	20.3	49.4	75.5	0	345
Human_11	2	52.3	0.24	19.4	45.1	69.0	0	330
Human_11	3	52.3	0.24	27.2	88.8	135.9	0	463
Human_11	4	52.3	0.24	28.3	96.1	147.0	1	482
Human_11	5	52.3	0.24	18.3	40.0	61.1	0	311
Human_11	6	52.3	0.24	39.6	187.9	287.4	1	673
Human_12	1	65.8	0.24	18.8	42.3	51.5	0	320
Human_12	2	65.8	0.24	42.0	211.4	257.0	1	714
Human_13	1	88.5	0.251	20.6	53.5	48.3	1	359
Human_13	2	88.5	0.251	20.4	52.3	47.2	0	355
Human_13	3	88.5	0.251	20.2	51.0	46.1	0	351
Human_13	4	88.5	0.251	29.5	109.4	98.9	1	514
Human_13	5	88.5	0.251	21.4	57.7	52.2	0	373
Human_13	6	88.5	0.251	29.5	109.5	99.0	0	514
Human_13	7	88.5	0.251	40.5	206.3	186.5	1	706
Human_13	8	88.5	0.251	29.5	109.5	99.0	0	514
Human_13	9	88.5	0.251	40.5	205.5	185.8	1	704

https://doi.org/10.52202/080042-0007

Specimen	Test	Specimen Mass (kg)	Impactor mass (kg)	Impact Velocity (m/s)	Impact Energy (J)	Reference Scaled Energy (J)	Rib Fracture (0/1)	Scaled Rifle Velocity (m/s)
Human_14	1	95.3	0.251	21.7	59.1	49.6	0	378
Human_14	2	95.3	0.251	30.2	114.5	96.1	0	526
Human_14	3	95.3	0.251	39.6	196.8	165.2	1	689
Human_14	4	95.3	0.251	20.1	50.7	42.6	0	350
Human_14	5	95.3	0.251	30.1	113.7	95.4	1	524
Human_14	6	95.3	0.251	20.5	52.7	44.3	0	357
Human_14	7	95.3	0.251	29.0	105.5	88.6	0	505
Human_14	8	95.3	0.251	38.7	188.0	157.8	0	674
Human_14	9	95.3	0.251	30.2	114.5	96.1	0	526
Human_14	10	95.3	0.251	39.1	191.9	161.1	0	680
Human_15	1	95.3	0.262	20.8	56.4	47.4	0	369
Human_15	2	95.3	0.262	30.0	118.3	99.3	0	534
Human_15	3	95.3	0.262	39.8	207.2	174.0	1	707
Human_15	4	95.3	0.262	20.0	52.2	43.8	0	355
Human_15	5	95.3	0.262	30.7	123.7	103.8	0	546
Human_15	6	95.3	0.262	39.7	206.8	173.6	1	706
Human_15	7	95.3	0.262	30.9	125.4	105.3	0	550
Human_15	8	95.3	0.262	40.1	210.8	177.0	0	713
Human_15	9	95.3	0.262	29.3	112.7	94.6	0	522
Human_15	10	95.3	0.262	40.4	214.2	179.8	0	719
Human_16	1	64.0	0.266	20.5	55.9	69.9	1	367
Human_16	2	64.0	0.266	20.2	54.3	67.8	1	362
Human_16	3	64.0	0.266	20.2	54.3	67.8	0	362
Human_16	4	64.0	0.266	30.1	120.5	150.6	1	539
Human_16	5	64.0	0.266	19.9	52.7	65.8	0	357
Human_16	6	64.0	0.266	29.8	118.1	147.6	0	534
Human_16	7	64.0	0.266	39.2	204.4	255.5	1	702
Human_17	1	74.8	0.276	26.6	97.6	104.4	1	485
Human_17	2	74.8	0.276	25.2	87.6	93.7	1	460
Human_17	3	74.8	0.276	32.8	148.5	158.8	1	599
Human_17	4	74.8	0.276	20.7	59.1	63.2	1	378
Human_18	1	85.3	0.276	19.0	49.8	46.7	0	347
Human_18	2	85.3	0.276	17.3	41.3	38.7	1	316
Human_18	3	85.3	0.276	17.2	40.8	38.3	0	314
Human_18	4	85.3	0.276	24.6	83.5	78.3	1	449

Appendix 2: Cadaveric Swine BABT Test Data

Specimen	Test	Specimen Mass (kg)	Impactor mass (kg)	Impact Velocity (m/s)	Impact Energy (J)	Reference Scaled Energy (J)	Rib Fracture (0/1)	Scaled Rifle Velocity (m/s)
Swine_01	1	37.6	0.22	15.3	25.7	54.8	0	249
Swine_01	2	37.6	0.22	15.6	26.8	57.0	0	254
Swine_02	1	45.4	0.22	12.7	17.7	31.2	0	207
Swine_02	2	45.4	0.22	25.1	69.4	122.2	0	409
Swine_03	1	42.6	0.22	16.8	31.0	58.3	0	274
Swine_03	2	42.6	0.22	14.8	24.0	45.1	0	241
Swine_04	1	37.2	0.22	15.5	26.6	57.1	1	253
Swine_04	2	37.2	0.22	31.5	109.1	234.6	1	513
Swine_05	1	35.8	0.22	17.6	34.0	76.1	1	287
Swine_05	2	35.8	0.22	18.0	35.6	79.6	1	293
Swine_06	1	39.5	0.22	20.1	44.6	90.3	0	328
Swine_06	2	39.5	0.22	24.1	63.9	129.5	1	393
Swine_07	1	35.38	0.22	16.6	30.3	68.6	1	271
Swine_07	2	35.38	0.22	27.2	81.1	183.3	1	442
Swine_08	1	44.45	0.22	20.9	47.9	86.1	1	340
Swine_08	2	44.45	0.22	19.9	43.6	78.6	1	325
Swine_09	1	43.5	0.22	17.7	34.5	63.4	1	288
Swine_09	2	43.5	0.22	14.4	22.9	42.2	1	235
Swine_10	1	41.3	0.22	16.8	31.2	60.4	0	274
Swine_10	2	41.3	0.22	13.4	19.7	38.1	0	218
Swine_11	1	43.3	0.29	19.9	57.4	106.1	0	372
Swine_11	2	43.3	0.29	22.2	71.1	131.4	0	414
Swine_11	3	43.3	0.36	24.6	108.5	200.4	0	512
Swine_11	4	43.3	0.36	41.8	315.1	582.2	1	872
Swine_12	1	52.1	0.348	23.8	98.8	151.7	0	488
Swine_12	2	52.1	0.348	22.2	86.0	132.0	0	456
Swine_12	3	52.1	0.348	31.5	172.3	264.6	1	645
Swine_12	4	52.1	0.348	31.5	172.3	264.6	1	645
Swine_13	1	45	0.24	21.5	55.3	98.3	0	365
Swine_13	2	45	0.24	19.6	46.1	81.9	0	333
Swine_13	3	45	0.24	40.9	200.4	356.3	1	696
Swine_13	4	45	0.24	52.3	328.2	583.5	1	890
Swine_14	1	28	0.231	19.9	45.7	130.7	1	332
Swine_14	2	28	0.231	19.5	43.9	125.5	0	326
Swine_14	3	28	0.231	19.4	43.5	124.2	0	324
Swine_14	4	28	0.231	26.0	78.1	223.1	0	434
Swine_15	1	69	0.262	33.8	149.7	173.5	1	601
Swine_15	2	69	0.262	41.6	226.7	262.8	1	740
Swine_15	3	69	0.262	19.3	48.8	56.6	0	343

https://doi.org/10.52202/080042-0007

Specimen	Test	Specimen Mass (kg)	Impactor mass (kg)	Impact Velocity (m/s)	Impact Energy (J)	Reference Scaled Energy (J)	Rib Fracture (0/1)	Scaled Rifle Velocity (m/s)
Swine_15	4	69	0.262	45.5	271.2	314.4	1	809
Swine_16	1	64	0.251	19.6	48.2	60.3	0	341
Swine_16	2	64	0.251	28.3	100.5	125.6	1	493
Swine_16	3	64	0.251	43.1	233.1	291.4	0	750
Swine_16	4	64	0.251	27.4	94.2	117.8	0	477

https://doi.org/10.52202/080042-0007

Loads Associated with Behind Helmet Blunt Trauma: Matched-Pair Load Sensing Headform Tests Correlated with Skull Fracture Severity

A. Iwaskiw[1], C. Howes[1], R. Hingorani[1], C. Bradfield[1], E. Mazuchowski[2], M. Clark[3], Q. Luong[1], D. Drewry[1]

[1]The Johns Hopkins Applied Physics Laboratory, 11100 Johns Hopkins Road, Laurel, Maryland 20723, alexander.iwaskiw@JHUAPL.edu
[2] Forensic Pathology Associates, Division of HNL Lab Medicine, 1255 S. Cedar Crest Blvd, Allentown, PA 18103
[3]U.S. Special Operations Command, 10 General Greene Avenue, Natick, MA 01760

Abstract. Events where ballistic rounds do not completely perforate combat helmets are hypothesised to cause blunt injury [1]. These events and subsequent injuries, termed behind helmet blunt trauma (BHBT), are caused by the rapidly-deforming backface of the helmet striking the wearer and are not well understood. Current and previous helmet testing methodologies and acceptance standards are based on the use and evaluation of the deformation left by the helmet backface in clay. Limitations in clay testing have led to interest in more repeatable mechanical surrogates, potentially for use in First Article and Lot Acceptance Testing. The Adaptable Testing and Load Assessment System (ATLAS) Headform was developed by the Johns Hopkins Applied Physics Laboratory [2] to provide physics-based measurements of backface loading from a live fire BHBT event. This novel, more repeatable, and user-friendly testing system would benefit from establishment of a relationship with human injury to inform testing standards. There are several examples of studies using post-mortem human surrogates (PMHS) to evaluate the injuries seen in these BHBT events [3, 4, 5]. To understand the mapping between physics-based measurements and injury probability, similar ballistic events including round type, velocity, helmet type, impact locations, and fit conditions should be tested on both PMHS and the ATLAS Headform system. This study compares sixty previously-conducted PMHS BHBT test results [5] with sixty-five ATLAS Headform results across a range of relevant ballistic doses and impact locations. These studies used the same type of helmet and fit conditions targeting a consistent standoff between the helmet and head, and live ammunition. These matched-pair tests were conducted to provide associated fracture lengths, injury severity, peak load, and force impulse for a similar BHBT event. The relationship between loading conditions and skull fracture can be used for comparison to similar predictive models for related injury types, human computational model validation, and future performance standard development.

1. INTRODUCTION

Combat helmets are designed to protect the wearer against many threats. As a crucial part of ballistic personal protective equipment, helmets are evaluated for their ability to stop complete penetration of specific ballistic threats while limiting the amount of helmet backface deformation (BFD). While the implications of stopping a round from completely penetrating are fairly clear, the implications of a round not fully penetrating the helmet and subsequently deforming are less understood. The rapid deformation of the backface of these helmets can impact the head and cause an injury termed behind helmet blunt trauma (BHBT). BHBT may include scalp contusions, scalp lacerations, skull fractures, and brain injury and has the potential to be fatal [5]. To mitigate this risk, ballistic tests are conducted on combat helmets mounted on headforms, but the relationship between BFD loading and injury is not well characterised. This limits the ability to create test standards that are relevant to reducing human injury.

Various headform testing has been conducted to evaluate BFD loading using live ammunition striking helmets. BFD loading is dependent on helmet impact location, suspension pad configuration, and projectile characteristics influencing deformation response. Two testing methods are favored, one being clay-backed headforms and the other being load cell-backed headforms. Clay headforms measure the deformation in clay caused by the helmet BFD and typically must not exceed a certain threshold [6]. These clay headforms suffer from a lack of repeatability stemming from the clay's less-controlled formulation and exacerbated by the specific geometrical constraints for helmet and headform interactions [7]. Conversely, load cell-backed headforms have been evaluated and are favored for their repeatability and temporal (dynamic) measured response for blunt ballistic testing, for example the Biokinetics Ballistic Load Sensing Headform [8], the Force Reaction Evaluation Device [9], and the load sensing headform developed by the researchers at the University of German Federal Armed Forces [10]. One of these load cell testing platforms is the Adaptable Testing and Load Assessment System headform (ATLAS) developed by the Johns Hopkins Applied Physics Laboratory (JHU/APL). The ATLAS is a

 https://doi.org/10.52202/080042-0008

newly developed test platform and has design features that allow for high-throughput armour performance testing [2]. There is great interest in gathering data on this headform associated with BHBT loading. While comparisons of headform results and computationally predicted injuries have been made [10], the direct relationship between BFD loading measured with mechanical headforms and actual experiments recreating human BHBT injury has not been established, hindering the ability to assess human injury risk. This results in non-injury-based helmet performance standards that may limit the optimisation of future helmet systems in protecting from BHBT. There is a clear need to study the phenomena of BHBT across a range of relevant loading conditions to establish a relationship between non-penetrating ballistic events and injury risk.

Previous studies have been conducted on postmortem human subjects (PMHS) to determine the relationship between loading to the skull and injuries [11, 12, 13]. Very few studies have been conducted with BHBT loading parameters [3, 4, 14], which are low mass and high velocity. In addition to the need for a more comprehensive PMHS BHBT injury study to determine the potential injuries from these events, having associated BFD loading parameters (e.g. clay deformation, peak load, etc.) with these injuries could form the basis for future helmet BHBT performance standards.

For this current study we leveraged the PMHS model, wearing light-weight helmets struck by live ammunition, to understand the relationship between BHBT loading and human injury (at four locations) and associated loading and armour performance measurements made by the ATLAS Headform. These relationships represent preliminary injury prediction models for BHBT and could be used to develop future helmet testing methods, performance standards, and more protective armour designs.

2. METHODS

2.1 Projectile and Helmet Combination

The same projectile and helmet combination was used for all tests in this study. The projectile chosen was a 124 grain, 9 mm full metal jacket, while striking velocities were intentionally varied ranging from 298-423 m/s to achieve research goals. The helmet was a modern lightweight helmet constructed primarily out of ultra-high molecular weight polyethylene. Each helmet was shot once at one of four locations, the crown, front, left side, and rear. Impact locations on the helmet were chosen in order to replicate the common test locations during helmet qualification testing. (Figure 27). Helmets had pads and suspensions systems installed according to the manufacturer's instructions. Pads included in the helmets were 1.27 cm thick made of expanded polypropylene, positioned behind each impact location. Due to the placement of the pads, the impact location on the crown of the helmet was between two pads and the impact location on the side of the helmet was offset (i.e., not-centred) with a pad behind. Both the rear and front impact locations were directly centred on a pad.

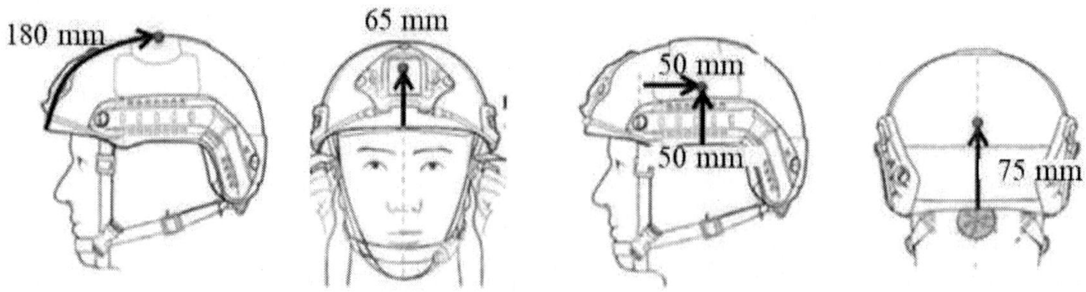

Figure 27: Helmet impact locations (red circle) for (left to right) crown, front, left side, and rear. Numbers (in mm) indicate distance measured across the surface of the helmet, along the direction indicated by the arrow

Tested locations for PMHS anatomical impacts were varied due to normal anatomical variation of skull bone and suture location as well as from the steps taken to get desired impact obliquity on the helmets (Figure 28). Crown impacts were on the mid-sagittal plane and posterior to the bregma (which is located at the intersection of the coronal and sagittal sutures). Front impacts were on the mid-sagittal plane and were approximately 10 cm above the most superior aspect of the orbits. The impact on the left side of the head was approximately 10 cm above the auditory meatus, perpendicular to the Frankfort plane on the parietal bone. The rear impact was on the midsagittal plane just below the lambdoid suture on the occipital bone.

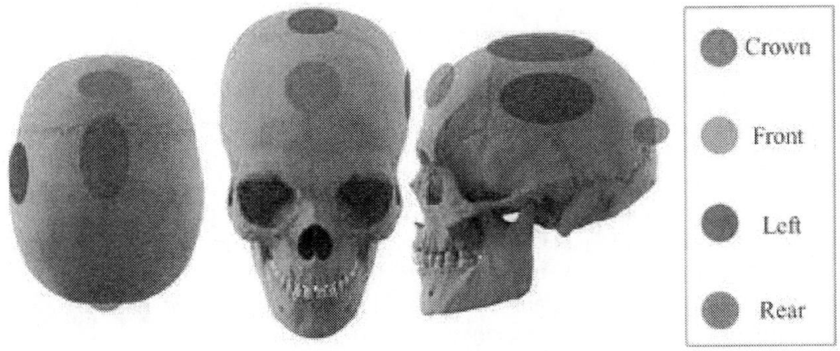

Figure 28: Approximate anatomical impact locations for the crown (red), front (green), left side (blue), and rear (purple)

2.2 PMHS Testing

2.2.1 PMHS Specimens

All PMHS specimens were fresh-frozen, thawed, un-embalmed head-neck complexes. The donor inclusion parameters and criteria that were targeted were age (18-61 years), gender (male), race (any), and absence of known medical history that may influence tissue response (e.g., absence of skull disease or head trauma). All specimens were comprised of intact skulls and external soft tissue (i.e., scalp and muscles), as the latter has been found to significantly influence skull fracture risk [3]. Since proper fitment of the helmet system being tested was necessary to remove additional variability in outcomes, specimens were selected with target specifications of head width (15.3 – 16.0 cm), head length (19.6 – 20.5 cm), and head circumference measured at the eyebrow level (56.0 – 59.0 cm) according to the fit documentation from the helmet manufacturer.

2.2.2 PMHS Specimen Preparation and Testing

Specimens were prepared by dissecting the lower cervical spine to expose the vertebral bodies to be potted in polymethyl methacrylate. Various electromechanical sensors were installed on the specimens, including strain gauge rosettes, acoustic emission sensors, intracranial pressure sensors, and 6 degree-of-freedom accelerometer packages. The results from the electromechanical sensors are not the focus of this paper and are not discussed further. Care was taken to install the sensors in a way that would not disrupt tissue at the impact site or introduce any loading of the BFD to potentially influence injury.

Prepared PMHS specimens were secured with a custom fixture securing the potted lower cervical neck to a rigid test frame. This custom fixture positioned the head-neck system in an inverted position, which used gravity to achieve more realistic specimen positioning due to the lack of active neck muscle tension.

After donning the helmet on the specimen, an internal helmet-surface to head standoff distance of 23 ± 0.5 mm was achieved for all shot locations. Standoff between the outside of the helmet shell to the surface of the head was measured using a FaroArm® (FARO Technologies, Lake Mary, FL) to remain consistent with the testing standard as well as between tests within this study. The thickness of the helmet shell was measured using precision calipers in order to determine proper standoff. Projectile velocity measurements were obtained using Oehler Research Model No. 57 (Oehler Research, Austin, TX) infrared screens with counter chronographs (universal counters, Hewlett-Packard model No. 53131A).

2.2.3 PMHS Forensic Evaluation

Following ballistic testing, the heads underwent Computed Tomography imaging with and without the helmet. Anatomical dissections were completed within two days of ballistic testing. For the anatomical

 https://doi.org/10.52202/080042-0008

assessments, an external examination of the scalp and impact location was completed before reflecting soft tissues to expose the skull surface. An internal examination was then conducted by opening the calvarium to inspect the inner skull layer and soft tissue structures (e.g., dura and brain). Detailed notes and photo-documentation were collected, specifically noting fracture type (e.g., linear, depressed), severity, location, and length. The length of the fractures in this study were obtained by measuring unique fracture lines on both the inner and outer table of the skull and combining them for a total fracture length metric. Complex and comminuted fractures were analysed after dissection through photo-documentation, measuring the boundaries of each piece and summing the inner and outer table fracture lengths. Images from the CT scan were reviewed and injuries documented. A medical report was provided for each test. Results from the anatomic dissection and CT scan were coded with Abbreviated Injury Scale (AIS) 2005 © updated 2008 [15] for each documented injury. No injuries to the dura mater or brain parenchyma were observed in this study. However due to the limitations of the PMHS model, injuries to these structures, including concussion and a more severe brain injury, may occur under similar conditions in a living individual. For each test, the maximum AIS was determined (MAIS). The dissections, review of CT scans, preparation of reports, and injury scoring were performed by a board-certified forensic pathologist with extensive trauma experience

2.3 ATLAS Headform Testing

The ATLAS Headform was developed to evaluate the ballistic performance of combat helmets as an alternative to the current clay-based approaches used to measure the potential for BHBT [2]. The ATLAS Headform is a platform that can accommodate different impact locations, helmet sizes, and helmet geometries via incorporation of a modular head structure. The system uses a piezoelectric force sensor with a measurement range up to 111 kN (Force Sensor Model 224C/FCS-DI IC, PCB Piezotronics), to provide a temporal measurement of impact force transmitted to the head. Importantly, as a result of the design of this system there is no need to reposition the stationary base, post, and load cell for different impact locations; rather, the head configuration changes around the load cell, as can be seen in Figure 29.

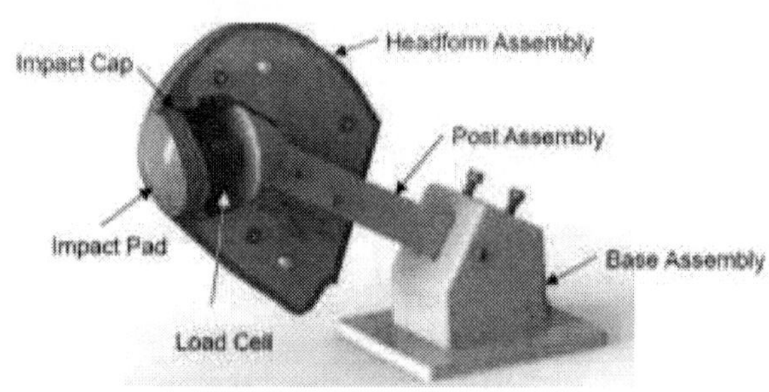

Figure 29: Schematic of the JHU/APL ATLAS Modular Headform system.

To swap configurations, the user removes the single-use neoprene impact pad, stainless steel impact cap, and polymer headform (2 pieces) and swaps them out with components for a different impact location or headform size. These modular components are held in place with magnets, alignment features, and/or friction and thus do not require mechanical fasteners. The ATLAS Headform was designed such that, when the helmet is seated properly on the headform, there is 23 mm of standoff between the helmet shell and headform and perfect alignment and zero-degree obliquity between the projectile, helmet and load cell.

For the ballistic test series, the ATLAS Headform load cell was connected to a signal conditioner (Model 482C05, PCB Piezotronics). The signal conditioner was connected to a high-rate data acquisition system (SIRIUS R4, Dewesoft, Trbovlje, Slovenia) that recorded load cell time history data. Data was collected at a 1 MHz sampling rate with a 5th order Bessel analog anti-aliasing filter with a 100 kHz cutoff frequency. No other digital filtering was applied to the data and force signal oscillations, potentially caused by system resonance and BHBT loading dynamics, were present in the force-time histories. All electronic systems were simultaneously triggered with the signal disruption of a frangible paper breakscreen disrupted by the incoming projectile approximately 1.5 m from the impact. Data was post-processed in MATLAB (MathWorks, Natick, MA). The data features were the peak force defined as the maximum of the signal, time to peak force defined as the duration between the trigger and peak force, and impulse computed by integrating the force-time history over 1 ms (Figure 30).

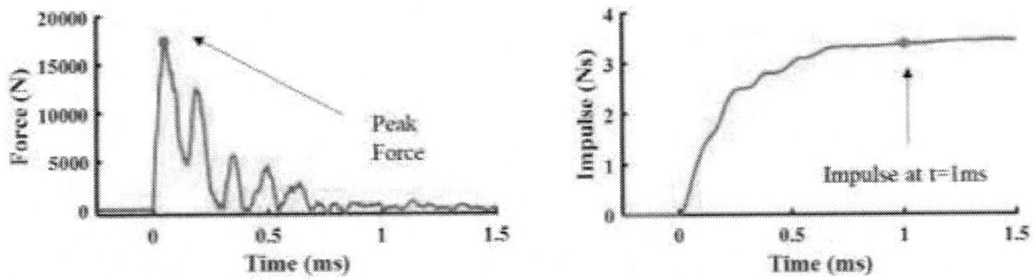

Figure 30: Example (left) force and (impulse) time histories and corresponding peak force and total impulse at 1 ms.

A sample size of four repeated tests were conducted at each of four impact locations and at each of four discrete impact velocities that were evenly spaced and spanned the range of impact velocities used for PMHS testing for a given location, for a total of 65 ATLAS Headform tests. Linear regressions were fit to the ballistic conditions versus loading results for each impact location, quantifying the relationship between ballistic conditions (e.g. striking energy, which was varied) to ATLAS peak load and impulse. The relationship between ballistic conditions and ATLAS headform results are not presented in this paper. However, these linear regressions all had high coefficient of determinations (e.g. R^2 values) with the lowest being 0.869. This higher coefficient of determination is influenced by the repeatability of the ATLAS headform system and the linear relationship between ballistic velocity and load response at the tested ballistic condition, which are deemed adequate for this ballistic testing application. These resulting relationships were used to evaluate the equivalent loading in the PMHS tests across the different studies and impact locations for specific testing ballistic velocities. These matched-pair load and impulse values were evaluated for their ability to predict injury in the PMHS data set.

3. RESULTS

The range and average results from PMHS injury data (i.e., MAIS and fracture length) are compared to loads evaluated at the ballistic conditions of each PMHS test (Table 14). One case from the rear test group had a disputed fracture and is not included in the results. Overall, assessed MAIS values were seen ranging from 1 – 4 and fracture lengths ranged from 0.0 (no fracture) to 142.0 cm. The associated matched-pair test loads ranged from 9.6 kN to 22.6 kN. The impulse ranged from 2.62 to 3.81 Ns.

For those tests with no fracture, the MAIS was 1 or 2 depending on the extent of the scalp injury. Total fracture length trended with MAIS. Complex and comminuted vault fractures have higher AIS severities, 3 and 4 respectively, and typically have increased fracture lengths compared to simple vault fractures (AIS severity of 2).

Injury results, organized by impact location and ATLAS derived metrics, can be found in Figure 31. Generally, increased loading corresponded with fracture and injury severity. However, there was a large region of mixed results, where peak loads and impulses seen with non-fracture cases were 17.3 kN and 3.34 Ns. Alternatively, the lowest peak loads and impulses for fracture cases were 10.9 kN and 2.67 Ns, respectively. The associated forces and impulse with crown fractures were greater than the loading associated with other location. The front impact location had the lowest loading associated with those tests.

Table 14. Average loading and injury outcomes organised by impact location

Impact location	Maximum AIS	Number of Outcomes	Average Peak ATLAS Load (kN)	Average ATLAS Impulse (Ns)	Average Total Fracture Length (cm)
Crown	1	4	15.9±1.57	3.21±0.14	0.0±0.0
	2	3	17.8±0.71	3.39±0.06	10.0±7.1
	3	2	19.0±0.96	3.49±0.09	41.0±11.3

https://doi.org/10.52202/080042-0008

	4	2	21.5±1.10	3.71±0.10	125.0±17.0
Front	1	7	11.1±1.00	2.75±0.09	0.0±0.0
	2	3	12.4±1.27	2.88±0.12	21.0±1.4
	3	8	14.4±2.42	3.05±0.22	38.0±17.4
	4	2	17.2±1.60	3.31±0.14	36.0±4.7
Left Side	1	7	13.1±1.37	2.91±0.15	0.0±0.0
	2	2	14.0±1.09	3.01±0.12	6.9±6.1
	3	4	15.6±1.44	3.18±0.16	44.9±41.4
	4	2	15.1±1.41	3.13±0.15	96.7±2.8
Rear	1	5	11.6±1.06	3.00±0.13	0.0±0.0
	2	1	14.9±0.00	3.39±0.00	3.5±0.0
	3	6	13.3±0.74	3.20±0.09	38.1±33.6
	4	1	13.1±0.00	3.17±0.00	56.2±0.0

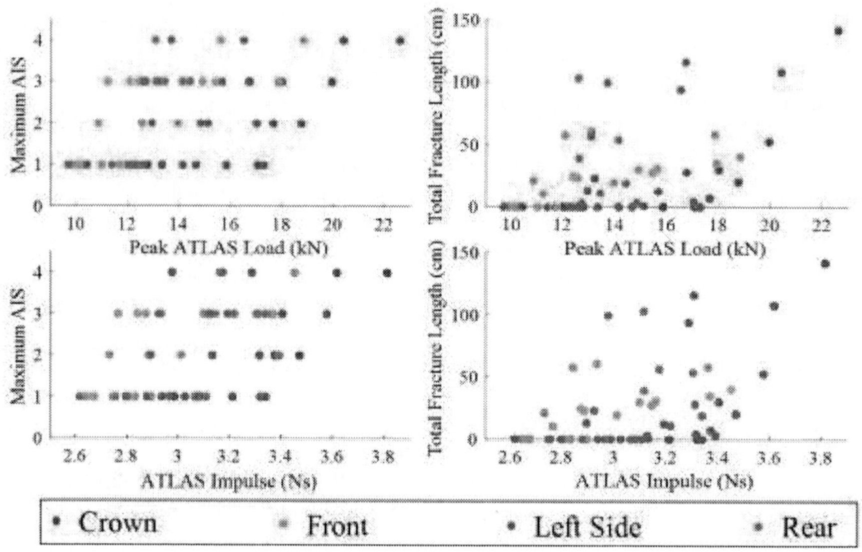

Figure 31: Scatter plots of the peak ATLAS Load (N) versus Maximum AIS, grouped by impact location (Left Top). Peak ATLAS Load versus total fracture length, grouped by different impact location (Right Top). ATLAS Impulse versus Maximum AIS, grouped by different impact location (Right Bottom)

Peak ATLAS load and impulse were fit, using a linear regression, to total fracture length for each impact location (Figure 32). Non-fracture cases were omitted from the regression in order to establish a predicted threshold of fracture (e.g intercept) and not weighing the regression with many non-fracture cases (n=24). All locations indicated a relationship of increasing load with respect to fracture length, aside from the rear location. The crown results are the only results that showed a coefficient of determination over 0.9. P-values for linear coefficient of ATLAS loading variable were lowest for the crown location (0.0009) and varied between 0.2427-0.4272 for the other locations.

https://doi.org/10.52202/080042-0008

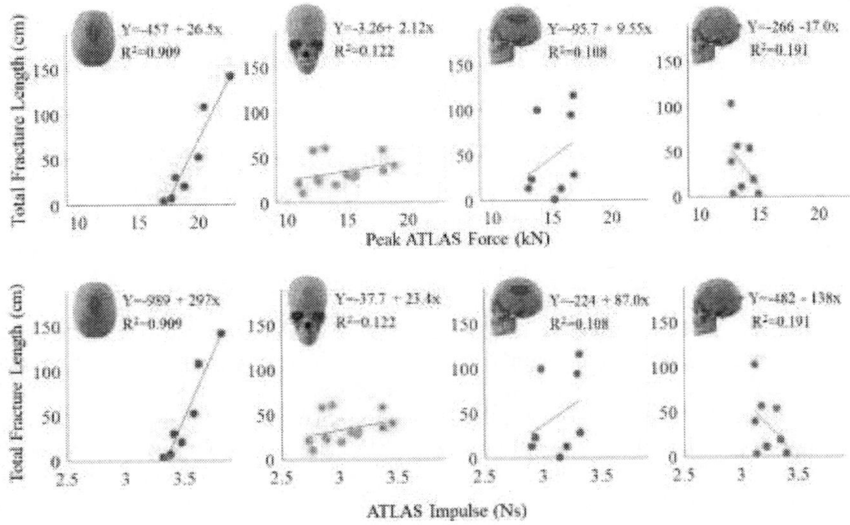

Figure 32: Linear regression of maximum ATLAS load and total fracture length for each impact location - crown, front, left side, rear from left to right (Top) Linear regression of maximum ATLAS impulse and total fracture length for each impact location - crown, front, left side, rear from left to right (Bottom)

4. DISCUSSION

4.1 ATLAS Loading

The main objective of this study was to provide a relationship between BFD loading and fracture outcomes using a matched pair testing methodology. In previous studies [16, 17], clay headform results showed large variations in clay deformations associated with a single ballistic and armour condition. For this reason, the ATLAS Headform was preferred for conducting these matched pair testing with the PMHS conditions.

The ATLAS Headform provides repeatable results across repeated conditions with the same ballistic conditions and same impact location. This was measured by the coefficient of variation, which is the ratio of the standard deviation of a repeated condition set with the mean of that set. The ATLAS headform tests had coefficients of variation ranging from 2.4%-11.1% and 0.2%-5.6% across all locations for peak load and impulse, respectively, and are a lower amount of variation to comparable clay tests. The lower coefficients of variation for the impulse measurements are due to the complete transfer of momentum of the ballistic round to the load cell in all tests. Peak loads have higher coefficients of variation, where varied pad interactions at different impact locations (e.g. loading directly over a pad for front and rear, or loading between pads for crown) may influence rise time and peak magnitudes. The repeatability in the results of the ATLAS Headform allows for more confidence in the associated loading metrics to human injury and fracture. The loading and injury results show a general trend of increasing dose resulting in increasing injury. The forces observed in this study are, on average, higher than fracture tolerances observed in literature from ballistic, blunt and motor vehicle crash loading [13]. It is hypothesised that fracture tolerances may be higher in BHBT due to the faster loading rate, shorter duration, and lower effective loading masses than in other loading scenarios [18]. The measured impulse values observed in these tests were generally in agreement with total ballistic momentum, however some tests saw higher measured impulse values than total ballistic momentum. This may be due to a number of factors influence the impulse delivered to the load cell such as helmet backface elastic rebound away from the load cell or due to the integration of post peak oscillations in the force-time history. It is noted that the ATLAS Headform steel impact cap and neoprene impact pad have a different stiffness than the human skull and scalp, likely resulting in a different force response than what a PMHS experiences during ballistics impact. While additional work is needed to evaluate the differences between the two, prior helmet testing resulted in similar residual backface deformation with the PMHS and the ATLAS headform, suggesting the dynamic helmet deformation interacts similarly with the human head and surrogate [19]. Ultimately, the ATLAS headform's steel impact cap and neoprene impact pad are integral

https://doi.org/10.52202/080042-0008

to its durability, which is critical to First Article and Lot Acceptance Testing that requires repeat evaluation.

4.2 PMHS Injury Models

While developing a relationship between AIS and BHBT loading can be helpful when understanding the implications of injury, the categorical nature of the AIS injury scores limits the ability to comprehensively model the data. While there is an implicit assumption that injury response has a monotonically increasing relationship with the loading, there is no knowledge as to the correspondence between monotonically increasing injury outcome categories. Practically, there is little to no information gained by characterising a lower threshold of injury (e.g. MAIS 1: scalp injury) when attempting to characterise the higher threshold of injury (e.g. MAIS 4: complex and comminuted fractures) other than the observation that higher thresholds of injury happen at higher loading conditions. These categorical relationships are best suited for binomial regressions and have been shown in other studies [20]. Additionally, the correlation of AIS to loading shows a large area of transition, where injury results from equivalent loading ranged from 1 to 4 maximum AIS results. This is due to potential influences such as geometrical uniqueness of impact locations with respect to one another, biological variance of fracture tolerance, and underlying anatomical features that may affect fracture and injury severity.

There are distinct advantages to using continuous variables like fracture length to create predictive models. These models strengths are their simplicity and ability characterise the relationship between the two variables with higher statistical power. Additionally, fracture length can be considered a biomechanical failure response, one that may be correlated to other human body models, such as finite element method models. Additional fracture based failure models can be leveraged to have a better theoretical understanding of the mechanicals involved in the loading and response. Future studies should further investigate the relationship between fracture length and measured load with the goal of creating these predictive relationships for the BHBT condition.

4.3 Impact Location Differences

There are differences in the relationships between BHBT loading and BHBT injury that are observed at different locations. Using a BHBT loading assessment method, such as the ATLAS Headform, allows for the combination of helmet, padding and threat interactions at each location to be evaluated together. This testing methodology should, ideally, measure loading equivalently at each location so the results of this data can be used to understand injury thresholds of each anatomical location. Ultimately, the differences observed across impact locations could provide insight on human injury risk.

The data from this current study indicates that the front location has lower fracture tolerances, or lower loads required to fracture the skull. This can be observed by noting that relatively lower forces and impulse are associated with higher fracture lengths and maximum AIS scores, as compared to other locations. In increasing order of fracture tolerance: the rear, side and crown have higher loads associated with similar injuries and fracture lengths. When evaluating skull fracture risk there is no consensus on the ordinal ranking of fracture tolerances of the skull as many studies have different loading methods as well as impact locations [21]. Therefore, it is hypothesised that the specific geometries of the PMHS skulls used in the study may have a great influence on the BHBT injuries and associated loads. Previous studies have utilised PMHS to evaluate the physical and mechanical properties of the skull [21]. These studies have found the flexural properties of the skull to be highly dependent on skull thickness. Additionally, it is known that the amount of flesh present at an impact location can influence loading [13], primarily by distributing load from the BFD, not unlike a helmet pad. It is important to identify the anatomical variations at the impact location to understand the potential influence on injury. There is no clear guidance established as it relates to skull and scalp thickness at different locations of the head [22, 23]. Using the CT scans, the scalp and skull thickness were characterised for these PMHS test impact locations [24]. For these specimen, the front and crown location had the lowest scalp thickness at 3.9 mm and 3.4 mm mean thickness, respectively. The scalp thicknesses of the left side and rear were the highest at 6.3 mm and 5.7 mm mean thickness, respectively. The mean skull thickness results for the impact locations in our study revealed higher values for the front (8.1 mm) and rear impact location (8.6 mm) than the crown (7.0 mm) and left side (6.3 mm). Variation of the skull thickness across the tests in this study were similar at each location. While the front impact location has relatively high skull thicknesses, it has the lowest scalp thicknesses of any impact location in our study. This may be a major factor in the fracture tolerance to BFD loading at this impact location. Additionally, proximity to the anatomical structures like the frontal sinus should also be considered when evaluation injury risk to locations on the frontal bone as they may also influence fracture patterns. Ultimately, different injury risk functions for different impact locations may have major implications on helmet performance standards and design, where one may have higher performance standards for vulnerable areas of the skull.

4.4 Study Limitations

The ATLAS Headform can measure peak forces and total impulse but does not capture the spatial distribution of loading that could affect injury results, due to the fact that is a single sensor in combination with an impact cap and skin pad. Melvin [25] describes a strain energy approach to skull fractures that highlights different dominating fracture mechanisms dependent on impactor type. Without an understanding of BHBT loading shape over time, it may be unable to predict the fracture tolerance due competing failure mechanisms. This might hinder the ability of setting a helmet performance standard based on load and impulse alone. For example, one might expect higher fracture risk with a "sharper" loading shape versus a "rounded" shape. However, it is recommended that further research be conducted to understand the relationship between shape characteristics and BHBT injury.

While a strength of the current study was the use of a single helmet type for all ballistic data collected with PMHS and load-sensing headforms to ensure strong matched-pair conditions, these injury outcomes are relevant for a single helmet and projectile type. Care must be taken to interpret these results in context of other helmet systems, particularly with different energy or BFD profiles engaging the head.

Finally, this study only evaluates four nominal impact locations with a limited amount of PMHS samples. The severity of outcomes seen in this study are dependent on the tolerance to failure as well as underlying anatomical features (e.g. blood vessels, sinuses) with respect to the impact location. Further studies and models should explore all potential impact locations before extrapolating the results to untested impact locations with additional samples to provide an increase in the robustness of the response predictions.

5. CONCLUSION

A total of sixty non-penetrating ballistic impacts against helmeted PMHS specimen have been completed in addition to sixty-five matched-pair ATLAS Headform tests. Helmets were tested at four locations (front, rear, crown, and side) with ballistic conditions to span injurious loading regimes that were then measured by the ATLAS in matched-pair tests to generate loads to be associated with the injuries. All tests sustained injury. Skull fractures types included linear, depressed, and comminuted and measured to obtain the length of the fractures of the inner and outer table of the skull. The data indicated that different impact locations have differing sensitivity to skull injury with the crown and front location showing the highest and lowest resistance to skull fracture of the locations in this study, respectively. Loads were generally higher than those seen in other head trauma studies, and the crown location was the only location with strong correlations between loading and injury (fracture length). With these preliminary relationships from loading on a testing platform (ATLAS Headform) and human injury, steps can be made to mature helmet testing standards to incorporate human injury risk. With better helmet testing methods and baseline injury data sets, future armour and testing standards can provide advantages to the warfighter, increasing survivability.

Acknowledgments

The authors would like to express their appreciation to the U.S. Special Operations Command for funding this work under Contract No. H92222-15-D-0004.

References

[1] A. W. Carroll and C. A. Sonderstrom, "A new nonpenetrating ballistic injury," *Annals of surgery*, vol. 188, no. 6, p. 753, 1978.

[2] M. M. Trexler, et al. "Development of a Modular Load-Sensing Headform," in *Personal Armour Systems Symposium*, Washington D.C., 2018.

[3] C. R. Bass, et al. "Helmet behind armor blunt trauma," in *Specialist Conference on Human Factors in Medicine*, Koblenz, Germany, 2003.

 https://doi.org/10.52202/080042-0008

[4] K. A. Rafaels, et al. "Injuries of the head from backface deformation of ballistic protective helmets under ballistic impact," *Journal of forensic sciences,* vol. 60, no. 1, pp. 219-225, 2015.

[5] C. Howes, et al. *Behind Helmet Blunt Trauma: A Potentially Significant Mechanism of Injury for Helmeted Warfighters,* 2022.

[6] J. L. Underwood, et al. *Technology Assessment Program NIJ Standard for Ballistic Helmets,* 1981.

[7] Y. Li, et al. "Review of Mechanisms and Research Methods for Blunt Ballistic Head Injury," *Journal of biomechanical engineering,* vol. 145, no. 1, 2023.

[8] L. Voo, et al. "Improved Repeatability and Reproducibility of the Ballistic Load Sensing Headform," in *Personal Armour System Symposium*, Amsterdam, 2016.

[9] J. Barnes-Warden and P. Fenne. "Developments in Ballistic Helmet Testing," in *Personal Armour Systems Symposium*, Amsterdam, 2016.

[10] R. Aranda, et al. "Integrated Assessment of Ballistic Backface Deflections on Combat Helmets and Prediction of Injury Risks," in *31st International Symposium on Ballistics*, Hyderabad, India, 2019.

[11] C. A. Weisenbach, et al. "Preliminary Investigation of Skull Fracture Patterns Using an Impactor Representative of Helmet Back-Face Deformation," *Military medicine,* vol. 183, pp. 287-293, 2018.

[12] D. C. Viano, et al. "Ballistic Impact to the Forehead, Zygoma, and Mandible: comparison of Human and Frangible Dummy Face Biomechanics," *Journal of Trauma and Acute Care Surgery,* vol. 56, no. 6, pp. 1305-1311, 2004.

[13] D. Raymond, et al. "Tolerance of the Skull to Blunt Ballistic Temporo-Parietal Impact," *Journal of biomechanics,* vol. 45, no. 12, pp. 2479-2485, 2009.

[14] C. J. Frietas, et al. "Dynamic Responses Due to Behind Helmet Blunt Trauma Measured with a Human Head Surrogate," *Int J Med Sci,* vol. 11, no. 5, pp. 409-425, 2014.

[15] T. A. Gennarelli and E. Wodzin, *Abbreviated injury scale 2005: updated 2008,* Russ Reeder, 2008.

[16] A. Miranda-Vicario, et al. "Experimental study of the deformation of a ballistic helmet impacted with pistol ammunition," *Composite Structures,* vol. 203, pp. 233-241, 2018.

[17] P. L. Gotts, et al. "Personal Armour Testing versus Small Arms Ammunition when the Test Standard Doesn't Fit," *Problemy Mechatroniki: uzbrojenie, lotnictwo, inżynieria bezpieczeństwa,* vol. 6, pp. 19-30, 2015.

[18] N. Yogananda and F. A. Pintar, "Biomechanics of temporo-parietal skull fracture," *Clinical Biomechanics,* vol. 19, no. 3, pp. 225-239, 2004.

[19] C. Howes, et al. *Effect of backing material on helmet backface deformation from partially penetrating ballistic impacts,* 2022.

[20] H. C. Cutcliffe, et al. *How few? Bayesian statistics in injury biomechanics,* 2012.

[21] J. H. McElhaney, et al. "Mechanisms of basilar skull fracture," vol. 12, no. 4, pp. 669-678, 1995.

[22] A. Moreira-Gonzalez, et al. "Calvarial thickness and its relation to cranial bone harvest," *Plastic and reconstructive surgery,* vol. 117, no. 6, pp. 1964-1971, May 2006.

[23] R. W. Young, et al. "Age changes in the thickness of the scalp in white males," *Human biology,* vol. 31, no. 1, pp. 74-79, 1959.

[24] A. S. Iwaskiw, et al. "Automated Measurement Technique for the Determination of Regional Skull and Scalp Constituent Thickness," *Biomedical sciences instrumentation,* vol. 51, pp. 165-172, 2015.

[25] J. W. Melvin and F. G. Evans, *A Strain Energy Approach to the Mechanics of Skull Fracture,* 1971.

Injury risk functions for behind armour blunt trauma based on clay backing cavity volume and depth

G. Pageau[1], S. Ouellet[2] and A. Bouamoul[2]

[1]*Calian Ltd., 2459 Route de la Bravoure, Québec, Qc, Canada*
[2]*DRDC Valcartier, 2459 Route de la Bravoure, Québec, Qc, Canada*

Abstract. A review of body armour behind armour blunt trauma (BABT) requirements and test protocols in national and international standards and specifications is presented. Back Face Deformation (BFD) measurement in a clay witness block remains used as a pass-fail tolerance criterion despite its limited medical basis, mainly due to the absence of a better validated and readily implementable testing method for body armour compliance certification. A review and meta-analysis of the extensive historical BABT data were conducted to determine if more relevant information and trends could be extracted. The emphasis was made on comparing the previously developed thoracic BABT Injury Risk Functions (IRF) based on tests with real bullets and armour, and with surrogate projectile replicating the interaction with the armour system, on live animal subjects (LAS) and post-mortem human subjects (PMHS). The test cases of Cooper and Bowen were found to be not fully exploited yet. The different injury scales and metrics used in previous studies have generated apparently diverging IRFs, potentially contributing to adopting widely different BFD tolerance levels in various body armour standards. Published data from reconstructions of BABT survivor field cases on soft armours were used to generate a new IRF called the VD^2 model, where the cavity volume (V) and the square of depth (D^2) in clay backings are combined as a blunt trauma dose parameter for predicting the risk of AIS-2/B injury. The same approach, using VD^2 as the injury predictor, was then applied to the data from reconstruction on clay blocks of the Cooper test cases on non-protected or bare porcine subjects and the Bowen test cases on bare canine subjects with their respective rigid impactors. The correlations obtained between kinetic energy, clay volume, and penetration depth were then used for generating the matched-pair VD^2 data related to the blunt trauma injury severity reported by Cooper and Bowen. The corresponding IRFs obtained for AIS-3 injury (Cooper) and AIS-6 (Bowen) are shown to follow a graded injury trend with the IRFs computed from Rafaels-Bir. The iso-injury-severity curves computed by plotting the data on a deformation-volume map are shown to be parallel with some logical upward offsets. For medium-size cavities (i.e., 125 cm^3), the 44 mm BFD limit can be associated with a 20% AIS-3 injury risk. Adopting a higher BFD limit (e.g., 50 mm) may be considered for clay cavities with smaller volumes but with caution. A lower BFD limit (e.g., 30 mm) would be needed for larger volume cavities to provide a more consistent level of BABT mitigation over the full range of cavity sizes. Until a more physically-based and medically-validated BABT assessment methodology is implemented, clay-based body armour specifications should consider adding the measurement of clay cavity volume, as currently prescribed by the VPAM standard but using laser scanning instead of the water filling method, as a complementary indicator of transmitted energy. However, additional validation of the VD^2 model with more recent and reliable data is needed before considering its potential adoption in body armour performance standards.

1. CURRENT STATUS OF CLAY-BASED BABT TEST METHODS

The role of body armour is to provide ballistic protection against bullet penetrating injuries and to mitigate the Behind Armour Blunt Trauma (BABT) to safe defined levels. BABTs are non-penetrating injuries resulting from the transmitted kinetic energy (KE) and Back Face Deformation (BFD) of the armour when impacted by a bullet. Ballistic resistance testing of a body armour system requires a backing material for replicating the as-worn condition while simulating the human body deformation resistance and acting as a recording medium for quantifying BABT with a measurable parameter. Multiple studies on BABT have been conducted over the last 45 years to understand BABT injury mechanisms better and develop proper assessment methods and relevant tolerance limits, which is a difficult task given all the critical parameters involved (Figure 1a). Due to its relatively low cost, long storage life, re-usability, and ease of implementation and shaping, clay-based BABT assessment has remained the only method specified for certifying body armour despite its limited medical basis and, more specifically, the uncertainty around the correlation between clay-based BFD measurements and BABT injury risks for both soft and hard armours.

Oil-based modelling clay also suffers from other important limitations related to its calibration procedure, temperature sensitivity, and the high variability and low reproducibility of the BFD measurements. Unlike human tissues, clay is naturally non-elastic with some elastic recovery with the residual deformation being slightly smaller than the transient one. The magnitude of the elastic recovery of clay is likely variable and related to the impact conditions. The maximum size of the permanent

https://doi.org/10.52202/080042-0009

residual cavity can then be captured, allowing the direct measurement of the maximum indentation/cavity depth. The maximum permanent BFD is the pass/fail criterion specified in current test standards.

To have the right consistency and pass calibration, clay must also be heated around 35 °C limiting the usage time (e.g., 45 min) at room temperature. Before ballistic testing, the consistency of clay is calibrated with a drop test which cannot be done at the same location on the clay box as the area where the armour sample will be positioned, which is another drawback of the method. The recent development of a temperature-insensitive clay ballistic grade [1], referred to as ARTIC, should make BFD measurements more repeatable and reliable without needing pre-heating and re-conditioning the clay block so that its calibration remains stable. The clay BFD methodology has been reviewed recently by the ASTM Subcommittee E54.04 [2-4] to improve its reproducibility and repeatability in supporting the NIJ-101.07 standard. For the more accurate measurement of maximum cavity depth for curved hard ballistic plates, the depth gauge and bridge caliper method, suitable for planar armour systems, has been replaced by laser surface profilometry in US DoD military standards [5] with a pre-scan of the curved clay reference plane and a post-scan of the clay cavity. Comparisons of the 3D solid models generated from the pre-scan of the original undisturbed clay surface and the post-impact surface are then made semi-automatically nowadays [6-8] with suitable point-cloud analysis software (e.g., *Geomagic*) to obtain the maximum cavity depth relative to the line of fire. Additional cavity shape parameters such as volume, cavity base and external surface areas (Figure 1b) can also be easily obtained. The complete 3D cavity profiles can be archived and used later in match-pair studies correlating blunt trauma severity. Clay being non-transparent, only post-impact static measurements can be made, which is another important limitation. For characterising the dynamic deflection-time response, an instrumented clay box with ultrasonic and pressure sensors [9] was developed to record the deformation, velocity and acceleration time history, which may be better related to the injury mechanisms but not captured with static BFD measurements where only magnitude and spatial distribution are recorded. Alternative test methods continue to be explored to replace clay-based testing, but despite promising progress, none is sufficiently mature and validated yet for implementation in a national standard.

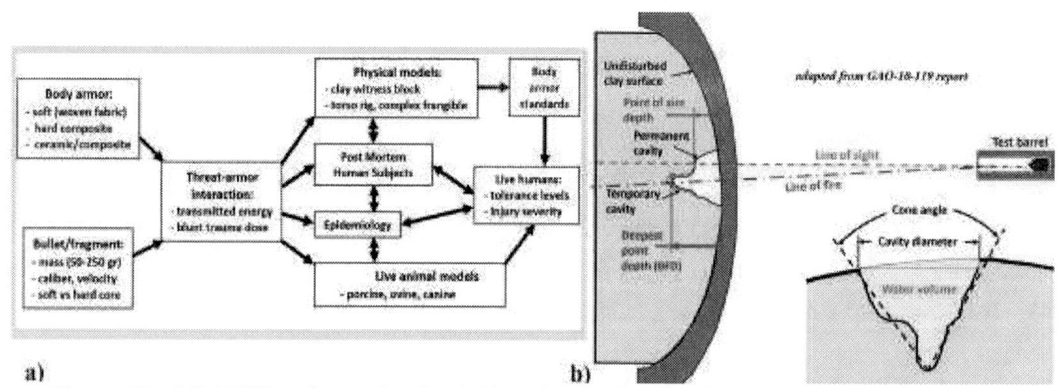

Figure 33: a) BABT key factors involved, b) cavity depth and volume measurement for curved surfaces

Table 1 compares the clay-based BFD test methods and requirements for commonly used law enforcement and military standards. NIJ-101.06 [10], the most widely used body armour standard, specifies a maximum BFD depth of 44 mm (80% upper tolerance limit and 95% confidence) as a pass/fail criterion. In the German VPAM BSW-APR standard [11], the BFD limit is adjusted as a function of the measured clay consistency, with a value of 44 mm for softer clay and 40 mm for harder clay. VPAM is the first standard to introduce the maximum cavity volume as an additional requirement. The UK Home Office [12] adopted a more conservative BFD limit of 30 mm for hard armour. To address soldier overload, lighter ballistic plates exploiting the latest material technologies and reducing torso coverage ratio are being procured and deployed for low-intensity threat environments [13]. For those plates, a BFD requirement of 58 mm was also adopted based on epidemiological data indicating no occurrence of severe thoracic BABT injury in recent military conflicts.

Furthermore, no soldier battlefield fatalities are known from the perforation of hard ballistic plates by projectiles they were designed for [14]. The 58 mm BFD limit is estimated not to affect soldiers' survivability significantly. In the Russian GOST standard [15], no BFD limit is specified for rifle projectiles, while for handguns, BFD is limited to 17 mm. Such a lower limit may indicate the desire to minimise blunt trauma and probably more incapacitation, ensuring that the wearer would remain functional during shooting incidents. The significant differences in the BFD tolerances levels between the national standards raise the question of which one should be used and if they really are incompatible and diverging that much.

Adopting a too-low and very conservative BFD limit is not necessarily better since it would impose a weight penalty hindering soldiers' mobility. A higher BFD limit may put soldiers at risk of severe BABT injuries. Establishing an optimal trade-off between armour weight, soldier mobility, and protection remains a non-trivial challenge. Scalable body armour systems are being developed, which allow the protection level to be adjusted to the perceived ballistic threat and take into account the acceptable level of injury severity from potential BABT overmatch ballistic threats. Improvement in ballistic materials often leads to lighter and more flexible armour systems having the same threat-stopping capability, but where the weight reduction may not be fully exploitable because BFD could become the design driver.

Table 15. Summary of back face deformation requirements in body armour standards

Body Armour Standard	Backing Type	Backing Material Consistency	Pass-Fail Criteria Measurement method
NIJ 101.06 USA, 2008	Roma Plastilina #1	1 kg sphere 2 m drop 19 ±3 mm range 19 ±2 mm average	All BFD ≤ 44 mm or All BFD ≤ 50 mm if 95 % confidence that 80 % of all BFD≤ 44 mm (one-sided CI)
NIJ draft 101.07 USA, 2018 ASTM E3004 ASTM E3068 ASTM E3086	Roma Plastilina #1 from Sculpture House Inc	1 kg sphere 2 m drop 19 ±2 mm all drops 1 kg cylinder 2 m drop 6.15 - 6.27 m/s velocity 25 ±3 mm	Same as NIJ 101.06 BFD measurement with Ø6.35mm probe tip depth gage
CAST-2017 UK, 2017	Roma Plastilina #1	1 kg sphere 2 m drop 19 ±3 mm range 19 ±2 mm average	Soft armour: BFD ≤ 44 mm Hard armour: - single shot BFD ≤ 30 mm - max mean BFD≤ 25 mm
VPAM BSW-APR 2006 Germany, 2009	Weible Plastilina	1 kg sphere 2 m drop 20.0 ± 2.0 mm Plasticity (P): 18-22 mm	BFD ≤ P + 22 mm Volume ≤ (0.134 x P-1.13) x 70 J Measured with water filling
ESAPI Rev. J US DoD, 2018	Not specified	1 kg cylinder 44.5 mm diameter hemispherical tip 2 m drop 25 ±3 mm	BFD ≤ 44 mm, laser scanning 1st shot: 90% UTL - 90% Confidence 2nd shot: 80% UTL - 90% Confidence
SPS Light Torso Plate PEO Soldier, 2020	Roma Plastilina #1	Same as ESAPI	BFD ≤ 58 mm Measured with laser scanning

2. REVIEW OF BABT INJURY RISK FUNCTIONS AND SCALES

The correlation between BFD and BABT severity is still limited and is still based on the ballistic tests conducted 45 years ago by Prather [16] with soft armour on goats from which lethality curves were drawn (Figure 2, left). Therefore, the 44 mm BFD limit was selected as the proposed threshold for soft armour corresponding to a 6% lethality risk. To address the limited medical basis issue, a number of complementary Injury Risk Functions or IRF have been developed with tests on PMHS and LAS. The main BABT IRFs developed based on clay BFD measurement and chest wall displacement [17-24] are shown in Figure 2, with some generated for soft armour and others for hard armour.

https://doi.org/10.52202/080042-0009

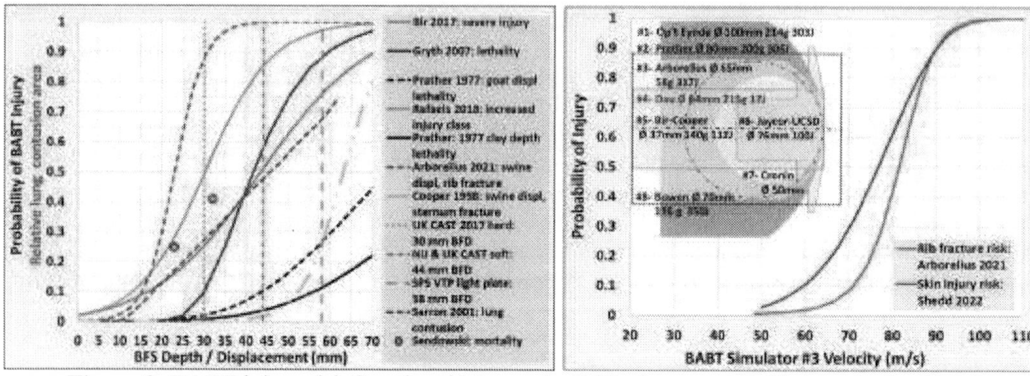

Figure 2: BABT injury risk functions/data: vs depth/displacement (left), vs velocity (right)

Several injury scales have been defined to provide a common basis for evaluating BABT injuries when conducting epidemiology studies and LAS/PMHS tests, which have led to the formulation of several IRFs which, at first, look diverging for the allowable safe BFD level. For the analysis of recreated field cases of law enforcement soft body armour, Bir and Rafaels have refined the AIS scale, which classifies the severity of individual injuries from grade 1 to 6 in terms of threat-to-life, by adding two classes, A and B (Table 2). This refinement was done to provide increased discrimination between minor and clinically insignificant injuries with no medical attention needed (class AIS-1A, green box) and those clinically significant since requiring wound care (class B: AIS-2/6, red boxes).

Table 2. AIS injury scale with Bir-Rafaels BABT complementary classes (A &B)

	Abbreviated Injury Scale (AIS-2015) **AIS code/severity & *typical injuries*** *Probability of death [44]*	**A: Clinically insignificant** **No wound care**	**B: Clinically significant** **Wound care needed**
1	**Minor:** ***1-2 rib fractures, skin/muscle contusion hematoma,*** *Probability of death: 0.2%*		
2	**Moderate:** ***3+ adjacent rib fractures, fractured sternum, 50% liver contusion*** *Probability of death: 1%*		
3	**Serious: not life-threatening:** ***4+ rib fractures, heart contusion*** *Probability of death: 2%*		
4	**Severe: threat to life but survivable:** ***displaced sternum, major contusions*** *Probability of death: 9%*		
5	**Critical: survival uncertain:** ***Bilateral flail chest, ruptured liver*** *Probability of death: 25%*		
6	**Fatal:** *Probability of death: 100%*		

Since BABT injuries have been found to have similar characteristics to those caused by the direct impact (i.e., target without armour) of non-penetrating projectiles used for Less Lethal Weapons (LLW), such projectiles and BABT rigid simulators (Figure 2, right) [25-31], specially designed to reproduce the backface deformation profile or force history, have been used in BABT studies to avoid using actual armours and bullets and provide better control of the variables involved when testing with PMHS and LAS. The IRF proposed recently by Arborelius [21] is shown in Figure 2 (right), where for an impact velocity of 82 m/s or 195 J, the risk of rib fracture would be 50%. Based on the IRF proposed by Shedd [32], scaled for the Arborelius BABT impactor using the Blunt Criterion, the probability of a skin and open wound injury at the same energy level would be 65%, which is coherent with BABT pathologies reported in previous studies.

3. DEVELOPMENT OF THE VD² IRF FROM THE DATA OF RECREATED FIELD CASES

https://doi.org/10.52202/080042-0009

Reconstructions of BABT survivor field cases have shown that clay cavity volume, although related to transmitted kinetic energy, is not a better injury predictor than cavity depth alone. Rafaels [19] developed an IRF based on the normalised cavity surface area/volume ratio to account for the empirical evidence that post-impact cavities with increased volume and BFD depth resulted in greater BABT, as illustrated in Figure 3, where cavity shape and volume can vary greatly depending on the armour type (soft/flexible vs stiff/hard) and projectile characteristics (velocity and core hardness) for the same BFD depth [33]. Projectile impacts on soft/flexible armour vests tend to produce more conical cavities than those on stiffer hard armour plates.

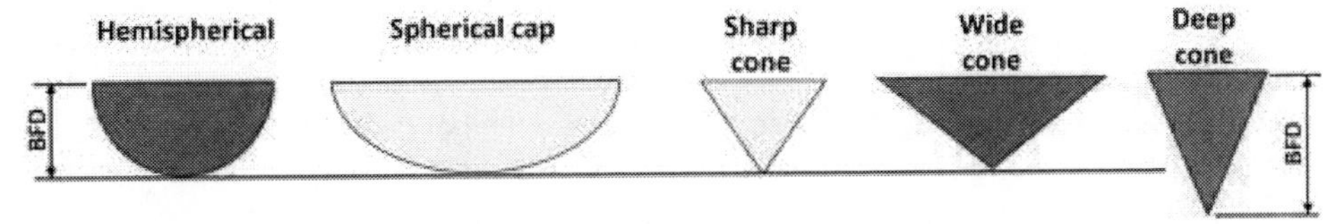

Figure 3: Comparison of backface cavity shapes

An analysis of the Bir-Rafaels data was conducted using a similar approach but only considering cavity depth and volume and not shape where the blunt trauma dose injury predictor selected is clay cavity Volume times Depth raised to exponent ***n***, i.e., $\boldsymbol{Vol{\cdot}BFD^2}$. This formulation is similar to the Gadd severity index [34] for head injuries, where the injury metric is based on acceleration and time with more weight on acceleration from the exponent 2.5, i.e., $\boldsymbol{a^{2.5}{\cdot}T}$. The proposed $\boldsymbol{Vol{\cdot}BFD^2}$ injury predictor does not, however, take time into account, which is one of its differences with the Livermore Conical Depression Factor (CDF) [35] where: $\boldsymbol{CDF = 3{\cdot}Vol{\cdot}BFD/T}$, where ***T*** is the time at which the cone/projectile velocity has slowed to 5% of its initial velocity and almost reached its maximum depth. Back face velocity ***V*** is known to be inversely proportional to the cone base surface area ***Ab*** of cone-shaped clay cavities [20], where ***Ab*** can be expressed as **3·Vol/BFD**, so $\boldsymbol{V \approx BFD/3{\cdot}Vol}$. The viscous injury criterion (***VC***) can then be expressed as $\boldsymbol{VC \approx BFD2/Vol}$, which is a different formulation where less weight would be given to cavity volume, which is somewhat counter-intuitive.

Logistic regressions of the BABT recreated test data were conducted with a Logit link function where the weighting factor "***n***" was varied from 1 to 3. An "***n***" value of 2 provided a degree of correlation similar to Rafaels' model [19], and the proposed model is referred to as $\boldsymbol{VD^2}$. Since the initial Logit function [36] gave a non-zero injury risk at an extremely low $\boldsymbol{Vol{\cdot}BFD^2}$ values, the data were re-analysed statistically by performing non-linear regressions with several sigmoidal functions. The best fit obtained with Equation 1a is shown graphically in Figure 4 (green curve), with the dotted lines representing the 95% confidence intervals. For example, a BABT dose of 715 cm^5 corresponds to a risk of a BABT injury of AIS-2/B of 50% (Equation 1b).

$$\textbf{a)} \quad P(AIS2/B) = 1 - e^{\left[-\left[Vol\cdot BFD^2/870.7\right]^{1.86}\right]} \qquad \textbf{b)}\ Vol \cdot BFD^2 = 715 \quad (1)$$

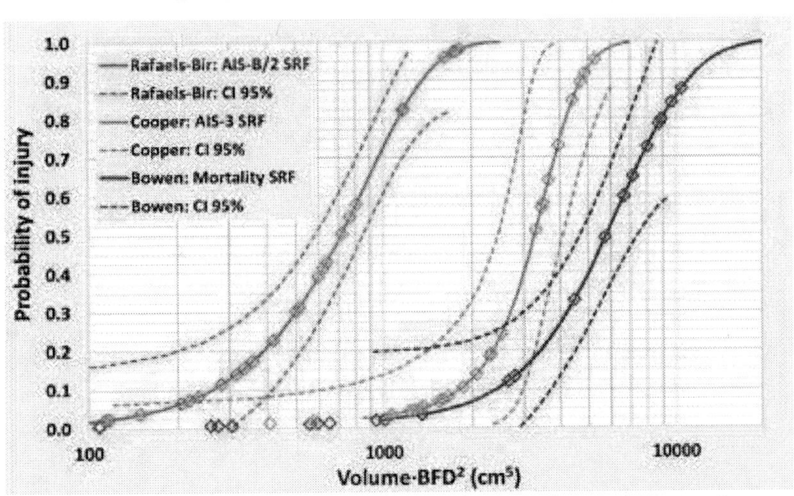

Figure 4: VD2 IRFs for Rafaels-Bir data, Cooper data, and Bowen data

https://doi.org/10.52202/080042-0009

4. CHARACTERISATION OF RP1 CLAY WITH RIGID BABT SURROGATE IMPACTORS

From the review of BABT historical data, the study of Cooper [29] with porcine subjects and the study from Bowen [30], with canine subjects were identified as a valuable source of test results with conditions resulting in higher BABT injury severities worth being analysed with the VD^2 model, but this required conducting reconstruction match-pair conditions with clay blocks and the same rigid impactors and velocities. The impact of rigid/blunt projectiles onto clay blocks, even though shown to reproduce chest wall velocity and injury severity when impacting unarmored animal bodies when adequately designed [21], has some limitations since the interaction mechanisms with clay are likely not entirely similar.

A series of tests were then conducted with the DRDC gas gun launcher, which included the 140-gram-37mm diameter cylinder of Cooper, also used by Bir later on, the 196-gram-70 mm diameter cylinder of Bowen as well as the BABT impactor of Arborelius [21], and a rigid version of the Lacrosse ball of Dau [26] (Figure 5a). Once impacted, the clay blocks were levelled and drawn off with a flat blade sliding on the reference side edges to remove the clay cavity lip.

The cavity depth was measured with a calibrated digital caliper, and the volume was measured as per the VPAM standard [11] with water using graduated syringes (Figure 5c) which work well with flat clay block but would not be adequate for curved armour as illustrated in Figure 1b). The plaster of Paris and silicone moulding methods could also be used for cavity volume measurement of flat armour samples but are much more laborious and less applicable for routine testing. The plaster or silicone cavity moulds (Figure 5b) help visualise and compare cavity shapes. Projectile velocity and orientation were measured using a *Photron* SA-Z high-speed camera. For the Bir impactor (140-gram, 40 m/s), the video data was analysed with the *Xcitex ProAnalyst* software to obtain the deformation time history of the projectile by tracking a reference line. The results obtained are presented in the four graphs of Figure 6. Based on the Bir-Wilhelm torso biofidelity corridors [25, 26] (Figure 6a), even though clay is not a tissue simulant, it does provide a similar match to ballistic gelatin, with both materials fitting within the male corridors initially and moving to the female corridors afterwards. The measurements of the permanent final depth were only 4% lower than the maximum transient deformation, which is less than the 30% recovery reported by Kinsler [36]. A linear fit (Figure 6b) provided a good correlation of cavity volume vs KE.

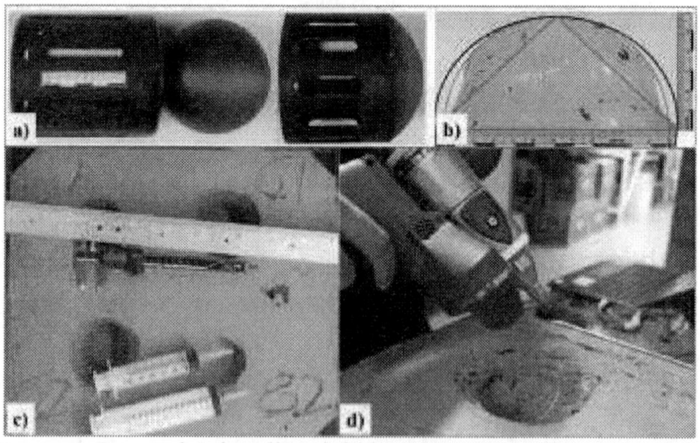

Figure 5: Dau and Arborelius rigid impactors (a), silicone cavity mould (b) Cavity volume measurement with water (c) and laser scanning (d)

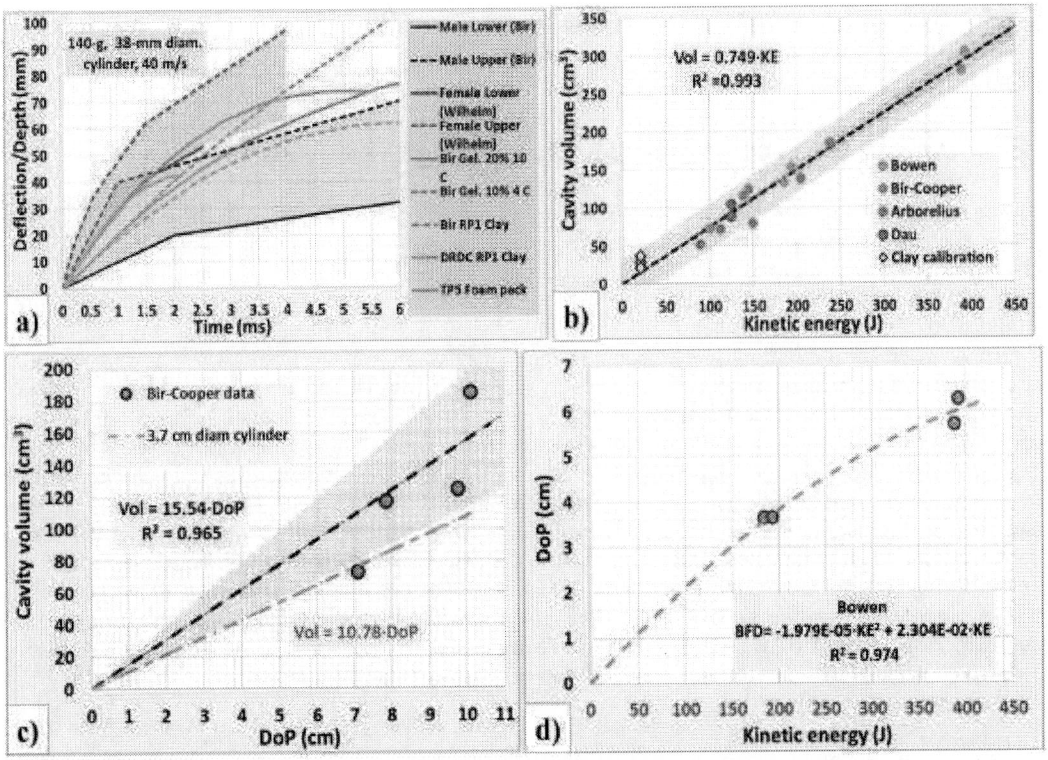

Figure 6: a) Bir-Wilhelm torso biofidelity corridors and RP1 cavity depth vs time for Bir impactor b) volume vs KE for 4 impactors, (c) volume vs DoP for Bir, (d) DoP vs KE for Bowen

5. APPLICATION OF THE VD² MODEL TO THE DATA OF COOPER

Cooper [29] conducted extensive live-fire tests on live anesthetised porcine specimens instrumented with pressure sensors where high-speed photography was used to measure chest wall displacement. The animals were impacted at the anterior mid-sternum with rigid projectiles with a diameter of 37 mm and masses of 140 and 380 grams striking at velocities between 20 and 72 m/sec. The 37mm-140-gram impactor from Cooper was used ulteriorly by Bir and Wilhelm [26, 27] for their studies with male and female PMHS. Wilhelm demonstrated that the impacts of this projectile onto bare 20% gelatin blocks provided a good match of the deflection time profile of actual soft armours on the same gelatin blocks. As shown in Figure 6a), the deflection-time response of RP1 clay and 20% gelatin are relatively close in the range of interest (i.e., up to 60 mm) for the 37mm-140 mm impactor. Comparative studies of soft armour on clay blocks and BABT torso rigs with membranes [38, 39] also show the same trend, i.e., that ***DoP*** in RP1 clay can be assumed to be close enough to the maximum chest wall displacement. For injury assessment, Cooper used a five (5) grades injury scale for rating cardiac injuries, while sternal injuries were rated as either no fracture or fracture with or without displacement. The injury scores obtained were plotted as a function of the chest wall displacement (P) ratio to the anteroposterior diameter (***AP***), i.e., ***P/AP***. A value of 24 cm was assumed for ***AP***, enabling the calculation of the chest wall displacement ***P*** for the data points included in Cooper's graph. The ***DoP*** for the match-pair tests were assumed to correspond to the chest displacement (***P***) values using the previously discussed assumptions. The injury grades from Cooper were translated to equivalent AIS numbers (Table 2), making possible comparisons with IRFs which are often based on the AIS scale. The volumes of the cavities for the match-pair tests in clay blocks with the same impactor were calculated using the equation given in Figure 6c), then allowing the blunt trauma doses ($Vol{\cdot}DoP^2$) computation for the 37 mm diameter impactor cases recreated experimentally with calibrated clay blocks. A logistic regression analysis with a Logit link function for the cases leading to AIS-3 and greater injuries gave Equation 2a, shown graphically in Figure 4 (light brown curve), with the dotted lines representing the 95% confidence intervals. Contrary to recreated soft armour field cases, the Logit fit did not display a non-zero injury for low blunt trauma doses, which is often a limitation when using logistic regressions for generating IRFs [40]. A BABT dose of 3280 cm^5 corresponds to a risk of a BABT injury of AIS-3 of 50%, as given by Equation 2b.

https://doi.org/10.52202/080042-0009

$$\text{a)} \quad P(AIS3) = 1/(1 + e^{(4.9192 - 0.0015)}) \qquad \text{b)}\ Vol \cdot BFD^2 = 3280 \quad (2)$$

6. APPLICATION OF THE VD² MODEL TO THE DATA OF BOWEN

The same approach used for the analysis of the data from Cooper was applied to the data from Bowen [31], which conducted experiments on live anesthetised canine specimens impacted onto the right lateral chest wall near the mid thorax by flat rigid cylinders with diameters of 70 mm and masses varying from 60 to 380 grams at velocities ranging from 20 to 90 m/s. Bowen rated the sustained blunt trauma injuries simply as a function of the number of rib fractures and time to death. The animals that survived were all sacrificed after 30 to 40 minutes. The canine specimens' weight ranged from 14.5 to 23.1 kg with no scaling applied to the data, which was a necessary simplification. The match-pair tests onto clay blocks were only performed with the 196-gram cylinders with a slightly smaller diameter (i.e., 68 mm instead of 70 mm) to fit the gas gun barrel readily available at the DRDC impact laboratory. The equation of Figure 6b was used to compute the cavities volume for the impact kinetic energies and velocities for the test conditions reported by Bowen. The resulting ***DoP*** in the clay blocks were calculated using the equation of Figure 6d. Assuming as previously that ***DoP*** can be used as a proxy for BFD, the corresponding blunt trauma doses ($Vol \cdot DoP^2$) were then calculated by performing non-linear regression analysis of the cases resulting in mortality (AIS-6) with a wide range of sigmoidal functions. The best fit was obtained with Equation 3a, shown in Figure 4 (red curve), with the dotted lines representing the 95% confidence intervals. A BABT dose of 5800 cm^5 corresponds to a 50% risk of AIS-6 injuries (Equation 3b), which is likely conservative since the data was not scaled to a 70 kg man.

$$\text{a)} \quad P = 1/\left[1 + e^{[-0.000356\cdot(29638 + Vol\cdot BFD^2]^{208387}}\right] \qquad \text{b)}\ Vol \cdot BFD^2 = 5800 \quad (3)$$

7. BABT SEVERITY MAP WITH POTENTIAL PASS-FAIL CRITERION

The three IRFs developed in sections 3, 5 and 6, using the VD² model for the Rafaels-Bir AIS-2B injuries from recreated field cases, the Cooper AIS-3 and Bowen AIS-6 BABT injuries are illustrated graphically in the injury map of Figure 7. They are shown to follow the same trends and be logically spaced with a larger upward offset between the 50% AIS-2/B and the 50% AIS-3 curves than between the 50% AIS-3 and the 50% AIS-6 curves. The solid black line represents the 50% risk of AIS-2B, the solid blue line the 50% risk of AIS-3 injuries, and the solid red line the 50% risk of AIS-6 injuries. The dotted lines, as indicated in the legend of Figure 7, were generated for the probabilities of injuries of 10%, 20%, and 70%. It can be seen that the iso-BABT severity curves generated follow the same trends where for the 50% injury risk, the allowable BFD is inversely proportional to the square of the cavity volume. This trend is coherent with the initial premise that for a given BFD, clay cavities with larger volume will lead to increased BABT severity and reversely that for a given cavity volume, impacts producing deeper cavities in clay will cause more severe BABT injuries.

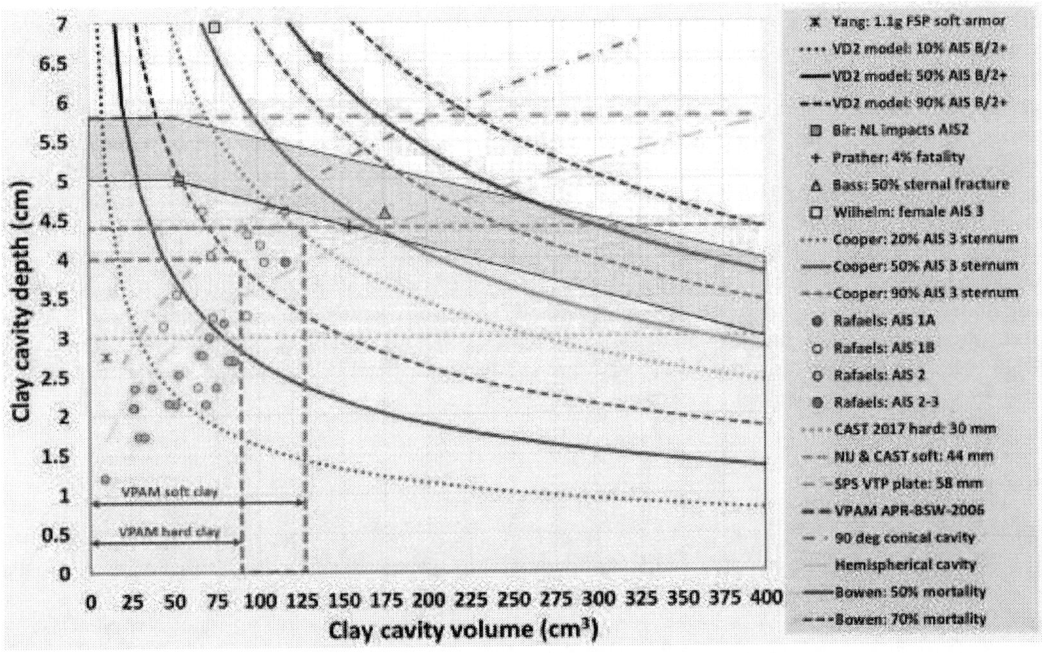

Figure 7: Cavity depth vs volume graph with iso-BABT severity curves derived from VD^2 IRFs

The BABT survivor data used for generating the AIS-2B IRF is also plotted in the injury map and is shown to be well distributed on both sides of the 50% injury risk curve. Relevant data points from other BABT studies [41, 42] not included for generating the three new IRFs are also shown to be in general concordance with the iso-BABT severity curves. The body armour BFD tolerance levels of Table 1 are also plotted in Figure 7, where it can be seen that the range of volumes and maximum BFD allowed by the VPAM standard (purple rectangular box) fits in the map centre between the 50% AIS-2/B and AIS-3 curves, and between the lines for hemispherical and 90° conical cavities.

The UK CAST, 30 mm limit, is shown to be highly conservative for small clay cavity volumes and more applicable to larger volume cavities. For cavity volumes greater than 75 cm^3, the 58 mm BFD limit is predicted to allow for severe to serious and potentially critical injuries. A good assessment of the likely direct fire threat to be encountered during the mission will be needed such that the body armour degree of ballistic perforation resistance and BABT injury mitigation adequately match the anticipated threat severity. A potential acceptance criterion is illustrated in Figure 7 as the green shaded area, where for cavity volumes below 50 cm^3, a maximum BFD of 50 mm would be allowed and then declining linearly to a maximum BFD of 30 mm at a cavity volume of 400 cm^3 following the Cooper 20% AIS-3 injury risk curve. The pink area in Figure 7 illustrates another trade-off option relying more on increased mobility from the lighter armour systems than BABT injury protection for soldier survivability, where the BFD limit was increased to 58 mm for low cavity volumes and 40 mm for high cavity volumes.

The impact velocities specified in body armour standards correspond to engagements at point blank, with standard barrel lengths and impacts occurring at zero obliquities, a worst-case situation rarely arising in operational theatres. The employment ranges of the heavier calibre weapons (e.g., 7.62x54mm sub-machine guns and marksman rifles) are also typically longer than those used with assault rifles such as the 5.45x39mm calibre (AK-74) more often used nowadays by Russia [43] or the 7.62x39mm AK47 rifles which are less of a concern for BABT overmatch due to the lower projectile mass (AK47 and AK 74) and velocity (AK47). When updating body armour standards, it may be worth adding BABT overmatch threat classes covering larger calibre sniper rifles (e.g., 338 Lapua) where the BFD assessment would be done at the typical engagement range (e.g., 500 m). For such threats, more prevalent in theatre nowadays, using anti-trauma plates worn in conjunction with plate inserts may be a worthwhile option

8. SUMMARY AND CONCLUSIONS

Based on the data from reconstructed soft armour field cases, the VD^2 model provided a good prediction of the BABT injury risks. Given the assumptions made, similar trends and levels of correlation were also obtained for the IRFs computed for the rigid impactors of Cooper and Bowen, where the original impact

https://doi.org/10.52202/080042-0009

conditions were recreated onto calibrated clay blocks. A linear correlation was obtained between cavity volume and transmitted kinetic energy indicating that $\boldsymbol{Vol{\cdot}BFD^2}$ could be considered as a weighted energy-based BABT criterion. The injury map produced was shown to be helpful in defining body armour acceptance criteria considering both cavity maximum depth and volume. It was shown that the current 44 mm BFD limit could be increased up to 50 mm, but only for impacts with cavities volumes smaller than 75 cm^3. A BFD limit of 58 mm should be used cautiously for impacts resulting in larger cavity volumes where the more stringent BFD limit of the UK CAST standard (30 mm) would procure BABT injury mitigation at a safer level. More experimental data with PMHS and LAS resulting in higher severity injuries are needed to quantify better the BABT dose effect and the development of better-suited medically-based IRFs. Improved laboratory assessment methodologies should continue to be developed in parallel and implemented in body armour standards when fully validated. Measuring clay cavity shape and volume using laser scanning should be recommended in body armour specifications to complement BFD measurement, as this would help discriminate between the BABT mitigation potential of two armour systems having the same maximum BFD.

References

[1] R. Mrozek, et al, Developing a room temperature replacement for Roma Plastilina #1 as a ballistic backing material, Proc. PASS 2020 Conference, Copenhagen, Denmark, 2021.

[2] ASTM E3086-20, Standard Practice for Creating Appliques for Use in Testing of Nonplanar Soft Body Armor Designed for Females, ASTM International, 2020.

[3] ASTM E3068-20, Standard Test Method for Contact Measurement of Backface Deformation in Clay Backing During Body Armor Testing, ASTM International, 2020.

[4] ASTM E3004-22, Standard Specification for Preparation and Verification of Clay Blocks Used in Ballistic-resistance Testing of Torso Body Armor, ASTM International, 2022.

[5] US DoD, Enhanced Small Arms Protective Insert (ESAPI), CO/PD 04-19 Rev. J, 2018,

[6] US Government Accountability Office, Warfighter Support Independent Expert Assessment of Body Armor Test Results and Procedures Needed Before Fielding, report GAO-10-119, 2009.

[7] F.D.P. Balbino et al., Systematic method for evaluating the performance of 3D optical scanners by structured light projection applied to ballistic vests tests, measurement, 190, 2022.

[8] V.D. Lee, et al, Design and calibration of an artifact for evaluating laser scanning articulating arm CMMs used for measuring complex non-concurrent surfaces, Proc. 30th ASPE, Austin, TX, 2015.

[9] E.P. Carton and Y.S. Khoe, Development and use of an instrumented alternative to the clay box, Proc. PASS 2020 conference, Copenhagen, Denmark, 2021.

[10] National Institute of Justice NIJ, Ballistic Resistance of Body Armor, NIJ Standard 0101.06, U.S. Department of Justice, Washington, DC, 2008, b) NIJ Draft Standard 0101.07, 2018.

[11] VPAM BSW 2006, Ballistic Protective Vests - Requirements, Classifications and Testing Methods, May 2009, https://www.vpam.eu/wp-content/uploads/2022/12/20090514_II_BSW2006_ENG.pdf

[12] T. Payne, et al, UK Home Office, Body Armour Standard, CAST report number 012/17, July 2017.

[13] SE. Miller and C.V. Ferguson, PM Soldier Protective Equipment, Joint Advanced Planning Brief to Industry Body Armor Session, USA, 2020.

[14] D.J. Carr, I.M. Horsfall and C. Malbon, Is behind armour blunt trauma a real threat to users of body armour? A systematic review, Journal of the Royal Army Medical Corps, 162, 2013.

[15] J. Ganor, Body armor and light ballistic armor materials and systems, Adept Press, 2021.

[16] R.N. Prather, et al, Backface Signatures of Soft Body Armors and the Associated Trauma Effects, U.S. Army Aberdeen Proving Ground, MD, Report No. ARCSL-TR-77055, 1977.

[17] C. Bir, et al, Behind Armor Blunt Trauma: Recreation of Field Cases for the Assessment of Backface Signature Testing, Proc. Int. Symp. on Ballistics, Long Beach, CA, 2017.

[18] C. Bir, C., K.A. Rafaels and K.L. Loftis, Behind Armour Blunt Trauma Injuries Sustained by Law Enforcement Officer, Proc. Int. IRCOBI Conference, Florence, Italy, 2019.

[19] K. A. Rafaels, et al, The Relationship between the Shape of Backface Deformation and Behind-Armour Blunt Trauma", Proc. PASS 2020 Conference, Copenhagen, Denmark, Oct. 2021.

[20] K.A. Rafaels, K.L. Loftis and C.A. Bir, Can Clay Tell Us More Than Deformation?, Proc. PASS 2018 Conference, Washington D.C, USA, 2018.

[21] U. P. Arborelius, et al, Pulmonary hypoxia and venous admixture correlate linearly to the kinetic energy from porcine high velocity projectile BABT, Exp. Lung Research., 47, 2021.

[22] J.C. Sarron, et al, Displacements of thoracic wall after 7.62mm non penetrating impact, Presentation to the NATO BABT Task Group, Copenhagen, 2001.

https://doi.org/10.52202/080042-0009

[23] D. Gryth, Hemodynamic, Respiratory and Neurophysiological Reactions after High-Velocity Behind Armor Blunt Trauma, PhD Thesis, Karolinska Institutet, Stockholm, Sweden, 2007.
[24] I. Sendowski, et al, Effets arrières des gilets de protection en mode non transfixiant: recherche de normes de tolérance", Report DRET 93/1349 J; restricted, Armées: DGA; 1994.
[25] J. Op 't Eynde, et al, Behind armor blunt trauma (BABT) indenter simulating high velocity impacts from rifle rounds on hard body armor, Proc. PASS 2020 Conf., Copenhagen, Denmark, 2021.
[26] N. Dau, Development of a biomechanical surrogate for the evaluation of commotio cordis protection, PhD Thesis, Wayne State University, 2012.
[27] M. Wilhelm, A biomechanical assessment of female body amor, PhD thesis Wayne State Un., 2003.
[28] C. Bir, The evaluation of blunt ballistic impacts of the thorax, PhD thesis Wayne State Un., 2000.
[29] G. Cooper, et al, Prediction of chest wall displacement and heart injury from impact characteristics of a non-penetrating projectile, Proc. Int. IRCOBI Conference, Salon de Provence, France, 1981.
[30] I.G. Bowen, et al, Biophysical mechanisms and scaling procedures applicable in assessing responses of the thorax energised by air-blast or by non-penetrating missiles, DASA-1857,1966.
[31] W. Shen, Y. Niu and J.H. Stuhmiller, Biomechanically Based Criteria for Rib Fractures Induced by High-Speed Impact, The Journal of Trauma (58) 3, March 2005.
[32] D.F. Shedd, et al, The risk of skin injury caused by high-rate blunt impacts to the human thorax, Human Factors Mech Eng for Defence and Safety, 2022.
[33] K.A. Rafaels, et al, Comparing the Backface Deformation Behavior between Soft and Hard Body Armors, Proc. Int. Symp. on Ballistics, Long Beach, CA, 2017.
[34] Gadd, C.W., Use of Weighted-Impulse Criterion for Establishing Injury Hazard, Proc. 10th Stapp Car Crash Conf., SAE, NM, USA, 1966.
[35] C.A. Honodel, Soft Armor Test Matrix 1 – Final Report, Lawrence Livermore Laboratory, Report UCRL-51677, Sept. 1974, https://www.ojp.gov/pdffiles1/Digitization/19963NCJRS.pdf
[36] G. Pageau, S. Ouellet and A. Bouamoul, Behind Armor Blunt Trauma Injuries Assessment with Clay Backing: A Meta-Analysis of Injury Risk Functions and a Proposed Cavity Volume-Depth Model, Proc. IRCOBI Conference, Porto, Portugal, Sept. 2022.
[37] R. Kinsler, "Benefits of gelation vs clay for analysing behind-armor effects in body armor testing", Proc. PASS 2008 Conference, Brussels, Belgium, 2008.
[38] K. Hewins, et al, Ballistic blunt trauma assessment methodology validation, Proc. PASS 2012 Conference, Nuremburg, Germany, 2012.
[39] P. Jain, et al, Effect of various backing materials on BABT in ballistic evaluation of body armour, Proc. Int. Symp. on Ballistics, Hyderabad, India, 2019.
[40] L. Voo, et al, Guidance Document for Human Injury Probability Curves (HIPCs) Development, DEVCOM report DAC-TR-2020-041, 2020.
[41] C. Bass, C., R. Salzar, R., et al, Injury Risk in Behind Armor Blunt Thoracic Trauma, Int. J. Occup. Saf. Ergonomy (JOSE), 12(4), 2006.
[42] Y. Yang, Study on Ballistic Performance of Hybrid Soft Body Armour, PhD Thesis, UK, 2015.
[43] Conflict Armament Research, Weapons of the war in Ukraine: a three-year investigation of weapon supplies into Donetsk and Luhansk London.
[44] M.S. Ulman and R.L. Stalnaker, Evaluation of the AIS as a measure of probability of death, Proc. Int. IRCOBI Conference, Zurich, Switzerland, 1986.

 https://doi.org/10.52202/080042-0009

Behind armour effects for overmatch threats

A. Azevedo[1], J. Dhaenens[1], A. Miranda-Vicario[1] and F. Coghe[1]

[1] *Department of Weapon Systems & Ballistics, Royal Military Academy, Brussels, Belgium, ana.ferreira@dymasec.be*

Abstract. Body armour systems, such as bullet-resistant vests or fragment-resistant helmets, are tested extensively according to different norms and standards before being introduced into service, in order to ensure the systems can provide the required level of protection. Unfortunately, a number of issues have recently emerged with in-service body armour and personal equipment for which currently no clear answer exists. This includes, among other things, differences in ballistic performance that have been noted due to differences between how in-service body armour behaves in operational circumstances, and how it behaves under controlled laboratory conditions during evaluation testing. This work will more specifically focus on the influence of the torso body armour on the wounding capacity of the projectile in case of overmatching threats. A selection of four different projectiles was made, based on their availability and their overmatching capabilities for different types of ballistic vests. For the soft armour vest the 7.62 x 39 mm M43 and the 5.56 x 45 mm NATO Ball projectiles were used. For the ceramic plate in conjunction with the soft armour vest, the .338 AP Lapua Mag and the 7.62 x 51 mm AP8 projectiles were selected. The residual wounding potential of the overmatching threats is determined using the standard approach of evaluating the energy deposit of the projectile inside gelatine, next to assessing the integrity of the projectile. This work will help better understand the behind armour effects for overmatch threats for several combinations of projectiles and ballistic vests. The effects of high velocity rifle bullets on soft armour vests will be investigated as well as the effects of armour-piercing sniper ammunition on hard armour plates. The results will give an idea about which combination of ammunition and ballistic protection is worth investigating more thoroughly during further research. Combining all the results of the four different combinations, it is possible to conclude that the temporary cavity on the gelatine occurred sooner when ballistic protection is worn. This research confirms that cavitation occurs sooner when the body armour is worn. However, no clear answer can be found to the question whether wearing body armour causes more damage or not to the body in the case of an overmatching threat. Some factors influencing wound severity seem to be worse in case of protected gelatine blocks (45 x 15 x 15 cm), while others do not. The gelatine block is 20 % concentration. Nonetheless, this research came with some new insights into the behind armour effects of overmatch threats for ceramic plates in conjunction with soft ballistic armour vests. The behaviour of the two projectiles impacting the ballistic armour with the ceramic plate is different to the two projectiles impacting the soft vest directly. In summary, regardless of whether the ballistic vest is worn or not in the case of an overmatch threat, all shots can cause a lot of damage on the human body which would most likely result in a neutralized target if they hit a vital organ.

1. INTRODUCTION

Police officers and military personnel wear body armour to prevent injuries from different threats, including firearms. The purpose of armour is hence to protect its wearer from specific threats [1]. In the case of a ballistic protection, a ballistic vest typically absorbs the kinetic energy of a bullet or fragments it in order to stop the fragments. Unfortunately, a ballistic vest does not stop every bullet. When a projectile completely perforates a specific vest, one speaks about an overmatch threat for that type of vest. There has been only limited research caried out on the effect of rifle bullets on tissue after the perforation of ballistic armour. This is the reason why this additional research was deemed necessary. Breteau et al. produced some remarkable hypothesis in 1989. During test firing on animals, they found that entrance wounds were larger for animals protected by soft and hard body armour, than those observed without protection. For the former case, the scientists also remarked that the neck was shorter and the cavity occurred earlier in the body. As a conclusion, Breteau et al. hypothesised that wearing body armour would cause greater damage [2].

These experiments have given rise to many ideas, even some scientists that started to think about the usefulness of not wearing armour. Their reasoning is in line to the hypothesis that Breteau et al. stated. As projectiles seem to have the tendency to tumble earlier inside the body when they have to pierce a ballistic vest, they might cause more damage compared with not wearing a vest. The quicker a projectile is destabilized, the earlier it induces larger cavities. The moment at which the projectile tumbles (prompt or late), determines the size of the entrance wound and the magnitude of the internal cavity. If its longitudinal axis does not deviate from its trajectory, the bullet does not cause as much damage.

To give the wound ballistics world some insight about this rumour, several researchers performed test firings. Their research led to some of the following insights.

https://doi.org/10.52202/080042-0010

As a first deduction, scientists confirmed that the projectile is destabilized much faster after perforating the vest, causing it to tumble sooner. However, Lanthier [3] remarked that this does not necessarily mean that more damage is inflicted to the tissue. During his experiments, he noticed that the total size of the cavity was smaller when a 5.45 x 39 mm bullet first perforated a soft armour vest compared with an unprotected body. Nevertheless, Mabbot et al. discovered that "earlier cavitation in a human target could cause more disruption and damage to a more susceptible area" [4].Yet, the authors also found that the cavity decreases for some projectile-armour combinations when the targets are armoured and conclude that "overmatching cannot be generalised for all overmatching scenarios" [4].

Secondly, one of the experiments conducted by Missliwetz et al. showed that the diameter of the temporary cavity was larger for the shots where body armour was present. The tests were conducted with 5.56 x 45 mm ammunition on light Kevlar® vests. Nonetheless, they did not mention a potential reason for this observation [5].

Thirdly, when ballistic protection surrounds a body, projectiles seem to have the tendency to bounce back in the body when they impact on the back armour after passing through the body. That way, the projectile can damage more vital organs and the risk of infection increases [1].
In conclusion, unambiguous insights have yet to be found regarding behind armour effects for overmatch threats. Yet it is already clear that a lot depends on which type of vest is combined with which projectile.

2. EXPERIMENTAL SETUP

The research described here, considers two different ballistic personal armour systems: a soft armour vest (NIJ-0101.06 level IIIA) and an in-conjunction plate with armour vest (NIJ-0101.06 level IV) (Figure 37). The two projectiles used for the soft vest test were: 7.62 x 39 mm M43 and the 5.56 x 45 mm NATO Ball. Both projectiles are full metal jacket (FMJ) constructions with a steel core or penetrator. For the soft vest in combination with the ceramic plate (Setup 2), two different armour-piercing ammunition types were used: .338 AP Lapua Mag and the 7.62 x 51 mm AP8. Both projectiles consist of a brass jacket and a tungsten carbide core.

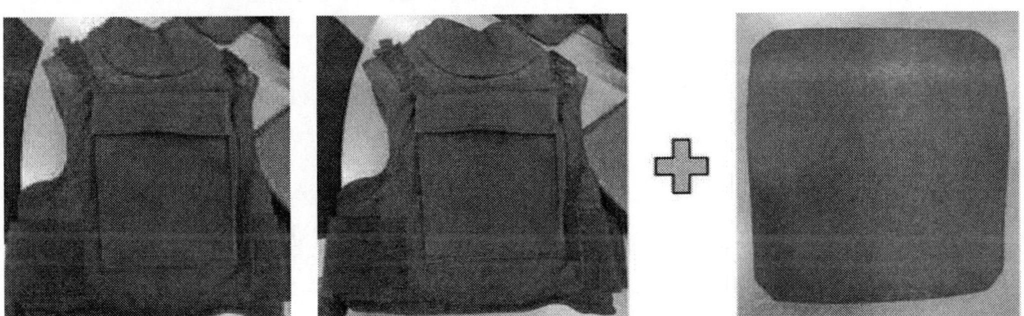

Figure 37. Left: soft armour vest; right: in conjunction vest (soft vest with a ceramic insert).

Figure 38 shows the schematic representation of the experimental tests. For these tests, a universal receiver with interchangeable barrel was used and the targets were placed at 15 m from the muzzle. A double velocity base was used to measure the projectile velocity. Two different target configurations were tested: the stand-alone gelatine block called Setup 1 and the two armour systems with the gelatine block on the back called Setup 2. For each type of projectile, three shots were performed.

Table 21 summarises all different experimental tests performed.

Table 21. Experimental tests

Projectile	Average Impact velocity (m/s) Setup 1	Target Setup 1	Impact velocity (m/s) Setup 2	Target Setup 2	Number of tests	Name

 https://doi.org/10.52202/080042-0010

7.62 x 39 mm M43	746	gelatine	740	Soft armour vest Level IIIA + gelatine	3	Combination 1
5.56 x 45 mm NATO Ball projectile	979	gelatine	981	Soft armour vest Level IIIA + gelatine	3	Combination 2
.338 AP	858	gelatine	859	In-conjunction plate + vest Level IV + gelatine	3	Combination 3
7.62 x 51 mm AP8	903	gelatine	903	In-conjunction plate + vest Level IV + gelatine	3	Combination 4

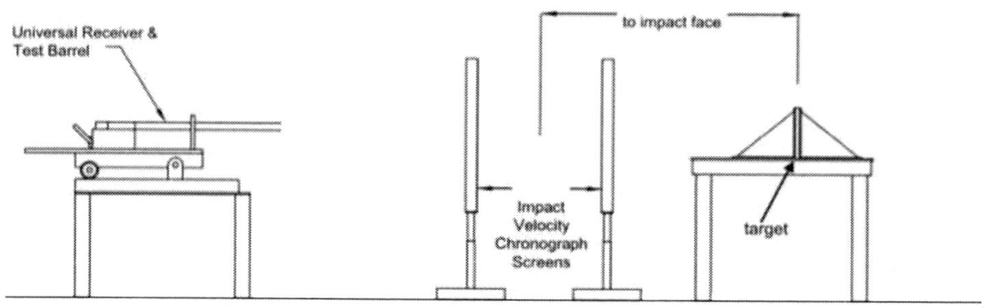

Figure 38. Schematic representation of the experimental tests.

3. RESULTS AND DISCUSSION

3.1 Quantitative analysis

For the quantitative comparison, two methods are used to obtain results related with the temporary cavity and the amount of kinetic energy that is transferred from the projectile to the gelatine block. The first method is a manual and labour-intensive practice, and is known as the Crack Length Method (CLM), whereas the second method is based on analysing the image recordings of the high-speed cameras using the PFV-software of the cameras.

3.1.1 Crack length method

By using the CLM, it is possible to estimate the transferred kinetic energy to the gelatine block. The method is based on measuring the total sum of the lengths of the cracks in different cross-sections of a block of gelatine. This Total Crack Length is proportional to the kinetic energy deposited by the projectile into the gelatine. The higher the Total Crack Length, the higher the transferred kinetic energy, consequently more tissue is damaged.

However, this method does not give a precise result in terms of kinetic energy. For example, poor visibility in the block can cause some cracks not to be taken into account. It is also possible that secondary cracks emerge from handling the block. Furthermore, the crack patterns are difficult to preserve since the gelatine deteriorates over time [6].

3.1.2 PFV-software

Due to the aforementioned issues with the CLM method, the transferred kinetic energy will also be calculated using the PFV-software, in order to confirm the accuracy of the results obtained by measurement of the different cracks.

In the software provided with the high-speed camera (Photron), different calculations can be performed, based on the images recorded from the high-speed cameras. First of all, the velocity of a projectile can be measured using the tracking function. This function enables the user to manually track the course of an object, in this case the bullet and its fragments, frame per frame. Knowing the pixel-to-mm ratio and the moment at which each frame was recorded, one can easily calculate the projectile's velocity. For each shot, the velocity of the respective projectile was calculated just before penetration of the ballistic gelatine and just after exiting the block, using this tracking method. Hence, knowing the entrance and exit velocity of a projectile allows calculation of the transferred kinetic energy. When using the same technique for each shot, it will allow correct comparison of the results between different settings. However, this method is not flawless either. Manually tracking the projectiles, can lead to small mistakes, resulting in diverging velocities. The PFV-software also allows to quantify the dimensions of the temporary cavity at a certain moment in time.

3.1.3 Analysis of the results

Combination 1: 7.62 x 39 mm M43 and soft armour vest

The first combination that is examined is a soft armour vest Level IIIA vest with a 7.62 x 39 mm M43 projectile. Plotting the crack length for each slice of gelatine (3 cm thickness each) on a graph, provides a graphical visualisation of the cavity within the gelatine block, for the 2 different setups (with and without gelatine). By comparing the graphs between the two setups, it is clear that the maximum dimensions of the cavity, represented by the maximum crack length for a slice, occur sooner for the test where the vest is in place.

In Figure 40, the curve without vest (dark grey colour), the maximum cavity takes place at a penetration depth of 21 cm. On the other hand, when the vest is in place (light grey colour) the maximum cavity occurs at 15 cm depth in the gelatine block. This result suggests that the projectile will start tumbling earlier after penetration of a Level IIIA soft armour. In terms of maximum values, the difference is only 17 mm between the 2 graphics. It is thus clear that a Level IIIA vest does not absorb a remarkable amount of the projectile's kinetic energy. In both cases, the projectile completely perforates the 45 cm long gelatine block. The average transferred kinetic energy of the shots fired in setup 2 is 3.5% higher than the average transferred kinetic energy in setup 1. This would mean that more damage is caused to the body when a 7.62 x 39 mm projectile has to perforate a Level IIIA body armour first (see Figure 39), although more data would be required in order to assess the statistical significance of this particular conclusion.

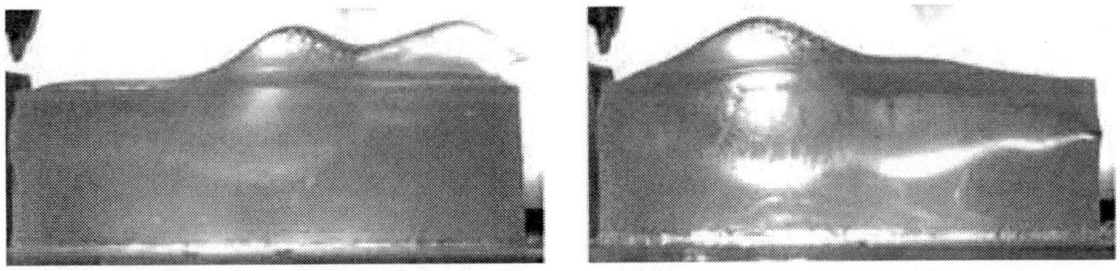

Figure 39. Left: 7.62 x 39 mm M43 on a gelatine block (Setup 1) - Right: 7.62 x 39 mm M43 on a gelatine block after penetration of a Level IIIA vest (Setup 2)

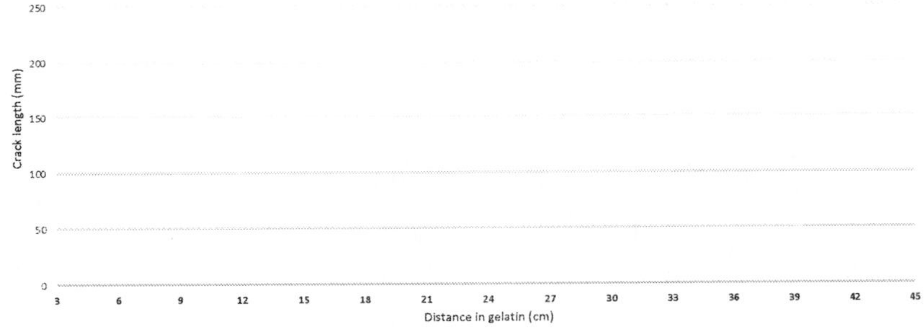

https://doi.org/10.52202/080042-0010

Figure 40. Crack length as a function of gelatine section (setup 1 – without vest and setup 2 – with vest)

When no body armour is in place, the projectile clearly has a longer stable phase, resulting in a longer neck in the ballistic gelatine. Due to this stable phase, the projectile loses less energy since it experiences less resistance. However, using the Crack Length Method (CLM), the opposite conclusion came out of the results. With a difference of 3% between both setups, according to the CLM, and a difference of 3.5% following the PFV-software calculations, it is not possible to state that one result is more significant than the other. Nonetheless, it is certainly clear that the Level IIIA body armour does not absorb a significant amount of the projectile's kinetic energy for the considered overmatch threat.

Combination 2: 5.56 x 45 mm NATO Ball and soft armour vest

In contradiction to the previous combination, here, the soft armour vest clearly absorbs part of the projectile's kinetic energy. Comparing the average Total Crack Length for the three shots in both setups (with and without vest), the Total Crack Length for the setup without vest is 273 mm higher than the value for the other setup. In this case, the soft armour vest does absorb a significant amount of kinetic energy when a 5.56 x 45 mm projectile penetrates it. This would thus intuitively result in less damage to the gelatine block.

For this projectile, the crack length measured in each slice of gelatine was again plotted on a graph (see Figure 41). Comparing the different curves for the two setups, it is obvious that the maximum dimensions of the cavity, represented by the maximum crack length for a slice, occurs sooner in setup 2 (with vest) than in setup 1 (without vest). As shown on the graphs below, the maximum cavity in setup 2 occurs at 15 cm in the gelatine block. In setup 1, this maximum cavity takes place at 21 cm. This result suggests once again that the projectile will also start tumbling earlier after perforation of the soft armour.

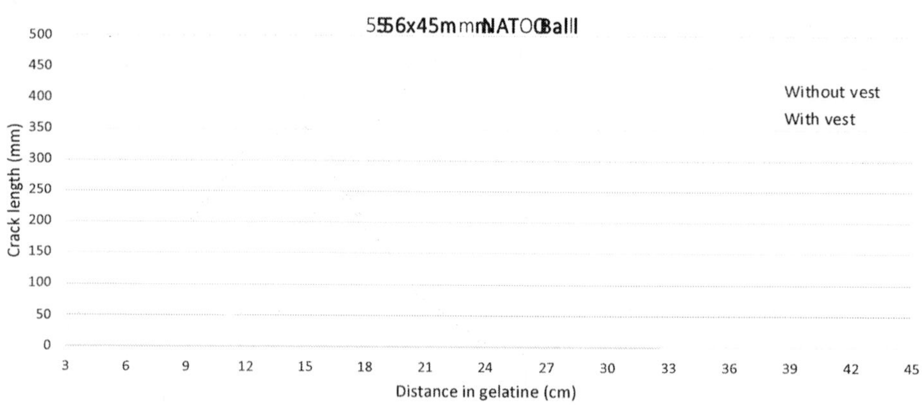

Figure 41. Crack length as a function of gelatine section (setup 1 – without vest and setup 2 – with vest)

Contradictory to the previous combination, the maximum crack length in a slice is significantly greater for setup 1, with a value of 456 mm. The tracking and measurement tools in the PFV-software confirm that wearing a vest will noticeably reduce the projectile velocity. The average entrance velocity of the bullet is 930 m/s for setup 1. For the other setup, the projectile loses on average 85 m/s when penetrating through the soft vest, hence, the projectile loses 17.4 % of its kinetic energy by penetrating the body armour. During its passage through the ballistic gelatine, the 5.56 projectile has the tendency to fragment. In both setups, the projectiles breaks into pieces in the ballistic gelatine.

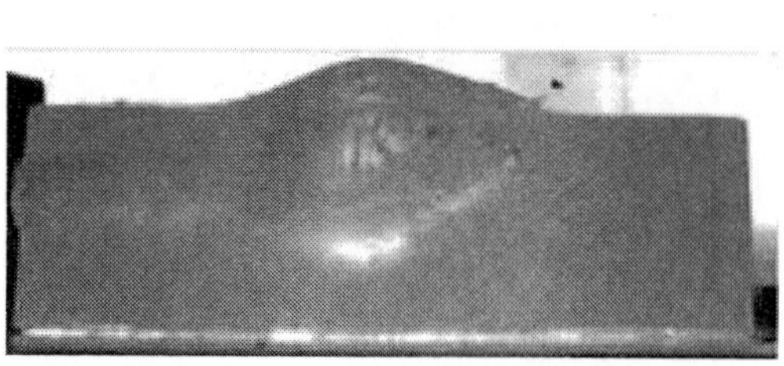

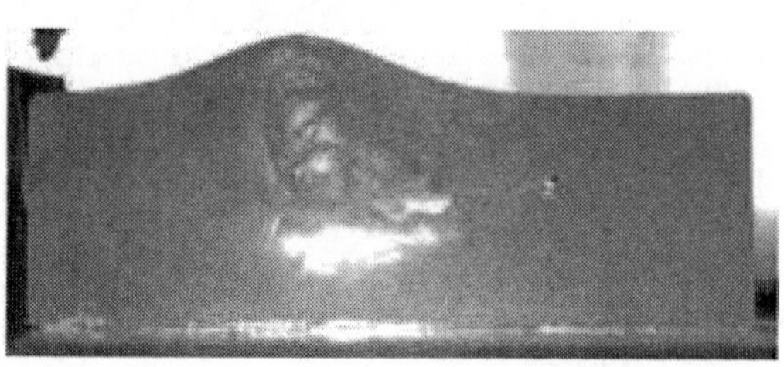

Figure 42. Left: 5.56 x 45 mm NATO Ball affecting a gelatine block - Right: 5.56 x 45 mm NATO Ball affecting a gelatine block after penetration of a Level IIIA vest

Looking at the images of Figure 42, the effect of the projectile's early tumbling is visible. The cavity on the right picture occurs earlier than the one on the left. Additionally, the dimensions of the cavity and neck in the Y and Z directions are similar for both settings. This means that the local damage is approximately the same with or without vest. In conclusion, the transferred kinetic energy is higher for the setup where the gelatine block is unprotected. A ballistic soft vest hence reduces the transferred kinetic energy of a 5.56 x 45 mm NATO Ball projectile.

Combination 3: .338 AP and soft vest with the ceramic insert

The impact on the body armour (now consisting of a ceramic plate combined with a soft vest) of a .338 AP projectile leads to an enormous amount of the projectile's kinetic energy to be absorbed by the body armour. The ceramic insert shatters the tungsten carbide core in such a way that the soft vest can absorb a large part of the remaining kinetic energy of the various fragments. The energy-absorbing capability of this in-conjunction system is nonetheless not high enough. There are still many fragments that pierce the vest and damage the gelatine block (Figure 43).

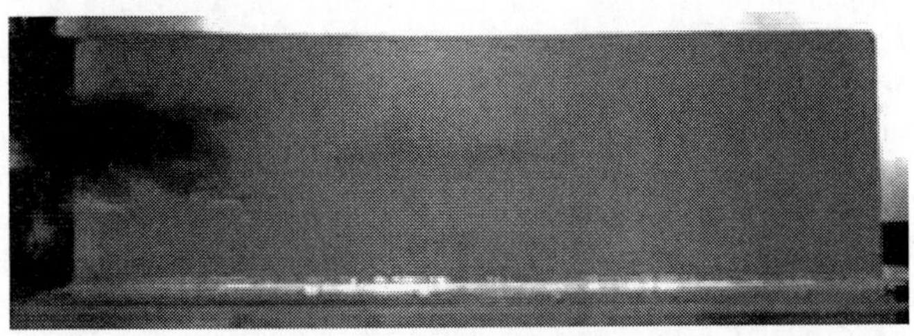

Figure 43. Fragmentation cloud of a .338 AP projectile entering the ballistic gelatine after perforation of an in-conjunction Level IV armour (Setup 2)

The average crack length measured in each slice is plotted in Figure 44 for both setups. Observing the graph, a clear difference is visible compared with the previous combinations. In this case, the maximum cavity in gelatine occurs directly at the entrance of the block for the setup 2 (with the soft vest in combination with the ceramic insert). For previous combinations, there was a more gradual build-up to this maximum value.

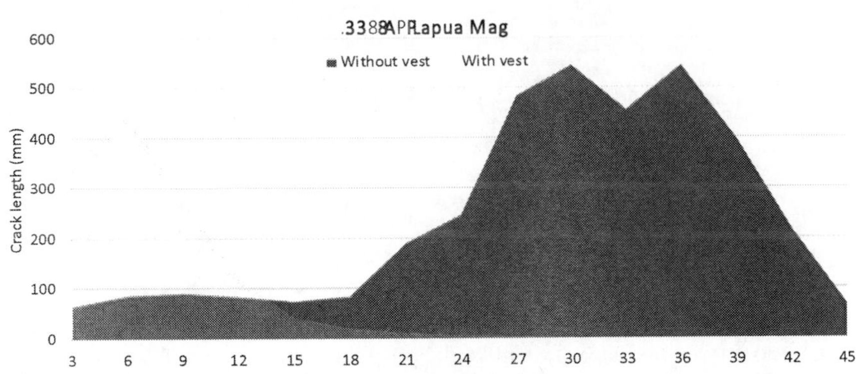

Figure 44. Crack length as a function of gelatine section (setup 1 – without vest and setup 2 – with vest)

Comparing the graph for the two setups, it is obvious that the maximum dimensions of the cavity, represented by the maximum crack length for a slice of gelatine, occurs sooner in setup 2 than in setup 1. Furthermore, it is also clear that the maximum crack length in setup 2 is lower than in setup 1. The difference between the location of the average cavity is however significant. The cavity that occurred for setup 2 is directly formed in the beginning of the block, while the cavity for setup 1 takes place in the second half of the ballistic gelatine. Notice that a difference in the location of the maximum cavity for both setups was also the case for previous combinations. The reason why in those cases the cavity

https://doi.org/10.52202/080042-0010

occurred sooner, lies in the tumbling of the projectile. However, for this combination, there is not any residual projectile (nor, as a consequence, any tumbling) visible after perforation of the in-conjunction armour. The projectile is completely fragmented and the particles propagating through the gelatine block consist of dust, ceramic fragments, projectile pieces and other materials. Figure 45 shows a snap-shot of the .338 AP penetrating the gelatine block, without and with armour system present respectively.

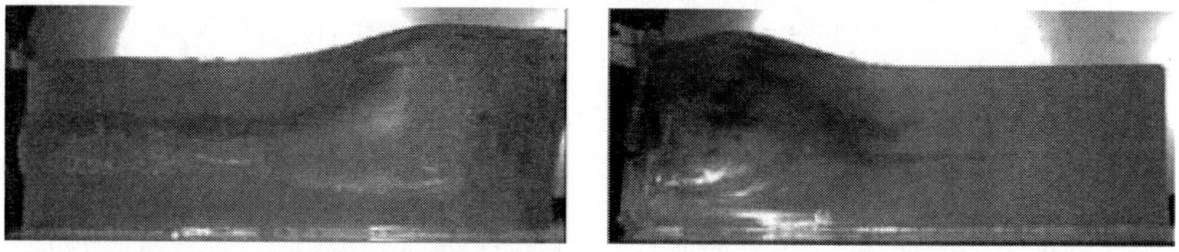

Figure 45. Left: .338 AP impacting a gelatine block (setup 1) - Right: .338 AP impacting a gelatine block after penetration of a Level IV vest (setup 2)

Combination 4: 7.62 x 51 mm AP8 and soft vest with the ceramic insert

Based on the comparison of the average Total Crack Length between both setups, it is possible to conclude that the soft vest in-conjunction with the ceramic insert again absorbs a lot of the projectile's kinetic energy, just as for the .338 AP. The Total Crack Length is 80 % lower in setup 2. As is the case for the previous combination, the ceramic insert shatters the tungsten carbide core in such a way that the soft vest can again absorb a large part of the remaining kinetic energy of the various fragments. The energy-absorbing capability of this in-conjunction system is nonetheless not high enough to protect the wearer. Some fragments are still able to perforate the vest and damage the gelatine block. Due to the fact that they enter the block in a fragmentation cloud, they leave a large entrance wound containing a lot of debris. The average crack length, measured in each slice is plotted in Figure 46.

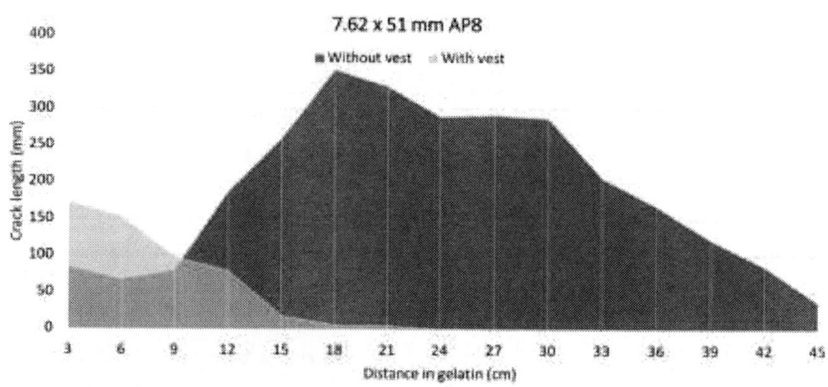

Figure 46. Crack length as a function of gelatine section (setup 1 – without vest and setup 2 – with vest)

Comparing the two setups, similar results as for the previous combination are observed. In this configuration it is clear that the maximum dimension of the cavity occurs sooner in setup 2 (with the vest) than in setup 1 (without the vest). Unlike the previous configuration, the maximum crack length found in setup 1 differs much more from the one in setup 2 (a drop between setups of 51% in terms of maximum crack length occurred). To determine the velocity of the fragmentation cloud, the most visible fragment was tracked in the PFV-software. Assuming that the velocity of the other fragments does not differ significantly, an estimation of the kinetic energy transferred to the block was made. The results are in line as the results obtained by the Crack Length Method. The average transferred kinetic energy in setup 2 is 85% lower than in setup 1, confirming that the Level IV protection vest absorbed a lot of the 7.62 x 51 mm AP8 projectile's kinetic energy.

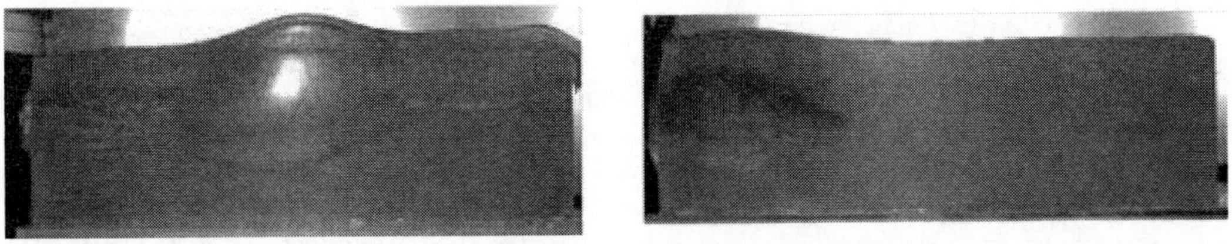

Figure 47. Left: 7.62x51 mm AP8 on a gelatine block (setup 1) - Right: 7.62x51 mm AP8 on a gelatine block after penetration of a Level IV vest (setup 2)

3.2 Qualitative analysis

3.2.1 Contamination caused by fragmentation

Fragments can occur in a lot of forms, dimensions and materials, diverging in origin. They commonly cause damage to the body just as any other projectile by shredding tissue and transferring their kinetic energy to surrounding materials. When wounds are contaminated, the fragments can complicate the treatment process, which can have severe consequences for the patient. If the person initially survived the primary tissue destruction, contamination is often considered as a major threat to that person's life. Nowadays, doctors are in many cases able to remove all the contaminated tissue while treating the wound [4]. However, this process can better be avoided in order not to waste crucial time treating the patient.
Every projectile can cause the created wound to be contaminated. As a bullet propagates through the air, fabric, skin or tissue, it can carry many different bacteria, which can be dispersed through the wound. If fragmentation occurs, there is a higher chance of contamination since the different fragments are from different sources. Some fragments originate from the bullet's core, others from the of the armour vest, or even from the perforated ceramic plate. Clearly, the more sources and different materials entering the body, the higher the chance of contamination. Fragmentation consequently leads to a higher infection risk, which is preferably avoided at all times.

Looking at the test results of the case where the soft vest is worn in combination with the hard plate, it is possible to conclude that fragmentation occurs more frequently after perforation of the body armour compared to when no body armour is worn. For combination 1 and 2 (being the 7.62 x 39 mm – Level IIIA and 5.56 x 45 mm – Level IIIA), no significant difference in fragmentation occurred. The 7.62 mm projectile did not fragment at all in either setup (unprotected gelatine and protected gelatine) and the 5.56 mm projectile has the tendency to fragment equally in both setups. This is shown in Figure 48 and Figure 49.

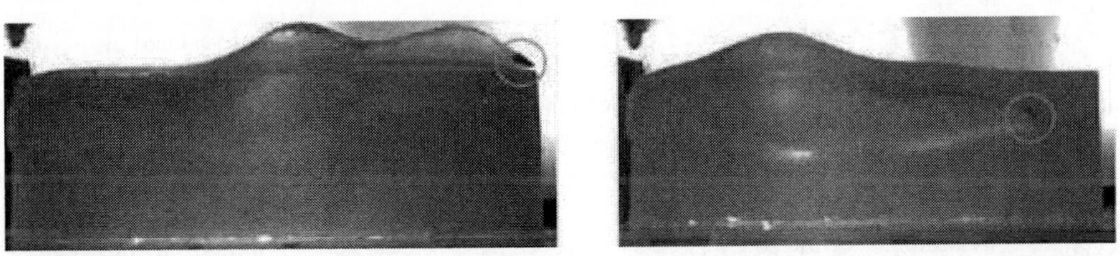

Figure 48. Left: 7.62 x 39 mm M43 projectile propagating through gelatine. No fragmentation occurred. - Right: 7.62 x 39 mm M43 projectile propagating through gelatine after piercing a Level IIIA soft armour. No fragmentation occurred.

Figure 49. Left: 5.56 x 45 mm NATO Ball projectile propagating through gelatine. 5 fragments can be distinguished. - Right: 5.56 x 45 mm NATO Ball projectile propagating through gelatine after piercing a Level IIIA soft armour. 5 fragments can again be distinguished.

https://doi.org/10.52202/080042-0010

The .338 AP ammunition fragmented remarkably when impacting the ballistic protection. After perforation, different particles of the projectile's tungsten core, ceramic tile (aluminon oxide) and fabrics of the carrier group together in a fragmentation cloud. Hence, this cloud consists of many different materials, increasing the risk of infection. In the case when no vest is placed in front of the gelatine, only limited fragmentation occurred. The fragments that are still formed obviously only originate from the bullet material. The 7.62 x 51 mm AP8 projectile shows similar behaviour as the .338 AP. In the former case, many fragments consisting of tungsten carbide particles, ceramics and fabrics group together in a fragmentation cloud. However, the dimensions of this cloud are smaller than for the one caused by the .338 AP bullet. Figure 50 and Figure 51 show the penetration of the Combination 3 and 4 for the 2 different setups with the fragmentation in detail.

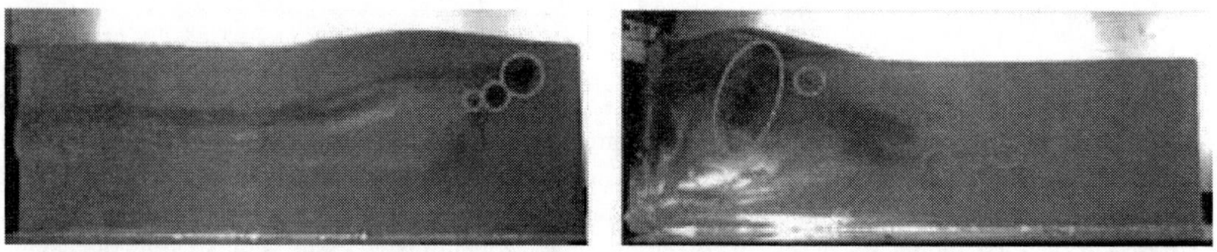

Figure 50. Left: .338 AP projectile propagating through gelatine. Only 3 fragments. - Right: .338 AP projectile propagating through gelatine after piercing a Level IV in-conjunction armour. 4 major fragments and an extensive debris cloud.

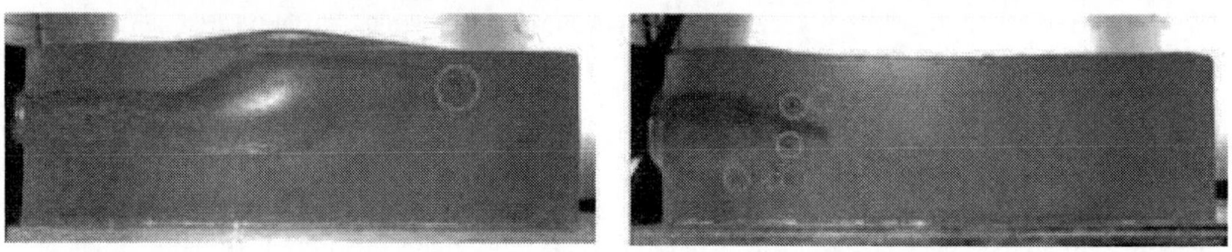

Figure 51. Left: 7.62 x 51 mm AP8 projectile propagating through gelatine. No fragmentation occurred. - Right: 7.62 x 51 mm AP8 projectile propagating through gelatine after piercing a Level IV in-conjunction armour. 6 major fragments and an extensive debris cloud.

4. CONCLUSIONS

The objective of this study was to retrieve data for four different combinations of vest and projectiles, in order to be able to investigate these combinations more thoroughly for overmatch conditions. Therefore, four different projectiles were shot at ballistic gelatine, with and without body armour protection. Two methods were used to retrieve data from the test results: the Crack Length Method and the measurement and tracking tools in the PFV-software. Combining all the results of the four combinations, it is possible to conclude that the maximum dimension of the cavity occurs sooner when a projectile first has to penetrate body armour. For the soft vests, this was because of the earlier tumbling of the projectile. For the tests with the in-conjunction armour, the cavities' dimensions shrunk significantly. However, cavitation occurs sooner.

The Crack Length Method showed that more energy was transferred to a protected body than to a bare block of gelatine. However, this was only the case for two combinations: the 7.62 x 39 mm M43 projectile, shot at a Level IIIA soft armour and the .338 AP ammunition, shot at a Level IV in-conjunction vest. Both the 5.56 x 45 mm NATO Ball and 7.62 x 51 mm AP8 projectiles still inflicted more damage when no body armour was worn compared with shots first perforating the armour systems.

Analysing the test results, not only the transferred energy has to be considered. The degree of fragmentation and the probability that the shot hits vital organs in the chest and abdomen region is also fundamental for this assessment. When projectiles fragment, the risk of infection in the affected tissue is seriously increased. Test firings showed that major fragmentation occurs in the case of a .338 AP projectile and a 7.62 x 51 mm AP8 projectile hitting an in-conjunction armour Level IV. After perforation of both the ceramic insert and the soft armour, all the particles grouped together in a fragmentation cloud of considerable kinetic energy. This fragmentation cloud affected the tissue greatly and thus caused a wide temporary cavity, now filled with debris. This increases the risk of infections, complicating the treatment process.

Because of the earlier cavitation, the greatest dimensions of the wound occur sooner in the body. Nevertheless, overall dimensions of the cavity and exit wounds seem to be smaller if the gelatine is protected.

In conclusion, this research confirms that cavitation occurs sooner if body armour is worn. However, no clear answer can be found to the question whether body armour causes more damage or not. Some factors influencing wound severity seem to be worse for protected gelatine blocks, while others do not. Nonetheless, this research offered some new insights in the behind armour effects of overmatched Level IV in-conjunction vests. The behaviour of the projectiles that hit this level of body armour is very different to projectiles hitting soft armour. Whether body armour protects its wearer or not, overmatching shots still cause a lot of damage, which would possibly result in an incapacitated target.

References

[1] A. Mabbott and D. J. Carr, "Effects of police body armour on overmatching ballistic injury," *Int J Legal Med*, vol. 134, no. 2, pp. 583–590, Mar. 2020, doi: 10.1007/s00414-019-02070-9.

[2] J. Breteau, M. L. Fackler, P. Martin, and I. Sendowski, "The Personal Protective Equipment Provided for Combatants: The Part Played by Wearing a Protection Vest in the Behaviour of Projectiles Wounding Outcomes," in *Proceedings of the 11th International Symposium of Ballistics*, Brussels, Belgium, 1989, vol. 3.

[3] J. M. Lanthier, M. J. Iremonger, E. A. Lewis, I. Horsfall, and P. L. Gotts, "Is the wounding potential of high velocity military bullets increased after perforation of textile body armour," in *Proceedings of the Personal Armour Systems Symposium*, The Hague, Netherlands, 2004, pp. 225–232.

[4] A. Mabbott, "The Overmatching of UK Police Body Armour.," PhD thesis, CRANFIELD UNIVERSITY, CRANFIELD (UK), 2015.

[5] J. Missliwetz, W. Denk, and I. Wieser, "Study on the Wound Ballistics of Fragmentation Protective Vests Following Penetration by Handgun and Assault Rifle Bullets," *Journal of Forensic Sciences*, vol. 40, no. 4, pp. 582–584, Jul. 1995.

[6] S. A. Bolliger, M. J. Thali, M. J. Bolliger, and B. P. Kneubuehl, "Gunshot energy transfer profile in ballistic gelatine, determined with computed tomography using the total crack length method," *Int J Legal Med*, vol. 124, no. 6, pp. 613–616, Nov. 2010, doi: 10.1007/s00414-010-0503-z.

 https://doi.org/10.52202/080042-0010

BHBT Impactor Classification Using Machine Learning

S. Magnan[1], E. Fournier[1], N. Shewchenko[1], and S. Ouellet[2]

[1]Biokinetics and Associates Limited, 2470 Don Reid Drive, Ottawa, ON, Canada, magnan@biokinetics.com
[2]DRDC Valcartier, 2459 Route de la Bravoure, Quebec, Qc, Canada

Abstract. A novel approach to improve injury risk assessment of behind-helmet blunt trauma (BHBT) events using machine learning is presented. Four rigid impactor profiles were fired at the ballistic load sensing headform (BLSH) using an air cannon to represent BHBT loading conditions. 14 key features were extracted from the seven-load cell array time histories recorded by the BLSH. No clear pattern emerged from the key features to easily identify which impactor corresponded to which test, therefore, a machine learning approach was used. The support vector machine (SVM) multinomial classifier was trained on a total of 48 shots, including combinations of four impactor face profiles, two test velocities, and three repeats at each of the two impact locations were provided as training/validation data. Cross-feature scaling was performed to prevent over- or under-fitting to specific features. The SVM accuracy was evaluated using stratified 12-fold cross-validation (leave-one-out cross-validation), where the model was found to have approximately 94% accuracy. Ballistic testing was then performed on infantry helmets mounted on the BLSH using 9mm FMJ and 64-grain FSP projectiles. The same 14 key features used to train the SVM on air cannon data were extracted from each ballistic event and fed into the model which predicted the equivalent impact profile. Equivalent testing was repeated on a clay-filled ballistic helmet to ascertain the actual deformation profile using the witness material which could then be compared against the profile predicted by the classifier. Finally, the predicted profiles and measured peak forces of the helmet testing were combined with the Allanson-Bailey BHBT injury risk curves to ascertain the probability of an AIS2+ injury for each event.

1. BACKGROUND

Combat helmets provide many protective benefits to the head including, in part, attenuation of blunt force impacts, resistance to penetration from ballistic projectiles and mitigation of behind-helmet blunt trauma (BHBT) from defeated projectile strikes. If not managed adequately, injuries to the head can occur and will vary with the helmet design, threat severity and human tolerance.

BHBT involves the transmission of forces from a defeated projectile strike resulting in local shell deformations and attenuation of the projectile's energy. The stiffness and strength of the shell, the amount of standoff from the head, and the use of impact liners can greatly affect the loads and resulting injury outcomes. For example, in the study of ballistic plate and ballistic helmet impacts, it was found that increasing the helmet stand-off or inserting an impact liner reduced intra-cranial pressures and skull fracture severity significantly during ballistic shell deformation [1], [2]. Additionally, helmets that exhibited larger shell deformations had greater contact areas with increased intra-cranial pressures and skull fracture severity. Skull fracture tolerance also varied with skull fracture mode (simple, comminuted) and severity (depressed, displaced) with dependence on load magnitude, distribution, rate, and the energy deposited into the anatomical structure. Observations of the injury mechanisms associated with some BHBT events have been documented in previous biomechanical studies:

1. Damage to the scalp with circular lacerations at the impact site (Bass, Boggess, Bush, & al., 2003), [1].
2. Cranial fractures with linear fractures radiating from and around the point of impact with the most severe cases resulting in comminuted fractures [4], (Bass, Boggess, Bush, & al., 2003), [1],. [2], [5].
3. Dural contusions from the dura separating from the bone at the impact site (Bass, Boggess, Bush, & al., 2003).

The risk of skull fracture under BHBT conditions has been linked to peak force metrics with the findings often limited to broad generalizations due to differences in test setups, test specimen variation, threat characteristics, data analysis metrics and test methods. Recent research by Allanson-Bailey [6] suggested that the risk of skull fracture is not only dependent on the peak transmitted force but also on the force distribution. As a result, characterizing the dynamic force and distribution may increase the specificity of the fracture risk assessments thereby improving the understanding of helmet design on protection including, for example, shell stiffness, helmet-head stand-off, local deformation shape, and impact liner interactions. For a comprehensive evaluation of the dynamic forces from shell deformation, combat helmets must be evaluated as a complete system in situ and must include the ballistic shell, impact

https://doi.org/10.52202/080042-0011

liner, and retention/sizing system, as applicable. To this end, instrumented headforms have been proposed to measure peak dynamic force but have limited ability to measure spatial force distribution [7], (Trexler, et al., 2018), [9], (Voo, Improved Repeatability and Reproducibility of the Ballistic Load Sensing Headform, 2016), [11], [12]. Alternatively, the Ballistic Load Sensing Headform (BLSH) [13] was assessed to be biofidelic under BHBT conditions (based on impacts to the temporoparietal location with a 38 mm diameter, 103-gram rigid impactor at 20 m/s and 35 m/s with comparison to force-deformation and force-time histories of PMHS corridors) [14] and has seven load sensors covered with compliant skin pad at specific cranial regions to measure peak dynamic force transmission and distribution, Figure 57. While the BLSH has been shown to be repeatable and applicable for limited BHBT conditions [13], [15], it is not known whether the spatial resolution of the BLSH sensing area is sufficient to properly characterize load distribution and skull fracture risks across a wide range of behind helmet loading conditions as proposed by Allanson-Bailey.

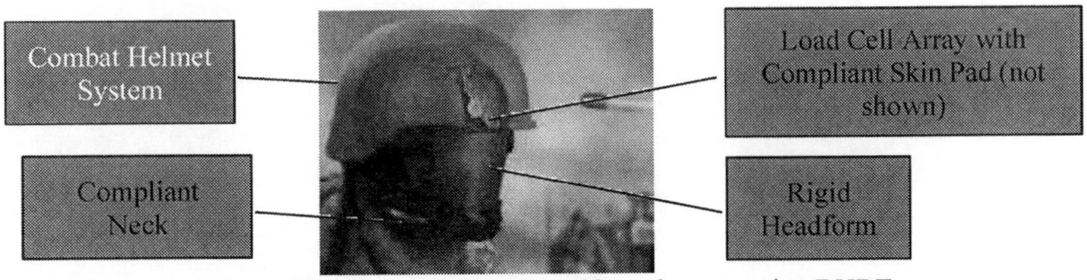

Figure 57: The Ballistic Load Sensing Headform for measuring BHBT.

The objective of the present work was to investigate whether the forces measured by the BLSH could be used to estimate the deformation profile of a helmet and, hence, skull fracture risk under the varying loading conditions defined by Allanson-Bailey. This was accomplished by comparing the measured force distribution to that obtained with direct impact to the BLSH with known projectile shapes and masses representing simulated behind-shell characteristics.

2. METHODOLOGY

The present study proposes a method for relating the BHBT force profile (i.e., time-histories) measured using the BLSH to skull fracture risk. Published rigid impactor test data from Allanson-Bailey, include injury risk as a function of strike velocity for several impactor geometries representing typical shapes, and hence, load distribution for BHBT conditions. Therefore, a link relating BHBT test data to the various rigid impactor test conditions is required for a comprehensive injury risk assessment. The approach described herein first requires the generation of a large set of air-cannon (rigid impactor) data at different velocities and impact positions for several impactor shapes, sizes, and masses. The data is recorded using the seven-load cell array of the BLSH to generate force-time histories for the impact event across the sensing area. Next, a machine learning classifier is trained and evaluated to predict the impactor type (i.e., class) based on key features extracted from the force-time histories. Subsequently, ballistics tests are performed on helmets to quantify the BHBT force-time histories using the BLSH. Finally, data from BLSH testing is inputted into the model to classify the deformation profile as being closest to one of the impactor shapes, which is then related to injury risk curves published by Allanson-Bailey.

Rigid Impactor Testing

For the current study, four impactor shapes (Table 23) used by Allanson-Bailey in the BSM test series were selected for direct impact on the left side load cell array of the BLSH headform. The four impactors selected for testing are. The shape of the impactors used by Allanson-Bailey were partly based on that used by Raymond [14] with additional impactors of varying face curvatures and loading area, all limited to 38 mm diameter for comparison to cadaveric data of Raymond. The progression of shell shape and size as it deforms and contacts the skull was thought to be well represented by the selected impactors.

Table 23 Characteristics of four projectiles selected for direct BLSH impacts.

Projectile	Flat 38 mm	Flat 20 mm	Curved 19 mm	Curved 50 mm
Mass (g)	103.2	104.0	105.4	103.1
Diameter (mm)	38	20	38	38

https://doi.org/10.52202/080042-0011

Length (mm)	74	74	74	74
Surface Radius (mm)	∞	∞	19	50
Projectile	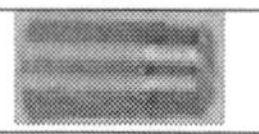	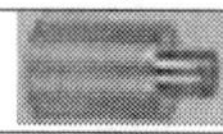	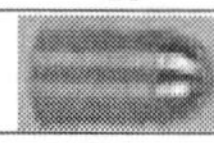	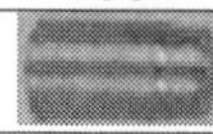

The projectiles were launched at the BLSH using an air cannon allowing for a free flight of the projectile before impact with venting of the muzzle backpressure. Testing was conducted at 15 m/s and 25 m/s (or approximately 1.56-2.60 N·s), corresponding to similar peak force values to those expected in the ballistic testing. The shots were centred on the load cell array (hex pad #1) or offset midway between the centre and upper load cells (hex pad #1 and #7) as shown in Figure 58. Each hex pad has an approximate nominal surface area of 440 mm^2 (420 mm^2 projected). The BLSH's load cell array was positioned 20 cm from the air cannon muzzle and aligned normally to the impactor trajectory.

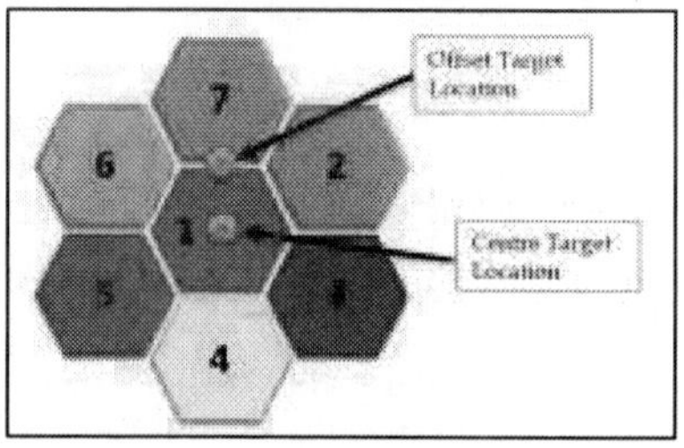

Figure 58. Targeted impact positions relative to the BLSH load cell array.

The load cell data was collected at 100 kHz using the BLSH's software with 10 kHz anti-aliasing analog hardware filters. Each channel was then digitally filtered using a phaseless Butterworth 4-pole low-pass filter with a corner frequency of 4,500 Hz, as per the BLSH's data collection protocol. The velocity of the projectile was measured at the exit of the barrel using a dual-beam IR light gate sampling at 80 MHz and triggering off the leading edge of the projectile. The BLSH skin pad covering the load cell array was inspected after every impact and was replaced when damage was observed.

BHBT testing

The air cannon tests on the BLSH aimed to develop an analysis methodology to predict the geometry of a striking impactor. The methodology could then be applied to non-perforating ballistic impacts on combat helmets to predict the resulting shell backface deformation profile. To gain insight into the shell deformations for the current study, ballistic tests were performed on the aramid combat helmets with a 9 mm 124-grain FMJ at 300 m/s (1.86 N·s) and 64-grain FSP projectiles at 400 m/s (1.66 N·s). Testing was conducted using the full shell/liner system on the BLSH to capture the loading profiles. Then, ballistic impacts to the shell with the liner removed were conducted to determine the backface deformation profile generated by the two projectiles. The 9 mm FMJ was selected to generate a flat or low curvature profile, and the 64 gr FSP was chosen to generate higher curvature profiles. To qualify the deformation profile, rigidly supported helmet shells were packed with clay (Roma Plastilina No. 1) behind the impact site, as seen in Figure 59, and were shot by the two threats. The resulting clay indentation provides a permanent record of the shell's maximum deformation and was then carefully removed from the helmet shell and cut along the mid-sagittal plane through the indentation to quantify the shape.

Figure 59. Helmet shell filled with clay witness material for qualification of BHBT.

Machine learning model

The problem of identifying the impactor shape based on a series of response variables is fundamentally a multinominal classification problem. For this application, Support Vector Machines (SVM) were selected due to the absence of a priori statistical information, the number of input parameters relative to the total number of points, the ability to manage multi-class supervised learning where all points have the same parameters (no missing data), and accessibility of open-source implementations. The SVM method is a supervised learning method, meaning, the correct classification must be included in the dataset for training and validation. The four classes included in the analysis were described in Table 23.

2.1.1 Feature Selection

For every test, full force-time histories were generated for the seven load cells in the array. The full data curves were not used to avoid overfitting given the limited data available. Instead, key features of each trace were extracted from the raw data for use in the SVM model. The following parameters, shown graphically and explained in Figure 60, were extracted from each event. The impact position was excluded because the classifier must distinguish the impactor profile independent of targeting accuracy and symmetries, and the velocity was excluded to not limit the model when the BHBT deformation velocity is unknown.

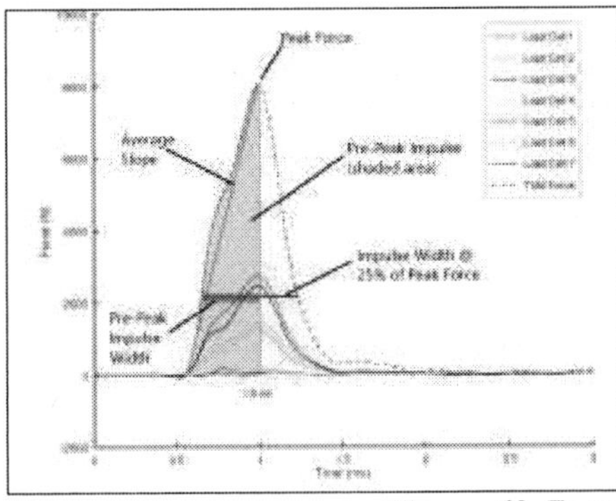

Peak force: Maximum sum of seven load cells.
LC Forces: Force of each load cells at time of peak force.
Loading start: First time exceeding 10 N total force.
Total loading time: From loading start to peak force
%Loading time: from X% peak force (rising) to peak force (at 10%, 25%, and 50%),
%Unloading time: from peak force (falling) to X% peak force (at 10%, 25%, and 50%),
Average slope: from impact start (10 N) to peak force,
Loading impulse: area under total load curve from loading start to peak force.
Average loading impulse: area under average slope

Figure 60. Features used in SVM training model.

2.1.1 Feature Scaling

Data for SVMs must be scaled to produce unbiased results. Typically, the scaling transforms each feature to have a minimum value of 0 and a maximum of 1. This is performed within each training group for cross-validation to prevent accidental information transfer (contamination) between groups. In most applications, parameters are independent of one another (i.e., the scaling is parameter specific). The BLSH features extracted from the time histories are not all independent. For example, a large impactor may increase the load on periphery load cells compared to a narrower impactor that only strikes a small area. This information may be lost if the parameters are scaled independently. Therefore, a scaling method which conserves the relative contribution of similar features was favoured in this application. This required scaling the parameters of a specific test using features extracted from the same test. A total of 14 parameters, shown in Table 24 were fed into the model.

Table 24 Scaling factors applied to features extracted from BLSH data.

Extracted Parameter	Scaled by	Min Scaled	Max Scaled
LC Forces (7)	Peak force	0.00	0.80
%Loading/unloading time (6)	Total loading time	0.17	1.92
Loading impulse (1)	Average loading impulse	0.68	1.29

2.1.2 Training Method

The Scikit-learn: Machine Learning in Python (version 1.2.1) was used to implement the SVM model and verification [16]. The "nu-SVC (classification)" formulation used in this analysis is an extension of SVMs that allows for multi-class problems [17]. The radial basis function kernel was used with nu set to 0.5. All other model parameters were set to the default values. Data was not augmented; rather, the load cell numbering was modified to automatically account for 12 rotational and line symmetry combinations.

https://doi.org/10.52202/080042-0011

2.1.3 Evaluation and Validation

In general, supervised learning involves training a model (e.g., SVM) on a portion of the data and evaluating the model against the remaining data that was held back from training. The performance of the classifier is determined by presenting the response parameters (forces, slopes, impulses, etc.) of the test data set, and comparing the actual class (i.e., impactor shape) against the class predicted by the model. For purposes of this analysis, the accuracy will be used to assess performance as this metric describes the ability of the SVM to predict multiple classes. The accuracy is the ratio of correctly identified objects to the total count of objects. A stratified 12-fold cross-validation was used to assess accuracy. This method first divides the data into groups of equal size with the same class distribution. Here, the fold count was equivalent to leave-one-out cross-validation for a multiclass problem. Therefore, the following process was followed 12 times: 44 of the 48 shots were used to train an SVM classifier, then four shots (one of each class) were used to test the model. The expected performance of the generalized model constructed using all 48 shots is the average accuracy of the individual models.

2.1.4 Application

The SVM model described in the previous section is a trained machine-learning classifier. The model can be applied to new data to predict the impact class. In the context of the present study, data collected in BHBT helmet tests using the BLSH are processed to extract key features and fed into the SVM as inputs. The model then predicts the impactor profile based on the training data. The peak force and impactor shape can then be related to injury risk using data published by Allanson-Bailey.

3. RESULTS AND DISCUSSION

A total of 48 air cannon rigid impactor shots were performed on a bare BLSH headform. Additionally, eight ballistic tests were performed on combat helmets: two were used to establish the BHBT profile on witness clay, and the remaining six used the BLSH to record the impact force-time histories. The shots were split evenly between the 64-grain FSP and the 9 mm FMJ projectiles. The SVM classifier was trained on the air cannon data and used to predict the BHBT impact profiles to correlate each event with injury risk.

Rigid impactor testing

For every air cannon test, time series data was generated for each of the seven load cells (e.g. Figure 60), additionally, representative force distribution maps were plotted to confirm targeting accuracy (Figure 61). The 20 mm flat projectile exhibited poor accuracy and the variability in targeting was assumed to be caused by the projectile pitching or yawing (weight distribution and aerodynamic effects) upon exit of the air cannon which then led to the centre of pressure, as measured by the BLSH, not being aligned with the targeted impact location. Targeting the other three projectiles, including the 38 mm flat projectile, was accurate and repeatable. The average peak total force and strike velocities for each of the test configurations are provided in **Error! Reference source not found.**.

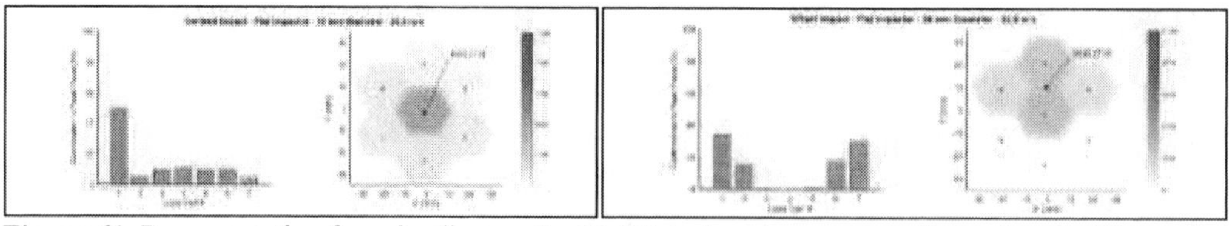

Figure 61. Representative force loading on the BLSH during air cannon testing using the 38 mm flat impactor for a centred impact (left) and an offset impact (right).

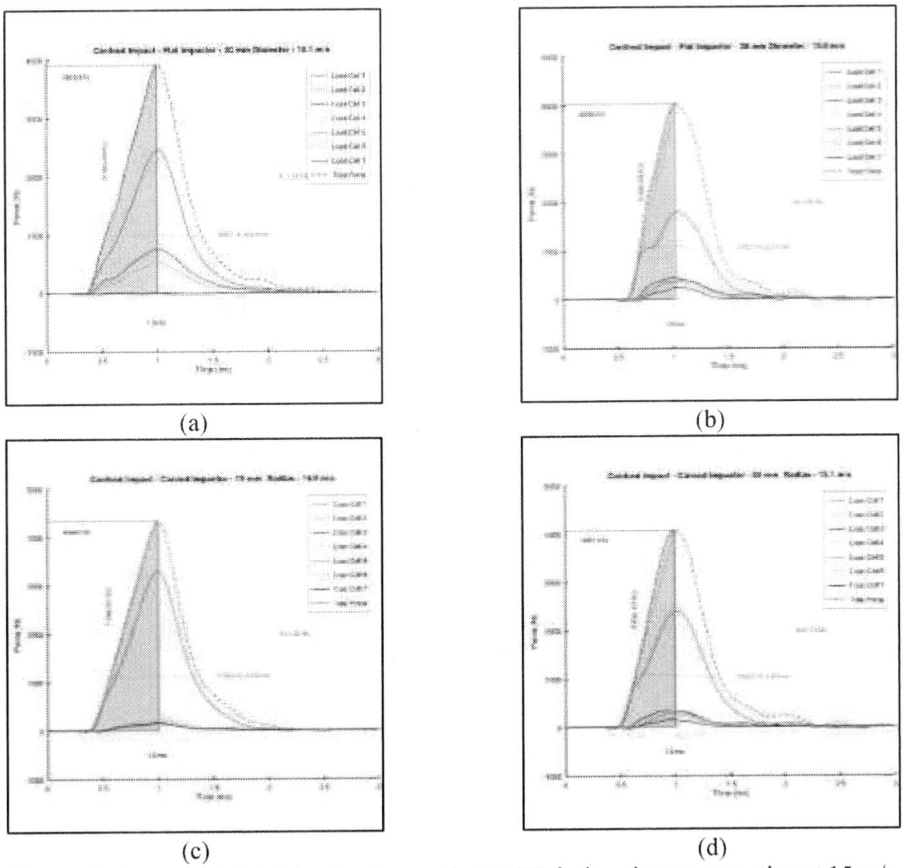

Figure 62. Representative force loading on the BLSH during air cannon testing at 15 m/s.

BHBT testing

The indentation in the clay backing from the 9 mm 124 gr FMJ and the 64-grain FSP helmeted clay impacts are shown in Figure 63 with the corresponding impactor shapes from Table 23 that best matched the indentations. Note that the shell deformations extend beyond the boundary of the projectile's body.

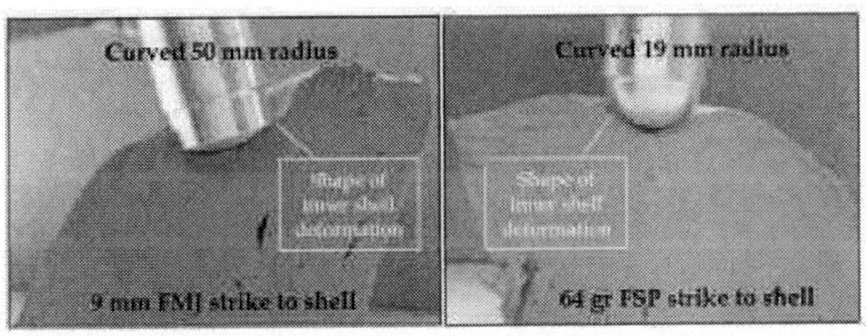

Figure 63. Clay indentations obtained from ballistic helmet strikes compared to rigid impactors used in the BLSH tests.

The curved 50 mm radius of the rigid impactor faces best matched the clay indentation from the backface deformation of the helmet shell for the 9 mm 124 gr FMJ bullet strike, whereas the curved 19 mm radius impactor best matched the indentation from the 64 gr FSP impact. Again, it is noted that the shell deformations extend beyond the outer body of the impactor.

Machine learning model

Before training the SVM model, a preliminary analysis of the BLSH data included a comparison of specific impact parameters for the different impactors. There were promising trends toward identifying the impactor shape based on the BLSH's load cell measurements, however, as there was no single parameter that could decisively identify the impactor shape, more sophisticated methods were required.

https://doi.org/10.52202/080042-0011

3.1.1 Evaluation and Validation

The procedure described from the implementation of SVM through testing and performance was applied to the BLSH test data to develop a method of identifying the impactor type based on the resultant BLSH force-time histories. Of the 48 shots studied using the SVM model with 12-fold cross-validation, the accuracy was 93.75% with two shots misidentified, Table 25. Therefore, given a set of experimental parameters, the SVM would be able to classify the closest equivalent impactor type during BHBT helmet testing with a high level of accuracy. A one-nearest-neighbour (1-NN) classifier was also trained as a baseline to establish predictive power using a trivial classification method. The 1-NN model, which is often included in published studies as a basis for comparison, had an accuracy of 85.42%. The SVM significantly outperformed the reference classifier with 3/48 misclassified events (compared to 7/48 for the 1-NN model). The high accuracy suggests that ML models can classify impactor shapes by detecting patterns that are not obvious to humans. Further research into alternate ML models is warranted. All testing was performed on the left side of the BLSH. Due to differences in skin pad and hex pad curvatures between the front, rear, and side impact sites, it is not known if the SVM model would be able to distinguish rigid impactor geometries at sites other than the one tested without additional training data.

Table 25. Binary classification confusion matrix.

Impactor Classification	Model Prediction			
	Flat 38 mm	**Flat 20 mm**	**Curved 50 mm**	**Curved 19 mm**
Flat 38 mm	11	-	1	-
Flat 20 mm	-	10	-	2
Curved 50 mm	-	-	12	-
Curved 19 mm	-	-	-	12

3.1.2 Application

The SVM classification model described in Section 6 above and trained using air cannon testing was applied to the BLSH data from the ballistic non-perforating behind helmet deformation impacts. The helmets with full suspension and retention system used for testing were fitted to the BLSH in the as-worn position by a soldier to achieve typical shell offsets. The BLSH load cell data was processed similarly to the BLSH air cannon test data and included the extraction of the 14 input parameters for the SVM. The SVM model classified the new BHBT data (three 9 mm FMJ and three 64-grain FSP) as belonging to the "Flat 38 mm" class. The injury risks are plotted in Figure 64 on Allanson-Bailey's injury risk curves for fracture, overlaid on the class predicted by the model.

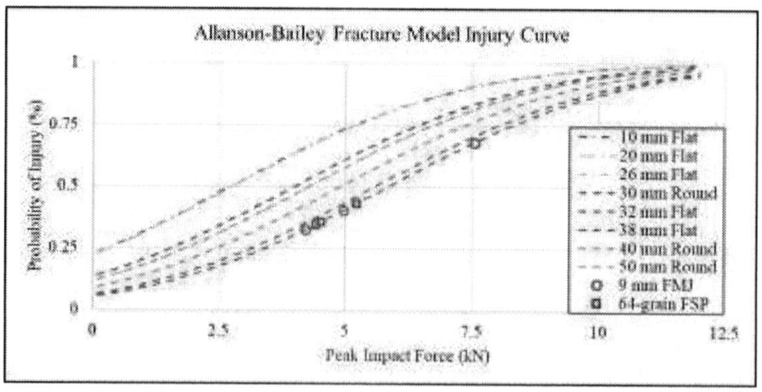

Figure 64. Clay indentations obtained from ballistic helmet strikes compared to impactors used in the BLSH tests.

Here, the boundary conditions between the BLSH/BHT tests and the clay-filled shell tests are inherently different (i.e., presence of liner/retention system and offset, rigidity of BLSH vs clay). It is therefore difficult to directly compare the BHBT profiles predicted by the SVM to the witness testing. Importantly, the projectiles and velocities selected for the analysis were demonstrated to produce different BHBT deformation profiles on the clay witness material.

The clay testing showed that BHBT profiles are different for the two projectiles, but they do not necessarily correspond to what they would be for the BLSH. It would be very useful to develop a method of measuring the actual BHBT impact profile with boundary conditions that match the BLSH to validate the SVM model. Alternate methods such as DIC on the inside surface of a helmet shell would likely provide similar deformation profiles as the clay witness due to similar boundary conditions. The dynamic helmet shell deformation is constrained by the contents (i.e., BLSH headform or operator skull). The SVM suggested that both projectiles produce loading most similar to the Flat 38 mm rigid impactor. This suggests one of the following scenarios:

1. Interactions between the helmet shell and BLSH headform produce similar BHBT profiles for the different projectiles as the maximum deformation is constrained.
2. The presence of a helmet liner/retention system with offset for BLSH testing produce a more distributed load than the clay-filled helmet tests with no liner or offset, thereby biasing the deformation profile. Similarly, the liner and comfort pads present during the BLSH tests may in fact be generating a wider backface profile.
3. The air-cannon data was not representative of the ballistic test conditions. The SVM should have produced different estimates of the impact profile for the two cases but did not because the closest case to both was not particularly representative.

Each of these potential explanations is potentially insightful and could further the understanding of BHBT. First, if the interactions between the helmet system and the headform, which is significantly less compliant, are critical to the proper assessment of BHBT, then an operator's skull also likely provides significantly more resistance than a clay witness. Second, if the presence of shell offset and helmet liner, which are designed to distribute load and provide additional protection, significantly changes the backface deformation profile, the approaches that aim to quantify the deformation of the inner shell surface (i.e., DIC) may have limited applicability to in-theatre events. Third, the fundamental underlying assumption relied upon in this analysis is that the air cannon impacts are representative of the helmet BHBT response during ballistic impacts. Characteristics of helmet BHBT responses have been published [6], (Voo, Improved Repeatability and Reproducibility of the Ballistic Load Sensing Headform, 2016), [14], [18] with varying characteristics which are likely due to the unique response of combat helmets to the specific threat and shot location, the varying stand-off distances between shell and head, and the helmet shell support conditions (e.g., edge clamped, air backed or supported by a liner). As a result, BHBT assessment studies will need to explore the range of responses that can lead to injury.

The *k*-fold cross-validation of the air-cannon model indicates that it is a strong model with high predictability for data similar to training data. SVM classifiers are known to be extremely sensitive to outliers (i.e., test data that is fundamentally different from training data) and unable to extrapolate beyond the training dataset as optimal hyperplanes may have high curvature outside the training bounds. Therefore, a fundamental question in the present study is whether the air-cannon training data conditions (constant cross-sectional loading) is representative of ballistic BHBT loading (decelerating end ballistics). The BHBT tests tend to have a much faster loading and a wider peak but similar maximum load. If further testing is performed using the same approach described herein, it would be beneficial to vary the rigid impactor masses and velocities to more closely match the peaks, slopes, and impulses seen in BLSH/BHBT testing. In theory, if the rigid impactors are designed to match the BHBT deformation and the mass is selected to represent the effective mass of the helmet shell and projectile, and the velocities are selected to represent the shell deformation speed, it may be possible to accurately represent ballistic events using air-cannon testing. The differences in loading curves, combined with the poor ability of SVM to extrapolate to new data not contained within the training data are critical limitations of this approach. By extension, if the air cannon data test conditions, based on elements of the Allanson-Bailey injury risk curves, are not representative of BHBT loading conditions, perhaps their relevance to BHBT injury severity ought to be questioned.

4. CONCLUSION AND RECOMMENDATIONS

Behind helmet blunt trauma is a potential threat when a helmet is struck by non-perforating ballistic projectiles where the resulting local shell deformation can impart significant loading to the head causing skull fracture. According to research conducted by Allanson-Bailey, in addition to the load magnitudes, the risk of skull fracture may also be dependent on the shape of the shell's backface deformation. The Ballistic Load Sensing Headform (BLSH) was used in a series of air cannon and ballistic tests to assess the headform loads and to estimate the profile of the shell's deformation based on characteristics of the headform load measurements. The direct load measurements with the BLSH's seven load cell array did

https://doi.org/10.52202/080042-0011

not have sufficient spatial resolution to distinguish the load profile of the impacting surface. That said, the characteristics of the resulting force-time data traces showed trends that may offer insight into the impacting surface's profile. A method was developed to combine multiple characteristics of the BLSH's response curves using a Support Vector Machine (SVM) to classify the response for different impactor shapes that were shot directly at the headform. The SVM was shown to be 94% effective at distinguishing between four different impactor shapes used by Allanson-Bailey.

In this study, significant assumptions regarding the applicability of air cannon testing to ballistic events were required. This may have resulted in BHBT events effectively being outliers that are not representative of the physical processes at play. Using a different ML approach that is more robust with respect to outlier sensitivity could help but the training data must still be representative of the test data. In theory, it may be possible to select a rigid impactor profile that is representative of the geometries seen in BHBT testing, tuning their mass to match the effective mass of the helmet shell and bullet, and matching inner shell deformation velocities. If these conditions are met, the applicability of air cannon testing to simulate ballistic BHBT events on the BLSH could be greatly improved. As helmet shell performance, stand-off and backings as well as threats change over time, the characteristics of BHBT simulating impactors may need to be revisited to better reflect current helmet technologies.

Additional limitations are noted with respect to the work of Allanson-Bailey with the use of a Bovine Scapula Model (BSM) as an analogue for fractures to the cranium [6]. While similarities were demonstrated with Raymond [14] when using a multi-parameter logistical regression model, limiting factors remain and are noted to include the scapula surface curvature, skin and bone thicknesses, effective mass, and mode of fracture with respect to the population being studied. Further limitations include the shape, projected area, mass, rigidity, and speed of the impactors used to represent the true dynamic shell deformations and interactions with the cranium for a range of helmet constructions. However, it should be recognized that while rigid impactors provide a first approximation of behind shell interactions, they are a valuable addition to help identify the contributing factors to injury by controlling the impact conditions compared to full helmet system tests with inherent greater variability. Ideally, rigid impactor conditions should span the range of expected responses of the helmet system in-situ for relevant estimation of the injury risk, as with the SVM classification approach being presented.

The discrepancy between the deformation profile seen in the clay witness testing and those predicted by the SVM may be a systemic artefact inherent to the comparison of different processes. The presence of the liner system that distributes force and interactions between the shell and a non-compliant headform may result in a different BHBT profile than when a shell is filled with clay. The objective of this study was to develop a method of predicting the impactor shape on the BLSH. This was achieved for air cannon testing, but it would be useful to have a method of validating the BHBT deformation profile on the BLSH to fully validate the approach.

An alternate development pipeline approach could be proposed, where BHBT tests are performed on clay or using DIC to quantify the deformation profiles. The machine learning model would then be trained and validated on ballistic tests thereby resolving any concerns regarding the applicability of air-cannon data to BHBT events. This method would be significantly more resource-intensive – from the cost of each helmet used in testing to build a dataset sufficiently large for machine learning to be used, to the time taken by technicians after every test to repair and recondition deformed clay. This approach would require a researcher to classify the deformation profile in clay or using DIC for each impact into one of a set of impact profile definitions. The tests would then be repeated on the BLSH to determine the load profiles for each test. Finally, every new BHBT test condition would be performed on the BLSH to assess the force distribution from the load cells, expected deformation profile from the ML classifier, and the injury risk from the combined peak force and deformation profile class using the Allanson-Bailey curves. It is fair to question the relevance of the Allanson-Bailey injury risk curves for BHBT testing due to differences in boundary and impact conditions, and test medium, however, until a more suitable dataset becomes available to the research community, this is arguably the most pertinent reference.

The approach described herein attempts to link BHBT data collected on the BLSH to published injury risk curves using a machine learning classifier based on rigid impactor tests performed using an air-cannon. Each step of the process required assuming the validity of certain aspects (air cannon testing to represent ballistic BHBT events, applicability of Allanson-Bailey injury risk curves to BHBT events, etc.) Addressing limitations identified in this study could improve the presented methodology and ability to link BHBT tests performed on the BLSH to injury risk using machine learning.

References

[1] J. Sarron, M. Dannawi, A. Faure, J. Caillou, J. Da Cunha and R. Robert, "Dynamic Effects of a 9 mm Missile on Cadaveric Skull Protected by Aramid, Polyethylene or Aluminium Plate: An

Experimental Study.," *The Journal of Trauma, Injury, Infection, and Critical Care,* vol. 57, no. 2, pp. 236-42, 2017.

[2] C. Freitas, J. Mathis, N. Scott, R. Bigger and J. MacKiewicz, "Dynamic Response due to Behind Helmet Blunt Trauma Measured with a Human Head Surrogate.," *International Journal of Medical Sciences,* vol. 11, no. 5, p. 409–425, 2014.

[3] C. Bass, B. Boggess, B. Bush and e. al., "Risk of Head Injury During Ballistic Loading of Helmets," Joint Applied Vehicle Technology / Human Factors & Medicine - NATO RTO Specialists' Meeting, Koblenz, Germany, 2003.

[4] J. C. Sarron, J. P. Caillou, J. Da Cunha, J. C. Allain and A. Trameçon, "Consequences of nonpenetrating projectile impact on a protected head: study of rear effects of protections," *Journal of Trauma and Acute Care Surgery,* vol. 49, no. 5, pp. 923-929, 2000.

[5] C. Freitas, W. Gray, J. Mathis and C. Weiss, "Behind Helmet Blunt Trauma Experiments with a Human Head Surrogate," in *Proceedings of the Personal Armour Safety Symposium*, Quebec City, Canada, 2010.

[6] L. S. Allanson-Bailey, "Improved Cranial Fracture Metrics for Assessing the Protective Benefits of Helmet Systems," The Open University, 2016.

[7] M. Bevan, J. Clark, J. Hrivnak, J. Herchek, V. Alphonse, Q. Luong and K. Sedberry, "Common Helmet Test System for Blast, Blunt, and Ballistic Testing," in *Proceedings of the Personal Armour Systems Symposium*, Copenhagen, Denmark, 2020.

[8] M. Trexler, Q. Luong, V. Alphonse, K. Perrino, M. Athey, S. Storck, J. Paulson, C. Bradfield, E. Gienger, Z. Xia, M. Bevan, C. Carneal, D. Colanto and B. Fasel, "Development of a Modular Load-Sensing Headform," in *Proceedings of the Personal Armour Systems Symposium*, Washington DC, USA, 2018.

[9] J. Barnes-Warden, "Developments in Ballistic Helmet Testing," in *Proceedings of the Personal Armour Systems Symposium*, Amsterdam, The Netherlands, 2016.

[10] L. Voo, "Improved Repeatability and Reproducibility of the Ballistic Load Sensing Headform," in *Proceedings of the Personal Armour Systems Symposium*, Amsterdam, The Netherlands, 2016.

[11] J. Barnes-Warden, "Ballistic Helmet Evaluation Developments," in *Proceedings of the Personal Armour Systems Symposium*, London, UK, 2014.

[12] S. Grobert, "Development of a Test Setup for an Integrated Assessment of Ballistic Backface Deflections on Combat Helmets," in *Proceedings of the Personal Armour Systems Symposium*, Amsterdam, The Neatherlands, 2016.

[13] B. Anctil, M. Philippens and N. Shewchenko, "Are the Sensor Based BABT test Method Suitable for Helmet and Body Armour Certification," in *Proceedings of the Personal Armour Systems Symposium*, Amsterdam, The Netherlands, 2016.

[14] D. Raymond, "Biomechanics of Blunt Ballistic Temporo-Parietal Head Impact," Wayne State University, Detroit, USA, 2008.

[15] M. Bolduc and B. Anctil, "Improved Test Methods for Better Protection, a BABT Protocol Proposal for STANAG 2920," in *Proceedings of the Personal Armour Systems Symposium*, Quebec City, Canada, 2010.

[16] F. Pedregosa, G. Varoquaux, A. Gramfort, V. Michel, B. Thirion, O. Grisel, M. Blondel, A. Müller, J. Nothman, G. Louppe, P. Prettenhofer, R. Weiss, V. Dubourg, J. Vanderplas, A. Passos, D. Cournapeau, M. Brucher, M. Perrot and E. Duchesnay, "Scikit-learn: Machine Learning in Python," *Journal of Machine Learning Research ,* vol. 12, pp. 2825-2830, 2011.

[17] J. C. Platt, "Probabilistic Outputs for Support Vector Machines and Comparisons to Regularized Likelihood Methods," *Advances in Large Margin Classifiers,* 2000.

[18] D. Hisley, J. Gurganus, Williams, S. Williams and A. Drysdale, "Experimental Methodology using Digital Image Correlation (DIC) to Assess Ballistic Helmet Blunt Trauma," Army Research Laboratory, Aberdeen Proving Ground, MD, 2010.

[19] M. Sturdivan, D. Viano and H. R. Champion, "Analysis of Injury Criteria to Assess Chest and Abdominal Injury Risks in Blunt and Ballistic Impacts," *The Journal of Trauma,* vol. 56, no. 3, pp. 651-63, 2004.

https://doi.org/10.52202/080042-0011

Comparison of dynamic and static backface deformation measures

K. Loftis[1], D. Barnes[2]

[1]*Army Futures Command, DEVCOM Analysis Center, Aberdeen Proving Ground, MD, USA, (kathryn.l.loftis2.civ@army.mil)*
[2]*Survice Engineering, Aberdeen, MD*

Abstract. The development of a medically-based behind-armour blunt trauma (BABT) injury criterion for evaluation of body armour (BA) performance remains a top priority for the U.S. Army and may lead to improved specifications and requirements for non-perforating BA impacts. Further, an updated criterion may open the design space for new materials and designs. This study investigated the backface dynamics and static residual deformation of hard plates impacted ballistically without complete penetration. High-speed video analysis of twelve gelatin-backed plate ballistics tests was used to characterise the dynamic backface response, and computed tomography (CT) analysis of the impacted plates were performed to obtain measurements of the static residual deformations. Hard plates with only a single curvature were used for this initial study to allow visualization of the backface deformation in a single view on high-speed camera and geometric (depth and area) and rate measurements were recorded. Further, a secondary backface deformation phenomenon was observed in high-speed video and compared to the dynamic deformation. Before and after the tests, the plates were CT scanned and analysis of static deformations was completed. Static backface deformations varied from 7.4 to 10.3 mm and maximum dynamic deformations varied from 29.7 to 38.0 mm. The secondary backface deformation observed in high-speed video varied from 4.1 to 13.7 mm. In this study, the static backface deformation did not correspond to the maximum dynamic backface deformation. The contradiction between static and dynamic backface deformation indicates that using static deformation as the sole indicator of BABT is insufficient. The observed secondary backface deformation was more consistent with the static measurements than dynamic. There was also a difference seen in deformation rates between initial backface deformations and maximum backface deformations. Additional testing and continued analysis should be conducted to gain more information regarding pertinent metrics for BABT injury risk.

1. INTRODUCTION

Current performance requirements for Army body armour are only loosely correlated to injury. The development of a medically based behind-armour blunt trauma (BABT) injury criterion for use as the basis for body armour performance requirements remains a top priority. Developing an injury criterion that can be used to develop specifications and requirements associated with non-perforating body armour impacts will significantly improve evaluation capabilities of military body armour. Previous experimental tests performed by multiple groups have investigated the loading characteristics behind armour, but it is currently unknown whether dynamic measures of deformation are related to static measures of deformation after an impact event (1-14). When analyzing body armour from real-world events, it is not possible to measure dynamic deformations, so it is currently unknown how well static measurements relate to BABT injury (15-17). Additionally, previous research has shown that backed materials deform differently than unbacked materials, which makes viewing and measurement of dynamic deformations more difficult than static measures. The goal of this article was to investigate static and dynamic measures of backed protective plates to understand if the two could be correlated for future behind armour blunt trauma studies.

2. METHODS

A hard plate was placed against a 20% gel block with a molded face so that the plate fit up against the gel with no gap or spacing. Both ends of the gel block were molded so that gel blocks could be used for two different tests (by turning it around for the second shot). No soft armour was used, as that would have prevented viewing of the full backface deformation. The test setup included two high-speed cameras orthogonal to the plate to capture pitch and yaw of the threat at impact, as well as capturing backface deformation shape and velocity on the back of the plate. Gel blocks were back-lit with diffused light banks to eliminate glare and intense regions of light to get clear views to measure backface deformation and velocity. Video frame rates were set at 80,000 frames per second (fps) to maintain appropriate field of view and image resolution while capturing the dynamic features of the test.

https://doi.org/10.52202/080042-0012

Twelve plates were tested in two cycles using two different threats (Table 1). A universal receiver gun was used to launch the threat, and the powder in the threat was hand-loaded to target the desired velocity. The prescribed threat velocity at impact was the same for both threats. Threats are listed as A or B, with A being the first one tested. All plates were pre- and post-scanned via computed tomography (CT). A preliminary analysis of the results was performed during the pause in testing to determine if any adjustments should be made. Following this review, it was determined that the results were as expected, so testing for round two was completed with the same setup and parameters as round one. Postshot analysis consisted of two unique methods for measuring static and dynamic backface deformation. Each method will be presented separately, with a comparison of the final results. A blind analysis of the postshot data was conducted so that results were not biased by the analysts.

Table 1. Test matrix for plates on gelatin

Threat	Tests
A	1, 2, 3
B	4, 5, 6
A	7, 8, 9
B	10, 11, 12

2.1 Residual Static Backface Deformation Measurement Method

Each plate was CT-scanned before and after testing. A standard protocol developed by DEVCOM Analysis Center was used, which includes scanning without extended Hounsfield units, with the plate lying on the CT bed and a radiographic grid behind the plate for reference. Preshot CT analysis, using the CT scout X-ray (XR), verified that the plates were undamaged and did not contain any defects prior to testing. Postshot CT analysis included determining the location of the centre of impact, and measuring overall plate thickness, maximum static deformation, and the radius of the extent of hard damage. All CT analysis was completed in Mimics version 23 (Materialize NV, Leuven, Belgium), with calculations completed in Excel 2016 (Microsoft, Redmond, WA). The method was adapted from previous plate analysis (15-17). For definition of the coordinate space for each plate, the top, bottom, left, and right corners were chosen and x, y, z coordinates were recorded. Using the coronal view from CT, the outermost slice showing the hard damage was chosen. Then a circle was defined around the hard damage using the 3-point method, as shown in Figure 1. The centre of this circle was then used to define the centre of damage for reference points from which maximum deformation and undeformed thickness were measured (B). This is possible with CT because the three views (axial, coronal, and sagittal) are linked. So, to gather the coordinates of the estimated undeformed front/back of the plate at the site of maximum deformation, the sagittal view was adjusted until the circle centre was in view. Then where the line between the top and bottom of the plate of the front surface within that view was intersected by the line along the maximum damage was chosen as the estimated undeformed plate front point (A). The same was repeated for recording the estimated undeformed plate back point (C). Along the same line, the maximum point of deformation was also recorded (D). The linear measurements were then computed using the distance formula between each set of points using the x, y, z, coordinates.

Figure 1. (left) Example coronal CT image of circle defining the damage in the plate, (right) Axial slice example diagram showing measurement points: A undamaged surface point, B internal damage centre, C undamaged back point, D maximum static deformation, where the yellow line defines the front plane of hard damage and the green line defines the centre of hard damage.

https://doi.org/10.52202/080042-0012

2.2 High-Speed Video Dynamic Backface Analysis Method

High-speed video was collected by two cameras (Photron FASTCAM SA4 model 500K-M3E), overhead and side views, aligned orthogonal to the gelatin block. The overhead-view camera focused on the superior surface of the plate while a side-view camera focused on the left lateral surface of the plate. Each camera was aligned such that the centre of the camera view was at the centre of the respective plate surfaces. The overhead-view camera used a Canon EF 50-mm lens with an f-stop of 9.9 and focus set at 3800 mm. The side-view camera used a Canon zoom EF 28–135-mm lens with an f-stop of 4.9 and focus set at 1100 mm. The lenses were controlled by Birger Engineering Interface software (v1.1.9). The frame rate and shutter speed of the cameras were set to 80,000 fps and 1/177,000 s, respectively, for all tests, with the exception of test one for which a frame rate and shutter speed of 72,000 fps and 1/98,000 s were used. Given these frame rates, the maximum resolution for each high-speed video was 192 × 192 pixels, with a viewing area of approximately 180 ×180 mm.

The viewing area was sufficient to capture the threat prior to contact with the plate as well as the full extent of backface deformation throughout the event. Two light banks comprised of 42 ERV halogen lamps, each rated at 340 W, were used to backlight the plate and gelatin during testing. High-speed video was captured for 0.5 s and was triggered in sync with the universal receiver. Prior to testing, a grid scale composed of white and black 1- by 1-inch squares was placed in each camera view and a single image was captured for determination of pixel dimensions.

Each high-speed video was exported as a TIF file. The TIFs were imported into MATLAB (version R2021a, Mathworks, MA, USA) for postprocessing and calculation of backface deformation metrics. Initially, the grid scale image was opened and viewed using the "imread" and "image" functions (Figure 2). The "colormap" function was used to apply a 256-bit gray scale to the grid image and each pixel was assigned a gray scale index value. The difference in index value between the white and black squares of the grid was used to measure the size of each grid square in both the vertical and horizontal axes of the image. Transitions between white and black were determined over 40 rows of pixels (horizontal direction of view) and 180 columns of pixels (vertical direction of view), and the average number of pixels between transitions formed the number of pixels-per-inch along the two axes (horizontal and vertical). The measurements were then converted from pixels-per-inch to pixels-per-millimeter and recorded for calculation of backface deformation.

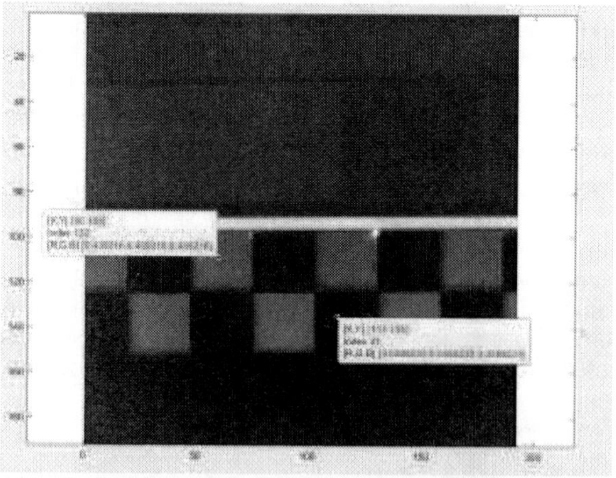

Figure 2. Gray-scale grid-scale image with the index of a pixel within a white square (index value 122) and index pixel within a black square (index value 31)

Next, a TIF displaying the threat during approach toward the front of the plate was opened and viewed in a similar manner as the image of the grid scale. The colourmap function was again used to apply a 256-bit gray scale to the image. With backlighting, the transition between solid material (i.e., the plate) and gel or air was a pixel index of 256, where the solid material is associated with a pixel index below 256, while gel and/or air had an associated pixel index greater than 256. The indexes were then used to determine the pixel location (row and column) of the rearmost portion of the plate along the shot line. This pixel was used as an initial origin for backface deformation during video analysis.

Finally, each TIF image of the high-speed video was opened using the imread function and a 192 × 192 matrix of gray scale pixel indices was created for each frame using the impixel function. These matrices were created for the first 200 frames of the video, as this sufficiently captured the full backface deformation. Comparison of the indices' values between the first frame and subsequent 199 frames were performed to identify pixels that were initially above an index value of 225 (gelatin) in the first frame and changed to an index value below 225 (plate) in subsequent frames. These changed pixels were assigned a "1" while all other pixels were assigned a "0". From this comparison, a second set of 199 binary matrices of 0s and 1s was created. An index value of 225 was chosen for this portion of the analysis because, although an index of 256 provided a clear distinction between solid material (the plate) and gelatin prior to initiation of backface deformation, it was observed that the gray scale index value of compressed gelatin fell slightly below 256. Therefore, to be certain that the backface deformation was accurately tracked as it compressed the gelatin, the threshold between the backface and gelatin was set at a pixel index value of 225. Inspection of initial frames of the video, prior to deformation occurring, indicated that using an index of 225 rather than 256 reduced the plate depth by only one pixel, if any, at any point along the backface.

2.3 Calculation of Dynamic Backface Deformation Metrics

The binary matrices from the overhead view camera were used to quantify geometric and rate-based metrics associated with dynamic backface deformation. The metrics calculated were depth of deformation, rate of deformation, area of deformation, and rate of change of the area of deformation (Table 2). Depth of deformation and rate of deformation were measureable along any desired vector extending posterior from the initial point of backface deformation. For this study, the apex of the backface deformation was identified frame by frame. Thus the maximum depth of deformation was measured over time, rather than assuming a set vector along which the metrics were measured. Figure 3 depicts an example video frame of backface deformation with the depth of deformation and area of deformation metrics highlighted. The rate of deformation was initially calculated on a frame-by-frame basis, where the rate for a given frame was the change in deformation from the prior frame, divided by the inverse of the video frame rate. The frame-by-frame deformation rate was observed to be highly variable, particularly as the rate of deformation decreased with increased depth. This variability is likely due limited resolution provided the high frame-rate requirements needed to capture initial deformation. Therefore, a moving average technique was used, which smoothed the frame-by-frame deformation rate by averaging each frame-by-frame rate with the prior and subsequent rates. Side-view camera videos were used to confirm the shape of the backface deformation but were not used for calculation of deformation metrics.

Table 2. Backface deformation metrics measured from analysis of high-speed video

Metric	**Description**	**Units**
Depth of deformation	Number of pixels identified as having transitioned from gelatin to backface along the apex of deformation (assigned "1" in the binary matrix), converted according to the grid scale measurements.	mm
Rate of deformation	Frame-by-frame change in depth of deformation divided by the video frame rate. The rate of deformation is smoothed by averaging the frame-by-frame rate with one prior frame and one subsequent frame.	mm/s
Area of deformation	Total number of pixels identified as having transitioned from gelatin to backface ("1" in the binary matrix) in the plane of the video, converted to area according to the grid scale measurements. Area of deformation does not represent a surface area of deformation, but rather a 2-D measurement of the area of the deformation within that specific plane.	mm^2
Rate of change of area of deformation	Frame-by-frame change in area of deformation within the transverse plane of the plate divided by the video frame rate. The rate of area of deformation is smoothed by averaging the frame-by-frame rate with one prior frame and one subsequent frame.	mm^2/s

https://doi.org/10.52202/080042-0012

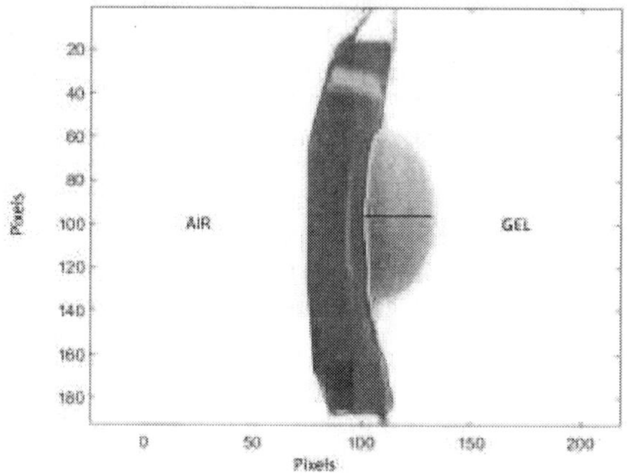

Figure 3. Exemplar high-speed video frame of dynamic backface deformation. The exemplar plate is represented in blue. The red line indicates the maximum depth of deformation while the total area of deformation is highlighted in yellow. For releasability, the front damage profile has been removed from the image.

2.4 Secondary Deformation Phenomenon from High-speed Video

Review of the overhead view high-speed video indicated that following the primary backface deformation, the backface appeared to return to its original form, and then deform again near the culmination of the video. Thus, a methodology was developed to quantify the observed secondary deformation phenomenon. Similar to the dynamic analysis, the imread function was utilised to import individual TIF files into Matlab. Once imported, the pixel values were utilised to manually identify the depth of the superior surface of the plate at two locations: (1) the point of threat impact (i.e. shot line) and (2) the right end of the plate. The former depth was utilised to identify static deformation of the backface relative to the initial depth of the plate along the shotline. The latter depth was utilised to aid in identifying change in the location of the superior surface of the plate within the camera view as the plate compressed the gelatin surface. This analysis was performed at increments of 250 frames from the first to final high-speed video frames. Once the potential plate shift was accounted for, backface deformation was identified in the final three incrementally-chosen frames and normalised to threat velocity. The prior frames were used to ensure that the depth of the superior surface of the plate was identifiable throughout the test.

3. RESULTS

All 12 plate tests on gelatin were completed successfully. Although threats A and B were meant to be striking the target at the same velocity, due to small variations in testing, threat A velocities were slightly lower than the overall mean while threat B were higher than the mean. For this reason, the results are presented as normalised according to incoming threat velocity, but due to releasability restrictions, actual velocities are not presented here. Of note however, the range in velocities overall was less than 16 m/s. To complete the normalization, all velocities were ordered and then normalised relative to the greatest velocity. A normalisation factor of 1.0 was assigned to that greatest velocity ($V_{greatest}$), while all other velocities (v_i) were assigned a normalisation factor (NF_i) greater than 1.0, according to Equation 1. Pertinent static and dynamic measurements were normalised based on these normalization factors.

$$NF_i = \frac{v_{greatest}}{v_i} \tag{1}$$

3.1 CT Static Deformation

For the static residual methodology, each plate was examined manually using Hounsfield unit values in the axial, coronal, and sagittal views of the postshot CT scan, with 3-D representation only used for visualization purposes. Figure 1 shows an example of the 3-D representation from CT and a view showing the coordinate points gathered.

A summary of the static residual plate deformation measurements is shown in Table 3. All results shown are normalised by threat velocity. The static postshot deformation was determined by subtracting the plate back surface of the post-impact plate from the back surface original plate at the impact location. To indicate the degree of plate damage, the postshot radius of damage was measured. This measurement was defined by fitting a circle around the damage within the plate and obtaining its radius. Results showed a statistically significant difference between the two sets of damage results, with the first set (A) having larger linear backface deformations but smaller radius of damage compared to set B.

Table 3. Static residual plate deformation measurements from CT

Test ID	Velocity Normalisation Factor	Normalised Post-shot Deformation (mm)	Normalised Post-shot Damage Radius (mm)
1 A	1.004	8.93	15.76
2 A	1.012	8.83	15.93
3 A	1.012	8.69	16.49
7 A	1.020	9.98	18.89
8 A	1.019	9.77	17.00
9 A	1.025	10.30	16.72
4 B	1.017	7.42	16.87
5 B	1.004	7.56	17.25
6 B	1.001	9.44	19.27
10 B	1.000	9.19	20.21
11 B	1.010	7.64	18.57
12 B	1.004	7.86	22.81
Total mean (SD)	NA	8.79 (0.99)	17.98 (2.06)
Mean (SD), A, B	NA	9.41 (0.68), 8.18 (0.89)	16.79 (1.13), 19.16 (2.17)
T-test (A vs B)	NA	$p = 0.045$	$p = 0.029$

3.2 High-Speed Video Analysis of Dynamic Deformation

Geometric and rate-based metrics were normalised and differences in metrics between threat types were determined using student t-tests (Tables 4 and 5). The maximum normalised depth of deformation varied from 29.7 to 38.0 mm with a mean of 32.2 mm for all tests (Table 4). The maximum depth of deformation tended to be greater for threat B than for threat A, but the difference in means was not significant ($p = 0.055$). The time of maximum depth, relative to initiation of deformation, varied from 1.31 to 1.75 ms with a mean of 1.56 ms for all tests. Maximum depth of deformation occurred in significantly less time for threat B than for threat A ($p = 0.001$). The area of deformation at maximum depth varied from 1898 to 2579 mm^2 with a mean of 2267 mm^2 for all tests. The area of deformation at maximum depth was not significantly different between threats A and B ($p = 0.481$).

https://doi.org/10.52202/080042-0012

Table 4. Backface deformation depth and area measurements normalised to threat velocity. Time of maximum depth was not normalised.

Test ID	Maximum Depth (mm)	Time of Maximum Depth (ms)	Deformation Area at Maximum Depth (mm^2)
1 A	31.7	1.75	2515
2 A	31.6	1.63	2282
3 A	29.7	1.64	2330
7 A	30.7	1.60	2036
8 A	29.7	1.63	2061
9 A	31.8	1.68	2069
4 B	32.6	1.41	2579
5 B	32.2	1.54	2556
6 B	30.3	1.46	2277
10 B	35.7	1.31	2077
11 B	38.0	1.56	2520
12 B	32.0	1.49	1898
Mean (SD), A, B	30.8 (1.0), 33.5 (2.8)	1.66 (0.05), 1.46 (0.09)	2216 (192), 2318 (284)
Student t-test	p = 0.055	p = 0.001	p = 0.481

Table 5. Rate of change of backface deformation measurements normalised to threat velocity

Test ID	Maximum Depth Rate (mm/ms)	Time of Maximum Depth Rate (ms)	Maximum Area Rate (mm^2/ms)	Time of Maximum Area Rate (ms)
1 A	205	0.014	4175	0.069
2 A	154	0.013	3800	0.063
3 A	103	0.025	3422	0.050
7 A	104	0.025	4739	0.038
8 A	103	0.038	3302	0.038
9 A	104	0.025	3272	0.050
4 B	129	0.038	4317	0.050
5 B	153	0.013	5174	0.063
6 B	102	0.038	3572	0.050
10 B	152	0.013	4431	0.050
11 B	154	0.013	4861	0.050
12 B	229	0.013	4353	0.038
Mean (SD), A, B	129 (43), 153 (42)	0.023 (0.009), 0.021 (0.013)	3785 (581), 4451 (546)	0.051 (0.013), 0.050 (0.008)
T-test	p = 0.345	p = 0.763	p = 0.068	p = 0.852

The maximum rate of change in deformation depth varied from 102 to 229 mm/ms with a mean of 141 mm/ms for all tests (Table 5). The maximum rate of change in deformation depth was not significantly different for threats A and B ($p = 0.345$). The maximum rate of change in the deformation area varied from 3272 to 5174 mm^2/ms with a mean of 4118 mm^2/ms for all tests. The maximum rate of change in deformation area tended to be greater for threat B than for threat A, although the difference between threats was not significant ($p = 0.068$). For each test, the depth rate maximum occurred within two to three frames after initiation of deformation, with the area rate maximum occurring one to two frames following the depth rate maximum. Therefore, the time of the depth rate maxima and area rate maxima did not vary based on threat type ($p = 0.763$ and $p = 0.852$, respectively).

The secondary deformation phenomenon was only observed in video for tests 7 through 12. The duration of high-speed video of tests 1 through 6 was sufficient to properly measure the secondary deformation. The mean depth of the secondary deformation observed in the final 3 frames of the video analysis are provided in Table 6. Mean secondary deformations were 9.8 mm and 5.7 mm for threats A and B, respectively. A student t-test of the secondary static results found that static deformation was not significantly different between threat types (p = 0.273).

Table 6. Secondary residual static deformation measurements from high-speed video for threats A and B normalised to threat velocity, where measurement was possible.

Test ID	Mean Secondary Backface Deformation (mm)
7 A	13.7
8 A	4.1
9 A	11.6
10 B	5.3
11 B	6.6
12 B	5.0
Mean (SD), A, B	9.8 (5.0), 5.7 (0.9)
T-test	p = 0.273

4. DISCUSSION

4.1 CT Analysis of Residual Deformation

The methodology for the CT analysis was developed initially to review returned theatre-damaged plates for BABT deformation and damage, and then relate their damage with injury (or lack thereof). In that scenario, there were no preshot CT scans to compare against. Therefore, for this initial analysis, only the postshot CT scans were used to estimate backface deformation measurements. Yet, in the future it is possible to reevaluate these measurements, comparing the preshot CT of each specific plate with their corresponding postshot CT. Furthermore, a comparison of the two techniques can be performed, including accuracy and ease of calculation. It is important to keep in mind how these measurements are to be used in the future and how applicable the measurement methodologies are in different testing and analysis situations (theatre-event analysis, research, etc.). Work is already in progress to compare these methodologies and expand the analysis.

After review of the CT scans, it should be noted that differences in plate design will result in different BABT characteristics and may also affect the methodology that is best suited for measuring backface deformations. These plates showed very clean, circular damage patterns, making the damage profiling more accurate and repeatable. In visual review of other plate designs with different threats, damage patterns varied greatly, with some resulting in such widespread cracking that this circle methodology would prove difficult. As BABT is investigated for links to injury risk, different plate designs and damage profiles will need to be incorporated to ensure widespread applicability of pertinent metrics.

There was a clear trend in the static deformation measurements from CT. It was possible to perform this analysis on all the plates, as the impacts were focused in the middle of the plates. This methodology would likely need to be revised for edge impacts where it was not possible to centre the damage and easily compare pre and post shot curvature. All plates showed some static deformation upon visual inspection and it was possible to view this deformation via the post-shot CT.

4.2 Video Analysis of Dynamic Deformation

Dynamic backface deformation metrics were successfully calculated from the over-head view camera videos. The curvature of the plates are such that the overhead view captured the depth of backface deformation from initiation to maximum. The full extent of dynamic deformation could not be observed in the side-view camera videos, thus the side-view videos were only used to confirm the rounded shape of the backface deformation. The dynamic backface deformation metrics provided in this study are from a 2-D analysis of the backface response. Therefore, the area of deformation does not represent the area of contact between the backface and the gelatin, but rather the total expansion of the deformed backface within the transverse plane of the plate. Creation of 3-D backface deformation metrics would require additional analysis.

Efforts were made to reduce the effect of parallax in the video analysis. The overhead and side-view cameras were carefully aligned to the centre of the surface of the plate, both vertically and

https://doi.org/10.52202/080042-0012

horizontally, and the cameras were kept orthogonal to the respective plate surfaces. Camera alignment and locations were identical for each test, and alignment of the plate and gelatin were consistent from test to test. When measuring the size of the grid scale (Figure 2), the horizontal lengths of the black and white squares, in pixels, were consistent across the length of the scale. There was no indication of decreased measurement of square size at the edge of the view relative to the centre, indicating limited parallax effect on video measurements. Therefore, any error in measurement of deformation due to parallax was less than that due to the resolution of deformation measurement and consistent from test to test.

The camera frame rate was selected to capture the dynamic nature of the backface deformation while allowing coverage of the entire area of deformation within each camera view with sufficient resolution. The goal of the current study was to create a deformation profile for the entire deformation event. This requires sufficient frame rate to capture high rate changes in deformation early in the event balanced with sufficient resolution to measure small variations in deformation geometry throughout the event. The chosen frame rate of 80,000 fps with a 192 × 192 pixel resolution resulted in a pixel size of approximately 1 mm^2. This pixel size was sufficient for capturing the geometry of the deformation while allowing for measurement of high rate changes to deformation over time intervals of 0.0125 ms.

4.3 Secondary Deformation Phenomenon from High-speed Video

The secondary backface deformation typically became quantifiable after 2000 high-speed video frames (0.025 seconds) and continued to be observed throughout the remainder of the video (approximately 3000 frames). Further, the depth of the secondary deformation remained consistent for the final video 500-750 frames. As shown in Table 6, the secondary backface deformation has a similar trend as the CT residual static deformation measurements, with greater secondary deformation for threat A than threat B. Thus, the secondary backface phenomenon may be the permanent final deformation observed in the CT analysis. However, this can not be confirmed at this time and further study is required to better understand the secondary deformation phenomenon. Most previous studies of backface deformation do not include data far enough after the event to look for this phenomenon in other testing. For future work, it is strongly advised to record data over a longer time frame to further investigate this finding. The phenomenon is likely material-dependent, so new testing of materials should record longer data to investigate further.

4.4 Comparison of Static and Dynamic Deformation Measurements

Figure 4 depicts the mean dynamic, static, and secondary backface deformations for all tests. Comparison of the results suggests that the generally accepted static CT depth measurements may not adequately describe the dynamic backface response and further investigation of alternative metrics may be warranted. Additionally, the secondary backface deformations more closely followed the trends seen for the static deformations from CT, but given the small number of cases where it was possible to measure this, more research is needed. Results are shown relative to threats A and B to simply show the reversal of maximum deformation between dynamic and static measurements. More research is needed to confirm the results of this small study and investigate a wider range of threats and velocities. Overall, these results show that there is little relationship between static and dynamic measures of deformation and this should be taken into account when investigating metrics for estimating injury from behind-armor effects.

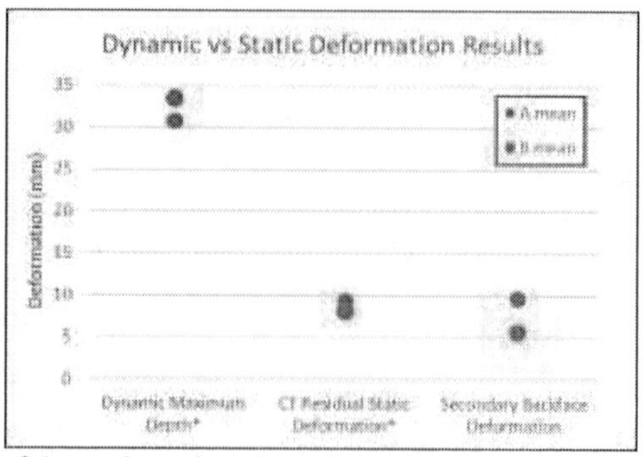

Figure 4. Comparison of dynamic and static deformation measurements for tests normalised to velocity, where * represents comparison results that were statistically significant at the p = 0.05 level.

5. CONCLUSION

The current study found that baseline post-CT hard-plate static deformation measurements trended opposite to the maximum depth of dynamic backface deformation observed in high-speed video. The tests with lesser static deformation had greater depth of dynamic deformation. Further, a secondary backface deformation phenomenon was observed in high-speed video, wherein the backface rebounded after the initial primary backface deformation event, and a secondary, smaller, backface deformation was observed. Like the static CT measurements, the depth of the secondary deformation trended opposite to the maximum dynamic deformation. The data indicate that static plate measures currently used to evaluate injury potential may be inadequate, and additional testing and continued analysis should be conducted to gain more information regarding pertinent metrics for BABT injury risk.

Acknowledgments

Funding support for this work was provided by the Army Futures Command Medical Research and Development Command through the Military Operational Medicine Research Program (MOMRP). We would also like to recognise the Army Research Laboratory Weapons and Material Research Directorate, with Dr. Karin Rafaels, for providing insightful experimental input for completing the testing and analysis.

References

[1] Rafaels, K. A., Loftis, K. L., Moholkar, N. M., & Bir, C. A. Comparing the backface deformation behavior between soft and hard body armours. 30th International Symposium on Ballistics. 2017.
[2] Rafaels, K. A., Loftis, K. L., & Bir, C. A. Can clay tell us more than deformation? Personal Armour Systems Symposium, 2018.
[3] Rafaels, K. A., Good, C. H., Loftis, K. L., Satapathy, S. S., Schuster, B. E., Zhang, T. G., McDonald, J. R., et al. Assessment of change in behind-armour blunt trauma (BABT) criteria (ARL-TR-8670). U.S. Army Research Laboratory, 2019.
[4] Cronin, D. S., Bustamante, M. C., Barker, J., Singh, S., Rafaels, K. A., & Bir, C. Assessment of thorax finite element model response for behind armour blunt trauma impact loading using an epidemiological database. Journal of Biomechanical Engineering, 143(3), 2021.
[5] Rafaels, K. A., Choi, K., Glasser, G., & Bir, C. A. Estimation of armour backface velocity. Personal Armour Systems Symposium, 2020.
[6] Lizins, M. E., Matheis, E., Loftis, K. L., & Rafaels, K. Investigating mortality of blunt and penetrating trauma to the torso. Military Health System Research Symposium, 2020.
[7] Rafaels, K. A., Gillich, P. J., & Moholkar, N. M. Importance of areal measurements for thoracic BABT test devices. Personal Armour Systems Symposium, 2016.
[8] Rafaels, K. A., Gillich, P. J., & Moholkar, N. M. Insights into evaluation tools for assessing thoracic behind-armour blunt trauma test devices. 28th International Symposium on Ballistics, 2014.
[9] Bir, C., Rafaels, K., & Loftis, K. Behind armour blunt trauma injuries sustained by law enforcement officers. IRCOBI, 2019.
[10] Bustamante, M., Barker, J., Rafaels, K., Bir, C., Singh, D., Sathananthan, P., & Cronin, D. Shell plate method of reconstructing of behind armour blunt trauma impact scenarios for soft armour using a detailed thorax model. IRCOBI, 2019.
[11] Cronin, D. S., Barker, J., Rafaels, K., & Bir, C. Injury risk assessment for behind armour blunt trauma impact conditions using thorax finite element model. World Congress of Biomechanics, 2018.
[12] Cronin, D., Rafaels, K., Bir, C., Barker, J., Singh, D., Bustamante, M., & Sathananthan, P. Reconstruction of behind armour blunt trauma impact scenarios for soft armour using a detailed thorax model. IRCOBI, 2018.
[13] Rafaels, K., Loftis, K., & Matheis, E. Correlating incoming threats with BABT skin injury. World Congress of Biomechanics, 2018.
[14] Good, C, H., Loftis, K. L., Schuster, B. E., Gillich, P. J., Eberius, N. L., & Ligda, J. P. Re-creation of ballistic impacts to enhanced small-arms protective insert (ESAPI) plates returned from theater (ARL-TR-8439). U.S. Army Research Laboratory, 2018.
[15] Good, C. H., Loftis, K. L., Gillich, P. J., & Eberius, N. L. Analysis of ballistic impacts to hard armour plates returned from theater (ARL-TR-8364). U.S. Army Research Laboratory, 2018.
[16] Loftis, K. L., Good, C. H., Schuster, B. E., & Gillich, P.J. Estimating obliquity of ballistic impacts from residual damage to hard armour plates. Personal Armour Systems Symposium. 2018.

 https://doi.org/10.52202/080042-0012

Development of a new thoracic surrogate for KENLW and BABT impacts

A. Andrei[1], C. Robbe[1], T. Penders[1], A. Papy[1]

[1]*Royal Military Academy - Department of Weapons Systems and Ballistics - 30 avenue de la Renaissance, 1000 Brussels, Belgium (anca.andrei@dymasec.be)*

Abstract. The studies regarding Kinetic Energy Non-Lethal Weapons (KENLW) and Behind Armour Blunt Trauma (BABT) have proven to be similar in terms of injury mechanism, and there are numerous attempts available in literature that demonstrate so. In order to be able to assess the injurious potential and effectiveness of KENLW, an injury criterion needs to be defined. This can be represented by one or more physical parameters that correlates with the injury severity of the impacted body region. For a certain value of the injury criterion and a given statistical probability, a distinction between a traumatic and non-traumatic event can be made. Regardless of the chosen criterion, the initial data is typically acquired from animal and Post Mortem Human Subjects (PMHS) testing. These are difficult to perform due to ethical, legal and practical considerations. Stating so, the use of surrogates, which are validated in terms of biofidelity, is imperative. Blunt Trauma Torso Rig (BTTR) is one the surrogates that is designed for BABT applications, which is also suitable for KENLW impacts. This system's injury assessment methodology is based on performing measurements of back-face displacement of a soft membrane and computing the $(VC)_{max}$. This represents the viscous criterion which is based on the viscous response of the human body and can be determined by the compression of the chest cavity and the velocity of compression. During the use of the BTTR, some difficulties regarding biomechanical results have been identified. A scaling factor needs to be applied in order to be able to correlate BTTR measurements with KENLW impacts observed in literature and biomechanical results published in NATO STANREC 4744: AEP-99. In order to be able to avoid altering the results obtained when testing by applying adjusting factors, the proposed study aims toward developing a new thoracic surrogate that will be both suited for KENLW and BABT applications. The proposed surrogate will be manufactured out of two types of polyurethane rubbers. The materials were chosen based on the material used for the BTTR, which is not stiff enough, the values of displacement are too high. In this paper, the next steps are followed. Firstly, sets of 50x50 cm membranes with different thicknesses are manufactured. Secondly, based on the corridors from NATO STANREC 4744: AEP-99 and those developed in literature, an optimization process is developed. The surrogates are calibrated in depth, in order to be able to accomplish validation in terms of biofidelity and standardisation. Thirdly, the calibrated version of the surrogates is tested under BABT conditions. Fourthly, conclusions and perspectives are drawn.

1. INTRODUCTION

The ability to assess and predict the injury potential, in cases of BABT solicitations, represents a major decisional factor when it comes to personal body armour. The phenomenon of BABT has been proven to be similar in terms of injury mechanism with KENLW impacts [1-5]. In order to be able to characterize the injury and wound potential of an impact, a correlation between one or more physical parameters and the injury severity of an impacted body region needs to be determined. Nowadays, this can only be accomplished by means of surrogate testing, due to ethical, legal and practical considerations of animal and Post Mortem Human Subjects (PMHS) testing. There are various surrogates that are suited for KENLW and BABT testing on the market. Due to practical and functional limitations, the need of creating a new thoracic surrogate emerged.

The objective of this paper is to develop a new thoracic surrogate, made out of polyurethane rubber with a flat design, that is easily manufactured in any laboratory setting, which is both suitable for KENLW and BABT testing. Following the course of this paper, the methodology of testing for both KENLW impacts and BABT solicitation, the manufacturing process of surrogates and the obtained results will be presented.

2. METHODS

2.1 Injury criterion and injury tolerance

In order to be able to assess the injurious potential and effectiveness of KENLW, an injury criterion needs to be defined. This can be represented by one or more physical parameters that correlates with the injury severity of the impacted body region. A distinction between a traumatic and a non-traumatic event

https://doi.org/10.52202/080042-0013

can be made based on the injury tolerance which is defined as a certain value of an injury criterion with a given statistical probability [6]. One of the most used injury criterion for dynamic thoracic impacts is the viscous criterion, $(VC)_{max}$, illustrated in Figure 1. This criterion is based on the viscous response of the human body and can be determined by the compression of the chest cavity and the velocity of compression [7]. The related injury tolerance is equal to $(VC)_{max}$=0.8 m/s. It corresponds to a 50% probability of observing 2 or more rib fractures or a sternum fracture [6, 7]. The purpose of the following section will be to measure displacement of the target as a function of time and the $(VC)_{max}$ for different KENLW and BABT loadings.

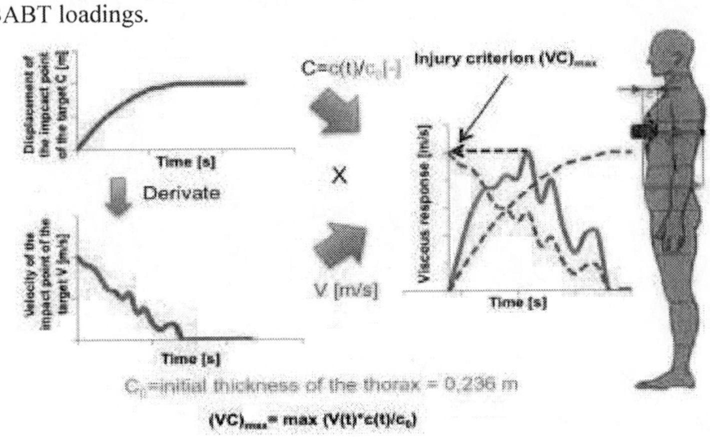

Figure 1. Illustration of the viscous criterion, $(VC)_{max}$

2.2 Experimental setup

2.2.1 KENLW

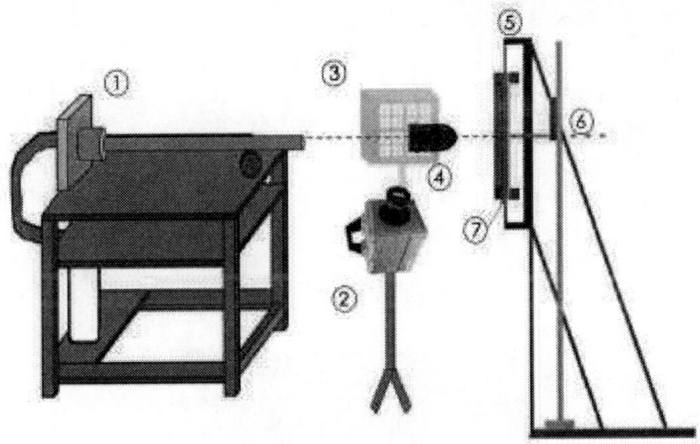

Figure 2. KENLW setup

The setup used for KENLW impacts is represented in Figure 2 and contains the following items:

1. Pneumatic launcher with interchangeable barrels, which allows the launching of any KENLW projectiles with a controlled velocity;
2. High-speed camera Photron SA-X2 High-speed Camera, used with a frame rate of 50 000 fps, allowing to perform velocity measurement and to visualise the impact. Previous studies have shown that the uncertainties of measurement using this setup remains lower than 1.5% [3,6];
3. Light spot used for optimising the high-speed camera images and measurements;
4. Projectile. 3 types of projectiles will be tested. One 40mm sponge grenade B&T Sir-X, composed of a plastic body and a deformable foam nose, and 2 types of 37 mm polyurethane. These are considered perfectly stiff during the tested impacts. The projectiles characteristics are presented on Table 2;
5. Frame. The frame provided by the BTTR has been adapted for holding the developed surrogate;
6. Wenglor PNBC006 laser used for measuring a single point backface displacement of the membrane with a measuring frequency of 30 kHz;

https://doi.org/10.52202/080042-0013

7. Surrogate.

Table 2. KENLW projectiles

Projectile	**B&T Sir-X**	**L5 Polyurethane projectile (BR1)**	**L5 Polyurethane projectile (BR2)**
Diameter [mm]	40	37	37
Mass [g]	32	134	30
	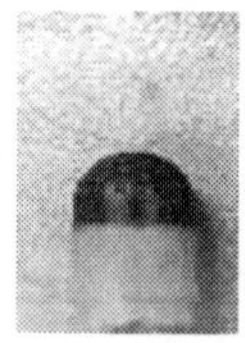		

2.2.2 BABT

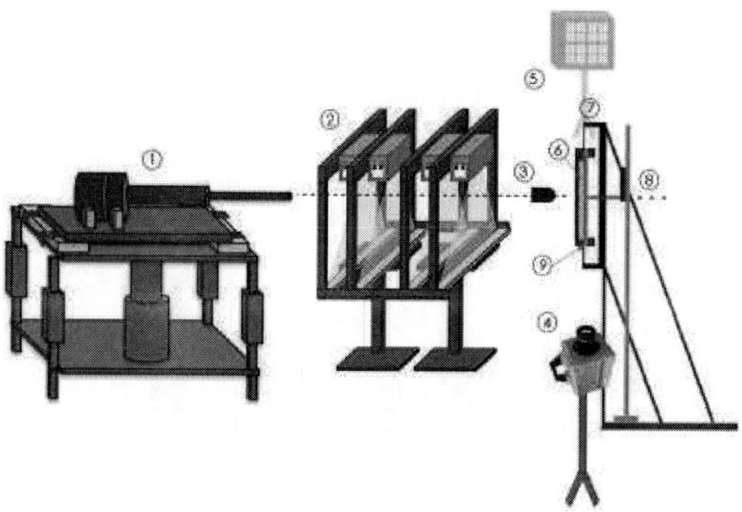

Figure 3. BABT setup

Figure 4. Placement of the ballistic protection. 2 steel frames are used to tighten the ballistic protection against the surrogate (behind the protection – not visible)

The setup for BABT testing is presented in Figure 3 and contains the following items:

1. Prototypa STZA 16M2 launcher;
2. Drello light screen LS9i3 velocity measurement system;
3. 9 mm Parabellum projectile;

4. High-speed camera Photron SA-X2 High-speed Camera, used with a frame rate of 50 000 fps, allowing to visualise the impact;
5. Light spot used for optimising the high-speed camera images and measurements;
6. Soft ballistic protection with 20 layers of Kevlar that is fixed against the surrogate (see Figure 4);
7. Adapted BTTR frame;
8. Mel M7L/100 laser used for measuring a single point backface displacement of the membrane with a measuring frequency of 10 kHz. This device is part of the BTTR system and, due to practical reasons, it was chosen to be used in the BABT testing, because its design has been adapted to sustain a hit in case of perforation (see Figure 5);
9. Surrogate.

Figure 5. Mel M7L/100 100 sensor on its support. The vertical direction of the sensor, the presence of the mirror and the metallic casing are designed so that an unlikely perforation only breaks the mirror.

2.3 Tested impact

Following the course of this paper, the proposed surrogate will be tested under two impact mechanisms, KENLW impacts and BABT.

In the context of KENLW, there will be 5 types of configurations for testing, presented in Table 1. These are determined by the necessity of validating the surrogate in terms of biofidelity, using the human response corridors determined by Wayne State University [7], and standardisation, using the corridors from NATO STANREC 4744: AEP-99 [8]. The results provided by the developed surrogate should be within the boundaries specified in Table 1 in terms of $(VC)_{max}$ values, and within the corridors specified in [7, 8] in terms of displacement as a function of time signals.

Table 1. KENLW testing configurations

Nr.crt.	Case	Projectile	Velocity [m/s]	$(VC)_{max}$ boundaries [m/s]
1	AEP 99	B&T Sir X	56 ± 2	0.28 - 0.32
2	AEP 99	B&T Sir X	86.5 ± 2.5	0.78 - 0.85
3	WSU Case A	L5, 134 g	20	0.24 - 0.51
4	WSU Case B	L5, 134 g	40	0.65 – 2.35
5	WSU Case C	L5, 30 g	60	0.14 – 0.60

Based on the results of the KENLW testing, an optimum regarding the depth of the proposed surrogate will be determined. The process of refinement of the surrogate's depth will be detailed in section 4.2. The optimal configuration will be tested under BABT.

BABT testing consists in impacting a soft ballistic protection with 9 mm Parabellum projectiles with velocities between 315 m/s and 335 m/s, in order to be able to avoid perforations. Due to practical reasons and availability of materials of the laboratory, the ballistic protection chosen in this matter is a 20 layer of Kevlar soft protection. The ballistic protection was preliminary tested to ensure no perforation at those velocities and was evaluated using plasticine testing, following the NIJ0101.06 standard, resulting in an indentation under 44 mm.

2.4 Surrogates

2.4.1 Existing surrogates

https://doi.org/10.52202/080042-0013

The process for determining the surrogate materials is based on existing surrogates on the market. These are the 3RBID (3 Rib Ballistic Impact Dummy) manufactured by Humanetics and the BTTR (Blunt Trauma Torso Rig) manufactured by Biokinetics.

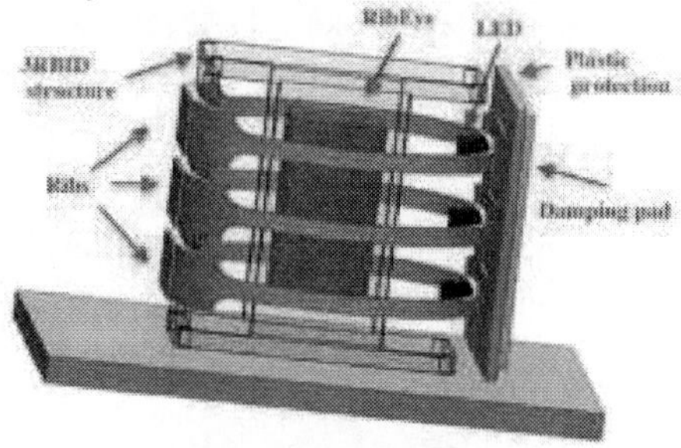

Figure 6. The 3RBID system

The design of the 3RBID, manufactured by Humanetics, presented in Figure 6, consists of three modified BioSID (Side Impact Test Dummy) ribs, with damping elements that allows a better fitting into the behaviour of the human thorax. Its measuring system performs 3D measurements of the displacement of the three ribs as a function of time, which can lead to computation of the (VC)max.

Figure 7. The BTTR system

The response of the system proved to be well correlated, in terms of biofidelity, with the biomechanical results presented by Wayne State University and KENLW impacts observed in literature [3]. Even though, in terms of practicality and reliability, the system proved to have some downsides. It is rather on the expensive side and due to its fragile design, in case of malfunctions, the repairing process seems to be time consuming and costly.

The BTTR, manufactured by Biokinetics, presented in Figure 7, consists of a cylinder shaped polyurethane membrane. The measurement system is represented by a Mel M7L laser, which performs measurements of a single point backface displacement of the membrane as a function of time, that allows determination of the $(VC)_{max}$.

Compared with results of the previous systems and NLW impacts observed in literature, the response of the BTTR, in terms of displacement, seems to be higher than desired. Due to this reason, in order to be able to correlate available results, the BTTR response needs to be scaled [4, 5].

2.4.2 Proposed surrogate

Based on the previously mentioned considerations the surrogate proposed in this paper is represented by a polyurethane rubber flat membrane with a determined thickness. It has a low level of complexity and is easily built within a ballistic laboratory. There are 2 types of polyurethane rubbers that are subjected to testing. These were based on the properties of the BTTR membrane material.

The membranes are manufactured by mixing two components in a mold at ambient temperature, extracting, and heating at 65°C. The mold is a 50x50x5 cm concrete plywood structure (Figure 8).

The optimum depth of the surrogate is determined through an iterative process of trial-and-error.

Firstly, sets of two membranes, out of each material, with an initial thickness is manufactured. Secondly, these will be tested in KENLW conditions and the $(VC)_{max}$ will be measured. Thirdly, the obtained results are compared with the reference data specified in section 2.3, and a new thickness is determined. The process will continue until response of the surrogates is validated through all the reference data.

Figure 8. Concrete plywood mold

The surrogate is secured using a metallic frame attached to the existing frame of the BTTR. The entire configuration is presented in Figure 9 for the KENLW configuration and Figure 4 for the BABT configuration.

Figure 9. BTTR frame with one of the two proposed surrogates

3. RESULTS AND DISCUSSION

3.1. KENLW results

In the current study, 2 polyurethane membrane materials are tested, referred to as M1 and M2. Under the process of material optimization, 7 thicknesses were tested, presented in Table 3. One sample was tested under each configuration.

Out of the two materials, M2 with a thickness of 3,07 cm, proved to reach satisfactory results, but there is still space for refinement. The optimal depth was determined based on KENLW tests performed on other three different depths.

The KENLW testing is composed out of 5 different testing configurations. The results presented in Table 4 and Figures 11, 12, 13 correspond to the M2 3.07mm configuration. The time-displacement graphs are presented and placed in contrast with the relevant corridors [7, 8] for each testing configuration and the moment in time when $(VC)_{max}$ takes place is indicated.

Table 3. Materials with different thicknesses

Material	Thickness (cm)
M1	1.84
	2.6
	2.82
M2	1.6
	2.81
	3.07
	3.29

https://doi.org/10.52202/080042-0013

Table 4. (VC)max values obtained under KENLW testing

Nr.crt.	Case	Projectile	Velocity [m/s]	$(VC)_{max}$ [m/s]	$t((VC)_{max})$ [ms]	$(VC)_{max}$ boundaries [m/s]
1	AEP 99	B&T Sir X	56 ± 2	0.36 - 0.4	0.9 - 1	0.28 - 0.32
2	AEP 99	B&T Sir X	86.5 ± 2.5	0.79 - 0.82	0.5 - 0.7	0.78 - 0.85
3	WSU Case A	L5, 134 g	20	0.25 - 0.31	0.3 - 1	0.24 - 0.51
4	WSU Case B	L5, 134 g	40	1.06 - 1.11	0.8 - 0.9	0.65 – 2.35
5	WSU Case C	L5, 30 g	60	0.47 - 0.5	0.4 - 0.6	0.14 – 0.60

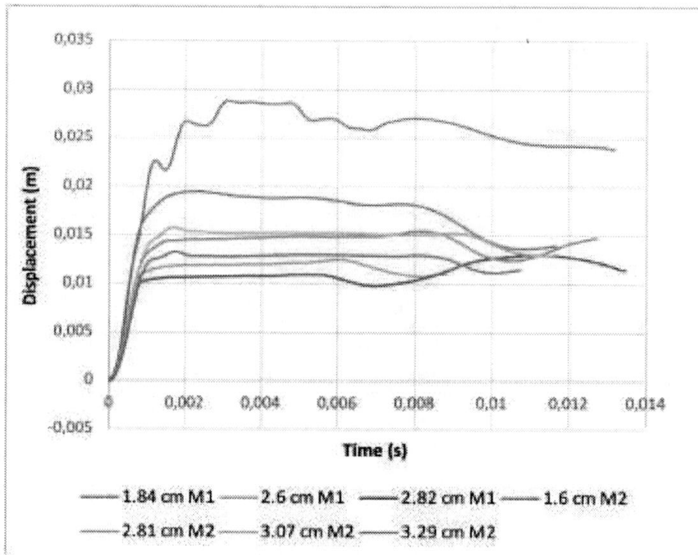

Figure 10. Time-displacement graphs of B&T Sir X impacts at 86,5 ± 2,5 m/s on 2 types of membranes at different thicknesses

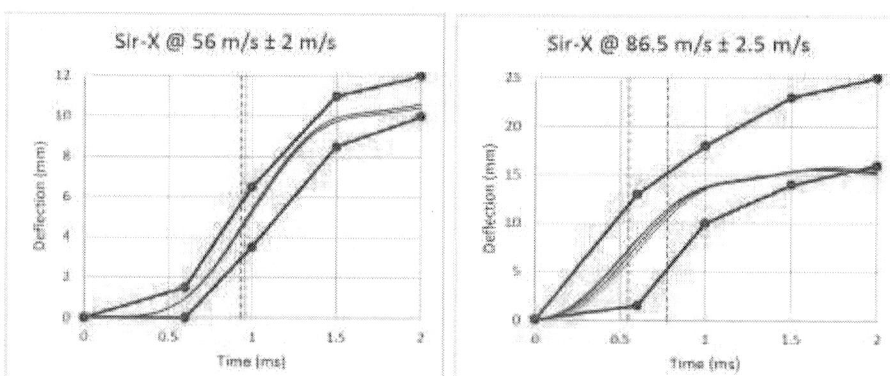

Figure 11. Testing under NATO STANREC 4744: AEP-99 conditions [8]

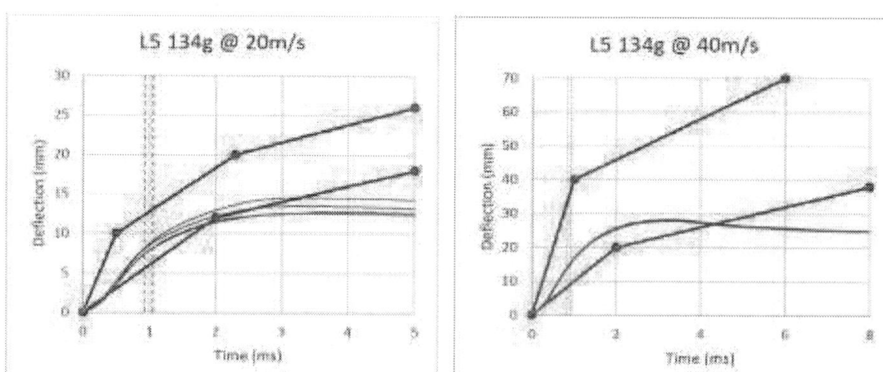

Figure 12. Testing under Wayne State University conditions, cases A and B [8]

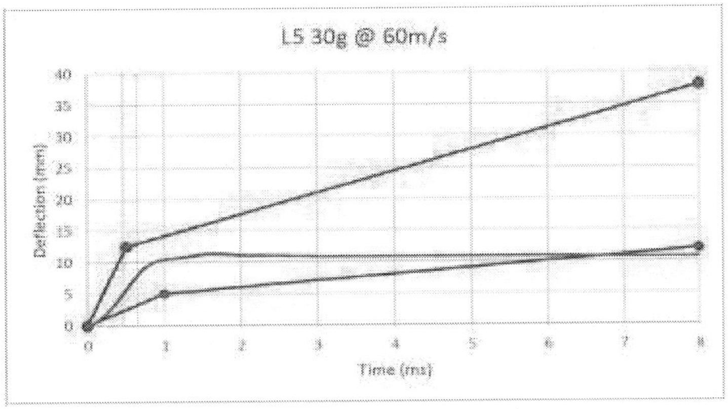

Figure 13. Testing under Wayne State University conditions, case C [7]

The time - deflection curves are very repeatable. For future development, new plates will be built in order to be able to validate the observed consistency with an increased number of shots.

In terms of biofidelity and standardisation, the results obtained are compatible with the corridors from NATO STANREC 4744: AEP-99 [8] and Wayne State University [7]. Even though, towards the end of the interval, they present a tendency to deviate outside the corridors, this doesn't represent an issue, due to the fact that the $(VC)_{max}$ occurs at a much earlier moment in time.

As the response fits the reference data properly, the injury predictions based on the response of the surrogate (based on values of $(VC)_{max}$ above or below 0.8 m/s) is also consistent with case reports from literature [9] and with other means of evaluation. [3-5, 7, 8].

3.2. BABT results

BABT is performed under a singular configuration and it consists of impacting the surrogate protected by a 20 layers of Kevlar protection with 9 mm Parabellum projectiles.

A concerning factor that occurred during the BABT testing was imposed by the mechanical constrictions of the frame. It is not in an optimised design and induces practical difficulties. As a future project to continue the development of the proposed surrogate, the authors are aiming towards developing a frame which allows moving the ballistic protection in plan and stand-off direction.

The obtained $(VC)_{max}$ values, presented in Table 5, are all below the value of 0.8 m/s, predicting a less than 50% probability to inflict an abovementioned injury. The injury prediction is therefore consistent with injury prediction using the 44 mm indent methodology [10]. The displacement-time measurements are presented on Figure 14. Due to the dispersion in velocities presented in Table 5 and to the more dispersive nature of BABT results, more dispersion is observed in the results as well.

Table 5. (VC)max values obtained under BABT testing

Nr.crt.	Velocity [m/s]	$(VC)_{max}$	$t((VC)_{max})$
1	318.79	0.36	0.53
2	324.16	0.54	0.39
3	321.73	0.7	0.28
4	322.88	0.61	0.41
5	331.9	0.39	0.6
6	334.31	0.53	0.31
7	330.63	0.53	0.33
8	326.46	0.56	0.72
9	328.45	0.56	0.48
10	331.81	0.49	0.43

https://doi.org/10.52202/080042-0013

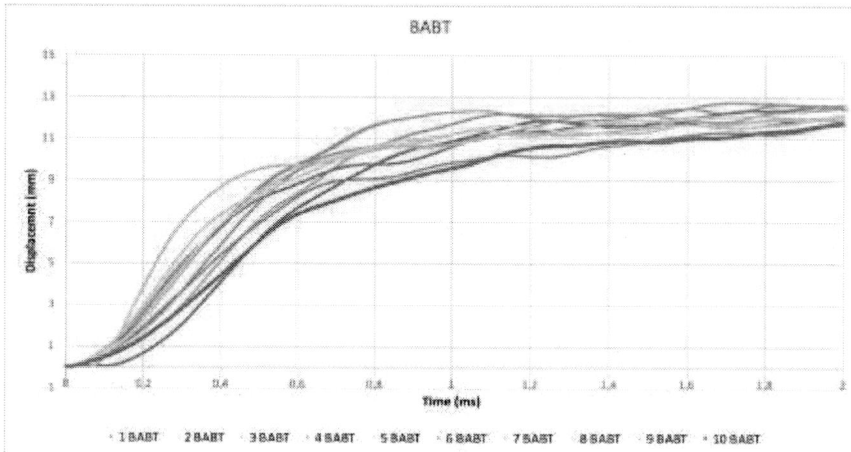

Figure 14. Time-displacement graphs of BABT solicitations

3.3. Comparison between KENLW and BABT

The dynamics of BABT loadings is similar in terms of time-displacement graphs to KENLW impacts (Figure 15). This is consistent with the literature [1-5].

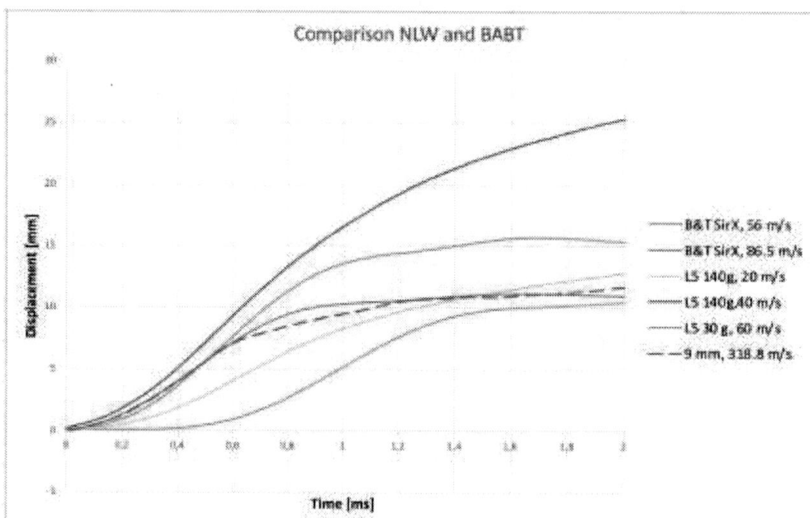

Figure 15. Time-displacement graphs of KENLW and BABT solicitations

4. CONCLUSIONS AND PERSPECTIVE

The topic of this article is to develop a surrogate which is suitable for KENLW and BABT impacts. The proposed surrogate consists of a polyurethane rubber flat membrane with a certain thickness. Its thickness is determined through a process of trial-and-error, until the targeted results are obtained. The surrogate is validated in terms of biofidelity using the human response corridors determined by Wayne State University[7], and standardisation, using the corridors from NATO STANREC 4744: AEP-99 [8]. Due to the possibility of adjusting the material and thickness, the mechanical response can be adapted to a certain degree in order to fit to the desired reference data. The optimal surrogate is tested under KENLW impacts and BABT solicitations and the results are promising. For KENLW testing, the results are consistent and present a good repeatability. The results obtained under BABT show more dispersion. The dynamics of both configurations is comparable, as already discussed in the literature [1-5]. These promising results remain to be validated on a longer test campaign, investigating the repeatability of results with different samples and the influence of the sample dimensions and aging. Practical improvements of the frame are also under consideration, in order to allow an easy plan direction movement of the protection between tests, and investigating the influence of the stand-off between the membrane and the protection.

References

[1] C. A. Bir and M. Wilhelm, "Female Body armour assessment: current methods and future techniques," PASS 2004 proceedings.

[2] K.Hewins, "Behind Armour Blunt Trauma: The Development of a Thoracic and Abdominal Simulator Rig," 2010.

[3] C. Robbe, N. Nsiampa, A. Papy, A. Oukara, and K. Meersman, "A new thoracic surrogate for assessing the impact of kinetic energy non-lethal projectiles," PASS 2014 proceedings.

[4] C. Robbe, N. Nsiampa, A. Papy, A. Oukara, and K. Meersman, "A new thoracic surrogate for assessing the impact of kinetic energy non-lethal projectiles," PASS 2016 proceedings.

[5] C. Robbe, N. Nsiampa, and A. Papy, "Sensitivity studies of the BTTR surrogate and comparison between NLW and BABT applications," PASS 2018 proceedings.

[6] C. Robbe, "Evaluation experimentale de l'impact thoracique des projectiles non-létaux - Experimental Evaluation of the thoracic impact of non-lethal projectiles," PhD thesis, Royal Military Academy (RMA) - Université de Liège (ULg), 2013.

[7] C. A. Bir, "The evaluation of blunt ballistic impacts of the thorax," PhD thesis, 2000.

[8] NATO, "STANREC 4744 Edition 3 - AEP 99 Thorax Injury Risk Assessment of NonLethal Projectiles - NATO Unclassified (available from your NATO National representative)," 2017.

[9] A. Andrei, C. Robbe, and A. Papy, "Literature review of case reports regarding NLW thoracic impacts," 11th Symposium on Non-Lethal Eeapons, 2023.

[10] K. Rice, M. Riley, and A. Forster, "Ballistic Resistance of Body Armor NIJ Standard-0101.06," U.S. Department of Justice, NCJ 223054, 2008.

https://doi.org/10.52202/080042-0013

Helmet Standoff Variation on Human Heads

Erika Matheis[1], Karin Rafaels

[1]Bennett Aerospace 1 GLENWOOD AVENUE, 5TH FLOOR RALEIGH, NORTH CAROLINA 27603 USA
erika.a.matheis.ctr@army.mil
410-306-4970

Abstract: Behind-helmet blunt trauma (BHBT) can occur when a threat is defeated by the helmet, but the helmet deformation exceeds the standoff distance between the helmet and the head. Previous research has demonstrated that effects are sensitive to standoff [1-5]. However, the standoffs that have been used in various assessments and studies are not consistent: 12.7 and 19.1mm to represent pad thickness [1], 22 mm from the ISO "J" headform [2]; and 14.9 mm in a study using the Ballistic Load Sensing Headform (BLSH) [3]. In a computational modeling study, large differences in injury risk were identified when the standoff was varied by 3 mm [2]. Therefore, assessments using different standoffs may not predict the same injury risk for identical helmet impacts. Consequently, helmet testing should examine fit and standoff more closely to ensure that unrealistic loading conditions are not being used in helmet evaluation tests. This study examines the helmet fit on human heads to provide insights on standoffs to inform laboratory testing methods. A single helmet geometry was fitted onto 25 human heads and scanned using computed tomography. Three-dimensional renderings of the helmet and head were created and analysed to determine the standoff between the helmet and the skin beneath. The standoff distribution for all 25 heads was normal with an average standoff of 24.1 ± 4.5 mm, with a range that spans 32.6 mm. Within each individual, the range was not as large as the whole population, but still spanned between 14.3 mm for the densest distribution and was as large as 27.6 mm for the widest distribution of standoffs. The median for 52% of the individuals was greater than the average standoff for the entire study, which means that more than half of the individual's measured standoffs were greater than the average. The wide range of standoffs measured between the head and helmet suggests that a single standoff for laboratory testing may not accurately represent the risk for injury behind helmets. Understanding the range and variance of standoff values in human heads may provide better insight for injury risk and potentially relevant standoffs to be used in test methods.

1. INTRODUCTION

With the increase in threats for law enforcement officers, ballistic head protection is suggested for scenarios when units are called upon to neutralise a situation [6]. Behind-helmet blunt trauma (BHBT) can occur in law enforcement tactical operations, where a threat is defeated by the helmet, but the deformation exceeds the standoff distance between the helmet and the head. This phenomenon is linked to some potentially serious injuries [4, 7-9]. Therefore, it is important to consider this injury mechanism in ballistic helmet test evaluations.

Standoff is important because it allows the helmet to dissipate more energy through deformation before contacting the head [2, 4, 10, 11]. In previous studies, there is not a consistent value for tested standoff, nor a consistent headform [1-3, 7, 12]. The different standoffs that have been used range from 12.5-25 mm [1, 7, 10, 12, 13]. Because of the lack of standardisation of standoff, there may be differences in the predicted injury effects through energy transfer. There are anthropometric differences of head shape within a human population and even some differences between standardised headforms. While headforms may be modelled after a human population [14], there are inherent differences. One headform used the "average skull" from a sample of 16 average-sized Western adults to determine its shape while the others have unknown origins of the original anthropometric dataset [15]. Some other headforms that are used for ballistic testing standards have no clear anthropometric origin and were developed for a specific test methodology [5]. Headforms may also have the same circumference measurement, but have different head breadth, length, and height measurements [14], which would change surface curvature and standoff values when helmeted. Some helmet studies have noted difficulties and performance differences with helmet fitting due to geometric differences [12, 16]. Previous research has recognised the importance of anthropometric differences and have characterised human head 3D measurements for helmet fitting and design [17-20], however standoff variation was not measured. The variance in standoff distances is unknown in a human population. Standoff variation is a known issue for helmet testing [5] and testing a single standoff may lead to unrealistic energy transfer predictions through BHBT for the

https://doi.org/10.52202/080042-0014

helmet-wearing population. This research investigates the range of standoffs within a human dataset to give insights for testing methods.

2. MATERIALS AND METHODS

2.1. Helmet Fitting and Image Acquisition

Twenty-five fresh-frozen post-mortem human subject (PMHS) heads, sectioned at the atlanto-occipital joint, were used in this study. Anthropometric measurements were taken to measure head length, breadth and circumference using calipers and tailor's tape. All specimens were fit into the same make, model, and size helmet. The suspension system was tightened so the helmet was level from side to side and stable when attempting to rock the helmet on the head. The front rim of the helmet was positioned just above the brow ridge. The head donned with the helmet was then CT-scanned using a defined protocol of 140 kVp, 250 mA, and a slice thickness of 0.625 mm.

The PMHS heads were purchased through a licensed and certified vendor of human tissue: Science Care, Inc., (21410 N 19th Ave, Suite 126, Phoenix, AZ 85027). Criteria for acceptable specimens included: male specimens without known previous or existing skull malformations and surgical procedures or interventions; fitting into the same size helmet; and approximately 50% in size for the male population. All handling and testing of the PMHS specimens were done in accordance with the Combat Capabilities Development Command (DEVCOM) Army Research Laboratory's (ARL) Policy for Use of Human Cadavers for Research, Development, Test, and Evaluation (RDT&E) under the guidance and oversight of the DEVCOM ARL Human Cadaver Review Board and the DEVCOM ARL Safety Office.

2.2. Image Segmentation and Post-processing

The images were analysed using Mimics (version 24, Materialise, Leuven, Belgium). The helmet was masked to a Housfield unit (HU) threshold of -200 to 200 HU; and holes were filled in where necessary (i.e. for equipment that may be mounted to the helmet or missing areas due to artifact). The soft tissue was masked to a threshold of -700 to 200 HU; and the skull was masked to the "bone" setting (226 HU to 3071 HU). The different masks were made into individual parts and then the parts were smoothed while compensating for shrinkage using 20 iterations with a smooth factor of 0.9. To improve rendering and reduce memory requirements, a triangle reduction of the parts was performed in edge mode with a 0.03 mm tolerance and 15-degree edge angle for 10 iterations. Finally, a wrapping surface was applied to the parts to filter out small inclusions and close any remaining small holes defined using the size of the new triangles, or the smallest detail, to be 1 mm and a gap-closing distance of 1 mm.

The STL parts were exported from Mimics to 3-matic (version 16) for standoff analysis. To define the helmet surface used for the standoff analysis, the helmet was sectioned to include only regions of interest. In this study, since the standoff is considered from the inside surface of the helmet, the helmet was sectioned to only include the inner surface closest to the head. As communication devices and ear protection can vary among users and ballistic helmets have varying degrees of coverage over the ears, this analysis did not include areas in and around the ears (Figure 1). Because some helmets have different brim and rim geometries, 20 mm of the helmet was removed at the front and back edge of the helmet to avoid issues with those components. The area around the attachment points of the retention or suspension system was also not included in this analysis since retention and suspension systems attachments vary. Once the final inner helmet surface area with only the regions of interest included, the final surface and part was reduced to 7,000-8,000 triangles. For the head, where there was an artifact or defect on the skin, the skin surface was locally smoothed. Once revised, the outer skin surface was defined as a part.

https://doi.org/10.52202/080042-0014

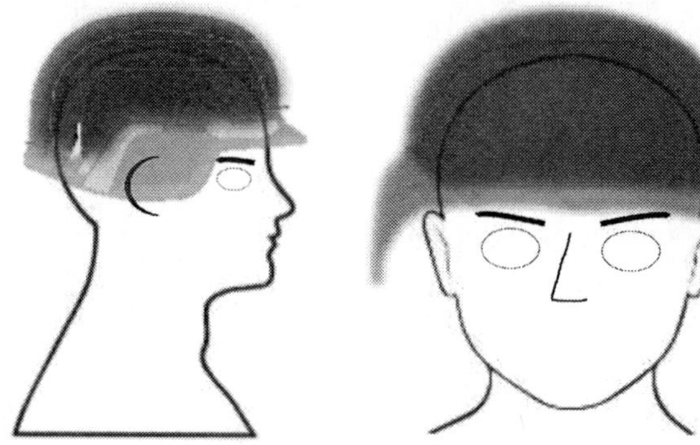

Figure 1: General geometry of a helmet with earflaps and a brim on head as worn (light gray), where the darker region (dark gray) was the approximate geometry assessed leaving out the earflaps and brim. The pink line represents a general inner surface of the helmet that was assessed for the differences between the head and helmet for the standoff measurement in the 3D viewing software.

To obtain the standoff between the helmet inner surface part and the skin outer surface part, the part comparison function in 3-matic was used. This function analyses the differences between two parts and outputs a point-cloud of coordinates and distances between parts. These calculated distances were then exported into JMP 14 (SAS, Cary, NC) for further analysis. The 25 different helmets and heads resulted in files with 3,200-6,000 standoff datapoints.

2.3. Data Analysis

Using JMP 14, descriptive statistics for all standoffs from each helmet and head combined were calculated, including the mean, standard deviation, median, quartiles, and range. A continuous best-fit density curve was generated where the Akaike Information Criterion (AICc) (Equation 1) was used to choose the best fit of a statistical model. This formula uses the number of estimated parameters (k) and the number of observations (n) used in the model. The lower the value, the better the fit when comparing multiple different options. There were a few standard options that JMP populates including comparisons with normal, normal mixtures, lognormal, Weibull, extreme value, Johnson, gamma, Sinh-Arcsinh, and exponential.

$$AICc = -2LogLikelihood + 2k + 2k(k+1)/(n-k-1) \quad (1)$$

Per specimen values for mean, standard deviation, median, quartiles and range were calculated for standoffs and head anthropometry. Individual distribution characteristics were explored through modality, kurtosis, skewness and density curve fit. A continuous best-fit curve was picked using the AIC for each individual to assess different characteristics across the dataset. Individual means were then also compared with each other to assess significant differences between individuals using a one-way analysis of variance (ANOVA) with post-hoc Tukey honest significant difference (HSD).

3. RESULTS

3.1. Total Population Standoff

Using all collected datapoints from the 25 heads, the standoff distribution was normal. In Figure 2, the total dataset looks like a normal distribution, however there are over 104,000 datapoints which may settle into a normal distribution because of the high density of datapoints across a range. The average standoff was 24.1 ± 4.5 mm with a median that matched the average (24.1 mm). Values from the 25-75% quartiles ranged from 21.1 mm to 27.1 mm, with only a 6 mm difference. The total range spanned 32.6 mm. When a best continuous-fit density curve was calculated for the total dataset, it fit a normal 3 mixture curve with an AICc of 609615 where other fitted curves tested were 204 greater than the chosen best fit. A lower AICc indicates a better fit of the model based on the likelihood probability. The normal 3 mixture model has 3 separate locations for the mean and 3 separate dispersions for the standard deviation, indicating that the numbers come from 3 separate populations. If you increase the parameter numbers to a fitted normal 25 mixture (the total population size) the AICc decreases by 557, indicating a better fit. This is to be expected because the more parameters included in a model, the better fit for the curve.

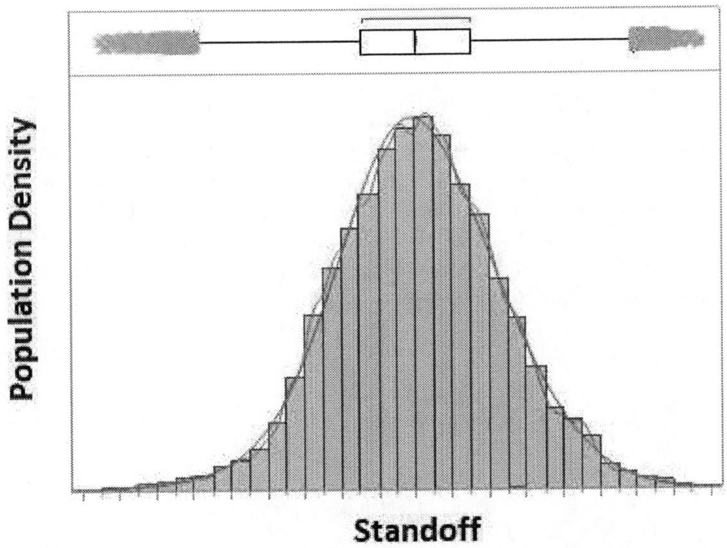

Figure 2: Total dataset distribution of standoffs. Red curve fit is a normal 3 mixture density curve and green is a normal 25 mixture density curve (matching population size). Boxplot at the top of the graph represents the spread of data where the centre line is the 50% median, the middle of the diamond is the sample mean with top and bottom indicating a 95% confidence interval, the whisker edges represent the furthest point within 1.5 times the interquartile range from the box, and the red bracket represents the densest part of the dataset that encompasses 50% of the data.

3.2. **Individual Standoff**

When comparing each individual distribution, 84% of the population had a unimodal distribution of standoffs and 16% of the population had bimodal distributions (Figure 3). The individuals with bimodal distributions (4 total) were included in the evaluation of skewness but were excluded from kurtosis analysis. There was variance in skewness for individuals, where 28% of individuals were less than 0.1 askew. The median for 52% of the individuals was greater than the average standoff for the entire study, which means that more than half of the individual's measured standoffs were greater than the average. Furthermore, the distributions of 68% of individuals was negatively skewed, indicating that the means of these individuals underestimate the most common standoffs seen in that individual. Evaluating the kurtosis of individual distributions identified only 2 individuals with a positive (leptokurtic) distribution, which suggests that these individuals had a higher likelihood of extreme standoffs. These two individuals also had the most extreme values of skewness in the positive direction (right skew), which may contribute to the positive kurtosis values. The other 90% of the individuals leaned toward a negative (platykurtic) distribution which denotes a flatter distribution with more standoffs centred around the mean.

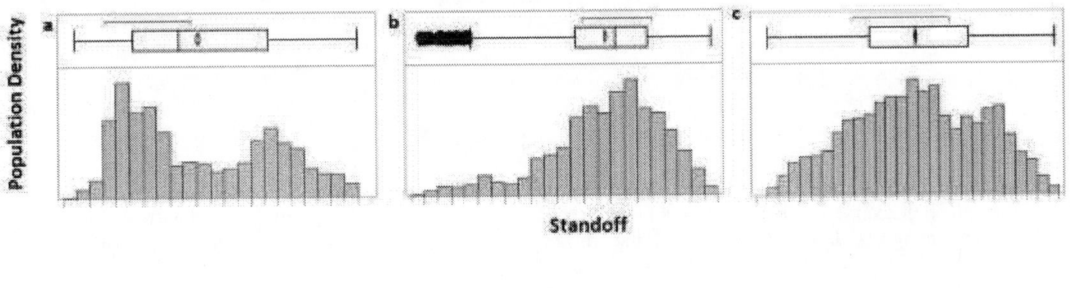

https://doi.org/10.52202/080042-0014

Figure 3:a) An example of a biomodal distribution with two clear peaks in data. b) An example of a negative skew (left skew) of datapoints indcating more collected data is more than the mean, and a leptokurtic distribution, with more extreme values. c) An example of a platykurtic distribution which is a flatter distribution. Boxplot at the top of the graph represents the spread of data where the centre line is the 50% median, the middle of the diamond is the sample mean with top and bottom indicating a 95% confidence interval, the whisker edges represent the furthest point within 1.5 times the interquartile range from the box, and the red bracket represents the densest part of the dataset that encompasses 50% of the data.

Continuous best-fit density curves were fit to each individual dataset by comparing the AICc values and the normal 3 mixture (8 parameters) had the best fit for 84% of the population, where 12% had a normal 2 mixture (5 parameters) for the best fit and only one participant (remaining 4%) had a Sinh-Arcsinh (SHASH) distribution (4 parameters) best fit (Figure 4). A normal 3 mixture curve fit represents a mixture of 3 different regions of more frequent standoffs (Figure 4), where a normal 2 mixture curve fit represents two regions of more frequent standoffs. These density curves indicate that 96% of the individuals evaluated in this study may have different regions around which standoff is distributed. A SHASH distribution identifies asymmetry and/or tails that are lighter than the normal as indicated by one individual and fits a single peak distribution curve.

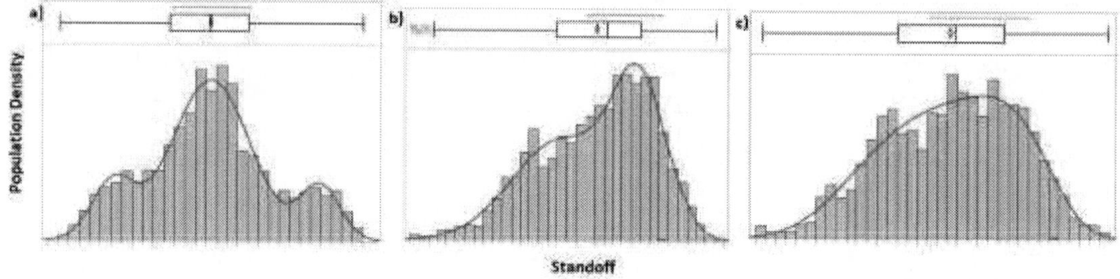

Figure 4: a) An example of a normal 3 mixture curve fit, which would represent a mixture of 3 different frequent standoffs as seen by the three peaks. b) An example of a normal 2 mixture curve fit, which would represent a mixture of 2 frequent standoffs as seen by the two peaks. c) An example of a SHASH distribution which identifies asymmetry and/or tails that are lighter than the normal. Boxplot at the top of the graph represents the spread of data where the centre line is the 50% median, the middle of the diamond is the sample mean with top and bottom indicating a 95% confidence interval, the whisker edges represent the furthest point within 1.5 times the interquartile range from the box, and the red bracket represents the densest part of the dataset that encompasses 50% of the data.

Within each individual, the range was not as large as the whole population, but still spanned between 14.3 mm for the densest distribution and was as large as 27.6 mm for the widest distribution of standoffs. The mean range of standoff within each individual was 19.62 ± 3.60 mm. The median range was less than the mean at 19.40 mm with 22.45 mm at 75% quartile and 16.70 mm at the 25% quartile. The range of standoff skewed toward a smaller range than the average (Figure 5).

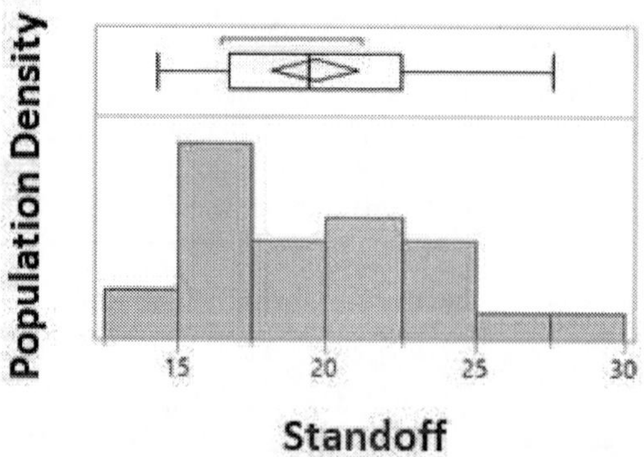

Figure 5: Range of standoff per individual in millimetres. Boxplot at the top of the graph represents the spread of data where the centre line is the 50% median, the middle of the diamond is the sample mean with top and bottom indicating a 95% confidence interval, the whisker edges represent the furthest point within 1.5 times the

interquartile range from the box, and the red bracket represents the densest part of the dataset that encompasses 50% of the data.

A mean for each individual was calculated and compared using a one-way ANOVA with Tukey HSD to understand if there were significant differences between standoffs in individuals. There were 10 significantly different ($p<0.0001$) groups of mean standoffs within the 25 individuals measured, where 4 individuals belonged to two groups (Figure 6). The range of the mean standoff values for individuals was 5.59 mm.

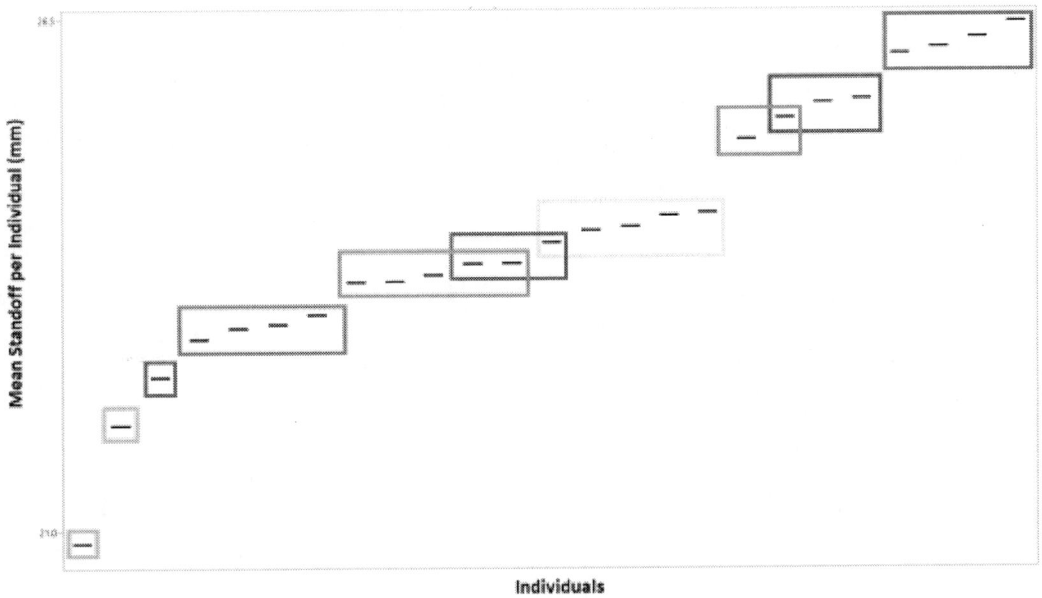

Figure 6: Mean standoff per individual in millimetres on Y-axis and each individual on X-axis, represented as a line, for comparison of statistically different groupings as highlighted by different color boxes.

3.3. Human Head Anthropometry

Anthropometric measurements were taken for length, breadth and circumference of each individual to understand the difference in human head anthropometry that is typically used for helmet fitting[5]. The range of head length was 167- 215 mm with an average of 187.72 ± 11.08 mm (Figure 7). The range of head breath was 119-170 mm with an average of 149 ± 12.91 mm. The range of head circumference was 538-591 mm with an average of 561.24 ± 15.32 mm. Calculated average eccentricity (length/breadth) was 1.27 ± 0.10 with a range from 1.07 to 1.45. These ranges may affect the standoff because of differences in helmet fit due to head shape. If the circumference is larger, then the expected standoff in that region would be less. Head length and breadth give a dimension of how circular or oblong the head shape would be at the measurement plane.

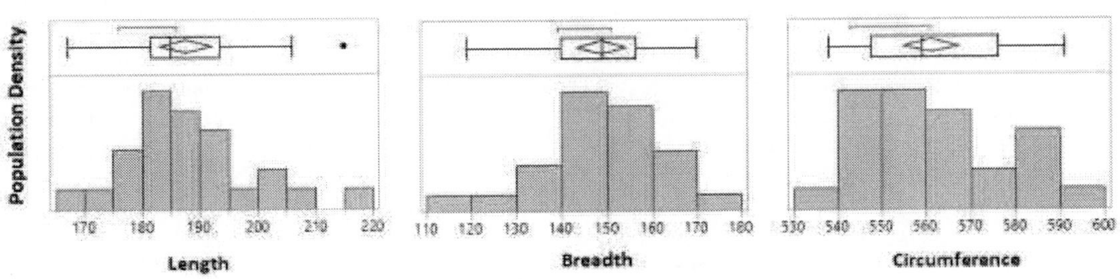

Figure 7: Distributions for head length, breadth and circumference in millimetres. Boxplot at the top of the graph represents the spread of data where the centre line is the 50% median, the middle of the

https://doi.org/10.52202/080042-0014

diamond is the sample mean with top and bottom indicating a 95% confidence interval, the whisker edges represent the furthest point within 1.5 times the interquartile range from the box, and the red bracket represents the densest part of the dataset that encompasses 50% of the data.

4. DISCUSSION

This study reported variation in standoff of helmeted human heads with a total dataset of over 104,000 standoff points for 25 individuals. It is to be noted that the individuals included in this study were from the United States and there may be differences in size and geometry depending on the region of origin [14, 18]. Furthermore, this study used a single helmet geometry, and other helmets might have different standoff distributions than the ones presented here. The standoffs that were measured in this study, in general, were larger than standoffs that have been previously reported in BHBT literature [2-5, 21]. The average reported standoff of the total dataset was 24.1 ± 4.5 mm. The individual averages ranged from 20.87 to 26.46 mm. While averages give an idea of a representative standoff value, it can be misleading to assume that the average standoff data encompasses the densest region of data as shown by the askew distributions and bimodal distributions of some of the human heads.

The reported ranges of the individual average standoffs and the 25%-75% quartile range for the total population from this study was around 6 mm. In the study of BHBT impacts by Deck et al, there was a distinct change in injury prediction from less serious at the 25 mm standoff to more serious at the 19 mm standoff, only with a difference in standoff of 6 mm [2]. Although these standoffs are smaller than the average standoffs identified in this study, this information indicates that standoff ranges as small as 6 mm can affect the predicted injury through energy transfer. In other words, for a single impact condition, the representative predictions from transferred energy for the population of potential wearers may not be captured by a single standoff.

There are some headforms that have been reverse engineered to have the same standoff at any location by matching helmet curvature to headform curvature [3]. This approach is reasonable if the purpose of the test is to only evaluate the material performance of the helmet but is not ideal for understanding the energy transfer from the helmet to head since it incorrectly assumes that a single standoff represents the fit of the helmet on the head. This approach also limits the application of the headform to helmets that may not have the same geometry as the helmet that the headform was designed for, resulting in more variation of future test results and an unequal comparison to previous test results.

There are other headforms that have been designed from human head anthropometric data [14, 15, 22, 23]. These headforms were designed for blunt impacts where energy transfer is correlated to the rigid-body motion of the head; therefore, the size parameters of these headforms prioritise factors associated with mass and moments of inertia. In other words, ensuring appropriate standoff in these headforms is not as important as ensuring proper kinematics. In BHBT impacts, the peak accelerations and bulk motion of the head occur hundreds and tens of thousands of microseconds after the peak loading [24], whereas standoff has been shown to play a significant role in energy transmission to the head [2-5]. Consequently, headforms to be used for BHBT evaluations should prioritize representing standoff over rigid-body motion.

Many helmets and headforms are sized using head circumference [5, 25]. However, in this study, the individual with the smallest circumference (and the other two anthropometric measurements) unexpectedly did not have the maximum recorded standoff value, though this individual was biased toward larger standoffs. Head circumference had a poor correlation ($R^2 = 0.15$) with mean standoff in this study, indicating that those with large head circumferences do not necessarily have smaller average standoffs (Figure 8). These findings suggest that circumference should not be the only component to consider when determining helmet fit.

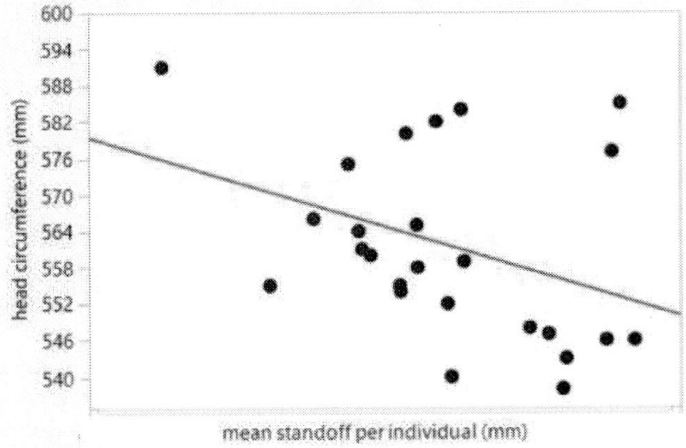

Figure 8: Comparison of head circumference in millimetres to mean standoff per individual in milimetres. The fit line (red) shows a poor correlation (R^2 value of 0.15) between mean standoff and head circumference.

Despite the same circumference of headforms, there are geometric differences that may affect standoff such as headform curvature, breadth, length and height from the reference plane [14]. The ratio of length and breadth, or eccentricity of the ellipse, describes how oblong or round a head may be and may provide insight into relevant head shapes for helmet fitting. Previous literature suggests that eccentricity be included in fit of helmets [17-19, 23] and consequently the standoff depending on shape. Eccentricity was calculated with the length and breadth values for each individual in this study, resulting in an average eccentricity of 1.27. Some existing headforms have a close-matching eccentricity value of 1.26 (Hybrid III 50th percentile), while others have different values such as 1.32 (NOCSAE and clay headform), 1.41 (ISO J), and 1.42 (DOT) [26, 27]. If implementing currently available headforms, the same helmet would inherently have different standoffs from the differences in eccentricity from these different headforms, potentially affecting the interpretation of the injury risk from results.

When exploring differences between individuals by fitting a density curve to the distribution, 92% had a platykurtic distribution and 96% of the population had a multi-peak distribution. The platykurtic distribution, or flatter distribution, suggests that extreme standoffs are not any less likely along the range of an individual. The multi-peak distribution implies that in a single individual there may be two or three distinct standoff distributions due to incompatibilities between the helmet and head geometries. For example, the four individuals with bimodal distributions had a shorter head height which led to the second peak from the larger standoffs measured in the crown region. This suggests that other anthropometric measurements including a parameter involving head height may be important to consider when fitting helmets and designing a headform for BHBT [17, 25, 28]. Additionally, only using a single-shape headform may not represent the fit of helmets on the soldier population since there are a wide array of head shapes [20, 28].

Behind helmet blunt trauma stems from the energy transfer of the helmet impacting the head after defeating the threat. The main purpose for helmets regardless of injury mechanism is to attenuate incoming energy, thereby, reducing or preventing injury. For BHBT specifically, some characterise the potential for head injury through recorded maximum depth on a clay headform [5, 27], measured energy transmission [1-3], measured force [29-31] or comparison to injuries sustained on PMHS [4, 9, 21, 32, 33]. For an accurate prediction of injury risk, an understanding of standoff variance is needed because of the effect on energy transfer.

Some helmet manufacturers choose to report the maximum depth of the ballistic transient deformation as a representative energy transfer metric because it is measured for body armour [34]; however, there is no reported correlation with head injury [5, 35]. Despite many issues that may affect the final depression in the clay headform [5], the intent is to statically capture the maximum displacement from a dynamic event to represent the differences in energy transfer behind the helmet. In the context of standoff, for a given impact condition, a larger standoff would decrease the measured maximum depth, indicating less energy transfer to the head. This reduction is due to the increased distance between the helmet and head allowing the helmet more space to dissipate the energy from the incoming threat before striking the head.

https://doi.org/10.52202/080042-0014

Through this type of testing, some specific BFS maximums have been suggested based on comparison to other quantitative metrics from previous tests, however it is unknown if the standoffs from the previous experimental tests using other headforms match the standoff on the clay headform [5, 35].

Some BHBT impact experiments use instrumented PMHS that measure similar engineering metrics to headforms and compare outputs in other laboratory settings [4, 7, 9]. Headform dynamic responses are mechanically different compared to PMHS [14], therefore the data from PMHS provides an integral component to understanding human injury. If there is a different relative standoff when testing helmeted PMHS, as shown by the increased overall standoff average in this study to previously tested values, then the associated results may not provide accurate predictions when tested on headforms which typically have smaller standoff values. In a previous study investigating the effect of standoff without considering the influence of pads or helmet support,, a difference of 1.5 mm in standoff, changed the predicted injury risk from 2% to 100% [2]. Since this study clearly shows that the range of variance within standoff values for a helmeted human is well beyond 1.5mm, it may be prudent to compare equal standoff values to better understand risk when using a headform because there is larger than 2 mm difference between the largest headform standoff to the average value of this study.

5. CONCLUSIONS

Law enforcement officers require head protection that minimises the injury risk of BHBT. The variation between previously tested standoffs makes cross comparison of injury results problematic due to differences in input energy from the defeated threat. Standoff variation is a known issue for helmet testing [5], even with standardised headform geometry. With the addition of human biovariablity in head geometry, the standoff could be larger or smaller at different locations which affects predicted outcomes. Testing a single standoff may lead to unrealistic predictions.

Acknowledgments

The research reported in this document was performed in connection with contract/instrument W15P7T-19-D-0126 with the Combat Capabilities Development Command (DEVCOM) Army Research Laboratory. The findings in this report are not to be construed as an official Department of the Army position unless so designated by other authorised documents. The views and conclusions contained in this document are those of Bennett Aerospace Inc. and the DEVCOM Army Research Laboratory.

References

1. Hisley, D.M., J.C. Gurganus, and A.W. Drysdale, *Experimental methodology using digital image correlation to assess ballistic helmet blunt trauma.* Journal of Applied Mechanics, 2011. **78**(5): p. 051022.
2. Deck, C., et al. *Assessment of behind armour blunt trauma risk for a combat helmet by a combined experimental and numerical method.* in *PASS2018-14th Personal Armour Systems Symposium 2018, Washington, DC, USA, 1-5 October 2018.* 2018.
3. Miranda-Vicario, A., P. Bravo, and F. Coghe, *Experimental study of the deformation of a ballistic helmet impacted with pistol ammunition.* Composite Structures, 2018. **203**: p. 233-241.
4. Sarron, J.-C., et al., *Dynamic Effects of a 9 mm Missile on Cadaveric Skull Protected by Aramid, Polyethylene or Aluminum Plate: An Experimental Study.* Journal of Trauma and Acute Care Surgery, 2004. **57**(2): p. 236-243.
5. Council, N.R., *Review of Department of Defense test protocols for combat helmets.* 2014: National Academies Press.
6. DiChiara, A. and M. Addonizio, *Law Enforcement Advanced Protection (LEAP) Requirements Focus Group Report.* 2007, Army Natick Soldier Research Development and Engineering Center, MA.
7. Rafaels, K.A., et al., *Injuries of the head from backface deformation of ballistic protective helmets under ballistic impact.* Journal of forensic sciences, 2015. **60**(1): p. 219-225.
8. Freitas, C.J., et al., *Dynamic Response Due to Behind Helmet Blunt Trauma Measured with a Human Head Surrogate.* International Journal of Medical Sciences, 2014. **11**(5): p. 409-425.

9. Sarron, J.C., et al., *Consequences of nonpenetrating projectile impact on a protected head: study of rear effects of protections.* J Trauma, 2000. **49**(5): p. 923-9.
10. Palomar, M., et al., *Relevant factors in the design of composite ballistic helmets.* Composite Structures, 2018. **201**: p. 49-61.
11. Azevedo, A., F. Coghe, and F. Teixeira-Dias. *Behind Helmet Blunt Trauma: Matching Measurement to Injury Using Finite Element Modelling.* in *Advanced Materials for Defense: Development, Analysis and Applications*. 2020. Springer.
12. Kulkarni, S., et al., *Ballistic helmets–their design, materials, and performance against traumatic brain injury.* Composite Structures, 2013. **101**: p. 313-331.
13. Bass, C., B. Boggess, and B. Bush. *Ballistic Helmet Backface Deformation Testing with a Dummy Subject [].* in *Personal Armour Systems Symposium (PASS)*. 2000.
14. Bonin, S.J., et al., *Dynamic Response and Residual Helmet Liner Crush Using Cadaver Heads and Standard Headforms.* Ann Biomed Eng, 2017. **45**(3): p. 656-667.
15. Becker, E.B., *Helmet development and standards.* Biomedical and Health Research, Frontiers in Head and Neck Trauma. California: IOS press, 1998.
16. Hopping, J.A., et al., *Methodologies for blunt trauma assessment in military helmets*. 2010, Project Manager Soldier Protection and Individual Equipment Haymarket,VA.
17. Guo, F., L. Wang, and D. Dong. *Human head 3D dimensions measurement for the design of helmets*. in *International Conference on Digital Human Modeling*. 2009. Springer.
18. Luximon, Y., R. Ball, and L. Justice, *The 3D Chinese head and face modeling.* Computer-Aided Design, 2012. **44**(1): p. 40-47.
19. Meunier, P., et al., *Helmet accommodation analysis using 3D laser scanning.* Applied Ergonomics, 2000. **31**(4): p. 361-369.
20. Zhang, J., et al., *Customize my helmet: A novel algorithmic approach based on 3D head prediction.* Computer-Aided Design, 2022. **150**: p. 103271.
21. Rafaels, K., *Behind Helmet Blunt Trauma Experimental Impacts Using Ballistic Helmets*. 2017, CCDC ARL.
22. Hubbard, R. and D. McLeod, *A basis for crash dummy skull and head geometry.* Human impact response: measurement and simulation, 1973: p. 129-152.
23. Gilchrist, A., N. Mills, and T. Khan, *Survey of head, helmet and headform sizes related to motorcycle helmet design.* Ergonomics, 1988. **31**(10): p. 1395-1412.
24. Council, N.R., *Testing of body armor materials: Phase III*. 2012: National Academies Press.
25. Ellena, T., et al., *The Helmet Fit Index–An intelligent tool for fit assessment and design customisation.* Applied ergonomics, 2016. **55**: p. 194-207.
26. Cobb, B.R., A.M. Zadnik, and S. Rowson, *Comparative analysis of helmeted impact response of Hybrid III and National Operating Committee on Standards for Athletic Equipment headforms.* Proceedings of the Institution of Mechanical Engineers, Part P: Journal of Sports Engineering and Technology, 2016. **230**(1): p. 50-60.
27. Justice, N.I.o., *NIJ Standard for Ballistic Helmets* 1981. p. 13.
28. Oudenhuijzen, A. and F. ter Haar, *Digital Head Avatars for combat helmet fit.* 2018.
29. Anctil, B., et al. *Evaluation of impact force measurement systems for assessing behind armour blunt trauma for undefeated ballistic helmets*. in *PASS conference proceedings, Edited by J Van Bree, The Hague, The Netherlands, September*. 2004.
30. Aranda, R., et al. *Integrated Assessment of Ballistic Backface Deflections on Combat Helmets and Prediction of Injury Risks*. in *31st International Symposium on Ballistics*. 2019.
31. Grobert, S., et al. *Comparative Investigations on Ballisitc Backface Deflections of Combat Helmets Using Force Measurement and Digital Image Correlation*. in *32nd International Symposium On Ballistics*. 2022.
32. Raymond, D., et al., *Tolerance of the skull to blunt ballistic temporo-parietal impact.* Journal of Biomechanics, 2009. **42**(15): p. 2479-2485.
33. Matheis, E., A. Eidsmore, and K. Rafaels, *Ballistic Factors of Behind Helmet Blunt Trauma Skull Fracture*, in *31st Annual International Ballistics Symposium*. 2019: Hyderabad, India.
34. Greene, M., et al. *The Next Revision of the NIJ Performance Standard for Ballistic Resistance of Body Armour, NIJ Standard 0101.07: Changes to Test Methods and Test Threats*. in *Personal Armour Systems Symposium PASS 2018 Proceeding Book*. 2018.
35. Young, L., et al., *When physics meets biology: low and high-velocity penetration, blunt impact, and blast injuries to the brain.* Frontiers in neurology, 2015. **6**: p. 89.

https://doi.org/10.52202/080042-0014

Test methodology for evaluating thoracic personal protective equipment against blast loading

J. Boutillier[1], S. De Mezzo[1], N. Prat[2], P. Magnan[1] and P. Naz[1]

[1]*French-German Research Institute of Saint-Louis (ISL), 5 rue du Général Cassagnou, Saint-Louis, 68301, France*
johanna.boutillier@isl.eu
[2]*Institut de Recherche Biomédicale des Armées, place Valérie André, Brétigny sur Orge, 91220, France*

Abstract. When designing new protective equipment for soldiers and law enforcement officers, the blast threat is not taken into account. The main focus is often on ballistic, stab and fragment protection. Primary blast injuries mainly concerned air-filled organs such as the lung and the gastrointestinal tract and studies have shown that some thoracic protective equipment (TPE) can worsen the level of injury.
An ISL anthropomorphic mannequin, called BOPMAN for Blast OverPressure MANnequin, was used to evaluate the efficiency of a soft ballistic TPE against blast threats of increasing intensities. Using the developed methodology, both qualitative (better or worse than) and quantitative (lung injury risk estimation) evaluations are possible.
Scenarios from 85g of C4 detonating at 3.8m from the mannequin to 4kg of C4 at 3m were performed unprotected and with a soft ballistic vest. Incident blast wave impulses from 17 kPa·ms to 237 kPa·ms were generated. Results show a near constant amplification factor of 1.35±0.20 on BOPMAN measurements with the vest compared to measurements unprotected. Estimated lung injury risks indicate that scenarios that should not generate lung injury when unprotected can be injurious with a soft thoracic protection. The percentage of increase of the lung weight ratio when equipped with a SBP are 0, 0.7, 2.7 and 18.7%. The augmentation of the ratio is due to pulmonary contusion and subsequent oedema. It was also noticed on high-speed videos that the TPE slaps the mannequin's chest when the shock wave arrived. The blast amplification observed could be the results of this slap caused by a small air-gap between the protection and the chest.

1. INTRODUCTION

The primary blast threat has not been considered in the development of protection systems to be used by soldiers and law enforcement personnel so far, mainly because no specification exists. The main focus is often on ballistic, stab and fragment protection. Nevertheless, air-filled organs such as the lung, ears and gastrointestinal tract are particularly susceptible to primary blast. So far, little is known about the efficiency of protective equipment against blast-induced thoracic damage. However, few studies have demonstrated that wearing low impedance thoracic protective equipment (TPE) worsens the level of blast-induced body injury, depending on the equipment used [1-5], although this finding seems to be inconsistent across studies [6-7]. However, placing a high density material (such as a ceramic plate) between the low impedance material and the incoming blast wave may help reducing blast-induced lung injury or mortality rate [2-3][8].

In order to correctly evaluate the performance of existing and future TPE against shock-waves produced by detonations of improvised explosive devices (IED), studies on thoracic models, especially mannequins, have recently emerged [9-13]. So far, the aim of these studies has been to demonstrate that the response of thoracic models is influenced by the TPE, although with this approach, one can only test if a protection system is better or worse than a reference system, without getting any information on the severity level of lung injury. Comparing the efficacy of different TPE using thoracic surrogates is a real progress in the process of designing optimal protections, but evaluating the level of protection they offer regarding the severity of lung injury would be more appropriate and informative. Indeed, such an evaluation could help find a good compromise between the weight of the systems and their ability to protect. More recently the ability of a new mannequin, called BOPMAN, to estimate the risk for lung injury in protected and unprotected soldiers was demonstrated [14]. The response of this mannequin was correlated with lung injury risk on 50kg swine.

In this study, the methodology developed with BOPMAN by Boutillier [14] for evaluating personal TPE against blast loading was applied to study the blast amplification behind a soft ballistic vest. The mannequin, unprotected and equipped with the vest was exposed to various short duration blast waves.

https://doi.org/10.52202/080042-0015

2. METHOD

2.1 Blast OverPressure MANnequin - BOPMAN

Figure 1 illustrates the anthropomorphic mannequin BOPMAN measuring 1.86 m for 78 kg. It is mostly made of solid polyethylene, with a specific instrumentation on the thoracic part, as shown in Figure 1C. The center of the thorax is not made of polyethylene but with a kind of drawer filled with silicone gel to represent the soft materials within the thorax. The thoracic part is equipped with:

- A pressure sensor (Kulite XT190M, 35 bar, USA) allowing the measurement of the reflected pressure on the thorax;
- A hydrophone (RESON TC4013, Denmark) placed in the silicone gel for the measurement of the internal pressure. It was located at the center of the gel block thanks to a thin plastic support, with the sensor tip located 1 cm behind the front wall of the thoracic part of the mannequin;
- A force sensor (B&K 8230, 22kN in compression and 2.2kN in traction, UK) at the rear part of the silicone gel block.

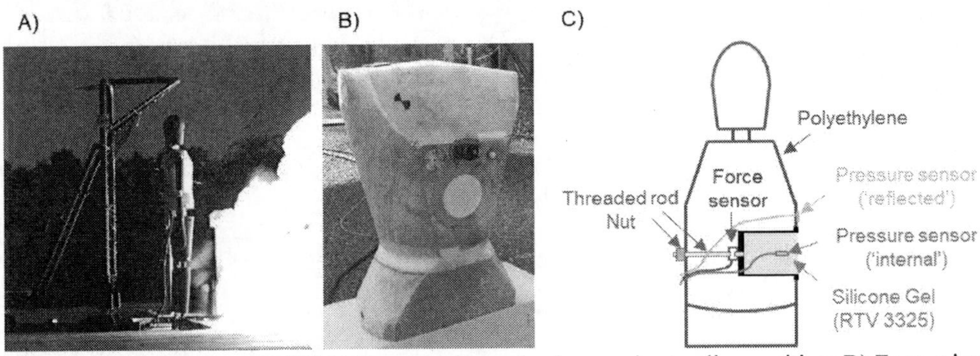

Figure 1. A) Illustration of BOPMAN exposed to a shock wave in standing position; B) Zoom view on the thorax; C) Schematic view of the thorax (side view) with details on the instrumentation

The response of this mannequin was previously correlated with lung injury risk on 50kg's swine [14]. The lung injury risk was estimated with the lung weight ratio RL/LL (RL for right lung, LL for left lung). In physiological condition, after exsanguination, this ratio is quite stable in swine. When this ratio changes, that means that fluid is trapped in or around the alveoli in one lung with the hypothesis that the exsanguination is equal between the 2 lungs, and that only one lung is damaged (that was verified during autopsy and previous unpublished histological data). The Axelsson Severity Score can also be determined, given a more descriptive aspect of the lung injury. All evaluations were done in a blinded manner (relatively to the experimental group). To fit with our animal model and experimental conditions, injury levels were determined as following, after macroscopic examination (both external and after slicing the lungs every centimeter): the lungs are graded ASS=0 for no injury, ASS = 1 for presence of surface petechiation, with no collection in the lung, ASS = 2 for presence of deep ecchymotic oedema with no "hematoma like" collection, ASS = 3 for large "hematoma like" involving less than 30% of the total volume or less than 50% of the surface on one of the slices, and ASS = 4 for large "hematoma like" involving more than 30% of the total volume or more than 50% of the surface on one of the slices. No medical imaging was available and used for the injury evaluation.

2.2 Tests with a soft ballistic thoracic protection

Four blast scenarios ranging from incident impulses of 17 kPa·ms to 237 kPa·ms were performed. For each scenario, BOPMAN thorax (27 kg with an height of 53 cm) is placed on a 15cm support at a given distance to a spherical explosive charge of C-4 suspended above the ground (height of burst around 20 cm). Quantity of C-4 and distance to the charge are determined to get the desired blast wave characteristics. Detail of the scenarios is given in Table 1. Results from exposing the whole BOPMAN or only its thorax are similar. For each scenario, reference tests were performed (thorax unprotected). Then, scenarios were reproduced while equipped with a soft ballistic thoracic protective vest. Three repetitions per scenarios and level of protection were performed for the reproducibility of the measurements. The experimental setup for the first three scenarios is illustrated in Figure 2. Experimental setup of scenario 4 is described in [14]. Exposing the standing BOPMAN to blast or only the thoracic

https://doi.org/10.52202/080042-0015

part does not change the response data. The soft ballistic pack (SBP) is composed of aramid layers, UHMWPE UD layers as well as a thin foam layer. Its weight is 4.7 kg.

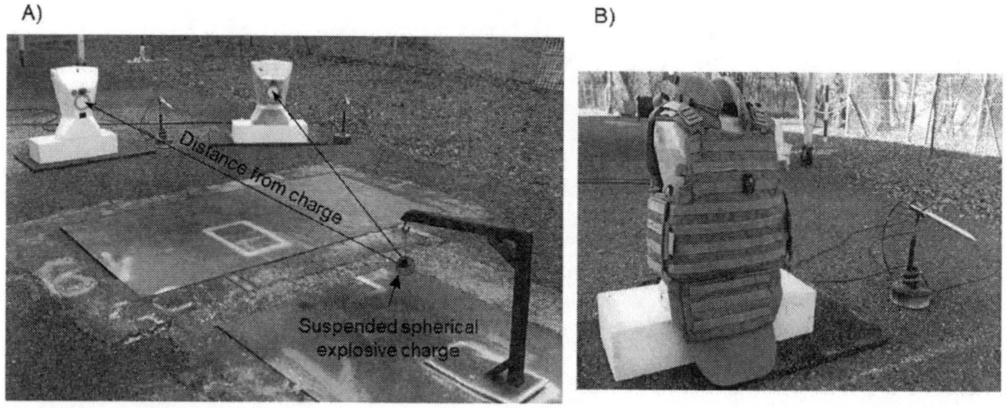

Figure 2. A) Experimental setup showing BOPMAN and pencil probe locations from the suspended explosive charge B) Illustration of BOPMAN thorax with the soft thoracic protective vest and the pelvis protection.

Table 1. Blast scenarios performed. The explosive charge was spherical and suspended 20cm above the ground.

Scenario	Mass of C-4, g	Distance from charge
1	85	3.8
2	500	3
3	800	2.3
4	4000	3

Every trial day begin with a test that serves to confirm the good response of BOPMAN by comparing the measurements with older data from a similar blast scenario. In addition to BOPMAN measurement, a pencil probe is placed near the mannequin at the same distance from the charge to measure the incident blast wave pressure profile. The height of the pencil probe sensitive part is 53 cm from the ground, similarly to the instrumented part of BOPMAN thorax. A FASTCAM Mini UX high speed camera at 10,000 frame per second was also used to visualize the vest movement under blast loading. This camera was placed at 25 m and orthogonally to the plane BOPMAN/explosive charge.

2.3 Data processing

All data were filtered with a 6th-order Bessel filter set at 90 kHz. Relevant metrics of interest were then computed for the pencil probe and BOPMAN (Table 2). The reflected and internal overpressure from BOPMAN are not presented as Boutillier [14] showed that those parameters are not relevant for protective system evaluation.

Table 2. Metrics of interest per sensor.

Sensor	**Metric of interest**
Pencil probe	Maximum positive incident pressure (ΔPi) Positive phase duration (T+) Maximum of incident impulse (ΔIi)
Reflected pressure sensor (BOPMAN)	Maximum of reflected impulse (ΔIr)
Internal pressure sensor (BOPMAN)	Maximum of internal impulse (ΔIint)
Force sensor (BOPMAN)	Maximum positive force (Force) Maximum of force impulse (ΔIforce)

2.4 Statistical analysis

Statistical analysis was done using Origin Pro software (OriginLab, United States). Metrics of interest were sorted based on thoracic protection level worn by BOPMAN and per scenario. If a normal distribution was observed, a one-way ANOVA test was performed to compare the mean values. Otherwise, a Kruskal-Wallis ANOVA test was performed. A $p < 0.05$ was considered significant.

3. RESULTS

3.1 Blast incident overpressure near the mannequin location

Blast incident characteristics for each scenario were comparable when the mannequin was unprotected or equipped with a soft ballistic protection, as indicated in Table 3. This table summarises mean and standard deviation (SD) of the metrics of interest from the pencil probe. P-values from paired student t-tests were also calculated. All p-values are above 0.05 (not statistically different), except for incident overpressure (ΔPi)from scenario 2 (p= 0.03) and scenario 4 (0.01). For scenario 4, this can be explained by the proximity of the targets with the fireball that can lead to disturbances of the shock wave. Moreover, the pencil probe is 50 cm from BOPMAN with the sensitive part looking upward and slightly rotated in the opposite direction of BOPMAN to avoid reflection from the thorax. From high speed video, the wave speed are around 375, 470, 607 and 770 m/s for scenario 1 to 4, respectively. The reflection off BOPMAN should then arrived 1.3, 1.1, 0.8 and 0.6 ms after the incident wave passage, affecting slightly the incident impulse, but not the incident overpressure. Shock reflection off BOPMAN cannot explains p-value on ΔPi from scenario 2. Nevertheless, data indicates that the blast pressure dose experienced by the mannequin and the vest were similar and, thus, allowed for a valid comparison across protection level tested.

Table 3. Incident pressure characteristics for the four blast scenarios. SBP: Soft Ballistic Pack

		ΔPi (kPa)	T+ (ms)	ΔIi (kPa ms)
Scenario 1	**Unprotected**	21.1 ± 0.6	1.89 ± 0.01	17.4 ± 0.0
	SBP	20.9 ± 0.3	1.89 ± 0.02	17.3 ± 0.2
Scenario 2	**Unprotected**	95.5 ± 8.0	2.33 ± 0.24	73.0 ± 3.2
	SBP	91.8 ± 2.4	2.24 ± 0.04	70.5 ± 1.8
Scenario 3	**Unprotected**	223.9 ± 14.9	1.72 ± 0.10	108.2 ± 2.6
	SBP	221.3 ± 11.6	1.67 ± 0.01	108.2 ± 2.2
Scenario 4	**Unprotected**	467.4 ± 48.6	2.28 ± 0.40	242.6 ± 22.6
	SBP	421.5 ± 37.6	2.28 ± 2.30	226.7 ± 7.8

Figure 3 illustrates an example of incident pressure and impulse profiles from unprotected and protected configurations for scenario 2 (500g at 3m).

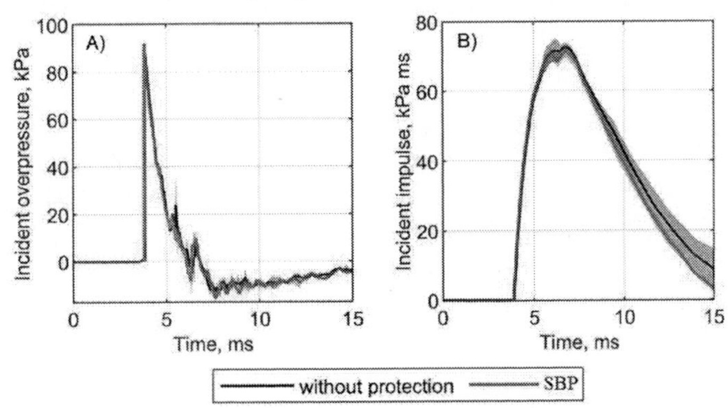

Figure 3. Example of incident pressure (A) and impulse (B) profiles from unprotected and protected (SBP) configurations (scenario 2, 500g at 3m)

https://doi.org/10.52202/080042-0015

3.2 Soft ballistic pack performance under blast loading

Figure 4 illustrates the force and the force impulse from BOPMAN unprotected and equipped with a SBP for scenario 2. An amplification of the maxima can be observed when equipped with the SBP compared to the unprotected scenario.

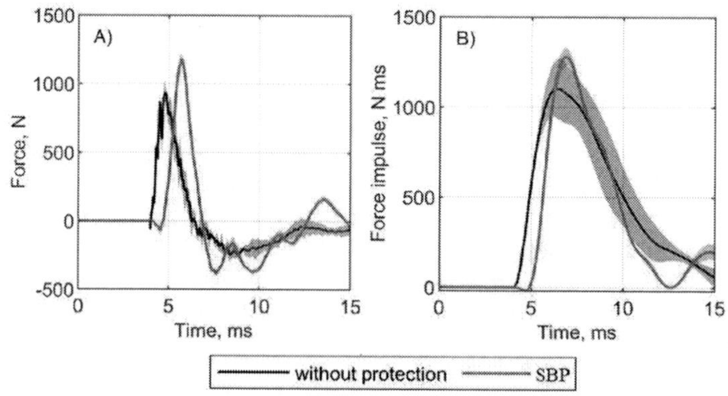

Figure 4. Example of force (A) and force impulse (B) profiles from unprotected (--) and protected (--) configurations (scenario 2, 500g at 3m)

The response of BOPMAN with the SBP differed significantly from the response unprotected for all tested scenarios (incident impulse up to 237kPa·ms, short duration wave). Figure 5 illustrates the comparison of the maximum values of the force and force impulse unprotected and with the ballistic vest for each scenario. BOPMAN lung injury threshold values are also plotted on the graph. It can be noticed that scenario 3 unprotected is close to/on the lung injury threshold while unprotected, which is in accordance with Bowen curves.

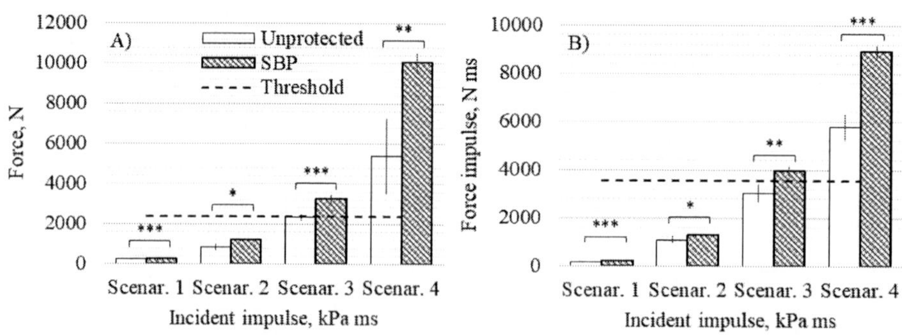

Figure 5. Comparison of the maximum values of the force and force impulse unprotected and with the ballistic vest for each scenario * p<0.05, ** p<0.01 and *** p<0.001

Figure 6 illustrates the mean and standard deviation of BOPMAN metrics ratio SBP/unprotected. The ratios SBP/unprotected from the force and the force impulse values were calculated. Then, mean values and standard deviation were calculated from previously obtained ratios. The mean ratio is statistically not different across scenarios ($p=0.587$). The amplification factor on BOPMAN metrics due to the wearing of a soft ballistic pack can be considered as constant for short duration blast wave of incident impulse up to 237 kPa·ms. The amplification factor is equal to 1.35 ± 0.20.

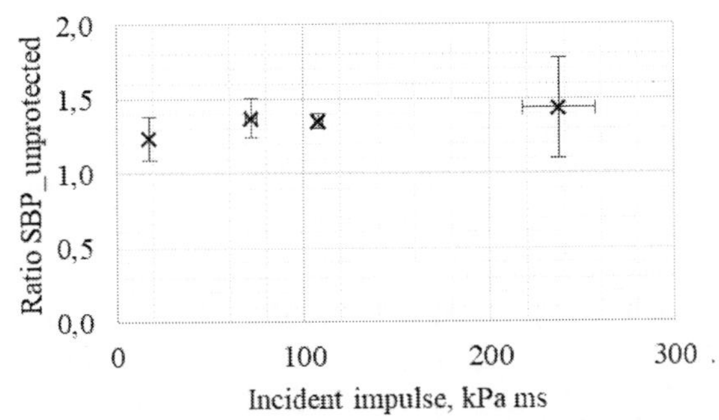

Figure 6. Mean and standard deviation of BOPMAN metrics ratio SBP/unprotected for each tested scenario. The ratio is plotted against the maximum of incident impulse.

It was also noticed with the high-speed video recorded that the SBP slaps the thorax when the shock wave hits it. Moreover, a double peak is seen on the internal pressure profiles when equipped with the protection. The more intense is the blast, the higher the inward movement of the protection will be. When the vest was positioned, there was (almost) no gap between the vest and the torso, but this small gap exists, leading to that displacement. SBP obeys to the successive compression and suction phases of the blast phenomenon. After this inward displacement, the negative phase of the incident blast wave led to the inflation of the protection with and without the pelvis protection attached to the vest. The movement of the SBP for the scenario 3 is illustrated in Figure 7.

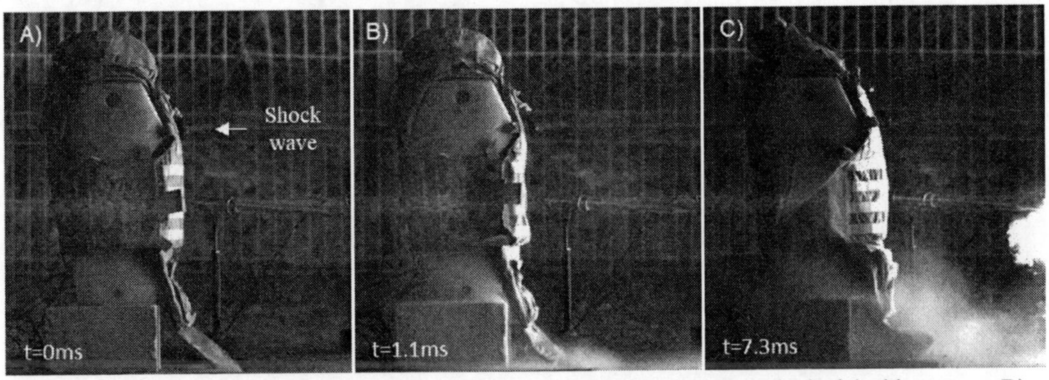

Figure 7. Movements of the SBP for the scenario 3. A) Vest before the arrival of the blast wave; B) Maximum compression of the vest; and C) Maximum expansion of the vest.

The addition of a ceramic plate was found to improve the performance of the TPE [14]. High-speed recording showed that the slap of the protection with a plate is slower than without it, which could be due to the addition of mass. Moreover, no expansion phase of this protection was visible, probably due to the rigidity of the protection with the ceramic plate.

3.3 Lung injury risk estimation

It was previously noticed a constant amplification factor on BOPMAN metrics while equipped with the SBP compared to unprotected scenarios. The constant amplification factor on BOPMAN metrics does not imply a constant amplification factor for lung injury level. As written in section 2.2, the ratio RL/LL was used to represent the extent of lung contusion. It can be calculated with equations (1) and (2) using the force or the force impulse measurement from BOPMAN. Those equations are slightly different from equations in [14] but are based on the same database. Only the chosen fitting equation is different.

$$\frac{RL}{LL} = 1.36 + 0.13 * e^{\frac{(Force-2348.2)}{4273.7}}, R^2 = 0.72 \quad (1)$$

$$\frac{RL}{LL} = 1.36 + 0.18 * e^{\frac{(\Delta I_{force}-3568.0)}{3653.7}}, R^2 = 0.77 \quad (2)$$

https://doi.org/10.52202/080042-0015

For information, during a blast event, lung injury was always on the lung facing the blast [14]. This can be proven by the left lung weight (LL) that was not statistically different across group (unprotected vs. equipped with a protection). The LL weight mean value is 158.6±34.0g.

Moreover, with RL/LL, the ASS (see section 2.2) can be calculated with the equation given in Fig 8. This metric give a more detailed description on the lung injury. Table 4 summarizes estimated RL/LL and ASS for each scenario and protection level. For the four tested scenarios, the percentage of increase of the ratio RL/LL when equipped with a SBP are 0, 0.7, 2.7 and 18.7%.

Table 4. Calculated RL/LL and ASS for all scenarios and protection level

	Scenario 1	Scenario 2	Scenario 3	Scenario 4
Unprotected	1.43 ± 0.01 ASS = 0	1.44 ± 0.01 ASS = 0	1.49 ± 0.02 ASS = 0	1.71 ± 0.09 ASS = 1
SBP	1.43 ± 0.01 ASS = 0	1.45 ± 0.01 ASS = 0	1.53 ± 0.02 ASS = 1	2.03 ± 0.14 ASS = 3

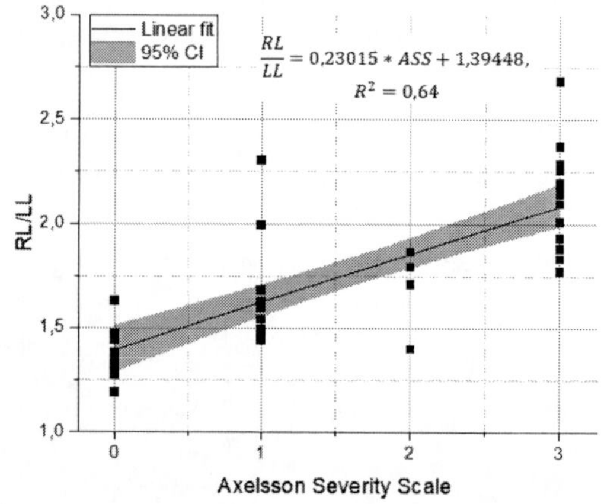

Figure 8. RL/LL as a function of the Axelsson Severity Score ASS from animal testing (data from [14]). The linear fit equation is written on the graph.

4. CONCLUSIONS

The aim of this study was to assess the feasibility of using an advanced surrogate system to evaluate blast amplification/attenuation behind a protective system. Therefore, the response of BOPMAN dummy is investigated while unprotected and equipped with a SBP. Incident blast wave impulses from 17 kPa·ms to 237 kPa·ms were generated by detonating spherical explosive charges of different masses at different distance from the target.

This study suggests that the test methodology used can detect relative differences in protection efficiency under blast loading. It was observed that for incident impulse up to 237 kPa·ms, the SBP amplifies the blast and so the lung injury risk. This latter was estimated with the RL/LL ratio and the ASS. BOPMAN metrics ratio (SBP over unprotected) was found to be roughly constant, at least over the incident impulse range tested, which does not imply a constant increase in lung injury level. This amplification factor is equal to 1.35 ± 0.20, while the percentage of increase of the ratio RL/LL when equipped with a SBP are 0, 0.7, 2.7 and 18.7%. It was also noticed with the high-speed video that the SBP slaps the thorax when the shock wave hits it. This movement is due to an inevitable small air gap between the vest and the thorax and the successive compression/suction phases as the shock front passed through it. The influence of the air gap on the amplification factor is unknown and should be studied.

While use of BOPMAN and the associated test methodology allowed for a comparison across different protection levels in terms of lung injury risk, some limitations were noted in this study. First, there is no standard to confirm the good positioning of the armor on the thorax. Nevertheless, a constant gap between the protection and the thorax was sought between the different experiments. Second, injury risk estimation is based on correlation performed at 237 kPa·ms with three levels of protection [14]. More data at other incident impulses are needed to increase the confidence on BOPMAN's ability to predict the lung injury risk in unprotected and protected configurations.

This study provides a first investigation of the use of BOPMAN to assess the effectiveness of TPE against blast loading based on lung injury risk. The results demonstrate that BOPMAN has potential to be used as a test mannequin to assess blast-induced lung injury risk. Additional works to investigate the sensitivity and reproducibility of BOPMAN to other blast exposures (e.g., severity levels, environments) and TPE are required to further validate this test device. Validation of a test device for the assessment of blast-induced lung injury is essential to properly evaluate protection systems for soldiers and law enforcement officers.

References

[1] Phillips Y.Y., Mundie T.G., Yelverton J.T. and Richmond D.R., Cloth Ballistic Vest Alters Response to Blast. J. Trauma, 1988; 28 (1); S149–S152.

[2] Cooper G.J. Townend D.J. Cater S.R. and Pearce B.P., The Role of Stress Waves in Thoracic Visceral Injury from Blast Loading: Modification of Stress Transmission by Foams and High-Density Materials. J. Biomech, 1991; 24 (5); 273–285.

[3] Cooper G.J., Protection of the Lung from Blast Overpressure by Thoracic Stress Wave Decouplers. J. Trauma, 1996; 40 (3); S105–S110.

[4] Gibson P.W., Amplification of Air Shock Waves by Textile Materials. J. Text Inst., 1994; 86 (1); 119–128.

[5] Prat N.J. Boutillier J. Cardona V. Monterspan F. Menini W. Mosnier J., et al., "Injury and Biomarkers from Pigs Following Isolated Open-Field Blast Exposure Depend on Ballistic protection Level", in Poster from the Military Health System Reseach Symposium, Kissimmee, Florida. 2019

[6] Long J.B. Bentley T.L. Wessner K.A. Cerone C. Sweeney S. and Bauman R.A., Blast Overpressure in Rats: Recreating a Battlefield Injury in the Laboratory. J. Neurotrauma, 2009; 26 (6); 827–840.

[7] Wood G.W. Panzer M.B. Shridharani J.K. Matthews K.A. Capehart B.P. Myers B.S. et al., Attenuation of Blast Pressure behind Ballistic Protective Vests. Inj. Prev., 2013; 19(1); 19–25.

[8] Sekine Y. Saitoh D. Yoshimura Y. Fujita M. Araki Y. Kobayashi Y. et al., Efficacy of Body Armor in protection against Blast Injuries Using a Swine Model in a Confined Space with a Blast Tube. Ann. Biomed. Eng., 2021; 49(10); 2944–2956.

[9] Jetté F.X. Dionne J.P. Williams K. Anctil B. and Makris A., Development of a Mannequin for Assessment of Blast Injuries and Lethality – Assessment of Protective Systems, in Proceedings of the MABS 18th Conference, Schneizlreuth, Germany. 2004.

[10] Bass C.R. Davis M. Rafaels K., A methodology for assessing blast protection in explosive ordnance disposal bomb suits. Int. J. Occupat. Safety Ergonom., 2005; 11(4) ; 347–361.

[11] Ouellet S. and Williams K., Characterisation of Defence Research and Development Canada's Mannequin for the Assessment of Blast Incapacitation and Lethality (DRDC MABIL), in Proceeding of the Personal Armour Systems Symposium, Brussels. 2008.

[12] Carboni M. DeCrisofano B. Segars R. and Maffeo M., Evaluation of a Blast Overpressure Test Device for Personal Protective Equipment, in Proceeding of the Personal Armour Systems Symposium, Quebec. 2010.

[13] Awoukeng-Goumtcha A. Thoral-Pierre K. Roth S., Toward a powerful tool for the prediction of human body trauma under extreme loadings through finite element simulations, in Proceeding of the Personal Armour Systems Symposium, Cambridge. 2014.

[14] Boutillier J. Cardona V. Magnan P. Ogier M. De Mezzo S. Montespan F. Menini W. Mosnier J. Naz P. and Prat N.J., A New Anthropomorphic Mannequin for Efficacy Evaluation of Thoracic Protective Equipment Against Blast Threats. Front. Bioeng. Biotechnol., 2022; 9:786881.

https://doi.org/10.52202/080042-0015

Shock Tube Size Considerations for Headborne Personal Protective Equipment: A Computational Sensitivity Study

M. Baker[1], R. Kumar[1], V. Alphonse[1], Q. Luong[1], D. Drewry[1], A. Iwaskiw[1]

[1]*The Johns Hopkins Applied Physics Laboratory, 11100 Johns Hopkins Road, Laurel, Maryland 20723, Alexander.iwaskiw@JHUAPL.edu*

Abstract: The study of blast-induced traumatic brain injury is becoming increasingly important to the design, manufacturing, and evaluation of headborne personal protective equipment (PPE). Existing testing methodologies rely either on live-fire testing or by recreating blasts using shock tubes to evaluate the blast overpressure attenuating performance of PPE. Previous studies have indicated that shock tube testing can introduce undesired anomalies when compared to live-fire testing [1, 2]. However, specific implications of the size and placement of the test article within shock tubes has seen less focus in literature [3, 4]. Due to the unique size and geometries of headborne PPE, it is important to elucidate both fundamental relationships between testing conditions and blast overpressure loading as well as testing effects of current shock tube systems.

This paper investigates how shock tube and test article size influences the blast overpressure loading using a computational fluid dynamics (CFD) analysis. The investigation used simplified geometries and dimensions to obtain fundamental relationships between geometry and blast overpressure loading. The study used a two-dimensional simulation that independently varied the shock tube cross-sectional dimensions and the circular test article size. The relationships between the shock tube sizes and test articles are comparable to similar shock tube and headform geometries seen in research [2, 5, 6, 7] and can be used to estimate realistic testing effects. Time histories of the blast overpressure signature were evaluated across all simulations. With these data, relationships can be developed and applied to the testing of various headborne PPE geometries to ensure that testing conditions are well understood. Ultimately, these data will inform standards for testing and evaluating PPE appropriately within various shock tube geometries and with a variety of headborne PPE sizes and shapes.

1. INTRODUCTION

Injuries due to blast are a sustained concern for the warfighter. A particularly insidious injury induced by blast, traumatic brain injury, has been observed in recent conflicts and projects to continue to be prevalent in combat [8]. Helmets and other headborne personal protective equipment (PPE) such as goggles, visors and face shields are potentially a very important tool in protecting the warfighter to these types of injuries. Currently there is no standard for evaluating blast-attenuating performance of headborne PPE [6], however progress is being made in developing a standard by utilizing laboratories employing shock tubes. These shock tubes re-create blast waves or shocks seen in live-fire for use in a repeatable testing evaluation methodology. Shock tubes have been known to create undesirable anomalies with respect to a live-fire blast, such as non-uniform shock waves, enhanced impulse exposure, and exit jet gassing. These anomalies have been studied using computation fluid dynamic (CFD) models and mitigation strategies have been integrated into the designs of newer shock tubes, such as tailoring shock generation through adjusting transition regions, changing gas composition for gas-driven shock tubes, and ensuring test articles are placed inside the tube vs outside [1]. While work has been done to generate a shock wave similar to live-fire, very few studies have been conducted to understand the implications of test article size with respect to shock tube size and its influence on testing and evaluation. There have been studies that have indicated that blockage ratio may play a role on pressure loading to a test article [3, 4] however there has not been a study to understand the factors that influence differences in pressure loading in a shock tube. With current geometrical constrains in place with existing shock tubes [2, 5, 6, 7] and headborne PPE, this study aims to evaluate the relationship between blockage ratio and other key metrics associated with pressure loading using CFD modeling. This CFD study spans realistic testing geometries using a simplified two-dimensional (2D) model (circle in a rectangular test region) to understand the effects of shock tube size with respect to test article size.

https://doi.org/10.52202/080042-0016

2. METHODS

2.1 Model selection and setup

Typically, the computational modeling of shock tube testing follows two potential approaches; 1) Fluid-structure interaction (FSI) which models the transient state of the solid test article using the finite element method (FEM) and models the aerodynamics using the finite volume method (FVM), often with reduced complexity to improve numerical stability, or 2) computational fluid dynamics (CFD) modeling which uses the FVM to perform complex modeling of the fluid dynamics and steady-state modeling of the solid test article. The former has the benefit of transient modeling of the fluid-structure interaction, which can predict test article kinematics and deformation – a desire for those interested in momentum transfer due to the shock front. The latter, while only being able to capture the steady-state of the test article, has a higher level of fidelity in the flow solution, and can capture more complex effects such as viscous drag on the test article. This analysis fundamentally is seeking to determine if the chosen metrics, which are derived from the dimensions of the shock tube and test article, can be used to properly describe a robust test configuration. Due to the lack of a need for kinematic results, and a particular interest in the shock wave interactions with the article and tube, a CFD-based analysis approach was chosen. The commercial solver *CFD++* from Metacomp Technologies was used for analysis as it has been validated against previous shock tube testing at the Johns Hopkins Applied Physics Laboratory (JHU/APL), and is sufficiently high fidelity for this analysis [1].

To prepare cases for simulation, a spatial representation (i.e. 2D grid consisting of connected nodes) of a shock tube was developed. Previous analysis leveraged the geometry of the advanced blast simulator at JHU/APL, the Blast Overpressure Simulation System (BOSS), to create a 2D shock tube simulation which would capture the driver and driven regions of the shock tube exactly through a 2D "slice" of the shock tube [1]. The grids used by this paper were simplifications of 2D grids used for the previous analysis and allowed for more specific tuning of the shock tube geometry to match the desired modeling conditions. Figure 18 contains images of these two grids. The simplifications include removing the driver and expansion region of the driven section used by the 2D BOSS model, and replacing them with a single high-pressure region upstream of the test section. As the simulation runs, the high-pressure regions results in a normal shock which arrives at the test article with a similar structure as the one present in the 2D BOSS model. This methodology was chosen to simplify the grid generation process, as varying the expansion region geometry to match the desired blockage ratio could impact the equivalence of the simplified grids. The decision to make these simplifications relies on the assumption that the simplified grid geometry can generate a similar shock at the test article to the 2D BOSS grid. To test this assumption, similar initial conditions were run for both shock tube configurations and the resultant shocks were compared. Both were shown to be planar and had similar overpressure signatures and wave durations.

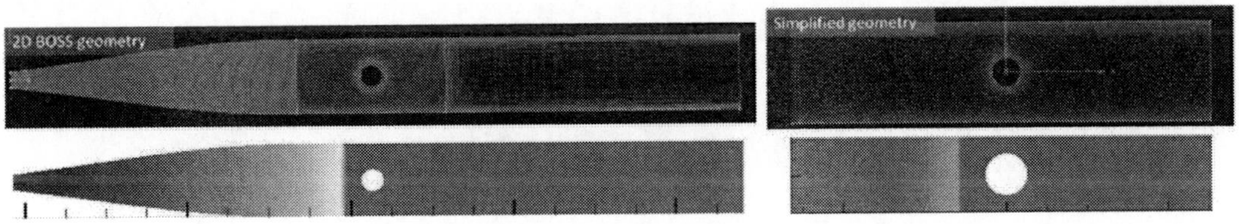

Figure 18. 2D BOSS shock tube geometry (Left) alongside the Simplified shock tube geometry (Right) and comparisons of 2D BOSS grid and simplified grid shock development

The simulation setup consists of 15,000 timesteps of 10^{-6} seconds each. This results in a simulation that models 15 milliseconds of flow, which is consistent with the timescale needed to run a full-scale BOSS shock tube test. All boundaries were treated as adiabatic walls with wall functions used to calculate values at the surface of the test article and tube walls. All flow solutions consider the viscous component of the flow, and the k-ε turbulence model was used. Finally, to match the conditions used in JHU/APL testing campaigns, the simulation was run using air as the fluid of choice.

2.2 Simulation test design

The purpose of this analysis was to leverage 2D CFD to determine if test article size and shock tube size influence the capability of shock tube configurations to simulate free-field blast profiles. To conform to the previous literature on shock tube test validation [1], it was important to consider blockage ratio to

https://doi.org/10.52202/080042-0016

describe the test configuration. Blockage ratio in the context of this analysis refers to the blockage of a circle, defined by the radius r, in a square cross section defined by shock tube side length L (Equation 1). Blockage ratio is varied by changing either the size of the test article (fixed tube case) or changing the width of the simplified shock tube grid (fixed article case) (Figure 19). In each test suite, one of these two variables was fixed to a representative geometry: in the fixed tube case, the tube width was fixed to 914 mm (the tube width of the BOSS); in the fixed article case, the circle diameter was fixed to 179 mm (an approximation of the cross-sectional area of a headform and large headborne PPE system). Varying these geometries can lead to equivalent blockage ratios, but do not necessarily lead to equivalent distances from the test article to the wall. For this reason, the minimum distance to the wall was recorded to identify cases where blockage ratio itself does not fully describe the test configuration (Equation 2). To create results that are representative to real world shock tube test configurations, shock tubes sizes seen in literature tubes [2, 5, 6, 7] were used. In addition to modeling those tube geometries, the bounds of the analysis were extended to capture both low and extremely high blockage cases.

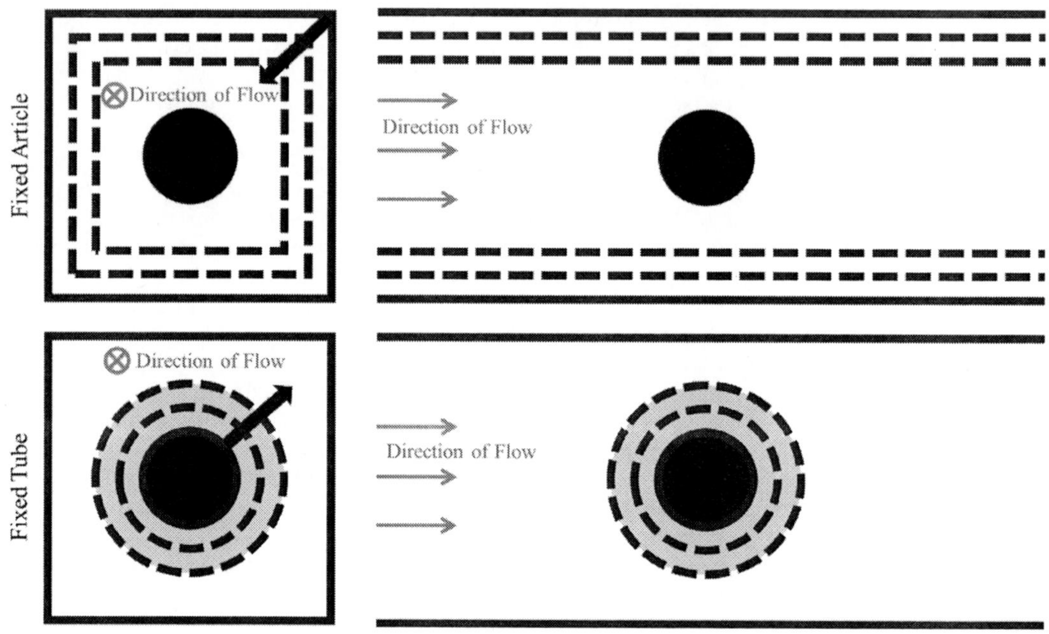

Figure 19. Depiction of the fixed article and fixed tube cases

$$Blockage\ Ratio = \frac{\pi r^2}{L^2} \tag{1}$$

$$Minimum\ distance\ to\ the\ wall = \frac{L}{2} - r \tag{2}$$

2.3 Data processing steps

The primary variable of interest from the CFD analysis is the pressure at the surface of the test article. The pressure recorded is the static pressure of the flowfield surrounding the article interpolated to the surface of the test article. The point where pressure is recorded is defined by the grid as described in section 2.1, where nodes act as "sensors" recording the simulation data at that specific point on the test article. The sensor locations were at 0°, 45°, and 90° degrees which provided potential for understanding the influence of distance to the wall to the sensor pressure measurement. Additionally, headforms used in headborne PPE testing typically have many sensors placed on the front hemisphere, locations selected in this study approximated these headform locations [6]. The pressure data at the surface is recorded at every timestep of the simulation, allowing the reconstruction of the full overpressure signature of the shock front on the test article. In addition to the pressure at the surface of the test article, the flowfield pressure, velocity, density, and temperature was recorded every 50 timesteps. These data are used to visualise the shock wave development within the shock tube to ensure it is normal, planar, and is traveling at an appropriate speed when compared to real world shock tube experiments. To record the article surface data, the 'data on boundaries' option in *CFD++* was used. This option allows for the user to extract pressure data at the surface of the test article at every timestep of the simulation.

In free-field blast test conditions, which can be created in live-fire testing, the structure of the blast overpressure at some distance from the blast source can be modeled as a Friedlander wave. The resulting Friedlander wave structure achieved in an empty shock tube is presented in Figure 20 beside experimental data obtained in the JHU/APL shock tube.

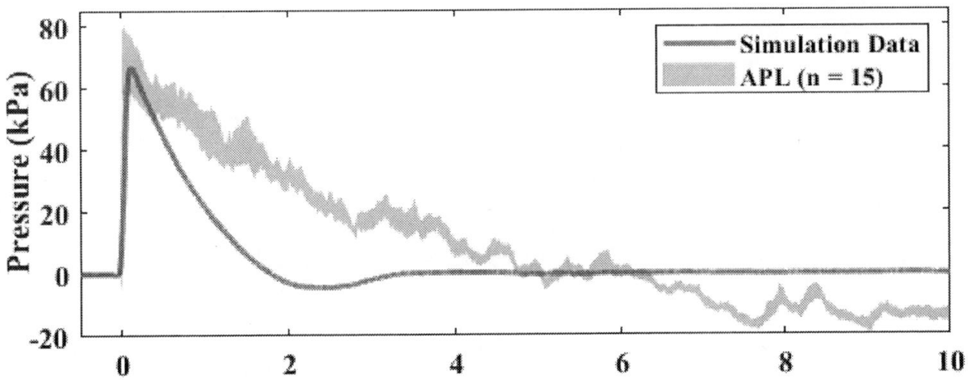

Figure 20. Friedlander wave pressure data from simulation and experimental static pressure response corridor (mean± 1 standard deviation) from 15 JHU/APL BOSS measured at the middle of the tube

The data collected is used to inform three main metrics that describe the effect of the shock on the test article: peak overpressure, positive phase duration, and impulse. Peak overpressure is determined by recording the pressure at every timestep of the simulation, subtracting from the pressure readings the standard atmospheric pressure, and determining the first point where the pressure was greater than zero (i.e. greater than standard atmospheric pressure). Positive phase duration is calculated by determining the first point after the peak overpressure is measured where the pressure returns to standard atmospheric conditions. Finally, positive phase impulse is calculated as the integral over time of the pressure acting on the sensor points of interest on the circular test article. This calculation was performed numerically during the post-processing of the computational results. Ten milliseconds of pressure data after the time of shock wave arrival was analyzed. The test article was placed in the shock tube such that the harmonic content of the wave would dissipate, leading to stable impulse measurements. After initial observation, it became clear that the reflected shockwave from the tube wall would be an important phenomenon to characterise, as the reflected peak (and subsequent reflections) occurred during the simulation. In order to characterise the extent of this reflected wave and the effects of the test configuration geometry on the test article pressure loading, two additional metrics are introduced: the ratio of peak pressures, and the second peak time of arrival. The ratio of peak pressures computes how strong the secondary shock is in relation to the strength of the initial shock. The second peak time of arrival denotes the amount of time that passed between the arrival of the initial shock and secondary shock. Both of these metrics offer additional insights into the impact both blockage ratio and minimum distance to the wall have on the pressure trace results.

3. RESULTS

The simulated flow characteristics of the shock through the 2D shock tube were produced every 50 timesteps, and were used to generate visual guides of the shock propagating through the shock tube (Figure 21) along with time histories of the static pressure response measured at three locations.

https://doi.org/10.52202/080042-0016

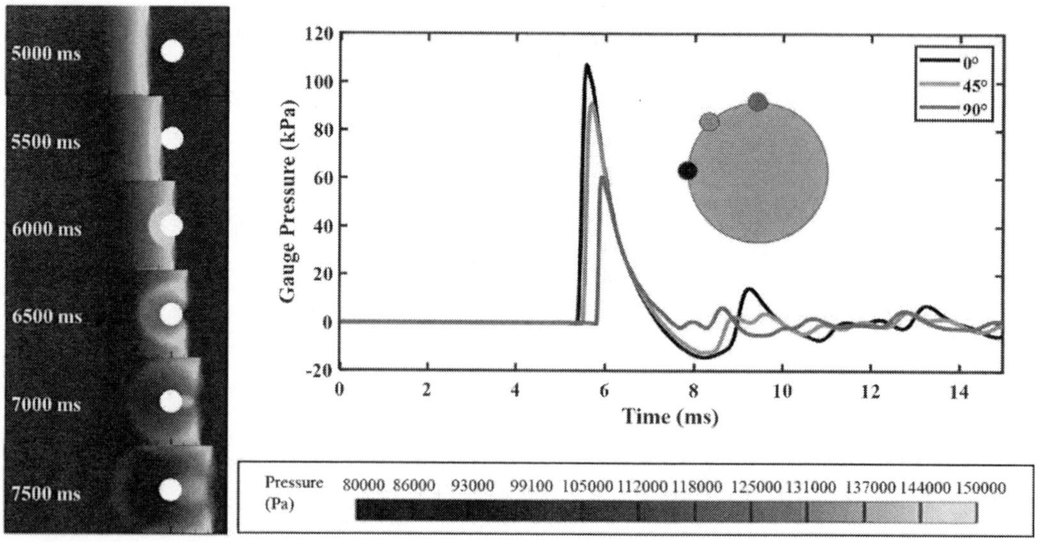

Figure 21. Evolution of blast overpressure for case with 914mm tube length and 116mm test article radius (left) and characteristic time histories of static pressure at 0°, 45° and 90° locations on the test article of a 5% blockage ratio case with a 914mm tube length and a 116mm test article radius (right)

Simulation solutions for low blockage ratio cases showed that the shock impact on the test article did not lead to significant impacts on the planarity and propagation of the shock, as shown in Figure 21. As blockage ratio increases to levels over 30%, a more significant impact on the flow surrounding the test article is observed (Figure 22). These impacts on the flow subsequently impact the recorded ratio of peak pressures, time of arrival of the secondary peak, and impulse experienced by the test article.

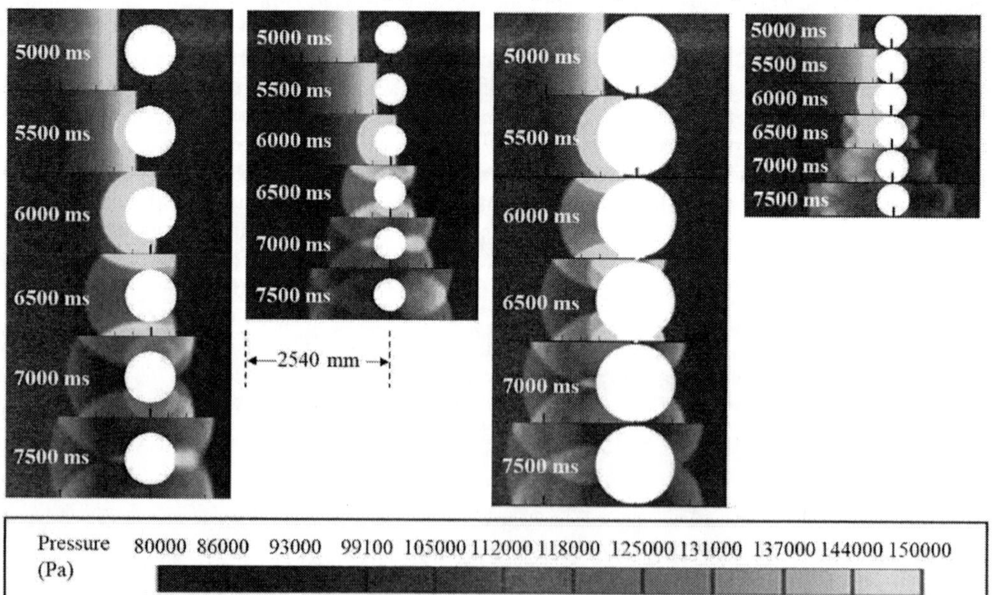

Figure 22. Evolution of blast overpressure for (left to right) 30% blockage ratio fixed tube case (with 914mm tube length and 283mm test article radius), 30% blockage ratio fixed article case (with 579mm tube length and 179mm test article radius), 70% blockage ratio fixed tube case (with 914mm tube length and 432mm test article radius), and 70% blockage ratio fixed article case (with 379mm tube length and 179mm test article radius). Plots of test article and tubes have consistent relative sizes with respect to one another.

Across both cases, an increase in the sensor angle leads to smaller impulse values. This trend holds until both blockage ratio and sensor angle are large, at which point simulation solutions become erratic. Additionally, for both fixed tube (Table 7) and fixed article (Table 8) cases it is observed that as blockage ratio increases the secondary peak time of arrival shortens significantly, even for the 0° sensor location. Relationships between the blockage ratio and impulse, for cases where both tube size and test article size are held constant, show nonlinear trends at higher blockage ratios (Figure 23) and deviations occur between fixed article and fixed tube cases.

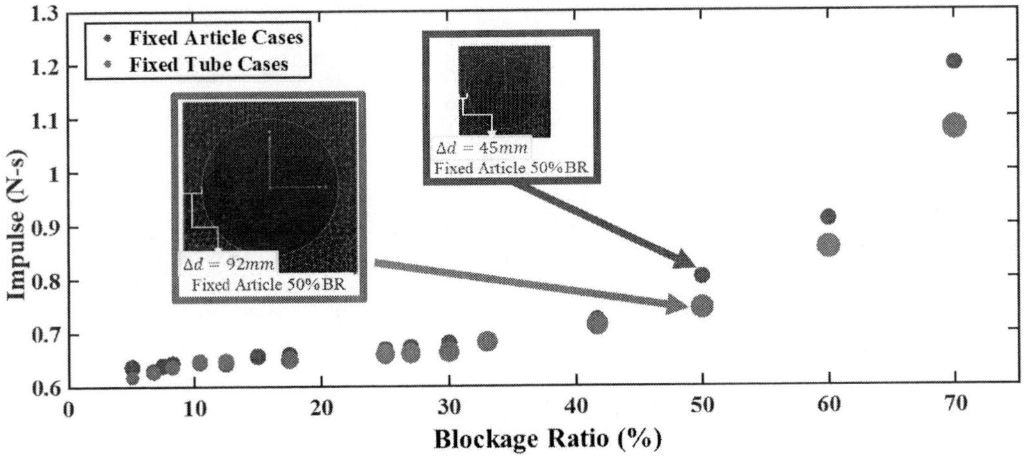

Figure 23. Relationship between blockage ratio and impulse (N-s). Size of circle marker correlates to test article size

Table 7. Fixed tube test matrix and results for 0°, 45°, and 90° sensor locations (omitted data represents anomalous data due to errors in the automated metric selection)

Run	Tube side length (mm)	Radius of the test article (mm)	Blockage Ratio %	Min Dist. To Wall (mm)	Second Peak Time of Arrival (ms)	Ratio of Peak Pressures	Impulse (N-s)	Second Peak Time of Arrival (ms)	Ratio of Peak Pressures	Impulse (N-s)	Second Peak Time of Arrival (ms)	Ratio of Peak Pressures	Impulse (N-s)
					0° sensor location			45° sensor location			90° sensor location		
1	914	116	5.1%	341	2.38	5.9%	0.75	2.10	7.2%	0.62	1.78	26.5%	0.43
2	914	134	6.8%	323	2.36	4.7%	0.76	2.00	6.0%	0.63	1.63	27.9%	0.45
3	914	148	8.3%	309	2.35	5.1%	0.75	1.96	8.2%	0.64	1.58	25.2%	0.44
4	914	167	10.4%	291	2.33	6.7%	0.76	1.89	10.5%	0.65	1.46	28.0%	0.44
5	914	182	12.5%	275	2.31	8.0%	0.77	1.84	13.8%	0.65	1.36	35.8%	0.45
6	914	216	17.5%	241	2.27	10.7%	0.8	1.70	20.1%	0.65	1.14	49.0%	0.46
7	914	258	25.0%	199	2.22	14.3%	0.81	1.53	29.5%	0.66	0.85	71.3%	0.47
8	914	268	27.0%	189	2.20	15.2%	0.82	1.51	29.8%	0.66	0.80	74.7%	0.48
9	914	283	30.0%	175	2.18	16.9%	0.82	1.44	38.1%	0.66	0.71	89.3%	0.48
10	914	296	33.0%	161	2.17	15.9%	0.82	1.38	35.1%	0.68	0.65	84.6%	0.47
11	914	333	42.0%	124	2.12	19.2%	0.84	1.25	46.3%	0.71	0.43	110.7%	0.47
12	914	365	50.0%	92	2.07	21.6%	0.86	1.12	56.5%	0.75	0.27	125.8%	0.43
13	914	400	60.0%	58	2.03	24.4%	0.96	1.01	68.1%	0.86	0.15	84.9%	0.29
14	914	432	70.0%	26	1.99	25.9%	1.16	0.88	78.3%	1.08	0.12	93.1%	

Table 8. Fixed article test matrix and results for 0°, 45°, and 90° sensor locations (omitted data represents anomalous data due to errors in the automated metric selection)

Run	Tube side length (mm)	Radius of the test article (mm)	Blockage Ratio %	Min Dist. To Wall (mm)	Second Peak Time of Arrival (ms)	Ratio of Peak Pressures	Impulse (N-s)	Second Peak Time of Arrival (ms)	Ratio of Peak Pressures	Impulse (N-s)	Second Peak Time of Arrival (ms)	Ratio of Peak Pressures	Impulse (N-s)
					0° sensor location			45° sensor location			90° sensor location		
15	1411	179	5.1%	526	3.68	13.4%	0.78	3.20	2.8%	0.64	2.72	10.5%	0.78
16	1222	179	6.8%	432	3.20	5.6%	0.77	2.68	0.0%	0.63	2.21	20.5%	0.77

https://doi.org/10.52202/080042-0016

17	1159	179	7.5%	400	3.00	3.2%	0.77	2.50	0.0%	0.64	2.04	22.7%	0.77
18	1105	179	8.3%	373	2.83	2.4%	0.78	2.34	1.3%	0.64	1.92	22.8%	0.78
19	983	179	10.4%	313	2.49	4.4%	0.79	2.00	8.5%	0.65	1.55	28.6%	0.79
20	898	179	12.5%	270	2.25	8.6%	0.77	1.76	15.8%	0.64	1.32	36.1%	0.77
21	820	179	15.0%	231	2.04	13.5%	0.79	1.55	22.0%	0.66	1.09	48.5%	0.79
22	759	179	17.5%	200	1.88	18.0%	0.79	1.41	29.0%	0.66	0.94	56.2%	0.79
23	635	179	25.0%	138	1.55	26.9%	0.8	1.08	43.3%	0.67	0.61	78.3%	0.42
24	611	179	27.0%	126	1.49	29.8%	0.81	1.02	46.6%	0.67	0.55	82.7%	0.4
25	579	179	30.0%	111	1.40	36.1%	0.81	0.92	58.1%	0.68	0.46	98.7%	0.38
26	553	179	33.0%	97	1.32	32.8%	0.81	0.85	52.3%	0.68	0.38	92.2%	0.36
27	492	179	42.0%	67	1.16	44.7%	0.85	0.68	71.7%	0.72	0.25	119.0%	0.26
28	449	179	50.0%	45	1.02	50.3%	0.92	0.56	84.0%	0.8	0.11	72.0%	0.13
29	410	179	60.0%	26	0.92	56.5%	1.00	0.46	96.58%	0.9	0.12	96.0%	
30	379	179	70.0%	11	0.84	63.1%	1.23	0.37	104.%	1.2	0.11	96.6%	0.08

4. DISCUSSION

4.1 Dominant pressure loading mechanism

The results of this study highlight key factors that affect pressure loading in shock tubes. Ritzel [7] considers two main mechanics when discussing short-duration moderate-magnitude (e.g. 200 kPa total pressure) shock waves causing blast-induced motion, which for the purposes of this study, can be a downstream result of pressure loading. These mechanisms are shock wave loading during the diffraction phase and drag loading during shock wave after-flow (or "blast wind"). Ritzel has determined that when a blast wavelength approaches the characteristic length (e.g. diameter of a sphere) that the diffraction phase dominates and there is little to no after-flow over the structure as the shock wave decays very rapidly, reducing the effects of both viscous and form drag loading. In this study, due to the blast wave length of roughly 2 meters, and test article diameters of 0.23-0.86 m, the assumption that there is no drag loading could not be made, however it is expected to be low with respect to the reflected shock pressure loading. When observing the pressure loading of the test article at 30% blockage, it can be seen that secondary shock waves reflect off of the shock tube walls and impart an additional pressure load on the test article while the original primary wave is still loading the test article (Figure 22). Complex wave superposition of test article and wall reflections can be observed creating high pressure loading regions after the wave has passed.

Higher blockage ratios show reflections that prevent a majority of the shock waves from traveling across the test articles (Figure 22). This results in lower pressures seen at the 90° sensor location at higher blockage ratios than the lower blockage ratios. While 70% blockage may be unrealistic for current shock tube designs, this reflection phenomenon may be important when considering test article and article placement in the shock tube as proximity to the wall may create similar complex reflection and choked flow phenomena.

4.2 Relationship between test article and shock tube sizes on response variables

This study aims to understand the relationship between shock tube dimensions, test article dimensions, and pressure loading. Tube size and test article size can vary independently and both have an effect on test articles in shock tubes. In order to further understand the effects of both variables on pressure loading, a commonly used scaled metric blockage ratio (%) can be evaluated. Equation 1 shows that blockage ratio is a metric derived from second order effects of both tube width and test article radius.

The time of arrival of the second wave, due to reflection, varies from 3.7 ms to 0.1 ms. The earliest a secondary wave arrives is at the 90° sensor location in both of the 70% blockage ratio cases. The latest the secondary wave arrives occurs in the fixed article case with 5% blockage ratio. The time of arrival of the second wave decreases (or occurs faster) as blockage ratio increases. Due to the fact that the shock wave used has a duration of 4 ms (and the experimental shock wave durations are even longer) it expected that reflections will be present in a recorded pressure at the surface of test articles that are of a relevant size, in real shock tubes, for headborne PPE testing. These reflected waves will be present in the pressure response, so the magnitude of the secondary wave should be understood. The range of ratio of the secondary peak to the first, which is a measure of the potential effect of the second wave, spans 0%-116%. Increasing blockage ratio increases the secondary peak ratio as expected. As minimum distance to the wall decreases the ratio of pressure increases, and is highest in the fixed article cases, where the tubes are the smallest. For some of these high blockage ratio cases, as well as when more complex test article geometries are considered, the simple relationship between the first and second peak magnitude may be insufficient in understanding the extent of influence of the reflections, so a method that integrates pressure over time, like impulse, could be better at explaining the entire effect.

When evaluating impulse for lower blockage ratios (5%-40%) one can identify a trend that exists in both fixed article and fixed tube cases (Figure 23). This trend implies that for lower blockage ratios, the impact of tube width and test article on the loading experienced by the test article is minimal and roughly linear. Moreover, it points to the validity of using blockage ratio to verify the quality of a shock tube configuration, so long as the blockage ratio is within acceptable bounds. Above 40% there is a larger divergence of impulse where this roughly linear relationship breaks down. For these large blockage ratios, cases with lower minimum distances to the side wall see much higher impulses. This may be due to the fact that reflections dominate this response, driven by the fact that cases with shorter distances to the wall have reflections that arrive sooner. For the case of a 445mm tube vs a 914mm tube (both having similar blockage ratios of 50%) the minimum distance to the wall doubles from 45mm to 92mm. It is recommended that further modeling be performed to fully understand the relationship between test article size and tube size, as describing the appropriateness of a test solely on blockage ratio may be insufficient.

4.3 Sensor location

Sensor location has a direct effect on the pressure loading across blockage ratios. For the 0° and 45° sensor locations, impulse and the ratio of peak pressures increase with increased blockage ratio. For the 90° sensor location, impulse decreases with blockage ratio while first and second peak ratio increases. The later trend resulting from the superposition of wave reflections driven by the low minimum distance to the wall. The decrease in impulse at higher blockage ratios is due to the choked flow phenomenon experienced by high blockage ratio test conditions. In these cases, the shock propagation over the top of the article is severely stifled, or entirely limited, by shocks reflecting off of the test article and shock tube wall.

4.4 Implications to headborne PPE testing and evaluation

When considering this research, it is important to think about how these results would affect headborne PPE testing evaluations and methodology. Previous studies have used classic aerodynamic blockage ratios as guidelines for headborne PPE testing methods. These blockage ratios of 5% may be unrealistic to achieve and are rooted in assumptions of quasi-static flow. These results demonstrate that the wavelengths of the short duration shock waves generated in shock tubes are similar in magnitude with the characteristic lengths of headborne PPE and headforms, so the rules that assume drag from flow are not directly applicable in these testing scenarios. Wortman [4] acknowledged that a less-stringent criteria of 20% blockage ratio was acceptable due to the rapidly decaying nature of these pressures, but a sensitivity study had never been run with relevant geometries typically present in PPE and headform shock tube testing. This study used shock tube sizes spanning 377 mm – 1401 mm, capturing and extending the typical sizes seen in literature (610 mm – 1220 mm from the cited reference tubes). The "fixed tube" case represents the JHU/APL's 914 mm x 914 mm square advanced blast simulator, with test article radii spanning 116 mm – 432 mm. The "fixed article" cases represent a rather large headborne PPE test condition to depict the worst-case headborne PPE that would be tested in a presently available shock tube.

These results show that shock tube testing with blockage ratios approaching 30%-40%, assuming representative geometries of headborne PPE, may be acceptable with only a 7-16% increase in impulse at the 0° and 45° sensor location. At larger blockage ratios, peak metrics such as impulse and magnitude of the second peak increase more rapidly, showing dramatically corrupted pressure loading, with respect to ideal Friedlander waves, versus the lower blockage ratios. At these large blockage ratio values (greater than 40%) minimum distance to the wall must also be considered as a limiting metric of the test configuration. While in some circumstances 30% - 40% blockage ratio can imply a valid testing configuration, the distances to the wall can play a role in enhanced pressure loading. Special care should be taken to maximise the distance between the test article and the side wall in configurations where blockage ratio alone is not beyond 30-40%. Practically, this means test articles should be placed in the middle of the shock tube, but this is not always possible. Additionally, to capture a more complete view of the loading present on a headform or PPE model, sensors should be placed near the point on the article closest to the wall to capture the reflected shock effects on the model.

https://doi.org/10.52202/080042-0016

4.5 Limitations

There are several limitations of this study. The durations of the shock wave generated in this simulation are less than what JHU/APL normally generates in shock tube experiments (Figure 20). Further work should be done to tune parameters in the simulation to modify the shock wave produced, as this could be used to match other relevant blast overpressure signatures seen in other shock tubes. However, the trends observed in the simulations study are relevant for the experimental conditions used in headborne PPE loading. In longer duration shock waves seen in some experiments, the effects of reflections on test article loading may be enhanced. There is a need to validate these results with headform data. It is recommended to continue to tune the initial conditions to create more representative shock waves for future validation. However, these results from the simplified shock tube simulations compared favorably with the previously validated 2D BOSS shock tube grid presented by Kumar et. al. [1]. This study placed a significant emphasis on understanding the impact the shock front has on the side of the test article closest to the driver section. There are additional wave effects on the opposite side of the test article that were not analysed as part of this study that should be considered in future work.

Additionally, the 2D simulations in this study lack the details of real headborne PPE testing such as facial features, head shape, and a helmet. Three dimensional (3D) simulations with those realistic geometries of a headform, like the JHU/APL Human Surrogate Head Model are recommended [6] to be modeled and compared to experimental data for proper validation. There may be more complex reflections that occur from these realistic 3D geometries as well. Another benefit of these realistic geometries is that a one-to-one comparison can be done between sensors on the headform and sensors in the 3D simulation.

This study applied the FVM as implemented in the commercial CFD solver, CFD++. For headborne PPE testing, the pressure loading will cause the kinematics seen in experimental testing and this is not explicitly modeled. However, by modeling with high fidelity the pressure interactions experienced by the test article, these results could provide reliable suggestions for the testing community as to the effects of these specific geometric conditions. Implementation of only the necessary physics, rather than evaluation using a multi-physics approach such as FSI, enables simulation of more complex fluid dynamics at a lower computational cost per simulation than other validated modeling methods implementing FSI approaches. For the exploration of shock tube design, these approaches may be sufficient; however, this method is limited in its ability to represent the kinematics as a separate response variable in the system. Currently, there are many proposed metrics in literature related to brain injury, including global kinematics-based and tissue deformation-based metrics. As this current model cannot account for the kinematic or deformation response of the test article, they do not directly allow analysis into brain injury, but it can be posited that understanding the changes in pressure loading on headforms during testing is an ideal variable in the evaluation of shock tube test configurations themselves.

5. CONCLUSION

The objectives of this study were to evaluate the relationship between shock tube size, test article size and pressure loading across ranges of relevant geometries in headborne PPE testing. Shock tube size and test article size were varied independently and both had an influence on changes in pressure loading as described by change in impulse, ratio of a peak pressures, and second peak time of arrival. Three locations on the test article were evaluated showing different relationships between the geometric variables and "sensor" locations. It was shown that sensor locations closest to the wall are the most sensitive to changes in blockage ratio. While blockage ratio was a good predictor of pressure loading for low to moderate blockage ratios, at high blockage ratios the minimum distance to the wall had significant impacts in the impulse and pressure responses experienced by the test article. Blockage ratios greater than the often cited 20% maximum recommended by Needham [3] were characterised and could be considered for testing, although blockage ratios above 40% seem to greatly affect the pressure loading. These results can better inform decisions in headborne PPE testing including how to avoid influences on pressure loading from higher blockage ratios or non-centred test articles. Ultimately, these data will help inform standards for testing and evaluating PPE within various shock tube geometries to ensure that blast testing and evaluation can be effectively leveraged to achieve greater research goals.

Acknowledgments

The authors would like to thank PEO Soldier Product Manager Soldier Protective Equipment for sponsoring this effort. Any opinions, findings and conclusions or recommendations expressed in this material are those of the author(s) and do not necessarily reflect the views of PEO Soldier or the NAVAL SEA SYSTEMS COMMAND (NAVSEA).

References

[1] R. Kumar and A. Nedungadi, "Using Gas-Driven Shock Tubes to Produce Blast Wave Signatures," *Frontiers in Neurology,* vol. 11, no. 90, 2020.

[2] A. Sundaramurthy, R. Gupta and N. Chandra, "Design considerations for compression gas driven shock tube to replicate field relevant primary blast condition," *ASME International Mechanical Engineering Congress and Exposition,* vol. 56215, 2013.

[3] E. Needham, D. Ritzel, G. Rule, S. Wiri and L. Young, "Blast testing issues and TBI: experimental models that lead to wrong conclusions," *Frontiers in Neurology,* vol. 6, no. 72, 2015.

[4] J. Wortman and R. Lottero, "Comparison of HULL Hydrocode Computations of Shock Tube Blockage Effects on Target Loading for Step Shocks and Rapidly-Decaying Shocks," Ballistic Research Laboratory, Aberdeen Proving Ground, MD, USA, 1982.

[5] A. Sundaramurthy et al., "A 3-D Finite-Element Minipig Model to Assess Brain Biomechanical Responses to Blast Exposure.," *Frontiers in Bioengineering and Biotechnology,* vol. 9, 2021.

[6] V. A. Alphonse et al., "Helmet Blast Attenuation Performance," in *Personal Armour Systems Symposium,* 2021.

[7] D. V. Ritzel et al., "Acceleration from short-duration blast.," *Shock Waves,* vol. 28, no. 1, pp. 101-114, 2018.

[8] P. Hayward, "Traumatic brain injury: the signature of modern conflicts.," *The Lancet Neurology*, vol. 7, no. 3, pp. 200-201, 2008.

https://doi.org/10.52202/080042-0016

Influence of torso protective equipment on intracorporeal shock wave behavior

H. Seeber[1], S. Grobert[2], D. Krentel[3] and T. Hauer[2]

[1]*Helmut Schmidt University – University of the Federal Armed Forces Germany Holstenhofweg 85, 22043 Hamburg, Germany,*
seeberh@hsu-hh.de
[2]*Bundeswehr Hospital Berlin – Department II, General and Thoracic Surgery, Scharnhorststraße 13, 10115 Berlin, Germany,*
[3]*German Federal Institute for Materials Research and Testing (BAM) Division 2.1 – Safety of Energy Carriers, Unter den Eichen 87, 12205 Berlin, Germany.*

Abstract. In the field of explosive reactions, there is a type of explosive effect that lacks a sufficient database and reproducible experiments regarding biomechanics. It concerns the primary explosive effect. It is defined as pure shock wave of the explosion. The physical behavior of the shock wave when interacting with different types of tissue and, in particular, the subsequent transitions of the shock wave, have barely been investigated. The transition of the shock wave into other materials is the focus of the research

Therefore, the aim of the investigations is the development of a multidisciplinary method to investigate shock wave behavior in various generic tissue simulants under the most reproducible conditions possible with realistic loads in an experimental test series with short set-up times. An autoclave is used to generate the pressure waves. A simplified torso model consisting of ballistic gelatin is used as a simulant.

In this paper, the influence of protective equipment on the pressure load in the tissue simulant is investigated. For this purpose, consecutive test setups are used. First, the behavior of ballistic gelatin as a tissue simulant is investigated. Then, the simplified torso model is covered with typical combat clothing consisting of four layers. Afterwards a currently used UHMWPE ballistic protective plate is placed in front of the simplified torso model. Finally, the combat clothing and the protective plate are examined in combination. Three cast-in pressure sensors are used as measuring devices, as well as an acceleration sensor attached to the protective plate.

The experiments show that the maximum overpressure in a model protected by combat clothing and the protective plate can be reduced by 95%. However, the propagation speed of the shock wave within the simplified torso model increases from 1535.5 m/s to 2204.5 m/s. This shows that even protective equipment, which is not primarily intended to protect against blast, offers a significant reduction in the pressure load in the protected area. On the one hand it is caused by the media transition from air to PE and the resulting higher reflection of the acceleration of the transmitted wave within the simulant. On the other hand, it is also reduced due to the damping and dispersion caused by the clothing layers.

1. Introduction

Blast injuries are among the most common injuries in military operations. Also, in civilian environments, more explosive threats are expected in the future due to emerging conflicts. While the effect of fragments, which is classified as secondary blast injury, could be minimized by police and military personnel's ballistic body protection systems, the effects of shock wave propagation in the body as part of the primary explosion trauma still remain a serious threat needing further research. Especially the highly dynamic pressure changes in the human body represent an area that has barely been investigated yet [1]. Dynamic pressure changes represent a relevant effect due to the reflection-related amplification of shock waves at organ-dermis interfaces and the compression and relief of the pressure wave at medium interfaces [2]. Various approaches have been used to investigate these aspects, like animal experiments on free field test sites or shock tube setups. Animal experiments have the disadvantage that measured values are interpreted with partly outdated or not validated limit values for overpressures (e.g. [3]). Whereas shock tube setups have limitations such as reflections, superimpositions, useable space and blocking [4]. Injury mechanisms and their effects have not yet been sufficiently elucidated for the torso and extremities [2].

Therefore, the intention is to use an experimental setup that allows for the generation of reproducible pressure waves under free-field conditions and the application of those waves to a sufficiently large body model. In a broader sense, the influence of combat clothing and personal protective equipment on the shock wave behavior of different tissue types is correspondingly less studied [5]. Therefore, this paper focuses on the behavior of the shock wave in a simplified first stage torso model after it has passed through personal protective equipment.

The paper is structured as follows. First, the experimental setup and materials, such as the simplified torso model, the protective plate and the clothing layers are introduced. Afterwards, the

https://doi.org/10.52202/080042-0017

measurement results are presented. Especially the pressure values in the torso model are considered. It concludes with a discussion of the results and an outlook of the further work.

2. Materials and Methods

2.1 Experimental Setup

The experimental setup used is shown in Figure 1 as a schematic illustration with distance specifications, as well as an overall view. In the following, the components of the experimental setup are explained related to the schematic illustration. The pressure wave is generated by a cylindrical autoclave, further called shock wave generator (SWG) (A). It has a volume of 0.065 m^3. It is filled with a stoichiometric acetylene-oxygen mixture under atmospheric conditions. Ignition takes place through an electronic detonator. The pressure wave escapes through a 12.5 cm diameter outlet, which is covered by an aluminum burst disc. The center of the outlet is 1.35 m above ground. Preliminary tests showed that the pressure wave propagates in a hemispherical shape. The simplified torso model (STM) is described in more detail in Section 2.2.1. (B). A high-speed camera is used to document the pressure wave propagation (C). The specification of the protective plate is described in Section 2.2.2. (D). The arrangement of the clothing layers is explained in Section 2.2.3. (E). The pencil probes (1, 4) are used to record the pressure values of the pressure wave in front of the STM. In addition, a runtime determination can be carried out. Kistler 6233AA0025 sensors are used for this purpose. Positions 6 - 8 are piezoelectric pressure sensors (three times PCB 132B38) casted into the STM. The orientation is side-on. The exact positioning is described in Section 2.2.1. Position 9 is an accelerometer (PCB 356A01), which is directly applied to the protective plate. The exact positioning can be found in Section 2.2.2. In general, a sample rate of 2 MHz is used.

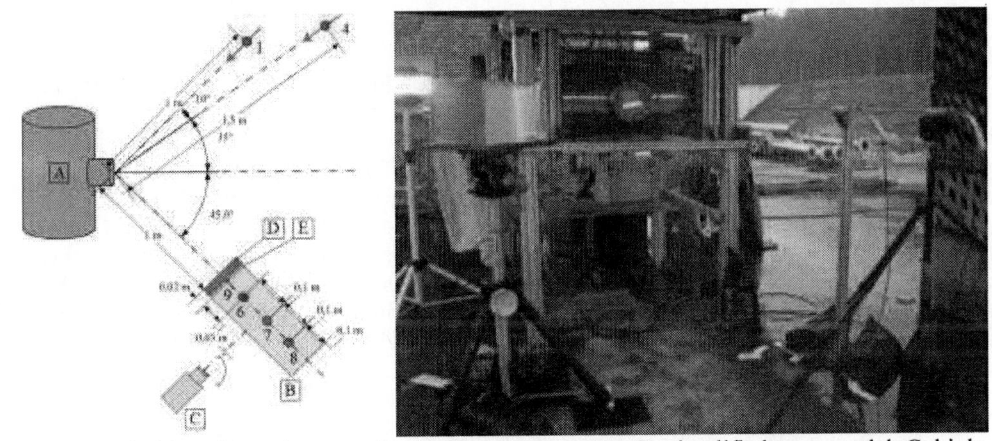

Figure 1. Left side: schematic setup. A: shock wave generator; B: simplified torso model; C: high-speed camera; D: protective plate; E: clothing; 1, 4: pencil probes; 6 - 8: internal pressure sensors, 9: accelerometer. Right side: measurement setup as an overview photo.

2.2 Materials

2.2.1 Simplified Torso Model

The simplified torso model (STM) consists of 17% (w/v) ballistic gelatin. Gelatin is produced according to the common procedures [6]. The used gelatin has a gel strength of 265 bloom. The mixing ratio of 17 w% is chosen based on measurements of the internal sound velocities. Ultrasonic transit time measurements are used to measure the internal sound velocity of gelatin samples with different mixing ratios. Human soft tissue, consisting of muscle and fat, is taken as the target value for the sound velocity. The individual sound velocities are calculated using their percentage distribution on the human body (limitations are known). For generic muscle tissue, a sound velocity of 1582.5 m/s and a distribution in the human body of 31.56 w% is chosen [7]. For fat tissue, a sound velocity of 1450 m/s and a distribution in the human body of 13.63 w% is chosen [7]. This leads to a calculated sound velocity for soft tissue of 1542.4 m/s. This value applies well to the calculated value of 17 w% gelatin. For the gelatin used in these experiments, an internal sound velocity of 1522 m/s is measured.

 https://doi.org/10.52202/080042-0017

The body model is in a rectangular shape with dimensions of 40 x 23 x 27 cm. In this test series, one side of the block is provided as a negative mold, which fits the protective plates. This convex shape allows full contact of the STM front surface with the protective plate, without an air gap between them. The STM is placed on a tripod with a table. The model is held on the tripod only by its weight. It is equipped with three cast-in piezoelectric pressure sensors (PCB 132B38). The orientation of the sensor is side-on. A sketch showing the exact positioning of the pressure sensors and a detail photo of the instrumented STM is displayed in Figure 2.

Figure 2. Left side: Sketch of the instrumented simplified torso model. Middle: Detail view of instrumented body model with protective plate attached. Right side: Sketch of the instrumented protective plate.

2.2.2 Protective Plate

In these trials, the focus is on torso protective equipment. Therefore, a chest protection plate from a plate carrier currently used by an army is used. This plate fulfills the test standard of protection class 6 (7,62 mm x 39, Fe-Core) of the Association of Test Laboratories for Attack Resistant Materials and Constructions (VPAM) [8]. It uses an ultra-high molecular weight polyethylene (UHMWPE) material as hard ballistic. Due to the stand-alone feature of the VPAM 6 plate, it has an integrated shock absorber layer. The plate is shaped concave in the direction of loading. The outer dimensions are 24 by 32 cm and the thickness is 1.9 cm. In the upper area, the plate has a trapezoid shape (Figure 2 r.s.).

The protective plate was mechanically clamped with the STM only (Figure 2 mid.). The contact pressure was adjusted so that it was comparable to the real operational conditions when worn with a real vest. Despite the concave shape of the plates, they lay flat on the face of the STM, since this face was cast as a negative mold. The accelerometer is posed directly onto the plate back face (Figure 2 r.s.). The position is related to the center of the tripod plate. The height of the accelerometer is 19 cm and the distance from the side edge is 12 cm.

2.2.3 Combat Clothing

Combat clothing is added over the STM. The area of the back side faces in the direction of the shock wave generator. This allows the most even and uniform textile layers possible. The structure of the clothing layers and their materials are stated in Table 1. In these trials where the protective plate has been added, it is placed between the undercoat rain protection and combat jacket. The layers are loosely arranged against each other. There is no mechanical clamping of the clothing to the tripod.

3. Experimental method

In this section, the experimental system used is presented. Table 2 lists the experiments carried out. The overall objective is to determine the influence of the protective equipment on the shock wave in the simplified torso model. For this purpose, each configuration with the STM is first considered individually. The influence of the combinations can then be identified with these comparative values.

Table 1. Layered structure of the used clothing.

Layer	1	2	3	4
Type	t-shirt	undercoat cold protection	undercoat rain protection	combat jacket with flame protection
Material	polyester	polyester / polyamide	polyamide	viscose fiber / aramid
Picture				

Table 2. Experimental systematic.

No.	Description/Configuration	Section
1.5	Gelatin, individual	4.1
1.6		
1.11	Gelatin + Clothing	4.2
1.12		
1.7	Gelatin + Protective Plate	4.3
1.8		
1.13	Gelatin + Protective Plate + Clothing	4.4
1.14		
1.15		
1.16		

4. Results

In this section, the results of the tests are presented in a respective sub-section. Specifics of the test configurations are discussed directly.

4.1 Gelatin, individual

In this experiment, the simplified torso model (STM) is placed individually on the tripod plate. The measured values for the tests can be seen in Table 3. The corresponding graphs of the external and internal overpressure can be seen in Figure 3.

The curve of the internal overpressure shows the typical pressure curve. Three distinctive peaks can be seen in the pressure curve of the front sensor. The first peak is the incident peak. The second peak represents a reflection peak from the base plate and the third peak represents a reflection peak from the following sensor.

Table 3. Measured and calculated (italic) values at tests No. 1.5 and 1.6.

No. /Unit	Max. outside overpressure at 1 m	Impulse duration	Pressure wave propagation velocity (Avg.)	Max. inside overpressure at the front sensor	Shockwave propagation velocity (Avg.)
	[kPa]	[ms]	[m/s]	[kPa]	[m/s]
1.5	92,1	*0,84*	*450,1*	83,9	*1539,9*
1.6	82,4	*1,17*	*459,6*	88,7	*1533,7*

https://doi.org/10.52202/080042-0017

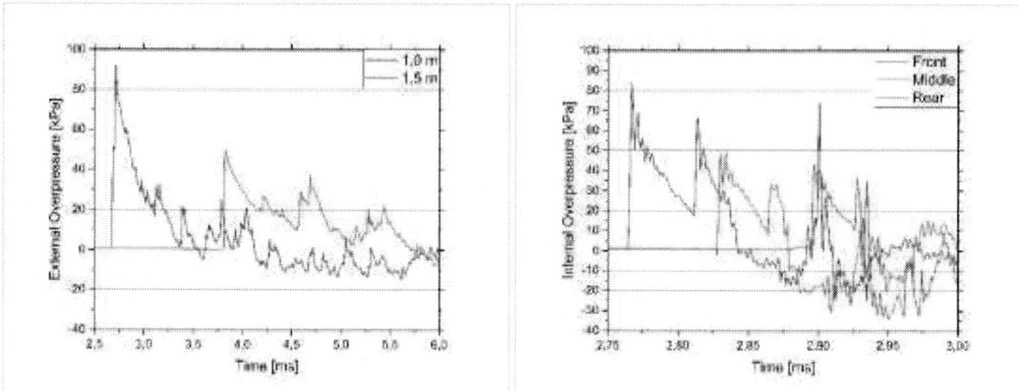

Figure 3. Left side: external overpressure. Right side: internal overpressure. (No.1.5)

4.2 Gelatin + Clothing

In this experiment, the STM is placed on the tripod plate. In addition, four layers of clothing are placed over the STM as shown in Table 1. The measured values for the test can be seen in Table 4. The corresponding diagrams of the external and internal overpressure can be seen in Figure 4 respectively in Figure 5. Due to the characteristic of the pressure curves, the runtime determination was not done by the peaks, but by the beginning of the pressure rise.

Table 4. Measured and calculated (italic) values at tests No. 1.11 and 1.12.

No. /Unit	Max. outside overpressure at 1 m	Impulse duration	Pressure wave propagation velocity (Avg.)	Max. inside overpressure at the front sensor	Shockwave propagation velocity (Avg.)
	[kPa]	[ms]	[m/s]	[kPa]	[m/s]
1.11	79,4	*1,19*	*471,5*	25,4	*1587,3*
1.12	71,8	*0,72*	*468,2*	15,1	*1587,3*

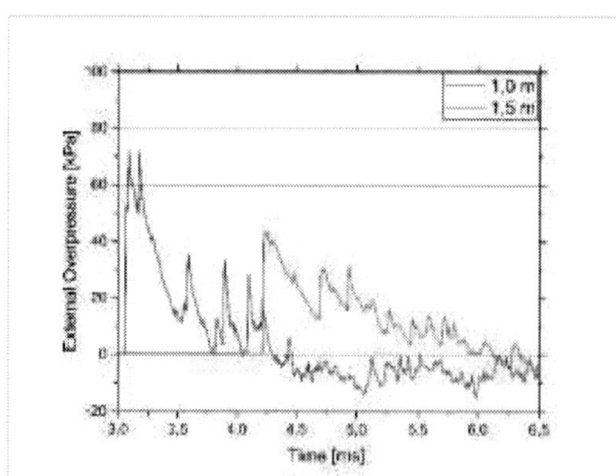

Figure 4. External overpressure (1.12).

https://doi.org/10.52202/080042-0017

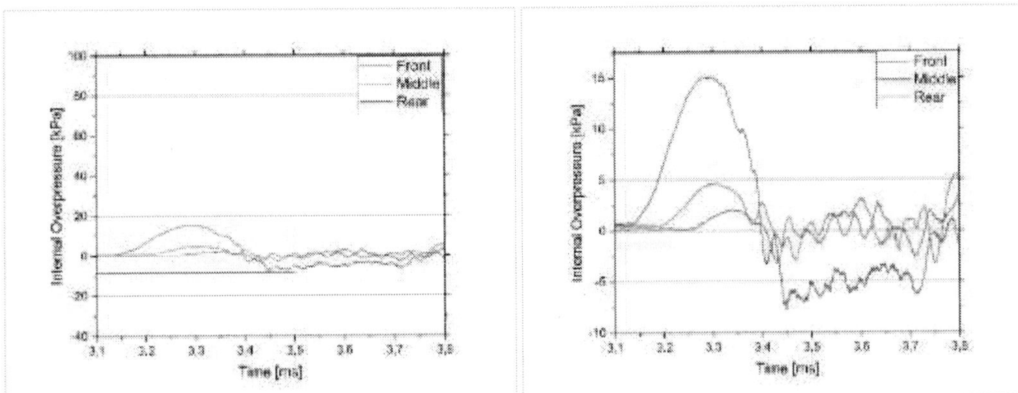

Figure 5. Left side: internal overpressure. Right side: internal overpressure, marked segment enlarged from the left graph. (1.12)

4.3 Gelatin + Protective Plate

In this experiment, the STM is placed on the tripod plate with a fastened protective plate in front of it. The measured values for the test can be seen in Table 5. The corresponding diagrams of internal overpressure can be seen in Figure 6 and of the acceleration in Figure 7.

With the measured acceleration values, there is a strong upswing of the entire STM starting at about 3.5 ms after ignition. The actual impact effect of the pressure wave takes place at 3.4 ms. Therefore, the 400 g acting at 3.4 ms are recorded. This time also correlates with the impact of the pressure wave on the pencil probe at 1 m distance. The swinging afterwards is caused by the free movement of the body model on the tripod plate. This does not represent an effect that can be found in real life.

Table 5. Measured and calculated (italic) values at tests No. 1.7 and 1.8.

No. /Unit	Max. outside overpressure at 1 m	Impulse duration	Pressure wave propagation velocity (Avg.)	Max. inside overpressure at the front sensor	Inside shockwave propagation velocity (Avg.)	Acceleration in propagation direction
	[kPa]	[ms]	[m/s]	[kPa]	[m/s]	[g]
1.7	95,6	*0,81*	*451,1*	16,3	*2325,6*	400
1.8	86,1	*0,99*	*468,2*	13,8	*2083,3*	371

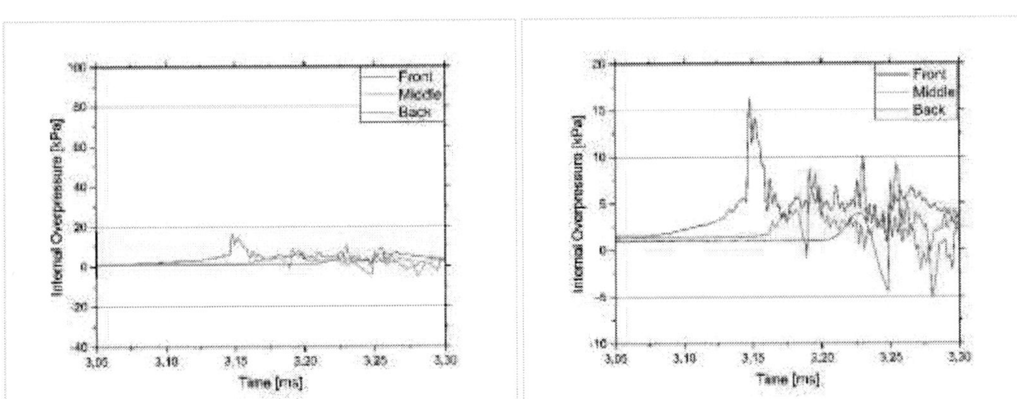

Figure 6. Left side: internal overpressure. Right side: internal overpressure, marked segment enlarged from the left graph. (1.7)

https://doi.org/10.52202/080042-0017

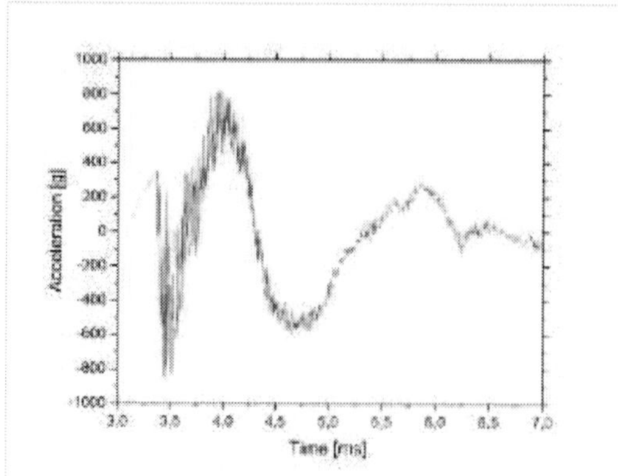

Figure 7. Acceleration (1.7).

4.4 Gelatin + Protective Plate + Clothing

In this experiment, the STM is placed on the tripod plate. It is covered with the clothing layers one to three (Table 1). Then the protective plate is fastened. Clothing layer four is then placed over it. For test numbers 1.13 and 1.14, the part of the clothing located under the tripod plate is fastened around the tripod. In test numbers 1.15 and 1.16, this part is freely movable. The measured values for the test can be seen in Table 6. The corresponding diagrams of internal overpressure can be seen in Figure 8 and of the acceleration in Figure 9.

It can be seen that lower accelerations are transferred to the STM in the tests where the clothing is fastened. A reason for this is the slap effect of the fourth cloth layer which can move freely in tests 1.15 and 1.16. Due to the characteristic of the pressure curves, the runtime determination was not done by the peaks, but by the beginning of the pressure rise. In test No. 1.15, the mean internal pressure sensor did not record any data, therefore a calculation of the internal propagation velocity was not possible. The error was due to a loose cable connection.

Table 6. Measured and calculated (italic) values at tests No. 1.13, 1.14, 1.15 and 1.16.

No. /Unit	Max. outside overpressure at 1 m	Impulse duration	Pressure wave propagation velocity (Avg.)	Max. inside overpressure at the front sensor	Shockwave propagation velocity (Avg.)	Acceleration in propagation direction
	[kPa]	[ms]	[m/s]	[kPa]	[m/s]	[g]
1.13	99,5	*0,68*	*461,9*	4,2	*1754,4*	364
1.14	89,1	*0,77*	*455,2*	3,5	*1904,8*	355
1.15	76,3	*0,71*	*474,4*	3,0	Def.	389
1.16	92,0	*1,25*	*457,7*	4,4	*1709,4*	422

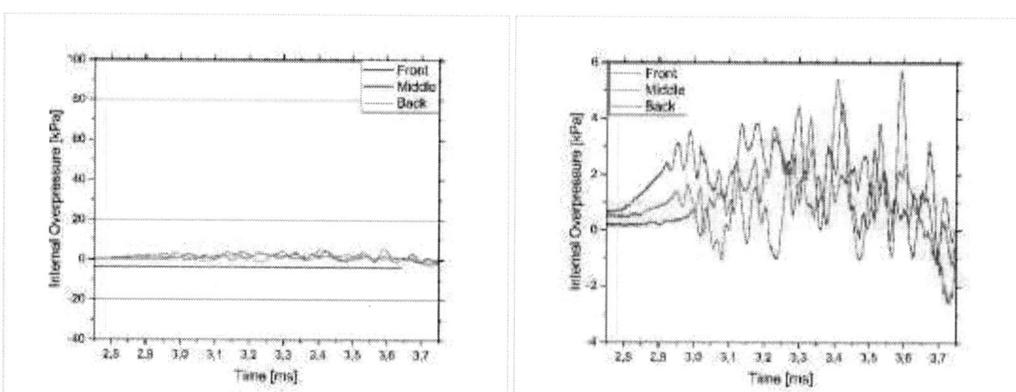

Figure 8. Left side: internal overpressure. Right side: internal overpressure, marked segment enlarged from the left graph. (1.16)

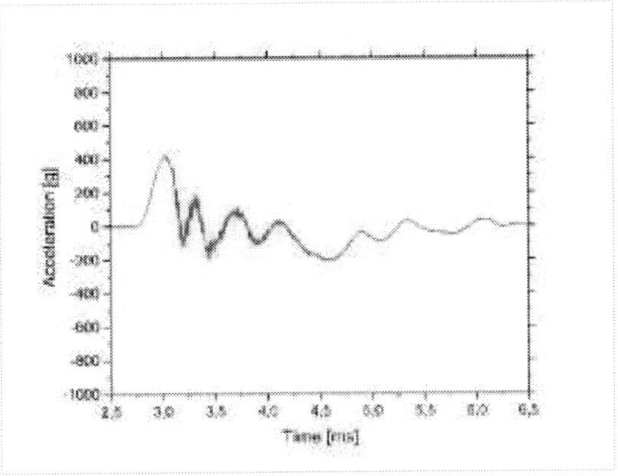

Figure 9. Acceleration (1.16).

5. Discussion

The external overpressure shows an ideal pressure curve (Figure 3 l.s.) [9]. A sudden increase followed by an almost exponential decrease in pressure. This is followed by the transition to negative overpressure and the leveling off around normal pressure. With optimum curve progression, 90 kPa overpressure is generated at a distance of 1 m in the test series. Lower maximum pressures (avg. 78 kPa) can be explained by a double peak at the maximum pressure peak (Figure 4). The impulse remained almost identical in both cases. On average, the impulse of the generated pressure wave is 37 kPa·s. The average impulse duration is 0.9 ms. Based on the time between the peaks of the pressure wave at the pencil probes at 1 m and 1.5 m distance, the average propagation velocity can be estimated. On average, the propagation velocity of the generated pressure wave is 458.72 m/s. Thus, the load case presented here remains below the level of a load case relevant for injury. It must of course be noted that there are no universal or validated injury thresholds [10]. However, the commonly used blast injury threshold curve of Bass et al. is used here as a guideline [3]. This results in a load case target value of 200 kPa at 2 ms duration. However, the characteristics of the pressure curve correspond to the expectations for a military explosive. The generated pressure wave can therefore be used for a general consideration of the effect of protective equipment on the internal shock wave, as it is a quantitative consideration.

For the internal pressures, it shows that through the clothing, the protective plate and then the combination of both, the pressure decreases (Table 7). Adding the plate and clothing reduces the maximum pressure to 5% of the pressure that occurs with an unprotected simplified torso model (STM). Furthermore, the respective protective equipment has an influence on the pressure curves and, thus, especially on the impulse. The adding of the clothing shows that the rapid increase in pressure is dampened. Almost bell curves are created (Figure 5). This shape of the curve, which almost corresponds to a mechanical excitation, is caused by the clothing layers, which act like a spring-damper system on the applied pressure wave. The respective maximum pressures are reduced. When the protective plate is added, very narrow peaks occur in the pressure curve (duration: 0,05 ms). These peaks show a rapid increase followed by a rapid decrease in pressure. The maximum pressure is lower. In both cases, the frequency of oscillation of the graph around normal pressure is lowered. So, one oscillation in the unprotected STM lasts 0.2 ms. In STM protected by clothing it lasts 0.7 ms. In the combination of clothing and a protective plate, the effects of both components occur in combination (Figure 8). The pressure sensors are not fully acceleration compensated inside the STM. Therefore, an acceleration of the sensors occurs, which is visible as oscillations of the pressure (< 1 kPa) (Figure 8).

Table 7. Avg. maximal internal overpressure at the first pressure sensor for every test configuration.

Configuration	Gelatin, indiv.	Gelatin + Clothing	Gelatin + Protective Plate	Gelatin + Protective Plate + Clothing
Internal peak overpressure (1st max.) (avg.) [kPa]	86,3	20,3	15,1	3,8

 https://doi.org/10.52202/080042-0017

Comparing the propagation speed of the shock wave in the unprotected STM with that in the one protected by the protective plate, an increase in the propagation speed of 669 m/s (44%) can be observed. This can be explained by means of impedance estimation. The following equation is used to estimate the material flow velocity [11]:

$$u_B = u_A \frac{2Z_A}{Z_A+Z_B} \quad (1)$$

Where u is the material flow velocity, Z is the acoustic impedance, subscript A is the origin material and subscript B is the coupling material. The material flow velocity is directly proportional to the propagation velocity. Therefore, it can be seen that at the transition from air to UHMWPE material, there is compression of the pressure wave. At the transition from UHMWPE material to gelatin, a decompression occurs. The decompression of the shock wave explains the increased velocity of propagation and the reduced maximum pressure (Table 5). The clothing layers also lead to an acceleration of the shock wave by 3% (Table 4). In combination with the protective plate, the clothing leads to absorption, since the propagation velocities in the combined setup are lower than in the setup with the protective plate by itself.

Considering the influence of acceleration, it can be seen that it has a maximum value of 385.5 g on average. The curve corresponds to that of a mechanical excitation and can thus be expected (Figure 7). Therefore, this is the acceleration that is induced by the pressure wave to the equipment and consequently to the STM. An effect of the clothing over the protective plate is not noticeable, as the excitation affects the entire tripod.

6. Conclusion and Future Work

The aim of this paper is to show the effect of protective equipment on the transmission and behavior of a shock wave in a simplified torso model. In summary, the combination of a UHMWPE protective plate and combat clothing has an attenuating but accelerating effect on the shock wave. Thus, the maximum overpressure is reduced by 95%. The impulse of the shock wave is reduced by 63%. The propagation speed is increased from 1532.50 m/s to 2204.55 m/s by the UHMWPE protective plate. The shock wave generator produces 90 kPa at a distance of 1m with an impulse duration of 0.9 ms in the configuration used during these experiments.

Also, it is important to consider the limitation that, in this case, the simplified torso model is protected over the entire frontal exposed area, which in reality is only a small part of the body. Moreover, the current version of the STM represents only a highly simplified version, which can be used to show the shock wave behavior in a well recognizable way. In addition, the load case is specifically relevant to highly exposed personnel, such as tactical door breaching operators, who are repeatedly exposed to comparable low-level blast. No injury-mechanical effectiveness can be assigned to the load cases generated for the tests. The reason for this is that there are many variables (some of which are unknown) that influence the effectiveness of injury mechanisms [2]. Nevertheless, the observations can be applied qualitatively.

Future objective will be to investigate the simplified torso model by means of injury-relevant relevant load cases in order to be able to make quantitative statements. Furthermore, the media transition of the shock wave within the model will be in the focus. Solids (bone simulants), air inclusions (hollow organs) and gelatin of different densities will be integrated into the simplified torso model. The detection of the shock wave is mainly done by means of pressure sensors. In addition, a general installation of the accelerometer is to be considered.

References

[1] Callaway, D. W, Burstein, J. L., Operational and Medical Management of Explosive and Blast Incidents. (Springer International Publishing, Cham, 2020); pp.19 – 35.

[2] Bull A.M.J., Clasper, J., Mahoney, P. F, Blast Injury Science and Engineering (Springer International Publishing, Cham, 2016); pp. 71 – 84.

[3] Bass, C. R., Rafaels, K. A., Salzar, R. S., Pulmonary injury risk assessment for short-duration blasts, The Journal of trauma, 65, 2008; pp. 604 – 615.

[4] Needham C.E. et al., Blast Testing Issues and TBI: Experimental Models That Lead to Wrong Conclusions, Frontiers in neurology, 6, 2015, No. 72.

[5] Hattingh, T. S., Skews, B. W., Experimental investigation of the interaction of shock waves with textiles. Shock Waves 11, 2001; pp. 115-123.
[6] Sellier K., Bolliger S., Kneubuehl B. P., Wundballistik Und Ihre Ballistischen Grundlagen (Springer, Berlin / Heidelberg, 2013); pp. 178 – 196.
[7] Azhari H., Basics of biomedical ultrasound for engineers (Wiley, Hoboken, N.J, 2010); pp. 313 – 314.
[8] Association of Test Laboratories for Attack Resistant Materials and Constructions (VPAM), General Basics of Ballistic Material, Design and Product Tests (Test Guideline: 2021-03).
[9] Sochet, I., Blast Effects, Physical Properties of Shock Waves (Springer International Publishing, Cham, 2018); pp. 37 – 57.
[10] Boutillier J., et al., A critical literature review on primary blast thorax injury and their outcomes, The journal of trauma and acute care surgery, 81, 2016, 371 – 379.
[11] Cooper P., Explosives engineering (Wiley-VCH New York, Chichester, Weinheim, Brisbane, Singapore, Toronto, 1996); pp. 203 – 223.

https://doi.org/10.52202/080042-0017

Blast load on operating personnel from shock grenade

A. Bjerke[1]

[1]*Norwegian Defence Research Establishment (FFI), P. O. Box 25, 2027 Kjeller, Norway, audun.bjerke@ffi.no*

Abstract. When using heavy weapon systems or explosives in military operations and training, blast waves can cause collateral blast exposure to the operating personnel. To best mitigate the effect of such exposures it is of interest to know how the positioning of the personnel will affect the risk of injury, both in terms of distance from the blast and how interior details like walls and doors affect the pressure build up in semi-enclosed rooms. The use of personal protective equipment is also a factor that will affect the risk of injury. In this research, we performed pressure measurements at selected points of interest of blast waves caused by the detonation of a stun grenade. We measured the pressure in several experimental setups; measurement of a free field wave, behind cover, in a trench system and usage in connected semi-enclosed rooms. From these trials, we estimated the risk of injury to the brain, ear and lungs from single point injury predictions models using the pressure history. We discuss the mitigating effect of personnel position during the blast exposure and how using protective equipment can mitigate the effect of the exposure. The results show that such grenades can cause significant injury and that mitigating measures are important to reduce the risk of injury. We also compared the obtained pressure values with results from numerical calculations using the non-linear explicit finite element solver IMPETUS Afea. Based on this comparison we discuss using numerical calculations as a supplement to measurements of actual events to assess the risk of collateral blast injury, which is useful when creating training procedures and operating doctrine.

1. INTRODUCTION

Military personnel are prone to injury when exposed to blast waves. During the Operation Enduring Freedom in Afghanistan, explosions was one of the main causes of injuries to participating Norwegian soldiers [1]. This is in line with experiences from other nations [2]. The risk of injury from blast exposure depends on several factors such as peak pressure, impulse, frequency and number of exposures. Research on blast injury has been ongoing for a several decades and recently there has been an increased effort on the effect of low-level blasts [3]. Studies have shown that this type of sub-concussive blasts can cause injuries, which are not immediately recognized but manifests itself as neurological disorders at a later time [4]. Blast exposure of this magnitude can be encountered when personnel are operating weapon systems or explosives in training [5] and this can cause the number of exposures and frequency to be higher than exposure to operational blasts.

In this research, we will investigate the blast exposure on the operating personnel when using stun grenades in typical training situations and estimate the risk using single point (SP) injury predictions models. We will then examine the use of numerical calculations for risk assessment without having to perform experiments of all relevant geometry configurations. Finally we discuss mitigating factors such as protective equipment in operational training.

2. METHODOLOGY

To investigate the blast exposure on operating personnel from a shock grenade we detonated a stun grenade with a net explosive weight of 116 g Comp B and recorded the pressure history at points of interest. These pressure measurements were then used to assess the risk of injury by applying various SP injury prediction models for the lungs, ears and the brain to determine whether the activity could cause injury.

2.1 Pressure measurements

To get a representative overview of the blast load for the personnel in various settings it was decided to perform pressure measurements in four different situations,

- a free field wave
- behind cover
- in a trench system
- in connected semi-enclosed rooms

https://doi.org/10.52202/080042-0018

2.1.1 *Free field wave*

Pressure measurements of the grenade as a free field wave are useful for verifying measurements when comparing to numerical calculations. The first step was therefore to measure the pressure in the case of a free field wave using piezoelectric pencil probes (PCB 137 A23). The pencil probes measured the pressure history at a distance of two, four and six meters from the grenade. The grenade and the pencil probes was placed on the same horizontal plane two meters above the ground with the pencil probes spaced out at 45 degrees to avoid interference with each other as shown in Figure 1. Using this setup and remotely detonating the grenade we could measure the passing pressure wave before reflections from the ground interfered with the wave pulse.

Figure 1. Experimental setup of the free field wave measurements

2.1.2 *Behind cover*

During throwing practice, the operating personnel takes cover from the blast wave by kneeling in a concrete booth after throwing the grenade at the training ground. It is thus of interest to measure the pressure history behind this cover to assess the risk of injury. In this experimental setup we switched to a piezoelectric sensor of the type PCB 102A05 which was fastened to the concrete wall in the booth, just below the ground level as illustrated in Figure 2. The membrane of the sensor was parallel to the ground to best record the side-on pressure. The ground from the booth have a slightly negative slope in the throwing direction and we measured grenades at a distance of two or four meters from the sensor. After remotely detonating the grenade we could record the passing blast exposure in the booth.

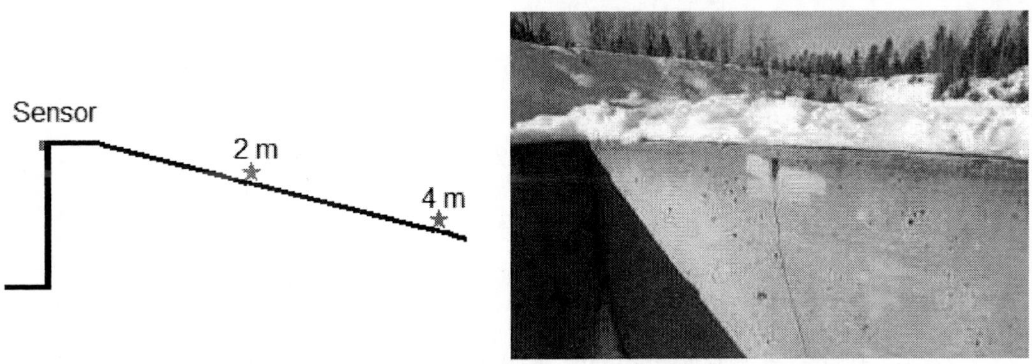

Figure 2. Experimental setup of the measurements behind cover

2.1.3 *Trench system*

Training with stun grenades in a trench system is another activity in which the operating personnel can experience blast wave exposure. The trench used consisted of concrete walls creating a canal with a width of one meter. At the time of the experiment, the height of the walls was around 115 cm, with the ground not being completely even due to a layer of snow. We put piezoelectric sensors (PCB 102A05) on the walls 40 cm from the top of the trench (approximately 75 cm above ground level) at positions shown in

https://doi.org/10.52202/080042-0018

Figure 3 with the membrane of the sensor parallel to the ground. Using remote detonation the grenade detonated at a distance of 150 cm from the intersection. We put sensor 1 and 2 at a distance of 50 cm from the intersection, whereas the distance between sensor 2 and 3 was 303 cm + 50 cm along the wall.

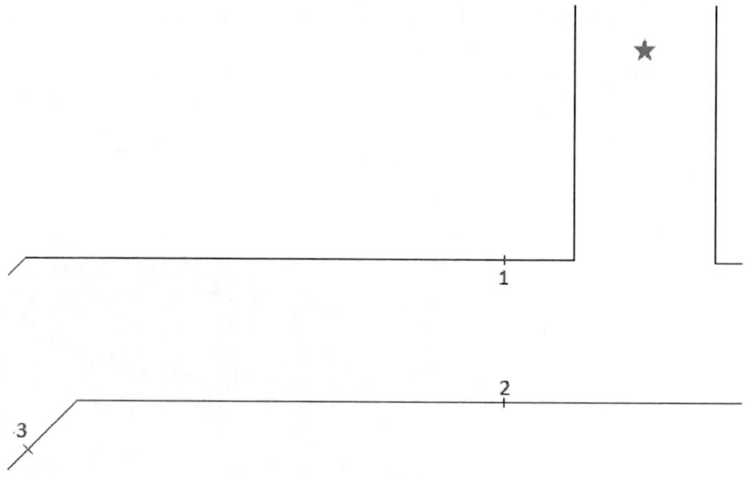

Figure 3. Experimental setup of the measurements in a trench system

2.1.4 Connected semi-enclosed rooms

The last setup was detonation of a grenade in connected semi-enclosed rooms. The position of the sensors and the grenade are as shown in Figure 4 with the measurements being in cm. The blue line indicates an open window and the open space is the door opening to the connected hallway. Sensors 1-3 are positioned 140 cm above the floor level in the hallway and sensor 4 is just below the window frame on the outside of the building. All sensor membranes were parallel to the ground. The height of the room was 238 cm with a total volume of 30.7 m^3.

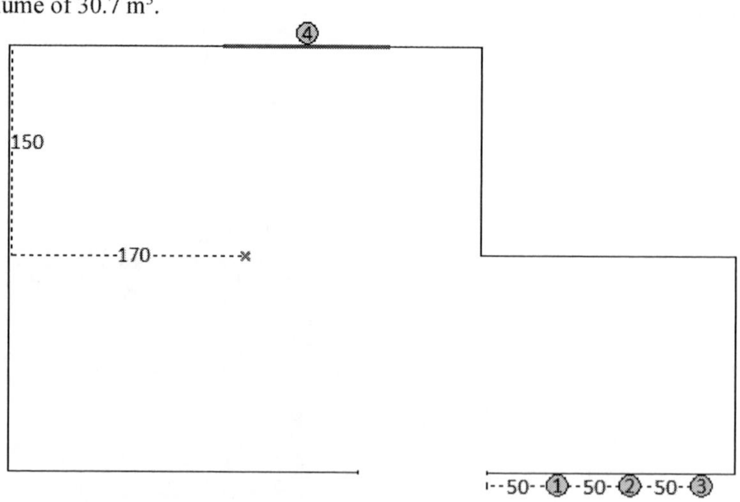

Figure 4. Sensor setup of the measurements in semi-enclosed rooms

2.2 Injury prediction models

After obtaining the pressure histories for the various experiments, we applied the collected data to different injury prediction models to assess the risk of injury for the operating personnel. The organs chosen were lungs, ears and the brain.

2.2.1 Lungs

When estimating the risk of injury to the lungs we used a modification of the blast injury model developed by Axelsson [6]. This model is a single degree of freedom (SDOF) system meant to describe the chest wall response of a human exposed to a given blast wave in a complex blast environment. It required pressure inputs from four gauges placed on a Blast Test Device (BTD) to calculate the chest wall velocities. Axelsson proposed that the average of the maximum of the four calculated velocities is a measure of injury called the Chest Wall Velocity Predictor (V). Based on experiments with sheep he created an Adjusted Severity of Injury Index (ASII) based on the level of injury. The correlation between injury level, ASII and V is shown in equation 1 and Table 1.

$$\text{ASII} = 0.124+0.117\text{V}^{1.205} \quad (1)$$

Table 1. Correlation between injury level, ASII and V

Injury level	ASII	V (m/s)
No injury	0.0-0.2	0.0-3.6
Trace to slight	0.2-1.0	3.6-7.5
Slight to moderate	0.3-1.9	4.3-9.8
Moderate to extensive	1.0-7.1	7.5-16.9
>50% Lethality	>3.6	>12.8

While the BTD-model requires data from four pressure gauges, a single point formula is defined by using the single point pressure in a given location as input to the Axelsson model. The value of V is then equal to the maximum chest velocity in the SP-model. The SP-model is much easier to work with numerically, as the ASII can be calculated in all locations with only one simulation. In contrast, with the BTD-model, the ASII can only be determined in one location at the time and a huge number of simulations (with different BTD locations) would have to be run to get an overview of the ASII as a function of position. The SP-model has earlier been shown to closely approximate the BTD-model for all kinds of scenarios, including short and long blast duration, free-field and near walls [7]. We will therefore be using the Axelsson SP-model as risk assessment for lung injury.

2.2.1 Ears

To evaluate the possibility of hearing injury we used the procedure for impulse noise defined in MIL-STD-1474D [8]. This procedure makes use of the peak sound pressure and B-duration calculated from the pressure history. The B-duration is the impulse duration to a point on the total blast wave where the rest of the exposure is at least 20 dB below the peak. The MIL-STD-1474D injury criterion for impulse noise is then a plot where the peak sound pressure (in dB) is plotted against the B-duration as shown in Figure 5. The three curves represent different amount of exposures a soldier can have during 24 hours.

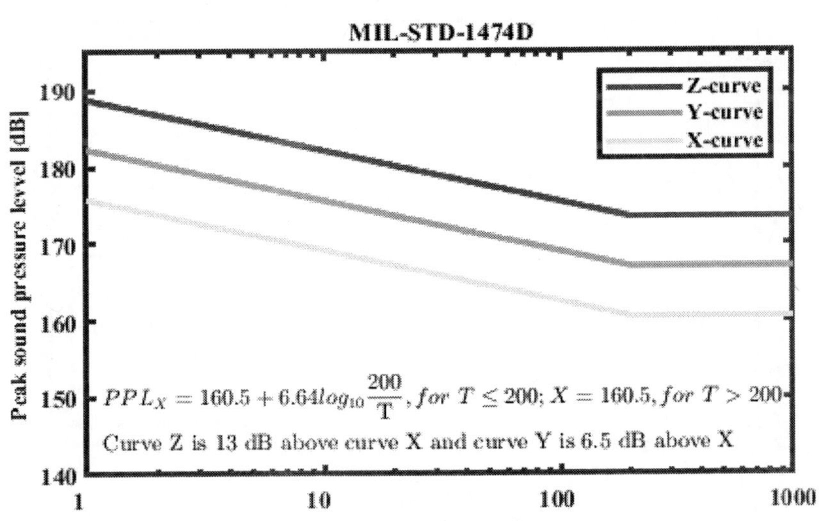

Figure 5. MIL-STD-1474D injury thresholds

https://doi.org/10.52202/080042-0018

2.2.1 Brain

The effect of blast overpressure on the brain is an ongoing area of research. The complexity of the matter spans across several subjects and currently there is a NATO Research Task Group within the Human Factors and Medicine (HFM) Panel to develop exposure guidelines for blast overpressure (HFM-338). In our work, we decided to apply a peak pressure limit of 4 psi (27.8 kPa) as an injury criterion. The historic rationale for this threshold is the risk of eardrum rupture on unprotected human ears [9] and thus not based on damage to the brain per se. However, it is relevant in our context as the 4 psi threshold for brain damage is used in most studies on low-level blasts and used by various range officers to calculate the minimum safe distance for explosive breaching [10].

3. RESULTS

The sampling rate of the pressure measurements was 204 800 Hz and all collected data was low-pass filtered using an 8th order Bessel filter with a cut frequency of 40 kHz. With the processed measurements and the previously defined injury criterions, we could estimate the risk of injury to the operating personnel from the blast exposure.

3.1 Free field wave

For the free field wave experiment, we detonated six grenades and from the recorded measurements, we extracted the peak pressure and calculated the B-duration and ASII values at two, four and six meters from the blast. The average of these values are tabulated in Table 2 and in Figure 6 we have plotted the peak sound pressure (in dB) and B-duration as given in MIL-STD-1474D.

Table 2. Averaged results from the free field measurements

Distance [m]	Peak pressure [kPa]	B-duration [ms]	$ASII_{SP}$
2	54.7	10.9	0.018
4	16.4	15.3	0.009
6	10.0	16.9	0.007

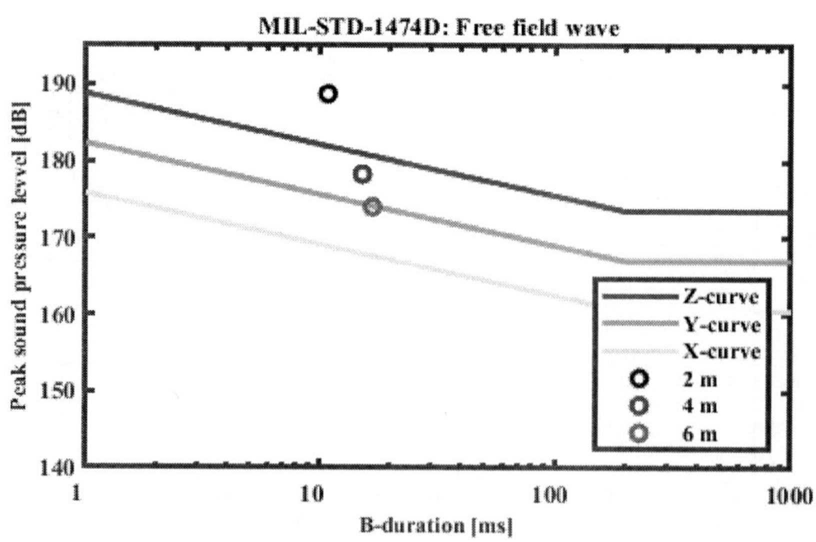

Figure 6. MIL-STD-1474D for 116g CompB in a free field wave

https://doi.org/10.52202/080042-0018

As we can see from Table 2 and Figure 6, the risk of injury from the blast exposure a stun grenade creates in the free field decreases below our injury criterions with increasing distance to the grenade. The risk of lung injury is insignificant, whereas the risk to the ears and brain can be significant if close enough.

3.2 Behind cover

In the setup behind cover, we measured grenades at two and four meter from the detonation. The values of interest are given in Table 3. In Figure 7 we have plotted the peak sound pressure (in dB) and B-duration as given in MIL-STD-1474D.

Table 3. Averaged results behind cover measurements

Distance [m]	Peak pressure [kPa]	B-duration [ms]	$ASII_{SP}$
2	20.1	7.1	0.012
4	7.0	26.3	0.007

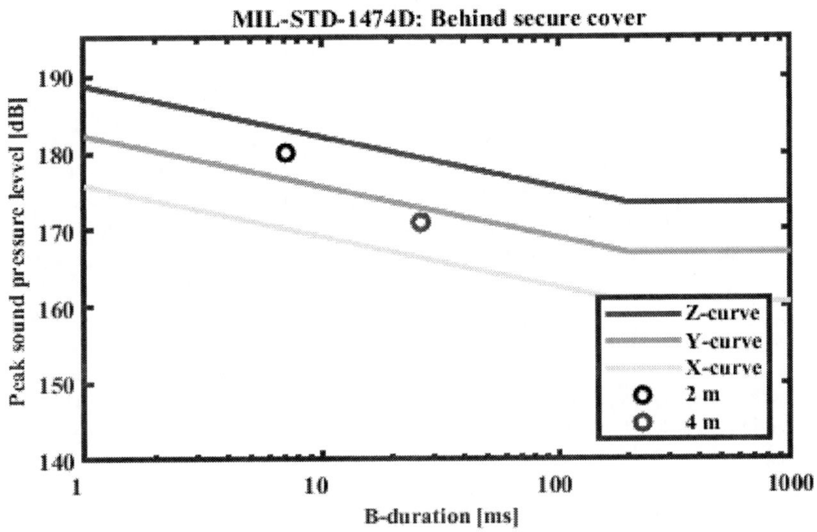

Figure 7. MIL-STD-1474D for 116g CompB behind cover

As we can see from the results, the risk of injury from the blast exposure when behind cover is insignificant when the distance to the grenade is at least 2 meters.

3.3 Trench system

In the trench system, we detonated three grenades and measured the pressures at the different positions outlined in Figure 3. The values of interest are tabulated in Table 4 and in Figure 8 we have plotted the peak sound pressure (in dB) and B-duration as given in MIL-STD-1474D.

Table 4. Averaged results in the trench system measurements

Sensor	Peak pressure [kPa]	B-duration [ms]	$ASII_{SP}$
1	19.5	33.9	0.012
2	80.8	9.1	0.018
3	4.7	87.2	0.008

https://doi.org/10.52202/080042-0018

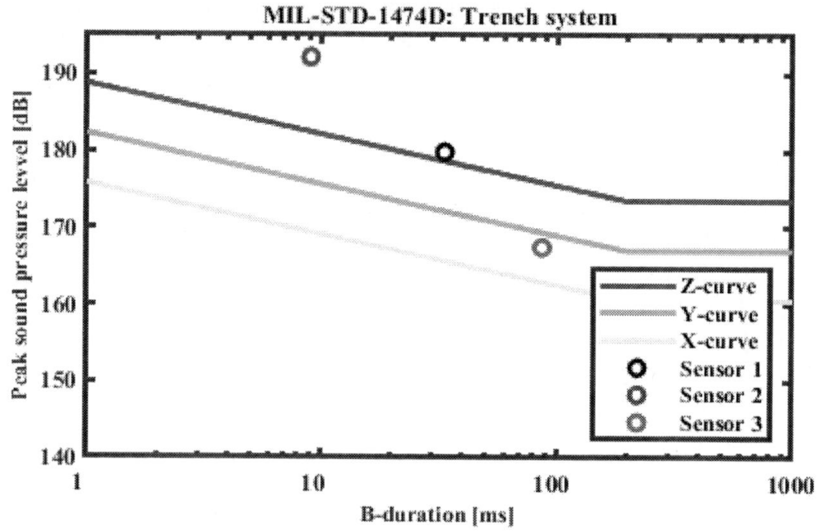

Figure 8. MIL-STD-1474D for 116g CompB in a trench system

As we can see from the results, the risk of injury from the blast exposure when operating a stun grenade in a trench can be significant. This comes due to the reflecting surfaces, which creates a peak pressure that can exceed the injury criterion for the brain. We also notice that the B-durations are quite long and will pose a risk to the soldiers hearing. It is thus of importance to ensure that the distance to the detonation is long enough both in terms of the distance in the channel in which the grenade resides and also the distance to the corner from the potential stacking position. The risk of lung injury is insignificant in this scenario.

3.4 Connected semi-enclosed rooms

In the setup of connected semi-enclosed rooms, we detonated three grenades. The averaged values of interest are tabulated in Table 5 and in Figure 9 we have plotted the peak sound pressure (in dB) and B-duration as given in MIL-STD-1474D.

Table 5. Averaged results in the trench system measurements

Sensor	Peak pressure [kPa]	B-duration [ms]	$ASII_{SP}$
0.5 m from door	24.7	373.4	0.009
1.0 m from door	19.0	491.6	0.009
1.5 m from door	16.0	469.1	0.009
Below window	37.6	114.0	0.014

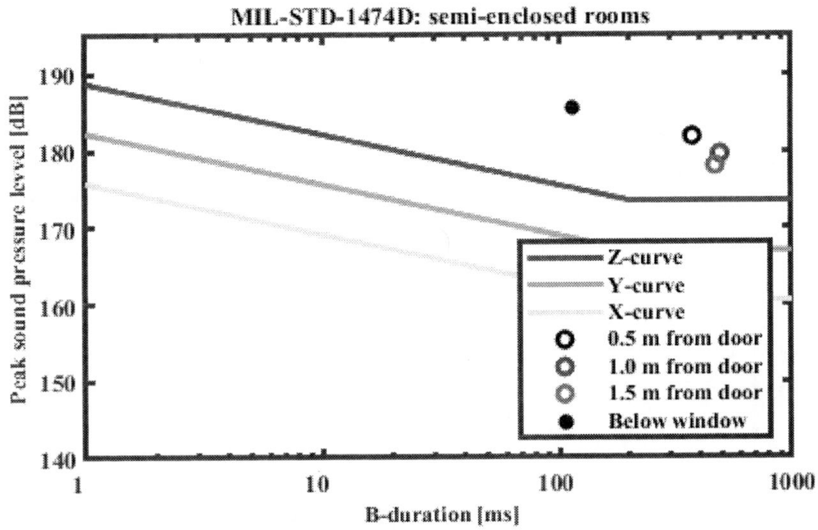

Figure 9. MIL-STD-1474D for 116g Comp B in connected semi-enclosed rooms

As we can see from the results, the risk of injury from the blast exposure when operating a stun grenade indoors can be significant. The reflecting surfaces in the geometry creates long B-durations will pose a risk to the hearing. The peak pressure outside the window is also exceeding the 4 psi threshold for the brain. It should also be stressed that the measurements done here is in the room connected to the room of detonation and the blast exposure inside the room of the grenade is even much higher. The risk of lung injury is insignificant in this scenario.

3.5 Summary

We have summarized the risk of injury in the various scenarios in Table 6. The risk of lung injury from a stun grenade is insignificant, but both hearing and the brain can be of risk to the operating personnel if close enough to the detonation. Especially when used in close proximity to reflecting surfaces that can cause buildup of the blast wave, which is the case when using the grenade in a trenches or indoors.

Table 6. Summary of the risk to operating personnel from stun grenades

Scenario	Lung	Ear	Brain
Free field	Insignificant	Yes (if close enough)	Yes (if close enough)
Behind cover	Insignificant	Insignificant (at 2 m)	Insignificant (at 2 m)
Trench system	Insignificant	Yes (caution required)	Yes (caution required)
Connected semi-Enclosed rooms	Insignificant	Yes (caution required)	Yes (caution required)

4. NUMERICAL CALCULATIONS

To create safe procedures for training it is of interest to know at which positions in a specific geometry that the blast exposure is at an acceptable risk level. Doing such experiments for each specific case is costly and if simulations can be employed this would be preferable. We used the numerical solver IMPETUS Afea to model the case of the free field wave and the semi-enclosed room and compared the simulation results to the measurements.

In the IMPETUS model the grenade is represented by a cylinder with height 5.94 cm, radius 1.97 cm and filled with Comp B using the IMPETUS particle model. We then defined output sensors as the same position as in the pressure measurements at two, four and six meters. Figure 10 shows the results of the Impetus simulation with cell sizes 0.02 m and 0.01 m compared to the measurements of the free

https://doi.org/10.52202/080042-0018

field waves with largest and lowest peak pressure at a distance of 2 meters from the detonation. As we can see the IMPETUS model have a similar development as the measurements, but slightly underpredicts the peak at two meters.

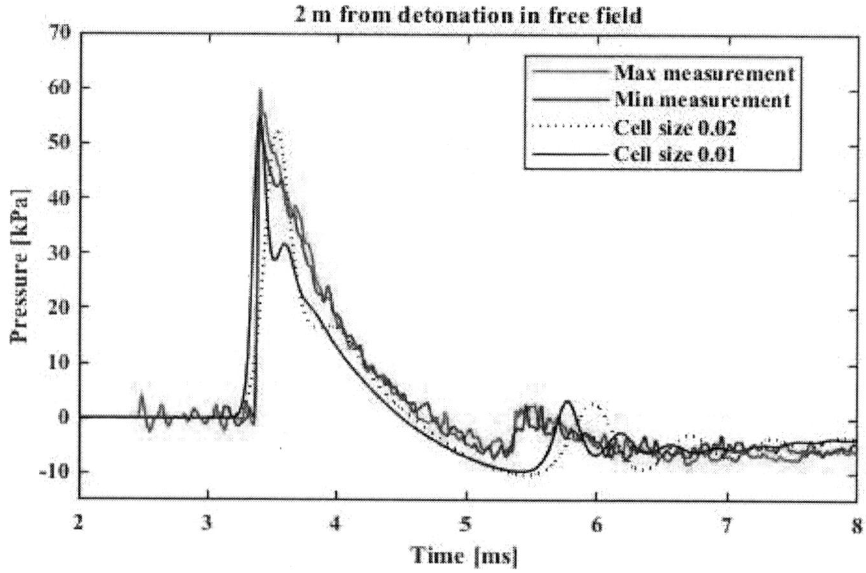

Figure 10. Comparison of IMPETUS and measurements of a free field wave

In the case of a detonation in a semi-enclosed room, we defined the roof, floor and walls using a rigid material and put this geometry inside an encompassing CFD domain filled with air. The total volume of the CFD-domain was 40 m^2. Again, we tried to simulate with cell sizes of 0.02 m and 0.01 m simulation with different cell size of the CFD domain to investigate the effect of the cell size.

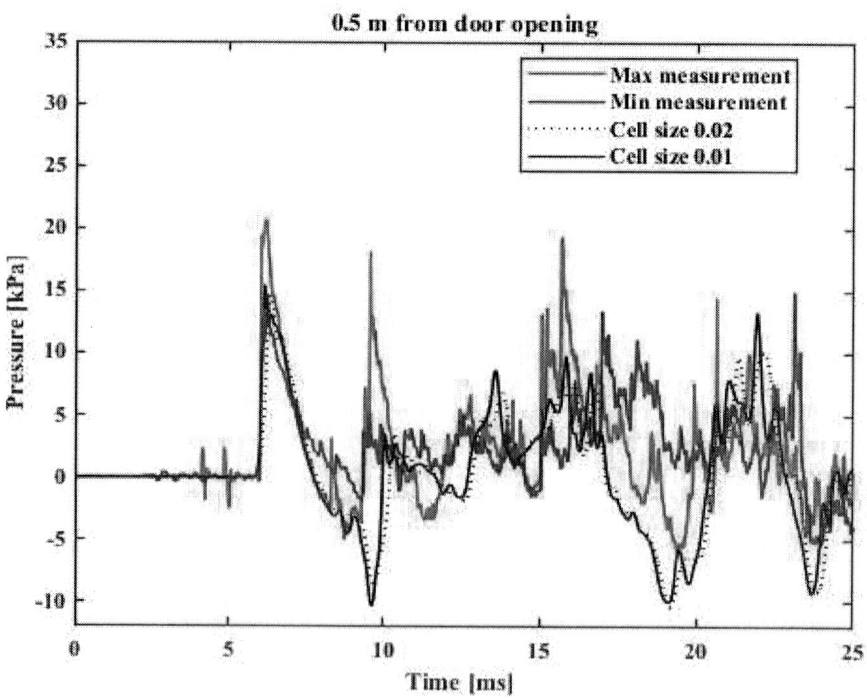

Figure 11. Comparison of IMPETUS and measurements in connected semi-enclosed room

https://doi.org/10.52202/080042-0018

Figure 11 shows the pressure history 0.5 meters from the door opening outside the room. As we can see, the simulation give a good estimate of the measurements. The small differences in measurements vs calculations can be due to several factors, such as inaccuracies in the measurement of the room, sensor positions, material properties, cell size or stochastic effects in measurements of detonation. In the case of cell size equal to 0.02 m the run time was just 4 minutes, whereas the cell size of 0.01 m gave a run time of 50 minutes. It is thus easy to obtain a good data foundation when developing training procedures that can maximize the training effect at an acceptable risk of injury.

5. DISCUSSION AND CONCLUSION

This work has shown that blast exposure from stun grenades can create a risk of injury for the operating personnel, especially when used in settings where there are surfaces that can reflect the blast wave. To mitigate this risk it is important to be aware of the blast exposure in the close proximity of the detonation. We have shown that employing a numerical tool as IMPETUS Alfea can provide calculations of the blast exposure to understand the risk of injury better, which is useful when creating training procedures for the personnel.

Another area of interest is the use of personal protective equipment (PPE) and how those can be designed to reduce the risk of injury from a given blast exposure. For instance, a review of the literature showed that wearing a helmet will, to some degree, protect against blast waves, given that the helmet is padded on the inside [11]. With the ability to use numerical calculations to get a good representation of the blast exposure of a given scenario, this data can in turn be used to assess the mitigating effect of varying PPE.

References

[1] Bøe H. J., Nordstrand A. E., Engen H. G., Gjerstad C. L., Bakker L. P., Thoresen S., Birkeland M. S. and Blix I. Afghanistanundersøkelsen 2020, Veteraners erfaringer, helse og livskvalitet etter tjeneste i Afghanistan 2001-2020. Oslo: Forsvaret, 2021.

[2] Owens B.D., Kragh J.F., Wenke J.C., Macaitis J., Wade C.E. and Holcomb J.B. Combat wounds in operation Iraqi freedom and operation enduring freedom. J Trauma, 2008; 64; 295–299.

[3] Belding J.N., Englert R.M., Fitzmaurice S., Jackson J.R., Koenig H.G., Hunter M.A., Thomsen C.J. and da Silva U.O. Potential Health and Performance Effects of High-Level and Low-Level Blast: A Scoping Review of Two Decades of Research. Front. Neurol, 2021; 12; 628782.

[4] Tate C.M., Wang K.K, Eonta S., Zhang Y., Carr W., Tortella F.C., Hayes R.L., and Kamimori G.H. Serum brain biomarker level, neurocognitive performance, and self-reported symptom changes in soldiers repeatedly exposed to low-level blast: a breacher pilot study. J Neurotrauma, 2013; 30; 1620–1630.

[5] Kamimori G.H., Reilly L.A., LaValle C.R. et al. Occupational overpressure exposure of breachers and military personnel. Shock Waves, 2017; 27; 837–47.

[6] Axelsson H., Yelverton J.T. Chest wall velocity as a predictor of nonauditory blast injury in a complex wave environment, J Trauma, 1996; 40; 31-37.

[7] Teland J.A., Review of blast injury prediction models, FFI/RAPPORT-2012/00539.

[8] Department of Defense. MIL-STD-1474D: Design criteria standard - Noise limits, 1997.

[9] Hirsch F.G. Effects of overpressure on the ear—a review. Ann N Y Acad Sci, 1968; 152; 147–162.

[10] Carr W., Dell K.C., Yanagi M.A. et al. Perspectives on repeated low-level blast and the measurement of neurotrauma in humans as an occupational exposure risk. Shock Waves, 2017; 27; 829–836.

[11] Teland J.A., Influence of helmet on blast propagation into the brain, literature survey, FFI/RAPPORT-2016/02314.

https://doi.org/10.52202/080042-0018

Computational Assessment of Headborne Equipment: Alteration of Head and Neck Biomechanics During Blast-Induced Accelerative Loading

M. Yates[1], M. Tumperi[1], E. Crane[1], G. Holt[1], L. Voo[1], V. Alphonse[1], D. Drewry[1], and Q. Luong[1]

[1] The Johns Hopkins University Applied Physics Laboratory, 11100 Johns Hopkins Road, Laurel, MD 20723, USA,
Melissa.Yates@jhuapl.edu

Abstract.
Headborne equipment (HBE), such as helmets, maxillofacial protection systems, and visual augmentation systems, alters the biomechanical state of the head and neck potentially influencing the inertial response and associated injury during exposure to blast overpressure. However, due to the complexity and substantial cost of experimentally recreating these events, parametric datasets covering a wide range of blast exposure and HBE combinations do not exist to evaluate injury outcomes. Human computational models can be applied to study the relative effect of changes in risk of injury based on the HBE combinations and assess if additional attention to risk of blast injury is necessary during HBE design. This work investigated the extent to which HBE may alter head and neck biomechanical response during blast-induced accelerative loading in order to help gain insight into the connection between blast loading and HBE combinations. A computational model of the head and neck was validated for kinematic response to impulse from blast overpressure exposure and a sensitivity study was conducted to evaluate the biomechanical effects due to headborne equipment selection, head orientation, and blast overpressure. The predicted biomechanical response data was then evaluated and then compared to established metrics and standards, the Head Injury Criterion (HIC), the Brain Injury Criterion (BrIC), and the Rotational Injury Criterion (RIC). These findings provide insight into the relative importance of assessing how current and future HBE systems influence biomechanics during a blast overpressure event, potentially informing equipment design as well as guidelines on HBE usage. These computational capabilities can provide insight for future testing and evaluation of HBE equipment prior to production of physical hardware. Further work is needed to experimentally capture these biomechanical effects in order to fully verify and validate the outcomes of this study.

1. INTRODUCTION

Combat helmet systems mitigate head injuries by absorbing the energy imparted by blunt impacts and arresting ballistic threats; however, current helmet performance assessments do not evaluate helmet performance for aspects related to blast exposure [1]. Blast exposure effects on the body are typically grouped into five distinct categories, with primary blast exposure induced by two biomechanical effects 1) overpressure loading and 2) inertial loading [3]. Prior efforts have investigated the effect of blast overpressure due to shock wave exposure on brain injury with a helmet system [4,5,6]. However, these studies focused on non-combat helmets, simplified pad systems, or did not evaluate the effect of changes in inertial loading on the head/neck system due to addition of headborne masses and geometries. The addition of helmets and other headborne equipment (HBE) change surface profile surrounding the head and in turn, results in an altered biomechanical state prior to threat exposure, as well as a change in impulse transferred to the head during the blast event. In order to understand if and how this altered biomechanical state affects injury risk, additional research is required. Characterization of the HBE performance during blast overpressure loading is necessary to understand if blast injury risk should be considered during helmet and HBE development, as well as to monitor the risks of augmenting personal protective equipment (PPE) to increase survivability; however, parametric datasets covering a wide range of blast exposure and HBE combinations do not exist to evaluate injury outcomes. *In silico* approaches can be applied to investigate blast-induced biomechanical alterations to the head/neck system and explore potential consequences in terms of injury. Numerous studies have developed numerical models to evaluate blunt injury risk for athletic and automotive applications [19], but a limited set of numerical models have been specifically developed to evaluate blast injury of the head and neck [2,20]. The purposes of this study are to (1) demonstrate a methodology that assesses how addition of HBE can influence the biomechanical response of the head and neck during blast-induced accelerative loading, and (2) evaluate results in the context of established head injury metrics to identify which metrics may warrant further study as potential tools for evaluation of HBE in the context of blast.

https://doi.org/10.52202/080042-0019

2. METHODS

A human head and neck finite element model (FEM), developed by The Johns Hopkins Applied Physics Laboratory (JHU/APL) was previously validated for blast overpressure loading using published brain displacement and rotation data was leveraged for this study [7]. During this effort, this existing model was refined and incorporated with a current-service combat helmet, fielded HBE surfaces, and blast wave propagation through an open-air environment. This computational setup is summarized in Table 26. These component models were then assembled into the configurations outlined in Table 26 and verified for pressure transmission and kinematic response when loaded by shock waves at incident pressures of 200 kPa (2kg TNT at 2.5m) and 400 kPa (2.27kg TNT at 2.0 m). A simulation dataset for sensitivity analysis was then developed and conducted to evaluate the biomechanical effects due to HBE. Simulations were conducted in an explicit FEM solver based on its established use for spatial and temporal characterization of head and neck biomechanics, as well as capability to simulate blast exposure effects.

Table 26. Summary of computational modeling components and range of study.

Model Component	Description
Helmet	FE model of a representative combat helmet
Mounted Equipment	A representative Night Vision Goggle (NVG) system with mounting hardware
Blast Environment	Blast wave propagation through air developed in LS-DYNA with evaluated incident pressures at 200 and 400 kPa
Human Anatomy	FE model head and cervical spine validated for dynamic motion during blast exposure.

2.1 Model Development and Validation

All blast simulations were conducted using ANSYS LS-DYNA (Livermore Software Technology Corporation, Livermore, CA). A hybridized approach with both Lagrangian and Eulerian approaches capable of material flow and solid deformation, named the Arbitrary Lagrangian-Eulerian (ALE) method, was selected for its appropriateness for evaluation of blast overpressure effects on deformable structures. Briefly, the head and helmet were positioned and oriented with respect to prior experimental tests conducted at JHU/APL with the face oriented toward the oncoming shock wave [7]. For HBE incorporation, geometries were generated by 3D laser surface scans and computed tomography (CT) scans of hardware provided by Program Executive Office Soldier (PEO Soldier) Product Manager Soldier Protective Equipment (PdM SPE). The attaching bracket of the HBE was positioned to be centered on the midsagittal plane and in contact with the brim and shell of the helmet. The night vision goggle (NVG) system was aligned with the eye. If a gap existed between the mount and NVG after positioning, small adjustments were made to ensure an airtight structure during blast exposure.

The head-neck FE model previously developed and validated by JHU/APL for use in evaluating the effect of blast loading was selected for this study [1]. This head model was meshed in TrueGrid to reflect the 50th percentile male and based on the ANSUR II and Visible Human geometries. As shown in Figure 1, the model consists of representations of the brain (336,000 hexahedral elements), the cerebral falx/tentorium (4,600 2D shell elements), skull cortex (45,000 2D shell elements), diploë (143,000 hexahedral elements), frontal and maxillary sinuses (22,600 hexahedral elements), and flesh (297,000 hexahedral elements). In the cervical spine, the ligaments between the individual cervical spine vertebral bodies were modeled as 2D spring elements. The interface between the skull and neck was refined to incorporate previously published rotational stiffness response for the Occipital-C2 joint.

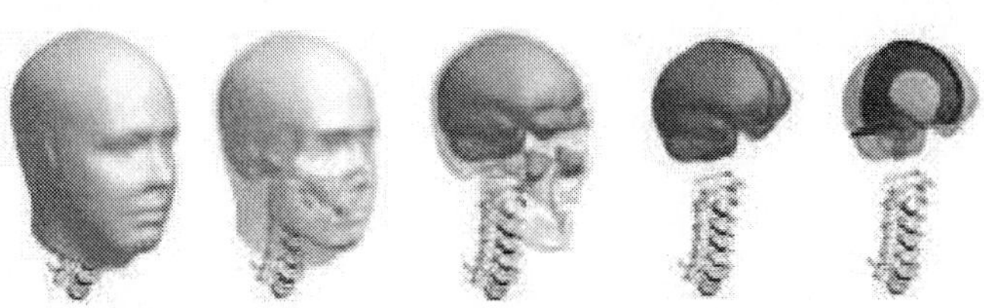

Figure 65. The head and neck FE model structures share nodal connections to enable pressure transmission. Key include flesh (tan), skull (white), vertebral bodies (white), intervertebral discs (black), cerebral spinal fluid (dark tan), sinuses (pink), brain (gray), cerebral falx/tentorium (red).

https://doi.org/10.52202/080042-0019

Dynamic validation of the human head and cervical spine model was conducted to evaluate kinematic motion in blast loading by simulating the experimental setup of Iwaskiw et. al. [7] and comparing kinematic results. Briefly, a series of front-facing short-duration dynamic overpressure were conducted on four postmortem human surrogates (PMHS) (male, ages 53-67) to characterize dynamic head and neck motion. These specimen were disarticulated in-between the first and second thoracic vertebral bodies, then a fixed boundary condition in the lower cervical spine was created by potting at the 6th and 7th vertebral bodies using poly(methyl methacrylate). Specimen were perfused to achieve physiological intracranial pressure. Authors collected head-neck kinematics, brain motion, and intracranial pressure. The pressure loading applied to the head model was determined using a computational fluid dynamics (CFD) model of the head outer surface as a surface grid embedded in a structured grid at the end of a modeled shock tube as described in the experimental setup. The pressure outputs at the skin surface grid were then mapped to the FE mesh of the flesh. ALE simulations were then run with a fixed sixth cervical vertebra (C6) of the spine to emulate experimental potting at the same location. The error between the mean and the 1st standard deviation of the experimental corridor and simulation predictions were then assessed using CORA (CORrelation and Analysis). CORA has been established as an objective method to evaluate finite element model validation based on error in signal shape, magnitude, and phase [17].

A representative combat helmet FE model was furnished by PEO Soldier PdM SPE and validated in work previously reported directly to PEO Soldier PdM SPE which quantified effects during blast loading. The validation data included axial compression test data for the individual pads of the helmet system and whole helmet axial compression tests at three strain rates with three-dimensional digital image correlation to quantify load displacement response of pads. This helmet model consisted of deformable representations of the helmet shell with edge trim (38044 hexahedral elements), retention system clips (3747 hexahedral elements), retention system straps (848 2D shell elements), and a pad-based suspension system (5248 hexahedral elements). The HBE models were developed from a combination of physical night vision goggle (NVG) and mounting hardware, and supplemented by available computer-aided design (CAD) models. Geometries of a helmet and suspension system were reverse engineered using hand measurements and callipers for initial and approximate shape development with CREO Parametric (PTC). Computed Tomography (CT)scans were then obtained using the X50 (North Star Imaging, Inc.) to complete geometry capture. CREO Simulation (PTC) was used to generate a tetrahedral mesh for each component with a minimum element size of 2.0 millimetres (mm). The meshed NVG was assembled relative to the helmeted head FEM model shown in Figure 66.

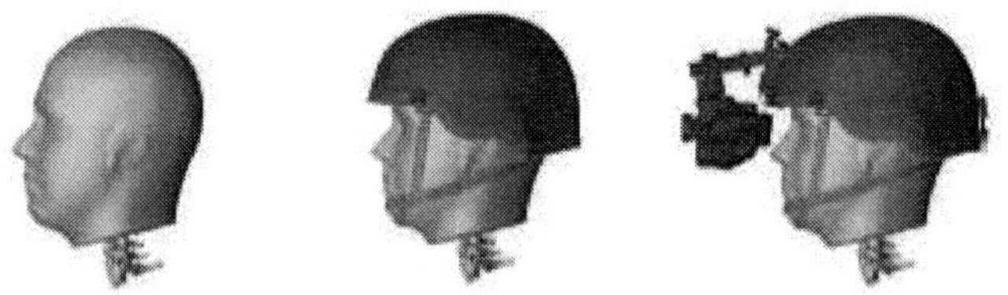

Figure 66. APL head-neck model without a combat helmet, with a combat helmet, and with NVG.

ANSYS LS-DYNA can simulate the blast event based on the empirical model outlined in TM5-855 US Army Handbook (ConWep) coded in the Load_Blast_Enhanced keyword and against ConWep and blast pencil-probe data from live fire blast testing [9,10]. This approach models the explosive event, including the charge, the expansion of the detonated explosive, and the near-field physical effects in air. The resulting pressures were then quantified at a distance where near-field effects are not substantial based on experimental data integrated into LS-DYNA. The nonlinearities of the shock wave propagation through the air domain was modelled using a polytrophic equation of state to initialize at atmospheric pressure, 101 kPa. A 100x100x100 mesh was developed using the Structured-ALE (S-ALE) solver. The centre of the blast was oriented to be in the same plane as the head centre of gravity (CG). After the impulse from blast wave was deemed sufficiently transferred (10ms) the ALE domain was removed.

2.2 Sensitivity Study

A simulation matrix was developed to assess the range of biomechanical responses under different conditions including presence of helmet, presence of HBE, incident pressure, and blast orientation. Table 27 below summarizes the configurations evaluated. All simulations were analyzed using LS-DYNA R11.0.0 129956 double precision massively parallel processing (MPP) solver. All simulations described in this matrix were run on a high performance computing system (HPC) running Linux CentOS v6.5.

Simulations were processed in Matlab (The Math Works, Inc.). Pressure propagation in the first 10ms was inspected to confirm sufficient time was permitted to transfer blast wave impulse to the head and that pressures had returned to ambient levels. The resulting head acceleration and rotation were extracted from the simulation and prepared using channel frequency class filters (CFC) 1000 for accelerations and CFC 180 (300 Hz) filter for velocity data, respectively, based on SAE J211, 8.4.1 and ISO 6487.

Table 27. Sensitivity study simulation matrix.

Model ID	Helmet Present	HBE Present	Incident Pressure (kPa)	Blast Direction
1	-	-	200	Frontal
2	-	-	400	Frontal
3	YES	-	200	Frontal
4	YES	-	400	Frontal
5	YES	YES	200	Frontal
6	YES	YES	400	Frontal

In order to assess the relative importance of sensitivity to these potential HBE configurations, the biomechanical data was compared to common metrics and standards in the head injury field. At the time of this work, the thresholds for blast injuries are an existing area of uncertainty and the field lacks a validated neurotrauma injury criteria specific to blast [8]. With the intent of motivating development of these blast specific criteria, this work proposed to place the effects of HBE in context where similarity in mechanisms between blast and blunt injury exists, with acknowledgement that future work will be necessary to develop a validated blast trauma criterion. Selection of injury criteria was based on evaluation of multiple kinematic measures and inclusion of a temporal component to account for the effect of duration on injury. The most commonly considered kinematic injury criteria were developed for blunt impacts. Peak acceleration is a common metric for short duration events such as helmet blunt impacts. For longer duration events such as those observed in automotive crash, a time component can be included such as in Head Injury Criterion (HIC) and Rotational Injury Criterion (RIC). In a primary blast event, high magnitude accelerations can occur over short durations without direct (blunt) impact.

Table 28. Duration of accelerative event is dependent on the loading environment.

Loading Environment	Duration of Accelerative Response
Blunt (helmet impact)	< 10 ms
Blunt (automotive crash)	15 – 30 ms
Blast	< 7 ms

Three injury metrics were selected based on their usage for blast-induced traumatic brain injury (bTBI) and their application feasibility in experimental test and evaluation of HBE using a human surrogate. The metrics are based on kinematic quantities measurable experimentally and include the HIC, the Brain Injury Criterion (BrIC), and RIC in Table 29. The 50% probability of mild traumatic brain injury (mTBI), as identified in the literature to provide context for meaningful injury scores.

Table 29. Summary of injury metrics considered in this study.

Metric	Biomechanical Data Applied	50% Probability of mTBI	Metric Calculation Formula
HIC	Linear Head Acceleration	265 [12]	$HIC = \max_{(t_1,t_2)}\left\{(t_2-t_1)\left[\frac{1}{t_2-t_1}\int_{t_1}^{t_2}\lvert a(t)\rvert dt\right]^{2.5}\right\}$
BrIC	Angular Head Velocity	0.59 [13]	$BrIC = \sqrt{\left(\frac{\omega_x}{\omega_{xcr}}\right)^2+\left(\frac{\omega_y}{\omega_{ycr}}\right)^2+\left(\frac{\omega_z}{\omega_{zcr}}\right)^2}$
RIC	Angular Head Acceleration	1.03×10^7 [14]	$RIC = \max_{(t_1,t_2)}\left\{(t_2-t_1)\left[\frac{1}{t_2-t_1}\int_{t_1}^{t_2}\lvert \alpha(t)\rvert dt\right]^{2.5}\right\}$

3. RESULTS

3.1 Model Validation

A critical aspect of any simulation study is that the applied model is grounded relative to experimental data or justifiable against accepted theory and literature where experimental data is insufficient. For this study, validation of global head kinematics in blast was targeted. The head CG response and cervical

https://doi.org/10.52202/080042-0019

spine FEM was compared to experimental head CG kinematics from the experimental shock tube tests at 400 and 550 kPa [7]. Results are shown in the Figure 4 below for Anterior/Posterior (A-P) translation and rotation in flexion/extension. CORA was used to assess the degree of validation as shown in Table 29. Summary of injury metrics considered in this study. While multiple degree of validation scales exist, for this work we consider an excellent degree of validation indicated by a CORA score of 0.94 or greater, good indicated as greater than 0.8, and fair as greater than 0.58 [18]. The cadaveric response has been shown to be more compliant than the *in vivo* response due to the compressive and stabilizing effect of the neck musculature [16]. While the present study does not model the musculature, the resulting response demonstrates stiffness with good agreement to the lower bound standard deviation of the cadaveric experiments. For upright head simulation of blast exposure, it is expected that the APL head-neck model stiffness values more similar to in vivo response.

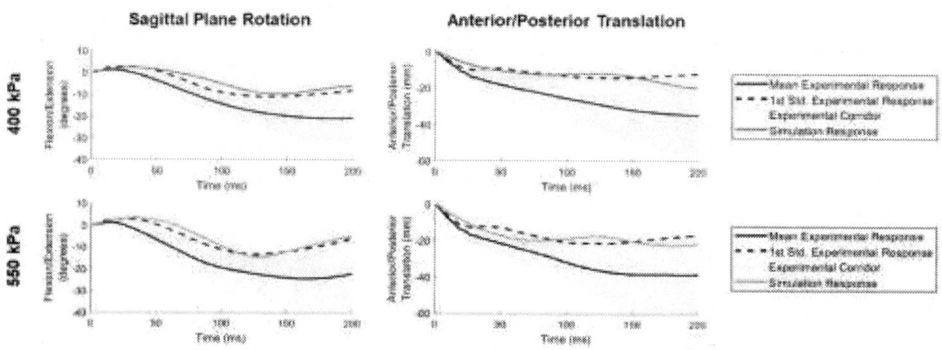

Figure 67. Model demonstrates response in line with stiff cadaveric response in rotation (left) and translation (right) to when exposed to experimental pressures at 400 kPa (top) and 500 kPa (bottom).

Table 30. Model achieves fair degree of validation relative to mean of experimental corridor and excellent when compared to 1st standard deviation of corridor

Incident Pressure	Mean Sagittal Response	1st Std. Dev. Sagittal Response	Mean A-P Response	1st Std. Dev. A-P Response
400 kPa	0.715	0.892	0.743	0.949
550 kPa	0.725	0.968	0.711	0.970

3.2 Headborne Equipment Effect on Head and Neck Biomechanics

The gross biomechanical effect of HBE on the head, when exposed to frontal blast, can be visualized in **Error! Reference source not found.** for each equipment configuration in the 200 and 400 kPa. The d istribution of the overpressure after contact with the human model can be observed in this figure, as well as the resulting head translation and rotation resulting from the blast exposure. The left column for each incident pressure depicts the FE model overlaid with a sagittal slice of the pressure contour map as the blast wave propagates at a single instance of time during the first millisecond. Right columns depict the final position of the head overlaid with the initial position at 200ms to show cumulative head displacement in each configuration. The addition of HBE increase overall head displacement in both blast exposure magnitudes. The biomechanical effect of adding HBE can be quantified by the head peak accelerations and velocities as well as their times at the peak (Table 31). In general, HBE increases head linear velocity magnitude and delay the time to its peak, but it does not increase the linear acceleration peak magnitude. The HBE effects on head rotational velocity and acceleration are more complex and quite different than the linear translational responses. HBE would increase rotational head velocity but could increase or decrease head rotational acceleration. It is worth noting that head rotational kinematic quantities peaked much later than the translational quantities which is an important characteristic of the human head supported by the anatomic multi-joint neck. The translational resultant acceleration peaks can be put in context relative to the acceleration threshold of 150g for combat helmet blunt impact test brain injury criteria as shown in Figure 6, as well as angular acceleration outcomes for flexion/extension and axial rotation. The peak accelerations observed at 200 kPa are consistently less than this test cutoff and the 400 kPa results are consistently greater. While resultant acceleration efficiently captures the peak acceleration event, it is necessary to discretize the rotational response quantities further. The primary rotational motion during a frontal blast event is flexion and extension; however, addition of an asymmetric NVG introduces extra head axial rotation which does increase with addition of HBE.

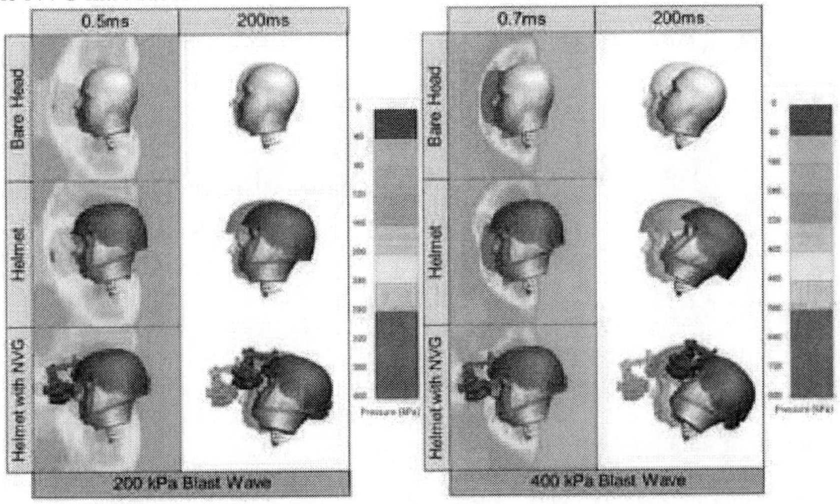

Figure 68. Snapshot of Simulation Results for 3configurations and 2 blast severities. The distribution of pressure after contact is altered by inclusion of the helmet and NVG surfaces.

Table 31. Peak and Time of Peak for Key Kinematic Quantities

Incident Pressure	Configuration	Peak Accel. (g)	Time of Peak (ms)	Peak Velocity (m/s)	Time of Peak (ms)
200 kPa	Bare Head	119.3	0.49	0.38	1.51
	Helmet	109.2	0.56	0.45	3.17
	Helmet with NVG	116.7	0.59	0.56	3.61
400 kPa	Bare Head	234.1	0.45	0.91	1.79
	Helmet	205.5	0.50	1.18	4.02
	Helmet with NVG	194.4	0.52	1.32	4.15
Incident Pressure	**Configuration**	**Peak Accel. (rad/s²)**	**Time of Peak (ms)**	**Peak Velocity (rad/s)**	**Time of Peak (ms)**
200 kPa	Bare Head	4137.2	30.24	1.98	111.03
	Helmet	6529.7	2.07	2.63	133.48
	Helmet with NVG	6177.5	2.04	4.04	16.37
400 kPa	Bare Head	8572.8	28.06	8.87	83.83
	Helmet	11045.9	1.97	9.16	198.58
	Helmet with NVG	11694.1	17.40	10.18	88.44

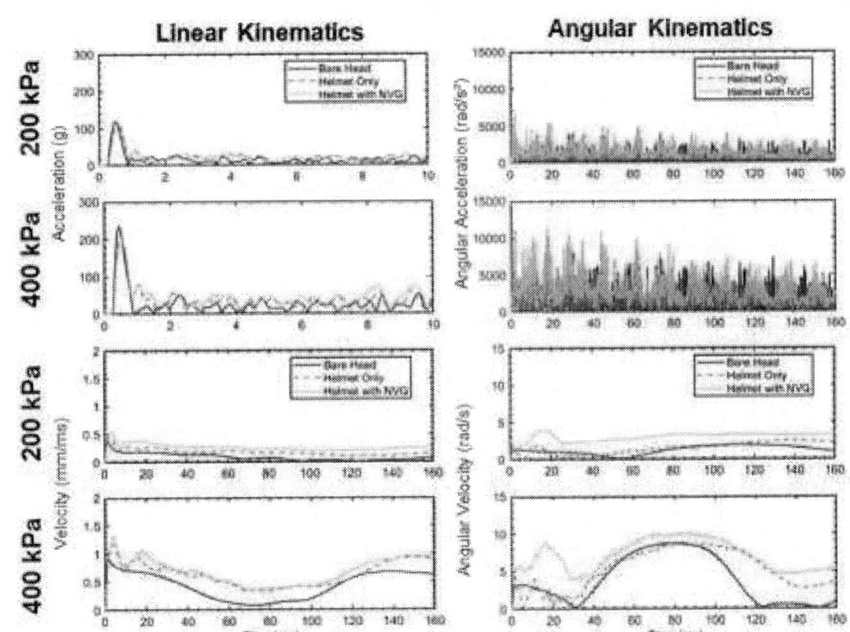

https://doi.org/10.52202/080042-0019

Figure 69. Resultant Kinematic response scales with increasing blast exposure for all kinematic measures. Linear acceleration peak occurs in first 10 milliseconds and is shown for reduced window.

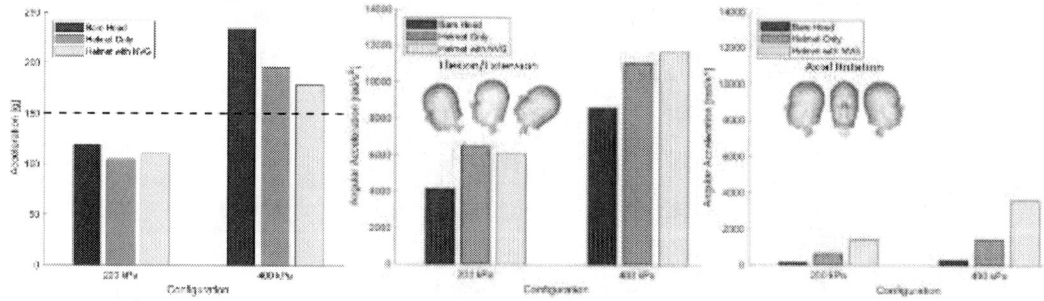

Figure 70. Linear acceleration results compared to acceleration threshold of 150g for combat helmet blunt impact threshold (left). Rotational acceleration results (middle) and axial rotation results (right).

3.3 Calculation of Common Injury Metrics & Standards

In order to relate biomechanical responses to risk of injury, human injury risk curves and metrics have been developed for numerous applications, including peak acceleration threshold metrics for short duration events such as helmet impacts in sports and criteria with temporal components for longer duration events such as automotive crash. In both cases, these metrics were validated for blunt events and have limitations when applied to blast due to differing mechanisms of trauma, though a growing number of metrics have been used to evaluate rotational motion and have been extrapolated to insults such as blast loading [22]. Comparisons with these metrics can motivate future investigations of validated blast metrics for HBE and other helmet mounted equipment if relevance can be established.

The kinematic injury metrics presented in Table 29 were calculated for all simulation conditions using the kinematic data presented in **Error! Reference source not found.**. Results are presented along with t he 50% mTBI threshold to provide context for the relative scales of each metrics. For HIC, a window of 15 ms was selected and applied to the resultant linear acceleration data. The HIC values ranged from 32 – 37 for the 200 kPa condition and 149 – 174 for the 400 kPa condition as shown in Figures 8. All values occur below the 50% mTBI threshold identified for blunt injury. The rotational kinematic metrics, BrIC and RIC are presented in Figure 71. BrIC scores range from 0.04 to 0.1 for the 200 kPa condition and 0.16 – 0.25 for the 400 kPa condition. All values occurred below the 50% mTBI threshold. RIC scores ranged from 0.17×10^7 to 0.56×10^7 for the 200 kPa condition and from 1.37×10^7 to 4.53×10^7 for the 400 kPa condition.

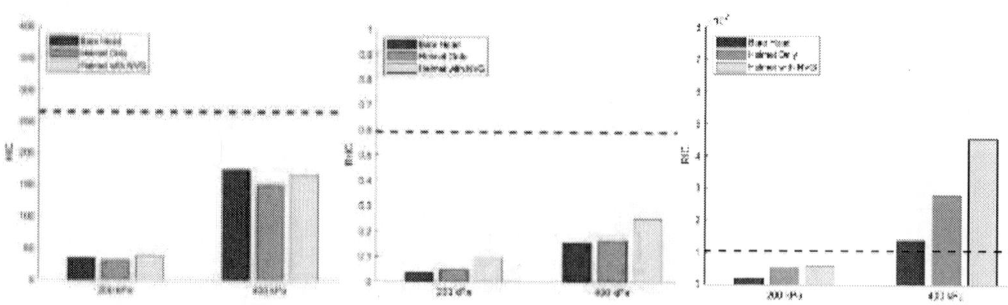

Figure 71. Model predictions of HIC score (left), BrIC score (middle), and RIC score (left). The dashed-black line is provided for context and represents the 50% probability threshold for mTBI

4. DISCUSSION

4.1 Headborne Equipment Effect on Head Kinematic Responses

This work serves as an investigation of whether blast acceleration with the addition of helmets and HBE could affect injury prediction, and therefore only one helmet and type of HBE were examined to make this assessment. The results indicate that there is a difference observed in predicted injury values when additional HBE were added, with introduction of additional axes of rotation when an asymmetric NVG was added. Future work should expand the type of HBE studied to provide a more comprehensive assessment for fielded equipment, as well as examine additional factors that might be affect the aerodynamics of the overpressure and shock wave transmission, such as shape, mass, and CG of HBE. Results demonstrate that total displacement and rotation of the head increases with both increasing blast severity and with addition of HBE. In the helmet with NVG case, we see additional axial rotation of the head due to the asymmetric design of the monocular NVG. For both translational and rotational kinematics, the peak resultant velocity and the time of peak predicted by the model increases both with exposure level and with addition of HBE. Considering all exposure levels and HBE configurations, translational velocity increases by an average of 35% with the addition of HBE and rotational velocity increases by an average of 38%. The percent increase in translational velocity is similar for both exposure levels; however, the percent change in rotational velocity is greater at the lower exposure, 68% for 200kPa, compared to the higher exposure, 9% at 400kPa. These results suggest that while the magnitude of rotational velocity increases at higher exposure levels, the effect of additional HBE is most evident at the lower exposure evaluated by this study.

The peak translational acceleration increased with more severe blast exposure and the addition of HBE decreased the magnitude of acceleration by 10%; however, addition of HBE did not clearly increase or decrease peak translational acceleration. The peak rotational acceleration increases with addition of HBE by an average of 43% and occurs at least 11ms earlier in all exposures. Rotational acceleration results can be further broken into flexion/extension rotation and axial rotation components. While addition of HBE increases flexion/extension rotation, addition of the NVG did not further notably change flexion/extension rotation; however, the NVG addition has a demonstrated effect on axial rotation, with addition of the helmet increasing average peak axial rotation by 383% and with helmet and NVG increasing by 1000%. The decrease in translational acceleration can be attributed in part to the increase in overall system inertia with addition of HBE mass, as well as translation of additional impulse from the blast into rotational acceleration due to addition of a moment from the helmet strap to the head resulting in increased flexion/extension rotation and introduction of an asymmetric cross-sectional area with addition of the NVG resulting in increased axial rotation. These results suggest that placement of additional HBE on the helmet introduces rotational kinematics with lower magnitude than the primary rotational direction. Future work should aim to validate models capable of assessing lateral blast exposure with high fidelity to further understand the effect asymmetric changes to the cross-section of HBE.

4.2 Application of Common Injury Metrics & Standards

This work aimed to study how the head and neck responds with the addition of HBE in the context of the kinematic assessments commonly applied to evaluate personal protective equipment, and to provide motivation for the future consideration of how HBE design could influence kinematic response of the head and neck during a blast event. Assessment of relative effect of HBE on head and neck kinematics was conducted with respect to existing standards and requirements, though these are validated primarily for blunt loading environments. For short duration blunt events, the 150g translational acceleration threshold is commonly applied for combat helmet blunt impact test based on risk of brain injury [15]. Prior work suggests that 275 kPa is an exposure level where instance of mTBI can be identified, and this factored into selection of one higher and one lower exposure level for this study [11]. The simulation results demonstrate translational acceleration peaks that agree with this prior study, with peak translational acceleration below 150g in the 200 kPa simulations and above 150g in the 400 kPa simulations; however, this cut off is not immediately informative of the significance of addition of HBE and this criterion cannot account for duration of the accelerative event which will likely influence injury due to inertial loading. Understanding of inertial events required evaluation of kinematic metrics and standards with a temporal component such as HIC, BrIC, and RIC.

Addition of HBE produced complex head response effects when exposed to a frontal blast wave. The overall head displacement and translational velocity would increase which may not correlate to increased brain injury risk based on the existing injury criteria. It does not increase the translational acceleration magnitude nor its integrated form for brain injury risk assessment. These results are expected from biomechanics perspective, as the added HBE increases the inertial resistance to acceleration without increasing the exposure cross-sectional area. Therefore, HBE would not expect to increase brain injury risk in frontal blast exposure if the head translational acceleration is the dominant factor. HIC measures the likelihood of head injury due to blunt impact, but has been applied more broadly to allow for comparison across insults including blast. This metric incorporates both the effect of head acceleration

 https://doi.org/10.52202/080042-0019

and the duration of this acceleration and can be limited in that severe, short duration accelerations may be assigned similar likelihoods as less-severe, long duration accelerations. Study results indicated less than 50% probability of mTBI for all exposures and HBE configurations and no substantial difference in HIC between Bare Head and with HBE. It has been observed experimentally that short duration (<7ms) peaks in excess of 1000g have the potential to occur. In this study, the filtered resultant accelerations indicate 100 – 250 g peaks with a duration of 1ms with oscillating peaks of 50g decaying across the remaining relevant window for HIC. Results show that HBE increases the head rotational velocity which could increase the risk of brain injury. BrIC considers both the duration and severity of an insult as a function of rotational kinematics, specifically rotational velocity. BrIC prediction increases slightly with addition of helmet and substantially with addition of helmet and NVG, suggesting that while BrIC as an assessment of injury may be limited, the magnitudes of change in angular velocity may be informative to the PPE design community when evaluating new HBE. Considering relative changes in BrIC, the bare head and HBE conditions where the surface geometry is systematic result in similar BrIC scores, while addition of asymmetric surfaces results in double the BrIC score. The RIC score demonstrated the most variation across HBE and mounted equipment conditions, with simulation results ranging from 0.17×10^7 to 4.53×10^7. The predicted RIC values are within the relevant range for brain injury based on the proposed criterion, with low exposures falling below the proposed threshold and high exposures above the proposed threshold. Additionally, addition of the helmet increases predicted RIC, which is further increased by addition of an NVG suggesting sensitivity to HBE design and configuration. The increase of head rotational acceleration and its associated brain injury criterion RIC due to HBE, could be an area of caution for PPE design and calls for further investigation. Continued effort is necessary to understand the mechanisms of blast injury and develop validated injury criteria. An area for further investigation could include analysis of the correlation between the head rotational acceleration or RIC with brain strain and/or strain rate which are based on brain tissue responses and potential tissue damage mechanisms.

One may notice that there appears to be a discrepancy of injury risk prediction between HIC/BrIC and RIC criteria. Both HIC and BrIC predicted low levels of injury risk while RIC predicted high levels of injury risk in our simulation results. This discrepancy is largely reflective of the injury criteria equation and the kinematic response characteristics of the head under our simulated blast exposure conditions. Both HIC is time integration of the acceleration signals (Table 4). The linear acceleration results from our study showed very short time pulse duration (typically <1 ms) and well below the common integration time period of 15 ms. Therefore, the peak acceleration magnitude is likely a more relevant injury criteria than HIC for this application. On the other hand, the high RIC results may suggest a greater role of head angular acceleration for brain injury in this application than the head angular velocity on which BrIC calculation is based.

4.3 Assumptions and Limitations

The assessment of head injury is a field of ongoing research. While numerous experimental and biomechanical studies have applied metrics such as HIC, BrIC, and RIC, as well as other tissue-based injury criteria to assess injury risk, the direct attribution of an injury mechanism has yet to be conclusively determined, in particular for more challenging assessments such as injuries occurring after the initial accelerative peak during blast loading. Additional research is needed to make this connection between blast loading and injury. In this study, injury metrics are calculated to provide context to the observed kinematic response. This assumes that the criteria are relevant for the loading based on prior use in the literature or sufficiently informative in providing context to warrant further study to confirm relevance to brain injury risk due to blast inertial loading. The hardware furnished at the time these results were generated during this study did not include fabric components or wires of the system. Higher fidelity HBE models would be necessary to enhance prediction specificity and may be a component of future work. Similarly, additional work is needed to fully validate the coupling of the NVG to the helmet and the helmet to the head. Ongoing work is assessing the stability of these systems, and this work could be incorporated into future studies to validate the coupling of this key load path. Finally, due to limited material data from NVG equipment suppliers, while helmet components have validated deformable material models, HBE components were modeled as rigid surfaces to enable pressure distribution, but cannot capture pressure transmission. This work could include validation with respect to whiplash experiments and incorporation of passive muscle response. Given the current model validation and the proposed future work, the model as applied in this study is sufficient for assessment of relative kinematic effects of HBE, but cannot provide an absolute injury risk assessment.

5. CONCLUSION AND FUTURE WORK

These findings provide insight into the relative importance of assessing how current HBE influences biomechanics during a blast overpressure event during the equipment design process. While there was an increase in blast-induced kinematic effects and a corresponding relative increase in injury metric score with HBE compared to the bare head condition, the addition of HBE showed reduced translation kinematics compared to the helmet-only condition. This demonstrates that there may be limitations in the assessment of risk of injury based on translational injury criteria such as HIC and that the addition of HBE is not a straightforward increase in injury risk. Rotational kinematics may offer more insight, specifically rotational acceleration, but assessment is limited by the need for ongoing fundamental research into the mechanisms of brain injury and further review of the proposed injury threshold. Investigation of HBE parameters such as mass properties and geometry, of helmets and other HBE may offer opportunities to tune designs to reduce head kinematics. This would aid in reducing tertiary injury, but also offer opportunity to reduce primary injury due to blast loading. These computational capabilities can be used to optimize headborne equipment design prior to production of physical hardware, as well as provide insight to end-users and decision makers regarding acquisition and usage of HBE.

Acknowledgments

The authors would like to thank PEO Soldier Product Manager Soldier Protective Equipment for sponsoring this effort. Any opinions, findings and conclusions or recommendations expressed in this material are those of the author(s) and do not necessarily reflect the views of PEO Soldier or the NAVAL SEA SYSTEMS COMMAND.

References

[1] Bradfield, C. Combat Helmet Suspension System Stiffness Influences Linear Head Acceleration and White Matter Tissue Strains: Implications for Future Helmet Design. Military Medicine; 2018.

[2] Tan, XG, Saunders, RN, & Bagchi, A. "Validation of a Full Porcine Finite Element Model for Blast Induced TBI Using a Coupled Eulerian-Lagrangian Approach." ASME IMECE 2017

[3] Merkle, A.C., J.C. Roberts, and A.C. Wickwire. Evaluation of the human surrogate test model response to ideal and complex blast loading conditions. PASS 2010

[4] Roberts, J.; Harrigan, T.; Ward, E.; Taylor, T.; Annett, M.; and Merkle,A. Sensitivity of a Human Head Finite Element Model to Anatomical Features. ASME IMECE Proceedings 2016.

[5] Rodríguez-Milllan, M.; Tan, L.B.; Tse, K. M.; Lee, H. P.; and Miguélez, M.H. Effect of Full Helmet Systems on Human Head Responses Under Blast Loading. Materials & Design; 2016.

[6] Zhang, L.; Makwana, R.; Sharma, S. Brain Response to Primary Blast Wave Using Validated Finite Element Models of Human Head and Advanced Combat Helmet. Frontiers in Neurology. 2013

[7] Iwaskiw, A; The measurement of intracranial pressure and brain displacement due to short-duration dynamic overpressure loading. Shock Waves. 2018.

[8] Young L, Rule GT, Bocchieri RT, Walilko TJ, Burns JM, Ling G, When Physics Meets Biology: Low and High-velocity Penetration, Blunt impact, and Blast Injuries to the Brain. 2015

[9] US Army TM5-855: Fundamentals of Protective Design For Conventional Weapons, 1986.

[10] Schwer, L.; Rigby, S. Reflected Secondary Shocks: Some Observations using Afterburning. In the Proceedings of the 12th European LS-DYNA Conference; 2017.

[11] Bauman R. An introductory characterization of a CCC relevant swine model of closed head injury resulting from exposure to explosive blast. J Neurotrauma. 2009

[12] King, A. I.; Yang, K. H.; Zhang, L.; Hardy, W. Is Head Injury Caused by Linear or Angular Acceleration? In the Proceedings of IRCOBI Conference; 2003

[13] Takhounts, E. G., Craig, M. J., Moorhouse, K., McFadden, J., & Hasija, V. Development of brain injury criteria (BrIC). Stapp car crash journal; 2013.

[14] Kimpara, H., Iwamoto, M. Mild Traumatic Brain Injury Predictors Based on Angular Accelerations During Impacts. Ann Biomed Eng; 2012.

[15] NRC, Review of Department of Defense Test Protocols for Combat Helmets. March 2014

[16] Panjabi MM, Miura T, Cripton PA, Wang JL, Nain AS, DuBois C. Development of a system for in vitro neck muscle force replication in whole cervical spine experiments. Spine 2001

[17] Gehre C, Gades H, Wernicke P. Objective Rating of Signals Using Test and Simulation Responses. Paper presented at: 21st ESV Conference.2009.

[18] ISO/TS 18571:2014€, "Road Vehicles – Objective Rating Metric for Non-ambiguous signals," International Organization for Standardization, Geneva, Switzerland, 2014.

[19] Miller LE, Urban JE, Stitzel JD. Validation performance comparison for finite element models of the human brain. Comput Methods Biomech Biomed Engin. 2017.

https://doi.org/10.52202/080042-0019

[20] Singh, D., Cronin, D. Multi-Scale Modeling of Head Kinematics and Brain Tissue Response to Blast Exposure. Ann Biomed Eng 2019

Effects of torso and head protection from blast overpressure on intracranial biomechanics

Pascal Magnan[1], **Aliénor Bardin**[2], Michael Ogier[2], Sébastien De Mezzo[1], Vénétia Cardona[2], Pierre Naz[1], Johanna Boutillier[1] and **Nicolas Prat**[2]

[1]*French-German Research Institute of Saint-Louis (ISL), 5 rue du Général Cassagnou, Saint-Louis, 68301, France*
pascal.magnan@isl.eu
[2]*Institut de Recherche Biomédicale des Armées, place Valérie André, Brétigny sur Orge, 91220, France*

Blast-related traumatic brain injuries (bTBI) become of major concern among active soldiers and veterans. The mechanisms underlying bTBI have been largely studied, but there are still diverging hypotheses in the literature. So far, these mechanisms have been divided into two categories: direct (effect of the shock wave on the brain) and indirect (remote effects from torso or acceleration of the head itself). The indirect components are the most controversial, even though they are supported by numerous animal studies. They include, among other phenomena, blast wave transfer to the brain through the vascular and cerebrospinal compartments. Since the mechanisms responsible for blast wave transfer from the thorax and abdomen to the brain are still unknown, it is difficult to determine if ballistic protection for the torso may protect soldiers from bTBI, or not. To begin understanding how adding torso ballistic protections would change brain overpressure kinetics during blast wave exposure, a representative large animal model was exposed to a blast overpressure load, important enough to cause a clinical response and generate a moderate to severe pulmonary and intestinal injuries. We used adult female Large-White swine equipped with sensors to collect data without affecting their physiology or increase risks of death. Animals were divided into three groups: 1) no torso or head protection against blast overpressure corresponding to the wearing of soft body armour protection alone, 2) torso protection only and 3) head protection only. Animals in the protected torso groups were fully protected by adding to the soft pack the ceramic plates at the torso and pelvic levels. The helmeted group was defined with the head enclosed in a rigid box whereas wearing the soft body armour only.
The blast threat used was characterized by a 460 ± 60 kPa side-on peak, with a 2.2 ± 0.4 ms peak duration and a 240 ± 40 kPa.ms impulse, which corresponds to a 50% risk of mortality on Bass' injury risk curve. Our results indicate that torso protection decreased the duration of blast overpressure in the oesophagus and reduced intracranial pressure, as compared to both the unprotected and soft armour protected groups. Head protection diminished the duration of the overpressure wave in the oesophagus and the maximal vascular and intracranial pressures and impulses, and increased the duration of vascular and intracranial overpressures. Taken together, these data give the first tendencies observed on large mammals resulting from direct blast exposure in free field which will be helpful for understanding of the mechanisms involved in the bTBI. Several avenues have been identified for future studies such as assessing the main characteristics of the threat involved, in order to update existing criteria or propose new ones and, ultimately, establish a strategy for developing new military armour that would protect soldiers against bTBI.

1. INTRODUCTION

Blast-related traumatic brain injury(bTBI) results from blast exposure during combat or training and are of major concern among the military. The estimated incidence of TBI in soldiers and veterans varies among studies [1-2]. BTBI incidence is likely to be underestimated, due to difficulties in reporting and documenting cases in combat theatre settings [3]. The main cause of military bTBI is the exposure to the blast event and its shock wave [1-2;7]. Sudden and transient increase in intracranial pressure during the travel of the shockwave and brain motion relative to the skull are expected to be the main causes of brain damage [8-9]. Few studies suggest that body protections can exacerbate the effects of the blast wave on the body, and particularly at the head level. Indeed, helmets [4] or ballistic vests [5, 6] can reinforce energy transfer of the blast wave. Therefore, military protections might need to be updated and optimized to be adapt to the blast threat.

Although there is no consensus about the mechanisms underlying bTBI, they have been divided into two main categories: direct and indirect [2-3;8;10-12]. Direct mechanisms include propagation of the blast wave through the cranium and orbital and/or oral openings [2;8;10;13]. The skull provides little protection against blast wave and skull flexure seems to be involved in its transmission to the cranial content [4;13-14]. Once the blast wave has passed the cranium, it may cause high-frequency contraction and relaxation of brain parenchyma and blood vessels, particularly resulting in damage of blood vessel walls and haemodynamic abnormalities [2]. *In silico* experiments suggest that the shockwave can also be transmitted to the brain from flexure of the skull, which creates localized regions of low pressure

https://doi.org/10.52202/080042-0020

waves [4]. Progression of the shockwave through orbital and/or oral openings has also been evidenced, and is associated with direct damage to ocular neurons. However, the involvement of this mechanism in TBI is controversial [7][8]

Indirect mechanisms are more controversial than direct mechanisms, but they have been supported by studies done in animal models [7;11-12]. Blast overpressure has also been reported to cause macroscopic translational and rotational acceleration of the brain, resulting in TBI similar to those caused by head impact [7]. Cernak et al. [11] proposed a mechanism in which the kinetic energy of the overpressure wave is transferred to the central nervous system and impacts brain tissue. Other proposed another mechanism in which the kinetic energy is transferred through the vascular system and cerebrospinal fluid [7;13;15]. The blast wave on the thorax would suddenly increase the pressure on the walls of large blood vessels, which would accelerate blood. With no valves to regulate blood ascension to brain, it would easily result in increased intracranial blood pressure leading to rupture of the capillaries [7;15].

Although they are controversial, indirect mechanisms are to be considered when developing military protections. Indeed, incidence of TBI has increased with the improvement of body armors and higher efficiency of thoracic protection [1]. It was proposed that soldiers wearing protections may get closer to the center of the blast and thus, may be exposed to higher blast levels. Alternatively, Kevlar vests may facilitate brain damage by increasing intrathoracic pressure [6;13]. The aim of our study was to test this latter hypothesis. For this purpose, a representative biological model (bodyweight, layout of organs, thickness of the chest wall which are as close as possible to the human body) was exposed to an explosive load. Adult female swine were used, with subgroups wearing different body armors conferring different levels of protection. Few swine were equipped with both the lower-level body protection and a helmet. Swine were equipped with sensors to collect holistic data relevant to blast overpressure kinetics, and wave transmission to the body, especially to the brain, without affecting its physiology or increasing risk of death.

2. METHODS

2.1 Threat characteristics and setup

The threat corresponded to a blast overpressure exposure under free-field conditions. A sphere of 4kg of Hexomax B2269B (Eurenco, France), moulded into a polystyrene shell, was placed on a cardboard cylinder at a corresponding height of burst of 39 cm above a concrete slab. Indeed, due to the distance from the charge fixed to 3 m range, the targets experienced the Mach stem regime of the blast, i.e. an ideal Friedlander wave, to simplify the threat profile and consequently our understanding (Figure 1A). The level was strong enough to cause sufficient pulmonary injuries, consistent with clinical and morphological changes. Sizing was chosen referring to the Bowen tolerance curves [16] revisited by Bass et al. [17] to achieve the 50% survival curve. Hence, a 400-kPa peak overpressure and 2.5-ms positive-phase duration were targeted. For comparison purposes, the targets were subjected to blast in pairs (Figure 1B). Each of them was accompanied at equal distance and height of a piezoelectric pencil probe 137A22 (PCB Piezotronics, United States) for recording the incident pressure-time history. The aim was to be able to follow both kinematics of the ribs and pressures which required invasive approach and short monitoring period, with, in the same time, an animal lightly instrumented monitored longer for others clinical and biochemical parameters trends.

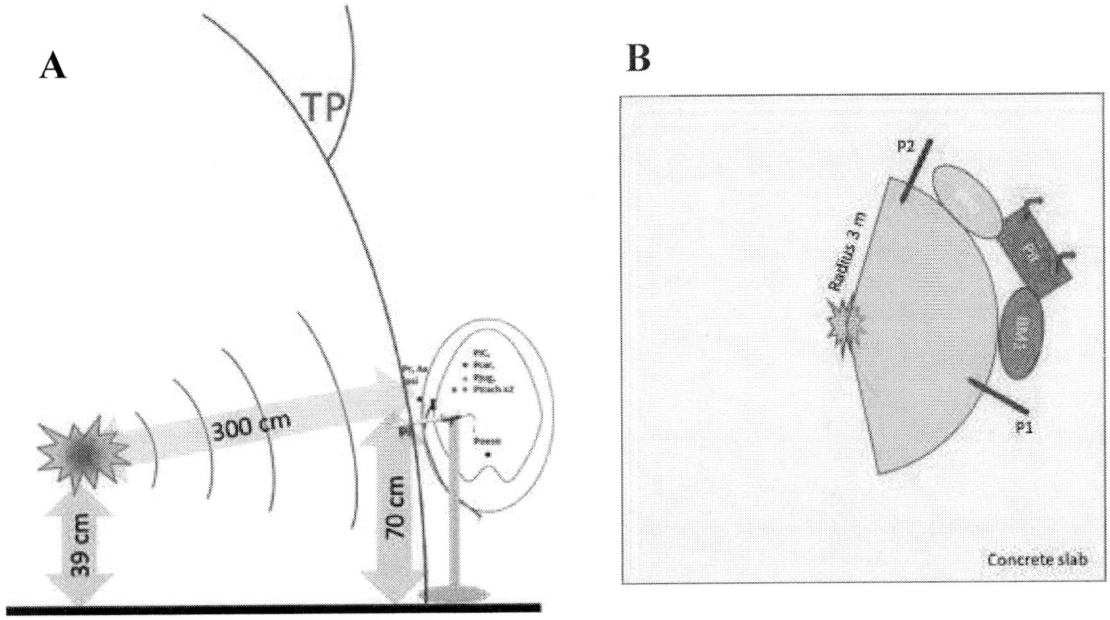

Figure 1. A): charge and targets distances (instrumented model and pencil probe for the reference incident pressure); the red line describes the Mach stem and its triple point passing around 2 m high. B): Overview of the scene near the pit: BM1 and BM2 exposed simultaneously with their reference incident pressure probes P1 and P2 at 3 m range from the charge

2.2 Model and instrumentation

Twenty-eight female Large-White swine (51 ± 4 kg weight; 118±6 cm length) were deeply anesthetized and prepared in accordance with the European directive 2010/63/EU on the protection of animals used for scientific purposes. They were adequately instrumented for cardio-respiratory monitoring with multi-parameter monitors (Propaq CS Monitor, Welch Allyn, United States) and Biopac modules (Biopac Systems, United States). Data were recorded on site from beginning at 15 minutes before blast initiation to 60 minutes after blast. Two distinct instrumentation protocols were used:

- strongly instrumented animals, for which a basic set of clinical parameters allowed to monitor cardiorespiratory functions and arterial blood pressure was used and reinforced by a set of more invasive instrumentation allowed to manage with physical data such as transient pressure surges through: large vessels (jugular vein or carotid artery), brain parenchyma (ICP), and with kinematics of the chest wall (acceleration, velocity, displacement of the rib). The monitoring time was limited to one hour on site after explosion.
- lightly instrumented animal model, for which the basic set of clinical parameters only was essential, allowed us to extent clinical monitoring period up to 6 hours after explosion in an operating room.

At the end of the sequences, animals were sacrificed by exsanguination under anesthetic overdose and autopsies were carried out for scoring cerebral, torso and abdominal injuries.

2.3 Protections and configurations tested

Table 1 below shows the distribution of the models according to their protection and the experiment duration. The level of instrumentation "S" for Strong and very complete instrumentation and "L" for the light and essential one, is also indicated and applies to every group.
In order to amplify the exposition of the torso to the blast overpressure effects, animals in the "unprotected" torso group were equipped with a soft body armour that has been reported to increase injury risks during exposure to blast overpressure compared with naked individual [6]. Targets were wrapped in a specifically designed thoracic protection either limited to the soft pack of the body armour

 https://doi.org/10.52202/080042-0020

defined as P2 and assimilated to the Th- entity, or including thoracic and pelvic ceramic plates in their respective pockets and defined as P3E-CERA and Th+ entity.
The helmeted condition, corresponding to the head encapsulated into a rigid aluminium box to prevent against direct blast loading, is defined as the H+ entity. The unprotected head is defined as H-.
Finally, three scenarios or groups are investigated: unprotected (Th-/H-), helmeted only (Th-/H+), or torso protected only (Th+/H-).

Table 33. Groups and their characteristics

Group	**Sub-Group**	Thoracic protection	Head protection	Instrumentation: S=strong; L=light	Number of cases	
					n	total
Th-/H-	**Th-/H- L**	P2	No	L	4	10
					2	
	Th-/H- S	P2	No	S	4	
Th-/H+	**Th-/H+ L**	P2	Yes	L	7	10
	Th-/H+ S	P2	Yes	S	3	
Th+/H-	**Th+/H- L**	P3E-CERA	No	L	3	8
					2	
	Th+/H- S	P3E-CERA	No	S	3	

2.4 Physical recordings

In addition with usual clinical instrumentation for hemodynamic measurement, cardio-respiratory and brain functions, other sensors were placed on the biological models to observe how the equipment affects the pressure transfer inside the body. The list of the sensors and their respective uses and filtering are presented in Table 34. Their locations are depicted in Figure 2.

Table 34. List of sensors, data collected and filter used

Sensors	**Parameters**	**IIR Filtering, type Bessel**		
		Frequency (kHz)	**Advanced parameters**	
Uniaxial accelerometer (PCB, 3501A)	Rib acceleration (screwed on #K8 ipsilateral right side)	1.65	4 poles	CFC1000
Hydrophone (Reson, TC4013)	Resultant pressure on the thorax	80	6 poles	Phase0/Begin/end
Pressure sensor (Kulite, XT190)	Reflected pressure on the jaw (ipsilateral right side)	80	6 poles	Phase0/Begin/end
Hydrophone (Reson, TC4013)	Intra-oesophageal pressure	10	6 poles	Phase0/Begin/end
Pressure sensor (Millar, MPR-500 Mikro-Tip®)	Intracranial pressure in the parenchyma	5	6 poles	Phase0/Begin/end
Pressure sensor (Millar, SPR-407)	Intravascular pressure from carotid and/or jugular	3	6 poles	Phase0/Begin/end
Pressure sensor (Millar, SPR-751 Mikro-Tip®)	Proximal and distal tracheal pressures	0.6	6 poles	Phase0/Begin/end

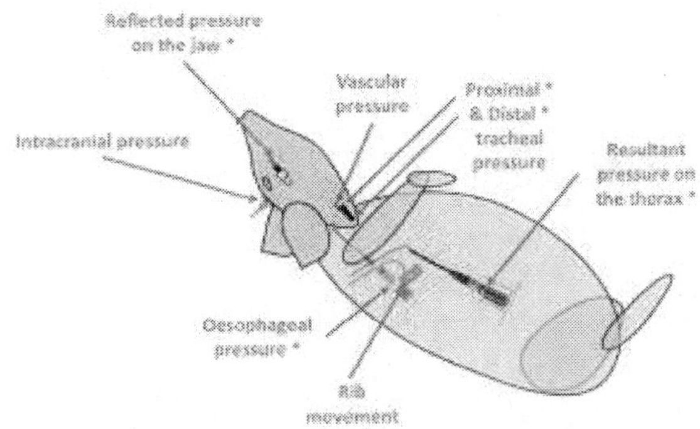

Figure 2. Sensor locations on the strongly instrumented animal model
(*: instrumentation for the lightly instrumented animal model).

Signals were sampled at 1MHz using a transient recorder TransCom (MF Instruments GmbH, Germany) from 1 second before explosion to 3 seconds after for all transient events, whereas a continuous basic sampling at 1 kHz was used from -15min to +60 min for other events. Raw data were then post-processed through their respective digital filters to suppress noise but keeping the characteristics of shockwave. The filter characteristics used in DIAdem (National Instruments) are reported in the table 2. The incident pressure was filtered in the same manner as for the thoracic resultant pressure and reflected pressure on the jaw.

The level of threat has been defined as the maximal peak pressure, the duration of overpressure or the impulse. Because of the lack of consensus, the three of them were considered in this study.

2.5 Statistical analyses

Statistical analyses were performed using JMP® (15.2.0). As previouslt described [18], the distribution normality was graphically checked and then the equality of variances were checked using Levene's test. Once the hypothesis of normal distribution and homoscedasticity were validated, statistical analyses was performed using the Tukey-Kramer's HSD (Honestly Significant Difference) test. If the homoscedasticity hypothesis was not validated, Welsh's test was used. Only the duration of threat overpressure, the threat impulse, the maximal distal tracheal pressure and acceleration of the ribs were compared using Welsh's test. The other parameters were compared among groups using the Tukey-Kramer's HSD test.

3. RESULTS

All graphs show results expressed in bars and not in kPa, keeping in mind that 1 bar = 100 kPa.

3.1 Blast parameters

Figure 3 illustrates the pressure-time profile and corresponding impulses during the last experimental trials. The incident pressure profile reached an average maximal value of 460 ± 60 kPa, with an overpressure duration of 2.2 ± 0.4 ms and an impulse of 240 ± 40 kPa.ms. No significant difference was observed in maximal pressure and overpressure duration between groups. Impulse was significantly higher for the Th-/H- group compared to the Th-/H+ group (+40 kPa.ms, $p= 0.0281$). No other significant impulse difference was observed between groups.

 https://doi.org/10.52202/080042-0020

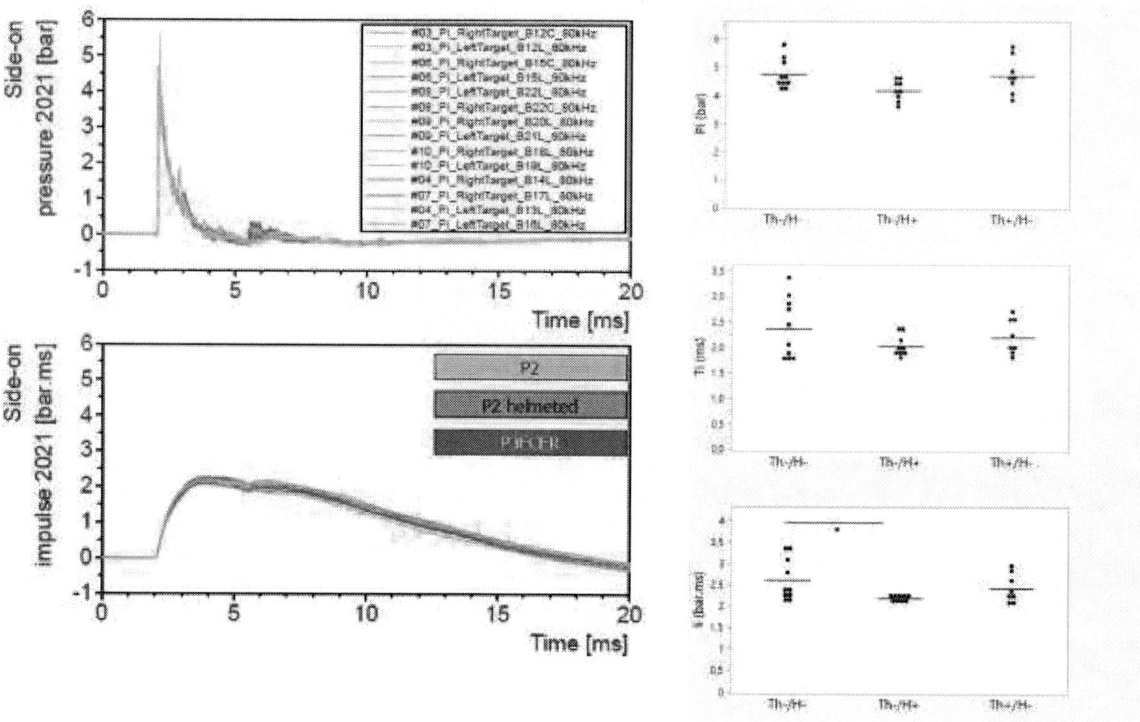

Figure 3. Blast kinetics for the 2021 test series (left); Pi: peak pressure, Ti: positive phase duration, Ii: impulse, *: $p<0.05$, describing the threat experienced by subgroups (right)

3.2 Screening on and through the body

- At the skin surface (exposed right side thorax/abdomen), the resultant pressure signals behind the armour reached a maximal value of 780 ± 360 kPa, with an overpressure duration of 1.7 ± 0.7 ms and an impulse value of 270 ± 120 kPa.ms. No significant difference was observed in maximal pressure, overpressure duration and impulse between groups, as described in Figure 4.

https://doi.org/10.52202/080042-0020

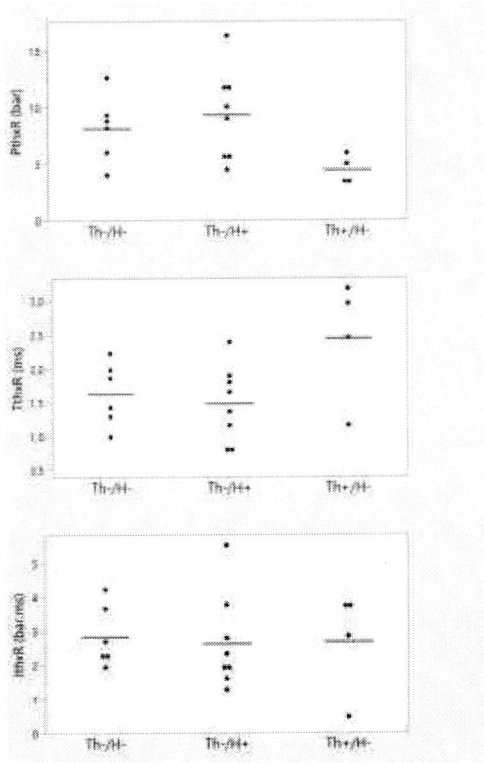

Figure 4. Resultant surge of pressure at the right thorax/abdominal skin surface: PthxR: peak pressure, TthxR: positive phase duration, IthxR corresponding impulse.

- At the jaw level of the exposed (right) side (Figure 5), the resultant pressure signals were significantly higher for Th-/H- and Th+/H- groups compared to the Th-/H+ group (+600 kPa, $p<0.001$ and +546 kPa, $p=0.0002$, respectively). No significant difference was observed between the other groups. The overpressure duration was significantly lower for theTh-/H- and Th+/H- groups compared to the Th-/H+ group (-6.72 ms, $p=0.0010$ and -6.52 ms, $p=0.0012$, respectively). No significant overpressure duration difference was observed between the other groups. Finally, the impulse value was significantly higher for the Th-/H- and Th+/H- groups compared to the Th-/H+ group (+213 kPa.ms, $p=0.00142$ and +266 kPa.ms, $p=0.0041$, respectively). No significant impulse difference was observed between the other groups ($p>0.15$).

https://doi.org/10.52202/080042-0020

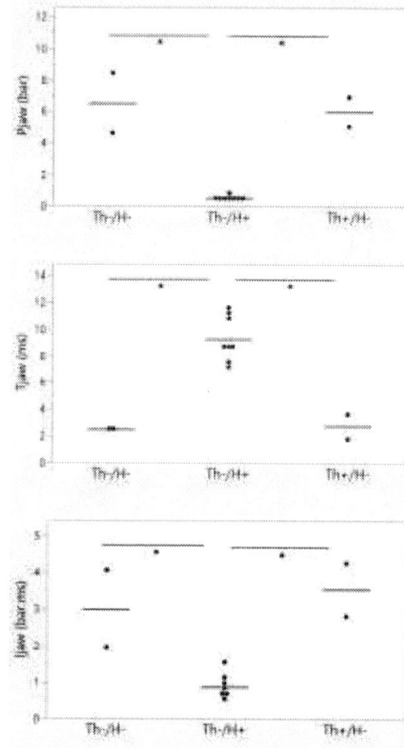

Figure 5. Resultant surge of pressure at the jaw surface: Pjaw: peak pressure, Tjaw: positive phase duration, Ijaw and corresponding impulse, *: $p<0.05$

- In terms of kinematics of the chest wall, the axial acceleration of the rib #K8 reached a maximal value of 25 ± 8 km.s^{-2}, with a maximal velocity of 7.3 ± 2.0 m.s^{-1} and a maximal displacement of 6.2 ± 4.3 mm. Values obtained from the double integration of the acceleration are questionable due to the inherent shift during the processing. No significant differences were observed in any of these parameters between groups.

- In the intrathoracic area, oesophageal pressure reached a maximal value of 240 ± 140 kPa, with an overpressure duration of 6.4 ± 1.0 ms and an impulse value of 300 ± 90 kPa.ms. Figure 6 depicts the surge of pressure propagating at the center of the thorax (oesophagus) while the chest wall is suddenly compressed. No significant difference was observed in maximal pressure between groups. The overpressure duration was significantly longer for the Th-/H- group compared to the Th-/H+ (+1.4 ms, p= 0.0021) and Th+/H- (+1.1 ms, p=0.0334) groups. No significant difference was observed between the Th+/H- and Th-/H+ groups (p=0.7070). The impulse value was significantly higher for the Th-/H+ group compared to the Th+/H- group (+130 kPa.ms, p=0.0137). No significant difference was observed between the other groups (p>0.1).

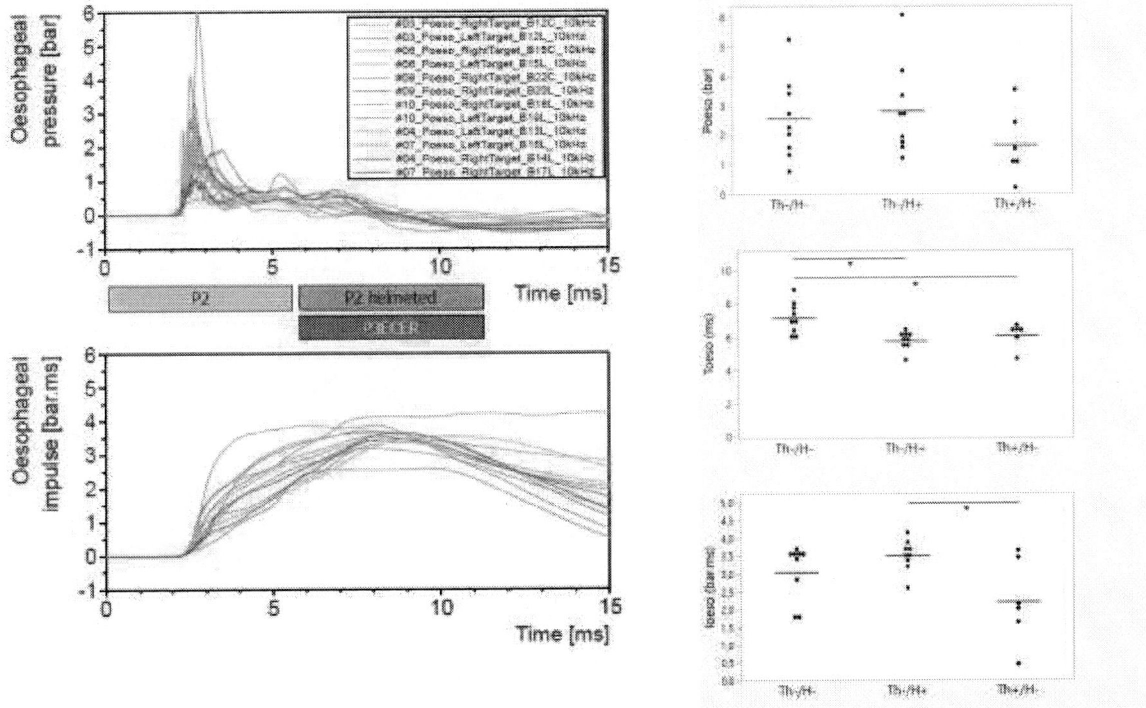

Figure 6. Pressure-time histories of the intrathoracic pressure for the 2021 series (left); Poeso: peak pressure, Toeso: positive phase duration, Ioeso: impulse, *: $p<0.05$, describing the resultant surge of pressure by subgroups (right)

- In the large vessels coming from or going to the head, the vascular pressure was measured at the jugular and/or carotid level (Figure 7). Jugular pressure was measured only for the Th-/H- and Th-/H+ groups. The maximal jugular pressure was significantly higher for the Th-/H- compared to the Th-/H+ group (90 ± 0.4 kPa vs +66 kPa, p= 0.0315). The overpressure duration was more important for Th-/H+ group compared to the Th-/H- group (4.8 ± 1.2 ms vs +2.2 ms, p=0.0004). The impulse value was higher in the Th-/H- group compared to the Th-/H+ group (1.0 ± 0.4 bar.ms vs+70 kPa.ms, p=0.0153). Carotid pressure was measured only for the Th-/H- and Th+/H- groups. Carotid pressure reached a maximal value of 160 ± 40 kPa, with an overpressure duration of 2.8 ± 0.6 ms and an impulse value of 127 ± 3 kPa.ms. No significant difference was observed between both groups.

https://doi.org/10.52202/080042-0020

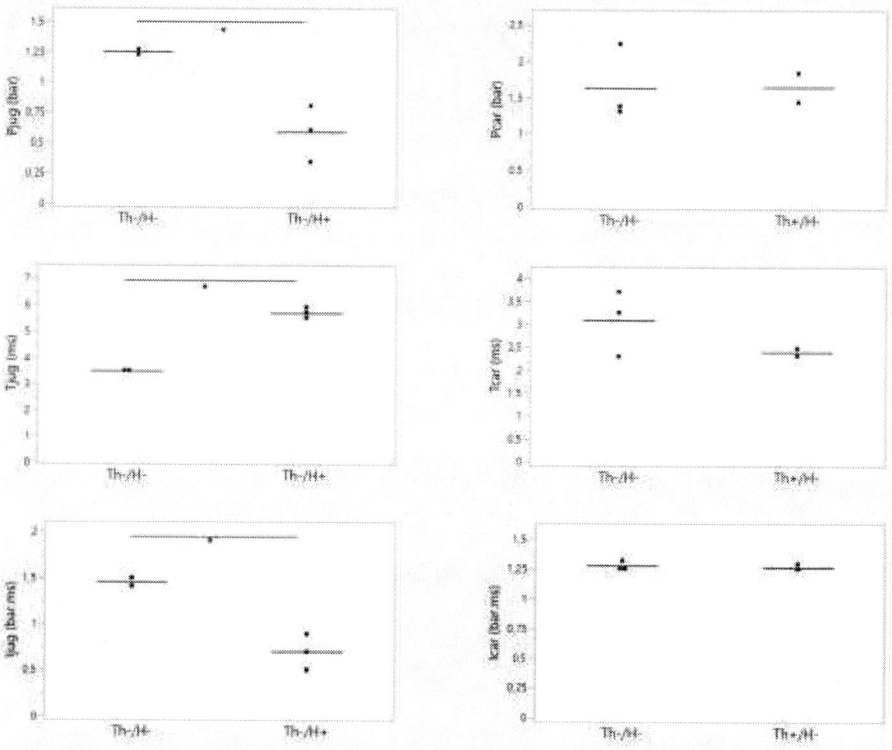

Figure 7. Resultant surge of pressure into the large vessels: jugular (left); carotid (right) in terms of peak pressure (top), positive phase duration (center) and corresponding impulse (bottom).

- Inside the brain parenchyma (Figure 8), the maximal intracranial pressure ICP reached significantly higher levels for the Th-/H- group compared to the Th-/H+ group (+186 kPa, p<0.0001) and for Th+/H- (+29 kPa, p=0.0206). ICP values were significantly higher for the Th+/H- group than for the Th-/H+ group (+157 kPa, p<0.0001). The overpressure duration was significantly longer for the Th-/H+ group than for the Th-/H- group (3.8 ± 2.5 ms +4.4 ms, p=0.0224) and for Th+/H- (+4.3 ms, p=0.0231). No significant difference was observed in duration of overpressure between Th+/H- and Th-/H- (p=0.9992). The impulse value was lower in the Th-/H+ group than in the Th-/H- group (-120 ± 40 kPa.ms vs -80 kPa.ms, p=0.0140) and than in the Th-/H- group (-60 kPa.ms, p=0.0348). No significant difference was observed in terms of impulse between the Th-/H- and Th+/H- groups (p=0.5524).

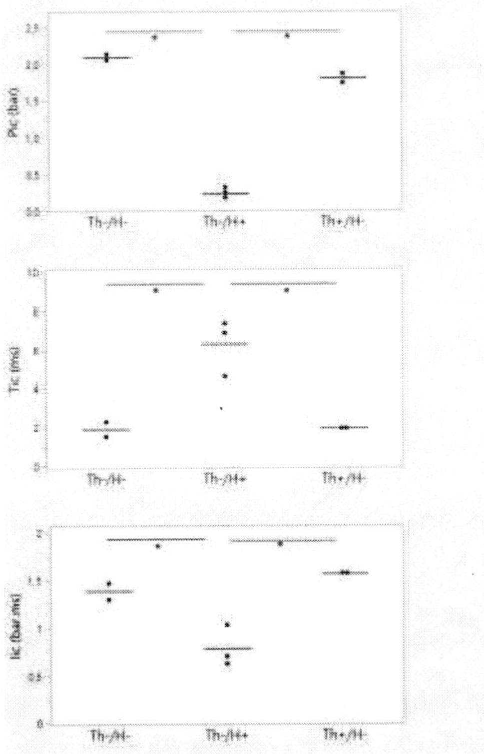

Figure 8. Resultant surge of pressure into the brain: Pic: peak pressure, Tic: positive phase duration, Iic: impulse, *: $p < 0.005$.

- In the trachea, the maximal proximal and distal pressures were not significantly different between groups ($p>0.7$). The delays in detection of the shockwave in the trachea were compared using a simple correlation analysis. The results showed they were related with a factor 1 ($p<0.0001$), but the constant was not significantly different from 0. Thus, it is not possible to determine which of the pressure sensors was hit first to deduce the wave direction. The difference of maximal pressures between both positions in the trachea was computed. Its average was 0 and no significant difference was observed between groups ($p>0.4$). These observations led to suppose there is no propagation of the shockwave in the caudal/cranial direction.

4. DISCUSSION

4.1 Blast threat characteristics

Statistical analysis on the data of the incident pressure showed the only significant difference was a higher impulse for the group Th-/H- than for the group Th-/H+ (+40 kPa, p=0.0463). As no other significant difference was observed, the threat was considered similar for all groups.

4.2 Effect of the different protections

None of the protection had a significant effect on the resultant pressure on the thorax and on the axial acceleration of the rib. Adding ceramic plates and head protection had no significant effect on the maximal pressure measured in the oesophagus whereas it significantly shortened the overpressure duration. Both effects seemed to be unrelated as no significant difference was observed between the Th-/H+ and Th+/H- groups. Removing ceramic plates and adding head protection (Th+/H- vs Th-/H+) significantly increased the oesophageal pressure impulse. However, the differences were not significant when removing ceramic plates (Th-/H-vs Th+/H-) or adding head protection (Th-/H+ vs Th-/H-) seemed to increase the impulse value.

https://doi.org/10.52202/080042-0020

So, there is a combined effect on the torso loading by the blast wave of adding the head protection and removing the ceramic plate.

Significant differences were observed in the pressure profile measured in the oesophagus whereas none were observed in the resultant pressure on the thorax and in the movement of the ribs. That could imply that complex mechanisms of reflection and propagation of shockwave are involved in the resultant pressure measured in the organs. Moreover, the protection equipment may have an effect on those mechanisms and not directly on the threat the body is exposed to.

Adding ceramic plates (Th+/H- vs Th-/H-) did not have any impact on the vascular pressure profile and on the duration and impulse of intracranial overpressure, but it significantly decreased the maximal intracranial pressure. Protecting the head (Th-/H+ vs Th-/H-) significantly decreased the maximal vascular and intracranial pressures and corresponding impulses while both vasculat and intracranial overpressure durations were significantly increased. The effect of protecting the head was stronger than the effect of adding ceramic plates as the differences between Th+/H- and Th-/H+ were similar to the differences between Th-/H- and Th-/H+ and opposite to the differences between Th+/H- and Th-/H-.

Our results showed that he maximal pressure peak was reached at the same time in the distal and proximal parts of the trachea, and had the same amplitude for both. This suggests that there is no propagation of the shockwave in the caudal/cranial direction.

In summary (Table 3), adding ceramic plates (Th+/H- vs Th-/H-) was associated with a shorter duration of overpressure in the oesophagus and a smaller maximal intracranial pressure. Protecting the head (Th-/H+ vs Th-/H-) was associated with a shorter duration of overpressure in the oesophagus, smaller maximal vascular and intracranial pressures and impulses and longer duration of vascular and intracranial overpressures.

Taken together, our results demonstrate that the addition of a protection to the thorax impacts the profile of the post-blast rise in intracranial and oesophageal pressure. This supports the results of previous studies showing brain consequences to chest exposure to blast [7;10-11;15].

Table 35. Summarised significant changes observed at the different sites when torso or head were protected

<table>
<tr><th rowspan="2">Group</th><th rowspan="2">Thoracic protection</th><th rowspan="2">Head protection</th><th colspan="7">Significant influence (Overpressure (P) / Duration (T) / Impulse (I))</th></tr>
<tr><th>Reflected pressure</th><th>Kinematics of the rib</th><th colspan="2">Oesophageal pressure</th><th>Vascular pressure</th><th>Intracranial pressure</th><th>Tracheal pressure</th></tr>
<tr><td>Th-/H-</td><td>No</td><td>No</td><td>Ref</td><td>Ref</td><td colspan="2">Ref</td><td>Ref</td><td>Ref</td><td>Ref</td></tr>
<tr><td>Th-/H+ vs Th-/H-</td><td>No</td><td>Yes</td><td colspan="2" rowspan="2">- / -</td><td>- / T↘ / I ↗</td><td rowspan="2">- / Combined or separate effects</td><td>P ↘ / T↗ / I↘</td><td>P ↘ / T↗ / I↘</td><td>-</td></tr>
<tr><td>Th+/H- vs Th-/H-</td><td>Yes</td><td>No</td><td>- / T↘ / I ↗</td><td>- / - / -</td><td>P ↘ / - / -</td><td>-</td></tr>
</table>

4.3 Propagation of the shockwave in organs

The results on the times of arrival and maximal pressures in the trachea suggests that the direction of the shockwave was mainly lateral to medial and not caudal to cranial. However, the consequences of adding thoracic or cranial protection led to think that the shockwave is partially transferred in the caudal/cranial direction. For both to be true, the only solution is for the shockwave to be transferred through fluids, such in blood vessels. Because fluids are less compressible than gas, it can be hypothesized that the shockwave transferred in caudal/cranial direction, based on the observations on the consequences of adding thoracic or cranial protection, is transferred through the vascular system. This is in agreement with mechanisms that have been described in the literature [3;7;13;15].

For two cases, both jugular and carotid pressures were measured, and in both cases, the carotid pressure was about 0.1 bar higher than the jugular pressure. This could be due to the cardiac valves allowing the blood to travel in one direction only, to the tissue nature (venous or arterial) and their mechanical characteristics. Further studies would be required to investigate the effect of blast on the cardiac system.

5. CONCLUSION

As evidenced in the literature [3;7;13;15] and in this study, when developing new military protection, it will be important to take into consideration the repercussions on local and distal body parts.
The first tendencies observed here on large mammals resulting from direct blast exposure in free field which will be helpful for understanding of the mechanisms involved in the bTBI. Indeed, a screen interfaced between threat and body can behave as a protector or a facilitator in relation to the injury. Our study showed a protective effect when hard plates were added in terms of reducing the duration of blast overpressure in the oesophagus and reducing maximum intracranial pressure. In helmeted animals, the chamber acts as a protector by reducing the duration of overpressure in the oesophagus, by reducing maximum vascular and intracranial pressures and impulses and by prolonging the duration of vascular and intracranial overpressures. In addition, it is not easy to protect the head effectively without redirecting blast waves to other parts of the body. Further studies will be needed to clarify this point and determine the strategy to be followed in developing protection, as a number of questions remain unanswered. For example, we need to determine which maximum pressure peaks, the duration of the overpressure and the pulses that present the greatest risk to organs.

Acknowledgments

This research was supported by a grant from the DGA (Direction Générale de l'Armement; upstream studies program PROTECOMB). The authors would like to thank the CETID (Centre d'Expertise des Techniques de l'Infrastructure de la Défense) for providing technical and pyrotechnical support and the proving ground.

References

[1.] Belmont, P. J., Goodman, G. P., Zacchilli, M., Posner, M., Evans, C., Owens, B. D. (2010). Incidence and epidemiology of combat injuries sustained during "the surge" portion of operation iraqi freedom by a U.S. army brigade combat team. *Journal of Trauma: Injury, Infection and Critical Care, 68*(1), 204–210.

[2] Kong, L.-Z., Zhang, R.-L., Hu, S.-H., & Lai, J.-B. (2022). Military traumatic brain injury: A challenge straddling neurology and psychiatry. *Military Medical Research, 9*(1), 2. doi:10.1186/s40779-021-00363-y

[3] Chapman, J. C., & Diaz-Arrastia, R. (2014). Military traumatic brain injury: A review. *Alzheimer's & Dementia, 10*, 97-104. doi:10.1016/j.jalz.2014.04.012

[4] Moss, W.C., King, M. J., & Blackman, E. G. (2009). Skull flexure from blast waves: a mechanism for brain injury with implications for helmet design. *Phys Rev Lett. ,4*;103(10):108702. doi: 10.1103/PhysRevLett.103.108702

[5] Philips, Y. Y., Mundie, T. G., Yelverton, J. T., & Richmond, D. R. (1988). Cloth ballistic vest alters response to blast. *J Trauma, 28(1 Suppl)*:S149-52. doi: 10.1097/00005373-198801001-00030

[6] Boutillier, J., Cardona, V., Magnan, P., Ogier, M., De Mezzo, S., Montespan, F., Menini, W., Mosnier, J., Naz, P., & Prat, N. J. (2022). A new anthropomorphic mannequin for efficacy evaluation of thoracic protective equipment against blast threats. *Front. Bioeng. Biotechnol., 9*:786881.

[7] Bryden, D. W., Tilghman, J. I., & Hinds II, S. R. (2019). Blast-related traumatic brain injury: Current concepts and research considerations. *Journal of Experimental Neuroscience, 13*, 1-13. doi:10.1177/1179069519872213

[8] Leonardi, A. D. C., Bir, C. A., Ritzel, D. V., & Vandevord, P. J. (2011) *J Neurotrauma 28*:85-94. doi: 10.1089/neu.2010.1324

[9] Huang, X., Xia, B., Chang, L., Liao, Z., Zhao, H., Zhang, L., & Cai, Z. (2023) *J Neurotrauma 40(7-8)*:788-790. doi: 10.1089/neu.2022.0229.correx

[10] Courtney, A., & Courtney, M. (2009). A Thoracic mechanism of mild traumatic brain injury due to blast pressure waves. *Medical Hypotheses, 72*(1), 76-83. doi:10.1016/j.mehy.2008.08.015

[11] Cernak, I., Wang, Z., Jiang, J., Bian, X., & Savic, J. (2001). Ultrastructural and functional characteristics of blast injury-induced neurotrauma. *The Journal of Trauma: Injury, Infection, and Critical Care, 50*(4), 695-706.

[12] Rubio, J. E., Skotak, M., Alay, E., Sundaramurthy, A., Subramaniam, D.R., Kote, V. B., Yeoh, S., & al. (2020). Does Blast Exposure to the Torso Cause a Blood Surge to the Brain?, *Frontiers in Bioengineering and Biotechnology, 17*;8:573647. doi: 10.3389/fbioe.2020.573647

https://doi.org/10.52202/080042-0020

[13] Chavko, M., Watanabe, T., Adeeb, S., Lankasky, J., Ahlers, S. T., & McCarron, R. M. (2011). Relationship between orientation to a blast and pressure wave propagation inside the rat brain. *Journal of Neuroscience Methods, 195*(1), 61-66. doi:10.1016/j.jneumeth.2010.11.019
[14] Bolander, R., Mathie, B., Bir, C., Ritzel, D., & Vandevord, P. (2011) Skull flexure as a contributing factor in the mechanism of injury in the rat when exposed to a shock wave. *Annals of Biomedical Engineering, 39(10)*: 2550–2559.
[15] Simard, J. M., Pampori, A., Keledjian, K., Tosun, C., Schwartzbauer, G., Ivanova, S., & Gerzanich, V. (2014). Exposure of the thorax to a sublethal blast wave causes a hydrodynamic pulse that leads to perivenular inflammation in the brain. *Journal of Neurotrauma, 31*(14), 1292-1304. doi:10.1089/neu.2013.3016
[16] Bowen, I. G., Fletcher, E. R., Richmond, D. R. (1968). Estimate of man's tolerance to the direct effects of Air Blast. *Defense Atomic Support Agency*, report DA-49-146-XY-372. Washington, DC: Defense Atomic Support Agency.
[17] Bass, C. R., Rafaels, K. A., & Salzar, R. S. (2008). Pulmonary injury risk assessment for short duration blasts. *J Trauma*, 65:604Y615.
[18] Morgan, C. J. (2017). Use of proper statistical techniques for research studies with small samples. *American Journal of Physiology - Lung Cellular and Molecular Physiology, 313*(5), L873-L877. doi:10.1152/ajplung.00238.2017

Hybrid III and THUMS Headforms Comparison for EOD Helmet Blast Mitigation Performance

Jean-Philippe Dionne, Ming Cheng, Jeffrey Levine, Aris Makris

Med-Eng Holdings ULC 2400 St. Laurent Blvd. Ottawa Canada
jean-philippe.dionne@safariland.com

Abstract. Historically, blast overpressure protection of Explosive Ordnance Disposal (EOD) Personal Protective Equipment (PPE) was assessed through full-scale blast experiments, with and without PPE. Existing injury criteria, sometimes from adjacent fields, were applied often without proper validation for blast scenarios. Over the last decade, advanced physical surrogates have been developed, focusing on the blast response of the human body, allowing for measurements that are more representative of an actual human head. Unfortunately, these advanced physical surrogates are complex and rarely available beyond the laboratories where they have been devised. Irrespective of the biofidelity of physical surrogates, blast testing of EOD PPE is challenging due to the severe threat and the inherent variability of blast. Thankfully, numerical models of the human body provide more insight on the body response and broaden the types of measurements reported. Importantly, the potential benefits of PPE can also be investigated with such human models, so long as the PPE itself is properly modelled (rate-dependent material properties, interaction with physical surrogates, sufficient resolution). The current study quantifies the performance of EOD helmets at mitigating blast overpressure, using data obtained from 1) blast experiments with physical Hybrid III mannequins, 2) computational simulation with a numerical Hybrid III model, and 3) computational simulation with a biofidelic human model. Numerical simulations of both the protected (EOD helmet) and unprotected cases revealed important differences between the two simulated scenarios as well as differences between experimental and numerical results for the Hybrid III case, when comparing common parameters. The computational biofidelic head models used also highlighted challenges in applying existing injury criteria since the exact locations within the head where parameters must be measured are not well-defined. This study is the most advanced numerical investigation to date of the performance of EOD PPE under representative blast loading, involving a human surrogate head.

1. INTRODUCTION

Explosive Ordnance Disposal (EOD) operators expect their EOD Personal Protective Equipment (PPE), also referred to as "bomb suits" (Figure 1) to provide protection from the blast overpressure threat. Unfortunately, there is currently no universally accepted quantitative test methodology for blast overpressure performance testing of bomb suits. Indeed, the US National Institute of Justice NIJ 0117.01 standard for public safety bomb suits [1], released in 2016, only addresses "blast integrity". This consists of observing the capacity of the bomb suit ensemble to resist a blast, from a purely qualitative standpoint. The NIJ rationale for the omission of quantitative requirements is that present research and data related to the effects of blast overpressure (e.g., blast head trauma, blast thoracic injury, blunt thoracic injury, blunt lower neck trauma, other neck injury, and blast ear injury) are limited. However, given that one of the main roles of bomb suits is to protect against blast overpressure, it remains highly relevant to quantify their blast overpressure mitigation performance, to ensure that end-users do not end up donning a poorly designed bomb suit, not providing sufficient blast overpressure protection.

Figure 1. Bomb suit for Explosive Ordnance Disposal

In the absence of a widely accepted quantitative blast overpressure standard, bomb suit manufacturers typically quantify PPE protection through percentage reductions in engineering variables measured on anthropomorphic mannequins, with and without bomb suit protection. The variables include head acceleration, as well as ear and chest overpressure, without any direct link to injury potential being provided. Dionne et al. [2] conducted a statistical analysis of the experimental blast overpressure test results related to these three variables. It was assumed that a reduction in engineering parameters measured on mannequins, must correlate with a reduction in blast injuries.

The Hybrid III mannequin mentioned in the NIJ standard [1] and by Dionne et al. [2] has only been validated for automotive crash tests. As such, its applicability for blast overpressure testing is of much debate. On the other hand, numerous human surrogates developed specifically for blast applications have been developed and tested over the last decade. The Warrior Injury Assessment Manikin (WIA Man)

 https://doi.org/10.52202/080042-0021

was developed for military vehicle under-belly blast testing [3]. This surrogate is aimed at quantifying vertical loading and human extremity response and is thus not suitable for bomb suit blast overpressure testing. Other suitable surrogates have also been developed by other groups, such as the Human Surrogate Head Model (HSHM) [4] and the Brain Injury Protection Evaluation Device (BIPED) [5]. However, these advanced blast surrogates tend to be expensive, possibly frangible, and are not standardized (at least not yet). Moreover, they are not readily available for purchase by industry, making their suitability for the severe EOD tests questionable. As a result, bomb suit manufacturers still rely on the Hybrid III mannequin (or equivalent) to characterize the protection performance of their products.

An alternative to blast overpressure experimental testing is to conduct numerical simulations. Indeed, computational modelling and simulation techniques have been used to study blast induced traumatic brain injury (TBI) and investigate the complex biomechanical and physiological factors leading to injury. For instance, Lockhart [6] implemented a rigid-body model (GEBOD) in LS-Dyna to compute the head response in blast scenarios. The head acceleration based HIC15 criterion was applied to investigate the effects of a PASGT helmet. Furthermore, a 2D sagittal biofidelic head model was used to explore overpressure distribution around the head. Unfortunately, no injury parameters at the brain tissue level were studied. Addressing this gap, Nyein [7] developed a 3D biofidelic human head model to investigate the effects of a military ACH helmet on the propagation of stress waves within the brain. Specifically, changes in intracranial pressure due to the ACH helmet were studied. More recently, Yu and Ghajari [8] implemented a high-fidelity human head model to study the effects of an ACH helmet worn with goggles on the head response to blast. It is reported that this protective gear led to increases in intracranial pressure (ICP), cerebrospinal fluid (CSF) cavitation, as well as brain strain and strain rate, compared to the unprotected case. The above studies all focused on military helmets aimed at protecting from blunt impacts and ballistic penetrations, not blast overpressure ingress.

Valverde-Marcos et al. [9] conducted an extensive study of the protective capability of an EOD helmet for small blasts, also using computational models. They used the HHFEM (Human Head Finite Element Model) model developed by J. Antona-Makoshi [10] and modelled an existing EOD helmet used by the Spanish police. In comparison with the unprotected case, the EOD helmet was found to delay the impact of the shockwave on the wearer's head and reduced the maximum head acceleration by 80% in all three cases simulated. Comparing to relevant published injury thresholds, they concluded that wearing an EOD helmet reduced the severity of injuries from a highly probable death (when unprotected) to a low probability of injury, of a mild and localized nature. It must be emphasized however that these findings were obtained through simulating relatively low explosive charges. In addition, the simulation model developed by Valverde-Marcos et al. [9] looked at the EOD helmet in isolation. Indeed, no interaction with an EOD suit was modelled.

In the present study, numerical simulations of the EOD helmet mitigation performance were conducted against the representative explosive charge described in the NIJ 0117.01 bomb suit standard. Even though far from biofidelic, a numerical Hybrid III head and neck model (Figure 2) [11] was used for a first set of numerical simulations, with the purpose of directly comparing with experimental results obtained with that same surrogate. For these simulations, an EOD helmet and an EOD suit (including blast-protecting collar) were modelled, both in terms of geometry and material properties.

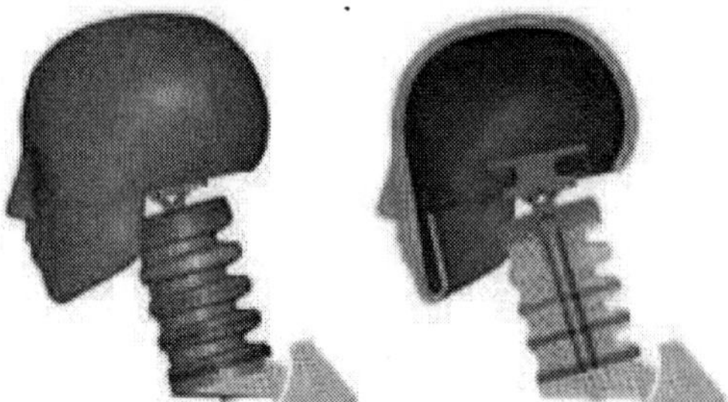

Figure 2. Hybrid III head and neck model [11]

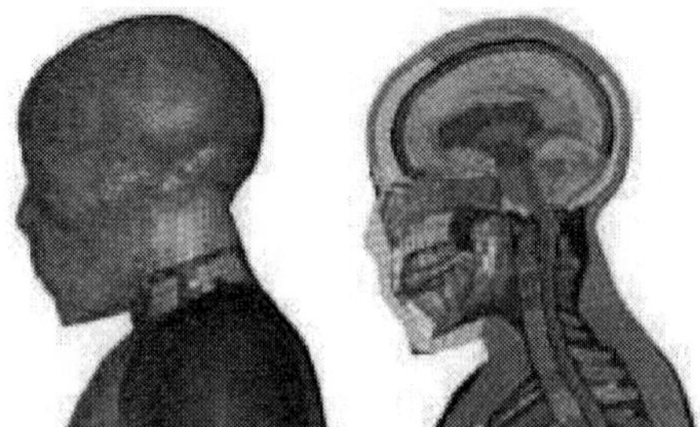

Figure 3. THUMS model [12], focusing on the head/neck portion

Simulations were conducted both in the protected (EOD helmet and suit) and unprotected scenarios. The work on the numerical Hybrid III model was funded in part by the US Army (2017-19) with an objective to get insight into the protection capabilities of EOD helmet protection concepts and validate the numerical EOD Helmet models. Similar simulations were then performed using a much more advanced head model (THUMS, Figure 3, [12]), also developed for the automotive industry, but featuring morphologically accurate details of the human head and brain. The rationale for using the THUMS head model was to quantify the response of actual brain tissues when subjected to blast, through parameters having the potential to be linked to injury mechanisms. It must be emphasized though, that to our knowledge, the THUMS model has not been validated for blast. Comparisons were made between predictions from the Hybrid III and THUMS numerical models, in the NIJ standard explosive scenario, but in three different orientations of the head surrogate with respect to the blast: 0° (directly facing), 45° (oblique) and 90° (sideways). To our knowledge, this is the first time EOD helmet response is simulated in orientations other than directly facing the blast.

2. METHODOLOGY

Simulations were generated using LS-Dyna, an advanced general-purpose multiphysics simulation software package developed by the Livermore Software Technology Corporation (LSTC) and owned by ANSYS. A 3D Finite Element Analysis (FEA) model was created for this study, which includes an air domain of 1200 mm by 840 mm by 1200 mm modelled using the Multi-Material Arbitrary Lagrangian Eulerian (MMALE) method.

To reduce computational time and increase the accuracy of the results, a technique that allows the mapping of results from 1D to 3D Eulerian domains has been employed [13]. A 1D spherical symmetry model with an element size of 1 mm was used to model the C4 explosive and its detonation. After the blast wave propagated to the boundary of the 3D air domain, the pressure distribution and particle velocity distribution were then exported to a binary map file. This map file was then used to initiate the 3D domain in the subsequent 3D simulation. The timestep was controlled by the LS-Dyna solver to achieve numerical stability. In Arbitrary Lagrangian Eulerian (ALE) simulations, the timestep is typically less than $1.0e^{-7}$ second, which might not be small enough to capture the real peak in blast waves, but still acceptable in the current context. The simulation data was generated at a rate of 1 MHz. Figure 4 demonstrates the 3D model schematic, where the blast wave comes from the left side.

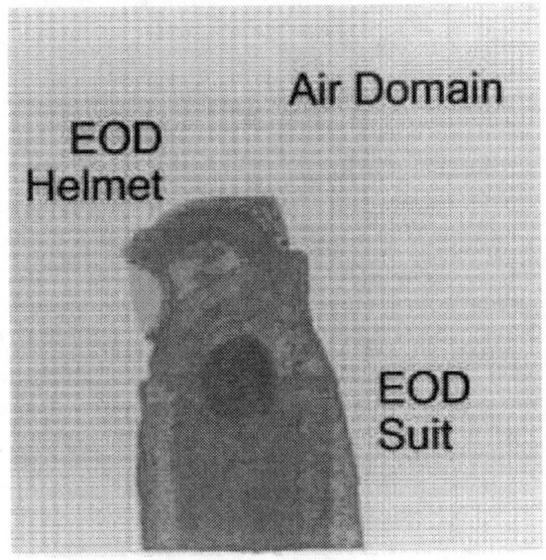

Figure 4. 3D model schematic of the simulation domain

In the 3D model, the Hybrid III mannequin head and neck model and the THUMS model were placed in the middle of the air domain, either with or without EOD protection. The Fluid-Structure Interaction (FSI) method was then applied to transfer the pressure from the air and explosive to the mannequin or the EOD helmet and suit in the protected case.

The EOD helmet model includes the helmet shell, the impact/comfort liner, the retention system, the face shield, and the housing for the electronics. These components were modelled with hexahedron elements using a Lagrange formation. Meshes were generated from CAD models of a Med-Eng developed EOD helmet version. The corresponding material models for these components were deformable. While the model constants were determined experimentally, the geometry of the EOD suit was scanned from a Med-Eng developed EOD suit, with its material properties estimated based on aramid fabric textiles. The purpose of this EOD suit in the simulation was to provide realistic surfaces to generate reflected waves that eventually influenced the head response. Based on the experimental studies conducted in the past, the reflected waves from the EOD suit, especially the collar, significantly modify the loading on the EOD helmet, and subsequently the head response [14].

All simulations were conducted to generate over 8 ms of data. This duration is sufficient to capture the original motion of the mannequin and PPE, given the absence of reflecting surfaces. Ground reflections would occur later and would induce response levels lower than for the original blast wave impact. Table 1 summarizes all simulations conducted (test matrix).

For both Hybrid III and THUMS cases, the global head acceleration was tracked. For the Hybrid III, an accelerometer positioned at the centre of gravity of the head is always used for the purpose of this measurement. On the other hand, the THUMS mannequin is deformable and no set location within the head model is dedicated for global head acceleration tracking. Ideally, the global THUMS head acceleration would have been obtained by computing the location of the centre of gravity at each time interval. But to simplify the calculations, a specific unique location, similar to the position of the accelerometer in the Hybrid III head, was determined as its centre of gravity. The head acceleration was tracked at that location throughout the event duration. With the head acceleration data available for both

https://doi.org/10.52202/080042-0021

head models, the Head Injury Criterion (HIC) was then calculated. In addition to the standard definition of the HIC15 (calculated over a maximum 15 ms duration), a version referred to as "HIC15d" was also calculated. The HIC15d is better suited when using head surrogates attached to mannequin bodies [15]. The equations for the HIC15 (free floating headform) and HIC15d (attached headform) are:

$$HIC15 = \frac{1}{(t_1 - t_2)} \int_{t_1}^{t_2} a_{res}^{2.5} \, (t_2 - t_1) dt \qquad HIC15d = 0.75446(HIC15) + 166.4$$

As the THUMS head model includes realistic human features, additional parameters such as the intracranial pressure (ICP), the cerebrospinal fluid pressure (P-CSF), as well as the cerebellum effective strain (ε_{eff}), were extracted. To determine the optimal location within the head to extract these measurements, a first simulation was first conducted, from which the approximate maximum location (for a given parameter) could be visually determined. A follow-on simulation then tracked the parameter at this selected location. The measurement locations thus varied according to the parameter being measured, and for each combination of orientation and protection configuration. Detailed results are presented below, for the three orientations with respect to the blast: front facing (0°), oblique (45°) and sideways (90°), for all variables of interest (head acceleration, intracranial pressure, cerebrospinal fluid pressure and cerebellum effective strain.

Table 1. Test matrix for all numerical simulations conducted

Headform	Protection	Orientation
Hybrid III	Unprotected	0°
		45°
		90°
	EOD Helmet	0°
		45°
		90°
THUMS	Unprotected	0°
		45°
		90°
	EOD Helmet	0°
		45°
		90°

3. RESULTS

3.1 Front facing (0°)

Figure 5 illustrates the numerical simulations conducted for all scenarios in the 0° orientation. Peak values for all parameters are listed in Table 2. Finally, detailed traces are provided in Figures 6 to 10.

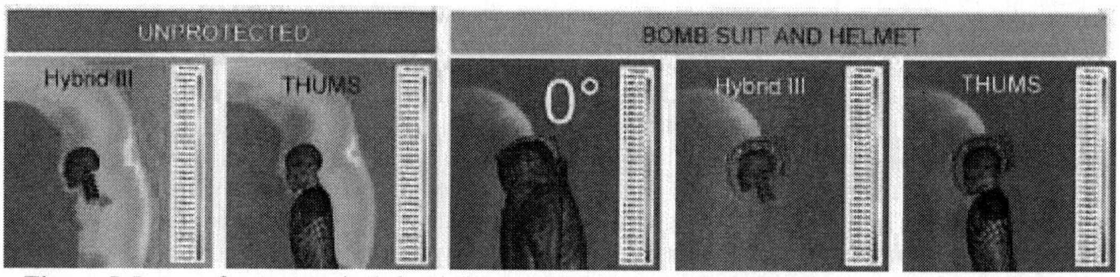

Figure 5. Images from numerical simulations at 0° (Hybrid III and THUMS, unprotected and EOD)

Table 2. Results (peak values and percentage reductions) obtained at 0° orientation

0°	Hybrid III			THUMS							
	a_{res} (g's)	HIC15	HIC15d	a_{res} (g's)	HIC15	HIC15d	ICP_{max} (kPa)	ICP_{min} (kPa)	$P\text{-}CSF_{max}$ (kPa)	$P\text{-}CSF_{min}$ (kPa)	ε_{eff}
Unprotected	2985	8648	6691	1967	18470	14100	2692	-2432	2894	-2627	0.15
EOD Helmet	130	258	361	305	266	367	210	-89	193	-123	0.20
% Reduction	95.6%	97.0%	94.6%	84.5%	98.6%	97.4%	92.2%	96.3%	93.3%	95.3%	-36.0%

https://doi.org/10.52202/080042-0021

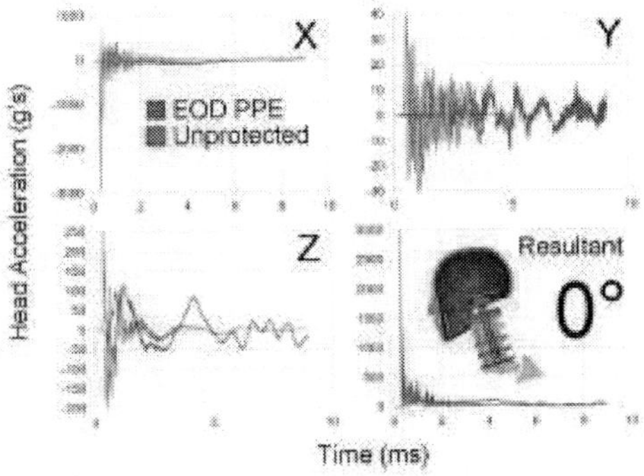

Figure 6. Hybrid III X, Y, Z, & Resultant head acceleration traces in the 0° orientation

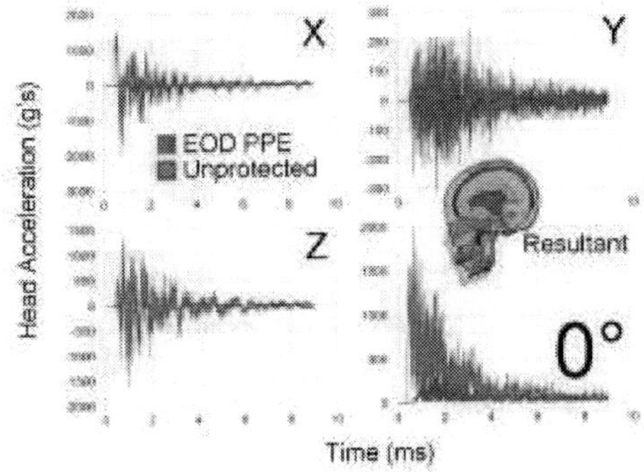

Figure 7. THUMS X, Y, Z, & Resultant head acceleration traces in the 0° orientation

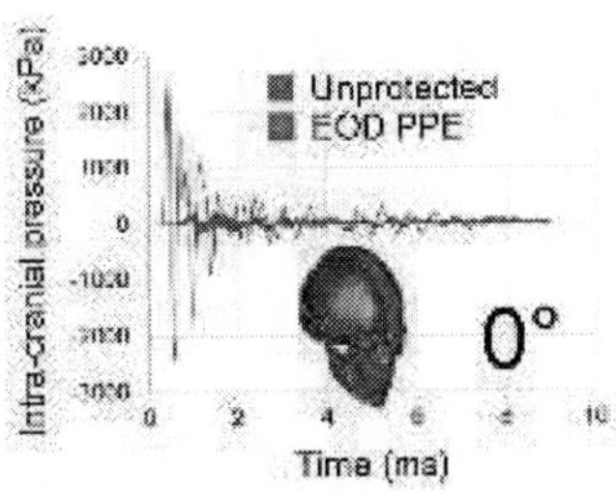

Figure 8. THUMS IC pressure in the 0° orientation

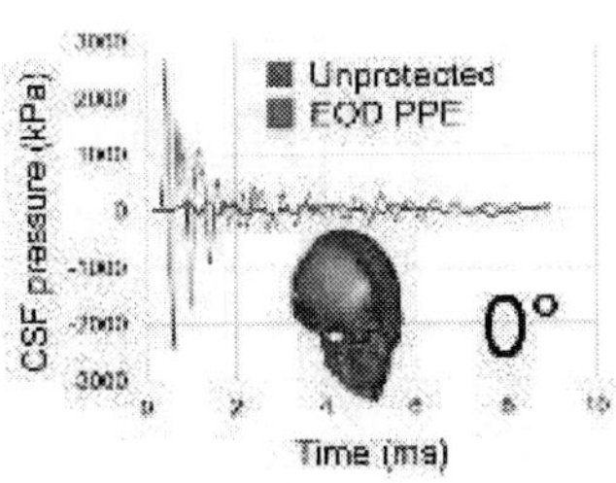

Figure 9. THUMS CSF pressure in the 0° orientation

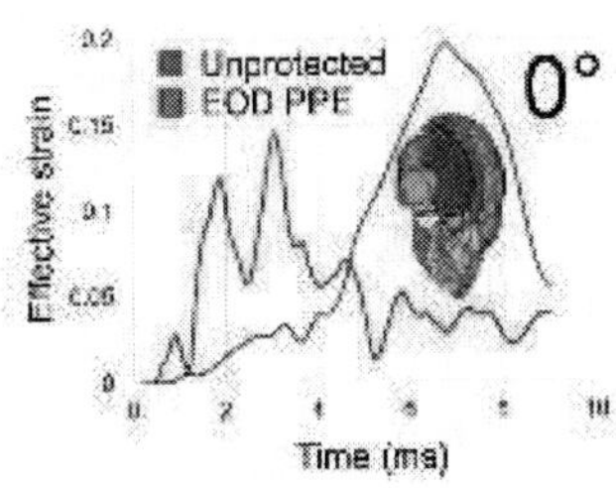

Figure 10. THUMS effective strain in the 0° orientation

3.2 Oblique (45°)

Figure 11 illustrates the numerical simulations conducted for all scenarios in the 45° orientation. Peak values for all parameters are listed in Table 3. Finally, detailed traces are provided in Figures 12 to 16.

Figure 11. Images from numerical simulations at 45° (Hybrid III and THUMS, unprotected and EOD)

Table 3: Results (peak values and percentage reductions) obtained at 45° orientation

45°	Hybrid III			THUMS							
	a_{max} (g's)	HIC15	HIC15d	a_{max} (g's)	HIC15	HIC15d	ICP_{max} (kPa)	ICP_{min} (kPa)	$P\text{-}CSF_{max}$ (kPa)	$P\text{-}CSF_{min}$ (kPa)	ε_{eff}
Unprotected	2670	7722	5992	1846	18950	14470	3092	-2053	3595	-2789	0.23
EOD Helmet	121	176	299	293	234	342.9	164	-99.5	154	-101	0.05
% Reduction	95.5%	97.7%	95.0%	84.1%	98.8%	97.6%	94.7%	95.2%	95.7%	96.4%	79.6%

https://doi.org/10.52202/080042-0021

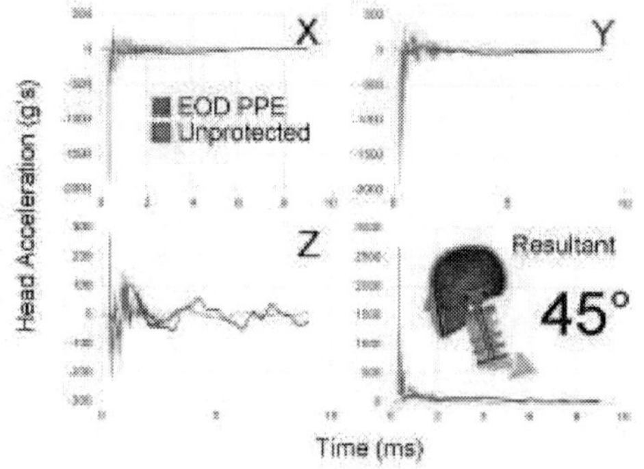

Figure 12. Hybrid III X, Y, Z, & Resultant head acceleration traces in the 45° orientation

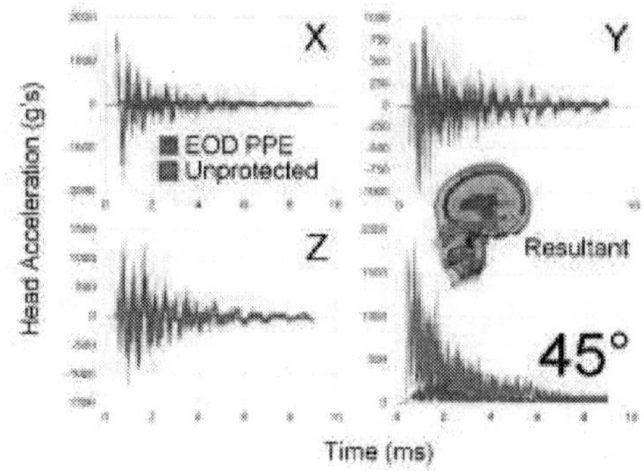

Figure 13. THUMS X, Y, Z, & Resultant head acceleration traces in the 45° orientation

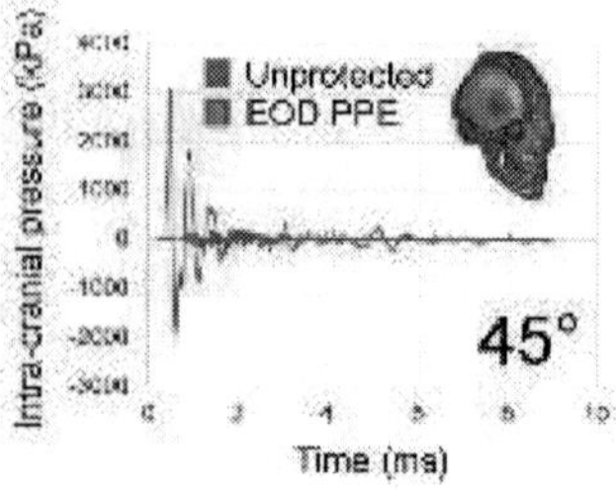

Figure 14. THUMS IC pressure in the 45° orientation

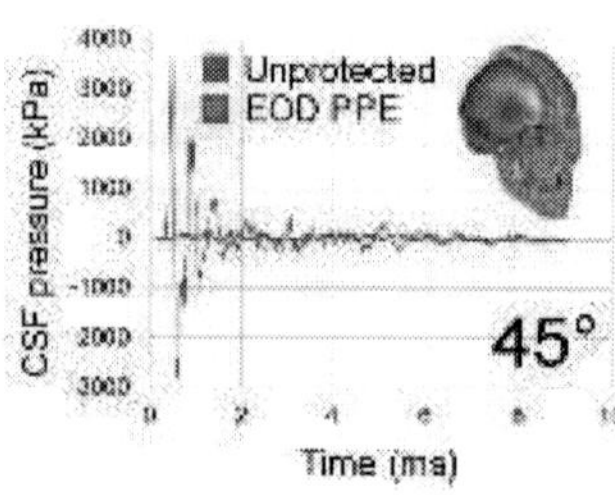

Figure 15. THUMS CSF pressure in the 45° orientation

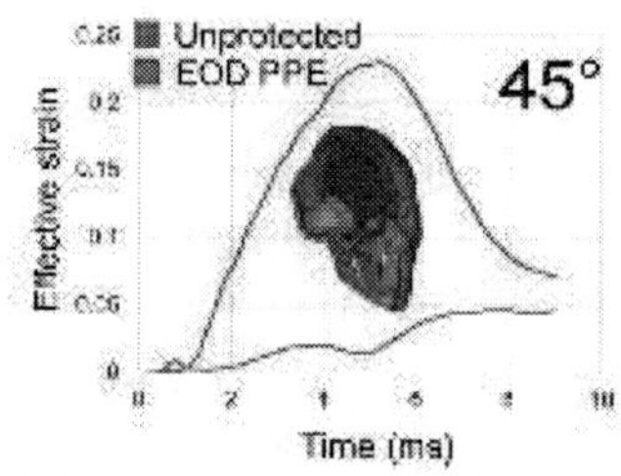

Figure 16. THUMS effective strain in the 45° orientation

3.3 Sideways (90°)

Figure 17 illustrates the numerical simulations conducted for all scenarios in the 90° orientation. Peak values for all parameters are listed in Table 4. Finally, detailed traces are provided in Figures 18 to 22.

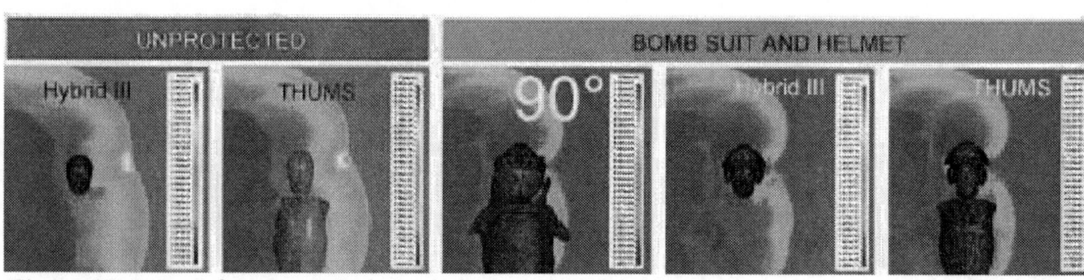

Figure 17: Images from numerical simulations at 90° (Hybrid III and THUMS, unprotected and EOD)

Table 4: Results (peak values and percentage reductions) obtained at 90° orientation

90°	Hybrid III			THUMS							
	a_{max} (g's)	HIC15	HIC15d	a_{max} (g's)	HIC15	HIC15d	ICP_{max} (kPa)	ICP_{min} (kPa)	$P\text{-}CSF_{max}$ (kPa)	$P\text{-}CSF_{min}$ (kPa)	ε_{eff}
Unprotected	3361	12400	9521	3233	43500	32980	4937	-2787	5500	-3391	0.11
EOD Helmet	173	171	296	533	944	803	205	-137	268	-190	0.10
% Reduction	94.9%	98.6%	96.9%	83.5%	98.1%	97.6%	95.9%	95.1%	95.1%	94.4%	10.1%

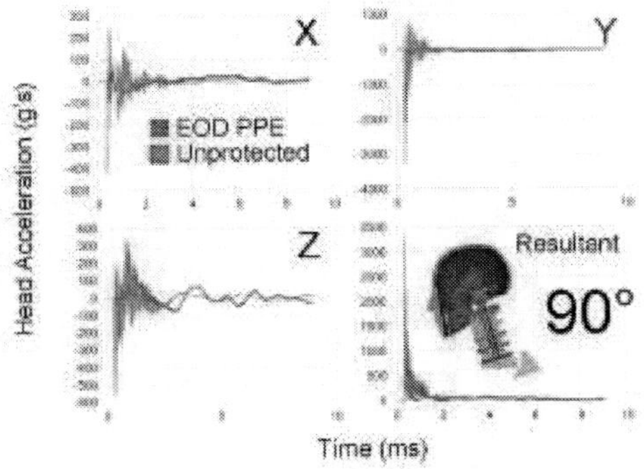

Figure 18. Hybrid III X, Y, Z, & Resultant head acceleration traces in the 90° orientation

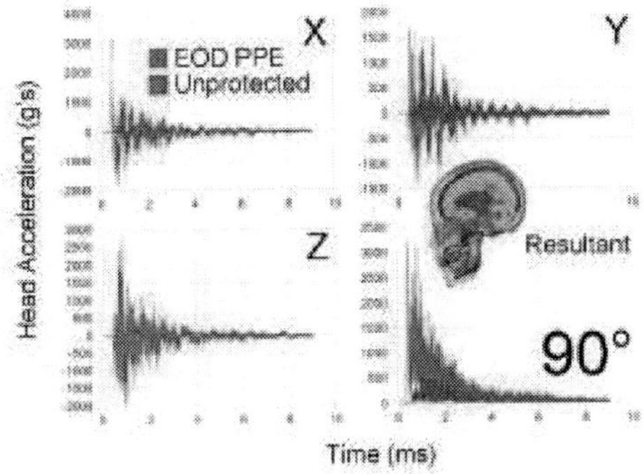

Figure 19. THUMS X, Y, Z, & Resultant head acceleration traces in the 90° orientation

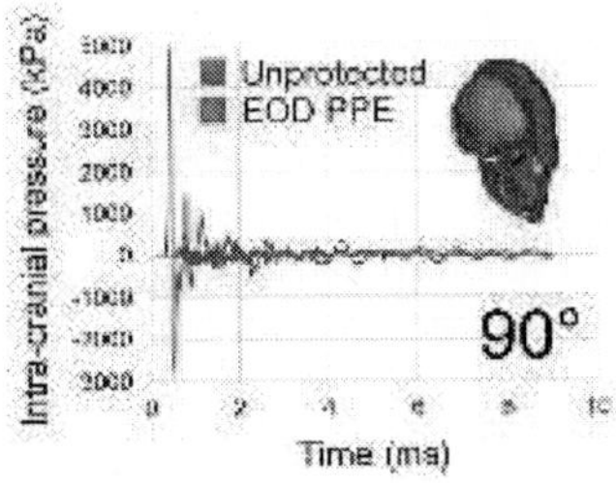

Figure 20. THUMS IC pressure in the 90° orientation

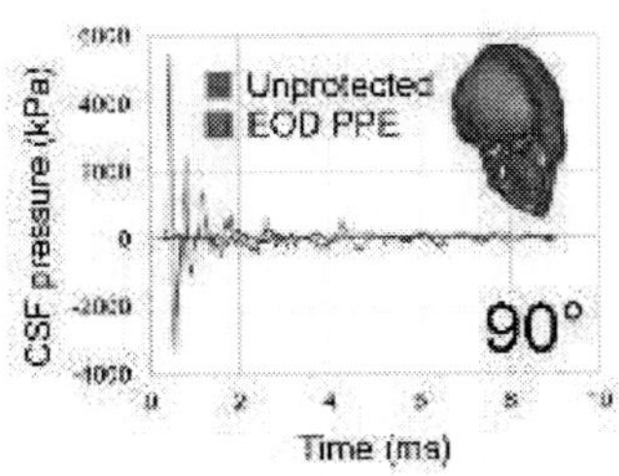

Figure 21. THUMS CSF pressure in the 90° orientation

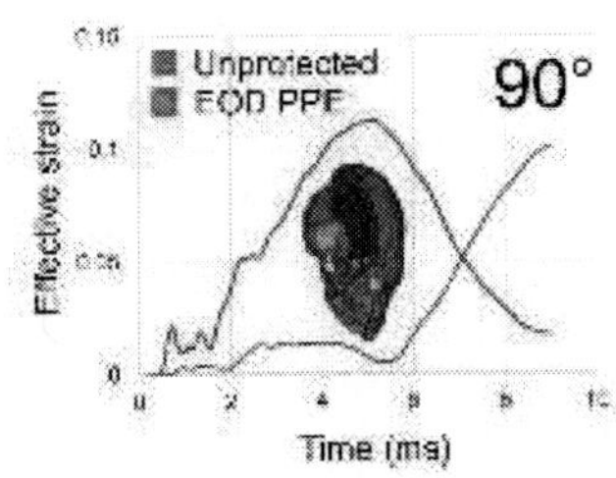

Figure 22. THUMS effective strain in the 90° orientation

4. DISCUSSION

Figure 23 tabulates the peak resultant head accelerations and HIC15d from the Hybrid III and THUMS models at all three mannequin orientations, with and without an EOD helmet. The results highlight that the PPE dramatically reduces the head acceleration and its derived HIC15d values, thus suggesting that the PPE provides effective protection to the EOD technicians. Moreover, this reduction is consistent across all three orientations for both Hybrid III and THUMS models. This insensitivity of results with respect to the orientation of the mannequin implies that the overall mass (inertia) of the head and helmet dominates over the kinematics of the head itself. Nevertheless, there are some notable differences in aerodynamic loading, with higher peak values for the 90° orientation, compared to the other two cases. This is due to the larger projected area inducing increased drag for the side (90°) exposure.

In terms of intracranial pressure (ICP), Figure 24 indicates that the *unprotected* peak pressures vary with orientation. While in the 0° and 45° cases, the maximum pressures are similar (approximately 2800 and 3000 kPa, respectively), a value nearing 5000 kPa was obtained in the 90° orientation. This large difference in the 90° case is due to the grey matter cerebrum not being spherically symmetric, implying that locations and values for the maximum ICP vary with the orientation. It must be kept in mind that peak acceleration values are very sensitive to the high frequency responses since the blast wave has a very short (almost zero) duration in the initial pulse around the peaking time. In contrast, the EOD helmet induced a significant reduction down to approximately 200 kPa for all cases, with differences in absolute pressures being modest across all orientations. The hard helmet shell and soft impact liner prevent a direct exposure of the head to the blast wave. The low ICP variations in the EOD case are thus a direct result of the interaction between the helmet and the head. Since the shock wave reflecting from the helmet outer surface remains at a similar level for all orientations (single blast charge and standoff), the orientation only exerts a small relative influence on the helmet kinematics. Consequently, the peak ICP values for the EOD case remained within a relatively limited range, likely below any injury threshold.

https://doi.org/10.52202/080042-0021

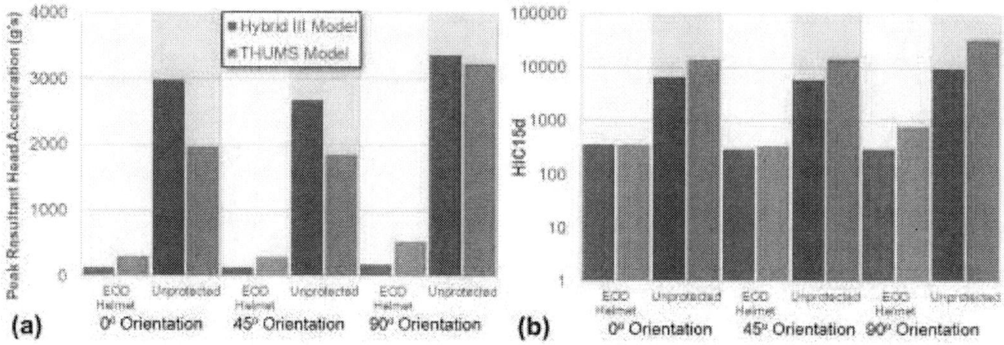

Figure 23. Comparison of the Hybrid III and THUMS models in terms of (a) peak head acceleration, and (b) HIC15d, in all three mannequin orientations (0°, 45°, and 90°) for both unprotected and protected (EOD) cases

Similar to the ICP, the cerebrospinal fluid pressure (P-CSF) also showed divergent peak pressures and locations based on the mannequin's orientation to the blast. Indeed, maximum CSF pressures of approximately 2900 kPa, 3800 kPa and 4900 kPa were noted in the 0°, 45°, and 90° orientations, respectively. Again, the EOD helmet reduced the pressure considerably down to approximately 200 kPa, in all three orientations. The reason for this phenomenon is the same as discussed above for the ICP.

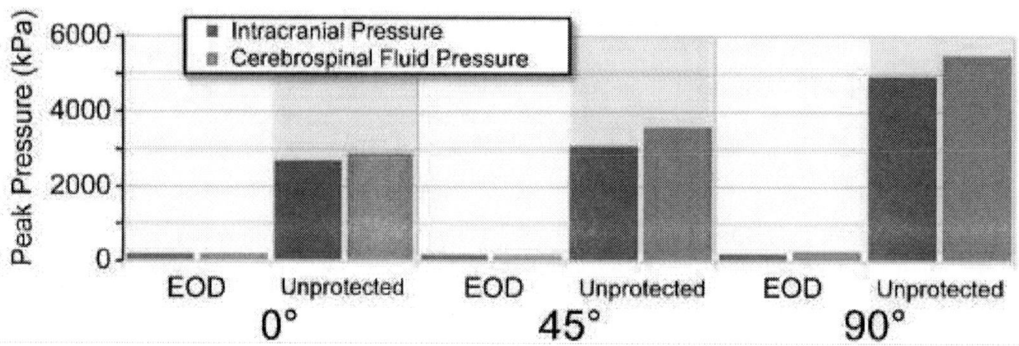

Figure 24. Comparison of intracranial and cerebrospinal fluid pressures with the THUMS model in all three mannequin orientations (0°, 45°, and 90°) for both unprotected and protected (EOD) cases

As for the cerebellum effective strain, the ability of the EOD helmet to reduce peak values is not consistent (Figure 25). Only in the 45° orientation was the strain reduction (80%) on par with those observed for the intracranial and cerebrospinal pressures. In the 90° orientation, the EOD helmet yielded a mere 11% reduction, compared to the unprotected. Finally, in the 0° orientation, the cerebellum effective strain was noted to *increase* by 36% with the EOD helmet. However, it should be noted that the time to reach maximum strain is much longer when wearing a helmet. Moreover, the rising rate of the cerebellum effective strain is dramatically reduced when introducing protection. Focusing on just peak strain might therefore not draw the complete picture. Unlike stress or pressure, the effective strain is a cumulative parameter and thus depends on the loading duration causing plastic deformation. As the cerebellum tissue has a low yield stress, accumulated plastic strain behaves completely different as a parameter, compared to the ICP and CSF pressure.

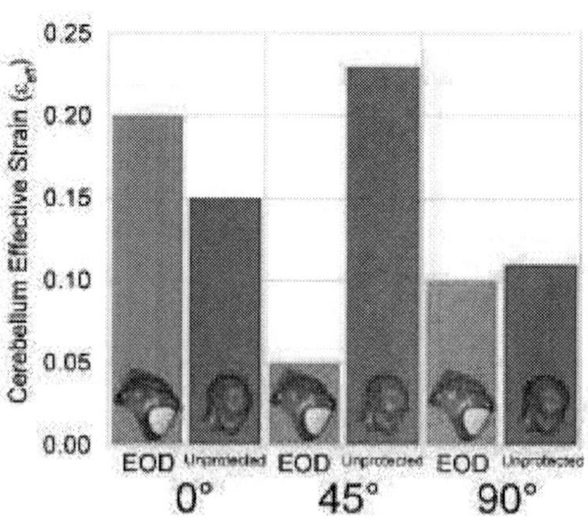

Figure 25. Comparison of cerebellum effective strain with the THUMS model for all three mannequin orientations (0°, 45°, and 90°) for both unprotected and protected (EOD) cases

4.1 Comparison with Experimental Results

As stated earlier, full-scale experimental blasts tests were also conducted with a physical 50th percentile Hybrid III mannequin for comparison with the modelling results. The experimental trials saw the mannequin placed in the same NIJ 0117.01 standard configuration [1], i.e., 60 cm away horizontally from a 0.567 kg C4 explosive at a 77 cm vertical height of burst. The experimental head accelerations were acquired at a sampling rate of 200 kHz with CFC1000 anti-aliasing filtering (1650 Hz cut-off frequency). Experimental trials were only conducted at a 0° orientation, in accordance with the NIJ standard requirements. Figure 26 compares the average and range of peak head acceleration and HIC15d for both the unprotected and protected (EOD) cases. Both numerical models (Hybrid III and THUMS) were found to overpredict the peak head acceleration and HIC15d values. Most simulated peak values nevertheless fall within the range of experimental data, with the exception of the head acceleration results in the EOD/THUMS case, and for both head models for the unprotected case.

The lower peak accelerations observed in the experimental data is likely due to filtering effects. Indeed, filtering at 1650 Hz leads to smoothing of the sharpest peaks in head acceleration signals. The HIC algorithm on the other hand, which incorporates the integration of the acceleration signal, is not as sensitive to filtering effects. As such, differences between experimental and simulated HIC15d values are not as significant, especially considering the scatter in experimental data. The high level of scatter in the experimental data is common to blast testing using Hybrid III mannequins, as previously reported by Dionne et al. [2].

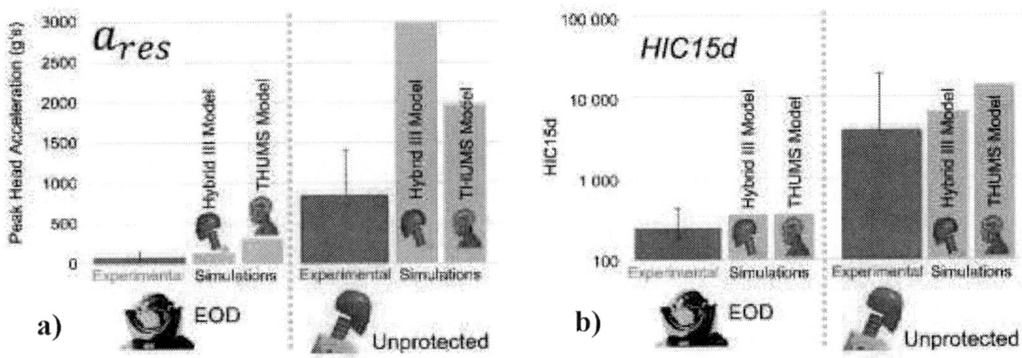

Figure 26. Average and range of the experimental (a) peak head accelerations, and (b) HIC15d, for both EOD helmet (green) and unprotected (red) experimental trials. Also denoted are the extracted values from the Hybrid III and THUMS simulations (grey)

5. CONCLUSION

The results obtained with the Hybrid III and THUMS numerical models do not show a very strong match with the overall experimental results (Figure 26), but among those, the protected (EOD) scenarios show a more favourable match, with the numerical results being more conservative. The use of numerical simulations to conservatively evaluate the effectiveness of EOD PPE (bomb suits) in configurations other than those tested experimentally is therefore promising, based on these results. In particular, the current study explored the effect of orientation, which is of interest to the EOD community, given that despite all efforts to follow standard operating procedures, EOD technicians do not always directly face an explosion. In particular, the 90° orientation for the EOD helmet differed from the other two cases, with higher acceleration and cerebrospinal fluid pressure values, thought to be due to a larger exposed surface area. For the Head Injury Criterion (HIC) though, much higher values were predicted for the more anatomically accurate THUMS model, compared to the Hybrid III, especially at 90° for the EOD helmet case, and in all directions for the other two orientations. The THUMS therefore yielded more conservative predictions for head injury, which is deemed preferable when designing protective equipment. Moreover, the THUMS advanced anatomical model allowed for the measurement of additional parameters (e.g., intracranial pressure, cerebrospinal fluid pressure, cerebellum effective strain) at various locations within the brain, helping to draw a more complete picture of the brain response under blast loading, and the role of blast protective helmets. Indeed, for all parameters measured, with the exception of the effective strain, substantial reductions (above 80% and often exceeding 95%) resulted from the presence of an EOD helmet.

https://doi.org/10.52202/080042-0021

Numerical simulations could also be conducted to investigate a wider range of explosive configurations (charge size, standoff, technician posture), at a much-reduced cost, compared to experimental trials. It is thus hoped that such numerical simulations could eventually guide and optimize EOD helmet design, when conducted in parallel with physical helmet development. However, further efforts will be required towards validating the numerical human body models and PPE models for blast, before numerical simulations can play a substantial role in the design of blast protective PPE.

References

[1] *U.S. National Institute of Justice (NIJ)*, "Public Safety Bomb Suit Standard", NIJ-0117.01, March 2016.
[2] Dionne, J.P., Levine, J., Makris, A., Investigating Bomb Suit Blast Overpressure Test Methodologies, in Homeland Security and Public Safety: Research, Applications, and Standards, ed. P. J. Mattson and J. L. Marshall (West Conshohocken, PA: ASTM International, 2019), 216–236. http://doi.org/10.1520/STP1614201800452
[3] Danelson, K., Watkins, L., Hendricks, J., Frounfelker, P., Pizzolato-Heine, K., Valentine, R., Loftis, K., Analysis of the Frequency and Mechanism of Injury to Warfighters in the Under-body Blast Environment, Stapp Car Crash Journal, Vol. 62 (November 2018), pp. 489-513.
[4] Merkle, A.C., Wing, I.D., Armiger, R.A., Carkhuff, B.G., Roberts, J.C., Development of a Human Head Physical Surrogate Model for Investigating Blast Injury, Proceedings from the ASME 2009 International Mechanical Engineering Congress and Exposition, November 13-19, 2009, Lake Buena Vista, Florida, USA.
[5] Ouellet, S., Phillippens, M., The multi-modal responses of a physical head model subjected to various blast exposure conditions, Shock Waves 28(10), January 2018, DOI: 10.1007/s00193-017-0771-3.
[6] Lockhart, P. A., Primary blast injury of the head: numerical prediction and evaluation of protection. A thesis presented to the University of Waterloo in fulfillment of the thesis requirement for the degree of Master of Applied Science in Mechanical Engineering, 2010.
[7] Nyein, M. K., Computational Modeling of Primary Blast Effects on the Human Brain. A dissertation submitted to the Department of Aeronautics and Astronautics in partial fulfillment of the requirements for the degree of Doctor of Philosophy in Aeronautics and Astronautics at the Massachusetts Institute of Technology in June 2013.
[8] Yu, X., Ghajari, M. Protective Performance of Helmets and Goggles in Mitigating Brain Biomechanical Response to Primary Blast Exposure. Ann Biomed Eng 50, 1579–1595 (2022). https://doi.org/10.1007/s10439-022-02936-x.
[9] Valverde-Marcos, B., Rubio, I., Antona-Makoshi, J., Chawla, A., Loya, J.A., Rodríguez-Millán, M., Numerical Analysis of EOD Helmet under Blast Load Events using Human Head Model, Applied Sciences 2020, 10, 8227; doi:10.3390/app10228227.
[10] Antona-Makoshi, J. Traumatic Brain Injuries: Animal Experiments and Numerical Simulations to Support the Development of a Brain Injury Criterion; Chalmers University of Technology: Gothenburg, Sweden, 2016; ISBN 978-91-628-9848-9.
[11] J. Sebastian Giudice, Kevin Kong, Adrian Caudillo, Sayak Mukherjee, Matthew B. Panzer, 2018. User Manual: Finite Element Models of Helmet Assessment Tools.
[12] Iwamoto M, Nakahira Y, Kimpara H. Development and Validation of the Total HUman Model for Safety (THUMS) Toward Further Understanding of Occupant Injury Mechanisms in Precrash and During Crash. Traffic Inj Prev. 2015;16 Suppl 1:S36-48. doi: 10.1080/15389588.2015.1015000. PMID: 26027974.
[13] N. Aquelet and M. Souli, "2D to 3D ALE Mapping," in 10th International LS-DYNA Users Conference, Dearborn, USA, 2008.
[14] Cheng, M., Levine, J., Dionne, J.P., Makris, A., Simulated head and neck responses with EOD helmet and collar protection, Proceedings from the 25th International Symposium on Military Aspects of Blast and Shock (MABS25), The Netherlands, 2018.
[15] U.S. Department of Transportation, National Highway Traffic Safety Administration, Laboratory Test Procedure for FMVSS No. 201U – Occupant Protection in Interior Impact – Upper Interior Head Impact Protection (TP-201U-02), January 2016.

Injuries and Operational Implications Caused By Behind Armour Blunt Trauma Across Various Impact Locations

C. Howes[1], E. Mazuchowski[2], A. Rohrer[3], M. Clark[4], Q. Luong[1], D. Drewry[1], A. Iwaskiw[1]

[1]The Johns Hopkins Applied Physics Laboratory, 11100 Johns Hopkins Road, Laurel, Maryland 20723, Caroline.Howes@jhuapl.edu
[2] Forensic Pathology Associates, Division of HNL Lab Medicine, 1255 S. Cedar Crest Blvd, Allentown, PA 18103
[3]Joint Trauma System, 3698 Chambers Pass, Joint Base San Antonio, Fort Sam Houston, TX 78234 and Armed Forces Medical Examiner System, Defense Health Agency, Dover Air Force Base, DE
[4]U.S. Special Operations Command, 10 General Greene Avenue, Natick, MA 01760

Abstract. Advances in materials development have enabled body armour plates to become lighter to optimise warfighter mobility and performance while maintaining or improving protection from penetrating ballistic injuries. These developments can worsen non-penetrating ballistic performance and cause injuries which are not currenrly well-understood. Behind armour blunt trauma (BABT) describes injuries caused by the deforming backface of body armour striking the wearer [1]. Non-penetrating ballistic standards for body armour are derived from a 1970s study of soft body armour tests on live goats [2]. Limitations of this study have led to interest in conducting experiments using post mortem human surrogates (PMHS) paired with live ammunition and hard armour towards the evaluation of the existing criteria. This pursuit of human-based injury data for BABT motivated a collection of studies characterising injuries from ballistic tests of personal protective equipment (PPE), specifically hard armour plates protecting the sternum, lower abdomen, and lateral ribcage. Thirteen impacts were conducted on the sternum, seven tests on the lower abdomen, and two tests, with tests ongoing, on the lateral ribcage. PMHS were prepared and physiological processes simulated to provide similar bodily conditions to a living individual. Each tested location resulted in characteristic injuries and collectively a new injury categorisation, the Projected Operational Injury Outcome (POIO), was created. The POIO was generated per test by a panel of board-certified physicians and projects a warfighter's status of returning to duty should they experience injuries as seen in the simulated tests. This provides a level of insight not currently provided existing metrics such as the Abbreviated Injury Scale's (AIS) Injury Severity Score (ISS). The studies aim to build knowledge of BABT towards a comprehensive understanding of the relationship between armour performance, possible injuries, and operational implications for various armour types and body regions.

1. INTRODUCTION

Warfighters are outfitted with personal protective equipment (PPE) to protect them from multiple threats they may be exposed to in an operational environment. This equipment includes body armour systems and helmets amongst other protective gear. While this gear protects the warfighter, it can hinder mobility and manoeuvrability throughout the mission. As such, advances in materials development strive to create systems with reduced weight and as a result, decreased load for the warfighter. These systems must conform to existing performance standards that all warfighter-borne PPE must meet. Current performance standards for ballistic PPE, including body armour, account for penetrating and non-penetrating ballistic events. Penetrating events generally occur when the ballistic conditions surpass the rated conditions of the armour. Non-penetrating events generally occur when the threat and velocity pairing is below the rated conditions. The current study focuses on such non-penetrating performance standards. Although the primary objective of defeating a threat is achieved, non-penetrating ballistic impacts can result in a complex trauma scheme, termed behind armour blunt trauma (BABT) [1]. A non-penetrating ballistic event to the torso that could cause BABT is depicted in Figure 34.

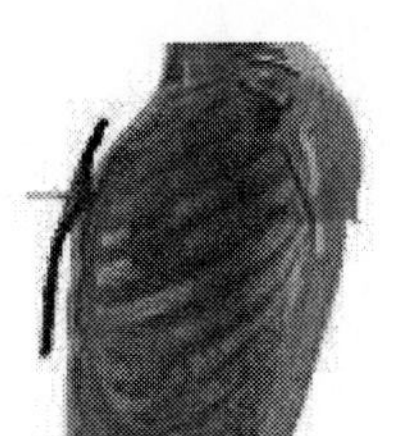

Figure 34. Depiction of BABT on torso anatomy

https://doi.org/10.52202/080042-0022

Non-penetrating BABT performance standards were originally derived from a study investigating the injury patterns generated by varying armour materials using live goats [2], [3]. The correlation between injury data, whether collected via live-animals or post mortem human surrogates (PMHS), and armour standards is achieved through matching the conditions used in such tests to those using a non-live backing medium. The specific medium used for these standards is Roma Plastilina No. 1 clay that backs the surface of the body armour plate during a live-fire ballistic test. Once the impacted plate is removed, the residual deformation left in clay can be measured [4]. The long-standing body armour standard in the U.S. Army was extrapolated from the injury data from these live-goat tests that was correlated to data from matched-paired clay tests. This standard requires that the fielded body armour systems cannot result in more than 44mm of residual deformation left in clay from a non-penetrating ballistic impact, and applies to any body armour plate regardless of region of coverage. This standard was not derived for use with high-velocity rifle rounds. The derivation of this standard from non-human testing has inspired several tests using PMHS, or cadavers, to better understand the injury implications of the 44mm standard [4], [5], [6]. The unknown relationship between the current standards and human injury, along with the desire to lighten the load to the warfighter, has resulted in many to consider relaxing the 44mm standard. Additionally, there exists a higher-level goal to seek knowledge of the injury mechanisms that cause such trauma by simultaneous collecting high-rate sensor measurements during these simulated events.

To further research within this space, the Johns Hopkins University Applied Physics Laboratory (JHU/APL) embarked on a multi-year effort, starting in 2018, to characterise injury outcomes from non-penetrating ballistic tests on hard armour plates for different body regions. The regions reported on within this larger study span the sternum, the lower abdomen, and the lateral ribcage. The studies have been conducted in pursuit of additional detail of BABT that can occur due to non-penetrating ballistic impacts. These studies investigated the injury results and patterns that occur throughout a range of striking velocities on an armour-threat pairing for each region of interest that could result in varied levels of residual deformation left in clay spanning below and above the 44mm standard.

The studies conducted by JHU/APL provide additional insights into the injury outcomes and mechanisms of BABT. The use of PMHS for research typically yields an older sample population that may interfere with obtaining applicable injury outcomes. However, this study utilised PMHS acquired with stringent inclusion criteria to provide results more relevant to the population of interest. Additionally, this study was conducted using live-fire ammunition, specifically the 7.61x51mm NATO round, with only one impact per armour system. The former is important as it replicates the loading conditions seen in-theatre. Collectively these attributes create a dataset that is invaluable to provide insight into BABT.

In the National Research Council's (NRC) 2012 report on the body armour testing, a key finding enumerated "the need for a robust and widely used ballistic trauma injury classification scale" [7]. In the present study, the authors used the results of the regimented and specialised dataset to respond to this finding to create a novel categorisation correlating the injury results observed on PMHS to a projected outcome that may occur on a living warfighter. This outcome was informed by a panel of, board-certified physicians knowledgeable of the military medical community, including what personnel and resources could be available should an injury occur due to BABT. The novel categorisation, termed the Projected Operational Injury Outcome (POIO), correlates armour performance metrics and the operational implications of the injuries resulting from BABT. Existing injury severity metrics, such as the Abbreviated Injury Scale (AIS) and affiliated metrics including the Injury Severity Score (ISS) and New Injury Severity Score (NISS), do not fill this gap when applied to PMHS data because they do not factor in considerations of the operational environment. This study enables the data-driven refinement of armour performance standards as well as foresight into the types of injuries that could be seen with increased backface deformation of armour systems.

2. METHODS

Each body region tested had methods tailored to the impact location to prioritise a realistic injury outcome. The specific regions studied within the greater study spanned the sternum, lower abdomen, and lateral ribcage. The general test locations and armour combinations as well as sample size are notated in Figure 35. The sternum tests investigated the response at a location on the midsagittal plane of the sternum between ribs three and four, and 13 tests were completed in total. The lower abdomen tests were conducted over either the pubic symphysis or the dome of the bladder, with seven tests conducted across both locations. The lateral ribcage tests were conducted directly over the most lateral aspect of rib nine either on the left or right side, with one test completed to date on each side. Overall, each region and impact location generally followed the same order of operations in preparation for and execution of testing.

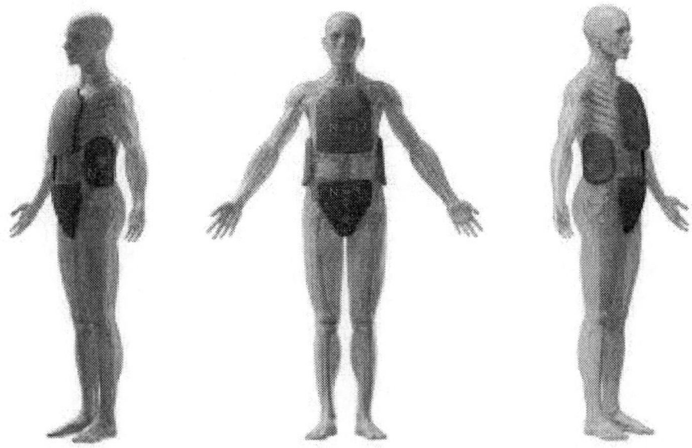

Figure 35. Anatomical locations of BABT impact and associated sample sizes annotated.

2.1 Specimen selection

For these studies, thawed, fresh-frozen, unembalmed cadaveric specimens were used to provide high-fidelity injury results in the context of a simulated physiological environment. All specimens were selected to fit stringent criteria that would allow the results to be more applicable to the relevant population, which in this case is a healthy adult male with good fitness. These criteria are detailed within

Table 16. Metrics screened for acceptance included a specimen's Dual X-Ray Absorptiometry (DEXA) score, Body Mass Index (BMI), and general medical history. DEXA and BMI values serve as an indicator of bone density and general health, which are important to align specimen quality to the relevant population and ensure that poor bone quality does not influence bony fracture results. Similarly, specimen cause of death (COD) and medical history were reviewed to ensure specimen did not have trauma or abnormalities related to the tested region. In addition to PMHS demographic information and medical history, anthropometry measurements were obtained prior to acceptance into the study. For the lower abdomen and lateral ribcage testing, measurements were taken to provide insights into where adipose tissue may be more prevalent, as both regions see increased amounts of adipose tissue [8].

Table 16. Specimen acceptance criteria

	Sternum	Lower Abdomen	Lateral Ribcage
Age (yrs)	18-65	18-65	18-65
Gender	Male	Male	Male
Stature (in)	65 – 73	65 – 76	65-77
Body Weight (lbs)	141 – 233	120 – 240	120 - 250
BMI (kg/m^3)	18.5 – 35	18 – 29	18 – 29
Specimen Type	Whole Body/Torso+Head	Whole Body/Torso+Head	Whole Body/Torso+Head +Proximal Extremities
DEXA BMD (Lumbar)	-1.0 < T-score <+2.5	-1.0 < T-score <+2.5	-1.0 < T-score <+2.5
Waist Circumference (cm)	Unspecified	Age < 40: < 39 inches (99 cm) Age > 40: < 36 inches (91cm)	Age < 40: < 39 inches (99 cm) Age > 40: < 36 inches (91cm)
Waist/Hip Ratio	Unspecified	Age < 40: < 0.96 Age > 40: < 0.92	Age < 40: < 0.96 Age > 40: < 0.92
Abdominal Circumference/Height Ratio	Unspecified	< 0.52	< 0.52

2.2 Specimen Preparation and Instrumentation

https://doi.org/10.52202/080042-0022

Specimen were thawed prior to preparation for the experimental test. Photos, planar x-ray, laparoscopic exams, and clinical computed tomography (CT) were leveraged throughout the process to document the specimen state. Pre-test CT scans (slice thickness of 0.625mm) enabled the documentation of existing disease processes and skeletal abnormalities prior to testing. Specimen preparation included simulating physiological processes including perfusion of the cardiovascular system, lung insufflation, and bladder-filling depending on the impact region. The details for each tested region are included in Table 17. For each impact location, the potential injury outcomes were evaluated to inform the selection of physiological processes to simulate. Lungs were insufflated for all of the testing locations to achieve realistic organ positioning within the abdomen and to simulate physiological lung pressurisation. Segments of the cardiovascular system were perfused for the sternum and lateral ribcage tests to create pressurised regions beneath the impact locations and were additionally injected with contrast to better visualise a cardiovascular injury. The bladder was filled for testing in the lower abdomen region to simulate worst-case loading to a fluid-filled sac.

Table 17. Simulated physiological processes or states for each impact region

Impact Region	Sternum	Lower Abdomen	Lateral Ribcage
Simulated Physiological Process or State	Lung Insufflation, Cardiovascular Perfusion	Filled Bladder (Volume)	Lung Insufflation, Cardiovascular Perfusion

2.3 Test Setup

All specimen were transferred to the same ballistics test facility for testing. Specimens were placed on a custom testing fixture to ensure orthogonal alignment of the impact location to the tested threat. The testing fixture also served to minimise bulk movement of the specimen. The testing fixtures for the each of the impact locations are shown in Figure 36. The specimen was placed on the test fixture and the physiological processes were simulated using active airway insufflation and vascular perfusion (with water) driven by down-regulated compressed air. The obliquity of the impact was controlled with the test fixture such that it simulated an impact straight on (zero-degree obliquity relative to the armour plate). Prior to fitting the body armour, a surface scan of the skin at the impact location was taken to supplement the understanding of the pre-impact condition. The scan was taken using a coordinate measurement machine (FARO Technologies, model FAROArm) was used to measure the armour standoff after the PPE was positioned to have a natural fit. The armour standoff is characterised by the depth of the air gap occurring between the skin surface at the impact location and the backface of the armour. Prior to the ballistic impact, an anterior-posterior and lateral planar x-ray were captured to document the pre-test condition. Still images of the test setup were taken pre-test. High-speed video was collected during the event to visualise the event. Immediately following the event, post-test photos and X-rays were taken, then the specimen was transported for a post-test CT scan.

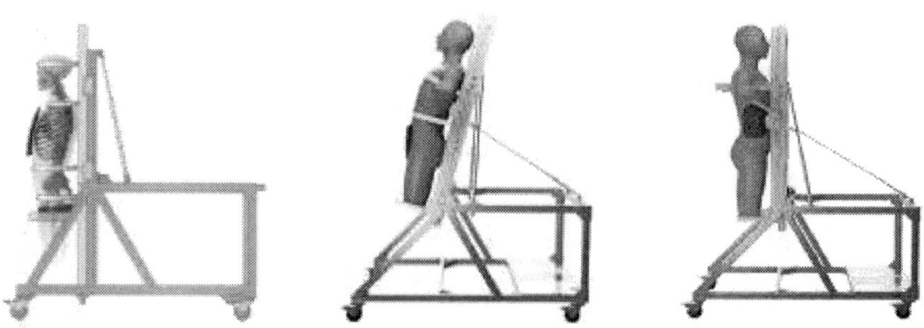

Figure 36. Fixture used for positioning specimen for sternum tests (Left), lower abdomen tests (Centre), and lateral ribcage tests (Right).

2.4 Post-test examination

Following the post-test CT scan, a comprehensive forensic examination was conducted on the specimen by a board-certified forensic pathologist. Sustained injuries were documented and photographed throughout the examination. Skeletal fractures were confirmed with CT scans, comparing the pre-test and post-test scans to ensure that the fracture was the result of the test event. An official post-test exam report, incorporating radiologic and examination findings, was prepared for each test subject. Documented injuries were coded using the Abbreviated Injury Scale (AIS) version 2015 by a Certified Abbreviated Injury Scale Specialist (CAISS) and reviewed by a board-certified forensic pathologist. An ISS was also generated. A pertinent metric also identified for each case was the Maximum AIS (MAIS) value, which is simply the maximum occurring AIS value for a given case's injuries.

2.5 Medical panels

Due to the lack of comparable injuries sustained in living invividuals, a medical panel was convened to provide additional insights into the operational impact of each simulated injury if it were to occur in a warfighter. The primary outcome of this panel was a categorisation of the injury severity coined the POIO. To support the medical panel, the study team compiled each test's data, which was called a 'case', to provide a comprehensive view of the injuries that occurred from the simulated BABT event. The team assembled packets of information containing all of the cases for a given region that included still images, CT scan snapshots, and AIS codes for each case. Additionally, specimen demographics (e.g., specimen age, weight, and cause of death) were included with each case.

Each medical panel was composed of a combination of board-certified forensic pathologists, trauma surgeons, and emergency medicine physicians. The panel reviewed the case information and deliberated on the projected operational outcome that could result if a living warfighter experienced the injuries that were observed in a simulated test with a PMHS. Their considerations included near-term impairment, long-term impairment, medical treatment, and rationale for assessment. The operational outcome, in this case, was a warfighter's ability to return to duty (RTD) or a potentially fatal injury. As part of their deliberation, the panel discussed what medical interventions were required to mitigate life-threatening injuries quickly and effectively. The defined categories of the POIO are detailed in Table 18.

Table 18. Projected Operational Injury Outcome (POIO) category definitions for body armour testing.

Level	Projected Operational Injury Outcome
I	RTD within 72 hours
II	RTD after 72 hours
III	RTD after 72 hours, potential for duty-limiting conditions
IV	Potentially fatal

In the case of the lower abdomen testing, additional deliberation occurred by the medical panel to determine the outcome dependent on the medical response time the hypothetical warfighter may experience. The three circumstances studied included a medical response time of less than one hour, less than 24 hours, or greater than 24 hours. This approach was taken due to the observation of bladder lacerations that could pose an infection risk that could escalate with increased time to treatment.

3. RESULTS

The results of each experimental study provided additional insight into the types of injuries that are characteristic of a BABT event. The subsequent sections provide additional detail on the injuries specific to each region. The test series were designed to vary injury outcome by input condition modulation, namely threat velocity, which was tested within the range of 580 to 945 metres per second. For the purposes of this paper the tests will be grouped by velocity range to anonymise specific conditions. Ranges are defined by Low (580-670 m/s), Mid (670-820 m/s) and High (820-945 m/s). Striking velocities vary within these ranges, and are not ordered within their group. For brevity, all of the AIS scores are not included for each test, but the ISS and MAIS scores are included.

3.1 Sternum

Sternum tests most often resulted in fractures of the sternum and ribs adjacent to the impact location at the sternum between ribs three and four. All tests had either a skin abrasion or laceration at the impact location, and in most cases, there was a sternum fracture occurring at the impact location. Rib fractures

https://doi.org/10.52202/080042-0022

occurred bilaterally and typically were centred around the impact location. Disruptions to the pleural cavity and mediastinal contents, that could result in hemomediastinum and pneumomediastinum in a living individual, occurred in severe tests. In some cases, there was an open chest wound along with a lung laceration to either side. Test results with injury scores including ISS, MAIS, and Projected Operational Injury Outcome are in
Table 19.

Table 19. Sternum testing injury results

Test	Velocity Range	ISS	MAIS	POIO	Skin Abrasion and Laceration	Sternum Fracture	Lung Laceration	Left Rib Fracture	Right Rib Fracture	Other Injuries
1	High	10	3	II	x	x		3-4	2-4, 7-9	
2	Mid	10	3	II	x	x		3-5	5	Hemomediastinum; Pneumomediastinum
3	High	10	3	II	x	x			2,4,5	
4	Low	1	1	I	x					
5	Mid	10	3	IV	x	x		2-5	3-5	Hemomediastinum; Pneumomediastinum
6	Mid	10	3	II	x	x		3-4	3-4	Ruptured pulmonary blebs of right and left lung; Superficial laceration of the epicardial adipose tissue of left ventricle
7	Low	10	3	II	x	x		4	2-5, 7-9	Ruptured pulmonary blebs of right lung
8	Low	5	2	II	x	x		4	4	
9	Low	26	5	III	x	x	Left	4-5	4-5	
10	High	26	5	IV	x	x	Right	3-4	2-4	Pneumothorax of right side
11	High	26	5	IV	x	x	Left	2-6	4-6	Laceration of the adipose tissue of the anterior mediastinum
12	High	26	5	IV	x	x		4-7	4-8	Laceration of the pericardial sac; Epicardial surface laceration spanning the left and right ventricles
13	High	26	5	III	x	x	Right		3-6	Multiple blebs on the right lung

3.2 Lower Abdomen

Lower abdomen testing injuries varied due to the different impact locations tested. Initial results at the pubic symphysis location were generally less severe with AIS values of 1. This did not warrant further exploration, so the impact location was moved to the anterior wall of the bladder to explore potentially more severe injury response. At the pubic symphysis, specific injuries spanned a superficial penetrating injury to the abdomen and a penis laceration. Impacts over the bladder resulted in skin abrasions, skin lacerations, rectus abdominus lacerations, and a bladder wall laceration. A seventh test was performed but was excluded from further analysis due to a complete penetration that occurred during the test that

rendered injuries indiscernible between non-penetrating effects of the armour and penetrating injuries. Test results with injury details are included in
Table 20**Error! Reference source not found.**.

A unique approach to the lower abdomen study was the designation of different injury outcome scores for different times to medical intervention. The timeframes of interest were less than one hour, less than 24 hours, and greater than 24 hours. Differences in scores across different timeframes occur due to likelihood of infection and pressing need for care, which was seen in the impacts centred over the bladder.

Table 20. Lower abdomen testing injury results.

Test	Impact Location	Velocity Range	ISS	MAIS	POIO <1 hr	POIO <24 hr	POIO >24 hr	Injury Description
1	Pubic Symphysis	High	1	1	II	II	II	Superficial abdominal penetrating injury; Penis laceration
2	Pubic Symphysis	High	1	1	II	II	II	Superficial abdominal penetrating injury
3	Bladder	High	9	3	III	III	IV	Skin abrasion; Skin laceration; Bladder laceration
4	Bladder	High	4	2	II	II	IV	Skin abrasion; Skin laceration; Rectus abdominus laceration
5	Bladder	High	4	2	III	III	IV	Skin laceration; Rectus abdominus laceration; Tissue avulsion
6	Bladder	Mid	5	2	I	I	I	Skin abrasion; Skin laceration; Bladder laceration, partial thickness

3.3 Lateral Ribcage

The lateral ribcage tests yielded both skeletal and soft tissue injuries. The test results varied when comparing tests conducted on the right and left side, which is expected due to the anatomic positioning of certain critical organs. On the right side, the injuries spanned ribs six through nine, with comminuted and displaced fractures. In addition, there was a large defect of the chest wall along with a large area of pulpification of the superior aspect of the right lobe of the liver. On the test impacting the left side, there were multiple comminuted and displaced rib fractures along with multiple pleural lacerations, a lung laceration, and hilar lacerations. Test results with injury details are included in Table 6. Tests on this region are ongoing.

Table 6. Lateral ribcage testing injury results.

Test	Impact Location	Velocity Range	ISS	MAIS	POIO	Injury Description
1	Right Rib 8	High	26	4	IV	Skin laceration; Fracture of right ribs 6-9; Defect of the right lateral chest wall; Area of pulpification of the right lobe of the liver
2	Left Rib 8	High	21	4	IV	Skin laceration; Fracture of left ribs 7-9; Pleural laceration; Lung laceration; Hilar lacerations

4. DISCUSSION

To characterise injury outcomes from non-penetrating ballistic tests of hard armour plates for different body regions, multiple tests were conducted targeting the sternum, the lower abdomen, and the lateral ribcage. These ongoing studies are in pursuit of additional detail of BABT that can occur due to non-penetrating ballistic impacts. These studies investigated the injury results and patterns that occur throughout a range of striking velocities on an armour-threat pairing for each region of interest that could result in varied levels of residual clay backface deformation spanning below and above the 44mm

https://doi.org/10.52202/080042-0022

standard. Each of the tested regions provided a range of different responses, however the conclusions span similar themes which are outlined below.

4.1 Merit of Projected Operational Injury Outcome (POIO)

The projected operational injury outcome score represents a different outcome than existing injury severity metrics that have traditionally been used to characterise polytrauma. AIS scores indicate injury severity, and when grouped together into an aggregate metric such as ISS and NISS, may be correlated to polytrauma severity [9]. AIS scores can be assigned to injuries, typically within a clinical setting, regardless of the cause or environment. The POIO incorporates the evaluators' knowledge of the injury results of a simulated BABT event and combines them with their medical experience to project the operational outcome should a warfighter encounter the same injuries in-theatre. The defined POIO categories are specifically designed to apply to a military operational environment due to the panelists' experience. While the POIO also factors in injury severity, it also incorporates injury treatments and interventions specifically required for a warfighter to return to duty. This uniquely accounts for the immediate interventions that a warfighter could receive in-theatre and any required monitoring or continued treatments that would support them to return in full capacity.

The merit of the categorisation has only been examined in the context of this study, elaborated further below, and warrants further examination as to its applicability to other studies. More broadly, the paradigm described of convening a medical panel with relevant experience to examine and project injury implications can be applied to other environments to achieve similarly insightful results. Thus, multiple categorisations are possible, but each categorisation should therefore be ubiquitously paired with its stated goal and context.

The differences between AIS and POIO in their definition and utility can be highlighted by a notable case per impacted region. Sternum test seven was assigned an ISS of 10, and a POIO of IV. Other cases that were assigned a POIO of IV, or potentially fatal, had ISS values of 26. The differentiator was the identification of the likely resulting hemomediastinum and pneuomomediastinum due to the location and severity of adjacent rib fractures. AIS scores, along with the ISS, does not convey the severity and implications of the polytrauma an individual could face if this injury was encountered in-theatre. Similarly, lower abdomen test five resulted in an ISS of four, and a POIO of II to V depending on time to treatment. The infection risk in a living individual, particularly over time, would grow and impart a significant risk to life if left without sufficient treatment. An ISS of four would not typically be deemed a significant risk to life. Lastly, lateral ribcage test two resulted in an ISS of 21 and a POIO of IV. Although the sample size is small, variation is already present with a distribution of ISS between 21-26 for both cases assigned a POIO of IV. Additional tests will likely further demonstrate deviations between trends of ISS and POIO for this region. This demonstrates the important insights that the medical panel was able to provide when viewing each case through the lens of warfighter operational outcome.

Another outcome of this study is the differences that were observed across regions. Assuming that the injuries produced within the PMHS tests occurred on a living warfighter, there are different implications for each region in terms of the Projected Operational Injury Outcome. A POIO metric of IV spans ISS values of 4 to 26 across all three tested regions, assuming the longest time to treatment for the lower abdomen injury. If assuming access to care is timely for lower abdomen tests, only the lateral ribcage and sternum tests produced a POIO value of IV with ISS values of 10 to 26. Collectively the results confirm, through the novel Projected Operational Injury Outcome categorisation, that injuries due to BABT across different regions are very complex and can have serious implications from an operational perspective of warfighters out of the fight.

4.2 Study Enablers

The main outcomes from each of the studies conducted are the injury results with affiliated metrics. These results enable the creation of statistical models that can predict an injury outcome given an input variable. An example of such an outcome is a binomial regression with an injury threshold that given a prescribed input condition predicts, with an affiliated confidence value, the likelihood that the threshold is surpassed. Additional models may be developed given the results of biomechanical sensor data that was collected during the tests.

One of the most significant results of the study spanning multiple regions is an improved understanding of how the current backface standard translates to different regions of the body. The use of the projected operational injury outcome could enable decision-makers in the armour space to assess the potential risk in fielding a given armour system. The injuries observed over the sternum and lateral ribcage were generally interpreted as more severe than those occurring in the lower abdomen region, though the regions have nuances to its input conditions. All armour systems tested resulted in injuries that were projected to take the warfighter out of the fight for a minimum of 72 hours, and some had the potential to be fatal.

Each armour system has a unique relationship between input conditions and backface deformation, but this study serves to shed light on the range of operational implications that a range of inputs could have on different systems. A working knowledge of the armour and threat space helps to contextualise the results as well understanding how different systems may respond to different threats and which are relevant. Ultimately, armour systems have an upper bound of input conditions that will result in non-penetrating impacts and those should be considered as additional context. Fundamentally, this study and each test conducted on each region was conducted in a controlled and methodical manner that closely simulates a BABT event, which is imperative to better understand the risks affiliated with BABT.

4.3 Study Limitations

While the conducted tests created a robust dataset, the research conducted has inherent limitations to be acknowledged.

4.3.1 PMHS model

The PMHS, or cadaveric, model is beneficial as it shares the anatomy of the human, but it lacks key qualities that influences the results. Cadaveric tissue lacks muscle tone, so it is not possible to understand the influence that actively contracted musculature may have on experimental results. Additionally, physiological processes, such as active blood flow and respiration are not present in cadaveric models. The test team took steps to simulate physiologic conditions by simulating these processes but they are not identical. Cadaveric models do not represent the inflammatory response to a traumatic event that could influence the outcome in an emergent scenario. The medical panel provides insight into the influence of the inflammatory response due to the representation of trauma surgeons and emergency medicine physicians, but it is still a limitation. Lastly, a well-understood limitation to cadaveric testing is the inherent variability that comes with studying humans.

4.3.2 Lack of generalisation

The tests previously described span a framework applicable across multiple body armour systems, but the injury results are not yet generalisable to other armour systems due to the multi-dimensional trade space spanning specimen qualities, armour metrics, and injury complexities. Additionally, most, if not all, tests conducted resulted in injuries which limits the applicability of these data to the population of interest. There are additional tests on isolated armour plates that may bolster the understanding of the influence of armour-specific qualities that may be more generalisable as well as tests conducted that result in more minor injuries. These tests include enhanced clay testing to understand the dynamic loading scheme as well as digital image correlation on unbacked armour systems. Both tests could provide a better understanding of the dynamics of an isolated event on an armour system that could bolster findings with PMHS.

5. CONCLUSION

This research study worked toward a comprehensive understanding of the relationship between armour performance and warfighter injury for an armour type configured to different body regions. A dataset representing a sample with controlled demographics, conducted with live ammunition, and hard armour was essential to gain a representative understanding of how BABT injuries may increasingly impact warfighters as there are advancements in armour materials. Each test on a new region further detailed the types of injuries that may occur during a non-penetrating ballistic event. Chest plates are most often studied in the literature as they are commonly used in both military and non-military applications, but peripheral armour, or armour of the side and lower abdomen, is lesser studied [10]. A conglomerate of tested regions provides a multitude of such insights that can add additional value to those making decisions about armour acceptance and risk management. The most substantial contribution to this study, along with its controlled attributes, was the medical foresight to create a link or correlation between cadaveric tests and probable living warfighter injury outcomes. The novel injury categorisation directly responds to the NRC's finding of a need for a trauma categorisation scale that is well-suited to ballistic trauma tailored to the military population [7]. This information is invaluable to the armour community to ensure that warfighters are properly protected and can perform their duties.

https://doi.org/10.52202/080042-0022

Acknowledgments

The authors would like to express their appreciation to the U.S. Special Operations Command and the US Army Combat Capabilities Development Command Soldier Center (DEVCOM Soldier Center) for funding this work under Contract No. is H92222-15-D-0004. In addition, they would like to express their appreciation to the medical examiners that contributed to the forensic examinations and medical panels for this work.

The views, opinions, and findings contained in this article are those of the authors and should not be construed as official or reflecting views of the Joint Trauma System, Armed Forces Medical Examiner System, Defense Health Agency or the Department of Defense.

References

[1] A. Iwaskiw, A. Wickwire, M. Vignos, C. Howes, N. Hahne, A. Injeian, N. Steiner, M. Bevan, E. Bar-Kochba, E. Mazuchowski, M. Clark, C. Carneal and D. Drewry, "Initial injury observations in postmortem human subjects: Non-penetrating ballistic impacts over the sternum with light- weight hard armour plates," in *Personal Armour Systems Symposium*, Copenhagen, 2021.

[2] N. Montarelli, C. E. Hawkins, M. A. Goldfarb and T. F. Ciurej, "Protective Garments for Public Officials," Defense Technical Information Center, Fort Belvoir, 1973.

[3] E. Hanlon and P. Gillich, "Origin of the 44-mm Behind-Amour Blunt Trauma Standard," *Mil Medicine,* vol. 177, no. 3, pp. 333-339, 1973.

[4] C. R. Bass, R. S. Salzar, S. R. Lucas, M. Davis, L. Donnelan, B. Folk, E. Sanderson and S. Waclawik, "Injury Risk in Behind Amor Blunt Thoracic Trauma," *International Journal of Occupational Safety and Ergonomics,* vol. 12, no. 4, pp. 429-442, 2006.

[5] M. DeMaio, "Closed chest trauma: Evaluation of the protected chest," in *NATO Task Group on Behind Armor Blunt Trauma*, Copenhagen, 2001.

[6] C. Bir, D. Viano and A. King, "Development of biomechanical response corridors of the thorax to blunt ballistic impacts," *J. Biomech,* vol. 37, no. 1, pp. 73-79, 2004.

[7] Committee on Testing of Body Armor Materials for Use by the U.S. Army--Phase III; Board on Army Science and Technology; Division on Engineering and Physical Sciences; National Research Council, "Testing of Body Armor Materials: Phase III," The National Academies Press, Washington, D.C., 2012.

[8] T. Rankinen, S. Y. Kim, L. Perusse, J. P. Despres and C. Bouchard, "The prediction of abdominal visceral fat level from body composition and anthropometry: ROC analysis," *Int. J. Obes.,* vol. 23, no. 8, pp. 801-809, 1999.

[9] C. S. Palmer, B. J. Gabbe and P. A. Cameron, "Defining major trauma using the 2008 Abbreviated Injury Scale," *Injury,* vol. 47, no. 1, pp. 109-115, 2016.

[10] N. Prat, F. Rongieras, J. C. Sarron, A. Miras and E. Voiglio, "Contemporary body armor: technical data, injuries, and limits," *Eur J Trauma Emerg Surg,* vol. 38, no. 2, pp. 95-105, 2012.

Bladed Weapon Assaults and Human Vulnerability

J. Barnes-Warden[1], and P. F. Mahoney[2]

[1] *Metropolitan Police Service, Physical Protection Group, 60 Albany St, London, NW1 4EE*
[2] *Centre for Blast Injury Studies, Imperial College London,* SW7 2AZ

Abstract. Within the UK, in the year between 1 July 2021 and the 30 June 2022 nearly fifty thousand (n=49,991) knife-enabled crime offences were recorded by police within England and Wales. Furthermore, over 22 percent (n=11,232) of these offences were committed within the Metropolitan Police District (MPD) [1]. Hence, bladed weapon assaults are a significant threat to frontline police officers and staff, particularly those patrolling within London.
Many bladed weapon assaults, and the associated injuries sustained, are survivable with swift medical intervention, however some injuries are so severe that death ensues. Whilst evaluating knife threats and anatomical vulnerability to such assaults, an opportunity arose to review post mortem (PM) reports for deaths caused by bladed weapons which occurred to citizens within the Greater London area. These fatal cases are of great value in understanding where critical and non-critical injuries have occurred and, therefore, are of benefit in evaluating the coverage of body armour, issued to frontline police officers and staff, and to determine where vulnerabilities exist.

This study reviews a large dataset (n=75) of PM reports of deaths due to bladed weapon assaults in 2019, within the MPD. It includes anatomically correct body maps to illustrate the location of the injuries and identifies the principal location of the fatal injury. It also summarises the cause of death. Statistics are presented with regards to the depth of the fatal stab wound, blade type, as well as an overview of the victims, such as age and gender.

In UK policing, Operation Hampshire is a recently introduced means for police officers and staff to report an assault, or having been subjected to hate crime, whilst on duty. This study presents data from Operation Hampshire associated with bladed weapon assaults. Hence, this second dataset outlines blade weapon injuries received during policing, as a useful comparison to the fatal injuries reported within PM reports.

The aim of this study is to analyse both datasets to determine the most probable location(s) of fatal and non-fatal stab wounds during attacks on police officers and staff with bladed weapons. Furthermore, this study also reviews current body armour designs, and proposes enhancements to protection schemes, to be worn by front line police officers and staff, to reduce their vulnerability.

. INTRODUCTION

Within the UK between 1 July 2021 and the 30 June 2022 nearly fifty thousand (n=49,991) knife-enabled crime offences were recorded by police within England and Wales [1]. Hence, a significant threat to frontline police officers and staff are knife assaults. Many bladed weapon assaults, and the associated injuries sustained, are survivable with swift medical intervention, however some injuries are so severe that death ensues. Whilst evaluating knife threats and anatomical vulnerability to such assaults, an opportunity arose to review post mortem (PM) reports for deaths caused by knife assaults which occurred to citizens within the Greater London area. These fatal cases are of great value in understanding how and where critical injuries have occurred and, therefore are of benefit to evaluating the coverage of body armour issued to police officers and staff.

This study reviews a large dataset of fatal weapons attacks (n=75) which occurred to citizens within the Greater London area, in 2019, by examining the post mortem (PM) pathological reports of murder victims. Examining these fatal cases is an invaluable resource when evaluating the coverage of body armour issued to frontline police officers and staff. Hence, this study evaluates the location and depth of injuries, cause of death and, if known, details of the weapon(s) and circumstances, and illustrates these injuries on anatomically correct, body maps.

Furthermore, records of assaults and hate crime on MPS police officers and staff, documented under Operation Hampshire, have been used to accompany the above data, by summarising non-fatal bladed weapon attacks during police operations.

The aim of this study is to appraise the most probable location(s) of fatal and non-fatal stab wounds and to review current body armour designs, worn by front line police officers and staff, to evaluate their vulnerability.

 https://doi.org/10.52202/080042-0023

. EXAMINING FATAL BLADE WEAPONS ASSAULTS ON CITIZENS WITHIN THE MPD

Preparing and applying best practice to handling the data

During the preparation and planning of this study the year chosen, to examine fatal bladed weapon assaults on citizens within the Metropolitan Police District (MPD), was 2019. This year was prior to the coronavirus global pandemic, therefore the dataset would not be influenced by government restrictions, such as national lockdowns of the UK population.

In 2019 there were 77 fatal stabbings of citizens within the MPD, of which 75 full PM reports were available. Best practice was applied when handling the data used within this study by:

1. the authors notifying the Senior Coroner at Westminster Coroner's Court;
2. applying Caldicott principles to the dataset throughout this study, including anonymization;
3. commissioning a Crown Prosecution Service (CPS) approved company (Evidential Ltd) to plot each bladed weapon assault on an anatomically correct model (similar to the professional avatar images used in court cases);
4. using Criminal Justice Secure eMail (CJSM) addresses to communicate;
5. retaining post mortem reports within the MPS electronic storage.

Illustrating the data

Four anatomically correct avatars, illustrating anterior, posterior, right and left lateral views, were used to create body maps to show the location of knife injuries received by the murder victims. The avatars were created using open source software Make Human, Version 1.2.0 (Mac IOS) [2] nominally for an average UK male [3]. Details of the Make Human avatar are male 100%; age 29; muscle 43.90%; weight 119.40% and height 173.01cm. This was combined with an internal anatomical male model from 3D 4 Medical - Essential Anatomy 5 (V5.0.8).

However the avatar is androgynous and illustrates wounds for the 75 male and female victims in the correct anatomical location. Hence, the body maps, in figures 1 to 4 below, illustrate the distribution of injuries, which include the fatal wound(s), as well as other injuries received during the assault. A caveat being, due to the high frequency, non-fatal defensive wounds to the hands and inconsequential wounds to the feet have not been illustrated.

Victims' background

Gender and age of victims

The gender and age of victims are summarised in table 1.

Table 1 – gender and age of victims of fatal bladed weapon assaults within the MPD in 2019

			Age range				
	Total of fatal stabbings in 2019	Fatal stabbings in dataset	<18	18 - 24	25 - 31	32 - 38	39+
Male	67 (87%)	66 (88%)	6 (8%)	27 (36%)	12 (16%)	11 (15%)	10 (13%)
Female	10 (13%)	9 (12%)	1 (1%)	2 (3%)	3 (4%)	0 (0%)	3 (4%)
Total	77 (100%)	75 (100%)	7 (9%)	29 (39%)	15 (20%)	11 (15%)	13 (17%)

Circumstances

The majority of the PM reports contain a brief description of the circumstances of the assault and often toxicology results are present. Although, we cannot know where details of the assault have not been passed onto the pathologist, the following percentages (in table 2 below) provide an indication of substance use and the events surrounding the fatalities in 2019.

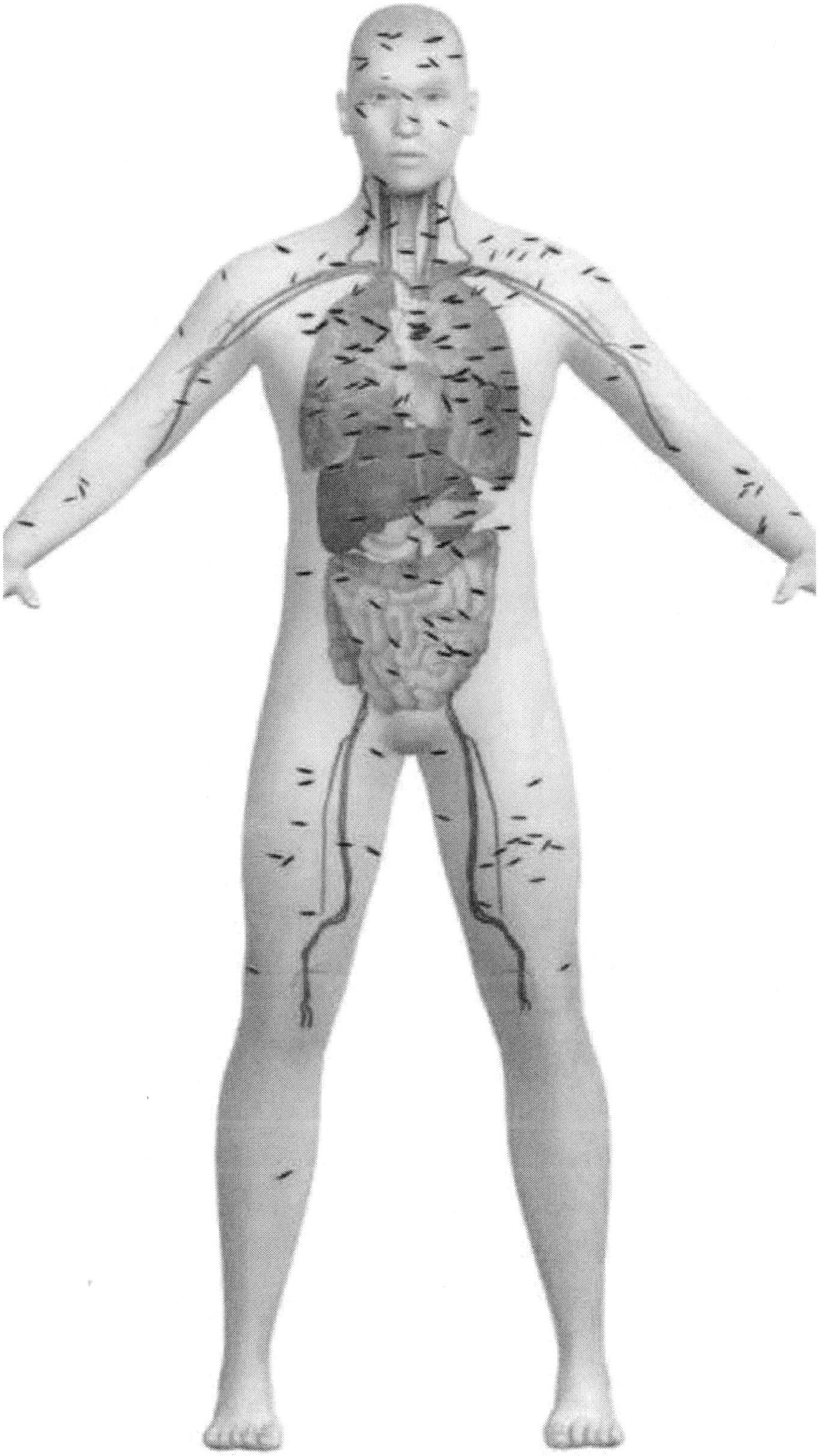

Figure 1: Anterior body map of injuries from fatal bladed weapon assaults

(Note: non-fatal, defensive wounds to the hands and inconsequential wounds to the feet, have not been illustrated).

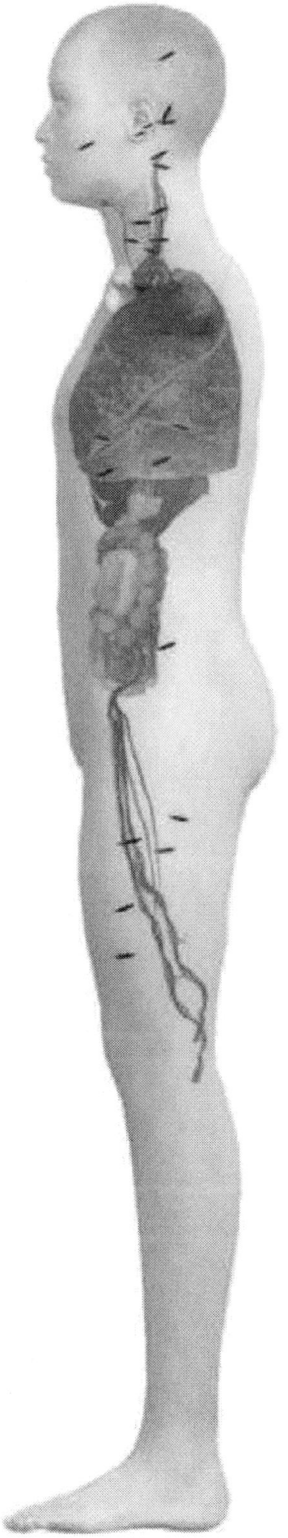

Figure 2: Left lateral body map of injuries from fatal bladed weapon assaults

https://doi.org/10.52202/080042-0023

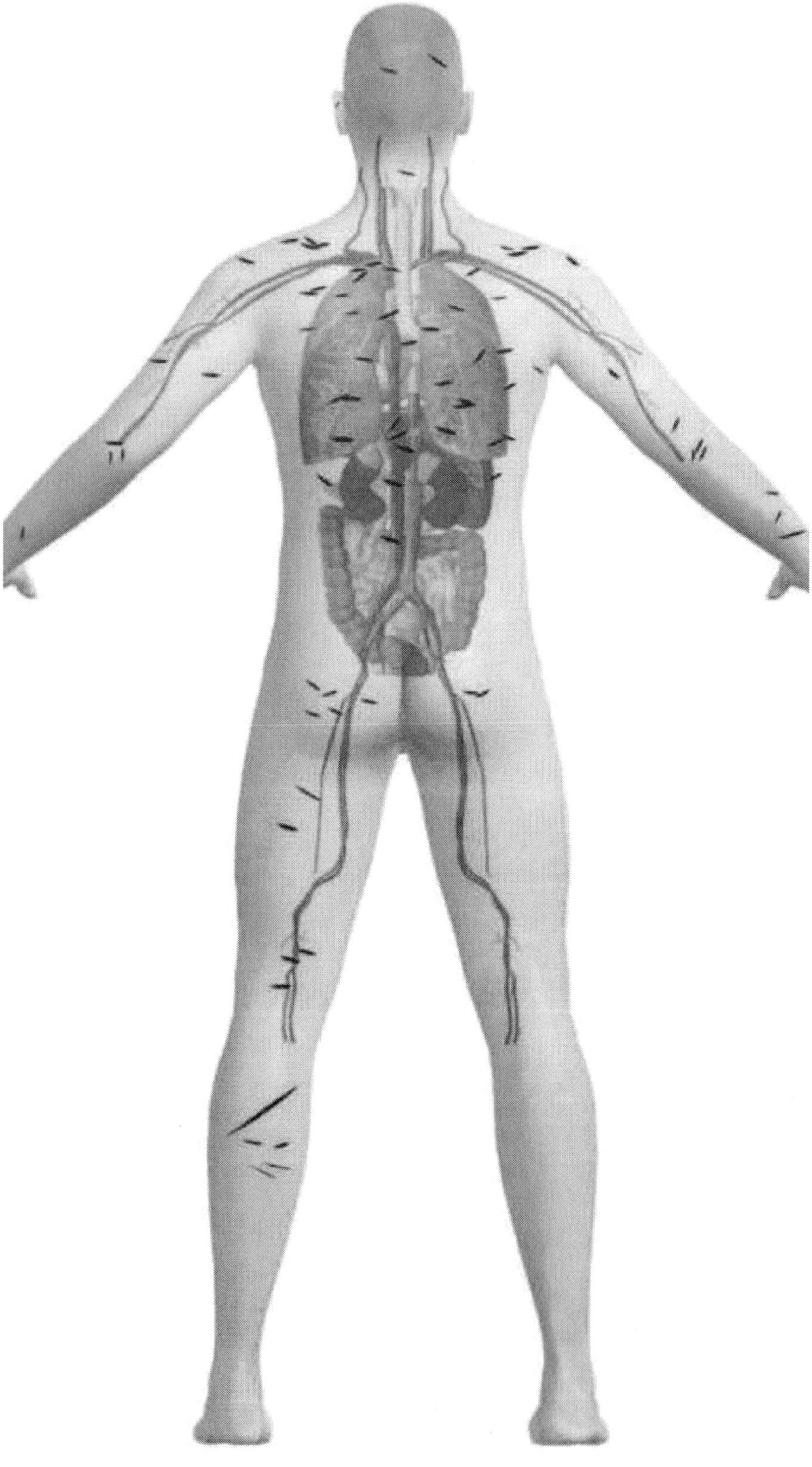

Figure 3: Posterior body map of injuries from fatal bladed weapon assaults

(Note: non-fatal, defensive wounds to the hands and inconsequential wounds to the feet, have not been illustrated).

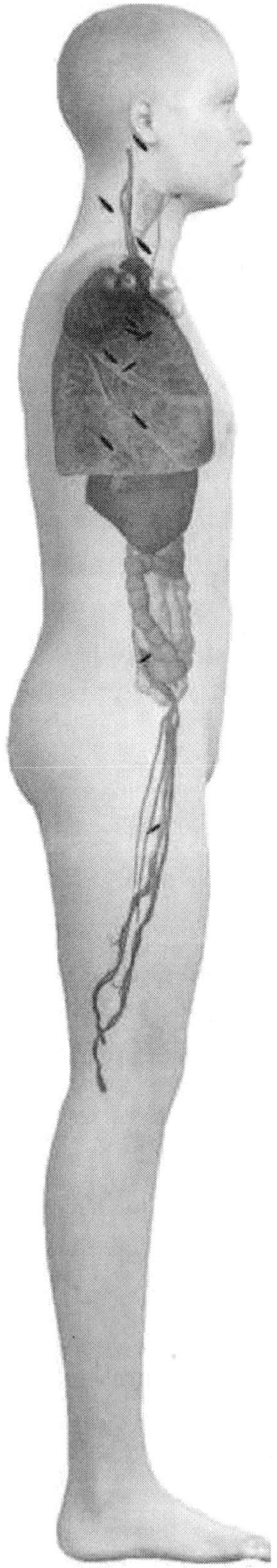

Figure 4: Right lateral body map of injuries from fatal bladed weapon assaults

Table 2 – circumstances of victims of fatal bladed weapon assaults within the MPD in 2019 (Note: if a victim has taken non-prescription drugs and was part of a gang they would be recorded in both categories)

		Circumstances				
Total of fatal stabbings in 2019	Fatal stabbings in dataset	Presence of non-prescription drugs	Presence of significant quantity of alcohol only	Multiple assailants or gangs	Domestic dispute	Terrorism
77	75	43 (includes 7 with alcohol)	5	29	8 (3 male; 5 female)	2

Principal anatomic locations of fatal stab wounds

Figure 5 illustrates a high level overview of the principal anatomical locations of fatal stab wound victims in 2019.

Two thirds of fatal stab wounds are located within the torso and nearly a quarter of fatal stab wounds are to the neck.

Within the dataset many victims received several bladed weapon injuries, clustered within a similar location, eg within their torso, which would include the fatal wound. However there are three victims, represented in figure 5, as having multiple sites of fatal stab wounds. The combined fatal wound locations are i) head and neck, ii) neck, chest and arm (brachial artery) and iii) torso and lower extremities (femoral artery and vein).

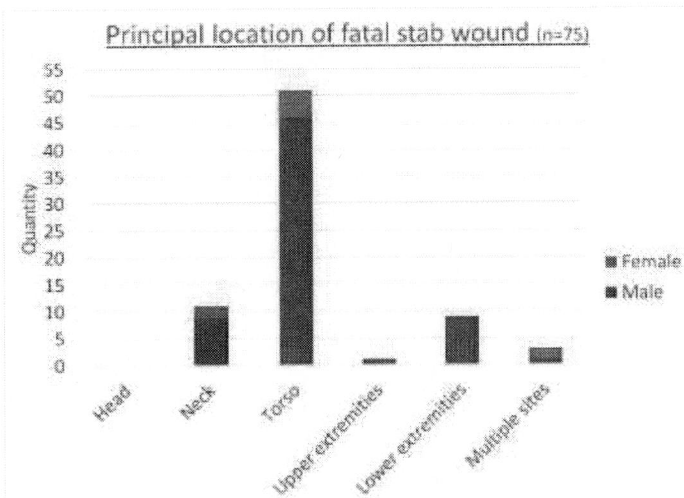

Figure 5 – Principal location of fatal bladed weapon injury

In over 90% of victims studied, in this review of bladed weapon assaults, their cause of death was ascribed to **loss of blood**.

Fatal damage within the torso

Looking further at the sites of fatal damage within the torso, the heart, lungs, aorta, inferior vena cava (IVC) and a combination of these, account for 74% of the sites within the torso.

Fatal damage to an organ(s) accounts for 60% of deaths, compared with fatal damage to a great vessel(s) accounting to 22%, in 6% of cases damage to both organ(s) and great vessel(s) are cited as the cause of death. Furthermore, in 12% of cases the wounds are complex, leading to extensive damage within the victim's torso.

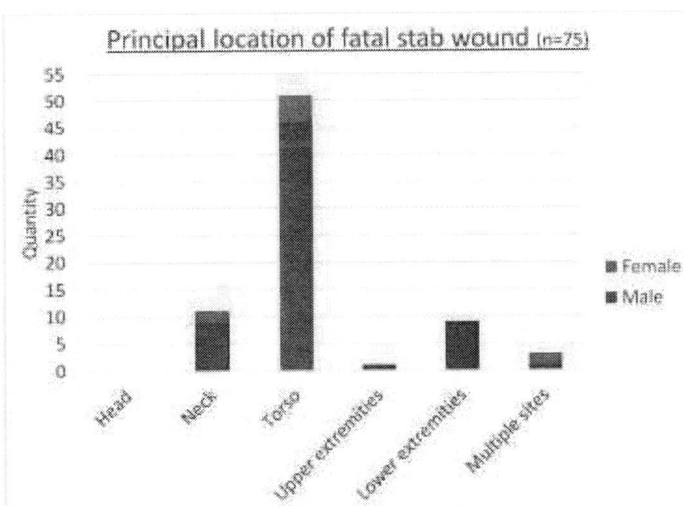

Figure 6 – Frequency of fatal damage to organs and great vessels within the torso

https://doi.org/10.52202/080042-0023

Fatal damage to the neck

There were 11 bladed weapon assaults to the neck which resulted in death, mainly due to critical damage to arteries and vessels of the victim. Eight of the victims had damage to their carotid artery and their internal or external jugular vein. In fifty percent of the above 8 victims their trachea was also damaged.

Two victims received fatal damage to their subclavian artery, which is near to the clavicle, however the point of entry of the knife was in the neck. One victim had fatal damage to their axillary vein.

Fatal damage within the upper and lower extremities

There was only one victim with an individual, fatal stab wound to the upper extremities which occurred to the victim's upper arm, cutting the brachial artery and damaging the brachial vein. Severing of the brachial artery also contributed to the death of another victim.

Nine of the victims of fatal stab assaults within the dataset were located in their lower extremities. Of these, six fatal wounds were attributed to severing the femoral artery and/or vein. Furthermore, two assaults were caused by fatal injury to the popliteal fossa (behind the knee where structures pass between the thigh and leg) and one death occurred due to a severed varicose vein.

Depth of fatal stab wound

Over 95% of fatal stab wounds were 50mm or greater in depth, as illustrated in figure 7. This is valuable knowledge for the design of protection schemes when combined with the depth of the critical organs, as reproduced from Breeze et al [4] in table 3.

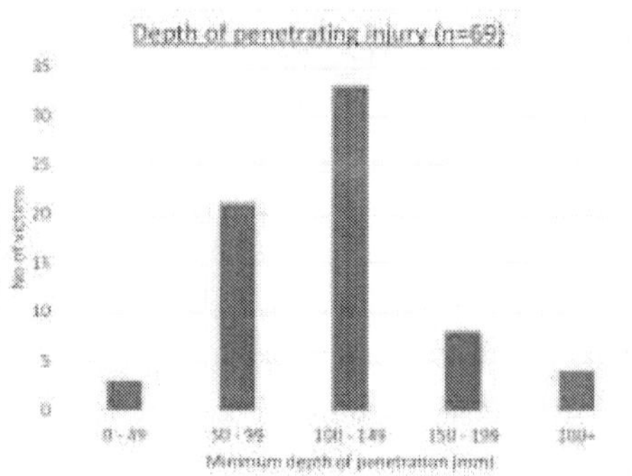

Figure 7 – depth of penetrating injury described within PM reports

Table 3 – Depth of organs reproduced from reference 4

Percentile	Entry point 1		Entry point 2		Entry point 3
	Skin to lung	Skin to heart	Skin to lung	Skin to heart	Skin to liver
1%	18	18	19	19	17
5%	19	20	23	23	19
25%	23	24	29	39	23
50%	25	27	33	49	26
75%	30	32	40	57	30
95%	37	40	51	65	37
99%	44	46	60	71	42
Mean	27	28	35	47	27

Type of bladed weapon

Within the 75 PM documents there are 42 reports which describe wounds solely from knives with a single cutting edge. Only six reports describe wounds solely from knives with a double cutting edge, and in one report both single and double bladed weapons are referred to.

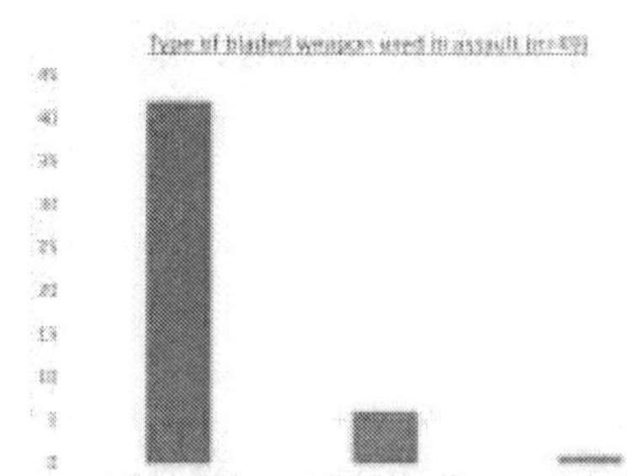

Figure 8 – type of bladed weapon described with PM reports

EXAMINING OPERATION HAMPSHIRE BLADED WEAPON ASSAULTS

Background to Operation Hampshire

Operation Hampshire was introduced into UK policing in 2016 as a means for police officers and staff to report an assault, or having been subjected to hate crime, whilst on duty. The principal data set is based on assaults within the Metropolitan Police Service (MPS) recorded from the beginning of January

https://doi.org/10.52202/080042-0023

2018 to the end of March 2022 [5]. Furthermore, detailed analysis of the location of injuries from a single year (2021) are presented [6].

Assaults recorded

Between January 2018 and March 2022 there were 30,763 Operation Hampshire victims reported, which averages at 603 per month. From this data set 467 victims, on average of 9 per month, were identified as victims of knife crime during the 51 months [5].

Severity of injury

In the full MPS data set (January 2018 to March 2022, inclusive) for Operation Hampshire 44% of MPS victims (n=13,548) sustained an injury. Of these victims 39% sustained minor injuries, 4% received moderate injuries, 1% was serious and one assault was fatal. This indicates that, in general, police officers and staff are more likely to endure minor injuries, rather than moderate, serious or fatal ones.

Table 4 – All MPS assaults reported to Operation Hampshire between January 2018 to March 2022 (inclusive) reproduced from reference 5

[illegible]	2018	2019	2020	2021	2022	Grand Total
Threats Only						
Offences	114	190	218	186	45	668
%	2%	2%	4%	3%	1%	2%
No Injury						
Offences	3107	3329	4458	4525	1118	16527
%	52%	55%	74%	75%	19%	54%
Minor						
Offences	2432	2743	3088	2923	694	11856
%	40%	44%	51%	47%	12%	39%
Moderate						
Offences	285	293	333	300	81	1294
%	5%	5%	6%	5%	1%	4%
Serious						
Offences	81	82	107	100	25	395
%	1%	1%	2%	2%	0%	1%
Fatal						
Offences	0	0	1	0	0	1
%	0%	0%	0%	0%	0%	0%
Grand Total	[illegible]	[illegible]	[illegible]	[illegible]	[illegible]	[illegible]

Table 5 – MPS knife crime victims reported to Operation Hampshire – January 2018 to March 2022 (inclusive) reproduced from reference 5

[illegible]	2018	2019	2020	2021	2022	Grand Total
Threats Only						
Offences	0	5	5	1	0	11
%	0%	4%	4%	1%	0%	2%
No Injury						
Offences	18	30	33	20	9	110
%	19%	26%	28%	17%	47%	24%
Minor						
Offences	56	68	61	64	9	259
%	60%	59%	51%	53%	47%	55%
Moderate						
Offences	14	8	11	27	1	61
%	15%	7%	9%	22%	5%	13%
Serious						
Offences	5	3	9	9	0	26
%	5%	3%	8%	7%	0%	6%
Fatal						
Offences	0	0	0	0	0	0
%	0%	0%	0%	0%	0%	0%
Grand Total	[illegible]	[illegible]	[illegible]	[illegible]	[illegible]	[illegible]

However, when comparing the above percentages, based on the full dataset, with injuries specifically sustained by knife crime victims a different distribution of injury severity emerges. Injuries were recorded in 74% of knife crime victims (n=346), with 55% of victims reporting minor injury, 13% with moderate injury, 6% with serious injury and no fatalities [5].

Location of injuries from knife assaults in 2021

In 2021 there were 121 knife crime assaults in which nine caused serious injury to the victims. Figure 9 is a Vitruvian map which illustrates the locations in which victims received serious injuries. Hence, victims received injuries to hands, wrists, arms, chest, abdomen, back, neck, jaw and forehead.

Furthermore, in one knife crime assault the police body armour was impacted, and in another knife assault struck the officers radio. Hence the location of both of these attacks were also to the torso [6].

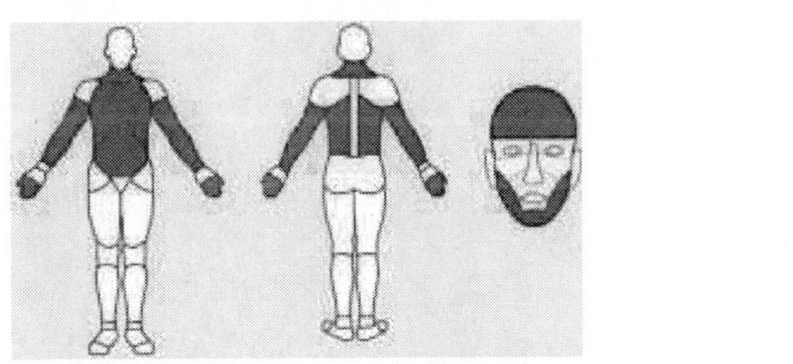

Figure 9 – Vitruvian map indicates the location of injuries recorded during 2021, reproduced from reference 6

DISCUSSION

The general distribution of wounds, described in the 75 post mortem reports from victims murdered in 2019, shown that there are approximately twice the quantity of stab wounds to the anterior body map, compared with the posterior. When analyzing the left and right lateral body maps, there are twice the number of stab wound on the left lateral body map, compared with the right, which is often attributed to more assailants being right-handed and thereby stabbing the victim on the left during a frontal attack [7].

https://doi.org/10.52202/080042-0023

Furthermore, when studying the positions of the wounds there is there is no cause to believe that certain anatomical locations have been targeted by the assailant.

When analysing the location of the fatal stab wound two thirds of fatal stab wounds are located within the torso and nearly a quarter of fatal stab wounds are to the neck. When determining the cause of death loss of blood was cited in over 90% of the victims.

Further scrutiny of the injuries to the torso revealed acute damage to one or multiple organs, great vessels or both. Victims with critical damage to their neck had injuries to their carotid artery, internal or external jugular vein, subclavian artery or axillary vein, typically the trachea was also damaged. Fatal damage to the upper and lower extremities was attributed to severing arteries and veins.

Data reported from Operation Hampshire between January 2018 and March 2022 (inclusive) shows that when police officers are subjected to knife assaults they are more likely to be injured and their injuries are more severe, compared with other assaults. Focusing specifically on assaults and injuries that occurred during 2021, the nine victims received serious injuries to hands, wrists, arms, chest, abdomen, back, neck, jaw and forehead (see figure 9). Although this is only a small sample of the data, these injuries are in similar locations when comparison with the body maps in figures 1 to 4 for murder victims. Even though police officers wear body armour for routine patrol duties, in 2021 two police officers received impacts to the torso (one to the body armour and one to the radio (worn around the shoulder region of the body armour)). In these assaults the assailant did not successfully target an unprotected location, however in both instances the impacts were close to the edge of the HO accredited armour panel.

In the UK police body armour for routine patrolling duties is a dual purpose armour, providing handgun and knife protection principally to the torso. (This armour scheme can be worn covertly or overtly, however the majority of the time it is worn in an overt cover). From the data in this study, the torso is the most vulnerable area and therefore the most important area to protect. However, due the inflexibility of materials used to construct armour schemes, "difficult to protect areas" such as around the arm, flank, lower abdomen, shoulders and around the neck are often left without protection as this can severely restrict the wearer's ability to range of motion and thermal comfort. Potential solutions for this issue could be achieved by i) a step change in the flexibility of armour materials or ii) seeking innovation in armour designs by engineering extra knife resistance in difficult to protect areas. The latter is the approach investigated and this concept is now known as Supplementary Knife Resistance (SKR).

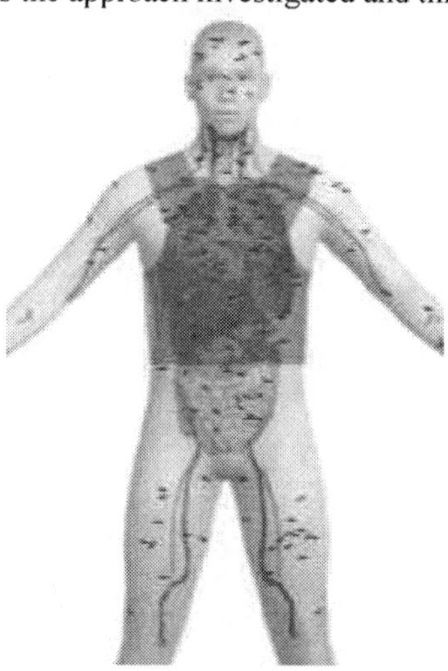

Figure 10 – anterior view of body map overlaid with HO accredited armour (blue panel) with SRK illustrated in red

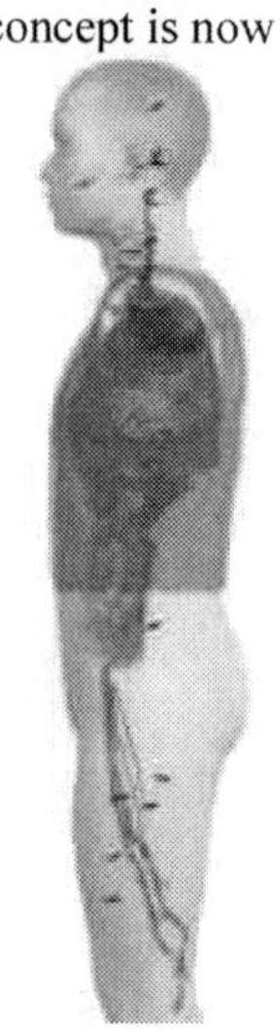

Figure 11 – left lateral view of body map overlaid with HO accredited armour (blue panel) with SRK illustrated in red

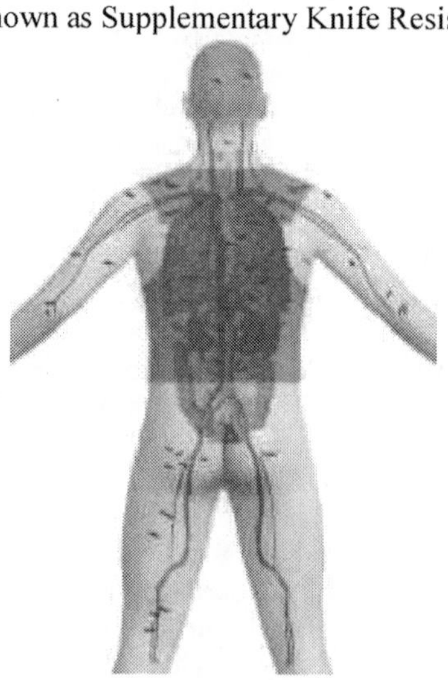

Figure 12 - posterior view of body map overlaid with HO accredited armour (blue panel) with SRK illustrated in red

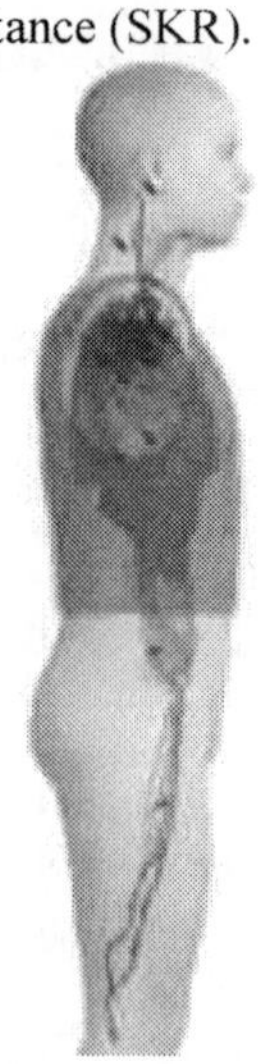

Figure 13 - right lateral view of body map overlaid with HO accredited armour (blue panel) with SRK illustrated in red

(Note: non-fatal, defensive wounds to the hands and the lower legs have not been illustrated).

This is a pragmatic approach to reducing vulnerability by increasing the area of knife resistance, using flexible materials, to augment the area of existing armour panels. In figures 10 to 13 above the

Home Office (HO) accredited armour panels are illustrated in blue, and the SKR is illustrated in red. Both HO armour panels and the SKR have been laid over the body maps (figures 1 to 4). By the addition of SKR there is nominally an increase of 10 to 15 percent in area of knife resistance.

It is imperative that SKR provides the appropriate balance between protection and enabling the officer to move. If the level of knife resistance is unable to achieve HO accreditation then ideally it should be suitable for high frequency knife threats. Furthermore, consideration of the depth of penetration of a knife through an armour scheme, with respect to depth of critical structures, must be considered.

Anatomically, the neck is recognized as a vulnerable and difficult area to protect area, particularly as the arteries and veins are close to the surface of the skin. Protection around the neck area would be advantageous from knife attacks, however there has been concern from the user with regards to a collar hindering movement. This requirements conflict is complex and a delicate balance is needed before the development of knife protection for the neck can be successfully introduced into future generations of routine patrol armour schemes.

There are other vulnerable areas such as the upper and lower extremities, containing the brachial or femoral artery, which are also not protected by police body armour. However police officers are provided with tourniquets and trained in their use in the event of injuries to the upper or lower extremities.

Finally protection schemes, such as body armour, are the final tier within the hierarchy of control, hence it is used in conjunction with police Public and Personal Safety Training (PPST) and equipment.

. CONCLUSIONS

This paper has presented the wounding patterns for 75 victims of bladed weapon assault in the MPD in 2019 in which:

- Two thirds of fatal stab wounds are located within the torso and nearly a quarter of fatal stab wounds are to the neck. In over 90% of victims the principal cause of death is loss of blood.

- Fatal stab wounds to the torso produced catastrophic damage to an organ or a great vessel or both. Fatal damage to the neck and the upper and lower extremities were due to severing an artery and/or vein.

- Over 95% of fatal stab wounds were 50mm or greater in depth;

Assaults to police officers and staff reported to Operation Hampshire showed that:

- police victims subjected to knife assaults are more likely to be injured, and their injuries are more severe, compared with other reported assaults. Hence the importance of PPST and body armour;

- nine victims in 2021 received serious injuries to hands, wrists, arms, chest, abdomen, back, neck, jaw and forehead;

- two police victims received impacts to the torso (one to their body armour and one to their radio (on the shoulder region of the body armour)). In these assaults the assailant did not successfully target an unprotected location, however in both instances the impacts were close to the edge of the HO accredited armour panel.

Finally, innovative solutions to reduce vulnerability, by engineering additional areas of knife resistance, can be achieved by enhancing armour designs in "difficult to protect areas". This is known as Supplementary Knife Resistance (SKR). In the future, the development of knife protection for the neck is an outstanding area of innovation for overtly worn, routine patrol police body armour.

Acknowledgements

We would like to convey our sincere thanks to John Foster (Forensic Operations), Joanna Konerska (Data Developer) and Inspector Stuart Kohring (Operation Hampshire) within the Metropolitan Police Service for providing access to data. We would also like to acknowledge the work of Gyles Denn at Evidential for the creation of the body maps.

https://doi.org/10.52202/080042-0023

References

[1] – Office for National Statistics. Crime in England and Wales: year ending December 2021. 2022. Available at:
https://www.ons.gov.uk/peoplepopulationandcommunity/crimeandjustice/bulletins/crimeinenglandandwales/yearendingdecember2021?_ga=2.168975204.45112.1655908206-296543442.1655908206#knife-or-sharp-instrument-offences
(Accessed 22/6/2022)

[2] Make Human. Make Human Community. 2023. Available at:
http://www.makehumancommunity.org/
(Accessed: 23 February 2023)

[3] Statista. Mean height of individuals in England from 1998-2021, by gender. 2023. Available at:
https://www.statista.com/statistics/332542/height-of-individuals-by-gender-in-england-uk/#:~:text=In%20the%20shown%20time-period%20the%20mean%20height%20of,centimeters%20in%202021%2C%20up%20from%20161%20in%201998.
(Accessed: 23 February 2023)

[4] Breeze J., Lewis E.A., Mahoney P.F. Minimum depths to essential structures in a UK military population using computed tomography: application to stab-resistant body armour. Int. J. Legal Med, 2020; **134**, 691–695. Available at: https://doi.org/10.1007/s00414-020-02256-6

[5] Konerska J. Knife crime and Operation Hampshire overview. Metropolitan Police Service. 2022.

[6] Kohring S. Assaults on Police with knives in 2021. Metropolitan Police Service. 2022.

[7] Bleetman A. Safety Standards for Police Body Armour. PhD thesis. University of Birmingham; 2000. Pt 7, p. 112.

Body Armour Comfort & Mobility Assessment

L. Boogh[1], C. Djololian[1]

[1]*DuPont Specialty Products Operations Sàrl, 154 Route du Nant-d'Avril, 1217 Meyrin, Geneva, Switzerland*
louis.boogh@dupont.com, christophe.djololian@dupont.com

Abstract. With the increased use of protective equipment, the notion of comfort & mobility becomes ever more important to reduce the strain imposed on the wearer. Considering the weight reduction of body armour made possible thanks to the development of increasingly protective material performances, comfort becomes a grail within reach when designing new protective body armours. This however requires capabilities to assess body armour comfort characteristics that are easier to implement than wear trials. Wear trials indeed allow the relative comfort of body armours to be characterized, but these are cumbersome to organize. Unfortunately, previous attempts to develop smaller scale laboratory test methods have proven to be unsatisfactory and shown to be rather poorly related to wear trial results. Various test methods have been explored and will be exposed to come up with reliable and accessible solutions that show good correlation with wear trial results. In this process, a variety of in-use situations, representative of the major discomfort and mobility constraints experienced by body armour users in the field, have been considered. Possibilities and limitations of lab-scale ballistic pack testing are discussed, mostly versus small scale individual body armour testing and wear trial testing. For all test methods explored, the human torso physiological characteristics in terms of shape and mobility have been at the core of the test method concepts and designs. Physiological criteria have also been used to define comfort and mobility criteria related to the discomfort in situations experienced by body armour users. Both 3D optical deformation measurement as well as mechanical test method results are presented. Based on this, a path forward is suggested for implementing new body armour assessment methods using accessible laboratory testing to assess flexibility of ballistic solutions in various situations. The presented methods allow a non-subjective numerical ranking of comfort criteria of body armours in good correlation with wear-trial results. It is also shown that when using various ballistic pack constructions with identical ballistic performances, significant impacts on comfort and mobility can be obtained. Novel solutions using new material offerings show that dramatically increased flexibility, mobility and comfort for the body armour wearer are possible when tools and criteria such as those presented here are used and taken into account in the selection criteria and solution design processes.

1. INTRODUCTION

1.1 The importance of comfort & mobility assessment

Body armour selection is to some extent a tradeoff between ballistic protection and wearability. As of today, the discomfort and mobility constraint of a ballistic vest, highly correlated to the lack of flexibility of a vest, are generally proportional to the level of ballistic protection it provides [1]; therefore, ergonomics, which is a combination of comfort and mobility capabilities, generally decreases as the protection level increases.

While ballistic performance of the material and design of a body armour can easily relate to the protection level, ergonomics, still play an important role in this area. An increased level of comfort, which limits discomfort and painful movement, enables a more systematic use of the vest while higher mobility, which limits movement restriction or energy spent during motion, provides more efficiency on the field, less exposure to risky situations and being less prone to musculoskeletal injuries.

Pope at al's [2] broader literature survey clearly demonstrated that body armour does have significant impacts on physical performance and biomechanical constraints on the wearer. On the other hand, they also determined that effects of body armour on marksmanship and physiological response had not yet been adequately ascertained. Their conclusion was that body armour should be carefully selected, taking into consideration the ergonomics and impact of the body armour on the perceived and measured exertion it induces as well as the impact on work capability, balance and stability.

Through the Functional Movement Screen (FMS) tool, in which scoring is based on different types of movements (involving overhead squat, hurdle step, in-line lunge, shoulder mobility, active straight leg raise, push-up, and rotary stability), different vests have shown significant differences regarding comfort and mobility properties, mostly on in-line lunge, shoulder stability and rotary stability [3, 4]. In addition to FMS results, time to execute certain movements during end-use situations (e.g. enter and exit of police cars) have been shown to be influenced by wearing ballistic vests [5].

https://doi.org/10.52202/080042-0024

Therefore, police officers appear to be tempted to remove their body armour on the job for discomfort reasons as recently verified and reported in a field survey [6]. This tendency would underline the fact that the wearing of body armours is not neutral to the well-being and efficiency of the wearers. Reasons to remove the body armour are mainly heat, similarly to the desire to remove a sweater in hot conditions, but in almost 50% of the cases, discomfort is considered as a main motivator to remove the protective gear. Weight still appeared as a factor, but only in 25% of the cases. In this survey, comfort came out as the most important factor users would consider for selecting their body armour.

Other studies made using military body armour, showed that reduced mobility could impact the movement and posture [7] and reduce the range of motion [8] even with moderate equipment weights. As a result, ergonomics, defined as a combination of comfort and mobility, is expected to not only impact immediate performance, but eventually, in the long run, also increase the risk of musculoskeletal problems.

Although the impact of ballistic vests on comfort and mobility has been established by wear trials in particular, no consensus toward representative and standardized test methods, which would be easier to set up and less subjective in nature, have yet been defined. Therefore, beyond ballistic performance, price and weight, one of the main purchasing criteria for the end-user remains challenging to be assessed in a standard manner. At the same time, body armour manufacturers have difficulties in developing comfortable solutions because of the lack of defined comfort criteria and related accessible test methods.

1.2 How comfort & mobility have been assessed so far?

1.2.1 Weight measurement

Comfort & mobility assessment has been considered through different perspectives these last decades. Only relying on weight initially in an attempt to determine load applied on the body as well as providing a first quantification level, it has quickly shown a lack of exhaustivity when it comes to defining comfort and mobility. In addition to weight, some approaches have been established to define material flexibility and enlarge comfort & mobility assessment. Three of them have been more commonly used in recent years.

1.2.2 Plunger pressure test

To assess flexibility of a material, various tests were measuring the resistance of a plunger pushing a ballistic pack through a hole in a support plate generally over a pre-defined displacement [9]. Although differences can be measured for different ballistic packs, it remains challenging to correlate these differences to real comfort & mobility feelings perceived during a wear trial. Through multiple end-users' feedbacks, ranking of materials obtained through this test did not correlate well with wear trials using the same ballistic packs. As a result, several initiatives have led to adjust and modify the test procedure to improve the correlation, but none have to our knowledge led to fully satisfactory results.

Obtaining a full correlation may in fact remain challenging as fundamentally the test principle is originally designed to determine the drapability of single layered fabrics on complex geometries [10], and not for assessing comfort and mobility provided by multi-layered constructions. As a result, the shape of the plunger and the punctual pressure it generates associated with the small displacements for which strength is measured (usually between 2 cm to 4 cm) does not realistically replicate ballistic pack deformations observed in practice when worn by law enforcement or military end-users.

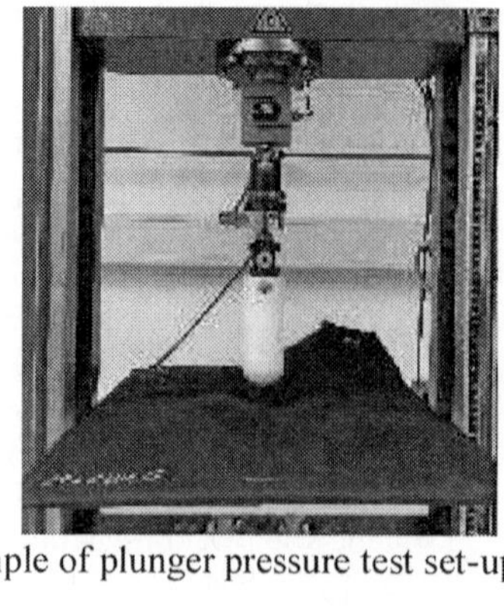

Figure 1. Example of plunger pressure test set-up used in DuPont

1.2.3 Pole or table edge test

Widely used for its ease of setting up, the pole or table edge test consists in measuring the angle formed between a ballistic pack and a horizontal axis. The ballistic pack is positioned on a pole or at the edge of a table. The part not being supported by the pole or the table bends under its own weight. In principle, the more open the folding angle formed by the pack with the horizontal axis is, the more flexible the solution will be [11].

While it can bring a first indication of the flexibility of a solution, results must be exploited cautiously as these do not correlate well with wear trials. The main obvious reason being that a body shape can be considered having a single main curvature but when in movement, it will impose body armour deformations along multiple deformation axes. This occurs when a subject bends forward to tie his shoelaces for example. The pole or table edge test only measures the material's capability to get deformed over one single axis and may thus eventually correspond to how well a body armour may fit while standing but will not correlate with deformations encountered when wearers are in movement. This limitation can be exemplified by the behavior of a sheet of paper that will bend readily on a single axis but will not accommodate to any multi-curvature bending.

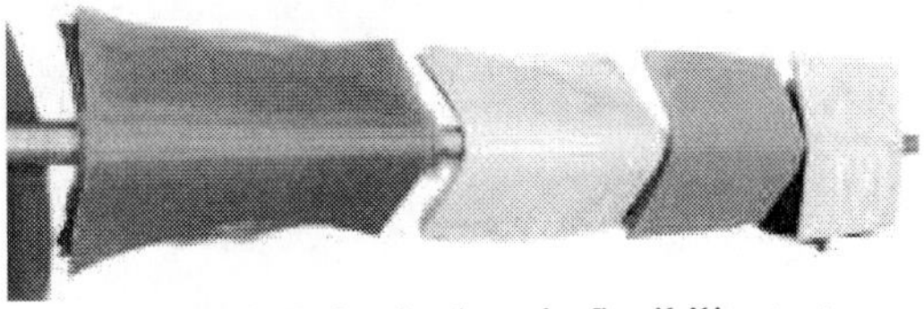

Figure 2. Pole single axis flexibility test

1.2.4 Wear trials

Largely used by law enforcement and military forces, wear trials are the most widespread and realistic ways to assess comfort today as it directly connects end-user perceptions during in-use situations. While methodologies are slightly different for law enforcement and military forces, the principle in essence remains the same. Assessing comfort & mobility in a qualitative and quantitative manner through subjective and objective measures.

For law enforcement, typical situations might involve getting in and out of a car, tying shoes, walking, running, handling weights, adopting a shooting position, or typing at the computer. Although the duration of the test can vary between one end-user and another, multiple days of testing are generally performed. During the test, a questionnaire is filled in by each wearer who provides a rating for each different task.

For military forces, obstacle courses are preferred to replicate field situations and constraints. As an example, the Load Effects Assessment Program (LEAP), used by different military forces has been developed to assess impact of equipment on mobility and combat tasks [12]. LEAP consists of a series of different obstacles to be overcome. The impact of the equipment on performance is generally measured through time spent to complete the entire course.

While these tests are highly relevant to define the in-use comfort and mobility impact of equipment, they suffer from several drawbacks. First, they can be quite demanding in terms of resources as multiple wearers are involved for a sustained period. Second, these tests are not fully standardized among the different law enforcement and military forces which makes it challenging for the industry to define a reference or representative procedure. As such, wear-trials in tenders are difficult to be used as a requirement that the equipment suppliers could implement to optimize their offerings.

1.3 How could comfort & mobility assessments evolve?

An ideal comfort & mobility assessment tool should address all the different weaknesses of the various testing methods underlined previously and be easily implemented preferably as a lab-scale test. Comfort & mobility being highly related to material flexibility, this material property would have to be assessed in different ways that correlate end-use situations. An important criteria though, is that the starting point of new tests aimed at correlating well with wear trials, should take into consideration situations met in day-to-day use, the associated movements, as well as the body shape of the wearer. It is likely that only by considering these two elements, test outcomes could eventually correlate satisfactorily with wear trials.

https://doi.org/10.52202/080042-0024

If such tests are eventually to be used as future standards for comfort assessment in tender processes, as well as for the development of optimized body armours by the industry, the test methods should be kept as simple and easy to set-up as possible. It is indeed important to ensure results can be replicated by end-users as well as by the rest of the value chain in their own facilities.

2. APPROACH TAKEN

2.1 Objectives

The aim of this study is to propose new appropriate laboratory test methods to assess comfort & mobility that would be in line with wear trial testing results. The target being to reproduce quantitatively differentiated and easy to interpret results.

2.2 Principle

The test methods are conceived not starting with existing material testing procedures but are developed based on body behavior, movement, and situations. As a reference, a specific and tailored wear trial has been led to assess the impact of different ballistic vests on comfort & mobility. This wear trial has enabled the identification of dominant wear constraints. Based on these, suitable prototype laboratory test methods are proposed to replicate the most impactful pains and mobility constraints underlined.

3. EXPERIMENT

Three different ballistic solutions have been considered in the frame of the wear trial with different levels of comfort. The ballistic solutions have been designed to pass a given identical ballistic threat and are thus compared at iso-performance, namely according to a NIJ06 level II requirement against 9 mm ammunition with a V50 of 500 m/s and a back face deformation < 44 mm. The three solutions, noted A, B and C are made using different ballistic protective materials and constructed as 100 % mono material assemblies with as a result, differentiated weights and rigidities.

The shape of the vest was identical for all samples and the ballistic packs were inserted into identical carriers. The shape used is representative of typical police ballistic body armours and was designed according to a NIJ medium size C3 design [1].

The wear trials have been led as a blind test with 8 male participants with a height ranging from 170 cm to 185 cm, a weight ranging from 70 kg to 85 kg and an age between 36 and 58 years old. Each wear trial had a duration of 60 to 120 minutes to allow for detailed feedbacks. Vests B, C and A have been tested in that order with the exact same order for each participant.

To define in-use situation scenarios to be included in the wear trial, feedbacks from law enforcement forces have been considered to replicate the most occurring standard situations to be as close as possible to real life experiences. Some situations were adjusted to enhance the potential impact of wearing a body armour, such as standing up from a lying position, starting from lying on the back rather than from lying on the front. Nine end-use situations have been considered to assess comfort and mobility behavior and were split as follow: office work, driving, tying shoes, standing still, handling weights (picking-up a bag from under a table), body search, walking/running, two-handed handgun shooting position, standing-up from a lying position on the back.

For each situation, a question related to comfort as well as a question related to mobility was asked. Each question got a qualitative feedback as well as a quantitative rating between 0 and 10. On the comfort scale, 0 to 2 referred to a very limited perceived discomfort, 3 to 6 referred to a definable level of discomfort, and 7 to 10 referred to a definable level of pain. On the mobility scale, 0 to 2 referred to a very limited impact on mobility or range of motion, 3 to 6 referred to a notable level of effort required to execute a task, and 7 to 10 referred to a restriction in the movement to be executed.

To provide an easier overview of the quantitative individual situation results, a sum of all the participant ratings has been made for each vest. The sums combine all different situations assessed for respectively the comfort and the mobility feedbacks, which allows for a relative performance rating of the vests. The results from vest A, the least comfortable one, have been taken as a reference with a total rating scaled to a value of 100.

The different ratings and feedbacks have in parallel served as a baseline to identify major pains and mobility issues and used to propose tailored test methods addressing the prominent issues. These test methods are described in the results and discussions chapter in their respective dedicated sections.

4. RESULTS & DISCUSSION

4.1 Wear trials results

4.1.1 The most challenging situations highlighted

The tailored wear trial has enabled the definition of several univocal statements. Overall, the most challenging situations which have been highlighted when it comes to the body armour impact on comfort are recovering from a lying position, tying shoes, handling weights, driving, and performing body searches as reported in Figure 5. These situations are often involving multiple body movements in parallel such as bending, squatting, and/or twisting. Vests A and B are the ones which have most impacted these situations and associated movements.

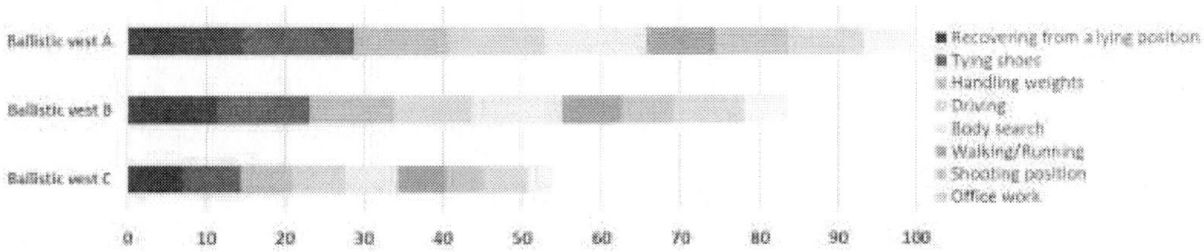

Figure 5. Wear trial discomfort assessment results for vests A, B and C. The relative impact of each situation on comfort is given by the width of each associated colored segment

From a mobility standpoint, differences between vests were in-line with the comfort assessment, with situations such as recovering from a lying position, driving, or tying shoes being reported as the most problematic ones, as described in Figure 6.

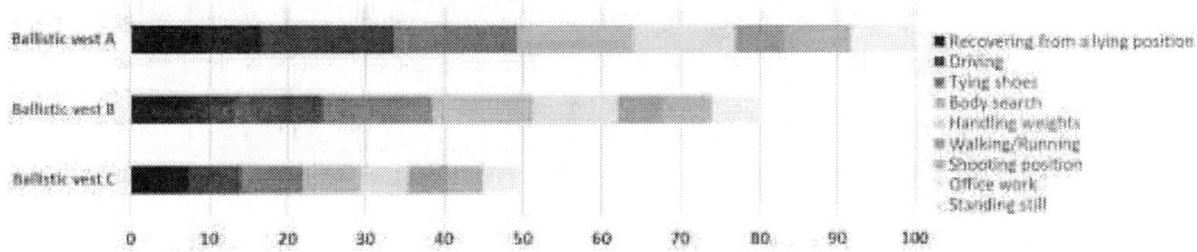

Figure 6. Mobility constraint assessment results for vests A, B and C. The relative impact of each situation on mobility is given by the width of each associated colored segment

4.1.2 Most recurrent pains and mobility constraints

For solutions A and B, most recurring were pains at the arm and shoulder locations due to the pressure exerted by the body armour edges affecting comfort. During the wear trial, these have been highlighted as number one in terms of intensity and occurrence in most of the situations although similar pains have been encountered at the lower abdomen and the neck. From a mobility standpoint, lower back and abdominal mobility constraints have been reported as problematic in executing multiple types of movements such as those described previously.

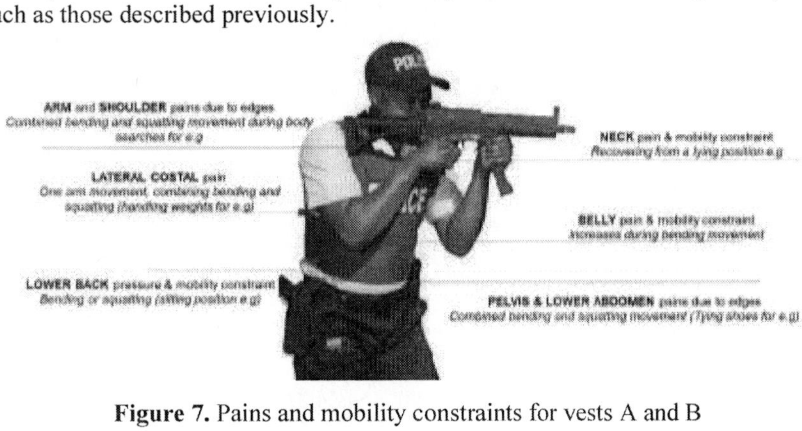

Figure 7. Pains and mobility constraints for vests A and B

https://doi.org/10.52202/080042-0024

4.2 Edge test design

4.2.1 Introduction

Wear trials have highlighted that the arm, shoulder, lower abdomen, and neck are locations of major pains encountered by the wearer. To provide a first quantification of this pain, it is proposed to replicate body parts and movement into an easy to set-up laboratory test configuration.

4.2.2 Description of the test

The arm is simulated by a semi-circular upper grip which is mounted on a tensile machine. The ballistic vest is replicated by a ballistic pack positioned with a radius corresponding to a typical body shape and upper-torso curvature. The lower grip holding the ballistic pack at its bottom edge is thus curved and allows for a free distance of the ballistic pack corresponding to approximately half the distance between arms, i.e. 20 cm. The semi-circular upper grip is shaped with a pulley profile to maintain the ballistic pack in place during the test. The pressure which is generated by the arm on the ballistic vest edge, in the vicinity of the shoulder-strap of the vest, is replicated by the force applied by the upper grip on the ballistic pack. When positioned at 90 degrees, the physiological mobility of the arm moving towards the front is typically estimated to be of 10 mm after contact with the body armour edge. The test is thus run as a compression test spanning over 10 mm, compressing the ballistic pack while pressing on its edge. Five successive tests have been run for each of the three ballistic packs. A preload of 5 N is applied before recording the force developed in Newton over a displacement of 10 mm.

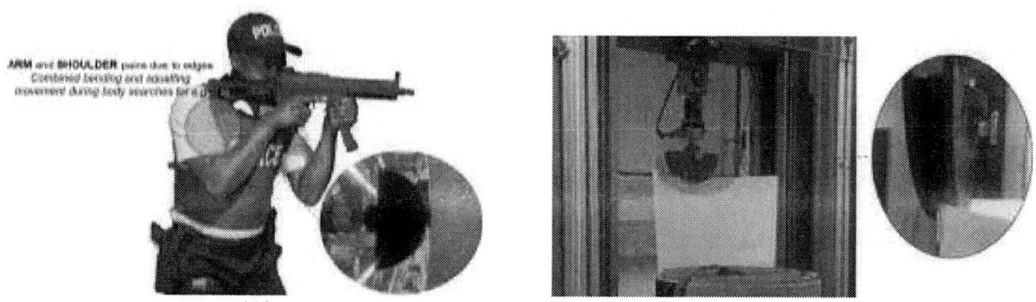

(A) (B)

Figure 8. (A) Edge test replicating pain occurring at the shoulder and (B) edge test set-up

4.2.3 Results of the Edge test

For certain types of material, it has been shown that the ballistic pack "softens" after multiple solicitations as shown in Figure 9 for packs A and B. The fifth measurement has been considered to represent stable conditions and are used for the analysis. Using the maximum force required to reach a 10 mm displacement thus shows clear differences between the three tested ballistic packs without ambiguity and with a ranking identical to the wear trial results. In the fifth measurement cycle, the maximum strength reached under a 10 mm displacement for ballistic pack A was 125 N, while it was of 53.7 N for ballistic pack B and 27.5 N for ballistic pack C.

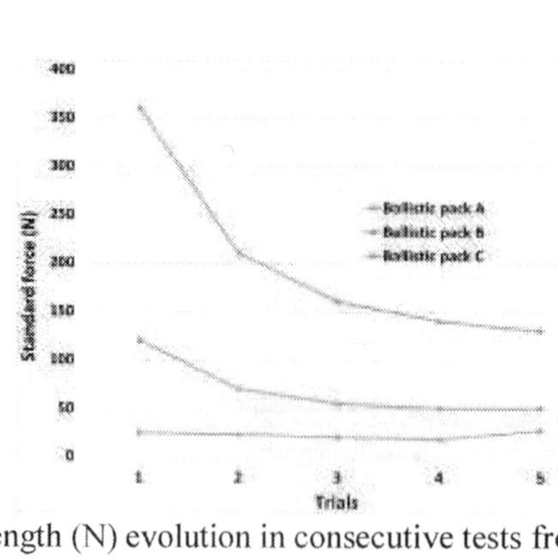

Figure 9. Maximum strength (N) evolution in consecutive tests from 0 to 10 mm displacement

https://doi.org/10.52202/080042-0024

4.2.4 Limitations and next steps

The test method established with a prototype set-up has proven to be easy to use and to be reproduceable even if minor practical optimizations are still in consideration. The procedure of the test, however, would benefit from additional care. In particular when defining if, and eventually how a physiological mobility limitation, such as a maximum strength that can be applied by the body before a pain threshold is reached, should be implemented as illustrated by the orange area in Figure 11.

For each of the three packs, five tests have been made to consider material fatigue and assess the tendency of the pack to get softer after multiple solicitations. It has not been clearly defined which maximum strength can be generated in real situations by the body and if the threshold leading to softening of the material can in fact be reached in-use. As an example, during the 1st trial, the compression force of ballistic pack B that led to softening of the material reached 120 N as shown in Figure 11. Further assessment needs to be made to understand if this level of strength can be applied by the body.

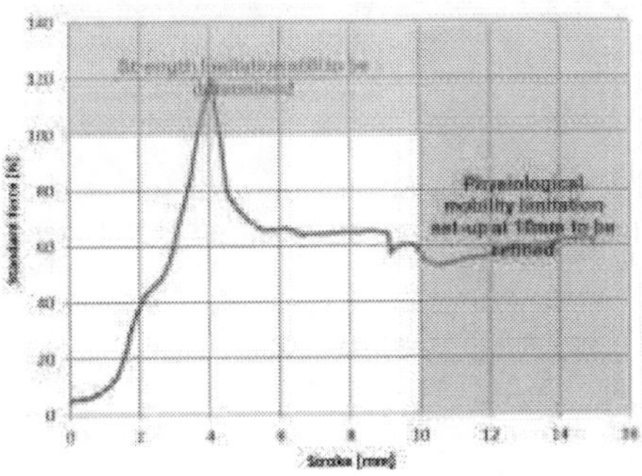

Figure 11. 1st measurement made with ballistic pack B showing limitation and future assessment work to be conducted

4.3 Bending test design

4.3.1 Introduction

Discomfort and mobility constraints are also related to the effort required to perform certain tasks and the constraints on free movement imposed by the rigidity of the body armour. A particular situation where such constraints become most evident is for example tying shoes or getting up from a lying position on the back. In these cases, the main deformation of the body armour and resistance to movement is the forward bending of the upper torso. Trying to stay as close as possible to the body and motion as for the edge test described above, a full-size torso manikin was built, able to fit a standard body armour. The aim is to allow a direct measurement of the resistance a body armour would impose on the upper torso during a forward bending movement.

4.3.2 Description of the test

The human upper torso is in fact not a rigid element as it will curve thanks to the flexibility of the spine in the back. In the front, however, the ribcage and sternum inhibit a regular bending deformation in the upper torso region. In the abdominal region, effective bending is also limited due to muscular tension developed during the effort. When schematizing the torso deflection at the front, one can consider two rigid parts with a main bending axis in the lower costal region slightly below mid-height of the upper torso as depicted in Figure 12(A). A prototype bending manikin was built following this schematic, with two cylindrical bodies articulated with an axis at two fifth of the total height. The lower part of the manikin is fixed to a table and the upper part of the manikin is pulled forward with a cable attached to the neck element illustrated in Figure 12(B). The cable is connected to a universal tensile tester machine load cell, allowing translation of the horizontal pull force on the rig to a vertical pull force on the load cell by passing over a pulley fixed to the tensile tester frame.

When bending forward from a straight to a flexed position, looking at your belt buckle for example, the lower costal bending axis (LCBA) plays a major role in the freedom of movement. There is nevertheless a maximum LCBA angle that can be developed. This was measured at about 33 to 35°, resulting in a forward movement at the neck of about 20 cm perpendicular to the lower torso axis as shown by the arrow in Figure 12(A). Body armours will cover the lower costal bending axis and require large scale deformations in this area. Since the body armour is curved around the torso along a vertical

 https://doi.org/10.52202/080042-0024

axis, forward bending will force the body armour into a double curvature. To allow for such deformations, the body armour needs to be able to conform to such a complex shape or be flexible enough to allow for folding to occur. The deformation pattern of two different vests was verified using a flexible and a rigid body armour with 3D speckle digital image correlation analysis (DIC) using an Aramis Pro set-up with a GOM software, which is illustrated in Figure 13. The flexible body armour showed an extensive deformation distributed over the entire front panel with the formation of folds in the lower costal bending axis region. The more rigid body armour was unable to deform in the center part along the original vertical body axis between the neck and the lower abdomen to adopt the imposed double curvature. As a result, the bending force reached a pain point at the neck before the physiological maximum forward displacement was reached.

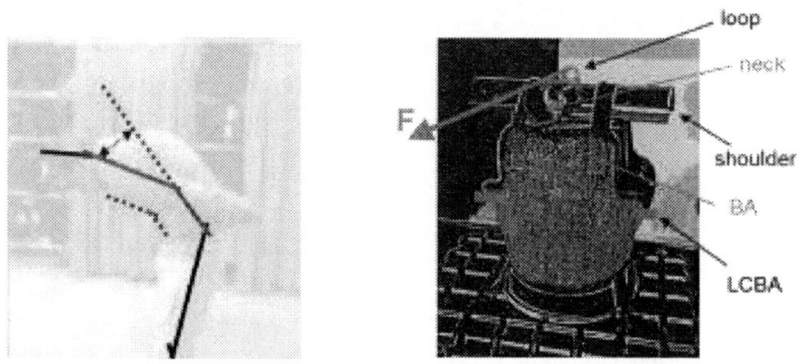

(A) (B)

Figure 12. (A) Human body main bending axes in a forward bent position. The upper torso is bending at a lower costal bending axis. (B) Prototype bending manikin with two cylindrical bodies articulated using the lower costal bending axis (LCBA).

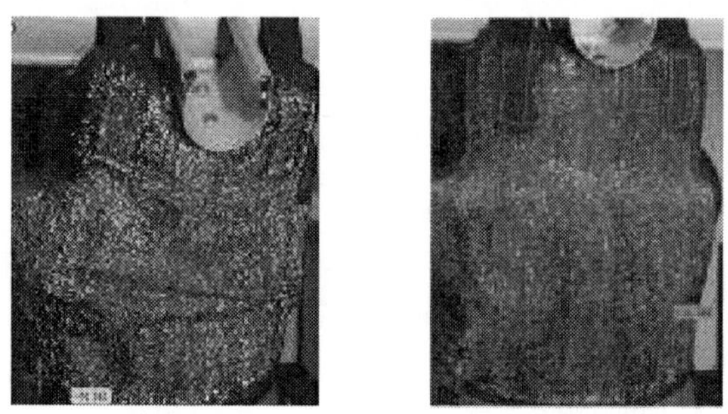
(A) (B)

Figure 13. DIC analysis of the body armour deformation: (A) a flexible body armour will deform similarly to a fabric producing folds, (B) a rigid body armour will not fold forwards and will exert high pressures at the neck and lower abdomen.

The bending test rig is used to determine the impact of the rigidity of a body armour on the discomfort induced by the cumulative required additional effort to perform certain tasks. For this measurement, the front panel of the body armour is fixed to the bending test rig. To avoid impact by relative movement, and thus, friction between the rig and the body armour panel, the upper and lower edges are strapped tightly to the test-rig. This allows measurement of force-displacement curves exclusively defined by the body armour's mechanical characteristics and resistance to bending into a double curvature shape. Testing was performed using an Instron Universal Tensile Tester equipped with a 2'000 N load cell at a crosshead speed of 100 mm/min and using a pre-load of 5 N. Pack mounting was done cautiously using reference positioning and tightening points. Force and displacement data were recorded using a dedicated Zwick software. Data were exported to Excel, treated and analyzed separately.

4.3.3 Results of the Bending test

During the first bending, the body armour packs undergo a first larger scale deformation resulting in "softening" of the panel. Softening however develops rapidly as of the second test cycle and the third solicitation has been used to compare the different materials. Figure 14 shows the distinct behavior in the force-displacement recording of the three ballistic pack constructions evaluated. Rather than testing over a maximum range of motion, it was chosen to limit the test to a neck displacement of 100 mm, i.e. half of the maximum displacement as determined previously which is considered as being more representative of the most common movements.

Double curvature bending of the ballistic protective body armour, as occurring during use, allows clear differentiation between ballistic pack constructions using different materials with varying rigidities. To quantify the impact a body armour may have on the wearer, one can compare forces at a given displacement illustrating the immediate response during solicitation in occupational specific tasks. For a longer-term impact eventually leading to exertion, the energy required to bend the ballistic pack can be compared. Results are shown in Figure 14 below.

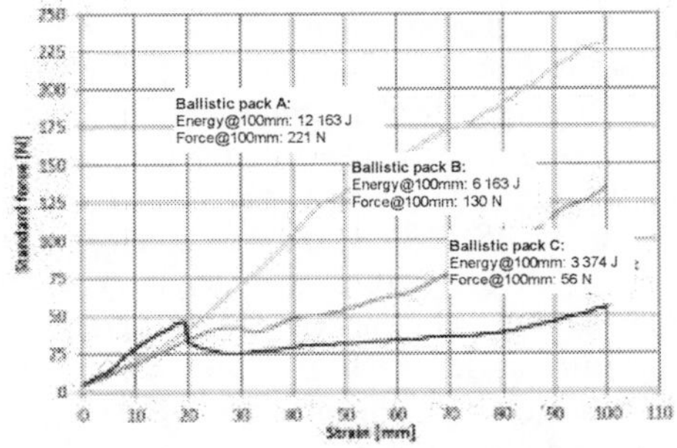

Figure 14. Force displacement curves of the third loadings of ballistic packs A, B and C

It can be seen that at a mid-range displacement compared to the body bending capacity, vests C and B require respectively about a quarter and half of both the force and energy needed to bend vest A. This difference is significant and would be expected to impact both immediate response during tasks as well as longer term cumulative exhaustion upon repeated or intense activities. This correlates very well with the wear-trial responses that also clearly differentiated the comfort and mobility perceptions of the three vests with similar magnitudes.

4.3.4 Limitations and Next Steps

Although the prototype bending manikin and test set-up appears to yield consistent results, it is of interest to attempt to simplify the test method. The current set-up would for example suffer from the fact that it can only be used on a dedicated set-up, not fully incorporated into the tensile tester. Nevertheless, to evaluate the impact of body armour ballistic pack design and material assemblies, the current evaluations indicate it is important to test body armours in a double curvature configuration. As a next step, a target is to develop and evaluate a simpler double curvature deformation test method that may be used directly on a universal tensile testing machine but would still respect physiologically defined body shape curvatures. This effort is in parallel pursued as a partnership with Hohenstein focused at developing a test method design starting from the Hyperbolic Paraboloid shape that replicates the body double curvature while bending. Mounted on a tensile machine, it would allow to measure the energy required to bend the ballistic pack into a double curvature in a simple way.

Figure 15. Illustration of a DuPont and Hohenstein prototype Double Curvature Compression (DCC) test method using a Hyperbolic Paraboloid shape to simulate the bent upper torso principal curvatures

5. CONCLUSION

Although the impact of ballistic vests on comfort and mobility has been identified by end-users and many wear trials, key factors involved in the perception of comfort and mobility have not been systematically defined and thus no consensus toward any representative or standardized test methods that would allow a rapid assessment of body armour comfort has been established. The approach taken to address this gap was to first, run a dedicated wear trial aiming at identifying key comfort related complaints and, to develop new test methods starting from human body and movement rather than using or adapting existing material test methods as has been attempted in the past. These new test methods, designed specifically to mimic the interaction between a body armour and the human body in movement, allow for a more

https://doi.org/10.52202/080042-0024

realistic assessment of in-use effects that wearing of body armour vests may have. The wear trials were led with a focus on the definition of comfort issues and mobility inhibition factors perceived by the wear-trial candidates. Two main factors came out as the most prominent affecting comfort and mobility. These were the pain provoked by the body armour edge pressing on the arm, abdomen or neck, and, the mobility constraint imposed by the rigidity of the ballistic pack on the torso. The origin of these issues arises mostly from the ballistic protective material assembly used within the ballistic pack inside of the body armour vest. The vest shape and design may help mitigate some issues but will not intrinsically solve these and may lead to compromising on the ballistic protection coverage. It was thus chosen to focus on the phenomena observed and the ballistic pack material's response for developing laboratory testing methods and procedures, allowing to anticipate body armour comfort and potential mobility issues before having to run wear trials that are long and complex to organize. Two test methods, which take into consideration the body shape of the wearer as well as situations met in day-to-day use and the associated movements, were developed. These are the Edge test and the Bending manikin test, both of which showed reproduceable results and were able to clearly differentiate the body armours tested in terms of comfort related responses. The obtained results were in-line with the wear trial feedbacks.

As a next step, design optimization for each of these tests is being pursued to further simplify implementation on standard universal tensile testing machines. The test procedures and analysis will also be refined taking into consideration further physiological parameters, such as pain thresholds or ranges of motion. The target is to ultimately enable and facilitate the accessibility of such tests and allow easy implementation by end-users as well as the value chain actors involved in supplying body armour solutions. Therefore, frequent exchanges and feedbacks from end-users and value chain partners will continue to be valued and taken into consideration to ensure relevance for all stakeholders interested in body armour comfort.

Acknowledgments

A special acknowledgement goes to Jeremie Maurin for the implementation of test design concepts, the assembly and manufacturing of test-rig elements, as well as for leading different testing campaigns. Nicolas Pont, Hakim Hoepffner, Philippe Caudoux and Michel Demolis are acknowledged for their support and technical contributions without which this work would not have been possible. Jan Beringer and Christine Lörcher from Hohenstein are acknowledged for their contribution in the development of the hyperbolic paraboloid test method. Antonio Recchia and Simeon Frei are acknowledged for all the learnings and knowledge shared in the 3D image analysis. Eoin Lynch and Adriano Mendonça are acknowledged for the insightful exchanges and relevant inputs. Ben Kennedy is warmly acknowledged for his support and interest in this project. Eoin Pierce and Connor Sheehan are gratefully acknowledged for reviewing this paper and the relevant feedbacks provided. Finally, Steven LaGanke and Joe Hovanec are acknowledged for their trust, sponsorship, and support of this project.

References

[1] Sabol W.J., Tillery C., Stoe D. and O'shea M., Body Armor Guide (National Institute of Justice, USA, 2014).

[2] Tomes C., Orr R.M., and Pope R., Annals of Occupational and Environmental Medicine, 2017; 29;14.

[3] Schram B., Orr R., Hinton B., Norris G. and Pope R., J of Bodywork and Movement Therapies, 2020; 24(3); 190-194.

[4] Tomes C.D. and Lewis M.D., Int J of Exerc Sci, 2019; 12(6); 536-546.

[5] Koblauch H., Zebis M.K., Jacobsen M.H., Haraldsson B.T., Klinge K.P., Alkjaer T., Bencke J. and Andersen L.L., Sensors, 2021; 21(5); 1-12 [1795].

[6] Avient Corp, US law enforcement survey, 2022; www.officer.com/on-the-street/apparel/uniform-fabrics/press-release/21289606/dsm-dyneema-dyneema-releases-us-law-enforcement-survey-results.

[7] Phillips M.P., Shapiro R. and Bazrgari B., Ergonomics, 2016; 59; 682–91.

[8] Lenton G., Aisbett B., Neesham-Smith D., Carvajal A. and Netto K., Ergonomics, 2015; 59(6); 808–16.

[9] Horsfall I., Champion S.M. and Watson C.H., Applied Ergonomics, 2005; 36; 283–292.

[10] Kissinger C., Mitschang P. and Neitzel M., Materials for Transportation Technology Vol 1 (Wiley-VCH Verlag GmbH, Germany, 2000), Winkler P.J. Ed.; pp. 176-182.

[11] ASTM D1388, Standard Test Method for Stiffness of Fabrics – Option A: Cantilever test, 2018.

[12] Karakolis T., Sinclair B.A., Kelly A., Terhaar P. and Bossi L.L.M., Human Factors, 2017; 59(4); 535-545.

Impact assessment of load bearing vests, combat armour and ventilated vest configurations on thermal strain

B.R.M. Kingma[1]

[1]TNO Netherlands Institute for Applied Scientific Research, Unit Defence, Safety and Security, Department of Human Performance, Soesterberg, The Netherlands
boris.kingma@tno.nl

Abstract. Dutch military personnel has access to a modular system of load bearing vests and combat armour that can be worn in combination with varying underlaying clothing. The equipment adds insulation and evaporative resistance, which may impact the environmental conditions that a military operation can be executed safely without risk of exertional heat stroke or other forms of heat illness. A ventilated vest worn under the load bearing vest may mitigate the heat burden. This study presents an assessment of the operational impact of load bearing vests and combat armour through measurements of thermal properties of clothing and equipment, combined with thermophysiological simulations. The insulation of the clothing and equipment configurations ranges between 0.68 clo and 1.03 clo; and the vapour permeability index ranges between 0.34 and 0.39. The impact of the equipment was larger on the warm climate clothing configuration than on the temperate clothing configuration. With inclusion of a ventilated vest worn under the load bearing vest the insulation ranges from 0.76 clo to 0.93 clo; and the vapour permeability index ranges between 0.41 and 0.46. The operational impact of the warm and temperate clothing configurations shows on average a 2°C difference in air temperature before core temperature reaches 38°C, and corresponds to an extended duration limit of exposure up to 4 hours. Finally, the ventilated vest on average compensated the added insulation range by −1°C for warm climate clothing and −2°C for temperate clothing configurations.

1. INTRODUCTION

Dutch military personnel has access to a modular system of load bearing vests and combat armour that can be worn in combination with varying underlaying clothing such as warm or temperate climate clothing. The equipment adds insulation and evaporative resistance, which may impact the environmental conditions that a military operation can be executed safely without risk of exertional heat stroke or other forms of heat illness.

This paper presents the outcome of a project for the NLD MoD concerning the assessment of the impact on thermal strain of load bearing vests and combat armour through measurements of thermal properties of clothing and equipment, combined with thermophysiological simulations [1]. Moreover the potential heat mitigating effect of an active ventilated vest, with channels to allow air flow, worn under the load bearing vest is incorporated [2].

The analysis provides a quantitative analysis to inform how protective clothing may influence heat strain and the potential effect size of wearing a ventilated vest. The information in this paper does by no means replace human experiments and should regarded as a sensitivity analysis to assess the impact on duration limit of exposure to warm environments; on top of that, the information can be used for time and cost-efficient design of experimental validation.

2. METHODS

Thermal manikin measurements of the thermal properties (clothing insulation and evaporative resistance) of load bearing vest and combat armour (equipment) configurations have been repeated for a warm climate clothing and a temperate climate clothing configuration. In total 18 configurations have been tested. Through this setup it is possible to study the effect of the interaction between the clothing and the equipment configuration on the thermal properties, for instance due to differences in compression of insulating air layers. Next, from a separate study the thermal properties of an active ventilated vest worn under the load bearing vest have been integrated into the results . Finally the operational impact has been estimated through simulation of heat balance with the ISO 7933 Predicted Heat Strain model [4]. The simulation encompassed a walking activity (6 km/h) without carried load for an unacclimatized person, and simulations were performed for a range of air temperature (20°C to 40°C) and relative humidity (10% to 100%). The output of the model is the time duration until core temperature reached 38°C. The

https://doi.org/10.52202/080042-0025

operational impact is defined as the difference in air temperature between configurations to reach core temperature equal to 38°C.

2.1 Clothing and equipment configurations

The clothing configurations consisted of a warm climate battle dress uniform and a temperate climate battle dress uniform. On top of both uniforms several configurations of equipment were added with increasing protection level. A brief description of the clothing and equipment configurations and their respective configuration number are shown in Table 1.

Table 5: clothing and equipment configurations and corresponding configuration number. The grey rows are shown in the results section.

	Load bearing vest	Ballistics	Warm climate battle dress uniform	Temperate climate battle dress uniform
Uniform			1	10
Uniform and backpack			2	11
Uniform,	Front Chest rig at back Hip belt	Soft	3	12
Uniform and backpack	Front Chest rig at back Hip belt	Soft	4	13
Uniform	Front and back	Soft and hard	5	14
Uniform		Protection vest	6	15
Uniform	Front and back	Protection vest	7	16
Uniform	Front and back Hip belt	Protection vest	8	17
Uniform	Front and back Hip belt Add-ons	Soft and hard Protection vest	9	18

2.2 Thermal properties of clothing and equipment configurations

In the determination of the effect of clothing and equipment on heat dissipation from the body to the environment, two main parameters are of interest: the total insulation (I_T [m^2KW^{-1}]) and the total evaporative resistance ($R_{e,T}$ [m^2kPaW^{-1}]) [3]. The insulation of clothing can also be expressed in 'clo', 1 clo corresponds to 0.155 m^2KW^{-1}. The total insulation consists of the intrinsic insulation of clothing and equipment including enclosed air layers (I_{cl} [m^2KW^{-1}]) and the insulation provided by the surrounding air layer (I_a [m^2KW^{-1}]) (**Error! Reference source not found.**).

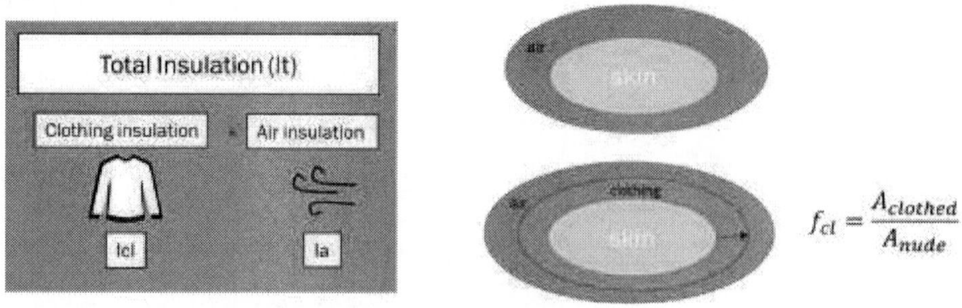

Figure 1: left, the total insulation consists of the intrinsic clothing insulation and insulation provided by the surrounding air layer. Right: clothing increases the surface area to exchange heat with the surrounding air layer. The increase in surface area is described by the clothing area factor f_{cl}.

The insulation values of clothing and the air layer do not simply add up to the total insulation (I_T [m^2KW^{-1}]). Clothing increases the surface area available for heat exchange with the environment; hence the insulation provided by air (expressed as m^2K/W) is inversely proportional to the increase of the surface area (**Error! Reference source not found.**, right). The reduction in insulation provided by air can be determined by correcting with the clothing area factor (f_{cl}: clothing area factor [-]); which is the ratio between the surface area of the outer clothing layer ($A_{clothed}$) and body skin (A_{nude}). Following the above the total insulation of a clothing configuration can be described as:

$$I_T = I_{cl} + \frac{I_a}{f_{cl}} \left[\frac{m^2 K}{W}\right] \quad (1)$$

The total evaporative resistance ($R_{e,T}$ [m^2kPaW^{-1}]) is described analogous to the total insulation of clothing and air and consists of a clothing-intrinsic evaporative resistance ($R_{e,cl}$ [m^2kPaW^{-1}]) and evaporative resistance provided by the air layer ($R_{e,a}$ [m^2kPaW^{-1}]): the latter again has to be corrected with the f_{cl}:

$$R_{e,T} = R_{e,cl} + \frac{R_{e,a}}{f_{cl}} \left[\frac{m^2 kPa}{W}\right] \quad (2)$$

From the total insulation and total evaporation resistance the permeability index is be calculated as follows:

$$im = \frac{I_T}{16.5\, R_{e,T}} \quad (3)$$

The value 16.5 [K kPa^{-1}] corresponds to the Lewis relation for evaporation at sea level and is a function of air pressure.

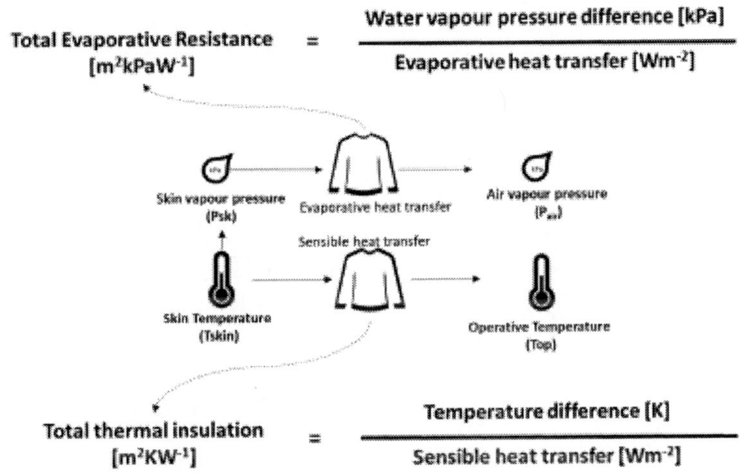

Figure 2: Total thermal insulation (I_T) is equal to the temperature difference between skin tissue and the surrounding air layer per unit of sensible heat transferring through the surface area. Sensible heat transfer is an all-purpose word for conductive, convective and radiative heat transfer. The total evaporative resistance ($R_{e,T}$) is equal to the vapour pressure difference between skin and the surrounding air layer per unit of heat equivalent to the evaporated water that is transferred over the surface.

The thermal properties are measured using a thermal manikin at Centexbel (Belgium). The thermal manikin consists of several segments for which the surface temperature can be regulated with a heat source (Figure 3). The thermal resistance per zone (i) consists of the resistance generated by the clothing (I_{cl}) [m^2KW^{-1}] and by the air layer (I_a) [m^2KW^{-1}]. The total resistance value for the whole body is determined according to the parallel method where the zones are weighted by the relative area:

$$I_T = \frac{A_{tot}}{\sum \frac{A_i}{I_i}}$$

 https://doi.org/10.52202/080042-0025

with A_{tot} [m^2] the total area, A_i [m^2] the local area, and I_i [m^2KW^{-1}] the local insulation value. For the evaporative resistance an analogous method is used.

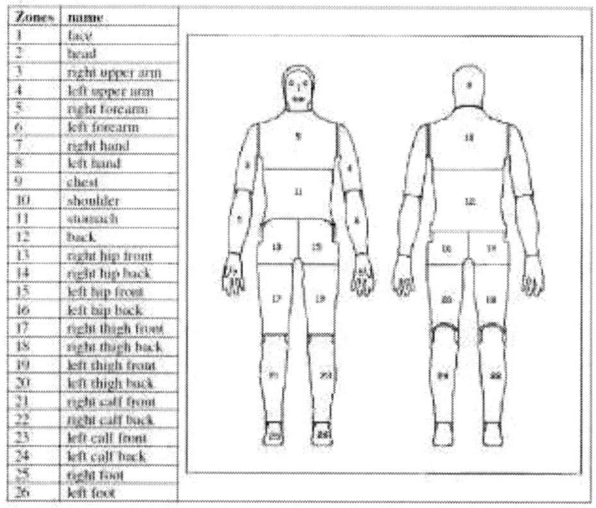

Figure 3: thermal manikin and division into segments.

The measured thermal properties are listed in Table 2. Moreover Table 2 also lists the thermal insulation values with ventilated vest over the torso. These values are estimated by substituting measured local values of the chest, shoulders, stomach and back (segments 9 through 12), with values measured with a ventilated vest system [2].

Table 6: clothing and equipment configuration thermal properties

Configuration	I_{cl} [m^2KW^{-1}]	im_{st} [-]	I_{cl} [m^2KW^{-1}]	im_{st} [-]
Warm Climate Clothing	Without ventilated vest	Without ventilated vest	With ventilated vest on torso	With ventilated vest on torso
1	0.105	0.39	0.118	0.46
3	0.122	0.38	0.122	0.44
9	0.140	0.37	0.131	0.43
Temperate Climate Clothing	Without ventilated vest	Without ventilated vest	With ventilated vest on torso	With ventilated vest on torso
10	0.137	0.36	0.136	0.43
12	0.147	0.36	0.141	0.43
18	0.159	0.35	0.145	0.41

2.3 Predicted heat strain according to ISO 7933

To gain insight into the operational impact of the clothing configurations, a simulation was performed using the model described in ISO 7933 (Predicted Heat Strain) [4]. The simulation performs a calculation of the physical heat balance to calculate the course of a person's core temperature and sweat loss given the weather (environment), personal characteristics, activity and clothing properties. The time (in minutes) to reach a core temperature above 38°C has been used, which corresponds to the standard duration limit value ($Dlim_{T,RH}$) as described in ISO 7933. Moreover, the duration limit of exposure difference between specific configurations ($\Delta Dlim_{T,RH,i,j}$, configuration i vs. configuration j) is determined by subtracting the corresponding values for each air temperature (T) and relative humidity (RH):

$$\Delta Dlim_{T,RH,i,j} = Dlim_{T,RH,j} - Dlim_{T,RH,i}$$

In order to provide an overview of the influence of temperature and air humidity, the simulations have been performed for a range of air temperature and air humidity; see Table 3 for details.

Table 7: input values for simulation of the Predicted Heat Strain model.

Type	Variable	Value or range	Unit
Environment	Air temperature	20 tot 40	°C
	Relative humidity	0 tot 100	%
	Air velocity	0.6	m/s
	Mean radiant temperature	5 °C above air temperature	°C
Person	Mass	80	kg
	Height	1.8	m
	Position	Standing	-
	Acclimatisation state	0 (not acclimatized to heat)	-
Activity	Metabolic heat production	415	W
	External work	0	W
	Walking speed	6	km h^{-1}
Clothing	Intrinsic clothing insulation	See Table 3 (column Icl)	m^2KW^{-1}
	Water permeability index	See Table 3 (column im_{st})	-
	Fraction covered by reflective clothing	0.54	-
	Emissivity of reflective clothing	0.97	-

3. RESULTS

The predicted time taken to reach a core temperature of 38°C during marching on an overcast day is shown in Figure 4 for a selection of the clothing and equipment configurations. The left column shows the results for warm climate clothing and the right column shows the results for temperate climate clothing. The dotted guides show how the results differ per configuration. Comparison of the top and bottom rows shows that the most insulating vs. least insulating configuration corresponds with an air temperature shift of about +2 °C for warm climate clothing and +1 °C for temperate climate clothing.

https://doi.org/10.52202/080042-0025

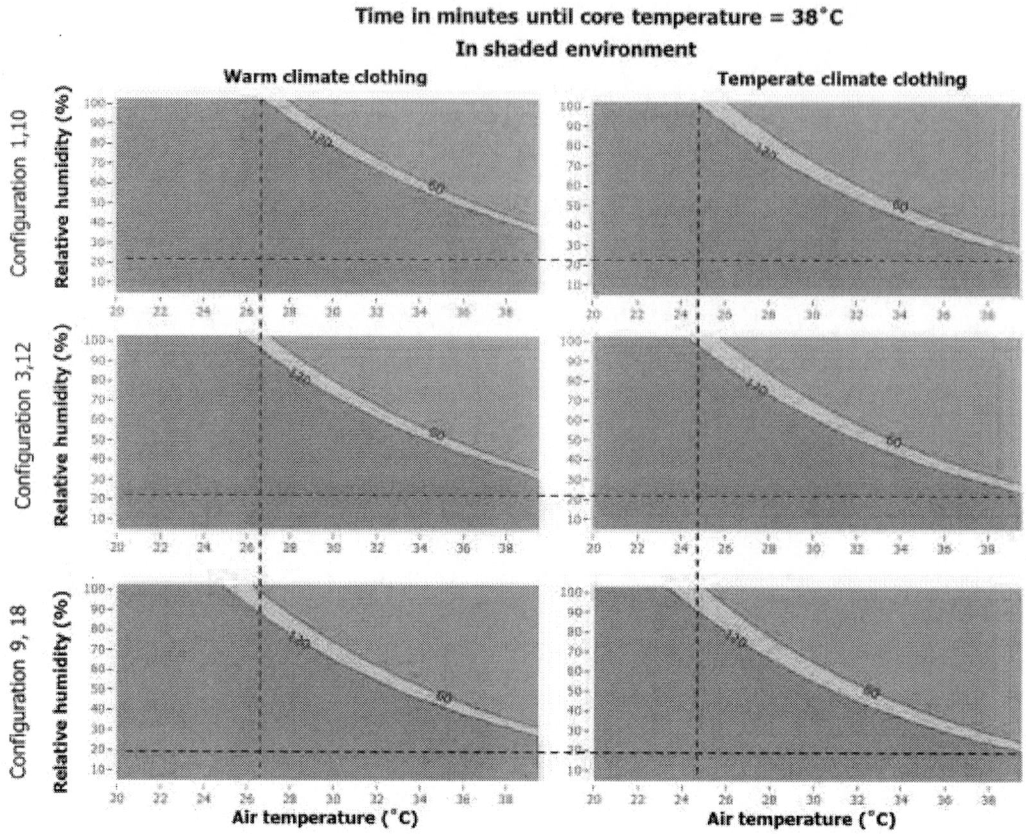

Figure 4: Duration in minutes until the core temperature reaches 38°C during marches (6km per hour). The green zone indicates that it takes longer than 2 hours to reach 38°C. The yellow zone indicates that it takes between 1 hour and 2 hours to reach 38 °C and the orange zone indicates that it takes less than 1 hour to reach 38 °C.

The temperature shift can lead to significant changes in duration limit of exposure, which is illustrated in Figure 5. For many combinations of air temperature and humidity the difference in duration limit of exposure to reach a core temperature of 38°C is less than an hour (shown as white area), however, the green zone depicts combinations of air temperature and humidity for which the time difference can increase to up to 4 hours. This means that the potential negative effect of added insulation on thermal strain is dynamic over the weather context. Interestingly, and actually adding to the complexity, the positioning of the green zone is not fixed and varies over clothing configurations. The analysis makes clear for human experiment design, the expected maximum effect size in thermal performance requires specific environmental conditions per clothing configuration. On top of that, not shown in Figure 5, but the positioning of the green zone is also dependent on the activity level (higher activity level means the green zone shifts to left), the prevailing wind speed (higher wind means the green zone shifts to right), and the presence of a radiant heat source (higher solar radiation means the green zone shifts to left).

Analogous to the comparison of clothing configurations operational effect on body temperature, also the potential effect of a ventilated vest on body temperature has been studied. The use of a ventilated vest can compensate 1°C for the warm climate clothing configurations to 2°C for the temperate climate conditions. The effect on extension of duration limit of exposure for selected clothing configurations is shown in Figure 6. Again the white zones correspond to less than 1 hour difference for the configurations to reach core body temperature of 38 °C; whereas in the green zone the difference becomes larger than 1 hour and can even increase to 4 hours. The analysis indicates that ventilated vests can be both crucial for the maintenance of operational capabilities in specific thermal environments, yet at the same time there are also many combinations of humidity and air temperature for which little difference in thermal strain can be expected.

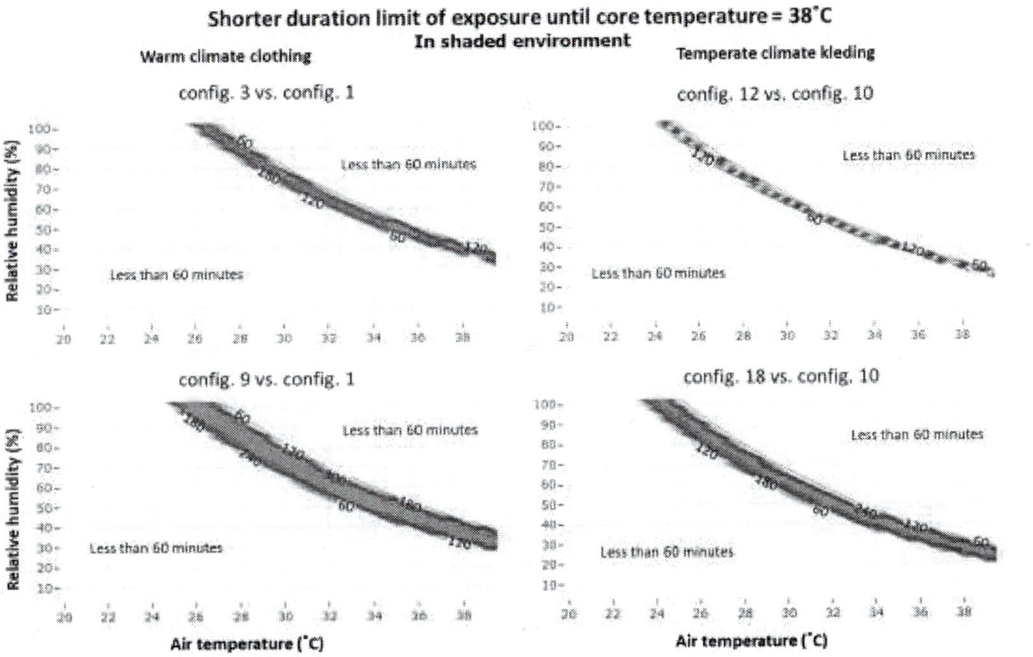

Figure 5: Difference in duration to reach 38˚C (in minutes) comparison for specific clothing and equipment configurations. Example: in the white area, the difference in time to reach 38°C is less than 1 hour for both configurations, within the green zone there is an expected difference in the duration before core temperature reaches 38 ˚C for both configurations.. The green area increases rapidly and can save up to 4 hours for specific combinations of temperature and humidity**.** The larger the difference in clothing properties, the larger the green zone.

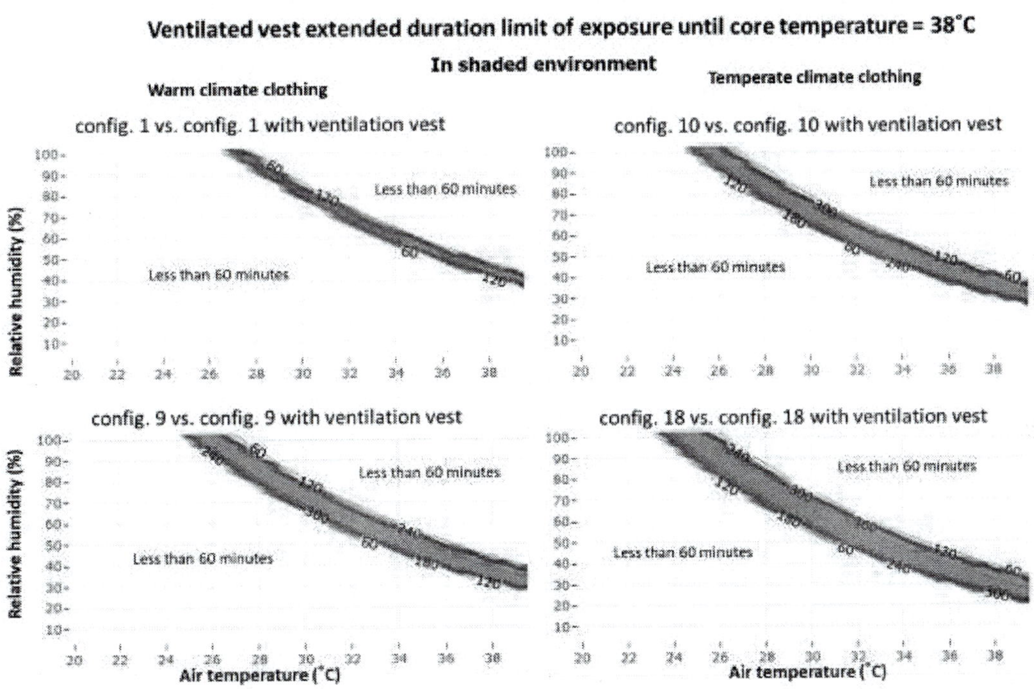

Figure 6: Extended duration to reach 38˚C (in minutes) when using a ventilated vest comparison for specific clothing and equipment configurations. In the white area, the difference is less than 1 hour. The green area increases rapidly and can save up to 4 hours for specific combinations of temperature and humidity**.**

https://doi.org/10.52202/080042-0025

4. DISCUSSION

The present analysis provides an overview of the impact of load bearing vests and combat armour on thermal strain. Moreover, the potential mitigating effect of a ventilated vest is quantified. In general terms highest insulating configuration has an impact that is comparable to that the air temperature is 2°C warmer. This result is comparable to NATO recommendations to add 2.8°C to wet bulb globe temperature thresholds when wearing combat armour [5]. Moreover, these 2°C can be operationally relevant as the duration limit of exposure can differ up to 4 hours for specific combinations of temperature and humidity (see Figure 5). A ventilated vest worn under the load bearing vest can potentially mitigate the thermal burden provided by the load bearing vest and extend the duration limit of exposure. The analyses show that the effect size of the added insulation or potential benefit of a ventilated vest on duration limit of exposure is highly variable over environmental conditions (see green zone in Figure 6), and can thereby inform decision support in an operational setting; or support the design of human experiments for the validation of the simulation results.

The comparative analyses between clothing configurations shown in Figure 5 indicate that it is perfectly possible to find little operational effect on the thermal burden for specific environmental conditions (white areas). However, in the green area's the difference in duration to reach a body core temperature of 38 °C can lead up to 4 hours. Analogous, Figure 6 indicates that the effect of a ventilated vest to extend the operational duration limit of exposure is also highly dependent on the exact environmental conditions. The width of the green zone, (see for example the panels for configuration 1 vs. configuration 18 in Figure 6), is apparently related to the corresponding difference in clothing insulation and evaporative resistance. The location of the green zone is further dependent on the prevailing wind speed, solar radiation, activity level and the acclimatisation state of the person.

The analysis of the heat load according to ISO 7933 is a model study and is in no way a substitute for experimental values. The results of the simulation can be interpreted as a sensitivity analysis to estimate the effect of the different insulation values on heat strain. The simulation in this report does not take into account biological variation and its impact on heat production and/or the ability to transfer heat to the environment. Furthermore, the simulation assumes one continuous activity level on a flat paved road. The increase in heat production due to the extra weight carried by the clothing and equipment configurations has not been included. This simplification may lead to an underestimation of the effect of the heat load. Next, part of the simulation performed with ISO 7933 Predicted Heat Strain model extends beyond the range of validity of humidity that the model claims to be valid (vapour pressure range between 0 and 4.5 m^2kPaW^{-1}); this means that for combinations of high humidity and air temperature (top right parts of Figure 4, Figure 5 and Figure 6) the model simulation validity is not defined. Since none of the operational differences are found in the top right corner of the corresponding Figures, the effect on the analysis interpretation is considered minimal.

5. CONCLUSION

The added insulation and evaporative resistance of protective equipment on military clothing configurations has been measured; and the operational impact has been simulated with the Predicted Heat Strain model (ISO 7933). Moreover the beneficial effect of a ventilated vest on thermal properties has been calculated. The most protective configuration shows an operational impact on average of 2°C air temperature during a marching activity, which may correspond to up to 4 hours decreased duration limit of exposure. A ventilated vest worn under the load bearing vest has the potential to compensate the added insulation by 1°C. The analyses show that the effect size of the added insulation or potential benefit of a ventilated vest on duration limit of exposure is highly variable over environmental conditions.

Acknowledgments

The support and discussions of H. Jager (KPU, NLD MoD) is greatly appreciated. This paper combines the outcomes of two projects performed for the NLD MoD and are reported in references [1] and [2].

References

[1] B. R. M. Kingma, "TNO21R12096 Thermische eigenschappen VOSS configuraties," Soesterberg, 2023.
[2] B. R. M. Kingma and J. Reinten, "TNO21M12094 RESPIRE II – LEWIS - Biophysical analysis of cooling power of a ventilated vest: focus on covered surface area and pre-wetting.," Soesterberg, 2022.
[3] I. ISO 9920, "ISO 9920 - Ergonomics of the thermal environment – estimation of the thermal insulation and evaporative resistance of a clothing ensemble," *Int. Stand. Organ.*, 2006, doi: 10.1086/261805.
[4] ISO, *7933: Ergonomics of the thermal environment - Analytical determination and interpretation of heat stress using calculation of the predicted heat strain*. 2004.
[5] NATO STO TG HFM-187, "Management of Heat and Cold Stress Guidance to NATO Medical Personnel, TR-HFM-187," 2013.

https://doi.org/10.52202/080042-0025

V_{50} Study of Light-Weight Body Armour Inserts under Angle Shot

K. Bolz[1], S. Hensellek, V. Acker, Dr. M. Veehmayer

[1]Rheinmetall Protection Systems GmbH, Im Rohnweiher 41, 53797 Lohmar, Germany
karl.bolz@rheinmetall.com

Abstract. Various ballistic tests are carried out on UHMW-PE inserts and plates at various angles from 0° to 30° NATO. The results show a decrease of the v_{50} by 14 % for plates with 12.7 mm thickness. For this purpose, several body armour inserts are cut along the shot axis after ballistic testing with the 7.62 x 39 PS in order to measure the pattern and to investigate the damage of the PE layers in more detail. It is observed that especially at the angular shots with 20 - 30° NATO the damage mechanisms in the UHMW-PE-layers behave differently. The thickness of the zone that is pierced without deformation is significantly thinner than in the case of shooting with 0° NATO. This can be particularly seen at the decreased deformation of the projectile. At the same time, increased shear stresses initiate delamination of the middle PE layers earlier. The projectile is deflected parallel to the UHMW-PE layers, so that it experiences less resistance, and more residual energy has to be absorbed by the layers close to the body. It was found out that this effect does not occur for plates with higher thickness. The v_{50} decreases just by 6 % for the same material with 1.1 mm higher thickness. The angle shot effect disappears for plates with thicknesses of 15.2 mm and 20.8 mm.

1. INTRODUCTION

Today's soldiers and law enforcement officers have an increasing need for body armour systems that provide the best possible ballistic protection, on the one hand. On the other hand, there is a huge demand of flexibility for any combat situation - as a driver, as a gunner, or in the field - in order to ensure the highest possible probability of survival as well as efficiency in the field. In order to cover various curvatures to provide an ergonomic shape of the hard ballistic, the issue of angular shots of UHMW-PE inserts is extremely important to consider. In this paper, investigations of the angular shots in the protection level VPAM Level 6 (7.62 x 39 PS) will be presented in order to better understand the mechanisms of the angular impact. In this context, the influence of the different thicknesses and PE materials will be considered in addition.

2. INFLUENCE OF ANGULAR SHOTS ON DAMAGE AFTER IMPACT

In recent years, an increasing number of bullet penetrations have been observed when testing UHMW-PE inserts at various angles although these inserts were already certified against shots at 0° NATO. For this reason, the resolution No. 23 was added to the VPAM to cover the additional requirement of angular shots [2].

In this test series, extensive ballistic tests of the lightweight PE inserts are carried out. On one hand, v_{50} velocities are obtained from UHMW-PE laminates to determine the angular dependence of the v_{50} (section 2.1). On the other hand, samples of a certified ultra-light weight body armour solution are cut to investigate both the deformation area of the laminate and the projectile (section 2.2). In the end, the influence of different PE materials and for each the response to an impact by 20° NATO is highlighted (section 2.3).

2.1 v_{50} Pre-Tests

The influence of the shot angle on the ballistic limit v_{50} is determined on the VPAM Level 6 certified body armour insert. To determine the v_{50}-values, 24 shots are performed for each of the firing angles 0° to 40° NATO and evaluated according to the Kneubuehl method [3].

Sample: UHMW-PE Laminate 1
Dimensions: 400 x 400 mm
Thickness: 12.7 mm
Number of shots per Plate: 8
Shot pattern: 120 mm triangle

https://doi.org/10.52202/080042-0026

In Figure 1, the results of various v_{50}-values are presented, as the v_{50} at 0° NATO is taken as 100 % - reference. The data of each v_{05} (5 % probability of penetration) and v_{95} are highlighted by the red and green graph. The v_{05} needs to be considered as well, since it is possible that the angle of attack has an influence not only on the v_{50}, but also on the distance between v_{50} and v_{05}. Furthermore, the v_{05} is close to the v_{stop} that is the crucial criterion for the certification in the end.
The v_{50}-values show a minimum at a shot angle of 20° NATO at which the value drops by 14 % compared to the v_{50} at 0° NATO. The results of v_{05} and v_{95} comply well with that as all values have a minimum at 20° NATO as well. An increase of the ballistic resistance due to a larger line-of-sight can only be observed from 40° NATO on.

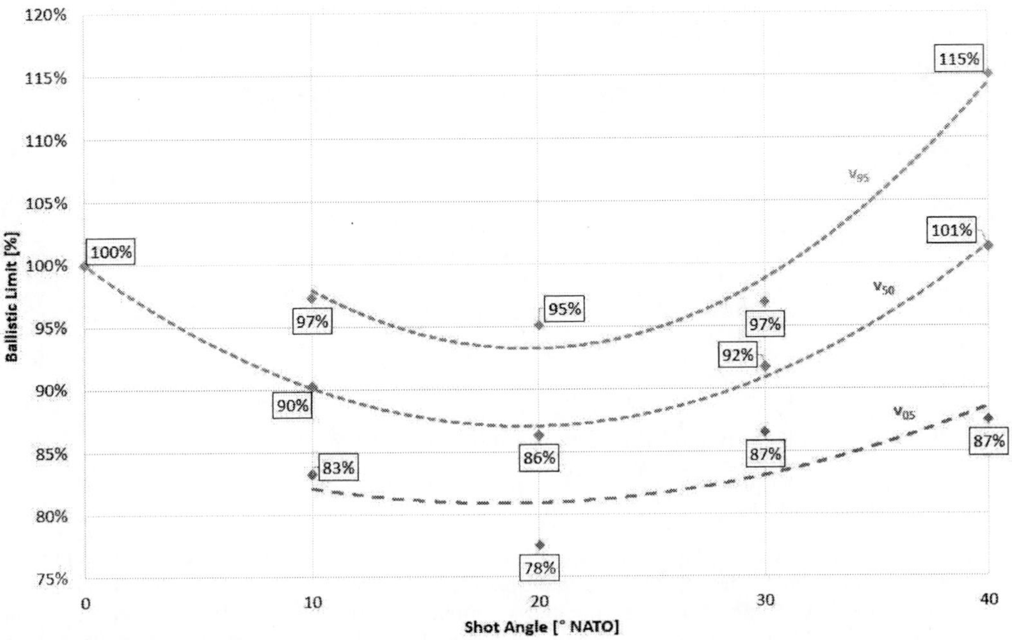

Figure 1: Laminate 1, Influence of the shot angle on v_{50}, v_{05}, and v_{95}; v_{50} at 0° NATO is 100 % - reference

In order to ensure the resistance against angle shots, a new, slightly thicker and heavier UHMW-PE insert of the same material have been developed.

Samples: UHMW-PE Laminate 2
Dimensions: 400 x 400 mm
Thickness: 13.8 mm
Protection Level: VPAM BSW Level 6 incl. angle shots
Number of shots per Plate: 8
Shot pattern: 120 mm triangle

Despite the increase in thickness and weight, the measured total v_{50}-value of Laminate 2 is approximately as high as the v_{50} of Laminate 1 at 0° NATO. In Figure 2, the test results of Laminate 2 are presented in the same manner as Laminate 1. The results show a minimum of the v_{50} at 20° NATO and a lower v_{50} at 10° NATO as well. However, the decrease of 6 % towards the v_{50} at 0° NATO is much weaker compared to Laminate 1. At 30° NATO, the v_{50} approximates the initial value. However, the course of the v_{95} and v_{05} deviates slightly from the courses of Laminate 1. V_{95} shows at 10° NATO the same value as for 0° NATO. The course of the v_{05} has even its minimum at 10° NATO.

https://doi.org/10.52202/080042-0026

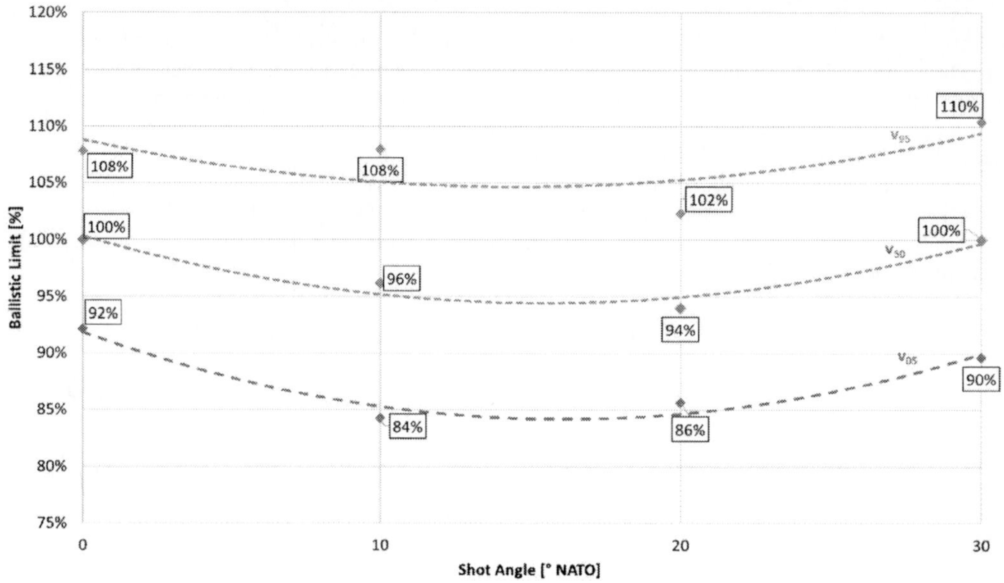

Figure 2: Laminate 2, Influence of the shot angle on v_{50}, v_{05}, and v_{95}; v_{50} at 0° NATO is 100 % - reference

2.2 Ballistic Tests and Cutting of ultra-light weight Torso Plates

2.2.1 Samples and ballistic Testing

To investigate the damage mechanisms in the PE after testing, an UHMW-PE ultra-light weight body armor insert is examined. The inserts have additional layers to reduce the trauma (BFS) and a textile cover. Therefore they have an increased thickness compared to the test samples in section 2.1. These inserts has been certified according to VPAM BSW Level 6 including the angular shots according to resolution No. 23. A scheme of the test setup is shown in Figure 3. Three shots on each insert were performed according to VPAM BSW Level 6. Table 1 gives an overview of the testing series. The tested samples are cut by water jet and the respective cross-section of the damaged area is examined.

Sample: UHMW-PE Ultra-light weight torso plate
Dimensions: 300 x 240 mm
Thickness: 18.5 mm (incl. trauma layer)
Weight: 0.98 kg
Number of shots per insert: 3
Shot pattern: 100 mm triangle
Protection Level: VPAM BSW Level 6

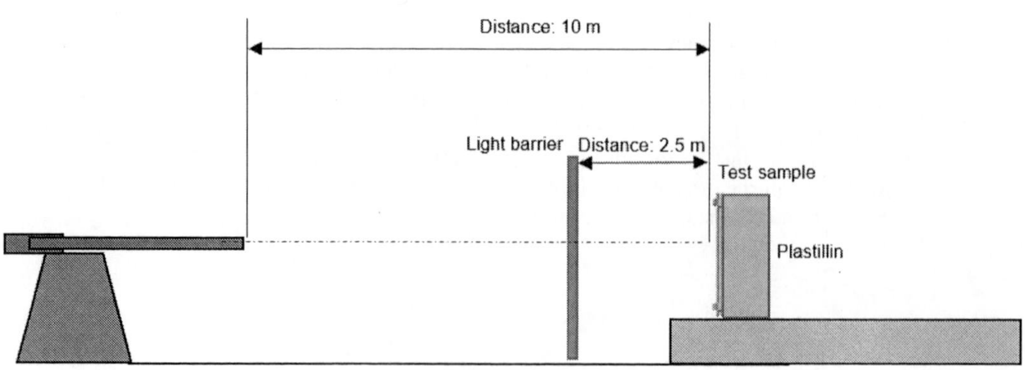

Figure 3: Schematic Test set-up of body armor inserts, according to VPAM BSW Annex 1 [1]

https://doi.org/10.52202/080042-0026

Table 1: Overview of samples

	Shot angle [NATO]	Temperature
1	10° / 10°	+20 °C
2	10° / 10°	-20°C
3	10° / 10°	+40 °C (95 % RH)
4	10° / 10°	+70 °C
5	20° / 20°	+20 °C
6	20° / 20°	-20°C
7	20° / 20°	+40 °C (95 % RH)
8	20° / 20°	+70 °C
9	30° / 30°	+20 °C
10	30° / 30°	-20°C
11	30° / 30°	+40 °C (95 % RH)
12	30° / 30°	+70 °C

2.2.2 *Results of cut Samples*

According to various studies [4] [5] concerning the damage in laminated composites, the damage area can be divided in three zones during penetration of a projectile in an UHMW-PE laminate.

1. Compression of the uppermost layers during penetration of the projectile, first layers fail due to shear stress **(Zone 1)**
2. Delamination of the middle layers initiated by (micro-) crack **(Zone 2)**. Two mechanisms lead to cracking:
   a. Transversal shear: Superposition of interlaminar shear stress and transverse normal stress leads to 45° incipient cracking
   b. Deflection of posterior layers, high tensile stresses, can lead to cracks
3. Compression / Deflection of the rear layers (tensile stress) **(Zone 3)**

Considering these mechanisms, the thicknesses of the three zones are measured on the tested samples as a function of the angle of impact. As an example, the measurement of the zones is shown in Figure 4.

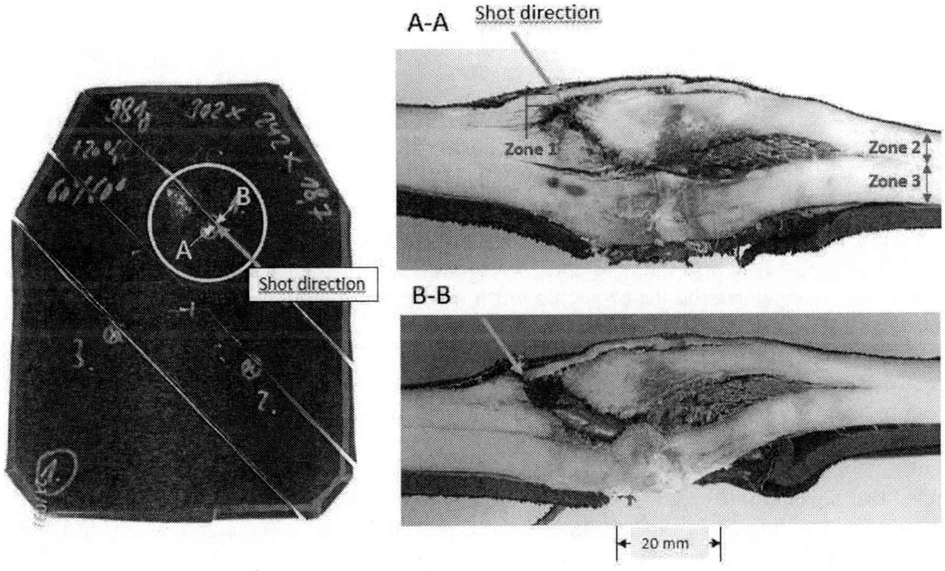

Figure 4: Example of an torso plate sample, tested at 30°/30° NATO (60°/60° VPAM) and 20 °C. Inserts were cut by waterjet, measurement of the zones

In Figure 5, the cross sections of the damaged area are shown as a function of the shot angle, while a sample tested at 0° NATO is added for reference. The pictures show that the projectile is deflected in the direction parallel to the laminate layers at 20° and 30° impact angles. As a result, a significantly wider delamination can be observed.

https://doi.org/10.52202/080042-0026

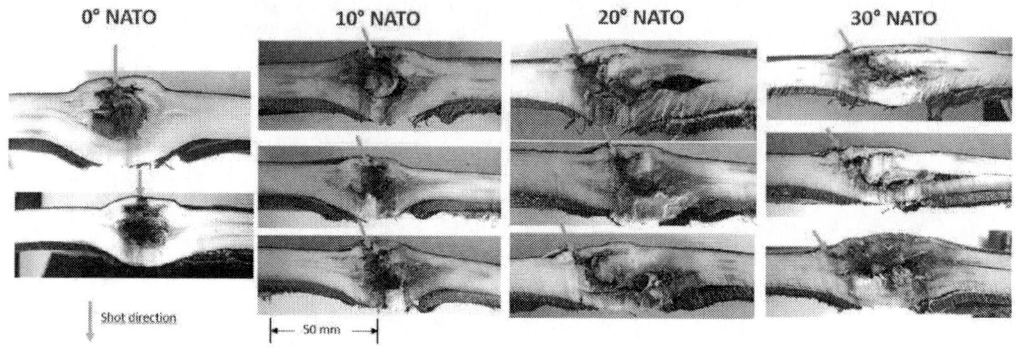

Figure 5: Cutting images of angle shot samples

The diagram in Figure 6 presents the relationship between the respective thickness of the three zones and the shot angle. It can be concluded that the angular shots lead to less compression, so a thinner zone 1, with a minimum at 20°. Through the deflection of the projectile the interlaminar shear stress is much higher. This leads to an earlier, larger and asymmetric delamination of zone 2.

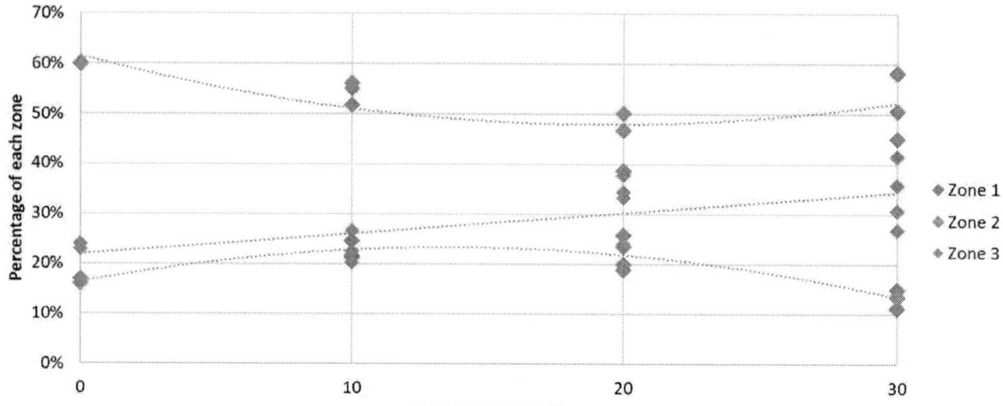

Figure 6: Thickness of zone 1-3 in depending on the shot angle

2.2.3 *Results of Projectile Deformation*

The different deformation of the projectiles is shown in Figure 7. Due to the angular impact at 20° and 30° NATO on the surface, less perpendicular forces and at the same time higher shear forces affect the tip of procectile. Consequently, the projectile mushrooms asymmetrically, which even leads to a lateral sharp edge at 20° NATO. This edge might facilitate the penetrating and piercing of the laminate.
To describe the deformation, the length as well as the diameter of the mushroomed tip is measured (Figure 8). Figure 9 presents these measurement results as a function of the shot angle. The greater the angle of impact, the less the projectile is shortened by the impact and the less the projectile mushrooms. However, at 10° NATO the measurements show a deviation of this correlation. The length after impact is slightly lower than at 0° NATO and the mushrooming is even slightly higher. This can possibly be explained that the additional shear forces at 10° NATO contributes to the shortening and mushrooming of the projectile. Taking this into consideration, the total forces is increased which lead to higher mushrooming and reduction of length of the projectile.
A dependence of the damage on the test temperature cannot be observed in this test series.

Figure 7: Projectiles after impact at different shot angles

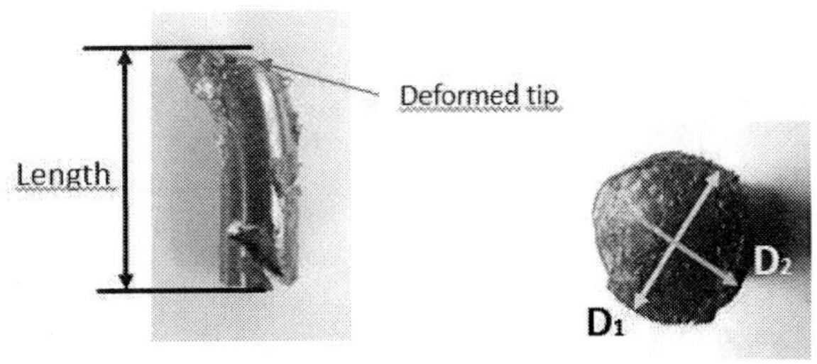

Figure 8: Measurement of length and diameters of the projectile after impact

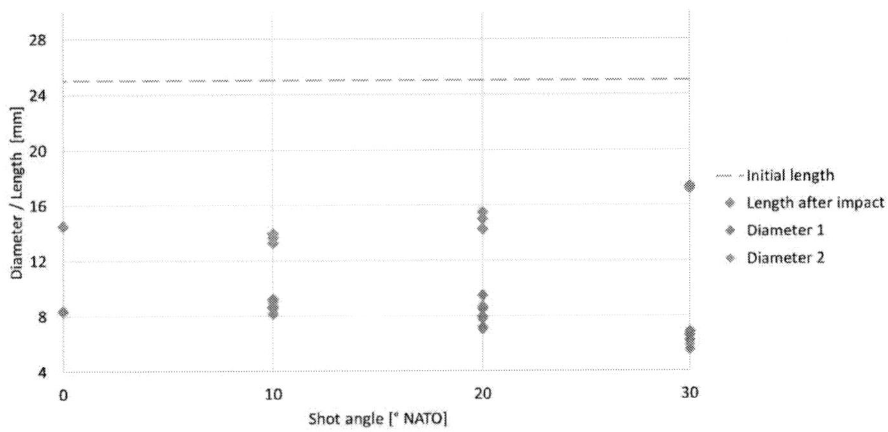

Figure 9: Angle dependence of diameter and length of the projectile after impact

2.3 Comparison of different UHMW-PE-Materials

Based on the results of the v_{50} pre-tests, it was determined that the shooting at 20° NATO is the most critical and has the highest probability of penetration. In order to compare the performance of two different materials, the respective statistical values v_{50} are determined. Applying the statistical evaluation according to the Kneubühl method [3] based on min. 16 shots, a sigmoid function can be generated for each material, which allows calculating the probability of penetration at different velocities. For this paper, the velocities v_{50} and v_{05} at 0° and 20° NATO are taken into consideration to investigate the shape of the sigmoidal curve in the context of the different UHMW-PE-materials. The two different materials have been expected to have very similar v_{50}-values.

Sample: UHWM-PE Laminate 3
Matrix: Rubber
Dimensions: 400 x 400 mm
Thickness: 15.2 mm
Number of shots per Plate: 8
Shot pattern: 120 mm triangles
Protection Level: VPAM BSW Level 6

Sample: UHWM-PE Laminate 4
Matrix: Polyurethane (PUR)
Dimensions: 400 x 400 mm
Thickness: 20.8 mm
Number of shots per Plate: 8
Shot pattern: 120 mm triangles
Protection Level: VPAM BSW Level 6

Figure 10 and Table 2 demonstrate the results of the v_{50}-tests taking the v_{50}-value of Laminate 3 as 100 % - reference into account. The two materials have the same v_{50}-value and show an increase of the v_{50} and

https://doi.org/10.52202/080042-0026

v_{05} at 20° NATO. At the same time, the sigmoidal curves have a steeper course at 20° NATO, which is outlined by the decrease of the standard deviation by 2 %. By comparing the two different materials, the PUR Laminate have a slightly higher increase (3 %) of the v_{50} and of the v_{05} (6 %).

Table 2: Results of v_{50}-testing of Laminate 3 and 4 at 0° and 20° NATO

	Laminate 3 (Rubber)		**Laminate 4 (PUR)**	
	0° / 0°	**20° / 20°**	**0° / 0°**	**20° / 20°**
v_{50}	100 %	102 %	100 %	103 %
v_{05}	91 %	95 %	91 %	97 %
σ	5 %	3 %	5 %	3 %

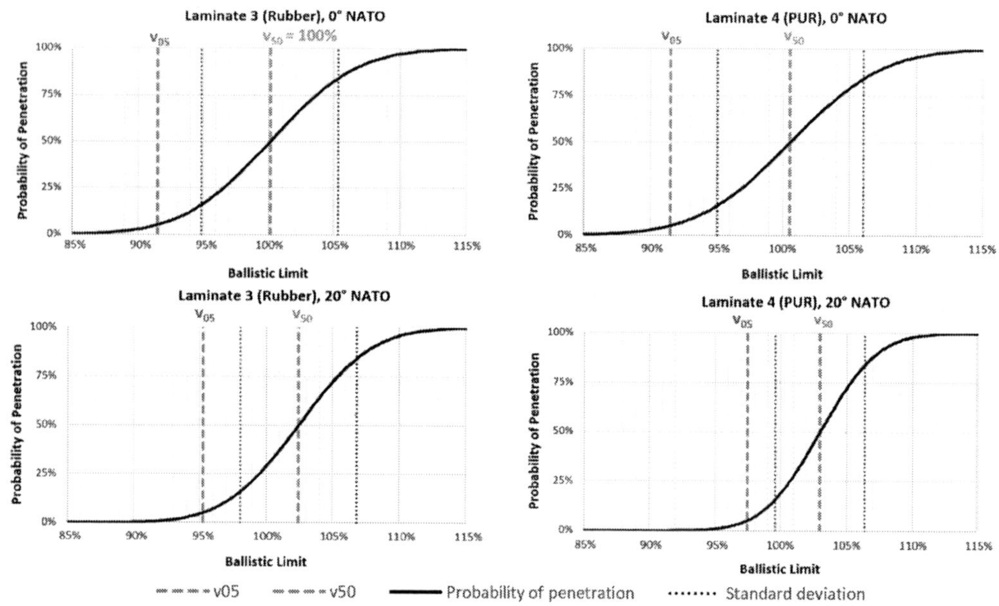

Figure 10: Sigmoidal Curves evaluated by Kneubuehl method of Laminate 3 and 4

3 DISCUSSION

The v_{50}-pretests of the lightweight Laminate 1 and 2 with thickness 12.7 mm and 13.8 mm show that the ballistic performance is the lowest at a shot angle of 20° NATO with a 14 %-decrease of the v_{50}. This can be explained by the different damage mechanisms due to the angular impact. Zone 1 decreases significantly so that the delamination initiates earlier and behaves asymmetrically. As the projectile is deflected nearly parallel to the layers it experiences much less resistance so there is a higher residual energy to be absorbed by zone 3. Additionally, the layers of zone 3 cannot deform in the same way and amount as under 0° NATO shooting, so that the energy absorption is even more impaired. Considering the projectile deformation, it can be concluded that the asymmetric mushrooming creating a sharp edge of the projectile tip amplifies this effect.

Regarding the thickness, both the results of Laminate 1 and 2 and of Laminate 3 and 4 indicate a dependency of the angle shot phenomenon with increasing thickness. By increasing the thickness of Laminate 1 it was found out that the susceptibility towards angle shots decreases. This is related to more layers of UHMW-PE layers which can stop the projectile in zone 3. Laminate 3 and 4 with thicknesses 15.2 mm and 20.8 mm do not show an dependency of the shot angle. There are enough layers in zone 3 to absorb the energy.

By comparing the two materials of Laminate 3 and 4, the v_{50} of the PUR matrix material increases by 2 % more than the rubber matrix material. This could be a hint that the PUR matrix material is more resistant to the angle shot effect. However, Laminate 4 has a higher thickness by approximately 5 mm than Laminate 3. As outlined above, the angle shot effect decreases with higher thickness. The effect of a higher line-of-sight under angle can possibly lead to this larger increase of the v_{50}.

4 CONCLUSIONS

- By testing various UHMW-PE laminates (VPAM BSW Level 6) it was found out that the v_{50} has a minimum at 20° NATO which is 14 % lower than at 0° NATO. For laminates with increased thickness, this phenomenon have a lower impact with just an decrease of 6 %.
- It was shown that the increased angles with 20-30° NATO lead to a thinner zone of the perforated initial layers of the PE laminate. The zone of delamination of the middle layers starts earlier and is larger. The projectile is deflected in direction parallel to the layers, so that it experiences less resistance by the PE layers.
- The projectiles themselves deform less and asymmetrically, which might even lead to a lateral sharp edge of the projectile.
- The angle shot weakness disappear at higher thickness at 15.2 mm and 20.8 mm.
- No clear evidence could be found that the material (PUR or Rubber) has an influence of the angular shot phenomenon.

References

[1] VPAM BSW 2006, Ballistische Schutzwesten Stand. 14.05.2009
[2] Beschlüsse zu VPAM-Richtlinien (Stand: 18.01.2023)
[3] Anschlussdokument zur VPAM-APR: Ballistischer Grenzwert v_{50} -Methode nach VPAM-KNB (Stand: 15.03.2021)
[4] Choi, H. Y., R. J. Downs, F. K. Chang, *A new approach toward understanding damage mechanisms and mechanics of laminated composites due to low-velocity impact* (1991): Part I. Experiments, Journal of Composite Materials, 25(8), 992-1011
[5] Scott, B., *The penetration of the compliant laminates by compact projectiles* (1999), Proceedings of 18th International Symposium on Ballistics, 1184-1991

https://doi.org/10.52202/080042-0026

Development and performance of an UHMWPE rifle helmet shell

M. van der Kamp[1], U. Heisserer[1], J. van Elburg[1], D. Louwers[1], J. Stolk[1]

[1]*Avient Protective Materials, PO Box 1163, 6160 BD Geleen, the Netherlands*
mirre.vanderkamp@avient.com

Abstract. Over the last years, body armour has become lighter as result of higher performing ultra-high-molecular-weight polyethylene (UHMWPE) materials [1]. These higher performing materials also enable protection against higher threat levels. Standard infantry helmet shells, as part of body armour for a dismounted soldier, are also becoming lighter. Developments on UHMWPE Uni-Directional (UD) laminates made it possible to produce light weight shells with mechanical stiffness and improved ballistic protection [2][3]. Most of the currently fielded helmets provide protection against fragments and handgun bullets. When aligning material innovations, processing, and moulding, helmets that stop rifle bullets (specifically 7.62x39mm with a mild steel core (MSC)) can be developed. Technological challenge was to not increase the weight of the helmet shell, compared to standard infantry helmets. In helmet processing, the double curved shape of the shell introduces two essential complications compared to flat plate processing. First, the originally flat UHMWPE UD laminates need to be formed into a double curved helmet shape without introducing defects like wrinkles [4][5]. Second, to obtain optimal ballistic performance, the curved shell needs to be homogenously pressurized at elevated temperature [6]. Due to the curvature of the shell this is technically challenging. This paper demonstrates experimentally the detrimental influence of wrinkling and inhomogeneous consolidation pressure on the ballistic properties of an UHMWPE rifle shell. It is explained, by using experiments and numerical deep draw simulations, how insights to overcome these processing challenges are gained. Finally, this article describes the rifle helmet test method applied, and presents ballistic performance of light weight helmet shells against 7.62x39mm MSC ammunition.

1. INTRODUCTION

Body armour is an essential component of personal protection for those who are at risk of injury or death from ballistic and fragmentation threats. Body armour has come a long way in terms of design and technology, and the latest trends reflect the ongoing efforts to improve its performance, comfort, and functionality. From lightweight materials and multi-hit capability to smart technology integration and concealability, the body armour industry is constantly evolving to meet the changing needs of its users. The introduction of new threats such as high velocity rifle bullets led to further developments in body armour technology. Especially for helmets, where weight is an important factor, defeating these more aggressive threats is a true challenge. Laing et al demonstrate [7] that reducing the helmet mass by 200g has recognized benefits with regards to user performance and decreased risk of chronic and traumatic injuries. Increasing the helmet mass to meet the higher threat levels is therefore undesired. The development of a rifle resistant helmet shell, meaning in this paper a shell that stops 7.62x39mm MSC rounds at muzzle velocity (720±10 m/s) at a weight corresponding to previously fielded frag and handgun resistant helmet shells, is therefore seen as a technology breakthrough, which required innovations in both ballistic resistant materials and helmet manufacturing.

The purpose of this paper is multifold. Main purpose is demonstrating to end-users the capabilities of current body armour materials to protect against rifle threats in helmet applications, more specifically against 7.62x39mm MSC. Secondly, it is meant to create awareness that the processing method of the helmet shell has a significant impact on the ballistic protection it provides. This is especially true for rifle helmets since the ballistic performance of UHMWPE UD laminates against 7.62x39mm MSC is more sensitive to imperfections than against other fragment or handgun threats. Finally, the quality of a helmet is difficult to assess without extensive testing but currently there is no consensus on how rifle helmets should be tested and what requirements should be met. By showing that rifle protection at low helmet shell weights is feasible, the authors would like to encourage end-users and industry to have the discussion on test methodology and performance requirements for rifle helmets.

The shells presented in this paper are made and tested at Avient Protective Material's R&D TechCenter in the Netherlands. Since Avient Protective Materials is not a commercial helmet manufacturer but the supplier of the Dyneema® brand ballistic UHMWPE UD laminate materials, the focus is on the ballistic performance of the unfinished ballistic shells against the 7.62x39mm MSC rifle threat. Other helmet requirements, for example related to stiffness or impact attenuation, are not discussed here. Also, potential influences of skin layers, coatings, retention, suspension, and inner pad system are not investigated.

https://doi.org/10.52202/080042-0027

2. RIFLE HELMET SHELL PROCESSING

2.1 Introduction to processing of a ballistic resistant shell

Combat helmets have evolved considerably over the years; from steel helmets in the early 1900's to aramid-based helmets around 1980 and light weight UHMWPE helmets from around 2010 [8][9][10][11]. Hand in hand with the evolution of materials also manufacturing methods have changed and are still evolving to get the most optimal performance out of the ballistic materials.

The traditional method developed for legacy aramid-based helmets shells is compression moulding of cut and dart ballistic materials in a heated matched metal tooling press. The dart pinwheel patterned sheets are aligned and staggered resulting in a uniform distribution of seams in the final helmet [10][11]. This can yield a high seam density, potentially lowering the protective performance. Although this method can also be used with UHMWPE UD laminates it is not the preferred method to manufacture helmets from UHMWPE UD laminate. Cutting the fibres introduces discontinuities limiting the spread of the energy away from the impact point.

A production method that circumvents the disadvantages of the cut and dart method is the deep-draw method. In deep-drawing, long fibre lengths are conserved, and high seam densities can be avoided. Deep drawing the UHMWPE UD laminate into a ballistic shell can be done in a single step, combining the forming and moulding process, using a hot press. This process is also often separated in a first step to produce a near net shape preform. This preform is then hot moulded in a second step after which the article obtains its final ballistic resistance. Although using a preform adds a manufacturing step it also brings many advantages. The repeatability of the process is typically very good, and the quality of the near net shape can be very high since optimal forming conditions are used which reduces artifacts. It also offers an opportunity to assess the quality of the forming process before investing in the actual moulding process. Furthermore, asset utilization can be optimized as the preform process is much faster than the consolidation step.

In the next sections two challenges in helmet shell manufacturing are highlighted, (i) avoiding wrinkling of the ballistic material during the forming step and (ii) applying homogeneous pressure to the double curved shell during hot moulding. And although both challenges are also valid for handgun and frag helmets, it will be shown that solving them is extremely critical for rifle helmets.

2.2 Wrinkling

Wrinkling is a common manufacturing defect observed in deep drawing. During drawing of the blank into the female cavity, excess material is drawn in and this excess material yields compressive forces in the circumferential direction when forced to follow the fixed circumference of the helmet shape. When the compressive forces exceed the buckling resistance of the sheet material, wrinkles are formed. Figure 83 shows the deformation and wrinkling during the deep draw process of a stack of UHMWPE UD laminates, as predicted from simulations using AniForm software [12]. In this case the Dyneema® HB311 UD laminate was characterized and the deep drawing of 50 sheets at room temperature was simulated. Figure 83 also shows pictures of experimental trials resembling the modelled case. The deformation and excessive amount of wrinkling predicted by the model is realistic. The wrinkling can be so severe that locally the amount of material is tripled by the presence of folds. This section will show that this challenge can be solved. Since it is not the goal of this paper to discuss modelling of the forming process, reference is given to [8][12][13][14][15][16].

Figure 83. Predicted (left) and experimentally observed (right) wrinkling during the deep draw process of a stack of Dyneema® HB311 consisting of 50 sheets

https://doi.org/10.52202/080042-0027

Wrinkling during deep drawing is not unique to UHMWPE shells; it will play a role in any synthetic fibrous sheet material or textile during deep drawing, and it is a well-known phenomenon in metal cup deep drawing and thermoforming of glass/carbon composites. Unfortunately, the presence of wrinkles is not only a cosmetic issue for ballistic shells, but it also lowers the ballistic performance. In [5] it is shown that out-of-plane fibre waviness had a significant deleterious effect on both the transverse wave velocity and the maximum energy absorption of the laminates before failure. Therefore, it is aimed to avoid the formation of wrinkles during the forming of the helmet near net shape.

Besides wrinkling there are two other deformation mechanisms that can be exploited to shape a flat sheet into a double curved object: Fibre extension and in plane trellis shear. The method utilized for the development of the rifle helmet as presented in this paper focuses on promoting trellis shear to avoid wrinkling. In an UHMWPE UD laminate construction, the fibre orientations are initially arranged perpendicular to each other, with the reinforcements aligned along the 0° and 90° directions. To accommodate complex curvatures the originally orthogonal fibre layers need to be sheared to adjust the angle between them in a scissoring fashion. This shearing can yield significant extension of the fabric if tensioned under the bias +/-45 fibre direction without the need to strain the fibres in axial direction, allowing the fabric to drape over a double curved surface without drawing in too much excessive material reducing the formation of wrinkles. Mechanistically there is a competition between trellis (in-plane) shear and out of plane wrinkling. The deep draw process should be optimized to favour shearing over wrinkling.

The resistance against in-plane shearing (angle change) of UHMWPE UD laminates is measured in a bias extension test. Figure 84 shows the test set up before and after deformation. In the pictures the fibres run in ±45-degree direction as indicated by the diagonal lines. There are no fibres spanning from clamp to clamp, so all the force needs to be transferred via fibre/matrix interaction and results are not reflecting fibre stiffness or strength. The left picture is the original configuration with orthogonal fibres, whereas in the picture to the right the angle between the fibres has changed from 90 degrees to about 30 degrees. This deformation range is relevant for the helmet shells presented in this paper. It can also be seen that the areal coverage of the material decreases when sheared but the UHMWPE UD laminate volume is conserved. This means the laminate thickness increases with local shear, with fibre filaments stacking on top of each other. The normalized thickness as function of shear angle is shown in Figure 84.

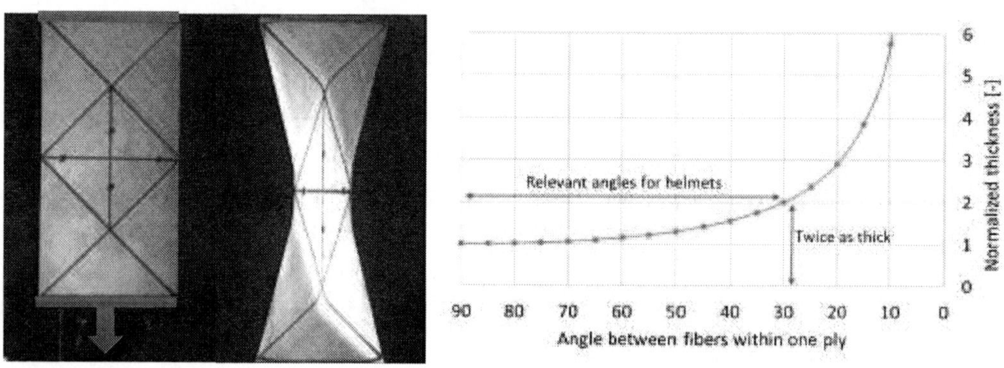

Figure 84. Bias extension test set-up (left), visual deformation during testing (middle) and thickness as function of shear angle normalized to the thickness of the unsheared material (right).

To obtain in-plane trellis shear, tensile forces in the cross-ply bias direction are needed. These forces can be obtained by utilizing a blank holder during deep drawing. The material flow into the die cavity is then controlled by frictional contact between blank holder, blank and die. Less material flow into the die cavity yields less wrinkles and requires more in-plane shear. This can be achieved by a higher blank holder force, and hence higher in-plane tension. However, if the blank holder forces restrict the material inflow into the die cavity too much this can yield material fracture.

An important factor in promoting in-plane shear is the materials resistance to shear. This can be reduced by heating the plies above the softening temperature of the resin. Figure 85 shows the force required to shear a single strip (10 cm width) of Dyneema® HB311 in a bias extension test at room temperature and at elevated temperature. The clamp displacement can be translated to shear angle in a nonlinear monotonic relation [17]. The forces required to change the angle between fibres are dominated by the properties of the resin and reduce if the material is heated up to 110 °C (383 K). The fibres themselves are not adversely affected at this temperature, and ballistic performance of them is preserved.

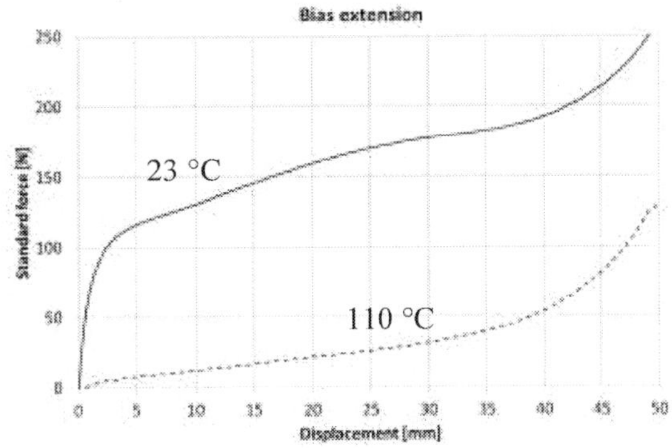

Figure 85. Force to shear single ply of Dyneema ® HB311 in bias extension test at room temperature 23 °C (296 K) (solid line) and at elevated temperature 110 °C (383 K) (dashed line)

To predict the effect of using a blank holder and a preheated blank the AniForm model is adapted to incorporate these changes. Figure 86 shows the results of a simulation predicting the deep draw process at 110 °C of 50 plies Dyneema® UD laminate using a blank holder ring that applies a force of 25 kN to the blank. Comparing the predicted deformation shown in Figure 86 with the deformation predicted and observed in Figure 83 (forming without blank holder at room temperature) it can be expected that preheating and using a blank holder with sufficient force helps significantly to reduce the amount of wrinkling defects.

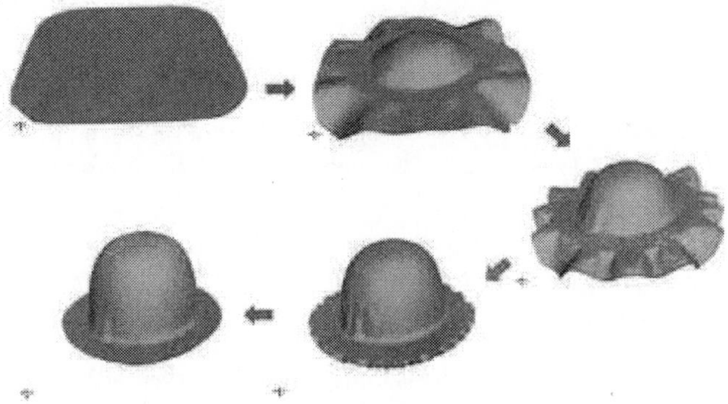

Figure 86. Simulation results predicting the deformation during the deep draw process using a blank holder and a preheated blank. For ease of modelling a simplified helmet geometry is used.

The preforms used in this study to obtain the results as presented in section 3.3 are manufactured using both a blank holder and a preheated blank. Pictures of the high quality near net shape preforms are shown in Figure 87 below.

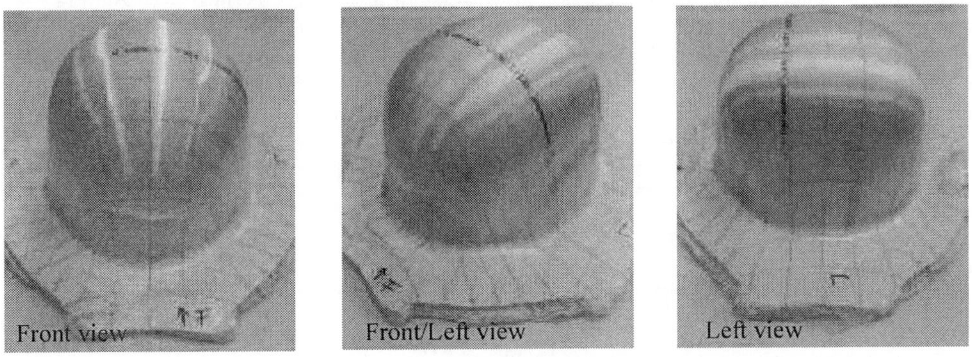

Figure 87. Pictures of high quality preform. Front view (left), preform 45 degrees turned to show front/left view (middle), preform 90 degrees turned to show left view (right)

https://doi.org/10.52202/080042-0027

2.3 Homogenous pressure

Final ballistic resistance is only obtained after the preforms are moulded using sufficient pressure and temperature. The influence of the consolidation pressure on the ballistic resistance is investigated in [6]. Higher moulding pressures yield higher ballistic performance. The relation between applied consolidation pressure and obtained performance is material and threat dependent. Figure 6 shows trendlines based on normalized Avient in-house data of the performance of Dyneema® HB311 against a 17-grain fragment simulating projectile (FSP) and a 7.62x39mm MSC rifle round as function of consolidation pressure determined on flat plates. The performance after consolidation using a pressure of 16.5 MPa is set to 100%. The ballistic performance against the 7.62x39mm MSC rifle round is more influenced by the applied amount of pressure than the ballistic performance against the FSP. This demonstrates that applying sufficient, and homogenous pressure is critical for the performance of rifle helmets.

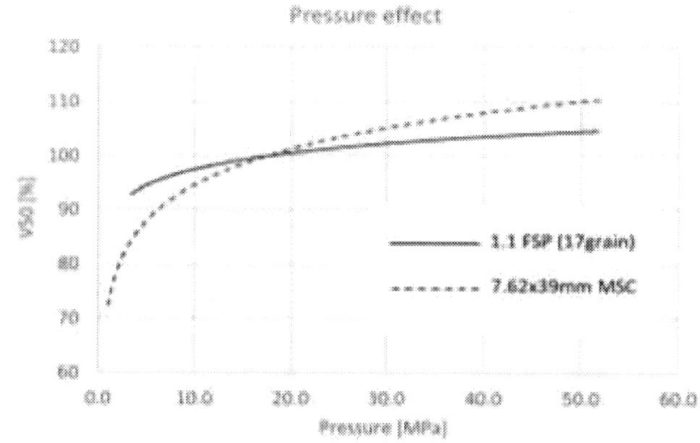

Figure 88. Performance of Dyneema® HB311 against a 17 grain FSP and a 7.62x39mm MSC rifle threat as function of consolidation pressure, normalized against the performance at a consolidation pressure of 16.5 MPa.

The strong dependence on pressure of the ballistic performance against the 7.62x39mm MSC rifle threat highlights that to obtain uniform ballistic performance over the complete helmet shell it is necessary to apply homogenous pressure to the ballistic material during moulding.

However, obtaining homogenous pressure in a double curved article using an axial press with matched metal tooling, a hardware configuration that is commonly used to produce ballistic shells, is not trivial. Underfilling the mould typically yields very high pressures at the crown region while overfilling yields undetermined high pressures at the circumference of the shells. Balancing the mould filling to obtain a uniform filling throughout the mould cavity is required to obtain uniform pressure across the part.

Since in-plane shear is used to deform the plies into the curved helmet shape, the plies are locally thickening. The amount of local thickening depends on the local shear angle as depicted in Figure 84. To obtain a uniform mould filling one must account for the various degrees of shearing, resulting in local thickening, by adapting ply geometries. Insight into how plies should be adapted to obtain a uniform filling can be obtained from modelling the forming process. Figure 89 shows the regions of the helmet shape with high in-plane shear angles. To obtain uniform thickness after forming one will have to reduce the amount of material at the locations experiencing high shear angles, compare also [4][16].

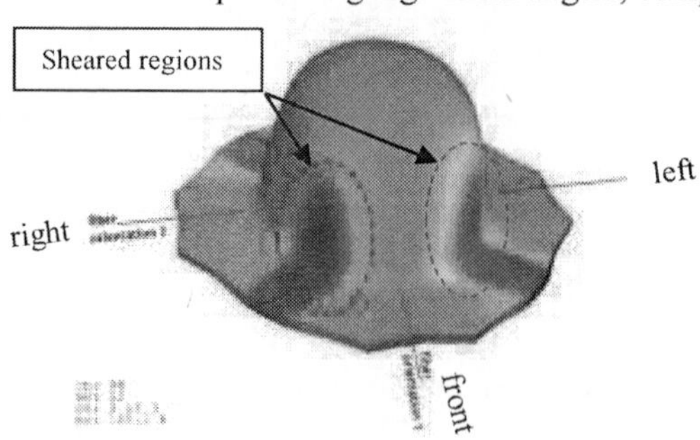

Figure 89. Simulation result highlighting the regions in a helmet with large in-plane shear

Besides the negative effect wrinkles have on the strain wave velocity [5] they also negatively affect the pressure distribution in the mould during consolidation. The process of wrinkling typically occurs in an uncontrolled manner making it non reproducible between helmets. This makes it difficult to adjust filler ply patterns to wrinkling. The local presence of folds or large out-of-plane wrinkles will yield relatively high local consolidation pressure hot spots while shielding lower areal density regions from receiving enough pressure. A high quality consistent preform is therefore essential to obtain a reproducible consolidation process with homogenous pressure.

3. RIFLE SHELL PERFORMANCE

3.1 Test methodology

The rifle shells are ballistically tested using the head form originally specified in NIJ 0106.01 [19] and modified to have clay filled slots in both the coronal and sagittal directions as later prescribed in the ACH standard [20]. Ballistic helmets covered by the NIJ 0106.01 standard are classified into various types, by level of performance, but the classification only includes handgun threats. So, although this standard is not written to include rifle rounds like 7.62x39mm MSC, this set-up is used as a basis in this study. To accommodate a good fit at the clay filled NIJ head form, a standard 7-pad inner system with a pad thickness of ¾ inch is attached to the inside of the helmet shell. A provisional strap is used to fixate the helmet onto the head form. Shot locations are chosen at the front and back respectively at 65 mm and 75 mm above the helmet lower edge in the mid-sagittal plane. At the left and right side of the helmet, the shot locations are chosen to be in the mid-coronal plane at a height of 120 mm from the helmet edge. The fit of the helmet on the clay filled head form and the shot locations are shown in Figure 90.

For perpendicular shots the head form is positioned such that the projectile impacts the shell perpendicular to its curved surface. When shooting the shells oblique[17], the impacting angles are adapted such that the impacting angle in the mid-sagittal (front/back) and mid-coronal plane (left/right) are 30 degrees towards the crown. Utilizing these rotational axes, in case of a complete penetration, the projectile ends up in the clay without damaging the metal parts of the head form. The standoff distance between the end of the barrel and the helmet is 15 m. The incoming speed of the projectile is measured using a chronograph light gate at 2 m from the impact location. The used projectiles are 7.62x39mm MSC rifle rounds and testing is started at muzzle velocity (720±10 m/s). Per shot location (front, back, left/right and crown) a V_{50}, i.e., the velocity at which 50% of the incoming projectiles are stopped, is determined using multiple identical helmet shells. The left and right shot locations are symmetric and the data on these locations is combined to determine one V_{50} for both sides. By determining a V_{50} per shot location instead of determining a V_{50} per helmet, as common in helmet testing against FSPs, the homogeneity of the ballistic performance across the helmet circumference is investigated. After a partial penetration at a certain location, the velocity at that location for the next helmet is increased. After a complete penetration the velocity for the next helmet is lowered at that location. The range of the step size is around 25-50 m/s. Per location the V_{50} is calculated by using an equal number of lowest velocity complete penetrations and highest velocity partial penetrations.

3.2 Sample information

The concept shells are produced using Dyneema® HB311 material. A high quality preform as shown in Figure 87 is manufactured prior to the shell being moulded using matched metal tooling in an axial 400 US Tons press using a set moulding temperature of 135 °C (408 K). The helmet moulds are designed to have a uniform cavity thickness. The shape of the helmets is a proprietary Avient design inspired by the ACH shape. The helmet shells are trimmed in a full-cut design and the mass of the shells is around 1200 grams for a size large. When trimmed in a high-cut design the mass of the shells is around 1000 grams for a size large.

[17] Testing under oblique angles can be critical for penetrators with sharp edges [18]

https://doi.org/10.52202/080042-0027

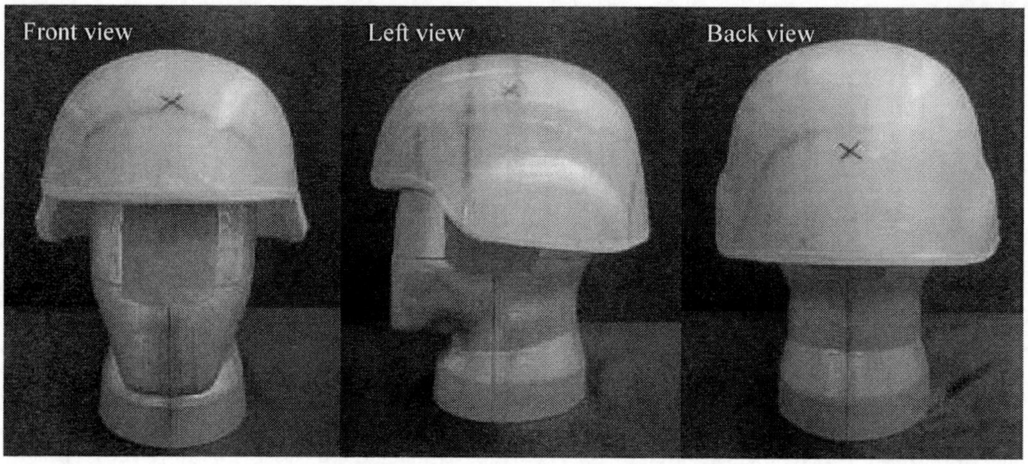

Figure 90. Rifle (7.62x39mm MSC) helmet shell on clay filled NIJ head form. Shot locations (front, side and back) indicated by the cross mark.

3.3 Results

For this investigation a total of 20 shells are tested. In most cases up to 4 shots per shell were possible, yielding a total of 74 fair shots. In some tests, the delamination after a partial penetration or the damage after a complete penetration affected the other shot locations and it was decided to not further test that particular helmet.

Figure 91 shows the calculated V_{50} per location on the shell for perpendicular shots (left) and oblique shots (right). The calculated V_{50} velocities are in between 805 m/s and 921 m/s. There were no complete penetrations at or below muzzle velocity.

The lowest V_{50}, at 805 m/s, is found at the back side. It is expected that the root cause of the slightly lower performance at the back side is local lack of sufficient pressure during moulding. The shape of the tested helmets is steep at the back making it more challenging to apply locally sufficient pressure using an axial press. Further optimization of the (filler) ply shapes would be needed to balance the mould filling to obtain more uniform pressure and more uniform ballistic performance.

The differences in V_{50} between the other locations (front, left/right and crown) are relatively small (statistically insignificant) and with a lowest complete penetration at 891 m/s and a highest stop at 925 m/s at those locations, the ballistic protection against 7.62x39mm MSC rifle threat is very good and consistent. The differences between the results shooting perpendicular and shooting oblique are too small to be statistically significantly different.

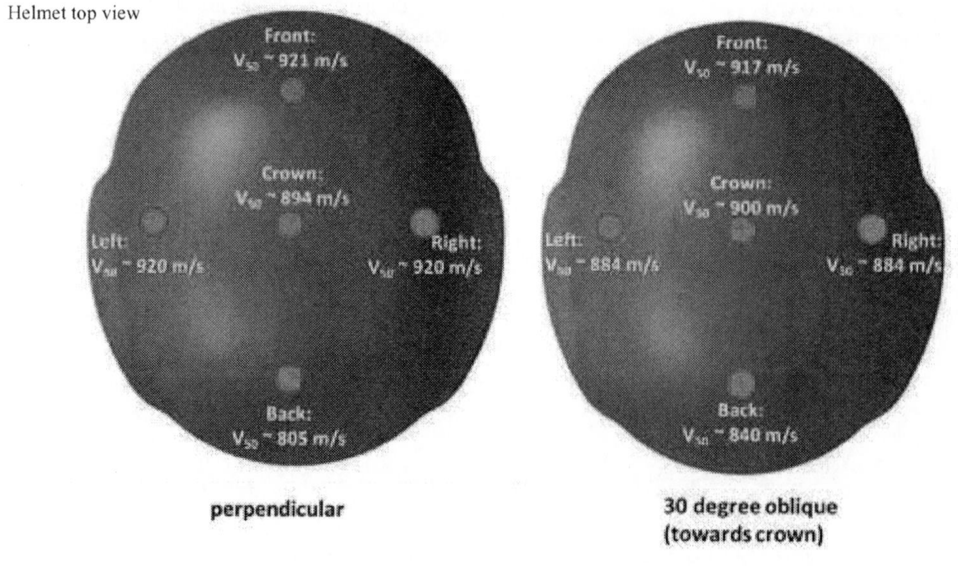

Figure 91. V_{50} against 7.62x39mm MSC per shot location on the shell (n=20 shells)

3.4 Experimental example of the effect of non-optimized processing

To emphasize the importance of a high quality preform, the effect of proceeding with a lower quality preform (i.e. reduced ballistic performance) is shown in this section. As a cautionary example a low quality preform is produced on purpose by reducing the blank holder forces and performing the shaping step at room temperature without pre-heating of the blank. Using the same product and same number of plies to obtain good preforms (Figure 87, and section 3.3), the outcome is different. Pictures of the on-purpose lower quality preform, including visible wrinkling are shown in Figure 92. The low quality preforms are moulded using the same hardware and process settings as the high quality preforms to obtain a consolidated shell.

The helmet is ballistically tested using the method described in section 3.1 and experienced complete penetrations in 3 out of 4 shot locations (front, back and right side) at muzzle velocity; only the shot at the left side was a stop. This result highlights how critical material processing is in developing a rifle helmet.

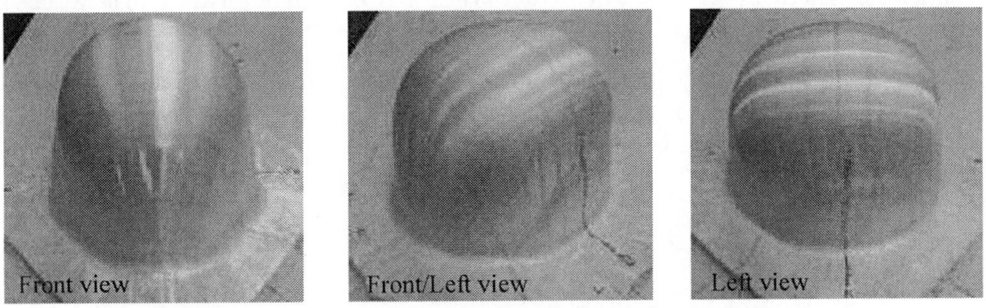

Figure 92. Low quality preform showing visible wrinkling defects

4. DISCUSSION

The work presented in this paper shows that using UHMWPE rifle performance is feasible, and that processing choices are key for performance against the 7.62x39mm MSC rifle threat. Processing defects introduced during manufacturing like wrinkling or lack of consolidation pressure, are not visible on the outside of a finished (painted) helmet. Even in consolidated unfinished shells, like the ones tested in this paper, it is difficult to visually detect potential processing defects and correlate them quantitatively to a potential drop in ballistic performance. Extensive ballistic testing will therefore be needed to check for uniform and consistent ballistic performance. And although the data presented here consists of V_{50} velocities, testing resistance to penetration (RTP) is likely more relevant.

The use of helmets that are certified or tested according to specific standards is important. For the ballistic testing presented in this paper choices were made on shot location, distance to the edge, obliquity angle and direction of oblique shots. Choices often prescribed by a standard.

Unfortunately, there is no public ballistic standard yet that describes a test method suitable for light weight rifle helmets. NIJ Standard 0106.01 Ballistic Helmets (including later ACH adjustments) does not include the 7.62x39mm MSC rifle threat and STANAG 2920 [21] focusses only on fragment testing. In parts of Europe, the VPAM HVN 2009 [22] is used in combination with the 7.62x39mm MSC rifle threat to obtain a certified VPAM6 rifle helmet. The residual energy requirement, less than 25 J deformation energy measured on a soap head form, as described in that standard is challenging for light weight rifle helmets, and work is ongoing on material development to address this.

This paper did not address the rifle helmet shell performance concerning back face deformation (BFD). The test method utilized here, using the NIJ 0106.01 based head form with clay filled slots, allows to measure the indent in the clay after a stop. The indentation in the clay of the head form, after a stop at muzzle velocity, has been measured for information only and is therefore also not included in the results section. The number of measurements is limited, and the range of measured indents varied greatly per location and condition. The lowest indent, 7.8 mm, was measured after an oblique shot at the right side. The highest indent, 32.0 mm, after a perpendicular shot at the helmet back location. It must be noted that these values are measured using non optimized unfinished helmet shells. It is likely that values are lower when shells are further optimized and the latest innovations on helmet finishing, pads and inner liners are included in the helmet system.

https://doi.org/10.52202/080042-0027

Besides the fact that the ballistic requirements in the standards currently are not developed for light weight rifle helmets, some even include secondary requirements that potentially have a negative impact on the uniformity of the ballistic performance of a rifle shell. Some standards for example include requirements on uniform shell thickness. The authors believe that uniform performance instead of uniform thickness should be the goal.

It is recognized that end-users prefer to implement helmet solutions that are certified or tested according to a specific standard and meeting certain requirements. The current lack of a suitable public standard for light weight rifle helmets holds back the application of such helmet in the field. Even though it's application will save lives.

5. CONCLUSIONS

Processing challenges particularly relevant in manufacturing of light weight rifle helmets, like wrinkling and inhomogeneous consolidation pressure, are discussed and insights that lead to solution routes to overcome these challenges are shared. It is shown that it is possible to produce light weight rifle protective shells. The unfinished ballistic UHMWPE shells presented in this paper weigh around 1200 grams and provide excellent ballistic protection against the 7.62x39mm MSC rifle threat. The current lack of a suitable public standard for light weight rifle helmets may hold back the end-user acceptance of such helmet in the field while it's application will save lives. This paper is therefore also intended to start the discussion on how a rifle helmet should be tested and what requirements should be met.

Acknowledgments

The authors acknowledge all the efforts from the researchers cited in this paper. Their insights are the foundation of this work. Also, without the efforts of the Avient Protective Materials colleagues working on the topic, who understand the relevance of in-house application development to show the capabilities of the Dyneema® materials in actual applications and to overcome application specific challenges, it would not have been feasible to present these results. Finally, a special word of thanks goes out to Jasper Maassen who produced all the shells shown in this paper.

References

[1] van der Werff, H.; Heisserer, U.; Lässig, T., Riedel, W., New protection levels of UHMWPE armor: From a hydrocode model of HB26 to new generation Dyneema® for armor applications, PASS 2014, Cambridge, 2014

[2] van Werff, H. & Heisserer, U. High Performance Ballistic Fibres: Ultra-High Molecular Weight Polyethylene (UHMWPE). in: Advanced Fibrous Composite Materials for Ballistic Protection, Woodhead, 2016

[3] Hazzard, M. K.; Trask, R. S.; Heisserer, U.; Kamp, M. v. d. & Hallett, S. R., Finite Element Modelling of Dyneema® Composites: From Quasi-Static Rates to Ballistic Impact, Composites Part A: Applied Science and Manufacturing, 2018, 115, 31 – 45.

[4] Dangora, L.; Mitchell, C.; Sherwood, J. & Parker, J. Deep-Drawing Forming Trials on a Cross-Ply Thermoplastic Lamina for Helmet Preform Manufacture. ASME. J. Manuf. Sci. Eng., 2016, 139,

[5] Parker, J. C.; Cyganik, J. E.; Kayhart, T. J.; Kubiak, D. & Sykes, R. C., The Effect of Out-of-plane Fiber Waviness on the Ballistic Performance of Ultra High Molecular Weight Polyethylene (UHMWPE) Laminates, 29th International Symposium on Ballistics, Edinburg, Scotland 2016, May 9th -13th, 2016

[6] Lässig, TR.; May, M.; Heisserer, U.; Riedel, W.; Bagusat, F.; van der Werff, H. & Hiermaier, SJ., Effect of consolidation pressure on the impact behavior of UHMWPE composites. Composites Part B, 2018, 147, 47 – 55
[7] S. Laing, J. Dutschke, S. Doecke and S. Davis; Human factors and traumatic injury considerations associated with small changes in combat helmet mass, PASS 2020, Copenhagen, 284-293
[8] S.G. Kulkarni; X.-L. Gao; S.E. Horner; J.Q. Zheng; N.V. David; Ballistic helmets – Their design, materials, and performance against traumatic brain injury, Composite Structures, 2013, 101, 313-331
[9] I.G. Crouch; J. Sandlin; S. Thomas, 5 - Polymers and fibre-reinforced plastics, Editor: Ian G. Crouch, In: The Science of Armour Materials, Woodhead Publishing, 2017, 203-268
[10] Liang, Y., Chen, X. & Soutis, C. Review on Manufacture of Military Composite Helmet. Appl Compos Mater 29, 305–323 (2022)
[11] Yongqiang Li; Hualin Fan; Xin-Lin Gao, Ballistic helmets: Recent advances in materials, protection mechanisms, performance, and head injury mitigation, Composites Part B: Engineering, 2022, 238, 109890
[12] AniForm virtual forming software, https://aniform.com/
[13] Cunniff, P. M., Parker, J., The Effect of Preform Shape on Ballistic Impact Performance, Coverage and Seam Density in Combat Helmets, 2008, 24th International Symposium on Ballistics, New Orleans, LA, September 22-26, 2008, 618-625
[14] Brands, D., Grouve, W., Wijskamp, S., & Akkerman, R. (2021). Intra-ply shear characterization of unidirectional fiber reinforced thermoplastic tape using the bias extension method. Paper presented at ESAFORM 2021. 24th International Conference on Material Forming, Liège, Belgium.
[15] Yeager M, Parker JC, Kubiak D, Shaw WS, Bogetti TA. Prediction of shear deformation and thickness gradients in deep-draw forming of ultra-high molecular weight polyethylene. Journal of Thermoplastic Composite Materials. 2022.
[16] Chen H, Wang J, Colin D, Li S, Drechsler K. Temperature-Dependent modelling of tension, in-Plane shear, and bending behaviour in non-isothermal thermo-Stamping process simulation of unidirectional UHMWPE fibre reinforced thermoplastic TPU composites. Journal of Thermoplastic Composite Materials. 2022.
[17] Boisse, P., Hamila, N., Guzman-Maldonado, E. et al. The bias-extension test for the analysis of in-plane shear properties of textile composite reinforcements and prepregs: a review. Int J Mater Form 10, 473–492 (2017).
[18] van der Werff, H.; Heisserer, U; Phoenix, S.L. Experiments and numerical simulations on oblique impact into pressed and unpressed targets consisting of cross-plied sheets of UHMWPE fiber in different polymer matrices, SSRN Journal, 2023, https://dx.doi.org/10.2139/ssrn.4373002
[19] Underwood, J.L. NIJ Standard-0106.01, Ballistic Helmets, December 1981
[20] CO/PD 05-04: HELMET, ADVANCED COMBAT (ACH), Oct 2007
[21] NATO Standardization Agency, "STANAG 2920 - Ballistic test method for personal armour materials and combat clothing", North Atlantic Treaty Organization (NATO), Jul. 2003
[22] VPAM, "VPAM HVN 2009 Durchschusshemmender Helm mit Visier und Nackenschutz 2009," Vereinigung der Prüfstellen für angriffshemmende. Materialien und Konstruktionen, 2017.

https://doi.org/10.52202/080042-0027

Extended Life Analysis (ELA) of Ceramic Plates

PL Gotts

Phil Gotts Consulting Ltd, 23 Thorney Rd, Capel St Mary, Ipswich, Suffolk, IP9 2HL, UK
phil@philgotts.com

Abstract. When any piece of personal armour exceeds the life of any warranty given, questions arise as to whether it can still be used or whether it should be replaced. In many cases armour that has reached the end of its warranty period, especially if that period is 10 or more years, can be considered to be obsolete, and perhaps should be considered for replacement on technology terms. However, some users will have armour which they are content to use for significantly longer than the warranty period, if they can be confident that it will still perform at a suitable level. Extended Life Analysis (ELA) is a method of testing which provides a numerical output for an armour system, which was probably originally specified with only a pass / fail type proof velocity criteria. Although not known as ELA at the time, this method was used extensively by UK MOD during the late 90s for specialised armour systems, for which potential replacement systems were not deemed to meet the wider requirements. This particular study, which is currently only into Year 2, consists of the ELA of a large batch of body armour plates which now date back as far as 2005 manufacturing dates. The user acknowledges that they are of an obsolete construction, albeit a very robust one, but due to their particular operations would prefer to continue to use them, rather than replace them. The plates spend most of their time being stored or in the back of vehicles and are only worn occasionally. The plates were originally specified to NIJ-0101.04 Level IV, and hence this is the standard used for the evaluation. The short annual test programme, consists of a small number of plates from which three separate results are acquired. For the ELA a V_{50} across the plates is achieved using 4, 6 or 8 shots. This value is compared with the values from previous years. From the same shot data, it is expected that there will be a number of proof shots, to prove that the plate still meets the NIJ-0101.04 level IV proof requirement. Finally, the user will probably never come up against a 30.06 AP M2 bullet, and so to provide them with some real world confidence, one of the plates will have three factory charge 7.62 x 39 mm PS ball fired into it. This paper outlines the latest test, and describe the successes, but also the limitations, of this approach to ELA of ceramic plates.

1. INTRODUCTION

Many items purchased by people in their everyday lives are supplied with a warranty. For example, in the UK, if a car is purchased new, it will be sold with a warranty covering a specific time, quite often 3 years. When the warranty expires there is the possibility of purchasing an extended warranty for a further period of time, but this is at the discretion of the purchaser. It is rare for items, such as cars, to be disposed of because they have reached the end of their warranty period. A car is a life-critical piece of equipment, so how does the owner know that it is still safe to use after the warranty has expired? In the UK, the car has to undergo an MOT test on an annual basis, from year 3 on, to allow it to still be used on the road, which also gives the owner, a level of confidence that it is safe to use. The MOT test is a predominantly visual test and should, in theory, be a non-destructive test.

Ideally personal armour should also be subjected to a non-destructive test, and there are methods that are used for ceramic-faced armour plates. However, these methods, such as x-ray and ultrasonics [1, 2], may identify damage such as ceramic cracks or composite delamination, but will not determine if the individual material components have degraded. Therefore, although they can investigate damage to the plate, they cannot determine if any of its performance is lost due to material degradation. Material degradation is caused by the aging process of the material and the effects of the environment over time. Various studies have considered both natural and accelerated aging assessments of typical ballistic fibres such as ultra-high molecular weight polyethylene (UHMWPE) or para-aramid, and some of this work was presented in 2012 by Bourget [3], Padovani [4] and Schaap [5].

When any piece of personal armour exceeds the life of its warranty, questions arise as to whether it can still be used or should be replaced. In many cases armour that has reached the end of its warranty period, especially if it is 10 or more years, is probably obsolete technology, and could be considered for replacement. Most items manufactured and sold with a warranty, are assumed to have a life much greater than the warranty period. However, personal armour is usually considered differently.

Despite the tendency for many users to consider that the warranty of personal armour equates to the expected life of the item, some users will have armour which they are content to use for significantly longer than the warranty period, if they can be confident that it will still perform at a suitable level.

Extended Life Analysis (ELA) is a method of testing, which provides a numerical output for an armour system, which was probably originally specified with only a pass / fail type proof velocity

https://doi.org/10.52202/080042-0028

criterion. Although not known as ELA at the time, this method was used extensively by UK MOD during the late 90s for specialised armour systems, for which potential replacement systems were not deemed to meet the wider requirements.

2. REASON FOR THIS STUDY

This particular study, which is currently only into Year 2, consists of the ELA of a large batch of body armour plates, which date back as far as 2005 manufacturing dates. The user acknowledges that they are of an obsolete construction, albeit a very robust one, but due to their particular operations would prefer to continue to use them, rather than go to the expense and effort of replacing them. These particular plates spend most of their time being stored, or in the back of vehicles, and are only worn very occasionally. They are also stored and transported in a climate where there can be a huge difference between day and night-time temperatures.

3. CERAMIC PLATES TO BE TESTED

The ceramic plates were originally specified to NIJ-0101.04[6] Level IV, and hence specific requirements of this standard are used during the evaluation. These requirements included the mounting of the plates for all tests, upon a conditioned, calibrated and formed Roma Plastilina Number 1 backing, the specific Level IV ammunition used, and the 30.06 AP M2 required proof velocities.

4. ELA TEST METHOD

The short annual test programme consists of a small number of plates, from which three separate results are acquired. For the ELA, a V_{50} across the plates is achieved using 4, 6 or 8 shots. This value is compared with the values from previous years. From the same shot data, it is expected that there will be a number of what can be considered to be proof shots, to prove that the plate still meets the NIJ-0101.04 level IV proof requirement. Finally, the user will probably never come up against a 30.06 AP M2 bullet, and so to provide them with some real-world confidence, one of the plates will have a relevant threat used to test them. This is in the form of three factory-charge 7.62 x 39 mm PS ball fired into it in a triangular pattern around the original 30.06 AP M2 centre shot.

In an ideal world the ELA would be conducted each year from the original manufacture date. However, the requirement to understand how an armour performs once it is out of warranty, only really becomes an issue in the minds of the user, once it is actually out of warranty, and in the case of this particular user, several years after it is out of warranty.

The testing was conducted in such a manner as to achieve three types of test results from the same series of plates:

1. Proof Test in the spirit of NIJ-0101.04
2. Extended Life Analysis (ELA) by V_{50}
3. Realistic Threat Proof Test

4.1 NIJ-0101.04 Proof Tests – Level IV Armour Plate

Any plate for which the impact velocity was within the proof velocity tolerance of 838 ± 9 m/s could be used as proof shots, whether they produce a partial penetration (PP) or a complete penetration (CP). Additionally, and in this case significantly more usefully, any shot whose velocity was greater than 847 m/s, for which the outcome was a partial penetration, could be included as a valid proof shot.

4.2 Extended Life Analysis V_{50}

The Extended Life Analysis V_{50} test consists of a single shot of the 30.06 AP M2 on each of the sample number of plates at velocities elevated above the required proof velocity, with the aim of achieving a 4-shot, 6-shot or 8-shot V_{50}, which can then be used for comparison with values from previous years. This then constitutes the Extended Life Analysis by V_{50}.

4.3 7.62 x 39 mm PS Ball

https://doi.org/10.52202/080042-0028

The most prolific, and hence realistic, threat in the theatre of operation of this user would be a Kalashnikov AK47 firing 7.62 x 39 mm PS ball. Therefore, to add some real-world confidence for the user, one of the plates which has already been impacted with a single 30.06 AP M2, is impacted with a further 3 shots of 7.62 x 39 mm PS ball, fired as full-charge factory rounds from a proof barrel.

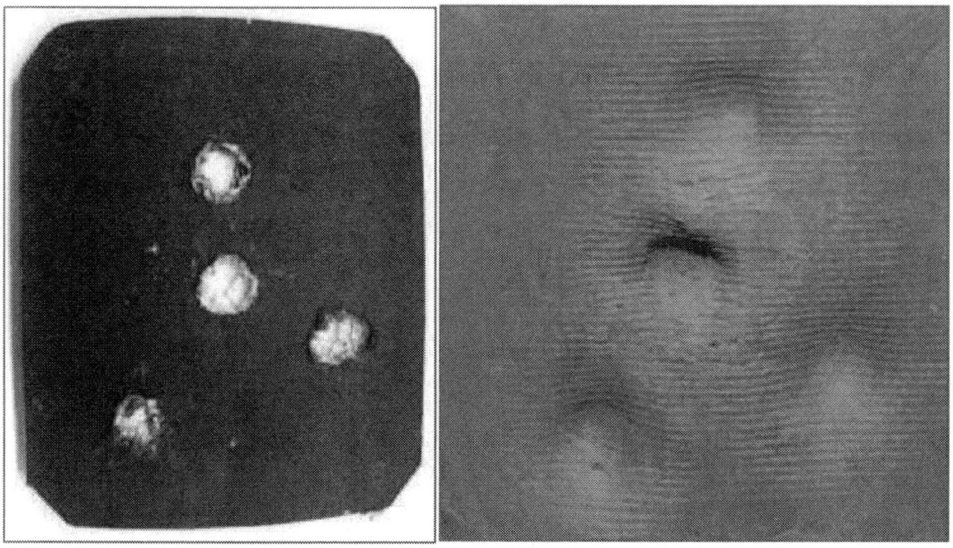

Figure 1. Plate with single shot of 30.06 AP M2 and 3 shots of 7.62 x 39 mm PS ball (left) and associated Plastilina backing with back-face signatures (right)

4.4 Test Method and Configuration

Each armour plate was mounted on a conditioned and calibrated box of Roma Plastilina Number 1.

The ammunition was charge-adjusted for each shot and fired from an appropriate proof barrel, mounted to a universal proof housing. The velocity was measured by optical sky-screens. The general trial configuration is as shown in Figure 2 below:

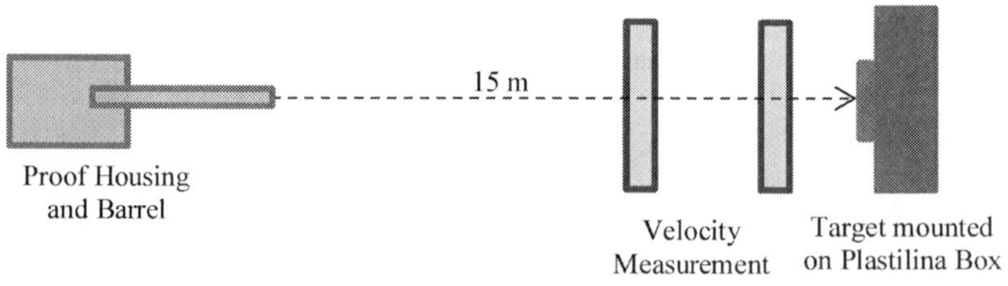

Figure 2. Plan View of General Trial Configuration

5. PLATE CONSTRUCTION

The ceramic-faced plates in this study have the advantage of being a very robust design. One of the first batch of plates delivered arrived without its black textile cover and hence it was easier to determine the construction of the plate. As previously stated, it is a robust design consisting of a thick (approximately 8 mm) alumina monolithic tile backed with a 9 mm glass-reinforced plastic (GRP) backing and all wrapped in GRP as a spall layer.

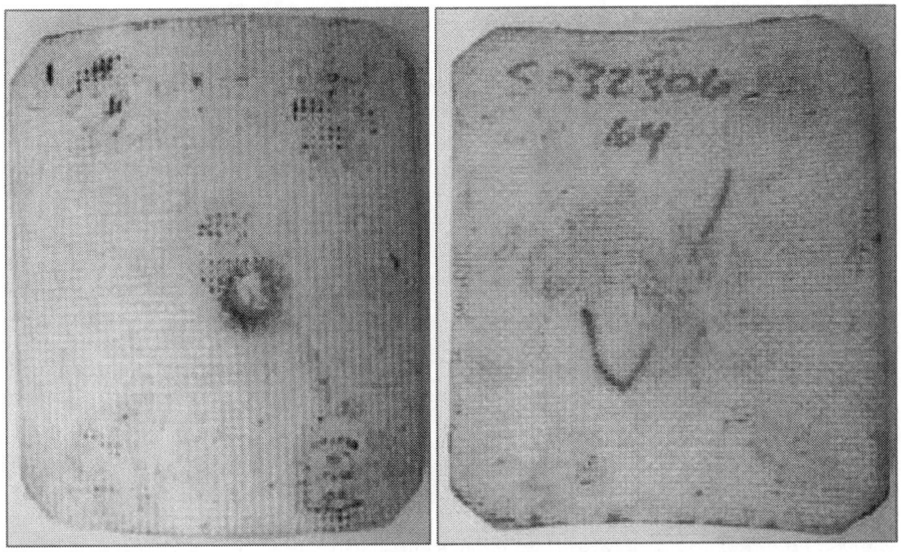

Figure 3. Example of Armour Plate without Black Nylon Cover

6. TEST RESULTS

6.1 Year 1 Test Results

The results of the Year 1 tests were not ideal, as despite very specific instructions, the user was not particularly careful about which plates they shipped for testing, and it turned out that there were 6 of one batch, 2 of another batch, a single plate of similar construction, and a single plate of a totally different geometry.

These constructions are outlined in Table 1 below:

Table 1. Construction of Level IV Armour Plates

Plate No	***Date of Manufacture***	***Construction***	***Plate Mass (kg)***
1	7/08	single curve – 8 mm alumina / 9 mm GRP	3.876
2	7/08	single curve – 8 mm alumina / 9 mm GRP	4.041
3	7/08	single curve – 8 mm alumina / 9 mm GRP	4.010
4	7/08	single curve – 8 mm alumina / 9 mm GRP	4.205
5	7/08	single curve – 8 mm alumina / 9 mm GRP	4.211
6	7/08	single curve – 8 mm alumina / 9 mm GRP	4.147
7	3/06	single curve – 9 mm alumina / 4.5 mm backing	3.825
8	3/06	single curve – 9 mm alumina / 4.5 mm backing	3.524
9	9/06	single curve – 10 mm alumina / 4 mm backing	3.880
10	3/06	flat – 9 mm alumina / 5 mm backing	3.861

6.1.1 NIJ-0101.04 Proof Tests – Level IV Armour Plate

For these 10 plates, a single shot of 30.06 AP M2 was fired centrally, at velocities at, or above, the required proof velocity. These velocities and the outcome are shown in Table 2. As the same shots are used for both the V_{50} calculation and the proof shots, many of the partial penetration proof shots are at velocities considerably higher than the requirements of the NIJ standard.

https://doi.org/10.52202/080042-0028

Table 2. NIJ-0101.04 Proof Results for Level IV Armour Plate

Plate No	*Ammunition*	*Shot No*	*Bullet Mass (g)*	*Velocity (m/s)*	*Outcome (CP/PP)*	*BFS (mm)*
1	30.06 AP M2	A1	10.6	835	PP	30.8
2		A2	10.6	874	PP	19.1
5		A5	10.6	878	PP	32.1
6		A6	10.6	887	PP	34.8
7		A7	10.6	853	PP	33.8
8		A8	10.6	867	CP	n/a
9		A9	10.6	840	PP	28.7
10		A10	10.6	841	PP	36.7

Plate numbers 3 and 4 were not used as proof shots as they produced complete penetrations, which were used in the V_{50} component of the test.

6.1.2 Extended Life Analysis V_{50} - Level IV Armour Plate

The Extended Life Analysis V_{50} was conducted on the six plates, which were manufactured in July 2008. The results are shown in Table 3 below:

Table 3. Extended Life Analysis Results for Level IV Armour Plate

Plate No	*Shot No*	*Bullet Mass (g)*	*Velocity (m/s)*	*Outcome (CP/PP)*	*BFS (mm)*	*Used for V_{50} (Y/N)*
1	A1	10.6	835	PP	30.8	N
2	A2	10.6	874	PP	19.1	N
3	A3	10.6	905	**CP**	n/a	Y
4	A4	10.6	889	**CP**	n/a	Y
5	A5	10.6	878	**PP**	32.1	Y
6	A6	10.6	887	**PP**	34.8	Y

Due to the limited number of plates of a single construction available, it was only possible to produce a 4-shot V_{50}. The 4-shot V_{50} achieved was **890 m/s** within a spread of 26 m/s, which is somewhat wider than would be ideal. This value may now be used as the basis for future extended life analysis tests.

6.1.3 7.62 x 39 mm PS Ball

Plate 6 was chosen for the additional 3 shots of 7.62 x 39 mm PS ball. The results of these shots are shown in Table 4.

Table 4. 7.62 x 39 mm PS ball (Factory 04) Proof Results for Level IV Armour Plate

Plate No	*Ammunition*	*Shot No*	*Bullet Mass (g)*	*Velocity (m/s)*	*Outcome (CP/PP)*	*BFS (mm)*
6	7.62 x 39 mm PS ball (04 66)	A12	8.0	723	PP	8.3
6		A13	8.0	726	PP	10.2
6		A14	8.0	717	PP	9.5

6.1.4 Conclusion of Year 1 Results

From the above results, obtained with a non-ideal mix of submitted plates, it was determined that for future years a quantity of 8 plates should be sufficient to achieve both the required V_{50} and suitably high proof shot velocities. Plates 7 and 8 would indicate a marginal proof shot pass, with that construction. Plates 9 and 10 achieved singe shot passes of the proof shot, for their respective constructions, although there is insufficient shot data to draw conclusions with any confidence.

6.2 Year 2 Results

This test was a much more controlled test. All 8 plates shipped were of the same batch, and hence the same construction, which was the same as the batch of 6 plates from Year 1. For Year 2, the construction and mass of the individual plates is as per the Table 5:

Table 5. Construction of Level IV Armour Plates

Plate No	*Date of Manufacture*	*Construction*	*Plate Mass (kg)*
1	07/08	single curve – 8 mm alumina / 9 mm GRP	4.065
2	07/08	single curve – 8 mm alumina / 9 mm GRP	4.121
3	07/08	single curve – 8 mm alumina / 9 mm GRP	4.172
4	07/08	single curve – 8 mm alumina / 9 mm GRP	4.085
5	07/08	single curve – 8 mm alumina / 9 mm GRP	3.717
6	07/08	single curve – 8 mm alumina / 9 mm GRP	3.946
7	07/08	single curve – 8 mm alumina / 9 mm GRP	4.092
8	07/08	single curve – 8 mm alumina / 9 mm GRP	3.994

6.2.1 NIJ-0101.04 Proof Tests – Level IV Armour Plate

For these 8 plates, a single shot of 30.06 AP M2 was fired at each at velocities at, or above, the required proof velocity. Those velocities considered suitable as proof shots are shown in Table 6 below:

Table 6. NIJ-0101.04 Proof Results for Level IV Armour Plate

Plate No	*Ammunition*	*Shot No*	*Bullet Mass (g)*	*Velocity (m/s)*	*Outcome (CP/PP)*	*BFS (mm)*
1	30.06 AP M2	1	10.6	901	PP	33.1
3		3	10.6	900	PP	25.7
4		4	10.6	906	PP	29.6
6		6	10.6	910	PP	36.9
8		8	10.6	910	PP	26.8

Plate numbers 2, 5 and 7 were not used as proof shots as they produced complete penetrations, which were used in the V_{50} component of the test.

6.2.2 Extended Life Analysis V_{50} - Level IV Armour Plate

The Extended Life Analysis was conducted on eight plates, with the results of six plates used to achieve a 6-shot V_{50}. The full set of 30.06 shot results are shown in Table 7 below.

Table 7. Extended Life Analysis Results for Level IV Armour Plate

Plate No	*Shot No*	*Bullet Mass (g)*	*Velocity (m/s)*	*Outcome (CP/PP)*	*BFS (mm)*	*Used for V_{50} (Y/N)*
1	1	10.6	901	PP	33.1	N
2	2	10.6	921	**CP**	n/a	Y
3	3	10.6	900	PP	25.7	N
4	4	10.6	906	**PP**	29.6	Y
5	5	10.6	917	**CP**	n/a	Y
6	6	10.6	910	**PP**	36.9	Y

https://doi.org/10.52202/080042-0028

7	7	10.6	914	**CP**	n/a	Y
8	8	10.6	910	**PP**	26.8	Y

The 6-shot V_{50} achieved was **913 m/s** within a velocity spread of 15 m/s. This value compares well with the March 2021 value of 890 m/s, and both values may now be used as the basis for future extended life analysis tests.

6.2.3 7.62 x 39 mm PS Ball

Plate 8 also had an additional 3 shots of factory charge 7.62 x 39 mm PS ball fired into it. The results of these shots are shown in Table 8.

Table 8. 7.62 x 39 mm PS ball (Factory 04) Proof Results for Level IV Armour Plate

Plate No	*Ammunition*	*Shot No*	*Bullet Mass (g)*	*Velocity (m/s)*	*Outcome (CP/PP)*	*BFS (mm)*
8	7.62 x 39 mm PS ball (04 66)	9	8.0	724	PP	15.8
8		10	8.0	736	PP	18.9
8		11	8.0	730	PP	17.6

6.3 NIJ-0101.04 Proof Tests over 2 Years

6.3.1 Year 1 Proof Results

Four shots above the proof velocity tolerance of 838 ± 9 m/s produced partial penetrations. The only complete penetrations were at 867 m/s and above. Therefore, there are no proof test failures of the armour plates tested at this time.

6.3.2 Year 2 Proof Results

Five shots above the proof velocity tolerance of 838 ± 9 m/s produced partial penetrations. The only complete penetrations were at 914 m/s and above. Therefore, there are no proof test failures of the armour plates tested at this time.

6.4 Extended Life Analysis (ELA) by V_{50} over 2 Years

6.4.1 Year 1 V_{50} Results

From the eight plates supplied, a V_{50} was obtained of 890 m/s, within a spread of 26 m/s. This value may be used as a basis for Extended Life Analysis in future years, as a means to extend the in-service life of the armour.

6.4.2 Year 2 V_{50} Results

From the eight plates supplied, a V_{50} was obtained of 913 m/s, within a spread of 15 m/s. This value, combined with the March 2021 result (890 m/s), may be used as a basis for Extended Life Analysis in future years, as a means to extend the in-service life of the armour.

6.5 7.62 x 39 mm PS Ball Results over 2 Years

6.5.1 Year 1 V_{50} Realistic Threat Results

The three additional shots of the 7.62 x 39 mm PS ball on plate 6, at velocities of 717 m/s and above, were easily defeated by the armour, leaving low back-face signatures.

6.5.2 Year 2 V_{50} Realistic Threat Results

The three additional shots of the 7.62 x 39 mm PS ball on plate 8, at velocities of 724 m/s and above, were easily defeated by the armour, leaving low back-face signatures.

7. DISCUSSION

In an ideal world, ELA should be considered as soon as an armour system is accepted for service, and the V_{50} should be conducted on the first production batch, as part of the initial acceptance testing. This would set the baseline performance value. Up until now this has not usually been done with ceramic armour plates, although it is common with items such as fragmentation vests, which are tested using fragment simulating projectiles (FSPs) and a V_{50} anyway.

For ceramic armour plates, accepting that the ELA is not started until the end of the warranty period, at the earliest, and that there is no baseline measurement to use, the 1st year should be considered as a learning experience, for both users and testers. Any V_{50} values obtained during the first year should be considered as indicative and should be used to inform the starting place for consecutive years.

ELA is, by its very nature, a destructive test, meaning that the inventory of armour plates diminishes by a number of plates each year. The programme is designed in such a way as to keep this annual plate reduction to a bare minimum. However, there will come a time when this testing is no longer viable, as the remaining number of plates will be too low to meet the requirements of the user. The number of plates requested for the ELA test is therefore kept to number, that many people may consider too low to obtain statistically significant results. For Year 1 this was 10 plates, which although compromised (due to the mixed batch), was deemed just adequate. For Year 2 it was deemed possible to reduce the number of plates required to 8 items. It is probably not possible to reduce that number any further, for a number of reasons. Predominantly, is the fact that these plates undergo a significant journey between the user base and the test house (with some challenging border customs to negotiate), so it seems prudent to remain with a sample size of 8. If this was reduced to 6, for example, any extras required due to issues during testing, would take a long time to arrive. This low number does mean that the 'proof' outcome is based upon much fewer shots than would normally be statistically required. However, due to the safety margin which still appears to exist with these plates, combined with the encouraging V_{50} results, this compromise is deemed acceptable, in this case. This acceptability is further increased by the fact that the realistic threat to the users is significantly lower than the armour specification.

It should be noted that the above discussion relates to these particular plates and the scenario in which they are used. In other cases, there could be some very good arguments to increase the sample size for testing. This could be, for example, if the first ELA test indicates that the plates are borderline in their performance. In such a case a larger sample size would provide the user with greater confidence, especially for the proof aspect of the tests. There may also be a scenario in which the initial specification of the plate is now borderline versus the current threat assessment. There could also be a situation where the inventory stock of plates includes many plates in excess to the operational requirement for numbers, and hence there is a greater resource for testing to call upon each year.

8. LESSONS LEARNT

For the first year, where the baseline is being set, the requirement for plate numbers is higher than it should be for subsequent years. The aim is to keep the number of plates used each time to a bare minimum, so as not to unnecessarily reduce the stock more than it needs to be. From the experience of this study, this is probably 8 plates. Ideally, plates would have an initial V_{50} test conducted when first manufactured, to provide a real baseline performance value. This would then allow the first year of ELA to be conducted with only 8 plates. In the absence of this initial production acceptance baseline, Year 1 of the ELA is considered to be that baseline, and hence the results for Year 1 should be considered indicative and used to inform Year 2 onwards.

It is critical that the user / supplier of the plates is meticulous in ensuring that the plates are from the same batch, and that there is evidence to prove it. In this study, the Year 1 plate submission was somewhat erratic, making it impossible to obtain a V_{50} with more than four shots from the same batch.

https://doi.org/10.52202/080042-0028

9. SUMMARY

When body armour upgrade plates exceed their warranty period, there is often a desire to gain confidence that they still perform as they should. ELA is a method, albeit a destructive one, which allows the year-on-year performance of the armour to be monitored in a numerical way.

The method of ELA used for this study allows for three different results: proof shots, V_{50} and reality check, from the same batch of 8 – 10 plates. V_{50} and proof values used will be the same shots in many cases, thus reducing the need for extra plates, and hence a greater reduction in the inventory of plates available to the user.

References

[1] Roberson C, Roberson G, Gotts P L, Hands G, Dulay D; ' Update on Practical Non-Destructive Testing Methods for In-Service QA of Ceramic Body Armour Plates' PASS 2006; Leeds; 2006

[2] Watson C, Ringrose T, Champion S, Horsfall I, Mallon B; 'An Assessment of the Ballistic performance of Contoured Protective Body Armour Plates'; PASS 2006; Leeds; 2006

[3] Bourget D, Withnall C, Palmer K, Rice S, Swann D; 'Aged Body Armour Testing: Further Results'; PASS 2012; Nurnberg; 2012

[4] Padovani M, Meulman J H, Louwers D; 'Effect of Real Aging on Ballistic Articles made of Dyneema® UD'; PASS 2012 Nurnberg; 2012

[5] Schaap, A A, Schotman, A H M, Bottger, C; 'Prediction of the Long-Term Performance of Twaron® Ballistic Vests by Accelerated Aging'; PASS 2012; Nurnberg

[6] NIJ-0101.04; 'Ballistic Resistance of Personal Body Armor'; US Dept of Justice; 28 Sep 1999

Glossary

AP	armour-piercing
ELA	Extended Life Analysis
NIJ	National Institute of Justice (USA)
V_{50}	the velocity at which, with a specified projectile and a specified armour system, the probability of perforation of 0.5

https://doi.org/10.52202/080042-0028

Generative Design of soft-armour

Y.S. Khoe[1]

[1]TNO PO Box 480 2501CL, The Hague Netherlands
yoeng_sin.khoe@tno.nl

Abstract. Computational models exist to quantify the performance of personal ballistic armour, where 'performance' refers to a variety of aspects of personal armour such as ballistic performance or human performance. Balancing the various aspects of performance is (and has always been) the quintessential element in body armour design due to limited load carrying capacity of a person. The Dutch organisation for *Toegepast-Natuurwetenschappelijk Onderzoek* (TNO) has combined these analytical models into a generative design algorithm to aid body armour design optimization by generating an array of potential designs within a given set of constraints. TNO foresees the use of this design tool to explore how body armour design changes with different performance requirements. Alternatively this design tool can be used to explore how a reduction in one aspect of performance requirements affects the other aspects of performance. Generative design of body armour is achieved through a Differential Evolution (DE) algorithm that combines a discretized approach to body armour parametrization with existing computational models of body armour performance. As a proof of concept, TNO implemented a detailed analysis to quantify protective performance, combined with simplified analyses to thermal, mobility and ergonomic constraints. The body armour is parametrized into squares of material that have varying areal density and can be positioned anywhere on the body. This allows (almost) freeform design of body armour. Protective performance is quantified using TNO's in house developed ICARUS-suite that evaluates four load cases of a fragmenting threat. The simplified approach to thermal, mobility and ergonomic constraints are constructed by assigning penalties to locations on the body where armour is placed. The algorithm evolves a group of designs over the course of a fixed number of generations. The result is a design that specifies body coverage and areal densities and shows the potential of the tool as a means to integrate and balance various aspects of body armour performance. This paper documents the development of the design tool and the various considerations that come into play when combining existing analytical tools in a generative algorithm.

THE PURPOSE OF GENERATIVE DESIGN TOOLS FOR PPE ACQUISITION SECTION

The Dutch MoD operates in a multitude of environments that vary in both climatic conditions and threats. Personal Protective Equipment (PPE), such as body armour, can be further optimized to maximize a soldier's performance and survivability when it is specialized to different environments (and soldier's tasks). TNO does not design PPE products nor does TNO do fabrication. What TNO does do is to support the Dutch ministry of defence in her quest to provide the men and women of the Dutch army with the protection that best suits the goals they are trying to achieve. TNO has a rich history (e.g. [1, 2, 3]) in the analysis and model development of soldier protection and soldier performance and we strive to combine this knowledge in a novel application to help the Dutch MoD in defining requirements for future soldier clothing and equipment.

The soldier system is human based, meaning that it inherits the human strengths, vulnerabilities, abilities and limits. The need to protect this valuable system is clear, as is the notion that any additional equipment burdens the soldier and its performance. However, quantifying this balance is a complex undertaking and has been a subject of ongoing research at TNO.

TNO is investing in the development of computer aided generative design tools based on machine learning. The purpose of this development path is to provide an optimization platform in which TNO's detailed performance models can be implemented, resulting in system optimization based on detailed submodels. A generative design tool can generate and evaluate designs and in this way autonomously explore the design space, a process that otherwise would have to be done manually. By exploring the design space and its resulting designs new insights can be gained to improve specification of system requirements for future systems.

This paper describes the optimization platform and the integration of different performance contributions. This proof-of-concept implements the Differential Evolution (DE) [4] optimization algorithm and TNO's Integrated Casualty Reduction Simulation (ICARUS) [5] tool, which scopes the application to soft-armour PPE against fragmenting threats.

MODEL DEVELOPMENT

Differential evolution

https://doi.org/10.52202/080042-0029

This generative design approach implements DE [4], an evolutionary type optimization algorithm. An evolutionary algorithm evolves a population consisting of individuals. The optimization loop (Figure 1) uses a parametrised design created by the DE. That design is sent to a submodel for evaluation to calculate the fitness of this design (i.e. the performance of the design). In the next generation parameters of random individuals (dubbed the parents) are combined (following the DE algorithm) and will evolve if the fitness of the new 'child' exceeds the fitness of the parents. Organized in a loop with a set number of generations, this mechanism drives the evolution that should result in increasingly better performing designs (see Figure 1). In essence this platform is an automated trial-and-error algorithm that autonomously that generates and evaluates possible designs.

TNO's implementation of the ICARUS in the DE algorithm is presented in brief in section 2.2. The armour parametrisation used in the DE is described in section 2.3. The implemented contributions to the fitness calculation are presented in chapter 3

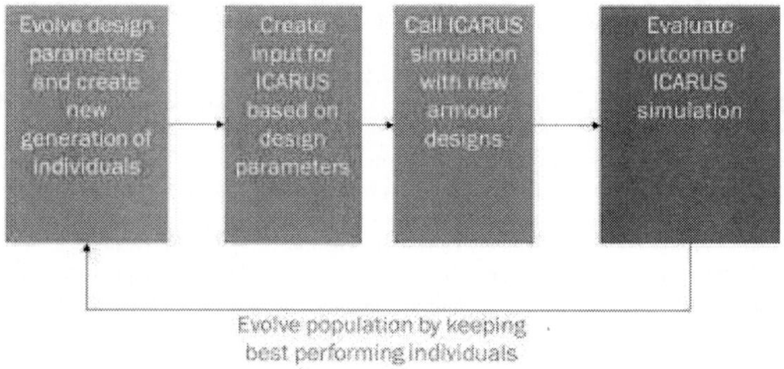

Figure 123 Implementation of ICARUS in the DE optimization loop

Evolutionary type algorithms require that all underlying submodels are evaluated for each individual in every generation. Hence, they are inherently slower than for example gradient based optimization techniques [6]. However, in the unknown design space for complex evolutionary algorithms they have the advantage that they are able to overcome local optima and don't require gradient information that is often not available when integrating existing simulation tools in optimization sequences.

Implementation of ICARUS in DE

ICARUS is a human vulnerability simulation tool that predicts the injury severity due to a fragmenting threat. This is achieved by (see **Figure 124**):

(1) modelling the fragment throw, ballistic trajectory and impact on the person
(2) modelling interaction with fragment protection using the Cunniff model [7]
(3) modelling penetration and retardation of fragments using ComputerMan [8]
(4) modelling injury severity using the Abbreviated Injury Scale (AIS) [9]

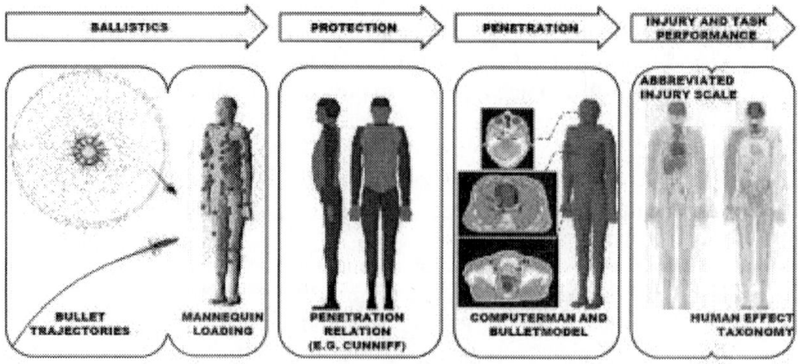

Figure 124 . Brief overview of main ICARUS modules to perform consequence analyses

Traditionally ICARUS has been employed in a forward manner where it is used to analyse the vulnerability of given PPE designs. By enabling automated design of PPE and quantifying the

https://doi.org/10.52202/080042-0029

vulnerability in a single metric the generative design loop can be closed, resulting in an automated system. The parametrised PPE design used in this demonstrator is described in section 0 and the evaluation procedure is described in section 0.

Armour parameterisation

The parametrised PPE must be compatible with the input format that is used in ICARUS and the underlying ComputerMan model. In ICARUS protection is included by calculating the velocity reduction of impacting fragments using the Cunniff model [7]. The area covered by protection is defined by selecting which voxels (the elementary building blocks of ComputerMan) are covered by protection.

Figure 125 shows all the voxels related to the outer skin of ComputerMan (front and back) folded out on a flat surface. This "skin projection view" is conveniently used to define which skin voxels are covered by protection by defining squares that encompass the voxels that are covered by protection.

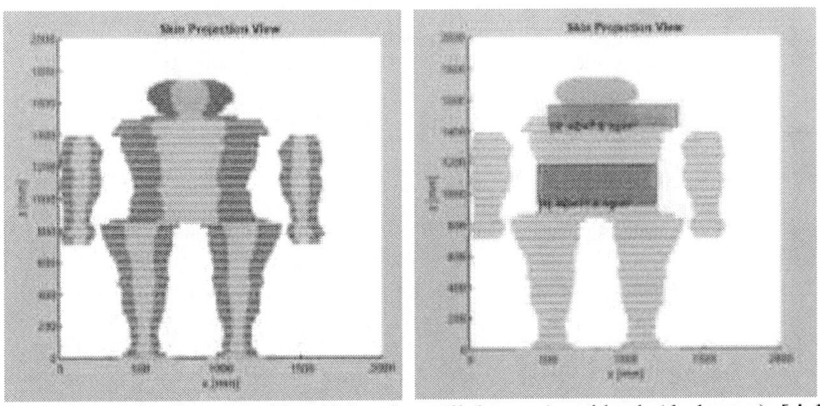

Figure 125. [left] Skin projection view identifying front (light grey) and back (dark grey). [right] Skin projection view with 2 armour squares generated by the DE algorithm. All voxels within the armour squares are covered by PPE.

Each armour square has 5 degrees of freedom: x-position, z-position, height, width and areal density. Currently only a single material is supported, which is the Kevlar KM2 material (see [7] for coefficients for the Cunniff model). Furthermore, the armour squares are restricted to the thorax region.

The DE uses a fixed number of armour squares and varies their degrees of freedom. **Figure 126** shows 4 armour squares, roughly representing a conventional soft-armour insert in an 'up-side-down-T' configuration. Armour squares can move and resize freely, which means that armour squares may result in a disjointed PPE design, or they may overlap. When overlap occurs, the areal density is combined and accumulated (i.e. it is treated as single thick protection, rather than multiple individual layers).

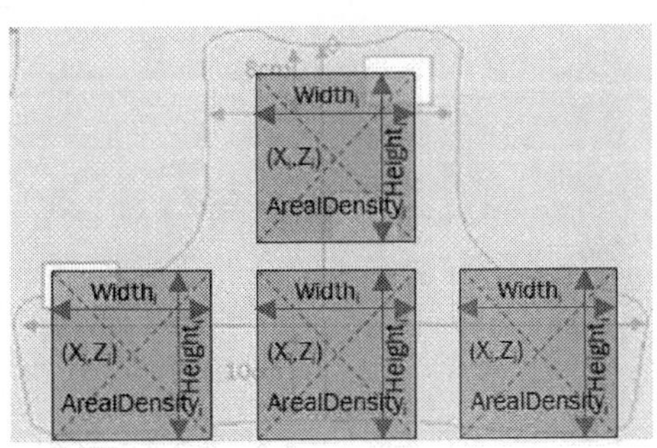

Figure 126 . Example with 4 armour squares

PERFORMANCE ASSESSMENT

Threat definition

ICARUS models injury due to a fragmenting threat, where each fragment is individually simulated and variations in the fragment throw are stochastically drawn. This discrete approach to simulation

https://doi.org/10.52202/080042-0029

potentially leads to, in an optimization loop, solutions that are optimized specifically to the used loading. To prevent this over-optimization, a variety of loading scenario's must be included in the simulations.

A generic 81mm mortar was selected that is placed at 4 orientations around ComputerMan (see **Figure 127**). Each position is simulated 5 times to obtain a spread in the loading that prevents over-optimization. Hence a total of 20 load cases are considered that are assumed to generate a representative loading.

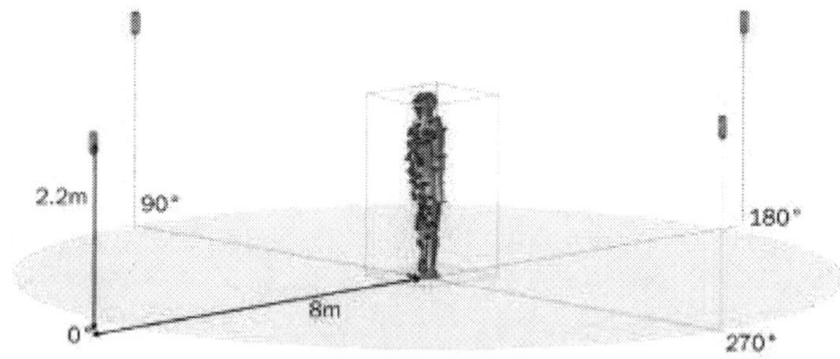

Figure 127 . Loading scenario for PPE optimization

Fitness function

The fitness (or performance) of a PPE design quantifies the outcome of a simulation (in this case ICARUS) in a single value. This fitness value follows 2 important guidelines:

(1) The optimization algorithm minimizes the fitness, thus a lower score must correspond to a better performing PPE design
(2) The fitness value must be normalized on a scale from -1 to 0, where -1 corresponds to the (best) upper limit of the achievable performance and 0 corresponds to the (worst) lower limit of the achievable performance

For the protection performance, the fitness is derived from the ability to reduce injury above a specified threshold. In example shown in **Figure 128**, the selected criteria is AIS>=5. The example shows that the PPE design realizes a reduction from 143 to 67 impacts (resulting in AIS>=5). The corresponding normalised fitness score is then -67/143=-0.469. If all fragments were to be stopped by a PPE the best score of 0/143=0 is achieved. If none of the fragments were to be stopped by a PPE the worst score of -143/143=-1 is achieved.

Criteria: AIS>=5	Unprotected	Protected
Nr. hits total	498	498
Nr. hits (AIS>=5)	143	67
Fitness score	-	-67/143 = -0.469

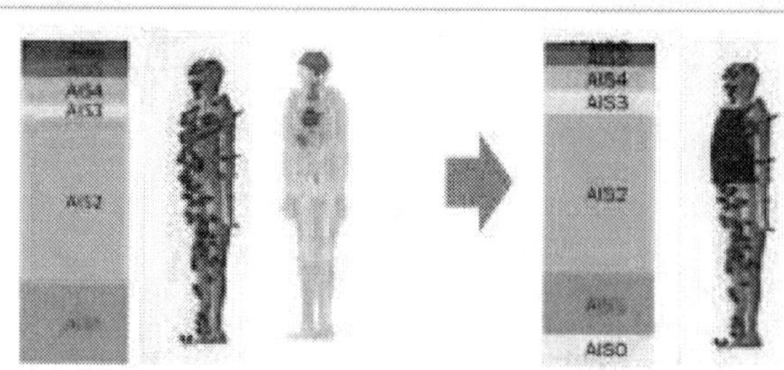

Figure 128. Example of fitness calculation for injury reduction

This fitness calculation is implemented such that the user can specify an injury criterion. In future versions, that are currently under development, we are developing a scoring system that takes injury reduction at all AIS levels into account.

Other performance contributors

An unrestricted optimization algorithm that is driven only by minimization of injury will always result in a design that will cover the complete body with the best (and usually heaviest) protection possible. In the future each contribution to the overall fitness will, like is the case with ICARUS simulation, come from detailed submodel analyses. In the presented proof-of-concept that interplay of detailed submodels is not yet available and a maximum mass for the PPE is prescribed.

Other contributions to performance have been included , but in a heavily simplified manner by applying for a weighted area approach. For example influence of thermophysiology is approximated by assigning more severe penalty values to areas that sweat more intensely [10] (see **Figure 129**), thereby giving the algorithm the incentive to avoid those areas.

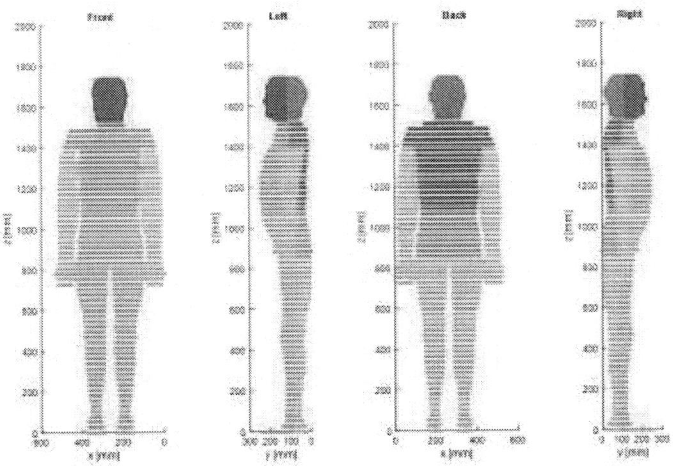

Figure 129. Penalty values assigned to sweat regions. Based on [10]

In addition to injury, the following performance contributions are implemented (as stated in a highly simplified manner): thermophysiology, thermosensitivity, ergonomics, stability.

Although the fitness score from each performance contributor is normalised there is subjectivity in the selection and combination of performance contributions. Using the normalised score per contributor implicitly assumes that each performance is weighted equally and is thus equally important. Secondly, for example thermophysiology and thermosensitivity address related (but not identical) performance contributions. By including both of them in the overall fitness, versus only 1 injury contributor, the optimization algorithm implicitly weighs thermal effects 2x as important. In the current implementation these considerations were addressed by manually assigning weighting factors to the contributors. Manual weighting factors are not preferred and in the next iteration of this optimisation algorithm a procedure is sought to objectively derive weighting factors.

RESULTS

A demonstrator was run, where the DE algorithm is allowed to evolve 8 armour squares to an optimal configuration, where the weighting factors were chosen to emphasize results from the vulnerability analysis. A population of 25 individuals are evolved over approximately 2000 generations , after which no more significant variations in the design are observed. The start, intermediate and final solutions are presented in **Figure 130**.

https://doi.org/10.52202/080042-0029

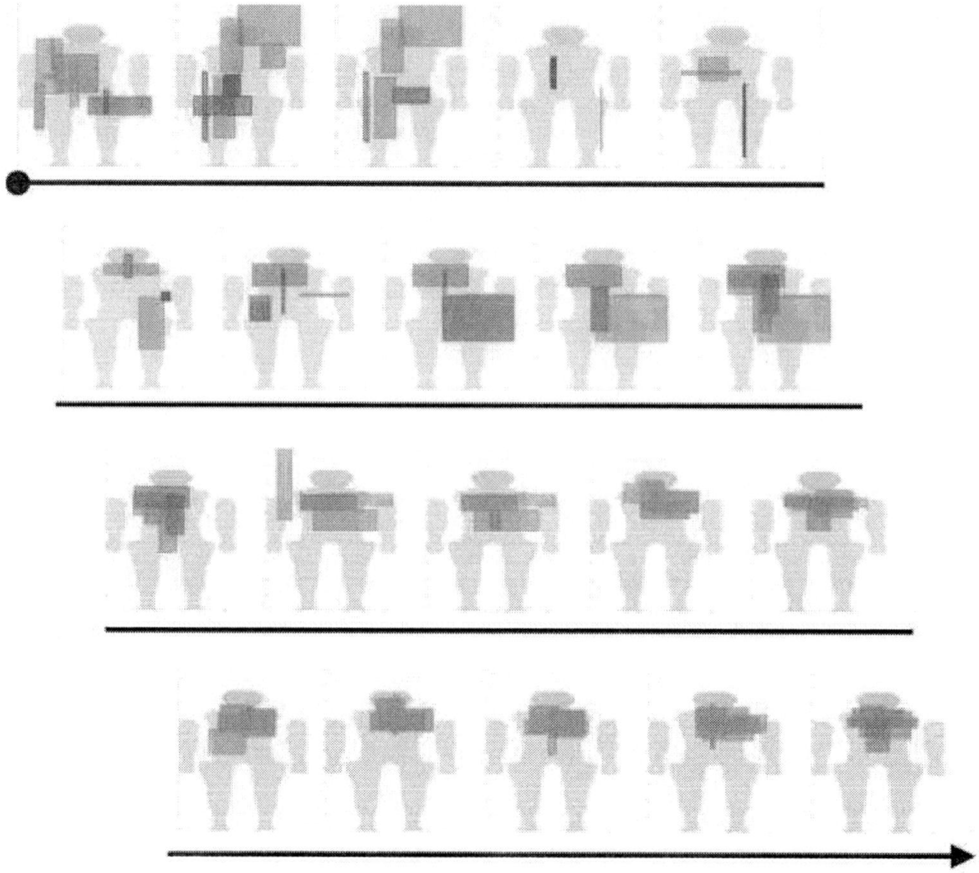

Figure 130 . Example result with 8 armour squares presented on the skin projection view. Colours represent areal density and range from green (low) to red (high)

The resulting PPE design has evolved to protect the upper thorax where most of the severe injuries (AIS>=5) can be found. Note that contrary to conventional up-side-down-T configuration, the found design more closely resembles a 'normal-T' configuration. This result can be explained because it maximizes protection of the requested AIS>=5 areas and is not rewarded for protection AIS4 zones (such as can be found in the abdomen. The horizontal upper-bar of the 'T' aims to protect the heart region (from impacts from all orientations). The vertical bar of the 'T' aims to protect the heart, aorta and nerves in the spinal region from impacts.

This demonstrates that the DE algorithm, coupled with ICARUS, can be used to generate logical designs that are in line with expectations. The design resulting from this exercise inspires the idea that in a high risk scenario, where high mobility is desired and only the most vital areas of the body are asked to be protected, an unconventional 'normal-T' design should be considered as an alternative.

CONCLUSION & FUTURE DEVELOPMENTS

In this paper the initial development of computer aided generative design algorithm is presented that can autonomously evolve PPE design. The underlying optimization algorithm is a differential evolution algorithm, coupled with TNO's ICARUS vulnerability assessment tool. The resulting tool shows that a outcome that is in line with expectations, but that still requires an subjective assessment of weighting factors

This development marks an important step in TNO's suite of tools, because it demonstrates that existing (analysis oriented) tools can be employed in an generative design algorithm. However, this generative design tool is not a step for TNO towards PPE design, but is a step to improve the ability to translate operational requirements to technical requirements for PPE. For example different PPE designs could be generated that prioritise different performance contributions. The outcomes will give insight in the dimensions and specifications that should be placed on PPE. We foresee application of this generative design tool early in the procurement phase, when defining the programme of requirements for new PPE.

The current generative design algorithm includes only protection against and injury due to fragment threat. The next iteration, which is currently under development, will introduce a number of new features: (1) introduce the ability to include different materials. (2) Expand the current implementation with TNO's Predicted Heat Strain (PHS) [11] model that models thermal strain due to physical activity and PPE. (3) Provide an procedure to objectively derive weighting factors. Here, the latter is of great importance, because it will allow for a more objective balance between the different performance contributions. The inclusion of PHS and ICARUS allows us to attempt to impose operational requirements on PPE (as opposed to artificial requirements such as weight) to see what designs are generated.

Despite all the so-called "intelligence" of generative design loops the results must always be carefully interpreted. As is the case for all simulations, the outcome is only as good as the underlying models and input. For example, at this moment the generative design loop does not consider aspects such as manufacturability or comfort, both of which are essential drivers of PPE design. Despite that shortcoming, explorations with this generative design tool will alleviate much of the trial-and-error effort that is otherwise manually performed and through its results it provides TNO and the MoD with insights into realistic and sensible technical requirements to impose on future PPE systems.

Acknowledgments

The author wishes to acknowledge the Dutch Ministry of Defence for funding this research, specifically the Kleding- en Persoonsgebonden Uitrusting Bedrijf (KPU-bedrijf)

References

[1] Y. Khoe, M. v. d. Jagt-Deutekom, T. Soullié and M. v. d. Horst, „Assessing the consequences on task performance due to the kinetic threat,", 14th Personal Armour Systems Symposium, Washington, DC, USA, 2018.

[2] T. Verhagen, H. Huisjes and M. v. d. Jagt-Deutekom, „Modern Personal Protection System Concept Assessment,", Personal Armour Systems Symposium, Nuremberg, Germany, 2012.

[3] Y. Khoe, „An integrated model approach to address the mobility versus vulnerability trade-off in soldier systems,", Personal Armour Systems Symposium, Cambridge, UK, 2014.

[4] R. Storn and K. Price, „Differential Evolution – A Simple and Efficient Heuristic for Global Optimization over Continuous Spaces," Journal of Global Optimization, p. 341–359, 1997.

[5] T. Verhagen, H. Huisjes and M. v. d. J. -. Deutekom, „Personal Protection System Concept Assessment,", Personal Armour Systems Symposium, Nuremberg, 2012.

[6] C. A. McCormick and M. R. Shepherd, „A comparison of algorithms for the vibroacoustic optimization of a beam: Gradient-based versus evolutionary," Proc. Mtgs. Acoust, vol. 42, nr. 1, 2020.

[7] P. Cunniff, „A design tool for the development of fragmentation protective body armour,", 18th International Symposium on Ballistics, San Antonio, TX, USA, 1999.

[8] R. Saucier and H. M. Kash III, „ComputerMan Model Description - ARL-TR-500," U.S. Army Research Laboratory, Aberdeen Proving Ground, MD, US, 1994.

[9] AIS Committee of the Association for the Advancement of Automotive Medicine, „The Abbreviated Injury Scale 2005 Revision," Association for the Advancement of Automotive Medicine, Barrington, 2005.

[10] C. Smith and G. Havenith, „Body mapping of sweating patters in male athletes in mild exercise-induced hyperthermia," Eur J Appl Physiol, vol. 111, pp. 1391-1404, 2011.

[11] M. Folkerts, A. Boshuizen, G. Gosselink, N. Gerrett, H. Daanen, C. Gao, J. Toftum, L. Nybo and B. Kingma, „Predicted and user perceived heat strain using the ClimApp mobile tool for individualized alert and advice," Climate Risk Management, vol. 34, p. 100381, 2021.

https://doi.org/10.52202/080042-0029

V_{50} instead of V_{proof} or alternative methodologies for highly protective EOD PPE

Jean-Philippe Dionne[1], Clint Hedge and Aris Makris

[1]*Med-Eng Holdings ULC 2400 St. Laurent Blvd. Ottawa Canada*
jean-philippe.dionne@safariland.com

Abstract. In 2015, AEP-2920 Edition A Version 1 [1] introduced the V_{proof} as a NATO personal armour ballistic test methodology. A given armour panel meets a V_{proof} rating against a specified fragment when at least 22 fragments shot at velocities in a range of V_{proof} to V_{proof} + 20 m/s all get stopped. In the context of Explosive Ordnance Disposal (EOD) Personal Protective Equipment (PPE), the suit areas most likely to be subjected to V_{proof} are those providing the highest level of protection. Historically, the reliance on measurements like V_{proof} for these areas arose from the technical challenge, for many ballistic laboratories, to reach high-enough velocities when firing the 1.1 g Fragment-Simulating Projectile (FSP), which is the most common fragment when testing EOD PPE fragmentation resistance. Prior to the release of the NATO V_{proof} methodology, and still today, some EOD PPE manufacturers and ballistic laboratories have applied unreliable customized test methodologies without statistical significance, to quantify the EOD PPE fragmentation protection performance. Among reasons for not applying the NATO V_{proof}, beyond the potential unawareness of the standard, is the need to fire a large quantity of fragments (22), which implies testing multiple complex and expensive protective panels. In addition, one must define, ahead of the test, a given V_{proof} classification to aim for. An alternative to V_{proof} for highly protective EOD panels, is the V_{50}, but using a heavier fragment. Contrary to V_{proof}, the V_{50}, which is a characteristic a given panel, does not heavily depend on a priori estimations of the performance (the starting velocity only slightly influences the V_{50}). For instance, the NIJ 0117.01 Standard for Public Safety Bomb Suits [2] mandates the use of the 13.4 g FSP for the chest and groin protective areas. Using such a heavy fragment results in lower V_{50} values compared to the 1.1 g case, which eliminates the challenges faced by ballistic laboratories in shooting 1.1 g fragments at very high velocities. In this paper, the statistical implications of selecting a V_{50} test with a heavy fragment as opposed to V_{proof}-like tests with a lighter fragment, are investigated through Monte Carlo simulations based on representative idealized materials. The results support the use of the 13.4 g to quantify the fragmentation protection performance of highly protective area of bomb suits, especially when facing limitations in costs or panels.

1. INTRODUCTION

Explosive Ordnance Disposal (EOD) suits (bomb suits) are not normally designed nor rated to stop bullets. As such, no backface deformation requirements are applicable to bomb suits. On the other hand, a bomb suit ensemble must stop explosively driven fragments originating from the explosive device itself, or those propelled by the force of the blast. As actual explosive devices vary greatly in size, components, construction and as the environment surrounding the explosive is also highly variable, fragmentation protection performance is characterized through standardized laboratory experiments. Such experiments involve the firing of identical and standardized representative fragments.

An ideal material would stop all fragments of a given type below and up to some characteristic velocity. Above that characteristic velocity, all fragments would completely penetrate the material (Figure 1). However, such "ideal" materials do not exist. Instead, in the real-world, there is a "zone of mixed results", where either complete stops or complete penetrations can be observed, when a specific fragment is fired (Figure 2).

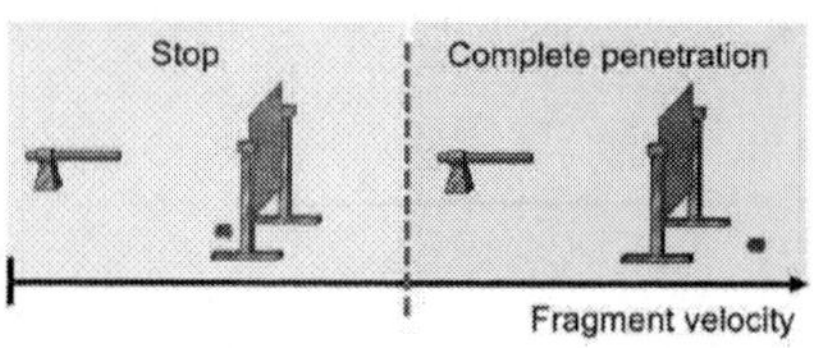

Figure 1. Idealized material

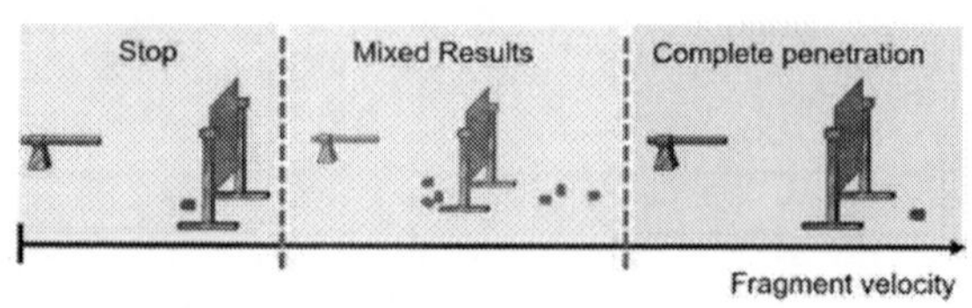

Figure 2. Real-world material, exhibiting a "zone of mixed results"

Binary logistic regression can be performed to illustrate this probabilistic characteristic of real-life materials, with the fragment velocity being used as the continuous explanatory variable and the penetration outcome being used as the categorical variable (Figure 3). At each velocity level, the resulting logistic regression curve from Figure 3, hereon referred to as the "S-curve", provides the probability of a complete penetration, for a specified fragment type. The S-curve is extensively used to completely

https://doi.org/10.52202/080042-0030

define the fragmentation protection characteristic of a material [3]. Unfortunately, accurately determining this curve requires extremely extensive testing. Such testing is generally not practical, other than in the limited context of scientific experiments. Instead, when it comes to characterizing materials, the emphasis is generally put on determining only one or a few points on the S-curve.

The V_{50} is the most widely studied point on the S-curve (Figure 3) and is considered by Cunniff [4] as being "among the most elegant performance metrics for armour systems". There exist well-defined methodologies for the determination of the V_{50} (e.g. MIL-STD-662 [5] and NATO STANAG 2920 [6]). These methodologies involve firing an equal number of fragments that completely penetrate a material, and fragments that get stopped, within a limited range of firing velocities. The V_{50} rating is then defined as the average of all velocities within the range. Many standardized fragments exist (Figure 4). But for evaluating bomb suit materials, the 1.1 g standardized fragment is the most widely used.

For the bomb suit areas requiring the highest levels of protection (red area in Figure 5), the V_{50} ratings with the 1.1 g fragment are typically of the order of 1800 m/s. Unfortunately, many ballistic laboratories experience difficulties in reliably and reproducibly firing 1.1 g fragments at such high velocities. As such, bomb suit manufacturers have often relied instead on means other than the V_{50} to quantify the level of fragmentation protection provided by the most protective area of their suits, against the 1.1 g fragment.

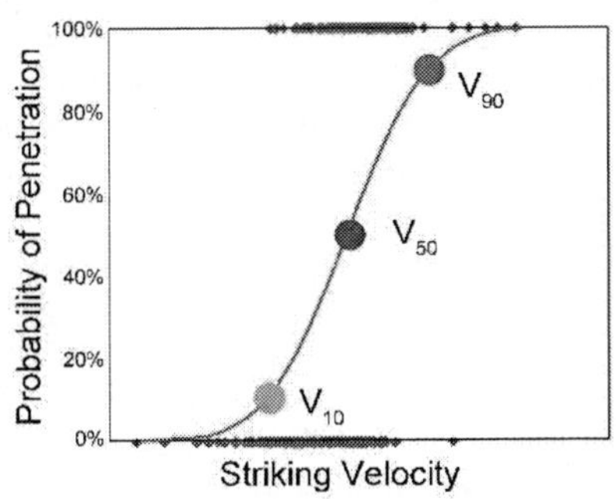

Figure 3. Generic logistic regression curve (S-curve) with three characteristic points highlighted

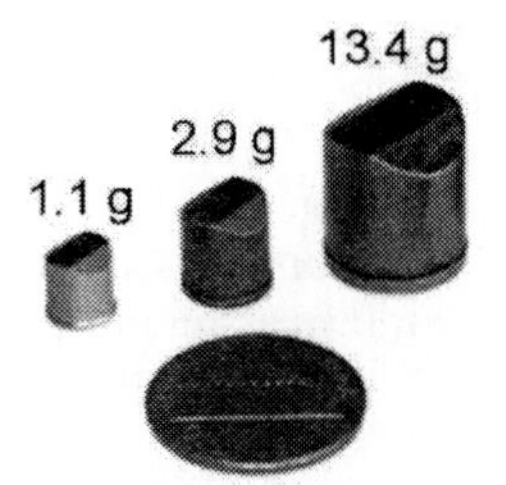

Figure 4. Fragment Simulating Projectiles (FSPs) used by NIJ [2]

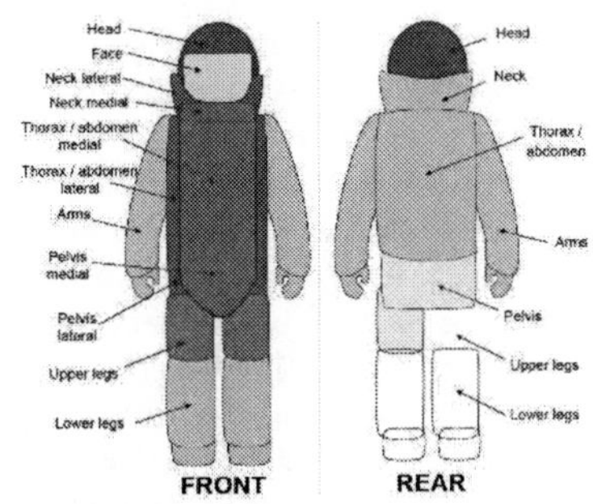

Figure 5: Bomb Suit protective areas as per NIJ [2]. Highly protective areas in red

An alternative to the V_{50}, when reaching the desired firing velocities becomes challenging, is to aim for another point in a lower range of the S-curve, such as the V_0. The V_0 represents the highest velocity at which no complete penetration is expected. The outdated NATO STANAG 2920 Ed. 2 [6] document provided a standardized method to estimate the V_0. This method implied firing fragments at velocities up to 1.5 times higher than the V_0 of a material (Figure 6). The V_0 was then estimated by extrapolating to zero the residual velocities arising from complete penetrations. This requirement to fire fragments at such high velocities made it even more difficult for test laboratories to characterize the protection performance of bomb suits with the 1.1 g fragment. As a workaround, some bomb suit manufacturers conduct customized tests involving a limited number of complete stops, close to the expected V_0. The V_0 is then estimated as the maximum value from these firing tests (Figure 7). However, such made up methodology is not based on a sound statistical approach and overestimates the protection performance. For the purpose of this paper, this custom methodology will be defined as $V_{0\text{-claimed}}$.

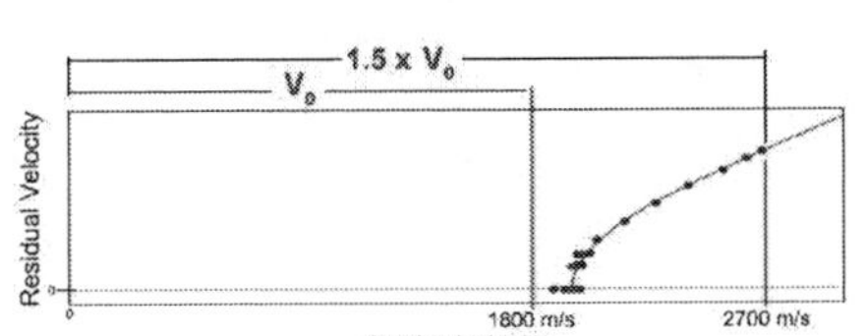

Figure 6. V_0 as per STANAG 2920 Ed. 2

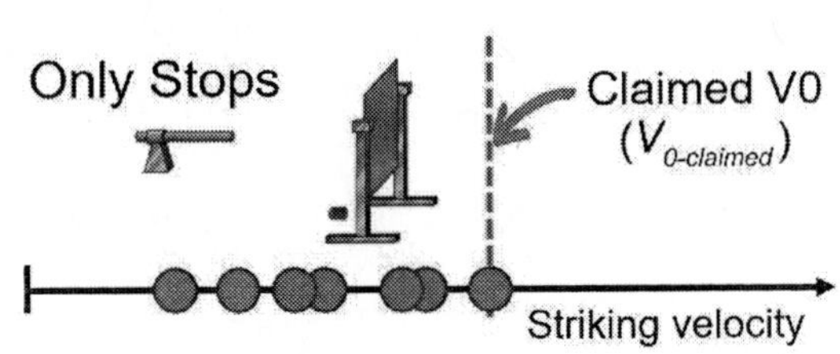

Figure 7. Customized methodology ($V_{0\text{-claimed}}$)

In 2015, NATO released a document entitled AEP-2920 Procedures for the Evaluation and Classification of Personal Armour, Edition A Version 1 [1]. This document includes a standardized approach to estimate a material performance rating similar in concept to the V_0 (velocity below which no penetration is expected), referred to as the V_{proof}. The V_{proof} involves testing in a lower range of the S-curve, compared to the V_0 from STANAG-2920 [6]. AEP-2920 defines the V_{proof} as a validation against

https://doi.org/10.52202/080042-0030

a specified projectile based on a statistical approach, where a defined number of projectiles are fired with a defined velocity at soft armour and/or hard armour, personal armour items, components or material samples. The classification is achieved when the said component has defeated the defined number of projectiles at the defined velocity. Bolduc et al. [7] provided a good overview of AEP-2920.

In this study, three test methodologies will be considered to characterize the fragmentation protection performance of materials: the V_{50}, the V_{proof}, as well as the made-up methodology referred to earlier as the $V_{0\text{-claimed}}$. Monte Carlo statistical simulations will be performed to characterize the performance of simulated materials (no laboratory experiments). Monte Carlo simulations of V_{50} tests have been conducted by Andres et al. [3], Cunniff [4], Eridon et al. [8] and Cheng et al. [9], among others. The approach adopted here will not deviate substantially from these previous studies. However, the focus here will be on the characterization of highly protective bomb suit areas. More specifically, the advantages, caveats, and biases of all three methods listed above will be presented, when applied to quantifying the protection performance of these highly protective armour panels.

2. METHODOLOGY

This study is based on statistical (Monte Carlo) computer simulations. No actual laboratory experiments were considered. Probabilities were used to estimate the firing velocity of any fragment given a target velocity, and to determine whether a complete penetration or a stop (partial penetration) was obtained as the fragment impacted the target material. For the firing velocity, a normal distribution centered around the target velocity was applied. A standard deviation equivalent to 2% of the target velocity was arbitrarily assigned, representative of deviations typically observed in ballistic laboratories (Figure 8). Andres et al. [3] also used a normal distribution with a similar standard deviation for the target velocity.

The S-curve of a material was determined assuming a normal distribution. A Weibull distribution, already discussed in the context of fragmentation testing [9, 10] or other distributions might have provided more realistic probabilities, especially in the low range of the S-curve. Indeed, a normal distribution provides for instance a non-zero probability of complete penetration even for negative firing velocities. Nevertheless, a normal distribution was selected to minimize the number of parameters (only the mean and the standard deviation are required). Andres et al [3] and Eridon et al. [8] have also assumed a normal (Gaussian) distribution when defining the S-curve of their idealized materials.

Two different idealized materials (Material A and Material B) were considered in this study. Both materials share the same V_{50} value but exhibit two different standard deviations. Specifically, a single V_{50} value of 1000 m/s was selected, while standard deviations of 20 m/s (Material A) and 100 m/s (Material B) were selected. Large enough differences in standard deviations were assigned to properly characterize the influence of S-curve steepness. The S-curves for Materials A and B (Figure 9) were used to determine the probability of complete penetration for any given firing velocity.

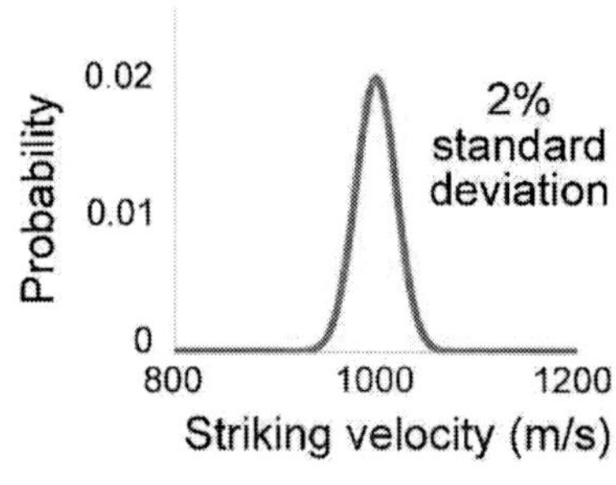

Figure 8. Distribution of striking velocities for a 1000 m/s targeted value, based on a 2% standard deviation

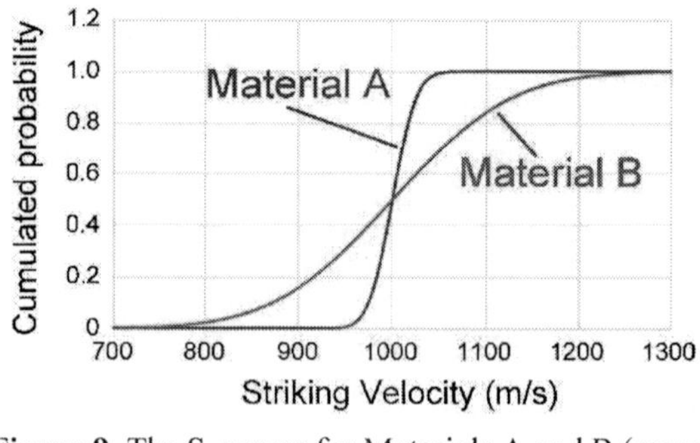

Figure 9. The S-curves for Materials A and B (same V_{50} of 1000 m/s, but standard deviations of 20 and 100 m/s respectively)

To determine the V_{50} of a material, the "up and down" procedure from STANAG-2920 was applied. According to this procedure, a first shot is fired at or near the expected V_{50} value. The following striking velocities are determined according to the decision algorithm illustrated in Figure 10. The procedure ends when either one of the four conditions listed in Table 1 is achieved. Figure 11 illustrates a representative test scenario, having led in that case to the determination of a V_{50} value based on the "3x3" condition from Table 1.

Table 1. STANAG 2920 V_{50} procedure – End conditions

Condition	Number of shots needed	Velocity range	V_{50} calculation
3x3	3 stops and 3 complete penetrations	within a range of 40 m/s	Average of the 6 velocities
4x4	4 stops and 4 complete penetrations	within a range of 60 m/s	Average of the 8 velocities
5x5	5 stops and 5 complete penetrations	within a range of 80 m/s	Average of the 10 velocities
Inconclusive	When none of three above conditions can be met. No V_{50} value can be determined		

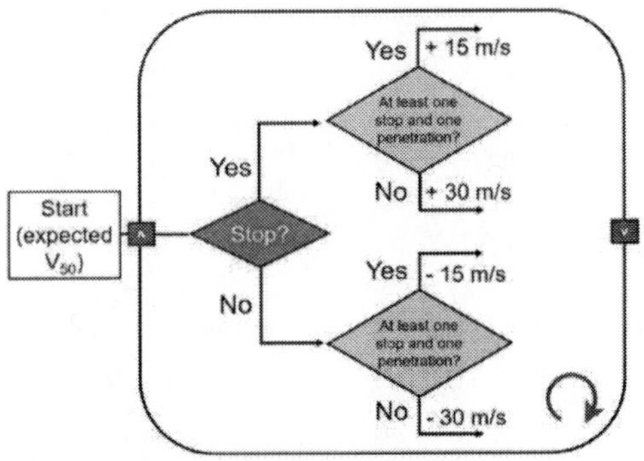

Figure 10. Up and down V_{50} procedure as per STANAG-2920. The suggested velocity changes are approximate

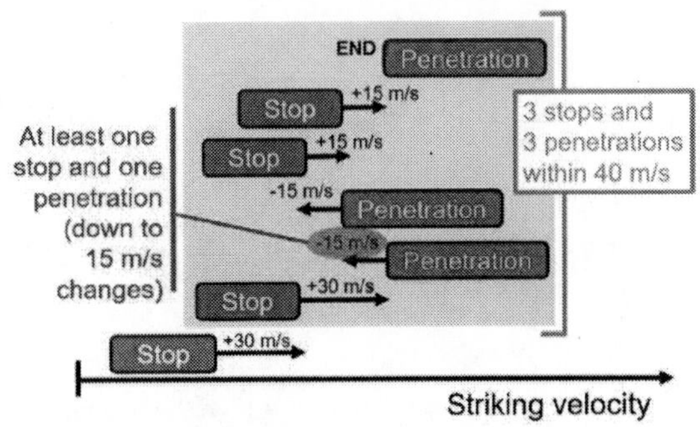

Figure 11. Example of a V_{50} scenario using the up and down procedure (from bottom to top)

To determine a V_{proof} value, AEP-2920 requires a total of 22 stops, fired within a range of the targeted V_{proof} value to 20 m/s above this target (Eridon et al. [8] provides the history and rationale behind this value of 22). Any single complete penetration within this range (or lower) results in a failed V_{proof} test. AEP-2920 does not mandate an actual procedure to determine the velocity at which fragments must be fired. But for the purpose of the current simulations, the target velocity was always in the middle of the sought range, i.e. the target V_{proof} plus 10 m/s. Variations in the actual firing velocity then resulted, from the assumed normal distribution illustrated in Figure 8. The procedure was interrupted either after 22 eligible consecutive stops were obtained, or as soon as a complete penetration was observed, within the desired velocity range. Figure 12 illustrates two representative V_{proof} test scenarios, having led to the two possible outcomes (pass or fail).

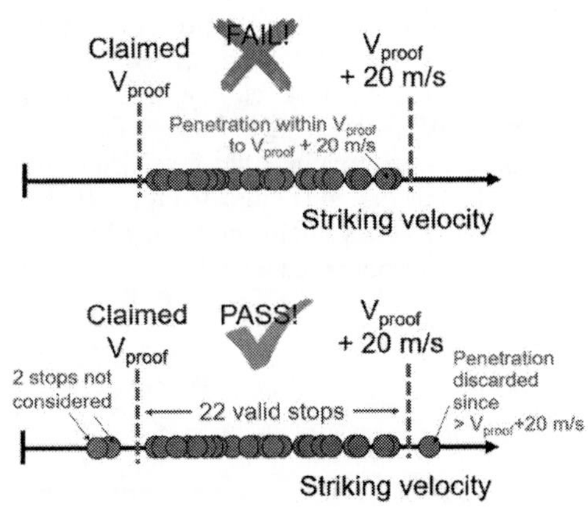

Figure 12. Two representative V_{proof} scenarios

As discussed earlier, some bomb suit manufacturers have sometimes relied on a customized methodology, not recognized by any standard, to quantify a performance characteristic meant to be similar in concept to a V_0 or V_{proof}. As this procedure overestimates the performance (as will be demonstrated), it was labelled as $V_{0\text{-claimed}}$ in the current study. Simulations were conducted whereby all shots were fired at approximately the targeted $V_{0\text{-claimed}}$ velocity. The actual firing velocity then varied according to the normal distribution illustrated in Figure 8. When a velocity exceeded the targeted $V_{0\text{-claimed}}$, the shot was excluded from the analysis, unless it corresponded to a stop. All other shots were considered. The procedure was repeated until a total of 8 accepted shots were fired. Any complete penetration within these 8 data points yielded a failed result. This number of shots (8) was selected based on anecdotal evidence collected over the years, obtained from open bid/tender processes for bomb suits.

https://doi.org/10.52202/080042-0030

3. RESULTS

When performing Monte Carlo statistical analyses, the number of individual simulations conducted must be large enough to obtain consistent average results. A sensitivity analysis was therefore first conducted to determine an acceptable number of simulations to conduct. To this end, V_{50} values were calculated for both materials, with the same starting point corresponding to the V_{50}. The average V_{50} and standard deviation values are plotted in Figures 13 and 14 respectively, as a function of the number of simulations. Variations in both parameters are minimal across all cases, and beyond 20,000 repetitions, the curves get fairly stable. Hence, for the remainder of this work, a total of 20,000 simulations were conducted for each experiment. This number far exceeds the number of Monte Carlo simulations conducted by Andres et al. [3] and Eridon et al. [8] (1000 simulations), but not as high as for Cheng et al. [9] (100,000).

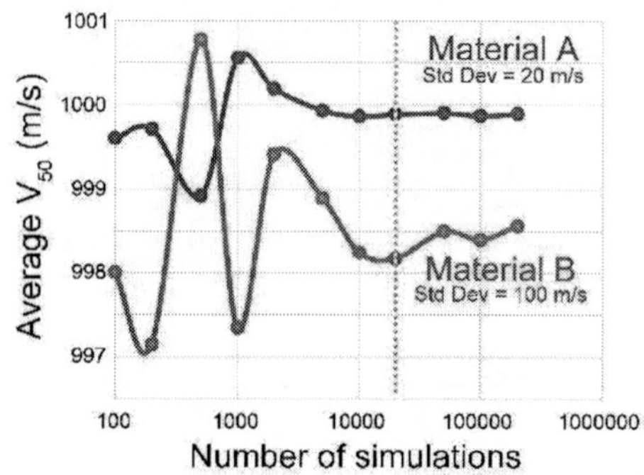

Figure 13. Average V_{50} as a function of the number of simulations

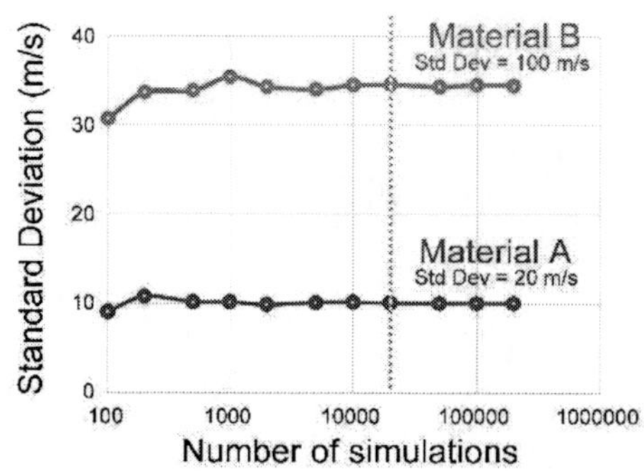

Figure 14. Average standard deviation as a function of the number of simulations

3.1 Effect of starting velocity on the V_{50}

While the STANAG-2920 V_{50} procedure is well defined, the start velocity is rather arbitrary. Indeed, the expected V_{50} is generally not known in advance. Andres et al. [3] and Cheng et al. [9] had already determined that the start velocity introduces a bias in the results (lower V_{50} with a lower start velocity, and higher V_{50} with a higher start velocity). To validate the model against these previous findings, simulations were conducted to investigate the effect of the start velocity on the V_{50} (Figure 15) and on the standard deviation (Figure 16) for both materials. The results indeed confirm the positive correlation between the start velocity and the simulated V_{50} value. The effect is much more pronounced for Material B, characterized by a wider zone of mixed results (less steep S-curve), compared to Material A.

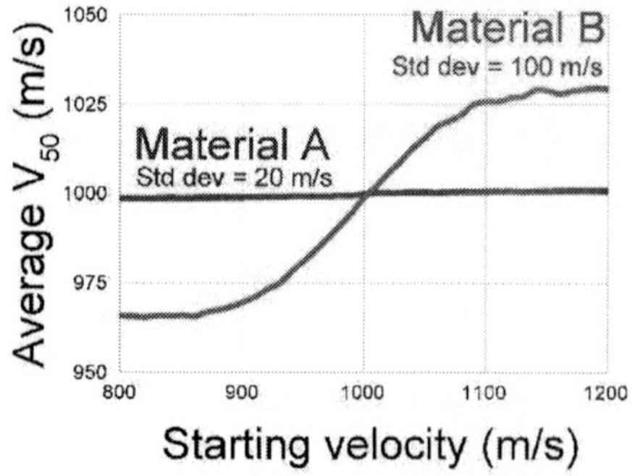

Figure 15. Average V_{50} as a function of the starting velocity

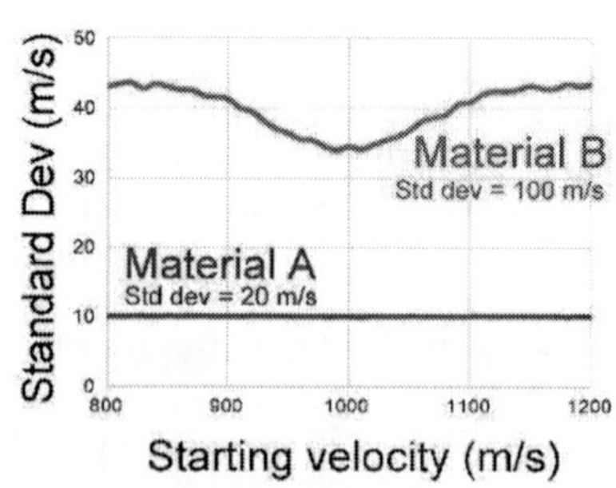

Figure 16. Average standard deviation as a function of the starting velocity

3.2 V_{50} determination in the NIJ 0117.01 standard for public safety bomb suits

The US National Institute of Justice NIJ 0117.01 standard for public safety bomb suits [2] includes V_{50} requirements for multiple protective areas against three specific fragment simulated projectiles. For every combination of area and fragment, a set of *three* V_{50} tests is required. A test panel is deemed to pass if: 1) the average of the three values exceeds the stated requirement, 2) no more than one single V_{50} test lies below the requirement and 3) if a test is below the minimum, it cannot be lower by more than 25 m/s. Table 2 summarizes the possible outcomes from a set of three individual NIJ V_{50} tests.

Table 2. NIJ 0117.01 V_{50} procedure – Possible outcomes

Average of 3 V_{50}s	# of results above the requirement	# of results >25 m/s below requirement	Result
> requirement	3	n/a	PASS
> requirement	2	0	PASS
> requirement	2	1	FAIL
> requirement	1	irrelevant	FAIL
< requirement	irrelevant	irrelevant	FAIL

The NIJ standard refers to the MIL-STD-662 test methodology, which is very similar to STANAG-2920. A similar requirement applies for the start velocity, which is near the expected V_{50} value. In the present study, the start velocity for the first of the three V_{50} tests was varied for investigation purposes. However, the start velocity for the second test was selected as the V_{50} obtained in the first test. And the start value for the third test was taken as the average of the first two V_{50} values. When tests were inconclusive, additional tests were performed to arrive at a set of three valid V_{50} values.

Figure 17 compares the V_{50} histograms obtained for Material A with a start velocity corresponding to the V_{50}, for the two cases of interest: a single V_{50} test (orange), and the NIJ scenario (green) involving the average of three tests. A tighter distribution is obtained in the NIJ case. Figure 18 then compares the V_{50} values obtained as a function of the start velocity for the first test when following the NIJ procedure (three tests) vs. the case with a single test, for Material B. Figure 18 demonstrates that the influence of the start velocity has been reduced in the NIJ scenario involving three tests instead of a single one.

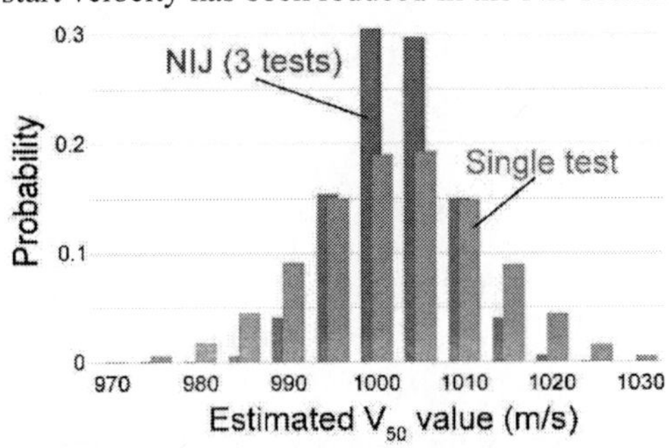

Figure 17. Reduction in V_{50} variability when using three tests as per the NIJ standard [2] (green) vs. a single test (orange) (Material A)

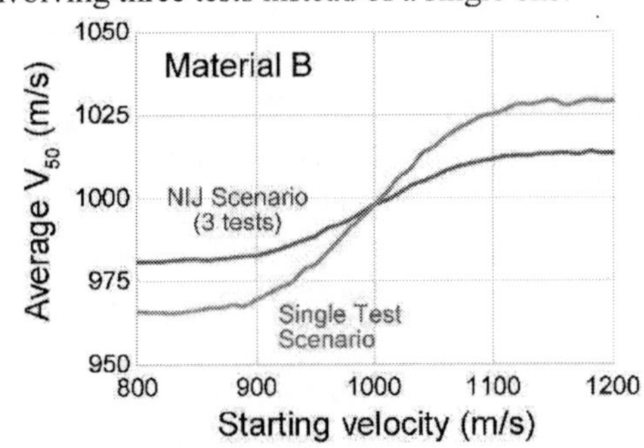

Figure 18. Reduction in V_{50} variability when using three tests as per the NIJ standard [2] (Material B)

Figure 19 investigates the probability of Materials A and B to meet different standard target velocities when tested according to the NIJ methodology. The horizontal axis displays the target NIJ velocity, while the vertical axis displays the probability of passing each of these target values. For both materials, the probability of meeting the true V_{50} of 1000 m/s (as per the S-curve) through an NIJ test is below 40%, which stresses the need to always include a buffer when making V_{50} claims. Specifically, when targeting a specific V_{50} requirement, a material with a higher "true" V_{50} (as per the S-curve) must be selected. Figure 20 complements Figure 19 by displaying the proportion of all 6 possible outcomes from Table 2, for the determination of the V_{50} as a function of the target NIJ value, for Material B.

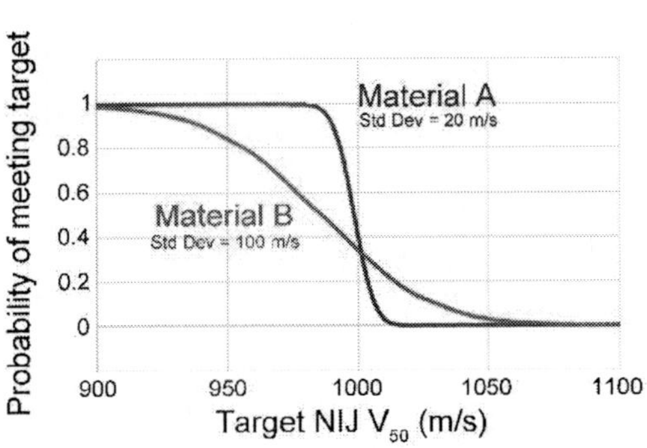

Figure 19. Probabilities for Materials A and B to meet NIJ target V_{50} ratings

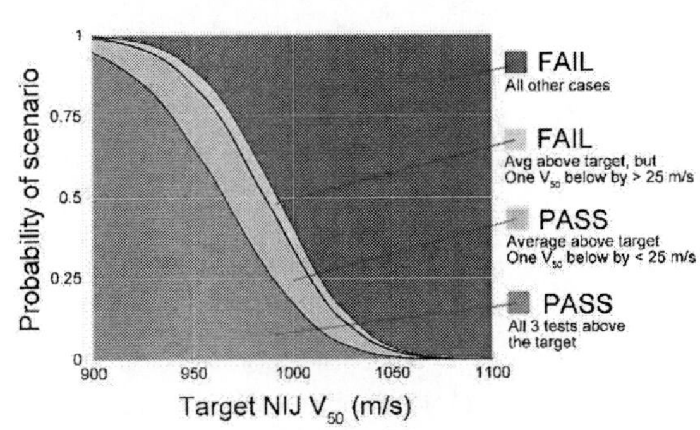

Figure 20. Proportion of all possible NIJ V_{50} outcomes for Material B

https://doi.org/10.52202/080042-0030

3.3 V_{proof} determination from AEP-2920

Monte Carlo simulations were conducted to estimate the probability of meeting a range of V_{proof} target values for a given confidence level (probability that the observed confidence interval contains the true V_{proof} value). Figure 21 provides these probabilities, highlighting the scenarios with a confidence level of 90% (940 m/s for Material A, 735 m/s for Material B). As a comparison, the V1 for these two materials are 954 m/s and 872 m/s respectively. The 90% confidence V_{proof} values therefore provide conservative assessments of the stopping capability of materials, below the expected V1 performance. A V_0, given the use of a normal distribution for the S-curves, could not be achieved through these simulations. Figure 22 shows the average number of shots required before a first complete penetration, as a function of V_{proof} target level, for both materials. For the 90% confidence level V_{proof} levels, the average required number of shots is 214 for Material A, and 202 for Material B, values much higher than the required number of stops for a V_{proof} (22). This being said, there is nevertheless a 10% chance of failing the V_{proof} test when testing at the 90% confidence level, meaning, 10% chance of a penetration within the first 22 shots.

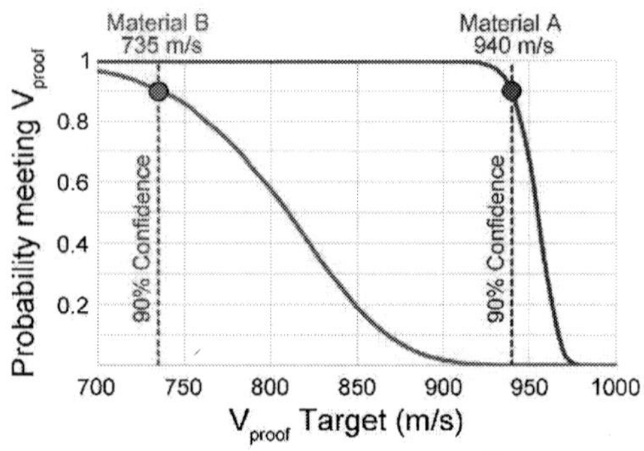

Figure 21. Probabilities for Materials A and B to meet V_{proof} target ratings, highlighting the 90% confidence case

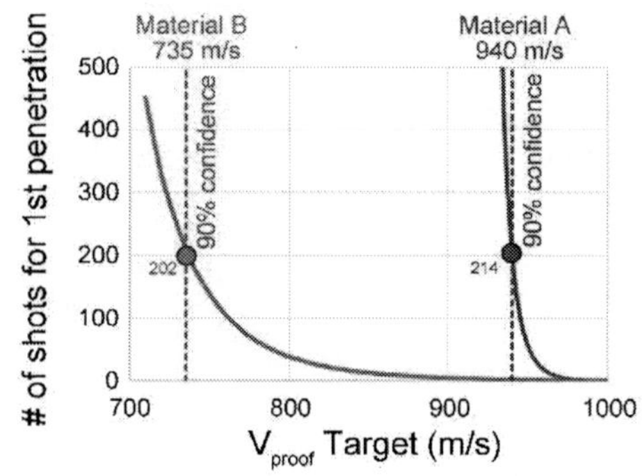

Figure 22. Average number of shots required for a first penetration, for Materials A and B. The 90% confidence level case is highlighted

3.4 $V_{0\text{-claimed}}$ determination (customized methodology)

Figure 23 shows the probability of meeting $V_{0\text{-claimed}}$ target values for Materials A and B. As highlighted before, the $V_{0\text{-claimed}}$ is a customized methodology not based on valid statistical grounds, that was used by some bomb suit manufacturers to *estimate* V_0 ratings. A $V_{0\text{-claimed}}$ experiment consists here of 8 stops (no complete penetration), with the highest stop being considered as the obtained $V_{0\text{-claimed}}$ rating. Figure 23 highlights the fact that for a $V_{0\text{-claimed}}$ target corresponding to the V_{50} of the material (1000 m/s here), there is a non-negligible chance (~20%) of meeting this target for Material A. This corresponds to estimating a V_0 exceeding the V_{50}, which does not make sense. Figure 24 further highlights the issue with the $V_{0\text{-claimed}}$ concept to estimate a V_0. In this figure, the $V_{0\text{-claimed}}$ and V_{proof} ratings for Material A are plotted with respect to the confidence level. The $V_{0\text{-claimed}}$ rating exceeds the V_{proof} rating by approximately 30 m/s across the entire range of confidence levels, which clearly indicates that the $V_{0\text{-claimed}}$, as its nickname suggests it, overestimates the protective capability of materials.

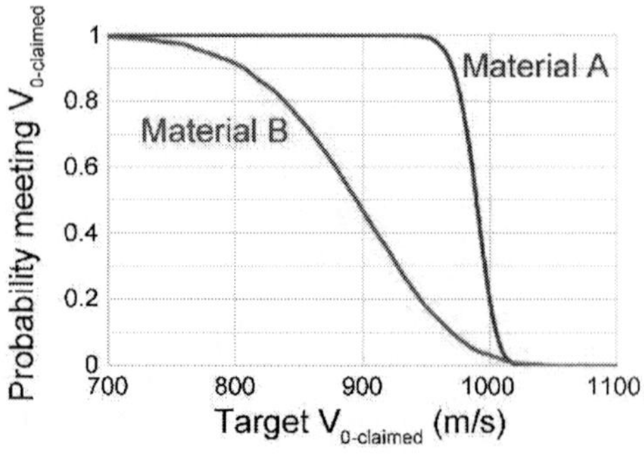

Figure 23. Probabilities for Materials A and B to meet $V_{0\text{-claimed}}$ target ratings

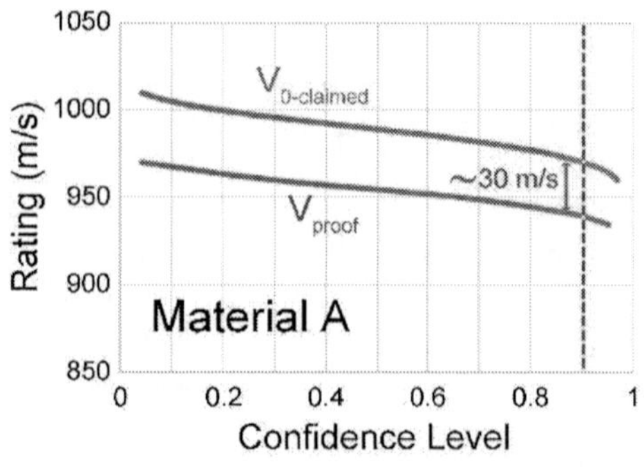

Figure 24. Comparison of the $V_{0\text{-claimed}}$ and V_{proof} ratings for Material A as a function of confidence level, highlighting the 90% confidence level case

4. DISCUSSION

As highlighted in this paper, the V_{50} methodology is effective at characterizing the protective capability of a material, using a standardized test method requiring only a limited number of strikes (typically between 10 to 15). The results are repeatable, as suggested in Figure 16 which shows relatively low standard deviations, relative to a V_{50} of 1000 m/s. The only arbitrary parameter in the V_{50} "up and down" methodology is the start velocity, as the expected V_{50} rating is often unknown. But the variations due to the start velocity can be minimized by applying the NIJ 0117.01 V_{50} test procedure (Figure 18), which requires three tests to be conducted on any given material/fragment combination. The start velocities for the second and third V_{50} tests can then be based on the previous results.

In general, the disadvantage of relying solely on the V_{50} to characterize a material is that only a single point on the S-curve is being quantified. But when applying the NIJ procedure, the requirement to have all three individual values exceeding the requirement (or only two, if the third one is no more than 25 m/s below the target) forces manufacturers to take into account the inherent variability of the materials. Indeed, as shown in Figure 19, Material B, which has the same V_{50} as Material A, needs an additional buffer when it comes to claiming NIJ ratings. For instance, the V_{50} for Material A that can be claimed with 90% confidence using the NIJ 0117.01 procedure is 992 m/s. For Material B, one can only claim 940 m/s. Conversely, to claim a specified V_{50} rating, one must select a material with a V_{50} as per the S-curve exceeding that rating, with a buffer increasing with the standard deviation (variability) of the material. It can therefore be inferred that the NIJ procedure takes the S-curve of the material into account.

It can thus be concluded that the V_{50} is a proper *characteristic* of a material, which can readily be determined without any prior knowledge about the material and based on a limited number of strikes and material samples.

The V_{proof} on the other hand, is *not* a characteristic of a material, unless it is associated with a specific confidence level (e.g., 90%), as also pointed out by Eridon et al. [8]. As an example, a material that meets a V_{proof} of 1800 m/s with 90% confidence can possibly meet a higher V_{proof} with a lower confidence level. It can also meet a lower V_{proof} of 1750 m/s. As such, if no prior knowledge on this material exists prior to testing for V_{proof}, and a test at 1750 m/s indicates a pass, there is no way to know whether the material could qualify for an even higher V_{proof} rating. An additional test conducted at 1750 m/s could even yield a fail, given the probabilistic nature of such tests. Finding the 90% confidence level V_{proof} experimentally for a given material can therefore be a daunting task, requiring an extremely large number of strikes (22 shots multiplied by the number of required iterations, which is much higher than the three times 10 to 15 shots required for the NIJ V_{50} scenario). Moreover, if a V_{proof} close to the 90% confidence value is claimed, it must be kept in mind that there is still a 10% chance that a single V_{proof} test at that velocity would yield a failed result. As such, the V_{proof} is not a *proper* characteristic of a material, the same way the V_{50} is.

In the case highlighted in the introduction, where the V_{50} of a given material against a specific fragment is too high for laboratories to consistently shoot at the required velocities, a heavier fragment should be selected to conduct a V_{50}, rather than relying on a V_{proof} test. This is exactly the approach adopted in the NIJ 0117.01 standard for public safety bomb suits. For the highly protective bomb suit chest and groin areas (red area in Figure 5), the NIJ standard mandates the use of the heavier 2.9 g and 13.4 g FSPs, as opposed to the 1.1 g, recognizing the difficulty in obtaining a proper V_{50} with such a light fragment, for these highly protective areas.

Moreover, the NIJ 0117.01 certification process mandates testing to be conducted at NIJ approved laboratories. This ensures that no liberty can be taken by bomb suit manufacturers in selecting the most favourable tests or picking a start velocity higher than necessary, hoping for a positive influence on the results. The conduct of V_{proof} tests on the other hand, is not governed nor controlled by NIJ, which could allow bomb suit manufacturers to "cherry pick" the best results (testing at various laboratories and selecting the most favourable results) or testing at velocities higher than the 90% confidence level, hoping for luck to be on their side. In addition, relying on a single test sample (*n=1*), which is typically the case when only a few shots are considered such as in the $V_{0\text{-claimed}}$ case, does not result in an acceptable statistical significance.

5. CONCLUSION

As also emphasized by Andres et al [3] and Eridon et al. [4], test cost and panel costs are always a limiting factor when it comes to characterizing the fragmentation protection performance of armour panels. The best fragmentation protection characteristic of a given material to aim for, when cost limitations are present, is the V_{50}. The V_{50} is a proper characteristic of a material, which can be obtained following a

https://doi.org/10.52202/080042-0030

limited number of strikes, even without prior knowledge of the material's performance. If obtaining a proper V_{50} becomes difficult, given the high velocities involved, a heavier fragment should be used. The NIJ 0117.01 standard for public safety bomb suits indeed mandates the heavier 2.9 g and 13.4 g FSPs to be used, as opposed to the lighter 1.1 g, for the highly protective chest and groin areas of a bomb suit. But unfortunately, a lot of requirements still specify the 1.1 g FSP for high velocity V_{50} and V_{proof} ratings. This situation highlights the need for procurement agencies, in addition to manufacturers and test laboratories, to be made aware of the issues highlighted in this paper.

Ideally, the fragmentation protective performance of a bomb suit should be evaluated in the context of the NIJ 0117.01 certification program, which requires testing at NIJ approved laboratories, imposes three V_{50} tests per combination of fragment and panel, and ensures that proper procedures are followed. Following such a certification process is much more desirable than adopting customized methodologies such as the $V_{0\text{-claimed}}$ introduced in this paper, which can inappropriately boost the apparent performance of protective panels.

Of significant importance is that manufacturers and test laboratories should not rely on customized test methodologies, such as the one referred to in this paper as the "$V_{0\text{-claimed}}$", involving a relatively low number of stops, and claiming the highest of the stops as an estimate for a V_0. Such a methodology, in addition to substantially overestimating the protective ratings (as seen on Figure 24) is not based on any statistical ground. Or, if a customized methodology is used, test reports should state it explicitly and avoid misleading references to other test methods like the STANAG 2920 V0, when clearly not followed.

A V_{proof} is only of relevance when stated as a *requirement* to meet specific threats, as opposed to defining a characteristic of a material. V_{proof} tests should only be conducted in conjunction with V_{50} tests or other more extensive tests, for the proper characterization of armour panels.

References

[1] AEP 2920 – Procedures for the Evaluation and Classification of Personal Armour, Bullet and Fragmentation Threats, Edition A, Version 1, 22 June 2015, NATO Standardization Office, Brussels

[2] *U.S. National Institute of Justice (NIJ)*, "Public Safety Bomb Suit Standard", NIJ-0117.01, March 2016.

[3] Andres, C., Boughers, W., An Analysis of V_{50} Ballistic Limit Results Adjusting 1st Shot Velocity, Step-Up Step-Down Increments, Truth Characteristics and Velocity Control Distributions, Technical Paper, Aberdeen Test Center (ATC), March 2012

[4] Cunniff, P.M., Assessment of the Ballistic Impact Performance of Variable Response Armor Systems, Proceedings of 24th International Symposium on Ballistics, New Orleans, LA, 2008

[5] MIL-STD-662F, Military Standard: V_{50} Ballistic Test for Armor (18 Dec 1997)

[6] STANAG 2920 PPS (Edition 2) – Ballistic Test Method for Personal Armour Materials and Combat Clothing, 31 July 2003, NATO Standardization Agency, Brussels

[7] Bolduc, M., Summary of Newly Ratified NATO Standard AEP 2920, Ed. A, V1, Proceedings from the 13th Personal Armour Systems Symposium (PASS), Amsterdam, Netherlands, Sept. 2016

[8] Eridon, J., Mishler, S., Ballistic Validation Test Statistics and Confidence Levels, Proceedings of the Ground Vehicle Systems Engineering and Technology Symposium (GVSETS), NDIA, Novi, MI, Aug 11-13, 2020

[9] Cheng, M., Dionne, J.P., Hedge, C., and Makris, A., Ballistic performance of personnel armour materials and evaluation methods, Proceedings from the 10th Personal Armour Systems Symposium (PASS), Québec City, Canada, Sept. 2010

[10] Maldague, M., Evaluation of some methods in order to determine V_{50}, Proceedings from the 9th Personal Armour Systems Symposium (PASS), Brussels, Belgium, Sept. 2008

V_{50} determination challenges for state-of-the-art body armour

M.J. van der Jagt-Deutekom and J.P.F. Broos

TNO, Ypenburgse Boslaan 2, The Hague, The Netherlands
marjolein.vanderjagt@tno.nl

Abstract. The material constructions in personal protective equipment (PPE) have shifted over the last decades. In soft armour ballistics the construction shifted from plain wave fabrics towards more use of Unidirectional (UD) sheets. Also, the use of Ultra High Molecular Weight Polyethylene (UHMWPE) fabric constructions increased significantly compared to aramid. Most state-of-the-art combat helmets are currently constructed from UHMWPE UD composite. All these changes were made to achieve weight reduction and increased ballistic protection for body armour equipment. The ballistic limit velocity (V_{50}) is widely used as a measure of the ballistic performance of a ballistic protective material or construction. This measure does depend on the standard or method used to determine the V_{50}. Four different standards for V_{50} determinations are discussed and compared to show that the ballistic limit velocity of a material is dependent on the test procedure, test requirement and the statistical analysis method used even if the threat, the mounting of the sample and the sample size were the same. Several challenges to accurately determine a V_{50} value have already been reported, such as the effect of start velocity and total number of shots. This paper focusses on effects observed due to the increasing ballistic performance: deformation of the rigid fragment simulating projectile, larger affected impact zone, and increasing Zone of Mixed Results (ZMR). Experimental data is given to support the effects. For all observed effects, solutions are proposed like using a hardened FSP and specifying a minimum shot-to-shot distance. The consequences of these challenges are discussed, including experimental challenges if high V_{50} values must be determined. It is questioned if the V50 value is always the consistent and reliable evaluation parameter to be used, especially when a large ZMR is observed. A solution could be, not to use the V_{50} value as a measure, but instead the percentage of perforations for one or a few specified velocities.

1. INTRODUCTION

The ballistic limit velocity is widely used as a measure of the ballistic performance of a ballistic protective material or construction. This is typically referred to as the V_{50} velocity, which is defined as the velocity for which the probability of perforation, or complete penetration (CP), of the sample by the chosen projectile is exactly 50%. This also means that the probability of the projectile being stopped by the armour, or a partial penetration (PP), will be 50%. The V_{50} is thus a statistical estimation of the ballistic performance of an armour.

While the definition of the V_{50} appears simple, the terminal ballistic event behind it is not. Terminal ballistics is about the behaviour and effects of a projectile when it hits and transfers its energy to a target. This is a complex interaction process. The interaction process depends on the armour and projectile construction and their material properties – quasi-static and dynamic. These determine the failure mechanism of the projectile and armour, such as ductile failure modes, shear failure modes and brittle failure modes. The failure mechanism of the projectile and armour material can differ from each other. The interaction process is also dependent on the striking velocity of the projectile. Assuming one main failure mechanism between a specific projectile – armour combination, the effect of the velocity can be illustrated with figure 1.

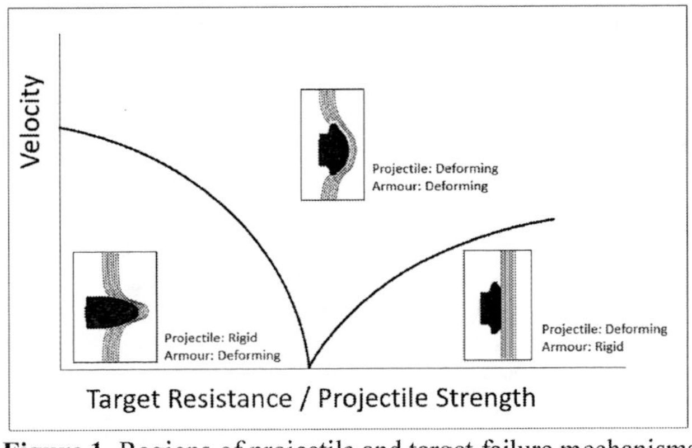

Figure 1. Regions of projectile and target failure mechanisms

https://doi.org/10.52202/080042-0031

Although the diagram in figure 1 has been derived to model long rod penetrator impact on metallic armour [1], it is illustrative to show the dependence between the striking velocity and the ratio between the target resistance and projectile strength. When the strength of the projectile material is much larger than the target resistance, the projectile remains rigid during the penetration process. For example, when a Fragment Simulating Projectile (FSP) impacts on a soft armour. When the target resistance is much higher than the projectile strength, the armour shows a rigid behaviour while the projectile deforms during the penetration process. The probability that both the target and the projectile deform during the impact and penetration process increases with increasing impact velocity.

The V_{50} value of an armour material is determined for different reasons. In the development stage it is to compare with competitive armour materials or with the minimum requirement. For acceptance testing it is used to check if the ballistic performance complies with the program of requirements. For lifespan control, the ballistic performance of used armour should be randomly checked to see if it still complies with the requirement. It can then also be used to compare the ballistic performance of the new and used armour. A significant decrease in V_{50} value is seen as a degradation of the ballistic performance. For all these evaluations, the V_{50} value must be determined consistently and reliably.

The material constructions in personal protective equipment (PPE) have shifted over the last decades. In soft armour ballistics the construction shifted from plain weave fabrics towards more use of Unidirectional (UD) sheets. Also, the use of Ultra High Molecular Weight Polyethylene (UHMWPE) fabric constructions increased significantly compared to aramid. Most of the state-of-the-art combat helmets are currently constructed from UHMWPE UD composite. All these changes were made to achieve weight reduction and increased ballistic protection for body armour equipment. Due to the increase in ballistic protection, the impact velocities needed to determine the V_{50} have also increased significantly.

The V_{50} is a measure of the ballistic resistance of an armour and is determined by performing a statistical analysis on the gathered ballistic data after a test. For a V_{50} determination, each result should be independent. This implies that in the tested velocity range, the armour material and the projectile should remain constant. That means not deform or deform in the same manner for the whole velocity range. It also means that if multiple shots are performed on the same sample, the distance between shots should be sufficient so that the armour at the next impact location behaves as if it was the first impact.

The procedure to determine a V_{50} value is described in several ballistic standards. All these standards give regulations on the experimental method and the V_{50} calculation method. These methods have all in common that the velocity varies around the zone that results in PP and CP and it is done for the perpendicular impact condition (0°NATO). However, these standards differ in the experimental assessment and in the statistical calculation method.

It is questioned if the V_{50} value is always the consistent and reliable evaluation parameter for all lightweight ballistic protection products, due to the challenges of the increased ballistic performance. This does depend on the standard or method used to determine the V_{50}. Therefore, first a summary of four different standards for V_{50} determinations are discussed and compared, before discussing the effects observed due to the increasing ballistic performance.

2. STANDARDS FOR V_{50} VALUE ESTIMATIONS

The NATO standard for the evaluation and classification of personal armour (STANAG 2920) [2] and the USA Department of Defense test method for V_{50} assessment (MIL-STD-662F) [3] both use the up-and-down firing method for the data acquisition for the V_{50} determination. This method is meant to converge to the average value with a limited number of shots. The V_{50} is calculated as the average of an equal number of highest PP velocities and the lowest CP velocities which occur within a specified velocity spread. The maximum allowable velocity span is dependent on the armour material and test conditions.

The Allied Engineering Publication (AEP), the underlying technical description of the STANAG2920, prescribes that the first round shall be loaded with an amount of propellant to give the projectile a velocity equivalent to the estimated V_{50} ballistic limit of the armour. MIL-STD-662F distinguishes between acceptance testing and other types of ballistics tests. For acceptance testing, the first round shall be loaded with a reference propellant charge so that the striking velocity is 23 to 30 m/s above the minimum required V_{50} as given by the appropriate specification. For other types of ballistics tests, it is the same as AEP-2920. So, for most ballistic tests, the starting velocity depends heavily on the given input before the execution of the ballistic test.

It has been shown previously that these established methods can be open to bias for some types of armour, like body armour, due to the chosen starting velocity and total number of shots [4]. Riley [4] concluded that the uncertainty in the estimated V_{50} will remain large when only a small number of test shots are used. Based on their results, 48 to 60 shots are necessary to reduce uncertainty. Both AEP-2920 and MIL-STD-662F use a lot less shots: respectively 6 to 14 and 4 to 10 depending on the bracket in which all velocities fall.

The methods described above only gives a V_{50} estimate. It is not able to determine a variance for the determination of any penetration probabilities. The AEP-2920 therefore recommends that an indication of the extent of the variability for a particular projectile and target material is given in the final report. For this, further impacts may be used after the first set of fair impacts that meets the criteria for the average V_{50} calculation. According to AEP-2920, firing shall continue until the three conditions mentioned below are fulfilled and the width of 95% confidence interval of the V_{50} is less than 4%:

- the highest velocity shall result in a CP,
- the smallest velocity shall result in a PP,
- there is a ZMR, which means that the lowest velocity producing a complete penetration shall be lower than the highest velocity producing a partial penetration is required.

The AEP-2920 mentions that a maximum of fourteen valid shots should be obtained to compute the V_{50} and standard deviation by means of the Probit method. The method does not give an indication what to do if not all the conditions are met after fourteen valid shots.

The standard of the Association of Test Laboratories for Attack Resistant Materials and Construction (VPAM) [5, 6] and Home Office [7] both describe a procedure and requirements for the V_{50} and the standard deviation calculation for bullet impact. However, the way they define it is quite different.

The VPAM APR2006 [5] does not prescribe a specific firing method or a starting velocity. It is assumed that the probability of penetration is a continuous, normal function of the impact velocity, based on the method from Kneubuhl [8] (KNB). The VPAM-KNB method replaces the probability function by the relative frequency. So, a classification of velocities in specific class ranges must be carried out (e.g. 5 or 10 m/s). From the results of a test firing, three areas can be identified: 1 - with only PP results, 2 – with PP and CP result (ZMR), and 3 – with only CP results. The firing is continued until it meets all the specified conditions:

- The minimal number of shots should be 16 (better 20 to 30)
- Every area must include at least 2 shots.
- Between two neighbouring partitions there can't be more than one empty class of velocity.

Given de minimal number of shots required, more than one PPE sample will be needed for one V_{50} determination. It is the authors experience that the VPAM-KNB method has a bias when the CP/PP ratio deviates significantly from 1. The V_{50} estimate is higher when the CP/PP ratio is low; based on significantly more PP results than CP results.

The Home Office body armour [7] standard uses Critical Perforation Analysis (CPA) software for the assessment of the velocity associated to a given statistical probability of body armour perforation. A minimum of 30 shots shall be performed with the test end conditions governed by the point at which the standard deviation of the V_{50} is below 10% of the mean. This condition shall be indicated by the CPA software. The advantage of this software is that it diminishes the influence of the operator. The software indicates which velocity should be used for each shot.

Helliker [9] gives more insight in the CPA method, which he used for the V_{50} determination against fragments. It is a tool using a Probit statistical analysis. Helliker mentions that the recommended number of shots is a minimum of 40 for fabric armour. The trial is divided into two phases of 12 and 28 shots, respectively. The first phase is a sighting phase to identify the "zone of mixed results" and to provide reassurance that the testing is in the area of interest. At the end of phase 1 a Probit model is fitted to the data from the first twelve shots. This model is used to estimate the V1, V20, V80 and V99 for the current data. The shots in the second phase are divided into seven sets of four shots. The velocities for each set are calculated per set of four.

The VPAM-KNB method and the CPA method specify the velocity at which 1% of shots are predicted to perforate the armour being tested, V_1, as well as the velocity at which 50% of shots are predicted to perforate the armour being tested, V_{50}. Main advantages of the CPA method are that it tries to capture the whole ZMR and that it limits the choices for the operator.

Besides the differences mentioned above, the different standards also prescribe minimal distances. Table 1 shows differences in the minimal distances between shots and from the edge. VPAM-APR does mention that the hits on the test specimen must be chosen in a way that there are no prior damages of previous shots around the point of impact, which could influence the result. VPAM-APR also mentions

https://doi.org/10.52202/080042-0031

that if the damage of the test specimen is too severe because of too many hits, the test must be continued using a further test specimen. MIL-STD-662F does specify a distance of at least two projectile diameters from any previous impact or disturbed area resulting from an impact. This all is very relative and susceptible on the judgement of the operator.

Table 1. Overview of requirements for distances mentioned in standards for body armour

	MIL-STD-662F [3]	AEP-2920 [2]	VPAM BSW [6]	Home Office body armour standard [7]
Minimal distance from edge	≥2 x projectile diameter	25 mm (50 mm from corner)	30 mm (75 mm from corner)	50 mm
Spacing between shots	≥2 x projectile diameter	≥ 65 mm or ≥10 x projectile diameter	≥ 75 mm	≥ 75 mm: undeformed panels

The ballistic limit velocity of a material depends on the test procedure, test requirement and the statistical analysis method used even if the threat, the mounting of the sample and the sample size were the same. Several challenges to accurately determine a V_{50} value have already been reported, such as the effect of start velocity and total number of shots. This paper focusses on the effects observed due to the increasing ballistic performance; deformation of the 1.1 g FSP fragment simulating projectiles, larger affected impact zone, increasing ZMR, and experimental challenges if high V_{50} values are to be determined.

3. DEFORMATION GAP

Modern fragment protective body armour is tested using fragment simulating projectiles (FSPs), not real fragments. The AEP-2920 standard defines chisel nosed FSPs (CN FSPs) as they provide repeatability, consistency, standardization and allow comparisons among armours. Previous work of Cant [10] shows no linear correlation between real fragments from a 81mm mortar shell and CN FSPs. As expected, CN FSPs behaved in a predictable manner, but did not accurately represent real fragments which behaved unpredictably due to the different shapes, sizes, and masses.

Due to the increase of ballistic protection, the impact velocities needed to determine the V_{50} are significantly increased. During ballistic limit testing of UHMWPE helmets against the 1.1 g FSP threat, the velocity increase is to such an extent that the 1.1 g CN FSP starts to deform at some point during the penetration process. This could mean that the interaction process changes within the ballistic limit velocity range, as illustrated in figure 2. At relatively high velocities the FSP deforms during the penetration process, creating a larger contact surface with the composite materials thereby engaging more fibres and thus becomes easier to arrest. At lower velocities, the FSP could however defeat the armour (complete penetration) because the impact energy is insufficient to deform the projectile. This could be quantified as a deformation-gap like the known shatter-gap phenomena. As described by AEP-2920, shatter gap can result in projectile/armour combinations having multiple ballistic limit (V_{50}) values. This is illustrated in figure 2 for the deformation-gap phenomenon as observed for FSP impact on UHMWPE composite helmets.

For three different UHMWPE helmets, FSPs were recovered after the test for a range of impact velocities. Figure 3 illustrates some recovered 1.1 g FSPs from one UHMWPE helmet shell for different ascending impact velocities. Two dimensions were measured: length and the maximum diameter of the chisel nose (see figure 4). Figure 5 shows the measured dimensions. It shows that the FSP starts to deform around an impact velocity of 550 m/s and that the amount of deformation not only depends on the impact velocity but also on the specific helmet shell construction.

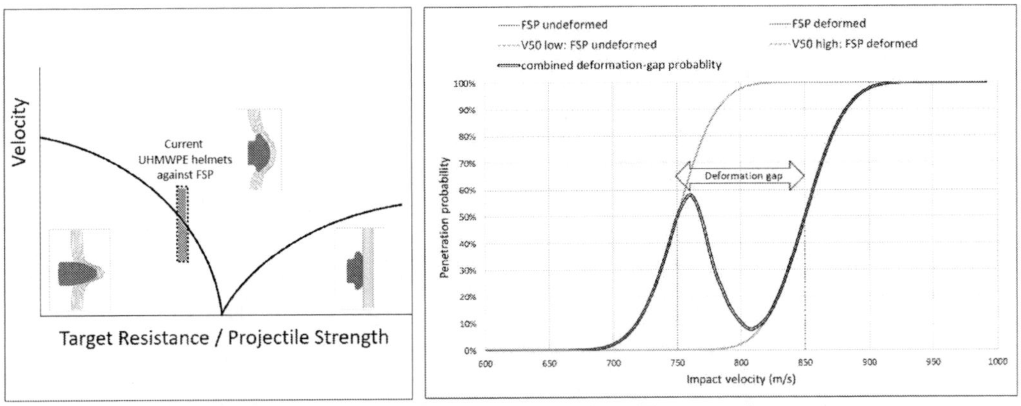

Figure 2. Schematic illustration of the deformation-gap phenomenon.
Left: Interaction process changes within ballistic limit velocity range.
Right: Example of the deformation perforation probability distribution of a deformed FSP V_{50}, undeformed FSP V_{50} and combined.

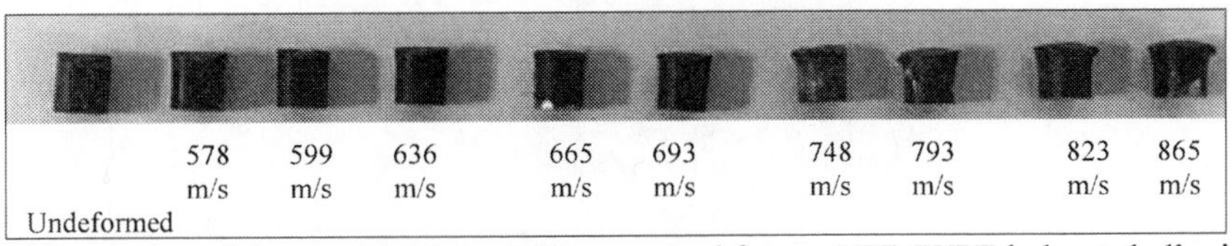

Figure 3. Illustration of the standard 1.1 g FSP recovered from a UHMWPE helmet shell with their corresponding impact velocities

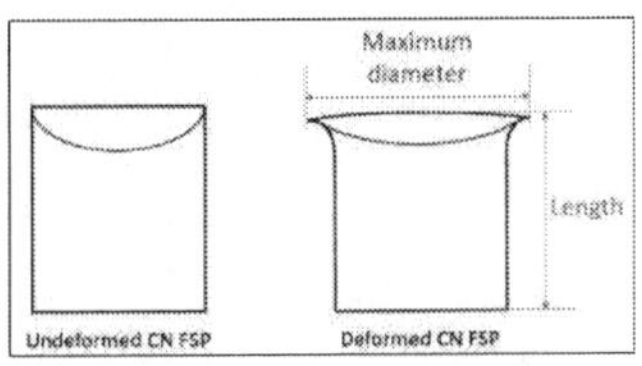

Figure 4. Schematic illustration of the two dimensions measured of the deformed FSP

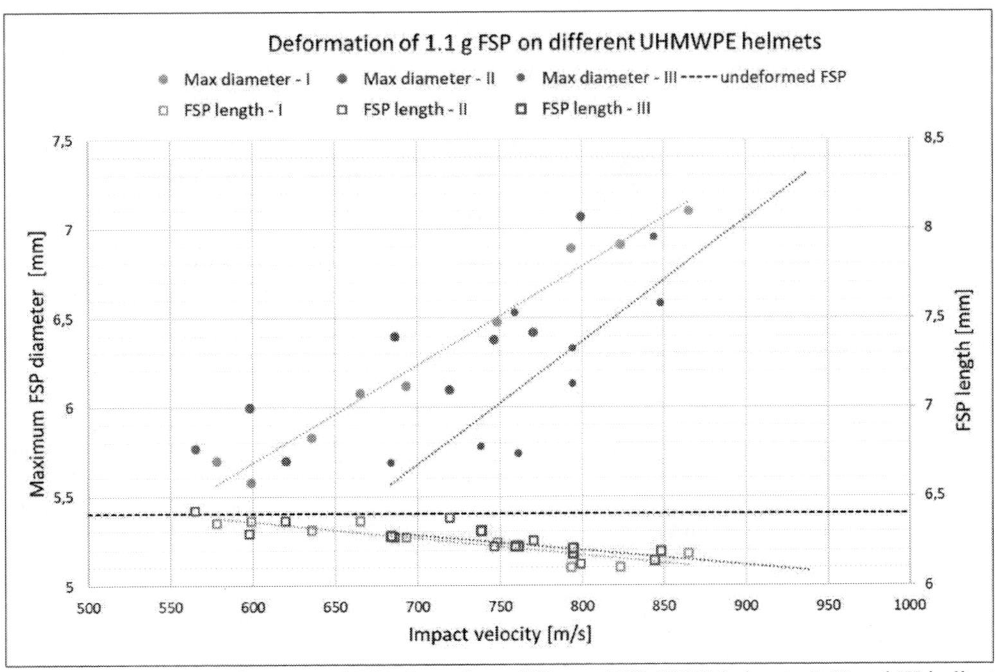

Figure 5. Dimensions of recovered FSP from three different UHMWPE helmets: I, II and III indicate from which helmet shell it was recovered

https://doi.org/10.52202/080042-0031

As mentioned before, the chisel nosed FSP's are defined to provide for repeatable and consistent comparisons between protective armour materials. It can be questioned if the current hardness of the FSP is still suitable for determining these increasing V_{50} results. The standard CN FSP has a hardness of 30 HRC. To explore the effect of the FSP hardness, V_{50} tests have been done with a hardened CN-FSP of 60 HRC. In these experiments the velocity was varied along the ZMR to strive to cover the whole ZMR. The V_{50} was estimated with the Probit method of AEP-2920. The data points and the Probit curves for both FSP types are given in figure 6. The two-coloured areas indicate the ZMR for the two FSP types. The partial penetrated hardened FSPs were recovered from the helmet shell (see figure 7) and showed no deformation for the whole velocity range tested. For the hardened FSP, the V_{50} decreases with 86 m/s compared to the standard FSP. The ZMR was comparable with standard FSP. This shows that the large ZMR is not only caused due to the deformation-gap, but also due to the inhomogeneity of the helmet shell.

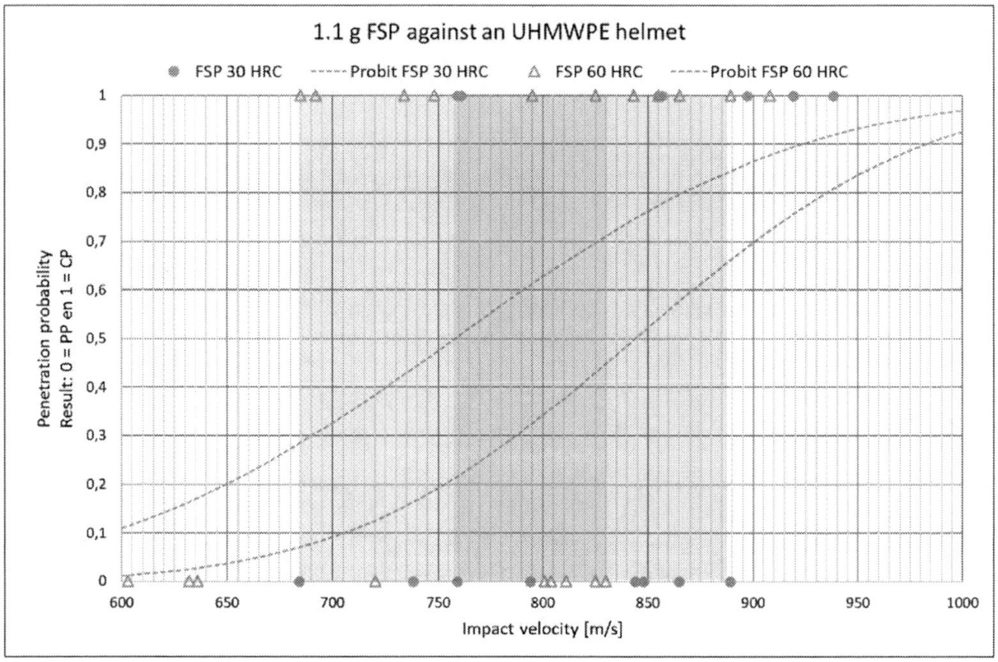

Figure 6. Penetration probability for the standard 1.1 g FSP of 30HRC and hardened to 60 HRC of a UHMWPE helmet: individual results and calculated Probit curve

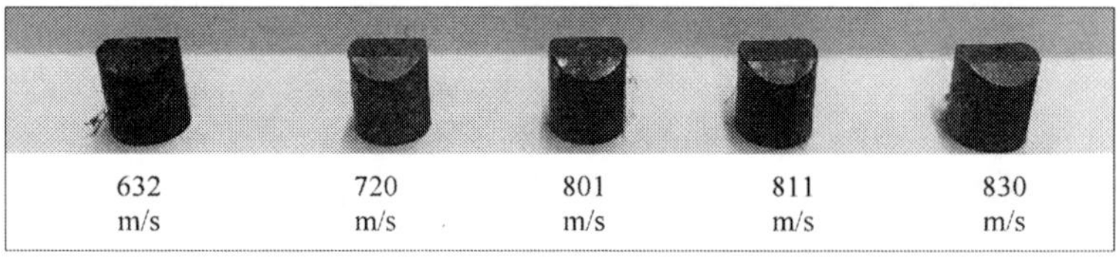

Figure 7. Illustration of the hardened 1.1 g FSP of 60 HRC recovered from a UHMWPE helmet shell for different impact velocities

Deformation gap can also be an issue for deforming bullets. When the V50 is determined in the velocity range where the bullet significantly deforms, it is likely to overlook the low velocity penetration probability of the undeformed bullet. This is also important to realize for V_{proof} classification of personal armour. In these tests the velocity remains constant, mostly around the muzzle velocity for the specific projectile. Such a test must ensure that at a confidence level of 90%, the probability of a partial penetration for the specified projectile at the velocity specified (V_{proof}) is higher than 90%.

4. SHOT SPACING

V_{50} testing standards assume that all impacts are independent. The hits on the test specimen must be chosen in such a way that there are no prior damages of previous shots around the point of impact, which could influence the result. It has been shown previously that these established V_{50} methods can be open to bias for body armour, due to the chosen starting velocity [4], total number of shots, and to the result of the previous shots on the test specimen [11]. Schaap [11] proposes an alternative test sequence by testing one velocity per panel to decouple the influence of stop-perforation history and bullet velocity. However, it is therefore important to consider the shot spacing.

In V_{50} testing of fabrics and composites, multiple shots are fired on a single piece of armour, while using a shot pattern that prevents hitting the same fibre twice. It is often implicitly assumed that by taking these precautions, individual shots in a V_{50} test do not affect the ballistic resistance of later shots and each shot can be regarded as independent. Analyses by van Es [12] of shot data with 9 mm FMJ DM41 on hard composite panels of Dyneema® HB26A showed that individual shots in the V_{50} test are not independent. The ballistic resistance of this material improves during a V_{50} test: The V_{50} of the third shot is higher than for the first shot. This was all done for one shot pattern of 8 shots per panel. Van Es concluded that it is recommended to limit the number of shots on a panel such that the ballistic resistance of the panel is not changed because of testing.

The effect of shot-to-shot distance and the effect of number of shots have been investigated with the 9 mm DM41 projectile against a hybrid soft armour of UHMWPE-UD and an aramid plain weave. A maximum of 6 shots per panel were performed with alternating multi hit pattern (based on VPAM-BSW pattern): the first 3 shots within a large equilateral triangle (150 mm) and the second 3 shots within a small equilateral triangle (75 mm) configuration as shown in figure 8 top left. Tests were done on 12 panels in total. The first 7 panels were tested with a constant velocity as recommended by Schaap [11] to scan for the whole ZMR. This was done in the velocity range 480 to 560 m/s with steps of 20 m/s. The results of the first five impact velocities are shown in figure 8. The two highest velocities resulted in CP on all six impact locations. Additional shots on the other 5 panels were with varying velocity to determine a Probit V_{50} per shot location. Figure 9 shows the Probit V_{50} for each shot and the Probit V_{50} for the two-triangle configuration. This shows that the V_{50} increases for the smaller shot-to-shot distance. A shot-to-shot distance of 75 mm is too small for the 9 mm threat; the material damage due to the previous shot does influence the subsequent shot and is thus not independent.

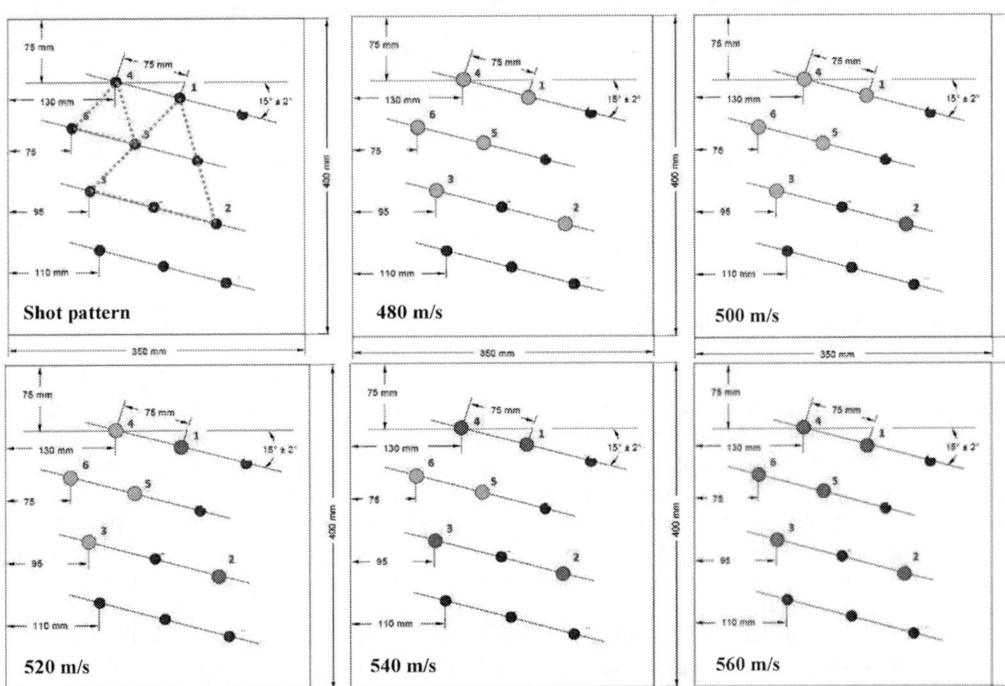

Figure 8. Schematic display of the hybrid soft armour after 9 mm DM41 impacts. Results for five different impact velocities per sample. Red dot = CP, green dot = PP. Top left: shot pattern.

https://doi.org/10.52202/080042-0031

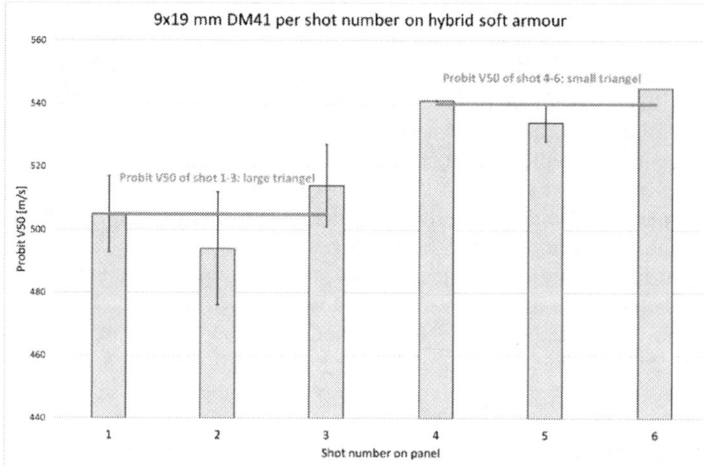

Figure 9. V_{50} results of 9 mm DM41 on hybrid soft armour with alternate shot pattern

5. DISCUSSION

The improvements in the material constructions in PPE resulted in thinner and lighter products with an increased ballistic protection. As discussed previously this can result in more projectile deformation and a larger affected impact zone. This all increases the complexity of the interaction process, which influences the ZMR. As illustrated in figure 10 the complexity increases when the projectile deforms during the penetration process and when the inhomogeneity of the armour material construction increases.

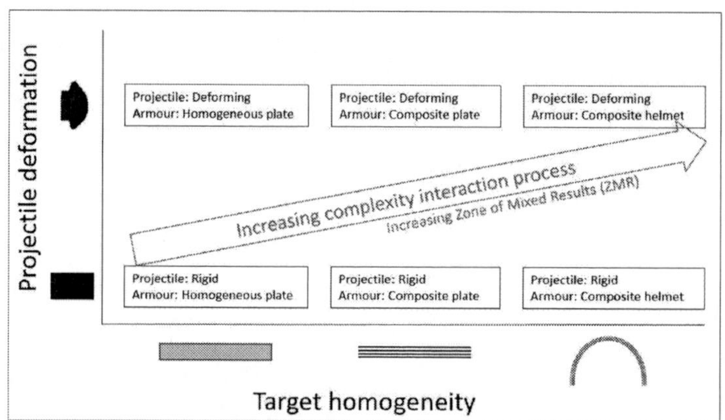

Figure 10. Illustration of the increasing complexity of the projectile target interaction process

A homogeneous hard armour with consistent thickness and material properties impacted by a non-deforming projectile usually has a small ZMR. This means that the probability of the projectile perforating the armour at a velocity slightly less than the V_{50} can be negligible. For soft armour panels and flat composite plates, the ZMR is a significant zone to be accounted for. This means that there is still a probability of the projectile perforating the armour at a velocity more than 50 m/s below the V_{50} [4]. The ZMR for UHMWPE helmets is much larger; there is still a probability of the projectile perforating the armour at a velocity more than 100 m/s below the V_{50}. For the current helmet testing this is partly due to deformation gap, but even with a hardened non deforming FSP, the ZMR is still significantly large with around 150 m/s (see figure 6).

Several factors contribute to the inhomogeneity of the helmet shell. The helmet shell varies around the surface in curvature, thickness, and laminate structure. In addition, there is also the variation in the applied production process, such as compression pressure and temperature and their distribution over the helmet shell. This would advocate to determine the ballistic performance of a helmet with a "one velocity" per helmet as previously proposed for hard composite plates [11] and soft armour panels [13]. For composite plates and soft armours, it is to account for the effect of the shot results. For the helmets, it is needed to account for the different ballistic performance over the shell surface. It can be questioned

https://doi.org/10.52202/080042-0031

if the V_{50} value is a representative value for helmets for assessing the ballistic performance, because it is clearly an average value for the whole shell.

If the V_{50} value is nevertheless desired, it is preferred to use a method least sensitive to the known bias factors and which covers the whole ZMR, like the proposed ballistic limit approach of Mauzac [13]:

- one-velocity-per-sample.
- velocities from approximately 0% to 100% CP.
- minimum of 6 test specimens per V_{50} (add more specimens if more velocities are needed).
- Probit method with confidence interval.
- In addition to [13], it is preferable to do the testing with an FSP with a hardness of 60 HRC to eliminate the effect of the deforming FSP.

A problem could be the substantial number of shots needed for statistical significance and accuracy of the test results. For soft armour this could be solved by optimizing the shot placement and order [13]. It should be investigated if and how this could be applicable for a helmet shell.

However, another problem will be the high velocities needed to achieve about 100% CP with the standard fragments. For the current fragment protective helmets, this means impact velocities of at least 1000 m/s are needed. Rifle helmets on the market already specify V_{50} values of more than 1000 m/s against the 1.1 g FSP. This means that impact velocities of at least 1400 – 1500 m/s are needed to achieve around 100% CP. These high impact velocities are experimentally possible, but it requires more sophisticated equipment than for "standard" V_{50} testing. At these high velocities more experimental variation will also occur with an FSP, like a larger absolute velocity variation and larger yaw. This all decreases the accuracy of the test results.

The problem of the high velocities could be mitigated by not using the V_{50} as a requirement or a measure, but by using the percentage of perforation for one or a few specified velocities. This method should still be based on the one-velocity-per helmet method. Instead of the V_{50} assessment, this method would focus more on the lower boundary of the ballistic limit, which is more relevant from a survivability point of view. Instead of a requirement for the V_{50}, a maximum percentage of CPs for the specified velocity/velocities are then given.

The above-mentioned solutions are also applicable for V_{50} determinations with bullets. However, V_{50} determinations with bullets are mostly done with impact velocities higher than the actual muzzle velocity, which gives even more restrictions. First, the number of independent impacts possible on a sample will decrease if the V_{50} increases. Second, for the penetration process to be comparable to the operational velocities, the bullet deformation during impact should not change significantly. When using higher velocities than normally used, it should also be considered that the bullet shape does not change during acceleration and flight. This requires expertise on internal and intermediate ballistics, to make sure it behaves the same as under normal operations. This could be achieved with an adjusted barrel and powder. Still, it is recommended that the bullet shape is checked with high-speed imaging before impact. In the effort to better define the V_{50}, the process around it is becoming increasingly complex.

6. SUMMARY

Nowadays, PPE equipment is thinner and lighter with increased V_{50} values. Even if the threat, the mounting of the sample and the sample size were the same, the V_{50} value is dependent on how it has been determined. Thus, dependent on the test procedure, test requirement, operator and the statistical analysis method used.

The increasing ballistic performance could result in more projectile deformation and a larger affected impact zone, which also affects the V_{50} determination. The increasing V_{50} values for helmets show the possibility of a deformation gap with the standard FSP. For velocities higher than 550 m/s, the 1.1 g FSP is staring to deform. Using hardened FSP's with 60 HRC could be a solution to eliminate the deformation gap observed for UHMWPE helmets. For 9 mm ball projectiles it is shown that the shot-to-shot distance of 75 mm leads to higher V_{50} values, because the material damage due to the previous shot influenced the subsequent shot and is therefore not independent.

The increasing complexity of the interaction process also increases the ZMR. For PPE with large ZMR results, like modern helmets, it is advised to use a one-velocity-per-sample method, because that is least sensitive to the known bias factors. Problem is the large number of shots needed to cover the whole ZMR. A solution could be, not to use the V_{50} value as a measure, but instead the percentage of perforations for one or a few specified velocities.

https://doi.org/10.52202/080042-0031

Acknowledgments

TNO would like to thank the Clothing & Personal Equipment Branch of the Defence Materiel Organisation of the Netherlands Ministry of Defence for funding part of this research.

References

[1] W-j. Song, et al., The effects of compressibility and strength on penetration of long rod and jet, Defence Technology 14 (2018) 99-108

[2] NATO standard AEP-2920, Procedures for the evaluation and classification of personal armour-bullets and fragmentation threats, Ed.A version 2, September 2016

[3] Department of Defense USA, Test method standard V_{50} ballistic test for armour, MIL-STD-662F, 18 December 1997

[4] M.A. Riley, D. Kirk, and A.L. Forster, Assessment of Uncertainty in Ballistic Response Estimates Obtained from Ballistic Limit Testing, Personal Armour Systems Symposium 2012, Cambridge, UK

[5] Vereinigung der Prüfstellen für angriffshemmende Materialien und Konstruktionen, Prüfrichtlinie "Allgemeine Prüfgrundlagen für ballistische Material-, Konstruktions- und Produktprüfungen", VPAM APR2006 Fassung 2, 30 November 2014

[6] Vereinigung der Prüfstellen für angriffshemmende Materialien und Konstruktionen, Prüfrichtlinie "Ballistische Schutzwesten", VPAM BSW 2006, 14 May2009

[7] T. Payne, S. O'Rourke, and C. Malbon, Home Office Body armour standard (2017), CAST publication number 012/17, July 2017

[8] B.P. Kneubuhl, Improved test procedure for body armour, Personal Armour Systems Symposium 1996, Colchester, UK, 3-6 September 1996

[9] M. Helliker, Degradation of fabrics used in military fragment protective clothing due to selected laundering regimes, PhD Thesis Cranfield University, May 2018

[10]D. S. Cant, et al. A comparison of fragment simulating projectiles and real fragments with respect to soft body armour performance, 29th International Symposium on Ballistics, Edinburgh, Scotland, UK, 2016

[11]A.A. Schaap, From Shooting Experiments to Ballistic Performance, Personal Armour Systems Symposium 2014, Cambridge, UK, 8-12 Sep 2014.

[12]M. van Es, Change of ballistic resistance of composite armour during V_{50} testing, Personal Armour Systems Symposium 2010, Quebec City, Canada,

[13]O. Mauzac, Practical and statistical aspects in the evaluation of the ballistic resistance of new and aged equipment, 3rd workshop on ageing effects in protective systems, components and materials, Saint Louis (ISL), 15 October 2019

https://doi.org/10.52202/080042-0031

Two ballistic test methods combined; residual energy method and digital image correlation (REM/DIC)

F. M. Creusen[1], E. P. Carton[1], E. van Zeijl[1], M. Vozarova[2], E. Neubauer[2]

[1] TNO Group Explosions, Ballistics and Protection, Ypenburgse Boslaan 1, 2496 ZA, The Hague, The Netherlands,
[2] RHP-Technology GmbH, Forschungs- und Technologiezentrum, A-2444 Seibersdorf, Austria.

Corresponding author: erik.carton@tno.nl

Abstract. In order to determine the ballistic efficiency of armour materials and armour systems several ballistic tests are available. In most tests only a binary test result is obtained; perforated or partially penetrated armour. TNO has used an alternative test method in which the material or armour system to be tested is always overmatched (sample perforated). The kinetic energy of the residual projectile is calculated from its mass and velocity. This kinetic energy is subtracted from the kinetic energy of the projectile before impact to obtain the projectile energy loss created during target perforation. This, so called, Residual Energy Method (REM) can be used on bare armour materials, as well as on complete armour systems and allows to determine changes in ballistic efficiency due to misuse, previous shots, aging, temperature changes, etc. When the target is not (yet) overmatched the response of the target can be determined using a digital image correlation (DIC) technique. For this method a speckle pattern is applied on the rear of the target or on a backing material. Two high-speed video cameras (each with a different angle towards the target) record the speckle pattern during and after projectile impact. These two video recordings are used in a post-processing software package that calculates the axial deflection and strain in the layer that holds the speckle pattern. From the time resolved data the distribution and history of the deflection, strains, velocities and strain-rates can be obtained. DIC data can be used to better understand the projectile-target interaction as well as for the validation of projectile-target interaction simulations. By combining these test methods, a useful test result is always obtained as for a perforating shot mainly the REM results are of importance, while for a stopped projectile the DIC results provide the relevant information.

1. INTRODUCTION

For many cases of projectile-target interactions typically the dynamic material properties of both the projectile materials and those of the target systems are unknown. Also, the failure mechanisms and failure loads/strains are unknown for the conditions that these materials experience during high-speed impact. This makes it hard to use (engineering and computer) models for the prediction of the projectile to target interaction, as these tools require these parameters as input variables for the calculation.

Materials and armour systems are generally ballistically tested. Such experiments may provide the depth of penetration (DoP) in a semi-infinite thick block, or a velocity of the projectile that has 50% chance of perforation (V_{50}). In both cases there is no role for the residual projectile, other than perhaps the (residual) velocity of it. Additionally, the DoP method was shown to suffer from a very large variation in test results [1].

Ballistic tests result in the interaction of two bodies colliding; the projectile and the target. In general, more attention in these tests is paid to the consequence for the target (dent or hole size, cracks, delamination, fragments, etc). Less attention is given to the residual projectile as it is normally not soft recovered. However, the residual projectile status (whether it is intact, deformed, broken or eroded) provides important information on the projectile-target interaction. Hence, in this work we will describe a soft recovery method for residual projectiles. This allows the residual projectile to be recovered after each test.

After measuring the mass and velocity of the residual projectile or rather the core of the armour piercing bullet (AP-core), its kinetic energy can be calculated and compared to that of the intact projectile (or AP-core) just before impact on the target. From the difference in mass, the amount of erosion of the projectile or core is obtained. Such information can be used for the ranking of targets and can also be used to analyse such data and learn about the projectile-target interaction. In order to determine the ballistic efficiency of ceramic tiles and armour systems the Residual Energy Method (REM) test has been used in the European Defense Agency category B projects CERAMBALL and CERAMBALL II. In

https://doi.org/10.52202/080042-0032

these projects most ceramic based armour systems are rated on their ballistic efficiency against Armour Piercing (AP) projectiles. Only the core of the projectile is considered as it is assumed that the projectile jacket is too soft to play a role in the penetration process. Targets consisting of a bare ceramic tile or ceramic based armour were impacted using regular AP projectiles (0.30" or 0.50" AP) with a constant impact velocity. A residual projectile catching device allowed to retrieve the residual projectile parts. However, this device used water (in a horizonal metal tube) for the soft recovery and this proofed to be a problem for many of the ballistic test centers involved.

Additionally, we can also learn from interactions in which the projectile is stopped. The impact of the projectile on thick targets frequently results in a dent formation and hence a large out-of-plane deformation of the target or backing material. The target accelerations, its top velocity and strain-rates are extremely high in high-speed impact tests and it has long been quite hard to perform measurements of such parameters. However, nowadays with the use of high-speed video and digital image correlation (DIC) software it is much easier to determine the distribution of such parameters over the rear surface of targets and backing materials.

By combining these two test methods, REM and DIC, useful information is gathered for each test; in case of target perforation the residual projectile is caught and weighted, while in case the projectile is stopped the DIC method allows to record many parameters of the target from the rear.

In this paper we will first focus on the residual energy method. The digital image correlation is described in a following chapter.

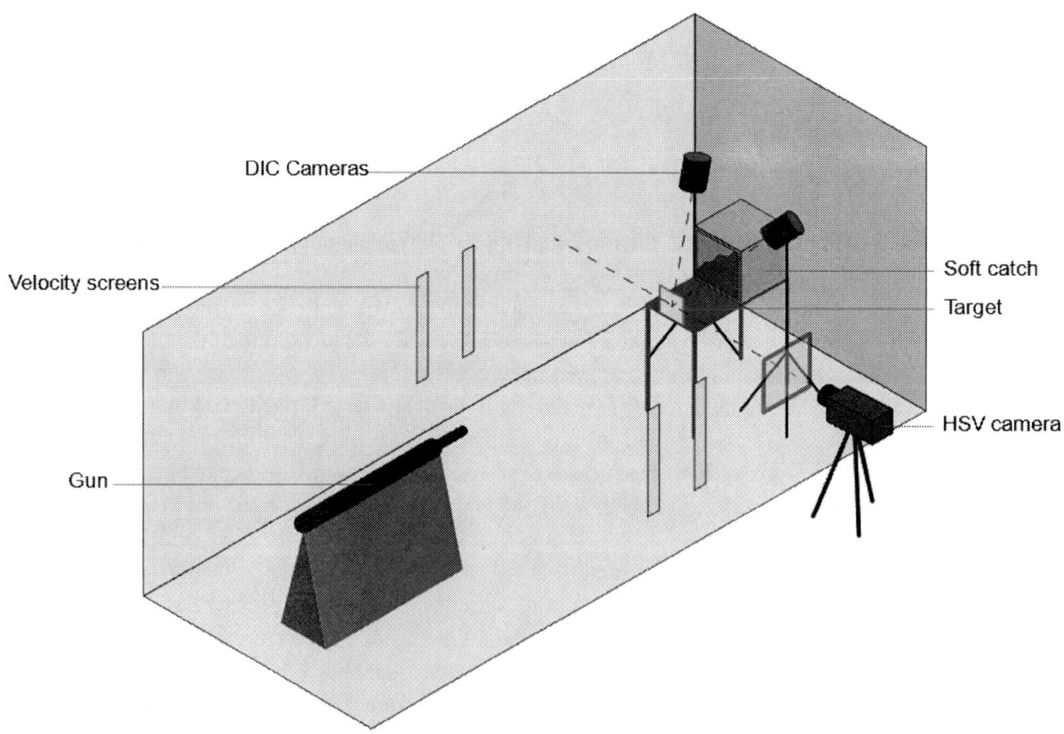

Figure 11. Test setup for the combined test method REM and DIC.

2. RESIDUAL ENERGY METHOD

The REM ballistic testing method more specifically determines the degree of erosion and deceleration of the projectile after target interaction compared to traditional testing methods. Figure 11 schematically shows the test set-up inside a ballistic testing range. Specifically for the REM test are the high-speed video camera, the projectile soft catch device and the velocity screens. The selected projectile is launched by a stationary gun. The projectile velocity is measured using the velocity screens (dividing the screen distance by the time-of-flight of the projectile between the screens). The high-speed video (HSV) camera and flashlight are triggered by the signal of a laser screen and starts recording after a pre-determined delay time. The camera is positioned outside of the shooting range and records the normal impact process in side-view through a transparent armour window. The frame rate used in the recordings should be no less than 50.000 frames per second and preferably between 200.000 and 500.000 fps. The optical lenses used should provide a clear image of the target (in side-view), as well as the fragment cloud (until 10 to 20 cm behind the target).

From these side-view recordings not only the impact on the tile, but also the fragment cloud behind the tile can be observed. This allows measurement of the velocity of the tip of the ceramic fragment cloud. The latter is assumed to be equal to that of the residual projectile. A scenario where an AP bullet interacts with a ceramic tile is schematically shown in Figure 12. The high-speed video results also allow the determination of the main dimensions of the fragment cloud.

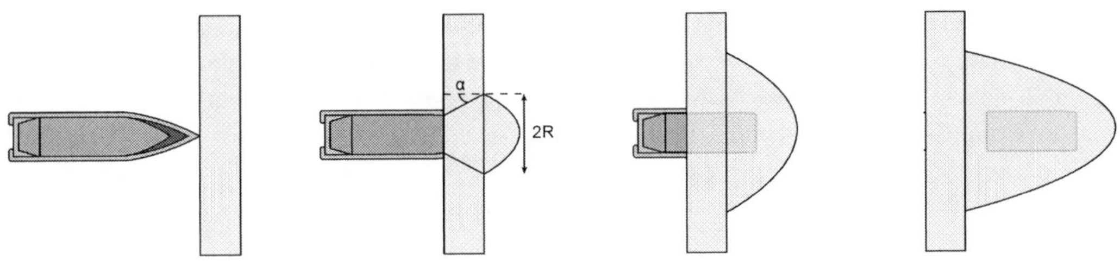

Figure 12. Schematic view of high-speed video images at various interaction times.

After the residual projectile has been caught, it should be recovered from the soft catch medium and weighted. Before weighting, the AP core fragments should be separated from any other particles that may have been entrapped, such as jacket parts and target particles. The residual velocity of the projectile can be obtained from the high-speed video recording. It is assumed that the velocity of the residual AP core is equal to that of the tip of the fragment cloud. The kinetic energy ($E_{kin,out}$) of the residual core is obtained through:

$$E_{kin,out} = 0.5 * m_{core,residual} * v_{cloud}^2 \quad (1)$$

Where $m_{core,residual}$ is the weighted mass of the residual projectile core fragments and v_{cloud}^2 is the measured velocity of the fragment cloud.

Based on the masses and velocities of the AP core before and after impact on the ceramic tiles the kinetic energy loss ΔE_{kin} of the AP core can be calculated:

$$\Delta E_{kin} = 0.5 * m_{core} * v_0^2 - 0.5 * m_{core,residual} * v_{cloud}^2 \quad (2)$$

Where m_{core} is the initial AP core mass and v_0 the impact velocity.

The material or armour system can be ranked using the fraction of energy loss of the AP core they provide.

$$Ballistic\ efficiency = \frac{\Delta E_{kin}}{E_{kin,0}} \quad (3)$$

https://doi.org/10.52202/080042-0032

Where $E_{kin,0}$ is the kinetic energy of the AP core before impact.

As the REM uses mass and velocity of the bullet before and after penetration, it can be applied to any armour system (a single plate or a complete armour system) as long as it is overmatched by the projectile. In this paper some experiments are presented considering bare ceramic tiles, as well as ceramic tiles with various backing materials using 7.62 AP bullets. Figure 3 shows an example of recovered core fragments of the 7.62 AP M2 bullet along with ceramic tile fragments that are formed during the projectile-target interaction. The AP core fragments have been caught using a revised projectile fragment catching device that is shown in Figure 4. In this new device, the earlier wet projectile capture method is replaced by a dry system, which is more friendly for the shooting range environment.

The new projectile fragment catcher makes use of a granular elastomeric material (granulated car tires) situated in a steel tube with a diameter of 33 cm and a length of 130 cm. The granulate filled tube is in horizontal position during a shot and positioned about 30 cm behind the target (this allows room for the high-speed video recordings directly behind the target). After a shot, the tube is rotated in vertical position and opened to allow the granulate to pour down into a container through a magnetic sieve. This sieve separates any ferro-magnetic particles from the granulate, hence the bullet fragments (steel and hard metal) are separated from the polymer and ceramic particles. The sieve section can be taken away and this allows the bullet fragments to be collected for each shot. After this separation the sieve is replaced in the set-up and after the metal tube is back in horizontal position. It is refilled with granulate using an industrial vacuum cleaner and hose. This avoids the experimenters having to fill the tube by hand after each shot, which saves a lot of work and time. It was proved that this system has a cycle time (time between shots) of less than 10 minutes.

Table 1 shows the results of a test series (using 7.62 AP P80 bullets) performed on 10 mm thick bare B_4C tiles that were made in various batches using starting powder of different particle sizes (0.5, 1.4, 2.5, 7 and 20 µm) at RHP Technology in Austria. From each batch 3 samples were tested to obtain an understanding of the variations in the REM test and allow an average ballistic efficiency to be determined for the ceramic (using equation 3). The results of the (average) ballistic efficiency of these tiles are shown graphically in figure 5. The B_4C particle size scale is logarithmic as this better shows the trend observed; the ballistic efficiency increases with decreasing particle size of the starting B_4C powder. However, below a particle size of about 1 µm the ballistic efficiency decreases rapidly and is again comparable to that of the coarsest starting powder with a particle size of 20 µm.

Figure 13. Residual fragments of the 7.62 AP core (in bag) and ceramic tile recovered from a REM test

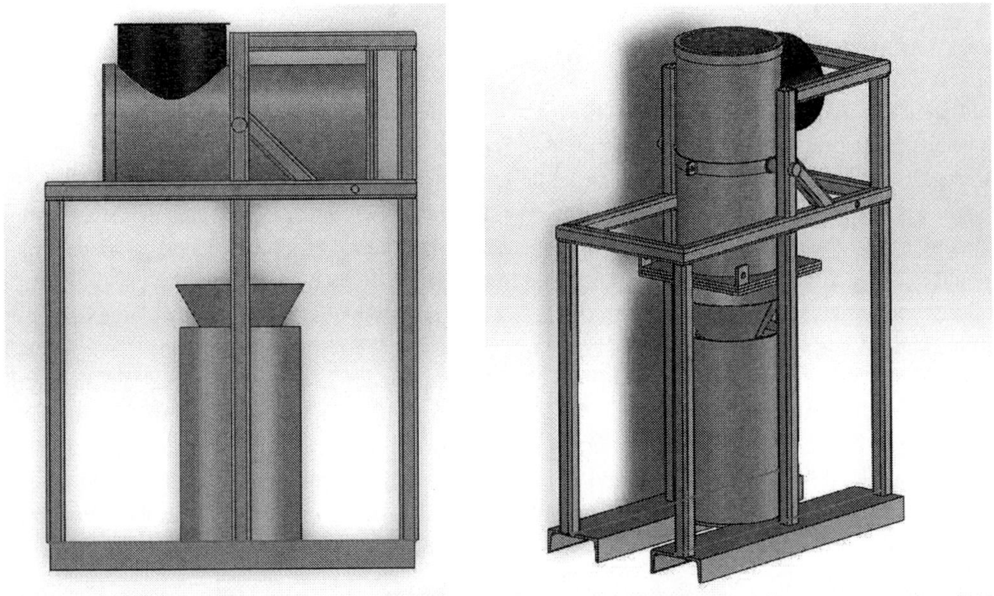

Figure 4. Residual projectile fragment catching device in shooting position (left) and after a test for separation of metal fragments using a magnetic sieve (right).

Table 1. Overview of REM test results for bare B_4C tiles

Sample Code	thickness [mm]	Composition	AD [kg/m2]	%TMD	Vin [m/s]	Vres [m/s]	Mres [gram]	E0 [J]	Eres [J]	ΔE [J]	BMEF [Jm2/kg]	BERatio [%]	average BE [%]
B11-01	9.70	B4C Grade 1 0.5 µm	24.5	99.2	820	631	3.29	1244	655	589	24	47	
B11-02	9.70	B4C Grade 1 0.5 µm	24.5	99.2	816	640	3.09	1232	633	599	24	48	
B11-04	9.70	B4C Grade 1 0.5 µm	24.5	99.2	819	638	2.94	1241	598	643	26	51	**48**
B12-02	9.80	B4C Grade 2 1.4 µm	24.6	98.2	827	566	2.49	1265	399	866	35	69	
B12-03	9.70	B4C Grade 2 1.4 µm	24.6	98.2	822	599	2.32	1250	416	834	34	66	
B12-04	9.70	B4C Grade 2 1.4 µm	24.6	98.2	818	664	2.84	1238	626	612	25	49	**61**
B13-02	9.90	B4C Grade 3 2.5 µm	24.8	98.80	534	-	2.84	528	-	-	-	-	
B13-03	9.90	B4C Grade 3 2.5 µm	24.8	98.80	823	616	2.89	1253	548	705	28	56	
B13-04	9.90	B4C Grade 3 2.5 µm	24.8	98.80	823	609	2.88	1253	534	719	29	57	**57**
B14-01	10.10	B4C Grade 4 7 µm	25.4	98.5	810	624	3.04	1214	592	622	24	49	
B14-02	10.10	B4C Grade 4 7 µm	25.4	98.5	823	594	2.88	1253	508	745	29	59	
B14-04	9.95	B4C Grade 4 7 µm	25.4	98.5	822	628	2.89	1250	570	680	27	54	**54**
B15-01	10.10	B4C Grade 5 20 µm	25	96.5	821	628	3.02	1247	596	651	26	52	
B15-03	10.00	B4C Grade 5 20 µm	25	96.5	822	637	2.91	1250	590	660	26	52	
B15-04	9.90	B4C Grade 5 20 µm	25	96.5	828	658	3.28	1268	710	558	22	44	**49**

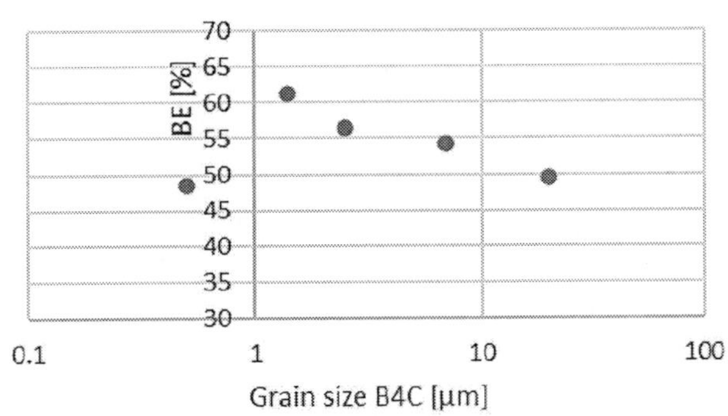

Figure 5. Graph showing the ballistic efficiency of B_4C tiles as a function of the particle size of the starting material (B_4C) powder as determined using a REM test series.

https://doi.org/10.52202/080042-0032

3. DIGITAL IMAGE CORRELATION

Digital Image Correlation is a method that allows the determination of motion of the rear side of a target over a wide area. In order to measure the complete target response, it is best when the projectile is stopped by the target, as this leaves the rear of the target largely intact during the interaction process. The DIC method makes use of two simultaneous (high-speed) video recordings of the target rear each at a separate viewing angle. The area to be recorded should have a large number of random discrete speckles with a high contrast to the target surface (speckle pattern). An example of such a pattern is given in Figure 6. From the stereographic high-speed video recordings, a post processing DIC software package is used to determine the out-of-plane deflection and velocity, strain distribution and strain rate history. Such parameter histories can be used to validate finite element simulations and provides the strain and strain rates experienced by armour materials during the interaction with an impacting projectile. For the DIC recordings the rear of the target should be illuminated with two lights behind and aimed at the target.

When the REM tests are performed in combination with digital image correlation, the use of flashlights should be avoided, as their intense light flash will overexpose the DIC images. Instead, a continuous light source is used for the high-speed video recordings.

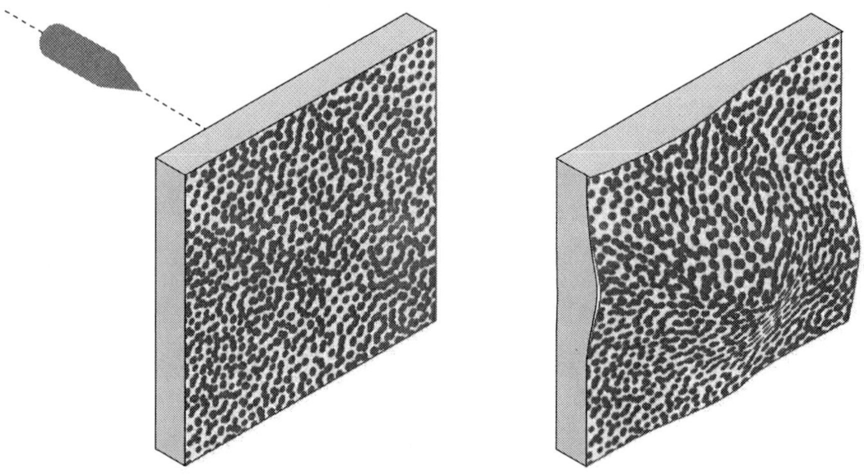

Figure 6. DIC speckle pattern target after a partial penetrated projectile interaction

Figure 6 also shows the interaction between a projectile and a speckled target. The reaction of the rear of the target in the second image occurs at approximately 0.1 ms after impact. The speckles are recorded by the DIC cameras at each video frame. The DIC software translates the speckle recordings into panel deflection, velocities, strains, and strain-rates of the rear of the target. The resulting time resolved data can be presented in graphs. Points on the target surface can be selected to prompt the required data from the software. An example of a deflection graph is presented in Figure 7. A ranking of ballistic efficiency could be based on a specific measured parameter by comparing various target responses using identical impact conditions. Prediction of blunt injuries can be done using the blunt criterion [2, 3, 4] or the viscous criterion [5, 6]. The maximal velocity of the backing is an important parameter in injury level calculations like behind armour blunt trauma (BABT).

The DIC data can also be applied to gain knowledge about the projectile-target interaction(s) and failure mechanisms that are involved, e.g., material properties largely influence the ballistic behaviour. From DIC data the mechanical behaviour of the material, like straining capability at high strain rates and the possibility to absorb energy by panel deflection, can be further investigated. Figure 8 and 9 show examples of the axial velocity and strain history, respectively, for a bullet (7.62 AP P80) hitting at 800 m/s on an 8 mm thick alumina tile with an 11 mm thick aramid backing material.

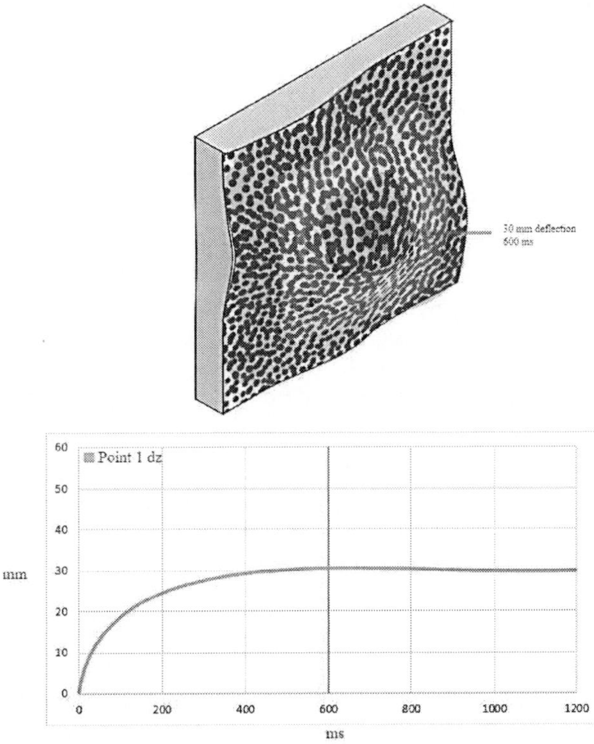

Figure 7. Example of DIC software image (top) with the corresponding graph (below) of the target displacement in time at the indicated position in the image in green.

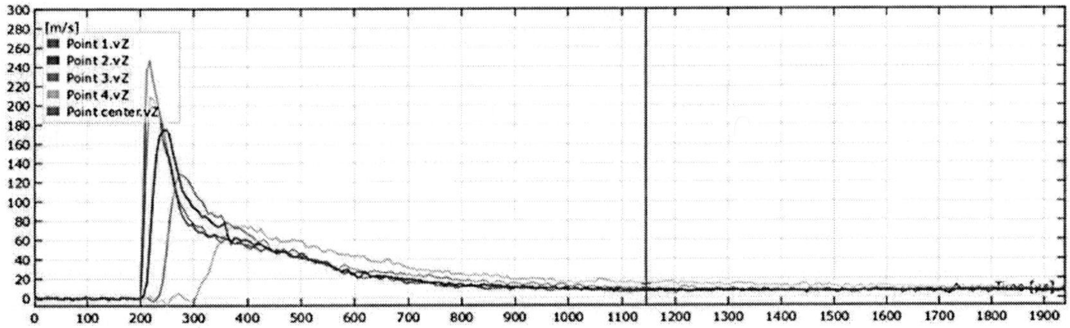

Figure 8. Example of a DIC result: velocity history of various points on an aramid backing plate; time scale is microseconds (µs).

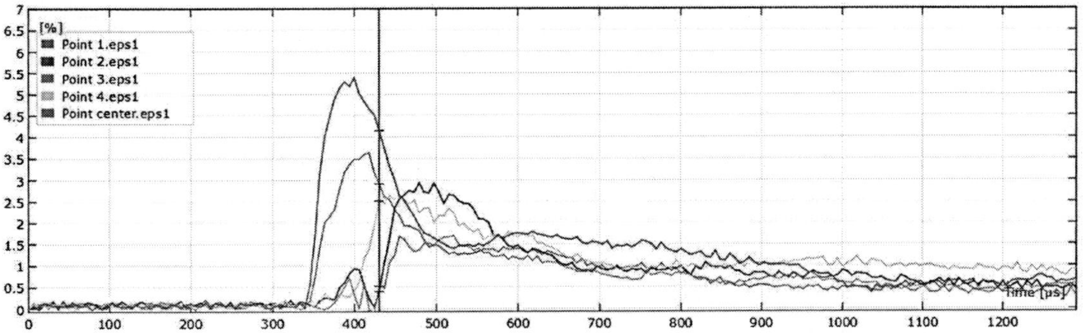

Figure 9. Example of a DIC result: strain history of various points on an aramid backing plate; time scale is microseconds (µs).

https://doi.org/10.52202/080042-0032

4. DISCUSSION

The REM test method is energy based and as such can be applied to any target material or armour system. Normally difficult to test armour materials like ceramics can be evaluated with this method, but the method could also be used to, for example, quantify the effect of damage or the aging of armour systems. The REM can only be used when the target is overmatched, as this allows the residual energy of the projectile to be determined, together with the status of the residual penetrator (intact, broken, deformed or eroded/shattered).

As an example, in this paper the effect of particle size of the starting powder used to sinter B_4C tiles is demonstrated. All tiles had an areal density of about 25 kg/m^2 and were tested using the same projectile (7.62 AP P80) and the same impact condition. This allows small differences between samples to be determined. The fact that the smallest particle size under-performed, may be explained by the large fraction of grain boundaries in such material. The crystalline order is likely disturbed leading to a large fraction of material with lower mechanical properties (compression strength and hardness).

With Digital Image Correlation in and out of pane deformation of the target rear can be measured (time and space resolved). The accelerations (velocity from zero to maximum) are major, where the highest velocity (240 m/s for the test displayed in Figure 8) is reached directly behind the impact point. Directly after reaching a maximal velocity the velocity reduces again, as more and more backing material is involved in the interaction. The other lines in Figure 8 correspond to the velocity history of positions on the backing further from the impact point. These points start to respond later in time and experience lower accelerations, while their peak velocity and velocity history is practically equal to that of the center point from that moment in time on (red line in Figure 8). This means that the points further away from the centre point adjust to the velocity of the centre point at these later response times. The fact that all moving points axially move at the same velocity, means that the shape of the dent is constant.

From the velocity history plot (Figure 8) also the acceleration of the local positions of the backing can be obtained. The impact point clearly experiences the highest acceleration as its peak velocity of 240 m/s is reached in about 10 microseconds (10^{-5} s) only. Points further away also experience large accelerations but have a lower peak velocity and reach that in a longer time frame. Hence, the axial acceleration is very high, yet decreases with increasing distance to the impact point.

Figure 9 shows the strain history plot of several points on the backing material. Also here, the peak strain value (5.5%) is obtained at the impact point (indicated in the graph as 'center point'). While points further away (points with higher number are further away from the center point), react somewhat later and experience ever lower peak strain values. Also here, the slope of the strain history graphs decreases with distance to the impact point, meaning that the strain rates are reducing with distance from the point of impact. The strain rate at the impact point is about 2700/s (4% in 15 microsecond), while that of point 4 is 1000/s (2.5% in 25 microsecond).

The DIC test method allows the determination of space and time resolved parameters. This can be used to validate FEM simulations and helps the researchers and engineers to understand and quantify the mechanisms involved in the projectile-target interaction. When the main deformation and failure mechanisms are identified and understood, we are in a better position to develop and fine tune materials and systems with better performance. Also, the REM test method proves to be valuable as it allows the quantification of the energy absorbed by (any) projectile-target interaction in which the target is perforated. Although no specific material properties are determined, the energy loss quantification forms a good starting point to understand the (main) mechanisms involved.

5. CONCLUSION

We performed ballistic tests using 7.62 AP bullets on bare ceramic tiles and ceramic based armour using a set-up that combines the residual energy method with digital image correlation (REM/DIC). Using this combined experimental set-up, useful results are obtained at each shot using any armour system. Shots that perforate the target can be used to calculate the energy lost by the projectile using the REM. Shots that are stopped by the target generate useful data from the DIC set-up; as the rear of the backing material remains largely intact, the speckle pattern remains available for recording by the two high-speed video cameras. This non-contact measuring method allows material behaviour to be measured during realistic impact conditions. High-speed DIC also provides measurements over a wide area with a high time resolution, this allows the important parameters to be determined both time and space resolved.

For the REM a new residual projectile fragment catching device was designed, built and used. It collects the residual projectile after each shot. This allows the residual projectile status and energy loss to be determined. It provides information on the projectile / target interaction process. Using a magnetic filter and an industrial vacuum cleaner, a cycle time of less than 10 minutes between shots was obtained.

Acknowledgments

We acknowledge our partners in the EDA project CERAMBALL II, especially the Dutch companies Cato Composites, Modified Materials and Allplast which made backing materials and bonded them to the ceramic tiles.

References

[1] Carton E.P., et al.," Round robin using the depth of penetration test method on an armour grade alumina", Defence Technology, 2019; Volume; 15 (6) 829-836 10.1016/j.dt.2019.07.014

[2] L. Sturdivan, D. Viano and H. Champion, "Analysis of injury criteria to assess chest and abdominal injury risks in blunt ballistic impacts," *J. Trauma,* vol. 56, pp. 651-663, 2004.

[3] D. Cronin, "Behind amour blunt truama evaluation using the blunt criterion," in *23rd International Symposium on Ballistics*, Tarragona, Spain, April 2007.

[4] K. A. Rafaels, K. L. Loftis and C. A. Bir, "Can Clay Tell Us More Than Deformation?," in *Personal Armor Systems Symposium*, Washington D.C., 2018.

[5] D. Viano and I. Lau, "A viscous tolerance criterion for soft tissue Injury assessment," *J. Biomechanics,* vol. 21, no. 5, pp. 387-399, 1988.

[6] National Research Council, Testing of Body Armor Materials: Phase III, Washington, DC: The National Academies Press, 2012.

https://doi.org/10.52202/080042-0032

The effect of backing methods on the measured ballistic performance of armour materials

G. James[1], **J. Keirl**[1]

[1]*Platform Systems Division, Dstl, Porton Down, Room CO12 Building 4, Salisbury, Wiltshire, SP4 0JQ, UK*
jjkeirl@dstl.gov.uk

Abstract. Backing materials are used during ballistic testing of armour materials to provide support and may additionally be used as an injury assessment method. For example, to record the back face deformation or assess penetrating injury risk from armour overmatch. Tests was conducted to investigate how different backing methods affected the measured ballistic performance of different protective materials. To characterise this potential effect, two ballistic materials/systems were chosen. The UK Tier 1 Pelvic Protection to represent a low Areal Density (AD) protection of 0.45 kg m^{-2} and a para-aramid woven pack, with total AD of 2.8 kg m^{-2} to represent a higher level of protection. A 20 shot V_{50} test was completed for each armour material on a selection of up to 14 different backing methods. These included Roma Plastilina® no. 1, 20% ballistic gelatin, foam based fragment packs, the AEP-2920 frame, the UK BABT rig (a silicone rubber based deforming element) and ex-vivo porcine tissue with armour overlaying the thorax and the thigh. The measured V_{50} performance on different backings was observed to vary from the baseline by up to -43% to +24%. Differences of -12% to +24% in the measured V_{50} performance were observed for backings that could be considered reasonable and/or common. Using a backing that does not sufficiently replicate an 'as worn' condition may result in Personal Protective Equipment that provides inappropriate protection: not protecting against the specified threat or producing unnecessary burden for the required protection, this therefore has implications for test methods and standards.

1. INTRODUCTION

1.1 Background

Backing materials are used during ballistic testing of armour materials to provide physical support. There are a variety of different backing methods in use across different nations, institutions and test specifications. Some of the backing methods enable additional measurements to be collected, for example to provide an assessment of the risk of penetrating injury during armour overmatch, or to record the back face deformation, which may be used as an additional pass/fail criterion or further analysed to provide an estimation of residual injury risk (e.g. with the UK Behind Armour Blunt Trauma (BABT) rig). Due to the differing requirements, it is not practical to enforce the use of a single backing method when assessing ballistic material performance. There is an embedded use of different backing methods across the ballistic material development and assessment community that complicates material performance comparisons, but provides comparison to legacy data.

It has previously been noted in the literature that conducting V_{50} testing (velocity at which the estimated probability of a complete penetration is 50%) of a given ballistic material using different backing materials can alter the measured performance [1-5]. Whilst this issue has been highlighted, there remains significant questions regarding which methods may be representative of 'as worn' performance or the degree to which methods may differ.

The focus of this paper is to investigate the influence on the measured V_{50} performance of an armour material on a selection of different backing methods and therefore, the resulting implications for choice of various test methods, comparison of results or considerations relating to optimisation of Personal Protective Equipment (PPE). The emphasis is on the potential changes to the measured ballistic response of an armour system, not the ability of a backing to assess any resultant injury in a person who may have been wearing the PPE when impacted. Two different soft armours designed for fragment protection were specifically considered.

The work described within this paper was instigated from the viewpoint of (injury) model development, which may be used in conjunction with an armour covering, not from a ballistic material characterisation or test standards position. This work supported the development of the TP5 fragment pack [1]. Therefore, the type of question being answered relates to providing the evidenced-based data to understand the caveats or limitations of the different backing methods on the resulting measured performance of the armour, given different test requirements.

https://doi.org/10.52202/080042-0033

1.2 Test standards and backing methods

The main test standard applicable to military body armour (at least in the UK) is the NATO AEP-2920 [6]. AEP-2920 has different requirements for soft armour target retention and support, depending if the testing is for a V_{50} assessment, V_{proof} validation (velocity at which the probability of a complete penetration is lower than a specified value for a given confidence value), using Fragment Simulating Projectiles (FSPs) or bullets. These recommended backing methods cover the use of a frame (air backing), layered foam, Back Face Signature (BFS) materials (for example Roma Plastilina® no. 1 ballistic grade (RP1), but other BFS materials are allowed) and suggestions for use of instrumented BABT assessment models to measure the dynamic response of the backing material. This single standard alone covers a multitude of different backing conditions.

Of relevance to the testing and advice on PPE provided by Dstl, is that for police body armour, covered by the Home Office Centre for Applied Science and Technology (CAST) Body Armour Standard [7]. The backing material for the CAST standard uses RP1 for proof tests of unformed armour, which is similar in requirements to the National Institute of Justice (NIJ) Standard-0101.06 [8]. The CAST Body Armour Standard also uses Plastazote® LD29 foam as a backing for critical perforation analysis [7].

The above standards focus mainly on a BFS material (namely RP1) to generate a BFS measurement to act as a pass/fail criteria in the ballistic assessment, in addition to the armour complete or partial penetration response. Dstl uses the UK MOD thoracic BABT rig [9] (shortened to 'BABT rig' in the remainder of this document) for some of its PPE research activities, as a model to measure the dynamic response of the 'body' wall from a silicone rubber based deforming element. This is done both in an instrumented form to understand the BABT risk, typically for hard armour, as well as un-instrumented, as a 'standard' backing method for V_{50} type assessments of soft armour during research activities. A significant issue with the use of the BABT rig for V_{50} assessments is that the relatively expensive custom moulded silicone elements require frequent replacement, even when no BABT measures are required.

The BABT rig or BFS materials may be suitable when assessing a ballistic material or PPE in a ballistic range setting. When trying to understand a material or system performance in an arena style blast trial (for fragmentation), there are a long list of practical issues, making them no longer suitable as a backing. In these types of scenarios, often a backing is required that can indicate if, or how many, complete penetrations of the armour material occurred. Often, some way of estimating the resulting severity of the overmatch is required. For this type of requirement, layered fragment witness packs are typically used.

Layered foam packs [1; 10], strawboard [11] or metal spaced packs [12] are examples of the types of layered fragment witness packs used for an arena style blast trial for fragmentation [13], which may be covered by a ballistic material or armour system in order to assess its effectiveness.

For Dstl, the situation has arisen where a different backing may be required for a V_{proof} or V_{50} test, for military or police, bullet or fragment, if it is for acceptance or research and different again to what might be required for an arena blast trial. Therefore, an understanding of how all these different backing methods influence the measured ballistic performance, in addition to other methods that may be used for specialist requirements or by other institutions, is critical to select a suitable test method and to allow robust conclusions to be drawn on the armour material or system performance.

1.3 Previous research

In an effort to reduce the time taken to reset armours between tests, the use of a foam backing in place of the frame was implemented for FSP V_{50} testing of different body armour systems in Reference [4]. The results for a 3.8 kg m^{-2} AD system showed no apparent differences in the measured V_{50} between the chosen low-density foam and frame.

Differences in the measured V_{50} performance between RP1 and the frame have been highlighted previously [1], showing increases in measured performance of up to 13% depending on the frame clamping conditions. An average difference of +8% was observed on the frame compared to RP1 for 3 different armour materials, two para-aramids and one Ultra-High Molecular Weight Polyethylene (UHMWPE) of 3.3 kg m^{-2}.

Testing by Nguyen, *et. al.* [2] showed that the specific material type was a factor to determine if a simple frame provided a measured V_{50} performance higher or lower than on 20% gelatin at 10°C for lightweight ballistic materials. For a 190 g cm^{-2} Twaron® weave, the measured V_{50} performance was around 55% higher on gelatin than on the simple frame across a number of different test fragments. A 610 g cm^{-2} Dyneema® knit showed an average measured V_{50} performance around 10% lower on 20% gelatin than on the simple frame [2].

https://doi.org/10.52202/080042-0033

Using 9 mm handgun rounds, References [5; 14] compared the V_{50} of a 3.5 kg m^{-2} soft armour system (Kevlar® 129) on RP1 and various foams (flat and curved). All the foams showed an increase in the measured V_{50} performance compared to the RP1, although the differences were not considered significant (maximum differences of around 5%).

1.4 Available backing materials and methods

In order to compare backings, the ideal would be to have a performance measure 'as worn' on a live person. Post Mortem Human Subjects (PMHS) or animals as a backing provide what is assessed to be the most realistic dynamic mechanical response, apart from a live person. Use of PMHS or live animals as backings for this work were not considered feasible due to ethical and practical limitations. The utility of different species of animal cadavers is debated for different applications within the wider wound ballistics literature. A goat thorax was selected as the baseline for comparison of methods in the 1970s to assess back face signature [15]. However, none of the armour backing materials/methods considered at that time matched the time-deformation response of the goat thorax [15].

The backing materials and methods selected for the current testing and reason for their use are summarised in Table 3.

Table 3. Summary of backing methods and reason for their selection

Backing	Reason(s) for inclusion
Ex-vivo porcine thorax	To give as close to 'as worn' conditions as possible. The porcine model (thorax) was previously chosen as a basis for UK BABT injury research as the mass and volume can be selected to match those of a human [16; 17]. Availability of fresh tissue from other ongoing trials, where it would otherwise have been disposed. Safety constraints concerning potential Transmissible Spongiform Encephalopathy (TSE) associated with goats and sheep from perforation of the head, spinal cord or abdomen [18][1].
Ex-vivo porcine thigh	To give as close to 'as worn' conditions as possible for items such as Pelvic Protection (PP) [19], designed to cover areas of the thigh.
Roma Plastilina® no. 1	A common backing material, specified in a number of standards[2].
Air backed frame	A common backing method, specified in AEP-2920.
ARTIC [20]	Development in the US of a ballistic-grade clay with predictable and controllable properties resulted in a material known as A Reusable, Temperature Insensitive Clay (ARTIC) [20]. This assessment supports the development and potential adoption of the backing method as an alternative to RP1.
UK BABT rig [9]	A common test method for armour research within Dstl.
20% gelatin at 10°C with a synthetic skin simulant [10]	Ballistic gelatin (20% concentration at 10°C) was used in early BABT research [21] and is a common model used for assessment of penetrating ballistic injury [10]. It may be convenient to use ballistic gelatin (or similar transparent gels) to assess injury risks from an overmatch of an armour system. Inclusion of the skin simulant layer was to help determine armour failure due to pencilling.
25% concentration SEBS [22; 23]	Poly(Styrene-b-Ethylene-co-Butylene-b-Styrene) triblock copolymer (SEBS) has been used as a ballistic [10; 23] and blunt [22; 24] assessment model, offering similar utility to gelatin without some of the practical limitations.
Strawboard, 3.8 mm thick, type D, 10 layer pack [11]	Used in UK for high velocity fragments or assessing injuries that could be lethal.
Metal spaced witness pack [12]	Legacy use in UK for high velocity fragments. The materials used were translated to the closest UK equivalent from the specification in Reference [12].

[1] Perforation of tissues within the head, spinal cord or abdomen may be unlikely when conducting shots against the thorax. However, as the testing was a V_{50} assessment, the trajectory and residual velocity of the shots that overmatched the armour could not be guaranteed.

[2] It is noted that the AEP-2920 and the CAST test standard specify the use of RP1 for bullets, but not FSPs.

MDFPIM V2.0 [10]	The Multiple Discrete Fragment Physical Injury Model (MDFPIM) V2.0 [10; 25] was developed by Dstl specifically for assessing risks from non-metallic fragments and low energy metal fragments[3]. The MDFPIM has been successfully used to compare effectiveness of different armour materials in buried Improvised Explosive Device (IED) arena trials and ballistic laboratory based multiple simultaneous fragment impacts.
TP5 fragment pack V1.0 [1]	A TTCP collaborative project was commissioned to develop a model to provide a backing that would provide representative boundary conditions to soft armour during ballistic impact testing, as well as additional metrics to correlate to injury risks in the event of an armour overmatch. Development of the model included matching blunt ballistic compliance to PMHS testing [1].
TP5 fragment pack V1.1 [1]	A variant of the TP5 pack with an updated skin material.
10 mm RHA	A 10mm thick Rolled Homologous Armour (RHA) plate was chosen to provide an indication of the result of using an extremely stiff backing material.

2. BALLISTIC TESTING

2.1 Backing materials preparation

For the ex-vivo porcine tissue, a single female large white pig was used, weighing approximately 60 kg. Shots were conducted on armour materials overlaying each rear thigh and on each side of the thorax. The shots for the thorax were completed randomly over ribs as well as the intercostal spaces. The testing started approximately 16 hours after the animal had been euthanized by a Schedule 1 method and was allowed to cool to room temperature (21±1°C) prior to testing.

For the purposes of the current testing, RP1 was used conforming to the CAST test standard [7], packed into trays of 420×350×100 mm and following the associated CAST ball drop calibration procedure[4]. If the required indentation depth was not achieved, the RP1 was reconditioned and re-tested. All shots were completed within 1 hour of calibration.

To facilitate direct comparison to RP1, ARTIC was packed into identical sized trays, and followed the identical calibration procedure (with the notable exception that all material was at room temperature, 21±1°C and not altered depending on the calibration result: it either passed and could be used, or failed and could not).

The 20% gelatin at 10°C was manufactured according to the 'Dstl 20% method', described in Appendix D.1 of Reference [10] and cast into blocks of 150×150×300 mm. Armour materials were held in contact with the skin simulant against one of the 150×300 mm gelatin faces. Impacts were completed within 30 minutes of removal of the gelatin from the conditioning cabinet.

The 25% SEBS by volume – mineral oil gel was made as follows:

- The required volume of mineral oil (Primol 352, an ExxonMobil product supplied by Univar) was heated to 100°C for 2 hours in metal trays.
- The SEBS powder (Kraton G1652, a Kraton Polymers product supplied by Univar), calculated to give 25% concentration by volume was gradually added whilst stirring continuously.
- The mixture was allowed to soak at 120°C for a minimum of 4 hours, with regular mixing (approximately 10 minute intervals).
- Once clear and free of bubbles, the mixture was transferred into glass moulds with internal dimensions 300×300×300 mm.
- Additional stirring was completed to release trapped bubbles introduced during the pour.
- The cabinet was programmed to gradually decrease in temperature (down to 20°C) over a period of approximately 24 hours.
- Once cool, the gel block was removed from the glass mould.

[3] The foam used in the MDFPIM has a density of 160±10 kg m^{-3}, compared to the foam as one of the backing options in AEP2920 at 40±5 kg m^{-3} and in the CAST Body Armour standard for critical perforation assessment at 29 kg m^{-3}.
[4] Three drops with a 63.5 mm steel sphere (1.043 kg), from a height of 2.00±0.02 m, 75 mm from an edge and 100 mm between indent centres. The mean depression depth of the three drops must lie between 19±2 mm with no single value outside of 19±3 mm.

https://doi.org/10.52202/080042-0033

2.2 Ballistic testing method

To characterise the effect of backing methods on the measured ballistic performance of soft armour materials, two ballistic materials/systems were chosen: one to represent a low Areal Density (AD) protection and one high AD protection. The low AD system was the UK Tier 1 Pelvic Protection [19] (NATO Stock Number 8420-99-873-0158), a three layer system based on antimicrobial undershorts with increased protection to vulnerable areas provided by two layers of knitted silk, with total AD of 0.45 kg m^{-2}. The Tier 1 PP was assessed with a 6 mm glass sphere (conforming to Reference [26]) as a representative threat for that armour system. The higher AD system (chosen to be more representative of soft armour to protect against metallic munition fragments) was a 20 layer para-aramid Kevlar® 640G woven pack, with total AD of 2.8 kg m^{-2}. The 20 layer Kevlar® pack was assessed with the 1.1 g Chisel Nosed (CN) FSP (G5 from AEP-2920 [6]).

A V_{50} assessment was conducted using a minimum of 20 fair shots for the different material/projectile and backing combinations. Not all backings were evaluated with every ballistic material due to limitations on available resources. The V_{50}s were calculated using a probit model within the statistical program R [27; 28]. This also enabled the 95% confidence intervals on the measurement to be calculated.

Shot spacing against the 20 layer Kevlar® 640G 2.8 kg m^{-2} pack followed Annex G.1 in Reference [6]; 63.5 mm spacing, along a line offset by 11° to avoid impacting previously damaged yarns. Shots were a minimum of 50 mm from any edge. For the Tier 1 PP, shot spacing was 50 mm. In all cases shots were a minimum of 50 mm from any edge of the backing method/material. The same shot spacing was applied to the backing material as to the armour, to avoid pre-damaged areas of the backing.

Testing was completed over multiple phases to the same method above, but utilising different compressed gas propulsion systems (air and helium), but with the same barrels. In each phase, velocity measurement was conducted using calibrated and cross-validated equipment.

For the 6 mm glass spheres, a 6.05 mm calibre, 300 mm length smooth bore barrel was used. For the 1.1 g FSP, a 800 mm length rifled 7.62 mm barrel and sabot were used to ensure stable flight (<3° yaw as per AEP-2920 [6]). A sabot stripper was placed between the barrel and target for shots with the 1.1 g FSP.

2.3 Ballistic Testing Results

A total of 22 separate V_{50} determinations were conducted with the various backing and armour material combinations (total of 459 fair shots). For security classification purposes, the V_{50}s and differences are reported as normalised values.

The 20 layer Kevlar® 640G 2.8 kg m^{-2} pack could not be assessed on the porcine tissue. A common baseline backing was desired to enable simultaneous comparison across the different backings for both armour materials. The frame was not considered a suitable baseline as the objective was to show methods to best replicate the 'as worn' performance. RP1 was selected as the baseline due to the lack of suitable data for an 'as worn' condition on both materials, not because RP1 was assumed to represent 'as worn' performance. The comparison is shown in Figure 14 with backing materials ordered in terms of increasing absolute average difference.

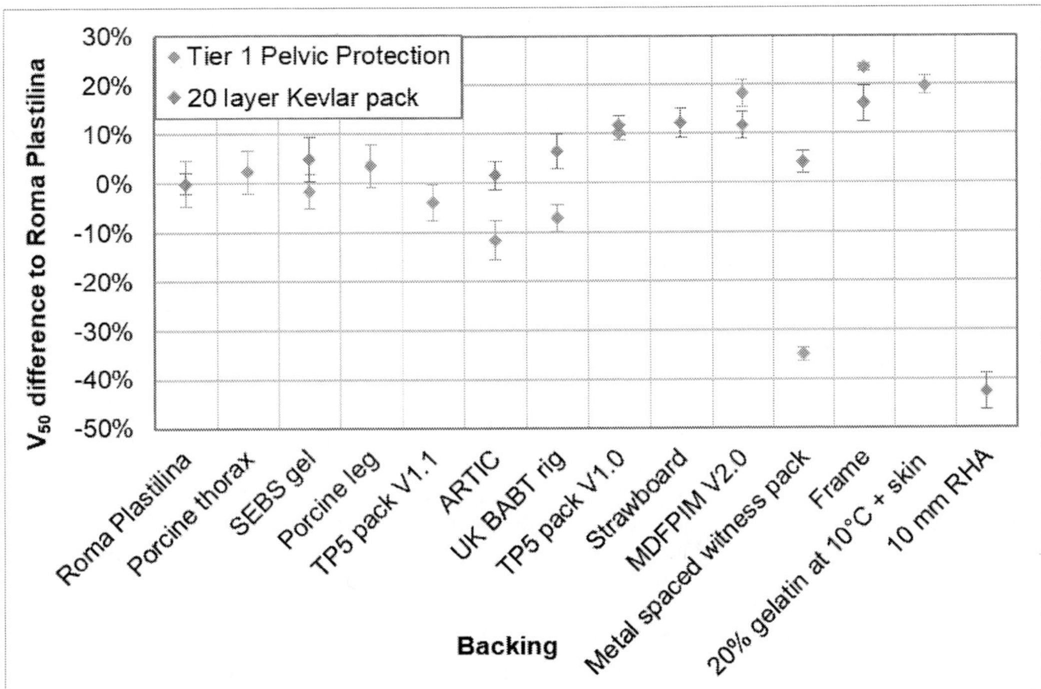

Figure 14. Measured V_{50} response of the Tier 1 PP and 20 layer Kevlar® 640G 2.8 kg m^{-2} pack, normalised to the response on a RP1, for different backing methods. Error bars show the 95% confidence interval.

3. DISCUSSION

3.1 Discussion of results

The 2.8 kg m^{-2} 20 layer Kevlar® pack showed differences in the measured V_{50} performance of up to 16% (+16/-0%) between common backing methods. The greatest differences observed were between the RP1 and the frame. The more flexible Tier 1 PP showed differences of -12% to +24% when compared relative to RP1, for backings that could be considered reasonable (all backings evaluated for the Tier 1 PP excluding the metal spaced witness pack), or -15% to +19% when the porcine thigh is used as the baseline. The difference in measured V_{50} performance between for the Tier 1 PP on RP1 and the frame was 24%.

A variation of 24% for the Tier 1 PP and 16% for the 20 layer Kevlar® pack between the measured V_{50} performance on RP1 and the frame is a much more significant source of 'error' in the armour system measured performance, than the ±2 m s^{-1} allowed for velocity system calibration within AEP-2920 [6] (around ±1% for the velocities used within this paper).

Using a stiffer backing, such as the metal spaced witness pack for the Tier 1 PP dramatically lowered the measured V_{50} response by 37% compared to the porcine thigh or by 35% when compared to RP1. The fact that the 2.8 kg m^{-2} 20 layer Kevlar® pack had a 4% increase in the measured V_{50} performance on the metal spaced witness pack compared to RP1 suggests a different loading condition between the rear face of the armour and front face of the backing compared to the Tier 1 PP. This highlights that the difference in the measured performance may not be consistent in sign or magnitude across different backings for different ballistic materials. This difference in sign in the measured V_{50} performance compared to RP1 was also observed when the UK BABT rig was used as a backing. However, in this instance, the magnitude of the difference was smaller (+6/-7%).

Of the 8 backing methods with measured V_{50} performance data for both the Tier 1 PP and Kevlar pack® (ignoring RP1 as it was used as the baseline), 2/7 backing methods showed the Tier 1 PP had a greater relative performance to the RP1 baseline than the Kevlar® pack. This showed the different backing methods did not consistently rank the materials based on the change in measured performance in the same order (to the same baseline).

Based on the data for the 2.8 kg m^{-2} 20 layer Kevlar® pack and Tier 1 PP: RP1, 25% SEBS gel, TP5 pack V1.0 and V1.1, ARTIC and the UK BABT rig are considered to give a reasonable

https://doi.org/10.52202/080042-0033

representation of an 'as worn' condition. This was based on <10% difference to the porcine leg for the Tier 1 PP and RP1 for the 2.8 kg m^{-2} 20 layer Kevlar® pack. The suitability of these backings may change for different armour materials or threats, or if compared to a more realistic 'as worn' backing condition.

The fact that the measured V_{50} performance for the Tier 1 PP on the porcine thorax and thigh were not statistically significantly different at the 95% confidence level suggests that a single, suitable backing method is likely to be appropriate for assessment of PPE designed to protect the thorax or thigh, at least for lightweight protection systems.

It is considered that the effect of the backing on the measured V_{50} performance of an armour is a factor of the backing material properties (stiffness, rate sensitivity and boundary conditions[5]) and loading conditions from the armour (back face deformation size and shape, rate, etc.). The armour loading conditions will be dependent on the threat (size, mass, velocity, deformation or fragmentation, etc.) as well as the ballistic material type, properties and construction.

3.3 Consequence for testing

The AEP-2920 test method drives towards two separate assessments on different backings to characterise an amour system (for fragment protection) in order to satisfy V_{proof} and V_{50} requirements. However, these two separate assessments are not comparable and do not appear to be mutually supportive. The reason for the use of the frame in AEP-2920 is likely for practical reasons (re-usable with no material wastage, efficient and inexpensive). However, given that different material types/constructions appear to potentially behave differently when air backed to a more realistic 'as worn' backing (some increase significantly in measured performance whilst others decrease), the AEP-2920 frame may not even be suitable to rank material performance. This situation is likely to be compounded with new materials, constructions, composites, etc..

One of the remaining benefits of the frame is the ability to conduct residual velocity (V_r) measurements when the armour is overmatched. However, if the measured performance of the material is increased in the region of 20% on the frame compared to an 'as worn' condition, then residual velocity measurements are also likely to be unrealistic of an 'as worn' condition. Residual velocity measurements can be collected with the use of alternative backings (gels or layered fragment packs) that are likely to provide more accurate performance measures of the armour under test, both in terms of V_{50} and V_r metrics.

These issues underpin the need to follow good modelling practice (of which the use of a backing material or frame for a ballistic test is a model): a fitness for purpose assessment should be conducted each time, before the model is used. Because it has been done that way in the past is not justification to do so again. However, this fitness for purpose assessment may be simple if the test requirement issued to the institution conducting the testing specifies a certain backing or test standard, but may be more in depth if left open. It also suggests that the requirements manager / staff setting the requirement should be provided suitable technical advice in order to specify an appropriate backing or test standard for their given scenario and requirements.

3.4 Limitations and way forward

Porcine skin is known to overestimate the ballistic resistance of PMHS skin [10]. There is a risk that using the porcine thigh or thorax may underestimate the measured V_{50} performance of a realistic 'as worn' condition. Within practical and ethical limitations, the assumption was accepted that the porcine thigh or thorax provided the closest model to the 'as worn' condition.

A limitation of a number of the backings (including the frame) is an inability to assess (realistic) failure due to pencilling. If a backing is used that has a validated skin perforation response, i.e. the TP5 fragment pack V1.0 or V1.1 [1], MDFPIM V2.0 [10], 20% gelatin at 10°C with a synthetic skin simulant [10], or a suitable animal model [10], this failure mode could be assessed directly. This is likely to be more critical for thin, flexible or low AD materials.

Where the spaced metal witness packs have been used to support UK MOD body armour tests (for assessment of ballistic post barrier risk to allow direct comparison to legacy data), the potential effect on the measured armour performance was presumed from the outcomes of phase one of this testing. This provided sufficient understanding on potential affects to the measured armour response to suggest that an additional layer of 25 mm polystyrene be placed between the armour and front layer of the pack. Whilst this arrangement was not specifically assessed in terms of the result on the measured V_{50}, it was

---

[5] Boundary conditions were not specifically addressed here, but have been demonstrated to influence measured V_{50} performance [1].

https://doi.org/10.52202/080042-0033

considered sufficient to mitigate any potential reduction in the measured armour performance for the testing in question.

The knitted silk construction of Tier 1 PP may be towards a worst-case material in terms of highlighting differences in measured V_{50} performance for different backing materials. Both the knit structure and silk yarns allow a large degree of deformation before failure and therefore the backing will play a significant role in how the material is allowed to deform. However, how other material types, constructions, or different threats; in particular handgun or rifle rounds that deform during impact with armour, or projectiles that cause a different failure mode in the armour, influence the effects of backings on the measured V_{50} performance is not known. This paper can be used to highlight the potential effects from different backings that need consideration when defining test standards or requirements, rather than as a method to relate or transfer performance measured on different backings.

Whilst this may appear critical of several backing methods, there is no evidence to say it is negligent to use them (and these backing methods have led to PPE that has saved lives on operations). This paper is aimed to identify areas for improvements and provide evidence. It is recognised that getting the required knowledge to build confidence in suitable backing method(s) will take time.

4. CONCLUSIONS

The backing material or method used during assessment of ballistic material performance (e.g. V_{50} or V_{proof}) can significantly affect the measured performance. Most importantly, the difference in the measured performance may not be consistent in sign or magnitude across different backings and ballistic materials. This effect is dependent on the backing used, as well as the amour (and threat) type.

Differences of -12% to +24% in the measured V_{50} performance were observed (relative to RP1) for what could be considered reasonable and/or common backing methods. This included an increased measured performance on the AEP-2920 frame of 16% for a 20 layer Kevlar® 640G 2.8 kg m^{-2} pack assessed with a 1.1 g FSP and an increase of 24% for the UK Tier 1 PP assessed with a 6 mm glass sphere, compared to the measured performance on Roma Plastilina® no. 1.

The measured V_{50} performance on the metal spaced witness pack showed the same model providing a difference of +4% to -35% to a baseline of RP1 for the materials assessed within this work.

Where possible, backing methods were compared to an 'as worn' performance, estimated by the measured performance on a porcine thigh or thorax. Comparison of the measured V_{50} performance on a porcine thigh and thorax indicated that a single, suitable backing method is likely to be appropriate for assessment of PPE designed to protect the thorax or thigh, at least for lightweight protection systems.

Selection of the backing material or method has the potential to have a much larger effect on the measured V_{50} performance of an armour material or PPE, than other potential sources of error already controlled within various test standards. This should be understood by the testers, users and requirements managers.

It appears challenging to justify the use of the frame as a backing method from an injury modelling perspective: it has potential to provide inaccurate measured performance outcomes, as well as potentially provide unreliable rankings of relative material performance compared to 'as worn' conditions. However, there may be other reasons for its use and there is no evidence to say it is negligent to use the frame providing the limitations and caveats are understood.

Considering the current testing alongside previous research, a different threat (such as 9 mm handgun rounds), a higher AD or stiffer soft armour system may reduce the measured differences in V_{50} performance between different backings.

There is no one backing material/method that should be used in preference to others – it will depend on the requirement and scenario. However, inappropriate choice of the backing method may lead to:

- PPE that does not provide as much protection as it would under 'as worn' conditions, i.e. does not protect against the specified threat.
- Development of PPE with unnecessary burden for the required protection.
- A different ranking performance of PPE/armour materials to their 'as worn' performance rankings.
- Increased resource required to meet specifications.

It is essential to follow good modelling practice (of which the use of a backing material for a ballistic assessment is classed as a model): a fitness for purpose assessment should be conducted each time, before the model is selected or used. The outcomes from this work can be used to support requirements managers or staff setting the requirements with suitable technical advice in order to specify an appropriate backing or test standard for their given scenario and requirements. This will help to prevent PPE with insufficient protection, overburden or wasted resource.

https://doi.org/10.52202/080042-0033

Acknowledgments

The authors would like to acknowledge the contribution of Chris Doona (US Soldier Center) for participation in phase 3 of this testing, support from TTCP for supply of materials (ARTIC and TP5 packs) and Andrew Sedman (Dstl) for the preparation of the SEBS gel.

References

[1] Ouellet, S., Pageau, G., and James, G. A new biofidelic backing for the evaluation of the ballistic performance of soft armour and lightweight protective fabrics. Personal Armour Systems Symposium, Copenhagen, Denmark 11th - 15th October 2021.

[2] Nguyen, T.-T.N., Tsukada, H., Breeze, J., and Masouros, S.D. The Critical Role of a Backing Material in Assessing the Performance of Soft Ballistic Protection. Hum. Factors Mech. Eng. Def. Saf., 2022; Vol 6(1), pp 13.

[3] Nguyen, T.-T., Meek, G., Breeze, J., and Masouros, S. Gelatine backing affects the performance of single-layer ballistic-resistant materials against blast fragments. Front. Bioeng., 2020; Vol 8, pp 744.

[4] Broos, J. and van der Jagt-Deutekom, M. Ballistic protection of fragment vest against IED threat. Personal Armour Systems Symposium, Royal Military Academy, Brussels, Belgium,2008.

[5] Anctil, B., Bayne, T., Bourget, D., Pageau, G., Binette, J.-S., Rice, K., and Toman, A. An Alternative to Plastilina for Evaluating the Performance of Body Armours. Personal Armour Systems Symposium, Royal Military Academy, Brussels, Belgium,2008.

[6] North Atlantic Treaty Organization (NATO). Procedures for the evaluation and classification of personal armour. Bullet and fragmentation threats, Allied Engineering Publication AEP-2920, Edition A, Version 2, 2016.

[7] Payne, T., O'Rourke, S., and Malbon, C. Body Armour Standard (2017), CAST Publication number: 012/17, 2017.

[8] Office of Law Enforcement Standards - United States of America. Ballistic Resistance of Body Armor NIJ Standard-0101.06. 2008.

[9] Tam, W., Rozant, O., Thoral-Pierre, K., Pope, D., Softley, I., Baker, L., Cook, W., Scaramuzzino, P., and Ferry, E. The UK Behind Armour Blunt Trauma Rig – 20 years on. Personal Armour Systems Symposium, Marine Establishment Amsterdam, Netherlands 19th - 23rd September 2016.

[10] James, G. Development of Models to assess penetrating injury from ballistic projectiles (PhD). Cranfield University, Cranfield Defence and Security. DSTL/PUB118063 V1.0. 2020.

[11] Defence Standard 93-59. Chipboard (For Use at Proof and Experimental Establishments), DEF STAN 93-59 Issue 3, 2005.

[12] Verlome, J., Szymczak, M., and Broos, J. Metallic witness packs for behind-armour debris characterization. Int. J. Impact Eng., 1999; Vol 22, pp 693-705.

[13] Weir, J., Shaw, B., Pizzolato-Heine, K., Ouellet, S., Martineau, L., McGuire, R., Mahoney, M., James, G., Hepper, A., Gillich, P., Gant, L., and Eberius, N. Fragment Characterisation & Threat Modelling - A Multinational Study to Re-Define & Represent the Fragment Threat. Personal Armour Systems Symposium, Copenhagen, Denmark 11th - 15th October 2021.

[14] Bosik, A., Hedge, C., Lightsey, S., and Pageau, G. Initial Findings on the Development of Test Procedures for Multi-Hit Testing Body Armour. Personal Armour Systems Symposium, The Hague, The Netherlands.

[15] Prather, R.N., Swann, C.L., and Hawkins, C.E. Backface Signatures of Soft Body Armors and the Associated Trauma Effects. 1977

[16] Cooper, G. and Taylor, D. Biophysics of impact injury to the chest and abdomen. BMJ Military Health, 1989; Vol 135(2), pp 58-67.

[17] Cannon, L. Behind armour blunt trauma-an emerging problem. J R Army Med Corps, 2001; Vol 147(1), pp 87-96.

[18] Advisory Committee on Dangerous Pathogens. BSE - Occupational guidance 2006.

[19] Lewis, E.A., Pigott, M., Randall, A., and Hepper, A. The development and introduction of ballistic protection of the external genitalia and perineum. J R Army Med Corps, 2013; Vol 159(suppl 1), pp i15-i17.

[20] Frreney, R. and Mrozek, R. The development of a room temperature clay backing material for the ballistic testing of body armor. Personal Armour Systems Symposium, Marine Establishment Amsterdam, The Netherlands 19th - 23rd September 2016.

[21] Metker, L.W., Prather, R.N., and Johnson, E.M. A method for determining backface signatures of soft body armors, EB-TR-75029, 1975.

[22] Bracq, A., Delille, R., Maréchal, C., Bourel, B., Lauro, F., Roth, S., and Mauzac, O. On the use of a SEBS polymer gel block as a new ballistic target to assess blunt ballistic impacts: application to a wide range of LLKE projectiles. Int. J. Impact Eng., 2021; Vol, pp 103874.

[23] Mrozek, R.A., Leighliter, B., Gold, C.S., Beringer, I.R., Jian, H.Y., VanLandingham, M.R., Moy, P., Foster, M.H., and Lenhart, J.L. The relationship between mechanical properties and ballistic penetration depth in a viscoelastic gel. J. Mech. Behav. Biomed. Mater., 2015; Vol 44, pp 109-120.

[24] Malbon, C. Female police body armour: improving comfort, protection, and testing. Cranfield University Cranfield Defence and Security. 2021.

[25] Witness Pack. Inventor: James, G. Application: 1918144.5. United Kingdom. Patent: GB 2595434 A, 01.12.2021.

[26] James, G. and Hepper, A. Ballistic simulation of fragmentation from buried improvised explosive devices. Personal Armour Systems Symposium, Robinson College, Cambridge, UK 8th - 12th September 2014.

[27] R Core Team. R: A language and environment for statistical computing. V2.15.2. Vienna, Austria. R Foundation for Statistical Computing, 2012.

[28] Kosmidis, I. brglm: Bias reduction in binary-response GLMs. (R package). 2007.

https://doi.org/10.52202/080042-0033

Testing light-weight personal protection impacted with sand particles

A. Miranda-Vicario[1], A. Azevedo[1], F. Coghe[1]

[1]*Royal Military Academy, Department of Weapon Systems and Ballistics Renaissance Avenue, 30 B-1000 / Brussels Belgium*
angel.miranda@dymasec.be
0244-14159

Abstract. Regarding the evolution of combat scenarios, it is necessary to comprehend the challenges provided by improvised explosive devices (IEDs), it is vital to investigate the sand that is ejected during an explosion. Due to the absence of primary components, the IEDs principal risks come from the explosion and secondary fragmentation. For dismounted soldiers, the explosion itself and the debris released could result in severe injuries to the exposed, unprotected body parts, primarily the limbs. More protective mass is not advantageous because the limbs need to be highly mobile. The addition of mass would reduce the ability to move which should be avoided. The Allied Engineering Publication (AEP) 2920, which focuses on primary fragments as they are the principal concern for the majority of explosives devices, is the reference standard for testing fragmentation threats. These primary fragments are represented by steel Fragment Simulating Projectiles (FSPs). Therefore, this threat is faster and denser when compared with a cloud of sand ejected from an explosion. Thus, a methodology tailored to IEDs must be developed or modified. The major goal of this project is to create a method for consistently releasing a cloud of sand, which will enable testing light personal protective equipment. Controlling the sand cloud's velocity, dispersion, and the ability to precisely measure these events as they occur, when the cloud hits the target, are essential. Secondly, it is crucial to be able to measure various targets' properties in order to research how well the sand grains are stopped by them. The AEP 94 "Skin penetration assessment of non-lethal projectiles" was used to evaluate the damage and potential skin wounds in order to assess the damage and possible skin wounds.

1. INTRODUCTION

The number of Improved Explosives Devices (IEDs) has increased. During the decade 2010-2020, there have been around 28,800 incidents involving IEDs, resulting in more than 35,000 casualties [1]. The main difference between IEDs and other explosive devices, such as shells or grenades, is the lack of primary fragmentation produced during the explosion. IEDs are often packed inside plastic containers, which limits the production of primary fragments. The dangers of IEDs are due to the explosion itself, but also the secondary debris (or fragmentation) from the soil surrounding the explosion site.

This secondary fragmentation can cause a variety of injuries to the unprotected parts of the human body, particularly the limbs. Due to the nature of the threat and the extremities, it is critical to increase the level of limb protection, without adding heavy armour that would severely restrict the user's movements.

The Allied Engineering Publication (AEP) 2920 [2] is the primary reference standard for testing fragmentation threats. 1.1 g Fragment Simulating Projectiles (FSPs) are typically required for armour and personnel protection. Other minor threats described in the standard, such as 0.16 g FSP or Right Circular Cylinder (RCC), are quite unstable in flight, making obtaining accurate results during testing difficult. Furthermore, as a reflection of typical primary fragments, all of these fragment-simulating projectiles are made of metal. These facts make it difficult to relate the test results to the actual protection provided by armour impacted with secondary fragmentation.

The study of the secondary fragmentation has been already addressed. Some systems have been developed to consistently launch projectiles with masses between 0.004 and 54 g, at velocities up to 1,600 m/s [3]. Several attempts have successfully developed different systems to launch different sand and fragments larger than 4 mm and 0.12 g [4, 5]. Smaller particles might be of special interest in sandy conditions or even to represent components of the IED itself [6]. The probability of hit by small (0.61 g) stones is similar to larger stones in a certain region (from 30° to 80° from the explosion point) [7], showing the necessity of being protected effectively for smaller threat.

For this reason, this work aims to expand the work in the literature and studies how to accelerate and eject sand and small debris particles at different light-weight fabrics, to obtain a methodology that is able to classify the fabrics. The main objective of this work is to test the ability of several light-weight armour systems to defeat a cloud of small debris and compare the results with the protection offered against FSPs.

https://doi.org/10.52202/080042-0034

2. METHODOLOGY

2.1 Test samples

The test samples are ten different fabrics of interest for comparison. These materials have been provided by different manufacturers and involve prototype fabrics and fabrics already used as uniform in different militaries. These light-weight fabrics have different areal densities and 2 different constructions, knitted and woven. Most of the samples are manufactured from aramid – viscose, while three of them use a high-performance ultra-high molecular weight polyethylene (UHMWPE) yarn (Table 4). The differences in areal density are mainly due to a tighter knit. An example of weave comparisons is shown in Figure 15.

Table 4. Tested samples

Sample	Material	Areal Density (g/m^2)	Fabric Construction
1	100 % UHMWPE	95	Knit
2	98 % UHMWPE	260	Knit
3	90 % UHMWPE	270	Knit
4	Aramid - Viscose	180	Knit
5	Aramid - Viscose	204	Woven
6	Aramid - Viscose	205	Woven
7	Aramid - Viscose	210	Woven
8	Aramid - Viscose	250	Woven
9	Aramid - Viscose	384	Woven

Figure 15. Comparison of fabrics: Sample 6 (left) and sample 8 (right)

2.2 Experimental set-up

The first part of this research is to estimate the ballistic resistance of the different fabrics when impacted with FSPs. For conducting the ballistic tests, a universal receiver with interchangeable barrel is used to fire the projectiles, and the target is positioned 5 m ahead of the muzzle. For the first series of tests, 1.1 g FSPs were used, with projectile velocity measured using optical chronographs. The target is the test fabric backed by 20% gelatine. Perforated fabric was assessed with the help of a high-speed camera, as the fragment might perforate the fabric but rebound out of the gelatine (Figure 16). During the testing the fabric was placed over the gelatine and pressed into a metal frame, which has an opening of 1,50 x 150 mm (Figure 17). This frame holds the fabric around its entire perimeter, with a single central shot for each test.

https://doi.org/10.52202/080042-0034

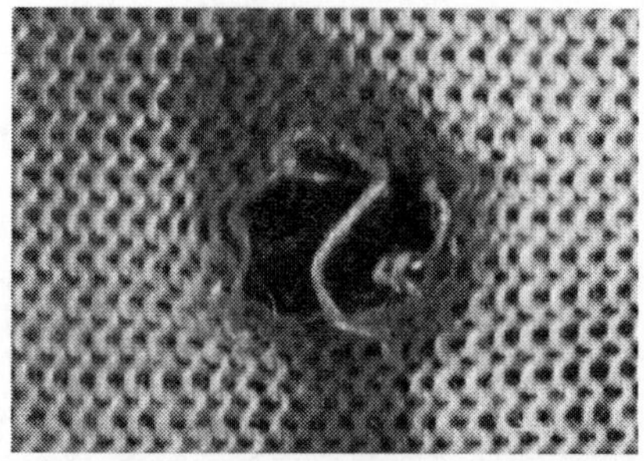

Figure 16. Example of hole in the fabric

Figure 17. Frame support

For the second round of tests, the ammunition used was a 12 gauge shell, filled with 6.37 g of sieved sand of between 1 and 2 mm (around 0.01 g) placed in a sabot. The sand remained in the cartridge after closing, and there were no issues regarding leaking of the sand outside the cartridge. These sizes were selected to provide a good compromise of mass and perforation effect without reaching the mass of an FSP. For that grain size it would need a complete ballistic or fragment protection. Small debris size was chosen to try to replicate small particles up to 2 mm size, a common size secondary fragment [6], as well as being a typical sand particle size in beach or desert environments. At this size, the sand ejected might find an easy way to get through the gaps of the fabrics. As shown in the example, there are gaps in the regular fabric of around 0.5 mm (Figure 4).

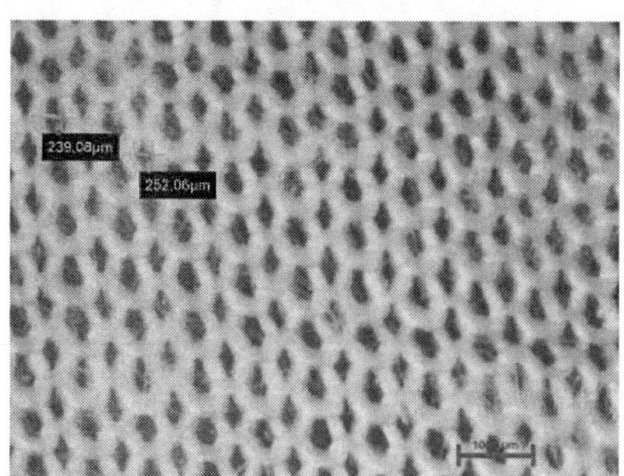

Figure 4. Gaps in the fabric (Sample 1)

The velocity of the particles was calculated from the recording of a high-speed camera placed orthogonally to the impact point. Five different particles were measured and averaged. The target is placed 3 m in front of the muzzle. The velocity of the cloud is controlled by varying this distance.

Different configurations of propellant have been tested and it is possible to achieve velocities up to 900 m/s at 2 m. The goal is to have a configuration able to fire sand in a controllable way from 300 m/s up to around 1,000 m/s. Below 300 m/s, there is very little interest as it has been observed that the particles in the target tend to rebound from the gelatine.

The test fabric is backed with natural chamois skin with an optimum thickness of 1.39 mm, a 6 mm closed cell foam and 20% gelatine (figure 5), as per AEP 94 [8]. Cameras were used to record the depth of perforation of the sand grains into the gelatine. The samples were held in place upon testing by stapling the fabric directly to the foam. This was enough to retain the fabric during testing.

It is possible to modify the impact velocity by modifying the distance between the muzzle and the target. This also modifies the density of the impacts in the back-face material. As the idea is to develop a quantitative method to rank the different materials, it should be possible to analyse the data and extract conclusions from the comparison between the different samples.

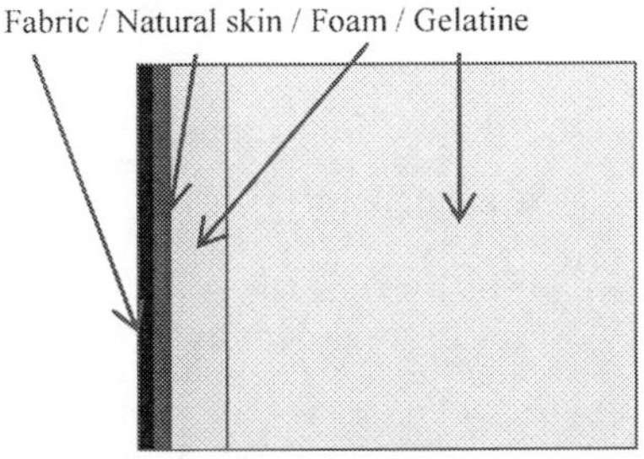

Figure 5. Lay-up of the test sample

It was not possible to always use the same backing material, as the material defined in the AEP 94 is no longer available, therefore available stocks were limited. An alternative solution is being sought in order to update the standard.

3. RESULTS AND DISCUSSION

3.1 FSP tests

The ballistic resistance of each sample backed with gelatine was estimated following the Probit method when impacting with FSPs (Table 5 and Figure 6). An average of 17 shots were fired per sample.

Table 5. Ballistic resistance of the samples impacted with FSP.

Sample	Material	Areal density (g/m^2)	Type of fabric	V_{50} (m/s)	σ (m/s)
1	100 % UHMWPE	95	Knit	137	11
2	98 % UHMWPE	260	Knit	194	0
3	90 % UHMWPE	270	Knit	161	15
4	Aramid - Viscose	180	Knit	92	5
5	Aramid - Viscose	204	Woven	92	3
6	Aramid - Viscose	205	Woven	94	0
7	Aramid - Viscose	210	Woven	96	0
8	Aramid - Viscose	250	Woven	97	2
9	Aramid - Viscose	384	Woven	106	2

https://doi.org/10.52202/080042-0034

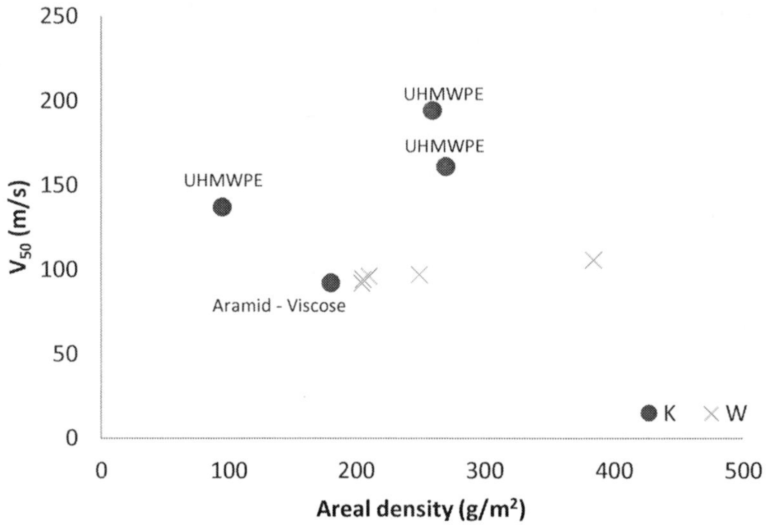

Figure 6. V_{50} versus areal density of the fabrics.

As shown in Figure 6; the higher the V_{50}, the higher the ballistic resistance of the sample. For woven fabrics, the ballistic resistance seems to increase slightly with the increment of the areal density for the considered range. This may happen because the level of the threat is too high for this type of material. Due to the low areal density of the structures, these samples are able to dissipate only limited quantities of kinetic energy. This may happen because of limitations on the response of the structure, e.g. in terms of deformation, to absorb the incoming kinetic energy. Adding a small quantity of material, even if it is almost double, may not have a visible effect in the protective characteristics of the fabric.

For knitted fabrics, it is more difficult to set conclusion due to the reduce number of samples. There seems to be an increase in resistance when increasing the areal density. The fabric is able to deform and dissipate significantly larger quantities of kinetic energy. The fabrics with polyethylene (Sample 1, 2 and 3) exhibit a higher performance than the regular yarns, as they have greater mechanical properties than the other woven sample.

3.2 Sand tests

When impacted with sand, the number of holes in the rear of the foam were counted, in order to differentiate the fabrics. This parameter was identified as a useful indicator to rank the different fabrics (Figure 7).

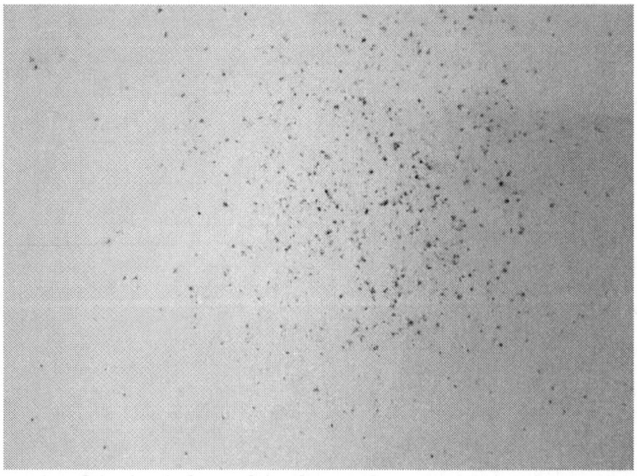

Figure 7. Example of the back face of the foam after a test (Sample size 305 x 225 mm).

Due to the inherent variability and dispersion of the ejection of different sand grains, there were differences in the density and velocity of the sand cloud. It is possible to observe a variation of around 10 % in the number of total holes in the rear of the foam for the test without samples, and a variation of around 20 % of the velocity of the sand cloud (Table 6). Results of the testing are shown in Table 6, Figure 8 and Figure 9. Currently, no study of variability has been performed, but it is possible to observe some differences in the behaviour of the fabrics. Only one shot per sample has been conducted.

Table 6. Ballistic results of the samples impacted with sand cloud.

Sample	Material	Weight (g/m^2)	Type of fabric	Impact velocity (m/s)	Holes
1	100 % UHMWPE	95	Knit	433	48
2	98 % UHMWPE	260	Knit	443	43
3	90 % UHMWPE	270	Knit	447	72
4	Aramid - Viscose	180	Knit	400	85
5	Aramid - Viscose	204	Woven	400	17
6	Aramid - Viscose	205	Woven	477	51
7	Aramid - Viscose	210	Woven	467	59
8	Aramid - Viscose	250	Woven	478	50
9	Aramid - Viscose	384	Woven	392	36
No sample	Backing material AEP94	-	-	470	83
No sample	Backing material AEP94	-	-	430	94

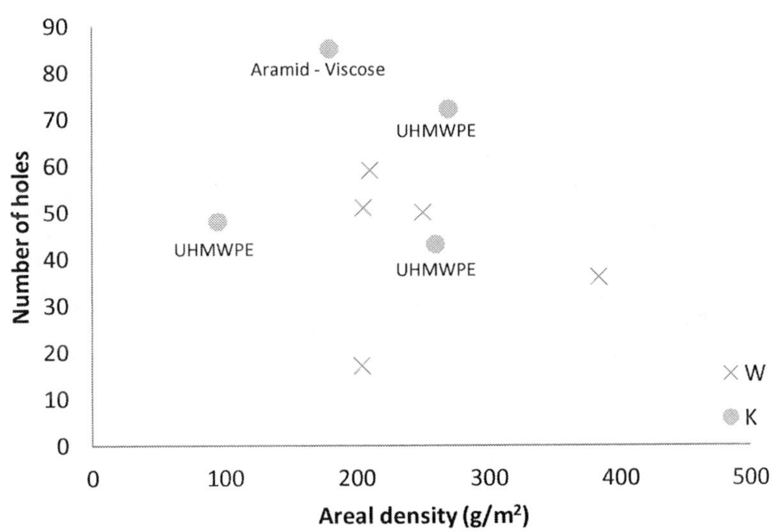

Figure 8. Number of holes in the foam regarding the areal density of the fabrics.

https://doi.org/10.52202/080042-0034

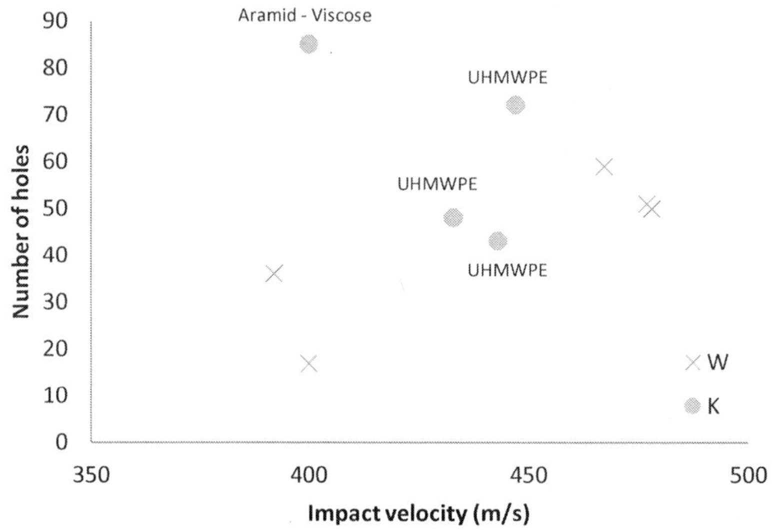

Figure 9. Number of holes in the foam regarding the impact velocity of the sand.

Despite the limited number of tests conducted, it is possible to observe some trends, which need to be confirmed with more testing. In Figure 8 and Figure 9, the lower the number of holes, the better the ballistic resistance of the sample, as there will be less sand debris perforating the fabric, causing potentials wounds. A criterion should be developed which relates the depth of penetration in the gelatine with the potential for wounding. The depth of penetration can be measured with the high-speed camera (Figure 10). The deepest sand impact is approximately 13 mm. Despite there being no clear wound criteria available to determine the possible damage upon the impact of sand, it seems plausible that it would be related to the number of impacts and the depth of the impacts. It has been previously postulated that the risk of skin perforation corresponds with impacts in excess of 24 J/cm^2. This value can be related to the number of impacts [9]. The depth of the impacts in gelatine may be related to an abrasion or more important wounds, but a reference needs to be established.

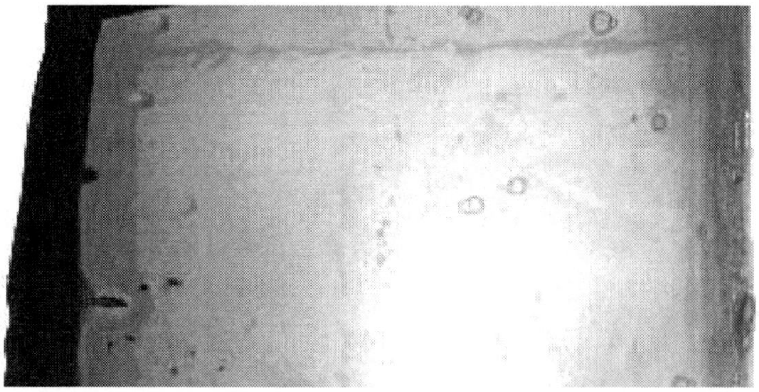

Figure 10. Sand impacts into a block of gelatine.

In Figure 8 and Figure 9, it is possible to observe the different points for all the different materials. This means that assessment of the holes in the fabric can lead to a ranking and allow the study of the best characteristic required to defeat this particular threat. Assessing the number of perforations is much easier with backing foam than in gelatine. These small perforations in gelatine tend to collapse, therefore the backing prescribed by the AEP 94, or a similar one, is of great importance.

Even though, it is difficult to obtain some conclusions due to the few samples shown, the number of holes increases with the impact velocity for knitted fabrics. There is still insufficient data for woven fabrics. The fabrics with polyethylene (Sample 1, 2 and 3) exhibit a higher performance than the regular yarns, as they suffer fewer perforations in comparison with the other woven sample.

It is noticeable the different behaviour of the samples when facing both samples. As an example, in Figure 6 and Figure 8, it is possible to observe that the knitted samples behaves better upon impact of the FSP's compared with the woven sample. But for the sand threat, knitted samples can only behave as good as the woven sample. This difference shows the importance of developing this technique to be able to study the different fabrics and improve them when facing small threats.

4. CONCLUSIONS

In this study, a methodology was proposed to test light-weight fabrics upon impact of secondary fragmentation in the form of sand debris ejected from an IED. Although a greater understanding of the mechanism of the perforation of multiple projectiles into a lightweight fabric is required, it is possible to identify some characteristics that would make this methodology suitable for classifying the fabrics. This methodology allows classification of the different fabrics, by counting the number of perforations in the foam. Trends observed include greater perforation with higher impact velocities and lower perforation with higher areal densities of fabrics. However, it is desirable to increase the control of the ejection of the sand, in order to enhance the repeatability in terms of velocity and dispersion of the sand cloud. The repeatability of a perforation pattern of a fabric when face a similar debris cloud should be studied.

Once the methodology has been refined, it will be possible to rank the fabrics and study in depth the different parameters that affect the ballistic resistance of a fabric, for this particular threat. It is not yet part of this work to establish the best characteristic of a fabric to defeat this type of threat. The difference of the results of the samples when facing the regular FSP's or a sand cloud shows a different behaviour that should be analysed developing this technique.

Acknowledgement

This research was done within the framework of the DYMASEC – Dynamic Material Behaviour and Security Applications – research pole of the Royal Military Academy, Brussels, Belgium, and funded by the Department for Scientific and Technological Research of Defense of the Belgian Armed Forced (study MSP 21 - 03).

References

[1] I. Overton, 'A decade of global IED harm reviewed', *AOAV*, Oct. 15, 2020. https://aoav.org.uk/2020/a-decade-of-global-ied-harm-reviewed/ (accessed Mar. 03, 2023).

[2] NATO, 'STANAG 2920 - Procedures for the evaluation and classification of personal armour-Bullets and fragmentation threats', Sep. 2016.

[3] J. Weir *et al.*, 'Fragment Characterisation & Threat Modelling - A Multinational Study to Re-Define & Represent the Fragment Threat', *Proc. Pers. Armour Syst. Symp. 2020.*

[4] G. Pageau and S. Ouellet, 'Development of soil ejecta surrogate projectiles for laboratory testing of lightweight protective materials', *Proc. Pers. Armour Syst. Symp. 2020.*

[5] M. J. Van den Jagt-Deutekom and T. A. T. Westerhof, 'Assessment of the secondary fragments from buried explosives', *Procedings Pers. Armour Syst. Symp. 2018.*

[6] B. Wilks, R. M. Morgan, and N. L. Rose, 'An experimental study addressing the use of geoforensic analysis for the exploitation of improvised explosive devices (IEDs)', *Forensic Sci. Int.*, vol. 278, pp. 52–67, Sep. 2017, doi: 10.1016/j.forsciint.2017.06.028.

[7] C. A. Couldrick, T. A. Ringrose, P. L. Gotts, and M. J. Iremonger, 'ASSESSMENT OF SECONDARY FRAGMENTS FROM BURIED AP BLAST MINES', *Procedings Pers. Armour Syst. Symp. 2004.*

[8] NATO, 'AEP- 94: Skin penetration assessment of Non-lethal projectiles', Jul. 2021.

[9] L. Koene, M. L. Van Essen, A. Oukara, and A. Papy, 'ASSESSMENT OF SKIN PENETRATION OF KINETIC NON-LETHAL PROJECTILES USING THE SURROGATE METHOD', *Proc. Int. Symp. Ballist. 2014.*

 https://doi.org/10.52202/080042-0034

Raising the Standards for Protective Equipment used by Public Order Police Officers

D. Longhurst[1], J. Horlick[2], B. Montgomery[2], C. Robinson[2], R. Schauf[3], R. Kinsler[4]

[1] *RTI, 3040 E. Cornwallis Road, Research Triangle Park, NC 27709 (dlonghurst@rti.org)*
[2] *National Institute of Standards and Technology (NIST), 100 Bureau Drive, Gaithersburg MD 20899, USA*
[3] *U.S. Capitol Police, 119 D Street NE, Washington, DC 20510*
[4] *Sydor Technologies, 105 Shell Cove Ct. Joppatowne, MD 21085*

Abstract. Law enforcement officers in the United States are facing increased acts of violence and aggression while protecting the First Amendment rights of citizens and communities to assemble in peaceful protests and demonstrations. Public order police officers may not know in advance what threats they will encounter during an incident, but their equipment must protect against the most likely and injurious hazards. A multi-discipline collaborative effort, initiated by the National Institute of Justice, was begun in 2017 to guide the development of performance standards for protective equipment worn by these officers, and the first equipment to be addressed was helmets. Public order officers defined their operational and functional requirements and described the threats and hazards they face. Those requirements and hazards were then considered by technical experts, researchers, manufacturers, and officers, working together through the ASTM International E54 Committee on Homeland Security Applications. The group adapted or developed new standard test methods to address officer head protection needs, which include protection against multiple blunt impacts with hard surfaces or thrown/launched objects, hand-swung penetrating weapons, slingshot projectiles, flammable liquids, and more. Those test methods have been incorporated into a specification for officer head protection that gives performance requirements and acceptance criteria. The two published standards are:

- ASTM E3343/E3343M – 22, *Test Methods for Nonballistic-resistant Helmets Worn by Law Enforcement and Corrections*
- ASTM E3342/E3342M – 22, *Specification for Nonballistic-resistant Helmets Specifically Designed to be Worn by Law Enforcement and Corrections Officers When Maintaining Order in Violent Situations*

These standards are being used by a newly formed *ASTM Verification Program for Law Enforcement Equipment*, which will help to raise the bar for protective equipment used by officers. This paper will introduce the officer-specified needs; detail the decisions, research, and testing upon which the standards were established; and provide a description of the ASTM Verification Program.

1. INTRODUCTION

United States (U.S.) law enforcement officers are facing increased acts of violence and aggression while protecting the First Amendment rights of citizens and communities to assemble in peaceful protests and demonstrations. These officers may not know the threats they will encounter until they are on scene, and their equipment must protect against the most likely and injurious hazards. The protective equipment officers wear includes helmets with face shields, supplemental eye protection, hearing protection, torso and limb protectors, gloves, and protective footwear.

2. NATIONAL INSTITUTE OF JUSTICE SPECIAL TECHNICAL COMMITTEE

In 2017, the National Institute of Justice (NIJ) was requested by public order police officers and their agencies to assist with the development of standards for personal protective equipment. A review of available standards revealed a gap in performance standards and test methods addressing the specific U.S. law enforcement requirements. NIJ is facilitating the development of baseline performance requirements, standardized test methods, and conformity assessment requirements for equipment used by U.S. public order police. An NIJ Special Technical Committee (STC) of public order police officers, stakeholder organization representatives, and technical experts from across the U.S. was convened and has been working collaboratively for three years. Officers defined their operational and functional requirements and described the threats and hazards they face, and using that information, technical experts worked together to address their needs through standard test methods and specifications.

https://doi.org/10.52202/080042-0035

Standards are being developed through the ASTM International E54 Committee on Homeland Security Applications, with the participation of U.S. law enforcement public order practitioners, testing laboratories, product certifiers, researchers, as well as manufacturers and industry.

3. IDENTIFYING OFFICER NEEDS AND REQUIREMENTS

As the starting point for developing standards, officers were asked to identify their needs and requirements in terms of threats, hazards, and other issues of concern. They provided a list of 75 threats ranked from frequently experienced to rarely experienced. Those threats were then separated into nine categories to indicate the type of hazard to the officer: blunt impact, puncture, cut, thermal and flame, biological, chemical, respiratory, and distraction. Appendix A, Table 1 contains a brief listing of these hazard categories with examples of hazards. It is emphasized that the listing gives examples of hazards within each category, but many hazards fall within multiple categories. For instance, slingshot projectiles may be blunt impact, cut, and/or puncture hazards. The officers also noted whether the threats were typically wielded by hand, swung by hand, thrown, or launched.

According to NIJ STC public order police representatives, head and face protection were the number one priority, and the most immediate and injurious threat to an officer during an event is blunt impact to the head and face. The impact and resultant injury may be caused by a launched, thrown, or swung blunt object. The swung blunt object may also penetrate due to embedded spikes or other protruding secondary threats.

The second threat of concern to the officer's head and face is harmful fluids, which may be of any configuration, caustic, biohazard, or toxic or it may be on fire. Specific concerns were raised about fluids running off the helmet shell onto skin or into eyes, pooling of flammable fluids on the shell, and the helmet components burning or melting when exposed to flaming materials.

Another threat of concern is high-powered, hand-held lasers that are frequently used against law enforcement officers to distract, disorient, or injure them. These lasers are inexpensive, easy to obtain, easy to conceal and carry, and easy to use. Some can cause temporary or permanent blindness with only momentary exposure. It is recommended personnel be equipped with eye protection against laser light at wavelengths of concern based on risk assessment.

Another threat used against officers is use of devices making extremely high decibel noises, primarily to distract and disorient officers but that can also damage hearing in a short amount of time.

Officers expressed additional needs beyond the threats listed above. The retention system holding the helmet to the head must be secure and easily released but have parts that allow for it to be snapped off in the event of a forcible removal by a protestor. The helmet will be worn for long periods of time, and the interior padding should be easy to clean.

4. PUBLIC ORDER POLICE HELMET STANDARDS

Because of its importance to officer safety, head protection was selected as the first item of protective equipment to be addressed by the STC. After identifying the hazards to be protected against, an effort was initiated to develop test methods and performance requirements for public order helmets, including face shields, and address those hazards. Technical experts, researchers, manufacturers, and officers, working through ASTM International's E54 Committee on Homeland Security Applications, began with a review of relevant existing public order head protection standards:

- *Protective Helmets – Test Methods,* BS EN 13087, 2000. [1]
- *Riot Helmets and Faceshield* Protection, CSA Z611-02, (Reaffirmed 2012). [2]
- *PSDB Protective Headwear Standard for UK Police*, HOSDB 21-04, 2004. [3]
- *NIJ Standard for Riot Helmets and Face* Shields, NIJ 0104.02, October 1984. [4]

A related resource reviewed by the group was a report done at the request of NIST: *Research Leading to Revised NIJ 0104.02 Standard for Riot Helmets and Face Shields*, Biokinetics Report R08-18B (Re-released July 2019). [5]

The review also included standards addressing blunt impact and eye protection for sports helmets, such as those for hockey and horseback riding, and industrial head protection.

The testing and performance requirements from each standard were analyzed and compared, and decisions were made to adapt existing test methods, where possible, and to develop new test methods as needed to address officer operational requirements and concerns. These test methods are included in ASTM E3343/E3343M, *Test Methods for Nonballistic-resistant Helmets Worn by Law Enforcement and*

https://doi.org/10.52202/080042-0035

Corrections [6], which provides a collection of test methods that may be used. The performance requirements and additional testing requirements are included in ASTM E3342/E3342M, *Specification for Nonballistic-resistant Helmets Specifically Designed to be Worn by Law Enforcement and Corrections Officers When Maintaining Order in Violent Situations* [7]. Appendix A, Table 2 provides a list of ASTM E3342/E3342M [7] performance requirements and associated test methods.

This paper focuses on three standard test methods of ASTM E3343/E3343M that were modified to address specific concerns of U.S. public order police: (1) protection against multiple blunt impacts in a single location on the helmet shell, (2) face shield impacts by thrown objects, and (3) face shield impacts by slingshot projectiles.

5. HELMET SHELL IMPACT ATTENUATION TESTING

Most existing public order helmet standards require a blunt impact attenuation test that simulates an officer being shoved or otherwise impacted that results in falling and hitting the head on a solid object, such as pavement or a curb. The test typically requires a single impact in each of several locations on the outer shell. The impact, at a specified energy, is usually achieved by dropping the helmet, mounted on a headform, onto an anvil of specified shape. The performance criterion is typically specified in terms of maximum linear acceleration, which predicts the maximum force acting on the head, and the pass/fail limit is typically 300g [5].

Officers have expressed concerns that their helmets must protect against multiple impacts that could occur during one event or over years of use. Helmets are not typically replaced following blunt impacts unless there is visible damage to the shell. The problem is that most currently available helmets use crushable foam inner materials (such as expanded polystyrene), and protection-reducing damage can occur with a single impact but not be visible on the outer helmet shell. These factors led to an obvious need to modify existing test methods to assess multiple locations on the helmet shell, with more than one impact in each location.

A reduction in the pass/fail acceleration limit was recommended in a NIST-funded research report to be 250g because it advances the protection offered by the helmet and respects the technology of modern energy absorbing materials and construction methods [5].

As a demonstration of the modified test methods, technical experts proposed that, according to the helmet impact attenuation method of ASTM E3343 [6], three impacts at 120 J should be done in each of five locations on the shell, and the three impacts should be done sequentially on a flat anvil, a triangular anvil, and a corner anvil. To evaluate how different helmets might perform, testing was done on five commonly used riot helmets, one ballistic-resistant helmet, and a football helmet, with the non-riot helmets providing points of comparison.

Figure 1 shows a graph of peak acceleration (g) for each impact on each helmet, with the pass/fail acceleration value of 250g shown.

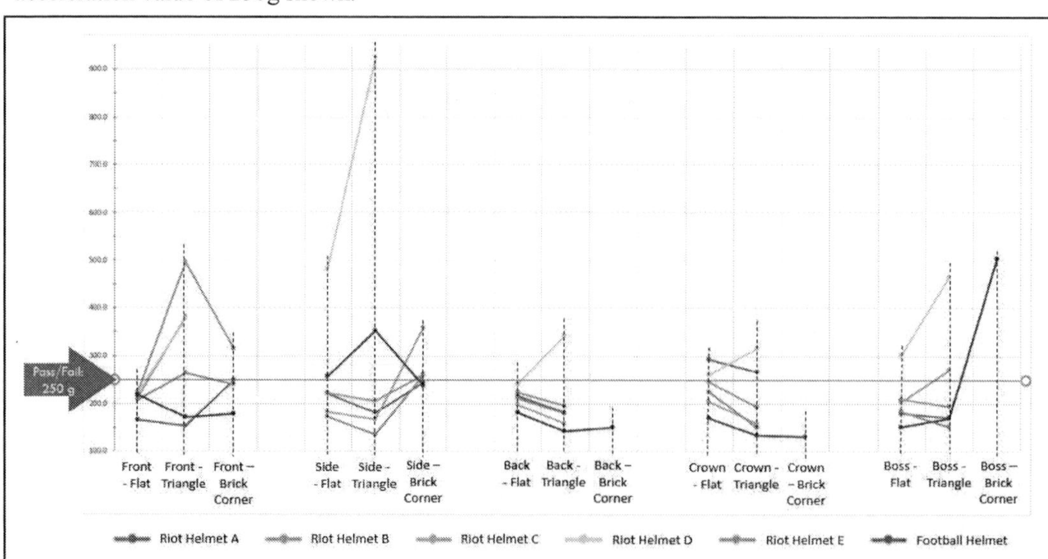

Figure 24. Graph of maximum recorded acceleration for each helmet impact by anvil

Observations based on these results shown in Figure 1 and examination of the impacted helmets are summarized below:

(1) Two impacts, once each with the flat and triangle anvils, in a single location do not appear to be too severe for the riot helmets tested, and a third impact may not be too severe, depending on the anvil chosen.

(2) The brick corner anvil penetrated many helmet shells as shown in Figure 1, brick corner impacts were omitted for some helmets due to concerns of damaging the test equipment. See Figure 2 for an example of shell puncture.

Because the test was too severe and not realistic, the task group made the decision to replace the brick corner anvil with the hemispherical anvil and conduct another round of testing that focused on anvils.

Figure 25. Brick corner anvil puncture

Subsequent testing was done on three helmets of a single model, rotating the order of the anvils and placing six impacts on each side of the sample (right and left). To obtain as many impacts as possible on three helmets, testing was done with side impacts only, based on the assumption that the sides are identical and would respond the same; other locations were not impacted because they are known to respond differently when impacted (as may be seen in Figure 1). See Appendix A, Table 3, for the impact locations and order of anvils.

Figure 3 provides two graphs of results showing the maximum recorded acceleration for each impact. Figure 3(a) shows impacts on the left side of three helmets using a single anvil. Six impacts were done; the first three are important for this test, and the second three were done for information. It can be seen from the first three impacts that the flat anvil impacts yield greater peak acceleration values. Figure 3(b) shows impacts on the right side of three helmets using an ordered sequence of three anvils. This figure indicates that impacts with the flat anvil yield greater acceleration values overall, regardless of the order of anvils. Evaluation of the test data led to the task group deciding to require three anvils (flat, triangle, and hemisphere) for each location and specify a different anvil order for each subsequent location on each helmet. The anvil order is listed in ASTM E3342/E3342M [7].

6. FACE SHIELD IMPACT AND DEFLECTION TESTING

Two existing public order helmet standards (NIJ 0104.02, PSDB 21/04) require procedures to test whether a known impact to face protection will cause deflection and contact to the wearer's face. Both procedures require that the helmet be mounted on a facially featured headform (positioned horizontally and nose facing up) with the face shield in the lowered position. Per these standards, assessing the face shield involves a single drop of a hemispherical impactor aligned with the headform nose, at a specified energy, onto the face shield. Contact between the headform nose and the face shield, via electrical connection, is determined during the impact.

Officers agreed with the above procedure for assessing face shield deflection; however, they also expressed their need to assess face shield integrity. The consensus of officers was that the face shield should be impacted in four locations, in the order listed: the nose, the upper edge center, the lower edge center, and at least one attachment point. After the final impact, the face shield integrity will be assessed, and the test result is considered a pass if each face shield tested does not contact the headform nose, has no visible cracks or splitting, is able to be raised and lowered, and remains fully attached. To support the inclusion of this test method and requirements, testing in accordance with ASTM E3343/E3343M [6] was done on three previously tested riot helmets. Deflection and impact testing revealed that some helmets showed no contact with the nose, while others did show contact; some helmets had visible structural damage, while others did not; and all helmet face shields remained functional and could be raised and lowered after impact. See Appendix A, Table 4, for the results of the test.

This testing supported the inclusion of the improved face shield deflection and impact test in the ASTM standard.

7. PROJECTILE TESTING OF FACE SHIELD

One of the concerns of public order police in the U.S. is protection against a projectile impact to the face shield, and the specific threat is a projectile fired from a wrist-supported slingshot. The *PSDB Protective Headwear Standard for UK Police* includes a procedure for assessing the face shield's ability to

https://doi.org/10.52202/080042-0035

withstand an impact from a low mass, high velocity projectile. While the general test was determined to be appropriate for U.S. purposes, the test projectile and its velocity (6 mm ball bearing at 200 m/s) were not appropriate because they are intended to address the threat UK police face from a projectile fired from an airsoft gun. Therefore, testing was required to determine a more suitable test threat and velocity.

NIST performed testing to determine a recommended projectile size, type, and velocity. Three commonly available wrist-supported slingshots were tested using five readily available slingshot ammunition types (See projectile details in Appendix A, Table 5).

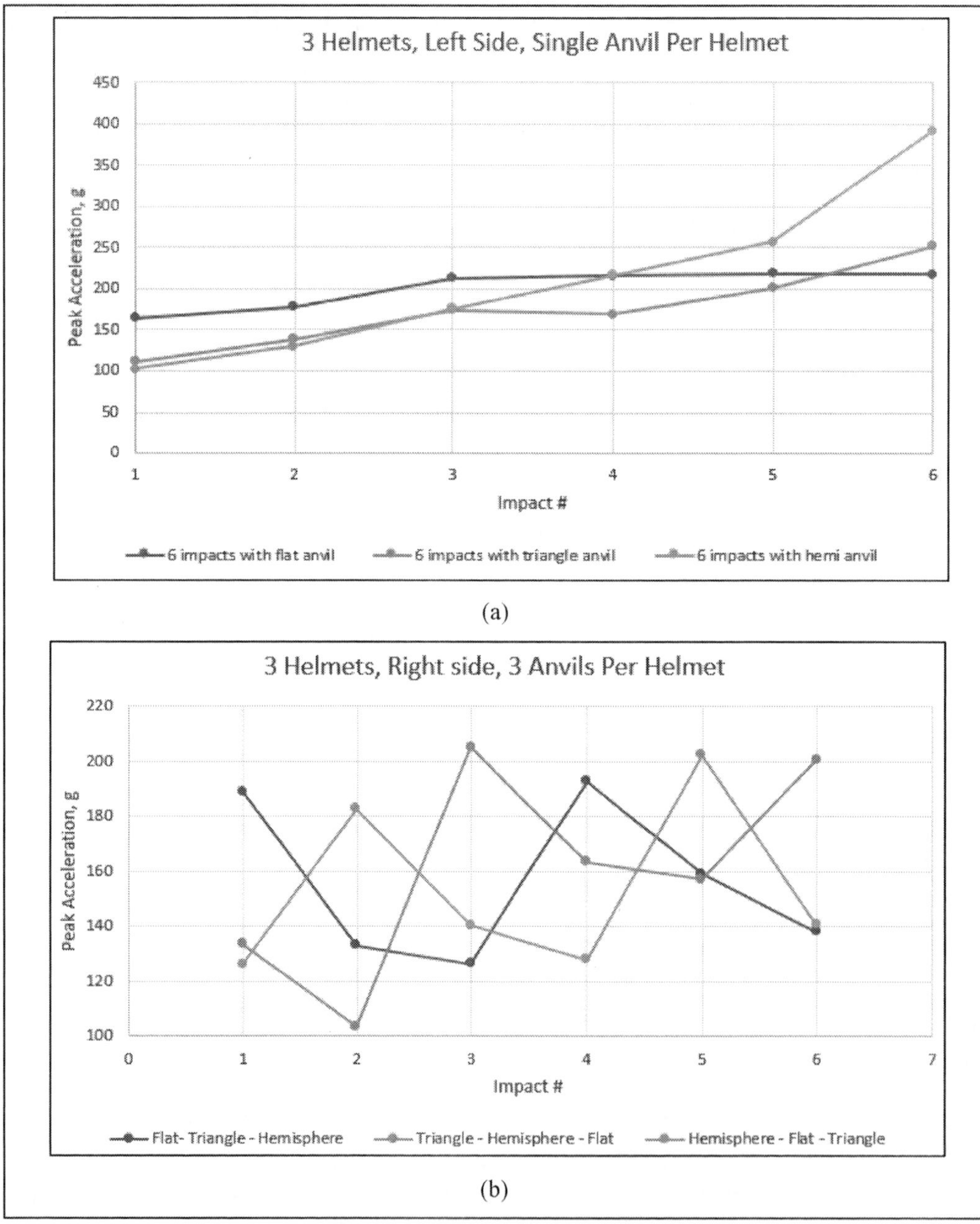

Figure 26. Test Results for Determination of Anvils and Order

For the test, the slingshot was mounted in a fixture, and the projectiles were shot through Oehler light screens to measure velocity (See Figure 4 for the test setup). Based on the abilities of several people to pull back and aim the slingshots, the draw length by the largest male, 81.3 cm, was selected as the draw length for the test. Five of each projectile were shot, and the average kinetic energy for each is shown in Figure 5. The ½-inch steel sphere achieved a maximum velocity of 47 m/s and delivered the greatest kinetic energy of all projectiles: 9.2 J. Based on this testing result, the ½-inch steel sphere was chosen as the projectile to require in the standard.

Figure 4. Two Views of Test Setup

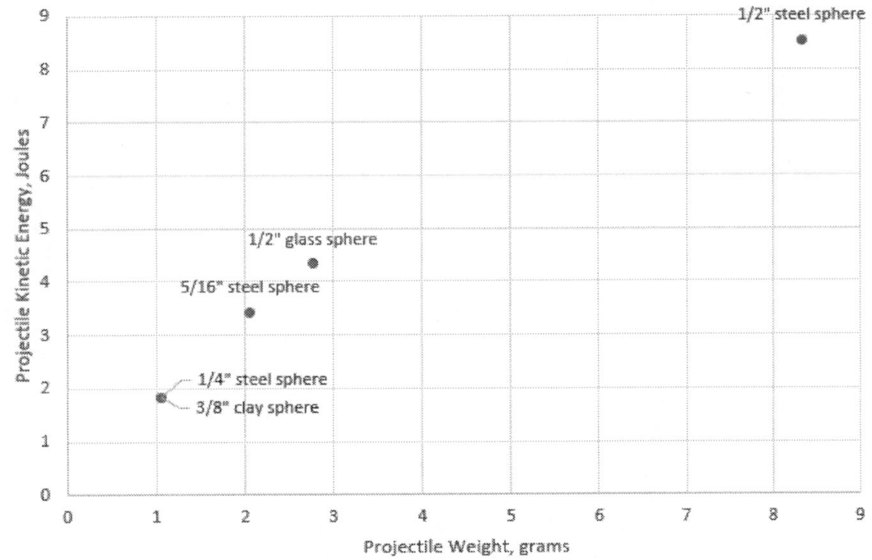

Figure 5. Projectile Mass vs. Kinetic Energy

The velocity was measured at different draw lengths as shown in Figure 6(a) and 6(b), and the relationship between draw length and velocity/k inetic energy can be seen. Both Figure 6(a) and 6(b) show extrapolated trendlines. This was done estimate the draw length of a larger adult male, size 40R, according to ASTM D6240 [8], having a length from wrist to opposite shoulder of 100.1 cm. Using that measurement (rounded up to102 cm (~ 40 inches)) to simulate one arm outstretched holding the slingshot, and the other hand at the opposite shoulder holding the projectile in the band (there is some hand length not taken into account) yields a corresponding velocity of 61 m/s (KE = 15.5 J). Adding in

https://doi.org/10.52202/080042-0035

a safety factor, the task group decided to set the required velocity for the ½-inch steel sphere at 65 m/s (KE = 19.2 J).

ASTM E3342/E3342M [7] requires testing with the "1/2-inch" steel sphere at 65 m/s. An impact is called a complete penetration if any part of the test projectile, or any part or fragment of the face shield, has damaged a witness panel such that the light from a light source can be seen through the witness panel. The test result shall be considered a pass if *(1)* each face shield shows no visible cracks or splitting and *(2)* the witness material has no complete penetrations.

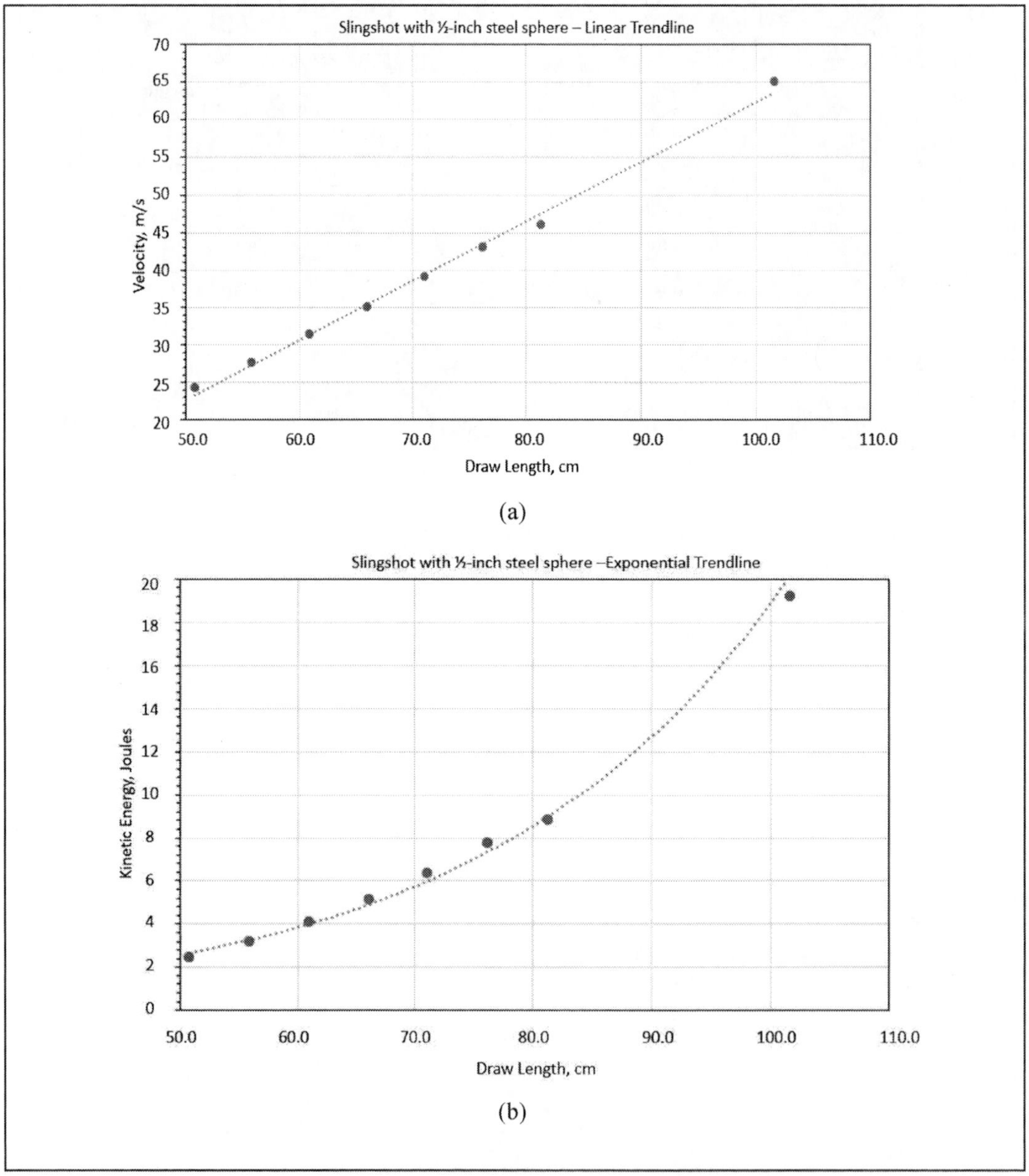

Figure 6. Graphs for Slingshot with ½-inch Steel Ball

8. ASTM VERIFICATION PROGRAM

The task group collaborating on these public order helmet standards recognized that published standards alone are not sufficient for improving the safety of law enforcement officers. There must be a method of conformity assessment to demonstrate that specified requirements are fulfilled, and it must provide both confidence in the helmet's performance but also be cost-effective for manufacturers and purchasers.

To meet this need, an ASTM Verification Program has been established to evaluate and verify that public order helmets meet the requirements of ASTM E3342/E3342M. The program is managed by the Safety Equipment Institute (SEI, an ASTM affiliate), which is an independent, third-party conformity assessment body, and requires testing by a laboratory accredited to ISO 17025 [9] with the relevant ASTM standards in their scope of accreditation. The laboratory will submit test reports to SEI for evaluation against the appropriate standard. Those products that are successfully verified will be included in an online listing of verified products (www.seinet.org), will receive authorization to have the ASTM-verified Mark placed on them (See Figure 7), and will undergo annual testing to assess continued compliance. Key benefits are listed below:

Figure 7. ASTM-verified Mark

- For purchasers, the program will greatly simplify the purchasing process by eliminating (or at least reducing) challenges caused by (1) unverified supplier claims of equipment performance; (2) incomplete, confusing, or misleading information about equipment performance; (3) and false advertising about equipment performance. A purchaser can require ASTM verification as a condition for purchasing a product and then check the online verified products list to see whether the product(s) being offered by a supplier has been verified.
- For manufacturers, the program will enable them to distinguish their ASTM-verified products from those that do not meet standards.
- For end users, the program will allow them to check their individual helmets to see whether the ASTM-verified Mark is present.

9. CONCLUSIONS

The work described in this paper began with identification of the needs and requirements of public order police officers in the U.S. The most pressing concern was protection of the head and face, which led to the effort to identify relevant existing test methods that could be applied as written or modified as needed. Fifteen existing test methods were determined to potentially be relevant, and three of those were modified to meet the officer-expressed needs, with the modifications supported by testing of commonly used helmets. ASTM E3342/E3342M specifies 11 performance requirements and test methods that the NIJ STC officers stated were their priorities (See Appendix A, Table 2 for a listing).

The ASTM Verification Program will help to ensure that U.S. public order police officers have access to helmets verified to meet ASTM E3342/E3342M and protect them against the threats they are likely to encounter during an incident or event.

10. ACKNOWLEDGMENTS

This work was funded by the National Institute of Justice (NIJ). Any opinions, findings and conclusions, or recommendations expressed in this material are those of the author(s) and do not necessarily reflect the views of NIJ. The authors acknowledge and appreciate the contributions of the NIJ Special Technical Committee.

REFERENCES

[1] *Protective helmets – Test methods*, British Standard 13087, 2000.
[2] *Riot Helmets and Faceshield* Protection, CSA Z611-02, (Reaffirmed 2012).
[3] *PSDB Protective Headwear Standard for UK Police*, Publication No. 21/04, 2004.
[4] *NIJ Standard for Riot Helmets and Face Shields,* NIJ 0104.02, October 1984.
[5] *Research Leading to Revised NIJ 0104.02 Standard for Riot Helmets and Face Shields*, Biokinetics Report R08-18B (Re-released July 2019).
[6] *Test Methods for Nonballistic-resistant Helmets Worn by Law Enforcement and Corrections*, ASTM E3343/E3343M-22.
[7] *Specification for Nonballistic-resistant Helmets Specifically Designed to be Worn by Law Enforcement and Corrections Officers When Maintaining Order in Violent Situations,* ASTM E3342/E3342M.

[8] *Standard Tables of Body Measurements for Mature Men, ages 35 and older, Sizes Thirty-Four to Fifty-Two (34 to 52) Short, Regular, and Tall*, ASTM D6240.
[9] International Standardization Organization/International Electrotechnical Commission (ISO/IEC) 17025, *General requirements for the competence of testing and calibration laboratories*.

Appendix A. Supporting Data Tables

Table 9. Categories and Examples of Hazards of Concern

Category of Hazard	Examples of Hazards
Blunt impact	Glass bottle, brick, rock, sign post, bike rack, rebar, mace, crowbar, bat, hammer, frozen soda can, slingshot projectile (e.g., marble, spark plug, ball bearing)
Puncture	Knife, sharpened dowel, bat with embedded spikes, improvised shiv, club
Cut	Glass, knife, box cutter, ax, machete, saber, bike chain, razor
Thermal/flaming	Flare, firework mortar, Molotov cocktail, flammable aerosols, fuel
Biological	Blood, urine, feces, saliva
Chemical & Respiratory	Pest repellant, hairspray, acetone, vinegar, bleach, ammonia, acids, lye, drain cleaner
Distraction	Visual: Paint, laser, strobe light, firework, mortar, Molotov cocktail Auditory: Yelling, loudspeaker, siren, air horn, whistle, trumpet, firework, mortar

Table 10. Performance Requirements and Associated Test Methods

ASTM E3342/E3342M Section	Associated Test Method	Purpose of Test: To assess
Section 7, Helmet Impact Attenuation	ASTM E3343/E3343M, Helmet Impact Attenuation Test	helmet's capability to attenuate an impact caused during a fall in which the head is hit on a flat, edged, or corner surface
Section 8, Helmet Shell Penetration Resistance	ASTM E3343/E3343M, Helmet Shell Penetration Resistance Test	helmet's ability to resist a thrown object, such as a brick
Section 9, Face Shield Deflection and Impact	ASTM E3343/E3343M, Face Shield Deflection and Impact Test Method	integrity of the face shield and its attachments and to assess whether a known impact to the face shield will cause deflection and contact to the wearer's face
Section 10, Face Shield Projectile Resistance	ASTM E3343/E3343M, Face Shield Projectile Resistance Test	face shield's ability to withstand an impact from low mass, moderate velocity projectiles, such as those launched from wrist-supported slingshots
Section 11, Flammable Liquid Trap	ASTM E3343/E3343M, Flammable Liquid Trap Test	whether there are liquid traps on the exterior of the helmet and whether the helmet is self-extinguishing within the defined period of time
Section 12, Liquid Penetration Resistance	ASTM E3343/E3343M, Liquid Penetration Resistance Test	helmet's ability to protect the wearer from contact with liquids
Section 13, Dynamic Retention System	ASTM E3343/E3343M, Dynamic Retention System Test	integrity of the retention system when subjected to a dynamic force as a drop weight delivers an impact load to the retention system
Section 14, Face Shield Optics	ANSI/ISEA Z87.1, Section 9.4, Refractive Power, Astigmatism and Resolving Power Tests, and, Section 9.5, Prismatic Power	whether the face shield distorts wearer's vision due to spherical and astigmatic aberration, and prismatic effects
Section 15, Accelerated Corrosion	ASTM E3343/E3343M, Accelerated Corrosion Test	ability of metallic components to resist corrosion

Section 16, Helmet Shell Spike Penetration	ASTM E3343/E3343M, Helmet Shell Spike Penetration Resistance Test	helmet's ability to resist a sharp weapon swung at the head, such as a board with an embedded nail

Table 11. Impact Testing Details

Impact Series	Helmet and Location	Order of Anvil Impact	
1	Helmet #1 – Left side	Impacts 1 through 6	All flat anvils
2	Helmet #1 – Right side	Impact 1, 4 Impact 2, 5 Impact 3, 6	Flat anvil Triangle anvil Hemisphere anvil
3	Helmet #2 – Left side	Impacts 1 - 6	All triangle anvils
4	Helmet #2 – Right side	Impact 1, 4 Impact 2, 5 Impact 3, 6	Triangle anvil Hemisphere anvil Flat anvil
5	Helmet #3 – Left side	Impacts 1 - 6	All hemisphere anvils
6	Helmet #3 – Right side	Impact 1, 4 Impact 2, 5 Impact 3, 6	Hemisphere anvil Flat anvil Triangle anvil

Table 12. Face Shield Deflection and Impact Testing Results

Impact No.	Impact Type	Sample	Riot Helmet B	Riot Helmet C	Riot Helmet E and Ballistic Helmet
1	Deflection	1	Contact to Nose	Contact to Nose	No Contact
2	Deflection	2	Contact to Nose	Contact to Nose	No Contact
3	Impact at nose	1	Visible Dent and crack	Visible Dent	No Visible Damage
7	Impact at nose	2	Visible Dent and crack	Visible Dent	No Visible Damage
4	Impact 2" below upper edge	1	Visible Dent	Visible Dent	No Visible Damage
8	Impact 2" below upper edge	2	Visible Dent	Visible Dent	No Visible Damage
5	Impact at chin	1	Visible Dent	Visible Dent and crack	No Visible Damage
9	Impact at chin	2	Visible Dent and crack	Visible Dent and crack	No Visible Damage
6	Impact within 1" of attachment point	1	No Visible Damage	No Visible Damage	No Visible Damage
10	Impact within 1" of attachment	2	Attachment pin broke; face shield fully functional	No Visible Damage	No Visible Damage

Table 13. Projectile Types, Weights, and Sizes

Projectile Type	Weight, grams	Diameter, mm
"3/8-inch" clay sphere	1.05	9.7
"¼-inch" steel sphere	1.06	6.4

https://doi.org/10.52202/080042-0035

"5/16-inch" steel sphere	2.06	7.9
"½-inch" glass sphere	2.77	13.2
"½-inch" steel sphere	8.33	12.7

NIJ Standards for Ballistic Resistance of Body Armor and Stab Resistance of Body Armor: New Developments

Mark E. Greene[1], Jeffrey Horlick[2], Daniel A. Longhurst[3], Lance L. Miller[3], Casandra Robinson[2] and Richard A. Sundstrom[3]

[1]*National Institute of Justice, 810 7th Street NW, Washington, DC, 20531, USA, mark.greene2@usdoj.gov*
[2]*National Institute of Standards and Technology, 100 Bureau Drive, Gaithersburg, MD 20899, USA*
[3]*RTI International, 3040 Cornwallis Road, Research Triangle Park, NC 27709, USA*

Abstract. The U.S. National Institute of Justice (NIJ) is revising their standards for body armor that protects law enforcement and corrections personnel against gunshots and stabbing weapons. The seventh revision to the NIJ standard for Ballistic Resistance of Body Armor, NIJ Standard 0101.07, includes many improvements that have been previously presented at the Personal Armour Systems Symposium. This paper focuses on new improvements to test methods for nonplanar armor designed for women and new test threats that take into account the changing landscape of policing and corrections in the U.S. Some improvements to the test methods for armor designed for women include new clay appliques to ensure better contact of panels with the clay backing material and new shot requirements to assess shaping features. Shot placement has also been reconfigured to explore potential vulnerabilities due to unique construction elements in the panel and nonzero angles of incidence in the proximity of edges. NIJ Standard 0101.07 requires testing with three rifle threats that were not previously required. One of the rifle threats new in NIJ Standard 0101.07 is a 7.62x39mm mild steel core (MSC) round, which is known to have variabilities in manufacturing and performance. A factory round was identified as an appropriate MSC test threat based on lot assessments and confirmed through a separate study conducted by an adjacent U.S. Government agency. The first revision to the NIJ standard for Stab Resistance of Body Armor, NIJ Standard 0115.01, has been extensively updated to include two protection categories, new test threats representing improvised weapons typically seen inside correctional facilities, updated test equipment, improved test procedures to better assess armor performance, and new methods for nonplanar armor designed for women. This paper will provide a overview of NIJ's body armor activities, describing the improvements to both standards, the changes to the NIJ Compliance Testing Program, and guidance for agencies, purchasers, and end users.

1. U.S. LAW ENFORCEMENT AND CORRECTIONS ARE DANGEROUS PROFESSIONS

Law enforcement and corrections are dangerous professions. The United States (U.S.) Bureau of Labor Statistics reports that in 2018 police and sheriffs' patrol officers in the U.S. experienced a fatality rate on the job of 13.7 in 100,000 officers—four times higher than the overall fatality rate on the job of 3.5 in 100,000 workers across all industries in the U.S. that year [1]. A majority of the accidental fatalities each year are traffic-related, while a majority of the felonious fatalities are due to assaults with firearms. Analysis of the Federal Bureau of Investigation's (FBI) Law Enforcement Officers Killed and Assaulted (LEOKA) statistics show that from 1987 through 2019, over 80,000 officers were assaulted with firearms. Furthermore, of the 1,923 officers feloniously killed in the line of duty by all means over that time span, 1,773 were feloniously killed by firearms. That equates to *over 92% of all felonious deaths* in the line of duty being due to firearms. Handguns alone accounted for at least 1,320 of those firearms fatalities [2].

Fewer law enforcement officers are killed by stabbing and cutting weapons than by firearms, but these assaults are also of concern to law enforcement. Analysis of LEOKA statistics show that from 1987 through 2019, over 38,000 officers were assaulted with knives with 20 officers feloniously killed [2]. Data to provide national estimates on assaults on correctional officers is harder to come by in the U.S., so much of what is known about the hazards in corrections is collected at the state and local levels in addition to anecdotal information. The Federal Bureau of Prisons in the U.S. publishes monthly serious assaults on its correctional staff [3], which totals several incidents per year, some of which involve stabbing or cutting weapons. For corrections officers, stabbing and cutting with inmate manufactured weapons are much greater concerns because those weapons are often intentionally contaminated with body fluids containing biohazards. While the assault may not result in immediate death, the long-term effects of infection can be deadly.

https://doi.org/10.52202/080042-0036

2. NIJ STANDARD 0101.07: BALLISTIC RESISTANCE OF BODY ARMOR

NIJ published its first performance standard for ballistic-resistant police body armor in 1972 [4]. The current revision is the seventh, NIJ Standard 0101.07, *Ballistic Resistance of Body Armor*, which includes improved test methods for female body armor and updated body armor protection levels that incorporate additional rifle threats faced by U.S. law enforcement, as discussed previously [5]. This revision takes into consideration the changing landscape of policing in the U.S. and implements lessons learned over the past decade of testing using NIJ Standard 0101.06, published in 2008 [6]. The improvements include more widespread stakeholder engagement, improved test methods and procedures, and updated test threats and protection levels, all of which result in better protection for officers wearing body armor.

NIJ Standard 0101.07 was developed with the guidance and input of a large group of end users and technical experts. Unlike previous revisions of the NIJ body armor standard which have been comprehensive, standalone documents, NIJ Standard 0101.07 references ten ASTM standards developed through the open and inclusive ASTM standards development process, with the participation and input from a wide range of stakeholders, including materials and equipment producers. These include standardized methods for laboratory measurements, ballistic testing, and data collection, among others [7].

In 2013, the U.S. Army, NIJ, and the National Institute of Standards and Technology (NIST) began a partnership to harmonize the standards and their implementation for ballistic-resistant vests. The federal agencies chose to work through ASTM's E54 Committee on Homeland Security Applications to develop standard test methods and practices for the purpose of improving and validating methods, increasing consistency among test laboratories, and ultimately increasing confidence in ballistic-resistant equipment. Incorporation of relevant ASTM standards into NIJ standards and U.S. Army requirements and testing documents affords the opportunity to harmonize laboratory test procedures and practices for both law enforcement and military ballistic-resistant armor and other ballistic-resistant equipment while allowing those end user communities ultimate control over product specifications, such as the specific threats against which their equipment must protect.

3. NEW NIJ STANDARD 0123.00 DEFINES BALLISTIC PROTECTION LEVELS AND TEST THREATS

NIJ's new test threats specification, NIJ Standard 0123.00, *Specification for NIJ Ballistic Protection Levels and Associated Test Threats*, is published as a companion to NIJ Standard 0101.07. NIJ Standard 0123.00 specifies the test threats—including projectiles and reference velocities—identified by U.S. law enforcement as representative of prevalent threats in the United States which will be used to test ballistic-resistant equipment for U.S. law enforcement applications. It is referenced by NIJ Standard 0101.07 for body armor and may be incorporated into future NIJ standards for ballistic-resistant helmets and ballistic-resistant shields. NIJ opted to develop a standalone specification of ballistic protection levels and associated test threats rather than specify the information directly in NIJ Standard 0101.07, as had been done in NIJ Standard 0101.06 and prior revisions. As a standalone specification, it may also enable testing of a variety of ballistic-resistant equipment, not just ballistic-resistant body armor, against contemporary U.S. law enforcement threats. However, NIJ Standard 0123.00 itself does not define any test methods.

The test threats in the inaugural version of NIJ Standard 0123.00 have been updated from section 2 of NIJ Standard 0101.06 to reflect the evolving threats faced by U.S. law enforcement end users, including a wider range and more severe ballistics threats, as shown in Table 1. The ballistic protection levels have been modified accordingly, with the protection level nomenclature also changed for better clarity and to reduce officer and end user confusion. Several new rifle test threats were added to the7.62x51mm M80, including 5.56mm M193, 5.56mm M855, and 7.62x39mm mild steel core (MSC), which is commonly seen by officers but is known to have huge variability in performance. That variability in the MSC round has driven the development of surrogate test round designs to ensure consistency in testing; however, NIJ Standard 0123.00 specifies a factory round until surrogate test round development activities are completed and validated, and the surrogate is commercially available.

NIJ Standard 0123.00 specifies a range of acceptable bullet dimensions, bullet mass, core dimensions, core mass, and core hardness for the factory 7.62x39mm MSC projectiles as well as audit procedures to assess ammunition lots to determine suitability for testing. NIJ previously engaged ammunition experts to investigate the possible solutions to define a 7.62x39mm MSC test round suitable to include in NIJ standards for the purposes of NIJ certification testing and Follow-up Inspection Testing (FIT) testing through its NIJ Compliance Testing Program (CTP) for ballistic-resistant body armor [8].

Physical characteristics of the ammunition as well as factors, such as availability of supply, were considered in the assessment of the various factory rounds available. The audit procedures were developed to conduct a multi-laboratory evaluation of the candidate ammunition, which performed consistently across the NIJ-approved laboratories. In addition, a factory round was identified as an appropriate MSC test threat based on lot assessments and confirmed through a separate study conducted by an adjacent U.S. Government agency.

Table 1. NIJ ballistic protection levels in NIJ Standard 0123.00.

NIJ Ballistic Protection Level	Test Threat	Ammunition Identifier	Reference Velocity
NIJ HG1	9mm Luger FMJ RN 124 grain	Remington #23558	1305 ft/s (398 m/s)
	.357 Magnum JSP 158 grain	Remington #22847	1430 ft/s (436 m/s)
NIJ HG2	9mm Luger FMJ RN 124 grain	Remington #23558	1470 ft/s (448 m/s)
	.44 Magnum JHP 240 grain	Speer #4453 or #4736	1430 ft/s (436 m/s)
NIJ RF1	7.62x51mm M80 ball NATO FMJ steel jacketed 147 +0/-3 grain	U.S. military supply or rounds meeting NATO specifications	2780 ft/s (847 m/s)
	7.62x39mm MSC ball ammunition Type 56 from Factory 31	Factory 31 Ammunition evaluated and meeting the requirements of NIJ 0123.00 Appendix A	2400 ft/s (732 m/s)
	5.56mm M193 56 +0/-2 grain	U.S. military supply or rounds meeting NATO specifications	3250 ft/s (990 m/s)
NIJ RF2	7.62x51mm M80 ball NATO FMJ steel jacketed 147 +0/-3 grain	US military supply or rounds meeting NATO specifications	2780 ft/s (847 m/s)
	7.62x39mm MSC ball ammunition Type 56 from Factory 31	Factory 31 Ammunition evaluated and meeting the requirements of NIJ 0123.00 Appendix A	2400 ft/s (732 m/s)
	5.56mm M193 56 +0/-2 grain	U.S. military supply or rounds meeting NATO specifications	3250 ft/s (990 m/s)
	5.56mm M855 61.8 ± 1.5 grain	U.S. military supply or rounds meeting NATO specifications	3115 ft/s (950 m/s)
NIJ RF3	30.06 M2 AP 165.7 +0/-7 grain	U.S. military supply or rounds meeting NATO specifications	2880 ft/s (878 m/s)

4. UPDATED LABORATORY PRACTICE FOR TESTING NONPLANAR BODY ARMOR DESIGNED FOR WOMEN OFFICERS

NIJ Standard 0101.07 includes many improvements that have been previously presented at the Personal Armour Systems Symposium [5]. Some improvements to the test methods for armor designed for women include new clay appliques to ensure better contact of panels with the clay backing material and new shot requirements to assess shaping features. Women comprised about 14% of full-time sworn officers employed by local police departments in the U.S. in 2020 [9]. Shot placement has also been reconfigured to exploit potential vulnerabilities due to unique construction elements in the panel and nonzero angles of incidence in the proximity of edges.

https://doi.org/10.52202/080042-0036

NIJ Standard 0101.07 initially referenced ASTM E3086, *Standard Practice for Creating Appliques for Use in Testing of Nonplanar Soft Body Armor Designed for Females*. This ASTM standard specified a procedure for creating appliques (e.g., build-up of clay) for use behind nonplanar soft armor panels and affixing the appliques to the clay block. The purpose was to specify critical parameters for creating appliques in order to improve consistency of the test setup between laboratories. The practice described a single applique shape applicable only to nonplanar, soft body armor designed for women. Implementation of this practice proved more challenging than expected, including difficulty creating the specific applique shapes described in ASTM E3086 and ensuring proper contact with the armor panel once mounted on the clay block, which required reconsideration of how to build up clay behind nonplanar soft armor panels.

A more simplified applique was developed to ensure that the panels are fully filled in with clay before mounting on the clay block. This applique is more monolithic in form with the general contours of a female torso in cross-section. It is created using one of two standardized mold sizes along with a procedure to shape its form once affixed to the clay block. The result is a better substrate to ballistically test nonplanar armor.

Figure. 1. Different views of the mold used to form the clay appliques for testing nonplanar soft armor in NIJ Standard 0101.07.

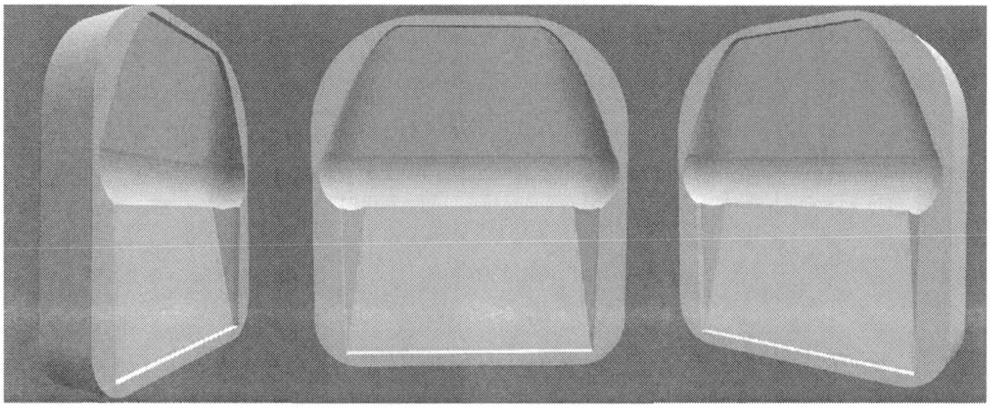

5. UPDATED P-BFD METHODS: ANGLED SHOTS ON SOFT ARMOR NEAR EDGES AND SHOTS ON THE CROWN OF MULTI-CURVED HARD PLATES

NIJ has updated perforation-backface deformation (P-BFD) testing to include an additional shot on soft armor panels. How soft armor responds to handgun projectiles striking very near the top edge of a front armor panel has been explored by an adjacent U.S. Government agency through experimental testing efforts. This involved mounting a ballistic vest with soft armor panels in an external carrier onto a model female torso made of molded ballistic gelatin. In this configuration, the top of the panel is naturally slanted back toward the torso in the carrier, creating an angle of obliquity between the armor panel the trajectory of the incoming bullet. Shots striking the top center edge at angles of obliquity in excess of approximately 40° have been demonstrated in some exploratory tests to not fully engage all layers of the armor panel and deflect off a middle layer into the neck region of the gelatin torso.

For planar soft armor, NIJ has added a shot located at the top center at the minimum shot-to-edge distances (2 in. or 3 in.) for the specific NIJ HG1 and NIJ HG2 threats at a 45° angle of incidence, which is achieved by rotating the clay block. For nonplanar soft armor, the built-up clay of the applique introduces an approximately 15° angle of obliquity by slanting the top of the armor panel back toward the clay block. The clay block is rotated an additional 30° angle of incidence to yield an overall 45° angle between the shot and armor surface. This new shot will provide minimum performance for soft armor for handgun projectiles striking that location.

NIJ has also reconfigured P-BFD testing on hard armor plates to include striking the crown on curved plates. The crown is defined as the location of the highest point of the strike face of the plate when the plate is lying horizontally on a flat surface, at the intersection of multiple different curvatures. The placement of a shot on the crown probes the performance of hard armor in a location that may be more vulnerable to penetration due to characteristics of the materials or construction methods used to manufacture plates. This shot location is consistent with testing conducted by the U.S. Army on hard armor to meet its specifications, bringing the NIJ standard into better alignment with DoD testing.

6. NIJ STANDARD 0115.01: STAB RESISTANCE OF BODY ARMOR

The first revision to NIJ Standard 0115.00, *Stab Resistance of Body Armor*, has been extensively updated to address law enforcement and corrections officer needs and requirements and to improve the standard based on lessons learned in the years since the previous publication [10]. NIJ published a draft of the new standard in 2020 through the *Federal Register* to request comments and input from the public, a practice commonly used by U.S. Government agencies to seek input on important guidance, policy, or regulations they plan to publish [11].

Stabbing is a major concern for officers working inside controlled-access facilities, such as jails, detention centers, and prisons. Inmates are well known to make improvised stabbing weapons from materials found in their environment and sharpened on concrete or other rough surfaces. Due to metal detectors and other detection methods at entrances and other key points, it is difficult to introduce weapons into a controlled-access facility, so firearms and commercial knives are not considered to be typical threats. Therefore, the revised standard NIJ Standard 0115.01, a *Stab Resistance of Body Armor*, includes two NIJ stab protection categories that are more descriptive of stab threats and the environments in which they are likely to be encountered: NIJ-STAB-Commercial and NIJ-STAB-Improvised.

The NIJ-STAB-Commercial threats address commercially made knives and spikes, typically encountered outside of controlled-access facilities or within the jail intake area. Within the commercial protection category, there are three test threats as shown in Table 2. These are the same threats specified in NIJ Standard 0115.00, but the test threat designator has been updated for clarity. The impact energy level for this test threat was derived from prior research [12]. The primary energy of 24 J corresponds to the 85th percentile of the population that was studied, and the 36 J energy corresponds to 1.5 times the primary energy value and is intended to ensure that the armor material performs in a linear fashion and does not suffer catastrophic failure at, or near, the primary energy level.

The NIJ-STAB-Improvised threats address improvised or inmate-made weapons, typically encountered inside controlled-access facilities, such as jails, detention centers, and prisons. Because improvised weapons are not as sharp or durable as commercial weapons, having an improvised weapon category will likely result in lighter-weight, more comfortable armor for corrections officers. Developing the parameters for improvised weapons required research to understand and analyze the types of improvised weapons found inside controlled-access facilities. NIJ funded an effort to characterize common improvised weapons and develop exemplars for testing [13, 14]. The research and development effort resulted in three test threats within the improvised protection category as shown in Table 2. The impact energy level for the improvised test threat was derived from the same study as for commercial weapons [11]. The primary impact energy of 43 J corresponds to the 96th percentile of the population that was studied, and the 65 J energy corresponds to 1.5 times the primary energy value.

NIJ requires that body armor within each stab protection category be tested with all three test threats. The performance requirements for resistance to penetration by commercial threats have been kept the same as "protection level 1" in NIJ 0115.000, namely: (1) 7 mm (0.28 in) at E1, for fair hits at angles of incidence of 0° and 45° and (2) 20 mm (0.79 in) at E2, for fair hits at angles of incidence of 0°. The penetration limit at E1 was determined through research indicating that internal injuries to organs would be extremely unlikely at 7 mm (0.28 in). The performance requirements for resistance to penetration by improvised threats are the same as for commercial threats, except that no penetration is allowed at the primary impact energy. This is intended to protect officers from exposure to infectious diseases when shanks and shivs are intentionally contaminated with body fluids or feces.

NIJ has added new testing requirements and procedures for effectively assessing shaped armor designed for women. Manufacturers are required to submit all test samples in NIJ template size NIJ-C-4, and female armor front panels must be submitted with cup sizes of B and E. The revised standard requires a build-up of backing material behind the front panel and impacts in specific locations to test potential weak point caused by stitching or other design features.

NIJ has improved all test procedures to better assess armor performance and to reduce interlaboratory variability in testing. Major improvements include an increased sample quantities from 3 to 16 samples and 16 more for shaped armor designed for females. Requirements were added for conditioning by submersion in water prior to impact testing and for sample panels to be rotated at several angles (0°, 45°, 90°, >30°). Multiple impact locations per sample panel were also added as were increased specificity for depth of penetration measurements and use of cut length for both commercial and improvised blades. To further reduce interlaboratory variability, NIJ increased the specificity in

https://doi.org/10.52202/080042-0036

requirements for the stab testing apparatus and had a new sabot designed and fabricated for the testing laboratories currently participating in the NIJ CTP.

Table 2. Stab protection categories, stab threats, and associated impact kinetic energies proposed for NIJ Standard 0115.01.

NIJ Stab Protection Category	Test Threat	Test Threat Designator	Impact Kinetic Energy, E1	Overtest Impact Kinetic Energy, E2
NIJ-STAB-Commercial	Commercial single-edged (SE) blade	Com-SE	24 ± 0.50 J (17.7 ± 0.36 ft-lbf)	36 ± 0.60 J (26.6 ± 0.44 ft-lbf)
	Commercial double-edged (DE) blade	Com-DE		
	Commercial spike (SP)	Com-SP		
NIJ-STAB-Improvised	Improvised single-edged (SE) blade	Imp-SE	43 ± 0.60 J (31.7 ± 0.44 ft-lbf)	65 ± 0.80 J (47.9 ± 0.59 ft-lbf)
	Improvised double-edged (DE) blade	Imp-DE		
	Improvised spike (SP)	Imp-SP		

7. NIJ COMPLIANCE TESTING PROGRAM

The NIJ CTP is a body armor certification program to provide U.S. law enforcement and correctional agencies and personnel confidence that the body armor they purchase and use performs according to minimum performance requirements to protect against common handgun and rifle threats as well as stab threats. Use of NIJ-certified body armor is ubiquitous among U.S. law enforcement and corrections agencies, and many agencies outside the U.S. make use of NIJ standards and NIJ-certified armor. Since the NIJ CTP began accepting armor submissions to meet the requirements of NIJ Standard 0101.06 in 2009, over 1,800 unique models of ballistic-resistant body armor have been submitted to the CTP for compliance testing through the end of 2022, with an overall failure rate of approximately 38% of models submitted for initial testing. Currently there are over 400 models of ballistic-resistant body armor listed on NIJ's Compliant Products List (CPL) [15,16].

Industry participation in the program by manufacturers is voluntary; however, it is recognized by manufacturers as the standard in body armor quality assurance. At the end of 2022, the NIJ CTP had 82 participants worldwide, with 127 manufacturing locations representing 73 locations in the continental U.S. and 54 manufacturing locations outside the U.S. The NIJ CTP also recognizes Body Armor Quality Management System Requirements called BA9000, which are optional quality assurance requirements to which body armor manufacturers can choose to have their manufacturing locations certified. There are 16 manufacturing locations certified to BA 9000 operated by 15 of the NIJ CTP participants, representing 228 models, or 55%, of the ballistic-resistant body armor currently listed on the CPL.

While NIJ will begin to certify body armor to the new NIJ Standard 0101.07 and will publish a new CPL for those armor models, it will also continue to accept armor for certification to NIJ Standard 0101.06 during a transitional period. NIJ will also continue to maintain its CPL for armor models compliant with NIJ Standard 0101.06 for a period of time and will continue to require FIT on these models. This will allow law enforcement agencies the time needed to transition their equipment as smoothly as possible over a reasonable amount of time. NIJ will also begin to certify body armor to the new NIJ Standard 0115.01 and will publish a new CPL for those armor models, it will also continue to accept armor for certification to NIJ Standard 0115.00 during a transitional period.

8. CONCLUSION

NIJ Standard 0101.07, *Ballistic Resistance of Body Armor*, includes improved test methods for female body armor and updated body armor protection levels that incorporate additional rifle threats faced by

U.S. law enforcement. NIJ Standard 0101.07 references ten ASTM standards, including standardized methods for laboratory measurements, ballistic testing, and data collection. NIJ's new test threats specification, NIJ Standard 0123.00, *Specification for NIJ Ballistic Protection Levels and Associated Test Threats*, specifies the test threats, including projectiles and reference velocities, identified by U.S. law enforcement as representative of prevalent threats in the U.S., including a 7.62x39mm MSC factory round. Improvements to the test methods for armor designed for women include new clay appliques to ensure better contact of panels with the clay backing material and new shot requirements to assess shaping features. NIJ has added a P-BFD shot located at the top center of soft armor panels at the minimum shot-to-edge distances to provide minimum performance for handgun projectiles striking that location and has reconfigured P-BFD testing on hard armor plates to include striking the crown on multi-curved plates.

The revised NIJ Standard 0115.01, *Stab Resistance of Body Armor*, includes two NIJ stab protection categories that are more descriptive of stab threats and the environments in which they are likely to be encountered: NIJ-STAB-Commercial and NIJ-STAB-Improvised. NIJ has added new testing requirements and procedures for effectively assessing shaped armor designed for women and has improved all test procedures to better assess armor performance and to reduce interlaboratory variability in testing. While NIJ will begin to certify body armor to the new NIJ Standard 0101.07 and NIJ Standard 0115.01 and will publish new CPLs for those armor models, it will also continue to accept armor for certification to NIJ Standard 0101.06 and NIJ Standard 0115.00 during a transitional period.

Acknowledgments

The authors would like to acknowledge the ASTM staff and the participants in the ASTM E54 working groups who contributed to the development of the ASTM standards referenced in this paper. The authors would like to acknowledge the participation of the NIJ Special Technical Committee on Body Armor for contributing to the development of the three new NIJ documents discussed in this paper. Individuals who provided public comments for the draft of the new standard as well as the new threat specification are acknowledged for their contributions to the revision process. Finally, the laboratory personnel and subject matter experts who have been consulted on various aspects of ballistic testing and evaluation of body armor and ammunition during the revision process are acknowledged for their insights and perspectives.

References

[1] "Injuries, Illnesses, and Fatalities: Fact Sheet – Police Officers 2018," Bureau of Labor Statistics, November 9, 2021, https://www.bls.gov/iif/factsheets/fatal-occupational-injuries-police-2018.htm.

[2] *Law Enforcement Officers Killed and Assaulted*, Federal Bureau of Investigation, https://ucr.fbi.gov/leoka.

[3] "Prison Safety," Federal Bureau of Prisons website, accessed March 10, 2023, https://www.bop.gov/about/statistics/statistics_prison_safety.jsp.

[4] NILECJ Standard 0101.00, *NILECJ Standard on the Ballistic Resistance of Police Body Armor*, National Institute of Law Enforcement and Criminal Justice, March 1972, https://www.ncjrs.gov/pdffiles1/Digitization/7037NCJRS.pdf.

[5] M.E. Greene, J. Horlick, D.A. Longhurst, L.L. Miller, C. Robinson, and R.A. Sundstrom, "NIJ Standard 0101.07, Ballistic Resistance of Body Armor: Changes and Challenges," *15th International Personal Armour Systems Symposium*, Copenhagen, Denmark, October 11-15, 2021.

[6] NIJ Standard 0101.06, *Ballistic Resistance of Body Armor*, National Institute of Justice, July 2008, https://www.ncjrs.gov/pdffiles1/nij/223054.pdf.

[7] "Subcommittee E54.04 on Public Safety Equipment: Matching Standards Under the Jurisdiction of E54.04 by Status," ASTM website, accessed March 7, 2023, https://www.astm.org/get-involved/technical-committees/committee-e54/subcommittee-e54/jurisdiction-e5404.

[8] Contra Threat Sciences, LLP, "7.62x39mm Mild Steel Core Ammunition," final report to the Justice Technology Information Center, April 29, 2019.

[9] S.E. Goodison, "Local Police Departments Personnel, 2020," Bureau of Justice Statistics, November 2022, https://bjs.ojp.gov/sites/g/files/xyckuh236/files/media/document/lpdp20.pdf.

[10] NIJ Standard 0115.00, *Stab Resistance of Personal Body Armor*, National Institute of Justice, September 2000, https://www.ncjrs.gov/pdffiles1/nij/183652.pdf.

 https://doi.org/10.52202/080042-0036

[11] "Request for Public Comment on Proposed Revision of NIJ Standard 0115.06, Ballistic Resistance of Body Armor," *Federal Register*, 85 FR 17600, March 30, 2020, https://www.federalregister.gov/d/2020-06516.
[12] I. Horsfall, P.D. Prosser, C.H. Watson, and S.M. Champion, "An Assessment of Human Performance in Stabbing," *Forensic Science International*, 102(2-3), 1999, https://doi.org/10.1016/S0379-0738(99)00055-9.
[13] J. Cavanaugh, C. Bir, and N. Rowley, "Characterization of Weapons used in Stab/Slash Attacks", report for NIJ Award 2011-DE-BX-K003, December 2015, https://www.ojp.gov/pdffiles1/nij/grants/249550.pdf.
[14] N. Shewchenko and D. Baines, "Characterization of Weapons Used in Stab Attacks – Phase II: Surrogate Development Final Report," report for NIJ Award 2011-DE-BX-K003, December 2015, https://www.ojp.gov/pdffiles1/nij/grants/249551.pdf.
[15] "Compliant Products List: Ballistic Resistant Body Armor," National Institute of Justice website, accessed March 8, 2023, https://nij.ojp.gov/topics/equipment-and-technology/body-armor/ballistic-resistant-armor.
[16] "Compliant Products Lists," Criminal Justice Testing and Evaluation Consortium website, accessed March 8, 2023, https://cjtec.org/compliance-testing-program/compliant-product-lists/.

An Alternative Method for Determining Penetration Limit Velocities Using Residual Velocity Data

K. Hohnecker[1] and C. Drake[2]

[1]DEVCOM Armaments Center, Picatinny Arsenal, NJ, USA
kenneth.h.hohnecker.civ@army.mil
[2]DEVCOM Armaments Center, Picatinny Arsenal, NJ, USA
christopher.g.drake4.civ@army.mil

Abstract. Limit velocities are the impact velocities at which a penetrator has a certain probability of perforating a given target. These limit velocities are often used as performance metrics to evaluate the effectiveness of targets (e.g., personal armour) at stopping a given penetrator. Limit velocities typically need to be determined experimentally, especially for new designs or concepts for which there is little or no pre-existing data. When evaluating protection against small arms, these limit velocity tests often employ an adaptive binary data gathering algorithm. One issue encountered when modelling binary data (perforation/no-perforation in this case) is that one needs a relatively large sample size to develop a model with reasonable confidence bounds (precision) due to the information-sparse nature of binary data. In recent years, the ability to capture the residual velocity of these penetrators after impacting the target has become more prevalent through the use of high-speed cameras or other modern instrumentation. The new methodology outlined by the authors in this paper demonstrates that the inclusion of this additional continuous data significantly improves both precision and efficiency with regard to the modelling of limit velocities. This paper will discuss the development of the equation for residual velocities that was sufficiently generic to apply to a wide range of penetrators and targets, while also remaining amenable to a tractable and computationally efficient statistical analysis. The authors go on to demonstrate the improvement to efficiency and precision using various Monte Carlo and re-sampling comparisons to traditional binary testing and modelling methods.

1. BACKGROUND

Ballistic penetration testing is an integral part of research and development, as well as demonstration testing, for various military commodities including personnel protection, ammunition, safety equipment, and weaponry. The binary nature of perforation testing can pose some problems with regard to the inherent inefficiency and variation of the data. There are also risks when trying to find the area of overlap from perforation to no-perforation (Zone of Mixed Results [ZMR]): if this region is missed in the test data, modelling the data has traditionally not been feasible. In most scenarios where continuous responses can be measured (e.g., velocity), useful estimates of relevant parameters including mean and standard deviation can be derived from relatively small sample sizes. With binary data, proportions are the relevant parameters, and more samples are generally required to estimate a proportion with similar precision. When trying to model a continuous response, like velocity as a function of another factor (i.e., propellant charge), interpolative Least Squares Regression models can be easily generated with a small sampling of points along the predictor input space. For binary data, more samples are required to produce a similarly precise model, with Binary Logistic Regression with a Logit link function being the preferred modelling method [1].

Adaptive test algorithms (e.g., Langlie [2], 3-Phase Optimal Design [3POD] [3]) have helped to improve efficiency with respect to lowering sample sizes, but sample size demands still surpass that of continuous responses. In cases where no residual continuous response exists, 3POD has been shown to be the preferred binary data collection method [4] and has seen increased use in United States Department of Defense applications since its development. However, when relevant continuous response metrics can be measured concurrently with a binary result, the authors posit that improvements in efficiency and precision can be obtained compared to Binary Logistic Regression and 3POD using the modelling technique described in this paper, and a new testing algorithm currently in development by the authors.

2. METHODOLOGY

2.1 Derivation of Residual Velocity as a Function of Striking Velocity

 https://doi.org/10.52202/080042-0037

The derivation begins by assuming that, for given target and penetrator materials and geometry, the resistive pressure (P) acting on the penetrator depends on the penetrator's instantaneous velocity inside the target (v), a strength parameter (S, units of force per area), and a characteristic velocity (v_c). Using standard non-dimensionalisation techniques, one can write that

$$\frac{P}{S} = g\left(\frac{v}{v_c}\right), \tag{1}$$

where $g(\cdot)$ is some non-negative, dimensionless function. Since P is a force per unit area, one can substitute Equation 1 into Newton's Second Law to obtain

$$\frac{dv}{dt} = -\frac{AS}{m} g\left(\frac{v}{v_c}\right), \tag{2}$$

where m is the mass of the penetrator, t is time, and A is the effective area of the penetrator. Equation 2 holds as long as A, S, and m remain constant throughout the impact event. This is not true in general, and one expects that the quantity $\frac{m}{AS}$ will depend on the initial conditions of the event (i.e., the striking velocity (v_s)) and how far into the event one is (i.e., t). The authors assume that the velocity of the penetrator is strictly decreasing in time so that there is a bijective mapping between t and v. Therefore,[6]

$$\frac{m}{AS} = q(v, v_s) \tag{3}$$

for some non-negative function $q(\cdot)$.

Since the authors are interested in the demarcation between perforation and non-perforation events, one must be able to model the dynamics near $v = 0$. The authors assume that $g(\cdot)$ is a smooth function, that $g(0) = 0$ (i.e., no retardation force when $v = 0$), and that it is strictly increasing (i.e., resistive pressure strictly increases with increasing velocity). A permissible candidate function for $g(\cdot)$ is therefore[7]

$$g\left(\frac{v}{v_c}\right) = \beta\left(\frac{v}{v_c}\right)^{\alpha}, \quad \alpha, \beta > 0. \tag{4}$$

Substituting Equations 3 and 4 into Equation 2 yields

$$\frac{dv}{dt} = -\frac{\beta\left(\frac{v}{v_c}\right)^{\alpha}}{q(v, v_s)}. \tag{5}$$

Separating variables, and integrating both sides from the beginning of the impact to when the penetrator comes to rest (assuming the target is thick enough) yields

$$t_f = \int_0^{v_s} \frac{q(v, v_s)\, dv}{\beta\left(\frac{v}{v_c}\right)^{\alpha}}, \tag{6}$$

where t_f is the total time elapsed. Since the factors in this integrand do not change signs, one can use the mean value theorem to obtain

$$t_f = \frac{v_c^{\alpha}\tilde{q}(v_s)}{\beta}\int_0^{v_s} v^{-\alpha} dv = \frac{v_c^{\alpha}\tilde{q}(v_s)}{(1-\alpha)\beta} v_s^{1-\alpha}, \; 0 < \alpha < 1 \tag{7}$$

[6] The authors chose to represent $\frac{m}{AS}$, and not $\frac{AS}{m}$, as a function since, for high striking velocities, significant erosion of the penetrator may occur such that m may approach 0.

[7] One reason for this choice is that $g(\cdot)$ cannot be represented by a power series near $v = 0$, as discussed later.

for some non-negative function $\tilde{q}(\cdot)$. Note that α must be constrained to less than unity to ensure that the penetrator stops in a finite amount of time.[8] With this additional constraint on α, take Equation 5, divide both sides by v, note that $\frac{dv}{dt}/v = \frac{dv}{dx}$, separate variables, and integrate across the thickness of the target that is perforated (T) to obtain

$$T = \frac{v_c^\alpha}{\beta}\int_{v_r}^{v_s} q\left(v, v_s\right) v^{1-\alpha} dv,\ v_r \geq 0,\ 0 < \alpha < 1 \quad (8)$$

where v_r is the residual velocity (i.e., exit velocity) of the penetrator.

The authors wished to be able to implement an estimation method to determine the parameters in Equation 8 that give the "best fit" to test data. These estimates could then be plugged back into Equation 8 to approximate limit velocities. Fitting the data is relatively straight forward if the penetrator perforates the target, since v_s, v_r, and T are all known or can be readily measured. Also, knowledge of the dependency of $q(\cdot)$ on v is no longer required, as the former can be pulled out of the integral (using the mean value theorem) and replaced with a function dependent only on v_s.

However, if perforation does not occur, then T will not be known. One must then measure T (which may be difficult, especially if the penetrator gets lodged in the target or if the target is littered with debris and fractures), and still one is left with determining the dependency of $q(\cdot)$ on v. Alternatively, one can throw out any test data where perforation does not occur (which is inefficient).

Therefore, this paper proposes the following method that ensures that all data is used in parameter estimation, without requiring the measuring of penetration depths. For non-perforating data, rather than setting $v_r = 0$ and measuring T, one leaves T set to the target thickness and extends Equation 8 to allow for *negative* values of v_r. The authors now abandon the definition of v_r as the physical residual velocity and instead think of it as a more abstract measure of the "unconsumed velocity" after penetrating a certain thickness of target. This definition still makes physical sense when the penetrator completely perforates a target. However, if complete perforation does not occur, and the penetrator only perforated a thickness T_1, then the amount of unconsumed velocity that is *lacking* (i.e., $-v_r$) to penetrate a second, adjacent target with thickness T_2 can be calculated from Equation 8 as[9]

$$T_2 = \frac{v_c^\alpha}{\beta}\int_{0}^{-v_r} q\left(-v, v_s\right) v^{1-\alpha} dv. \quad (9)$$

Therefore, if the thickness of the target that the penetrator actually perforates (T_1) is less than the total thickness (T), then one sets $T_2 = T - T_1$ and obtains an extension of Equation 8 to negative values of v_r, given by

$$T = \begin{cases} \frac{v_c^\alpha}{\beta}\int_{v_r}^{v_s} q\left(v, v_s\right) v^{1-\alpha} dv, & \frac{v_c^\alpha}{\beta}\int_{0}^{v_s} q\left(v, v_s\right) v^{1-\alpha} dv \geq T \\ \frac{v_c^\alpha}{\beta}\int_{0}^{v_s} q\left(v, v_s\right) v^{1-\alpha} dv + \frac{v_c^\alpha}{\beta}\int_{0}^{-v_r} q\left(-v, v_s\right) v^{1-\alpha} dv, & \text{otherwis} \end{cases}$$
$$= \frac{v_c^\alpha}{\beta}\int_{v_r}^{v_s} q\left(v, v_s\right) |v|^{1-\alpha} dv. \quad (10)$$

For a fixed T and fixed target/penetrator properties, v_r is strictly a function of v_s. Additionally, the factors of the integrand do not change signs; therefore, one can use the mean value theorem again to obtain

$$T = \frac{v_c^\alpha}{\beta} q_{eff}\left(v_s\right)\int_{v_r}^{v_s} |v|^{1-\alpha} dv, \quad (11)$$

where $q_{eff}(\cdot)$ is the effective value of $\frac{m}{AS}$ across the thickness of the target. The authors assume that $q_{eff}(\cdot)$ is an approximately linear function of v_s near some striking velocity of interest (v_τ), so that

---

[8] Had $g\left(\frac{v}{v_c}\right)$ been expressed as a power series expansion about $\frac{v}{v_c} = 0$ with $g(0) = 0$, then, near $v = 0$, α would have effectively been greater than or equal to unity, implying that the penetrator would not come to rest in a finite amount of time. Since this is not physically realistic, the authors did not permit any function for $g(\cdot)$ that could be represented by a power series near $v = 0$.

[9] Here the authors have substituted v_s in the upper integration limit of Equation 8 with $-v_r$, since this is now the "initial velocity" going into the second target. The lower integration limit of Equation 8 is set to zero as one is interested in how much unconsumed velocity is required to exactly penetrate the second target (i.e., when is the residual velocity out of the second target exactly zero). The authors have also included a minus sign in front of the v inside of $q(\cdot)$ to indicate that this is the extension of q to "negative velocities" given the true initial striking velocity of v_s. Note that $v_r < 0$ for $T_2 > 0$.

https://doi.org/10.52202/080042-0037

$$q_{eff}(v_s) \approx Q_* \left[1 - c\left(\frac{v_s}{v_\tau} - 1\right)\right], \tag{12}$$

where Q_* is the effective value of $\frac{m}{AS}$ across the target thickness at a striking velocity of v_τ and c is some constant. Since $q_{eff}(v_s)$ is non-negative, one also wants the approximation to be non-negative at all values of v_s under consideration. The authors assume that Q_* is not zero, therefore imposing the constraint

$$1 - c\left(\frac{v_s}{v_\tau} - 1\right) \geq 0 \tag{13}$$

Substituting Equation 12 into Equation 11 and solving for v_r yields

$$\begin{cases} v_r &= \text{sign}(x)\,|x|^{\frac{1}{2-\alpha}}, \\ x &= v_s^{2-\alpha} - \frac{T\beta(2-\alpha)}{v_c^{\alpha} Q_* \left[1-c\left(\frac{v_s}{v_\tau}-1\right)\right]} \end{cases} \tag{14}$$

Observe that v_c is just a constant with respect to a penetrator-target system and can be absorbed into β; therefore, v_c is essentially arbitrary. For convenience, then, the authors set $v_c = v_\tau$, where v_τ is now defined to be the penetrator striking velocity at which the target is perforated with probability τ (that is, the limit velocity associated with probability τ). Additionally, replace the expression $\frac{T\beta(2-\alpha)}{Q_*}$, which is non-negative and does not depend on v_s, with the variable u^2, so that Equation 14 simplifies to

$$\begin{cases} v_r &= \text{sign}(x)\,|x|^{\frac{1}{2-\alpha}}, \\ x &= v_s^{2-\alpha} - \frac{u^2}{v_\tau^{\alpha} \left[1-c\left(\frac{v_s}{v_\tau}-1\right)\right]} \end{cases} \tag{15}$$

Note that u has units of velocity.

Lastly, given the intended use-case of modelling v_r at a v_s near the limit velocity, the authors expect that v_r will be strictly increasing with respect to v_s (i.e., $\frac{dv_r}{dv_s} > 0$), which yields the constraint

$$1 - \left(\frac{v_s}{v_\tau}\right)^{\alpha} \frac{cu^2}{(2-\alpha)\, v_s v_\tau \left[1 - c\left(\frac{v_s}{v_\tau} - 1\right)\right]^2} > 0 \tag{16}$$

The authors thus arrived at their form for v_r as a function of v_s, which is given by Equation 15 and subject to $0 < \alpha < 1$, Equation 13, and Equation 16. Note that caution must be taken if Equation 15 is used to model v_r's beyond its intended purpose of finding limit velocities.

2.2 Probability Distribution of the Residual Velocity

The authors now wish to approximate the probability distribution of the residual velocity as a function of striking velocity. Upon examination of Equation 15, this paper proposes that the majority of shot-to-shot variation in residual velocity will result from the variation in u. A physical justification for this is that u contains variables related to the angle of attack of the penetrator at impact, the mass erosion of the penetrator, and the strength parameter of the penetrator/target interaction (via Q_*, T, and β), which the authors expect to be the more dominant stochastic processes in the impact event when compared to the "shape" of the velocity decay curve (defined by α) and the second order interactions with v_s (defined by c). A heuristic mathematical justification is that, for v_s's near the limit velocity of interest (v_τ) and after factoring out $v_s^{2-\alpha}$, α will be the exponent of a number close to unity, and the term containing c will be small compared to unity, so that the variations in either α or c will make small changes to the value of v_r. Thus, an approximate distribution for the random variable (r.v.) of the residual velocity (V_r) is given by:

$$\begin{cases} V_r \mathrel{\dot\sim} \text{sign}(X)\,|X|^{\frac{1}{2-\alpha}}, \\ X \mathrel{\dot\sim} v_s^{2-\alpha} - \frac{U^2}{v_\tau^{\alpha} \left[1-c\left(\frac{v_s}{v_\tau}-1\right)\right]} \end{cases} \tag{17}$$

where the capital letters X and U are used to denote r.v. representations of x and u, respectively.

One still needs to determine a probability distribution for U. Recall that U represents a product of non-negative r.v.'s; therefore, $\ln(U)$ is a sum of r.v.'s. The Lyapunov variant of the Central Limit Theorem can be used to show that the sum of independent (but not necessarily identical) r.v.'s asymptotically approaches a normal distribution as the number of r.v.'s increases, assuming the higher moments of the individual r.v.'s are not "much bigger" than their variance ([5], [6]). Thus, the authors assume that $\ln(U)$ is approximately normally distributed, or equivalently, that U is approximately log-normally distributed with parameters μ and σ, which are the mean and standard deviation, respectively, of $\ln(U)$.

The authors previously defined v_τ to be the striking velocity at which the penetrator will perforate the target with probability τ. In order for this to hold, one must have that

$$\tau = P(v_r > 0 \mid v_s = v_\tau) = P\left(v_\tau^{2-\alpha} - \frac{U^2}{v_\tau^\alpha} > 0\right) = P(U < v_\tau) = F_U(v_\tau) \quad (18)$$

where Equation 17 has been used, and where $F_U(\cdot)$ is the cumulative distribution function (CDF) of U. Because of the log-normal approximation of U, one then has that

$$\tau = \frac{1}{2}\left[1 + \operatorname{erf}\left(\frac{\ln v_\tau - \mu}{\sigma\sqrt{2}}\right)\right]$$
$$\Longrightarrow \mu(v_\tau, \sigma) = \ln v_\tau + \sigma\sqrt{2} \cdot \operatorname{erfc}^{-1}(2\tau), \quad (19)$$

where $\operatorname{erfc}^{-1}(\cdot)$ is the inverse of the complementary error function. Therefore, the parameter μ is not free but is in fact a function of v_τ and σ.

2.3 Maximum Likelihood Estimate of Limit Velocity

The authors chose to use the maximum likelihood estimate (MLE) to determine the "best fit" parameters for Equation 17 given test data. The main idea behind the MLE is to create a "likelihood" function (L) that is the joint probability density function (PDF) of the m observed values of V_r (denoted $v_{r,1}$, $v_{r,2}$, ..., $v_{r,m}$) for a given a set of parameters. This likelihood function is then maximised with respect to the model parameters [6]. In other words, these maximising parameters ensure that the modelled probability distribution has the highest probability of drawing the observed data. If one assumes that the tests are independent, then L is simply the product of the PDFs of V_r for each data point.

In order to construct a likelihood function for the test data, therefore, one must first derive the CDF and PDF of V_r. Let F_{Vr} be the CDF of V_r. With the help of Equation 17, one obtains, after some rearranging:

$$F_{Vr}(v_r; v_s, v_\tau, \alpha, c, \mathbb{U}) = P(V_r < v_r \mid v_s, v_\tau, \alpha, c, \mathbb{U}) = 1 - F_U[G(v_s, v_\tau, \alpha, c) M(v_r, v_s, \alpha)], \quad (20)$$

where

$$G(v_s, v_\tau, \alpha, c) = v_s\sqrt{\left(\frac{v_\tau}{v_s}\right)^\alpha \left[1 - c\left(\frac{v_s}{v_\tau} - 1\right)\right]}, \quad M(v_r, v_s, \alpha) = \sqrt{1 - \operatorname{sign}(v_r)\left(\frac{|v_r|}{v_s}\right)^{2-\alpha}}, \quad (21)$$

and $\mathbb{U}$ is, in general, the set of parameters that defines the distribution of U; for the authors' particular assumption that U is log-normally distributed, $\mathbb{U} = \sigma$.[10] The PDF of V_r is found by taking the derivative of $F_{Vr}(\cdot)$ with respect to v_r:

$$\therefore f_{V_r}(v_r; v_s, v_\tau, \alpha, c, \mathbb{U}) = \frac{(2-\alpha) G(v_s, v_\tau, \alpha, c)}{2 v_s M(v_r, v_s, \alpha)} \left(\frac{|v_r|}{v_s}\right)^{1-\alpha} f_U[G(v_s, v_\tau, \alpha, c) M(v_r, v_s, \alpha)]. \quad (22)$$

[10] Recall that μ is not a free parameter.

https://doi.org/10.52202/080042-0037

With Equation 22 derived, one can now construct the likelihood function for the parameters α, c, v_τ, and $\mathbb{U}$.

Here one again runs into an issue with their test data when perforation does not occur. If $v_{r,i} > 0$, then the PDF of V_r for the ith test instance is given by Equation 22. However, if the penetrator does not perforate, there is no means on knowing how "negative" $v_{r,i}$ is,[11] only that it is non-positive. To address this problem, the authors treat the V_r r.v.'s for non-perforating data points as Bernoulli r.v.'s, where $v_{r,i}$ has a probability p_i of being less than or equal to zero. One can calculate p_i directly from Equation 20. Thus, the likelihood function given test data $\vec{v_s} = (v_{s,1}, v_{s,2}, \dots, v_{s,m})^T$ and $\vec{v_r} = (v_{r,1}, v_{r,2}, \dots, v_{r,m})^T$ is:[12]

$$L(v_\tau, \alpha, c, \mathbb{U}; \vec{v}_s, \vec{v}_r) = \prod_{\substack{i=1 \\ v_{r,i}=0}}^{m} F_{V_r}(0; v_{s,i}, v_\tau, \alpha, c, \mathbb{U}) \cdot \prod_{\substack{i=1 \\ v_{r,i}>0}}^{m} f_{V_r}(v_{r,i}; v_{s,i}, v_\tau, \alpha, c, \mathbb{U}) \quad . \quad (23)$$

Thus, the authors seek to find the parameters that maximise L; that is, the MLE parameters.[13] The v_τ obtained as part of this maximising set of parameters (denoted $\hat{v}_\tau$) will therefore be the "best guess" of the limit velocity associated with probability τ.

2.4 Construction of Confidence Bounds Around $\hat{v}_\tau$.

To improve the usefulness of the approximation $\hat{v}_\tau$, one needs to construct upper and lower bounds on the estimate. Theoretically, the true v_τ will reside within this interval with some specified probability (often referred to in percentage as the confidence level [CL]).

Due to the complexity of the probability distribution of V_r (nonstandard distribution and, under the authors' assumptions, four unknown parameters: v_τ, α, c, and σ) and the sparsity of data (it is not uncommon for a penetration test to consist of only 10 - 15 data points), the authors have chosen to use parametric bootstrapping to estimate the confidence bounds. The general procedure to find the upper confidence bound is as follows (a similar approach can be used to find the lower confidence bound) [6]:

1. Compute the MLE parameters $(\hat{v}_\tau, \hat{\alpha}, \hat{c}, \hat{\sigma})$ using the approach discussed in Section 2.3. This set of parameters is called the alternative hypothesis (H_1).
2. Make a guess for the upper bound of v_τ (denoted $v_{\tau,u}$).
3. Compute the MLE for the remaining three parameters assuming $v_{\tau,u}$ is true. This set of parameters (including $v_{\tau,u}$) is called the null hypothesis (H_0).
4. Randomly simulate many repetitions of the penetration test (at the same striking velocities as the data) using Equation 17 and the H_0 parameters;[14] use the results to approximate the distribution of some test statistic. The authors propose using the likelihood ratio test statistic (λ_{LR}) with $v_\tau = v_{\tau,u}$ as the null hypothesis in computing λ_{LR}. This is therefore an approximation of the distribution of λ_{LR} assuming H_0 is true.
5. Using the simulated distribution, determine the (100 - CL_u)th percentile of λ_{LR} (denoted $\tilde{\lambda}_{LR}$), where CL_u is the confidence level of the upper bound.
6. Compute λ_{LR} using the H_1 and H_0 parameters (denote this as λ^*_{LR}). If $\tilde{\lambda}_{LR} > \lambda^*_{LR}$, increase the guess for $v_{\tau,u}$; otherwise, decrease the guess for $v_{\tau,u}$.
7. Repeat steps 3 - 6 until $\tilde{\lambda}_{LR} = \lambda^*_{LR}$, to within some tolerance.

3. EXPERIMENTAL TESTING AND NUMERICAL SIMULATIONS

As an initial exploration of the validity of the method proposed in this paper, two pre-existing sets of perforation data (with accompanying residual velocity data) were obtained. One data set was of a small arms projectile against a "soft" metallic plate, and the second set of data was of the same projectile against a "hard" metallic plate. The v_s vs. v_r data is graphically depicted in Figure 1. These metallic data

---

[11] Recall the generalised definition of v_r.

[12] Here, T is the transpose operator, not target thickness.

[13] For computational purposes, a common practice is to maximise the log-likelihood function ($\ln(L)$) instead of L. This transformation has the benefit of turning multiplications into summations.

[14] To be compatible with the test data, simulations of striking velocities that did not perforate during testing should be modelled as binary r.v.'s (i.e., perforation/no-perforation), while simulations of striking velocities that did penetrate during testing should be modelled as continuous r.v.'s. Also note that for the latter simulations it is possible (and acceptable) to generate negative v_r's.

sets were investigated first due to their immediate availability; however, the authors plan on performing a similar analysis on other target materials, such as personal armour, once funding and testing logistics can be arranged.

The first step in the authors' evaluation was to fit the entire population of the data using the method discussed in Section 2.3 for both data sets. Due to the large number of data points, they therefore assumed that these fitted parameters were reasonable approximations to the population parameters. Next, using the fitted parameters, the authors graphed the Q-Q plots of $\ln(u)$ (calculated using Equation 15) against a normal distribution, which are shown in Figure 2. The closer the dots lie on a straight line, the more the u's follow a log-normal distribution. Except for a few outliers near the tails, the distribution of the u's appear to closely resembles a log-normal distribution for both data sets.

The second step in the evaluation process was to measure the precision of the proposed residual velocity method compared to logistic regression, which is a commonly used method in the estimation of limit velocities in the small arms field. For each of the data sets, evenly spaced striking velocities were chosen that roughly spanned the data. Penetration "tests" were simulated by generating residual velocities using the "population" parameters calculated previously, in conjunction with Equation 17.[15] A limit velocity was then estimated for each test. This process was repeated many times and the relative mean square error (RMSE) was computed for each of these tests when compared to the "true" limit

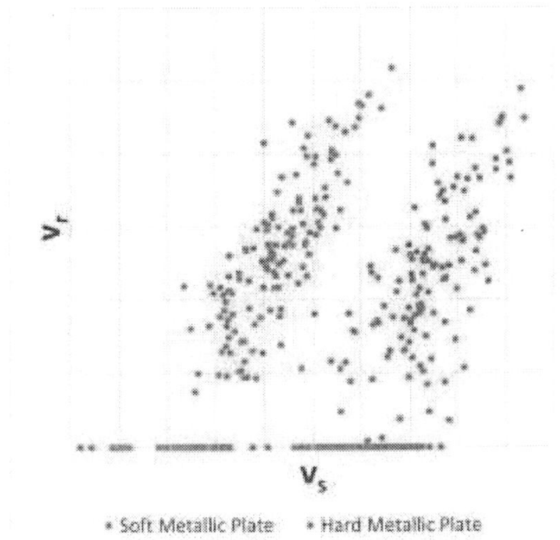

Figure 1. v_s vs. v_r curves of two penetration data sets

velocity. Figure 3 shows the plots of the RMSE when estimating the V10, V50, and V90 limit velocities as a function of the number of shots used in the simulated test. Observe that for V50 estimates for both the soft and hard plate, the proposed residual velocity method and logistic regression are comparably precise. However, when trying to estimate V10 or V90, the proposed method is significantly more precise than logistic regression, on the order of 1.5%. To look at it another way, the proposed method only requires 10 shots in a test to have the equivalent precision of a 25-shot test using logistic regression.

The final step was to evaluate the coverage of the confidence intervals[16] (CI) around the limit velocity estimates. That is, if one computes a 90% confidence interval, does the population limit velocity actually lie within the confidence interval with probability 0.9? Figure 4 shows the coverage of both the proposed method and logistic regression when computing confidence intervals for V10, V50, and V90 estimates when a 90% confidence interval is requested. Observe that logistic regression generally overshoots the desired coverage. At first, this result may seem to favor logistic regression; however, what this implies that logistic regression will, on average, construct confidence intervals that are larger than necessary. Figure 5 shows the average confidence interval widths for both the proposed method and logistic regression.

Note that logistic regression's confidence interval widths are on the order of one-and-a-half to two times as large as the proposed method. Consequently, when you evaluate data using logistic regression, you will have significantly less confidence in your results than if you evaluated the same data using the proposed method, potentially by a factor of 1.5-2.

[15] Assuming this distribution is justified based on the results of the Q-Q plots.

[16] Confidence interval is the interval bounded by the lower confidence bound and the upper confidence bound.

https://doi.org/10.52202/080042-0037

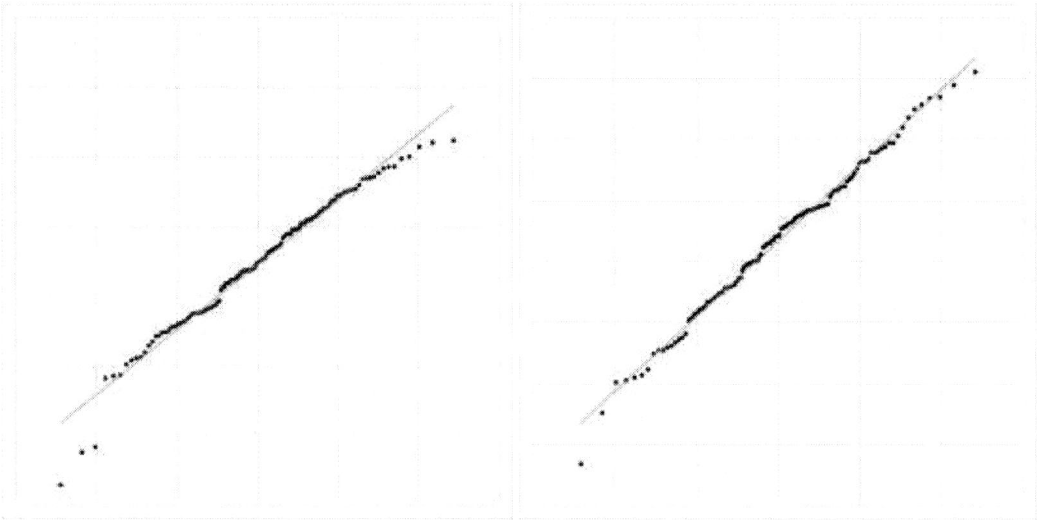

Figure 2. Q-Q Plots of ln(u) against a normal distribution for soft (left) and hard (right) metallic plate data

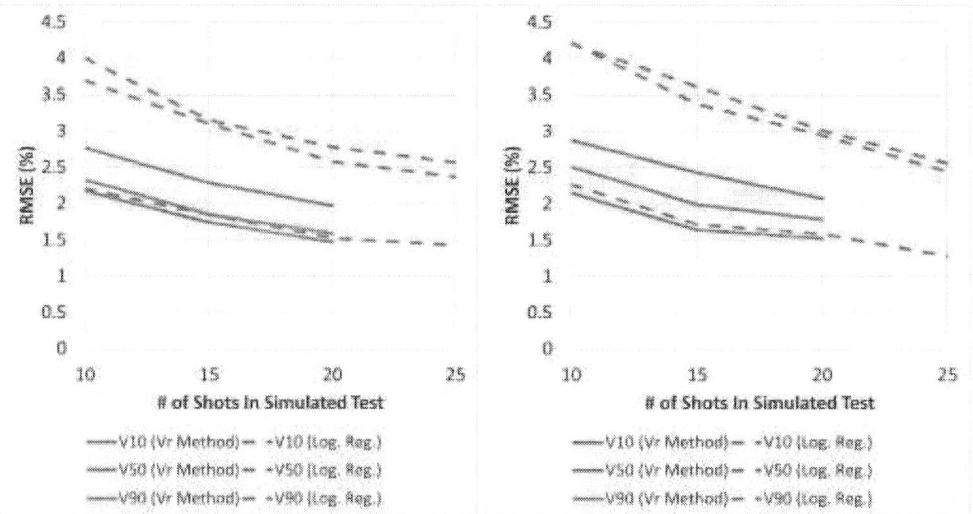

Figure 3. RMSE of limit velocity estimates vs. # of rounds in simulated tests for soft (left) and hard (right) metallic plate data

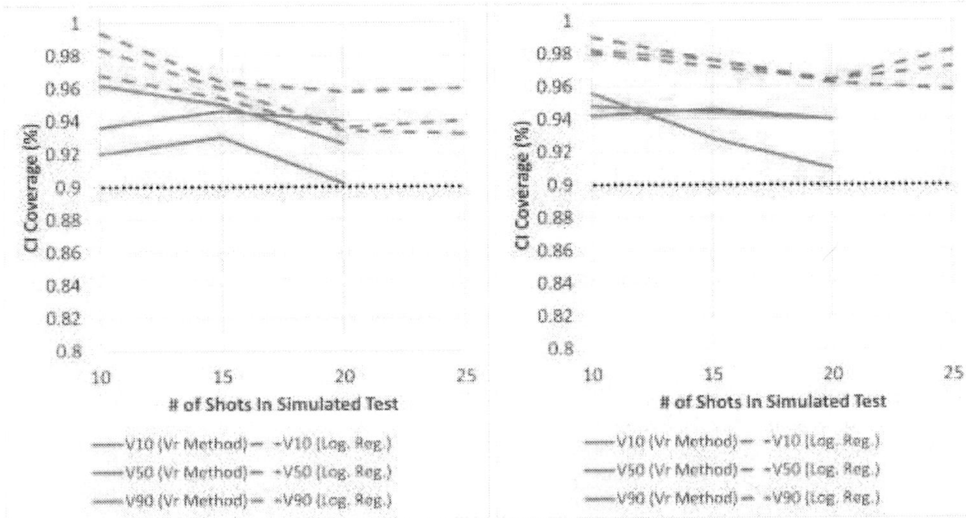

Figure 4. 90% CI coverage vs. # of rounds in simulated tests for soft (left) and hard (right) metallic plate data

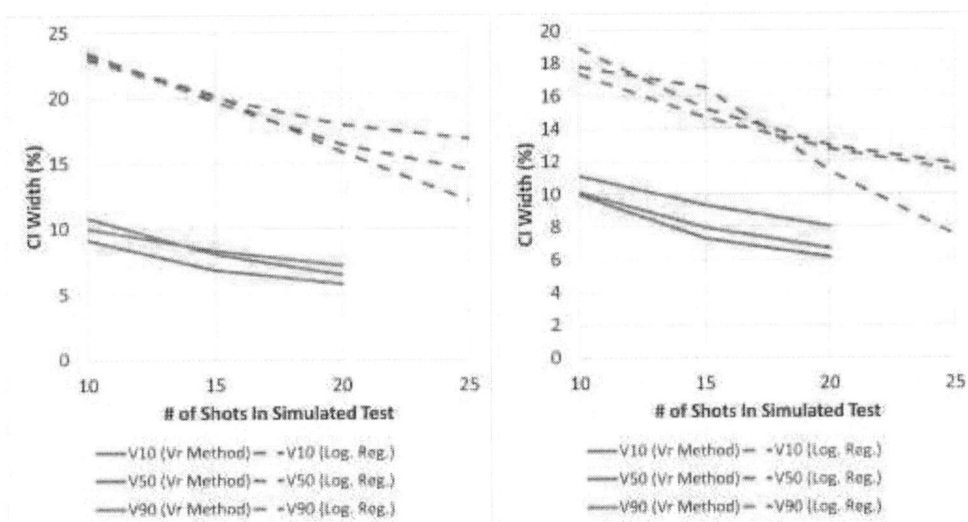

Figure 5. Avg. width of estimated 90% CI vs. # of rounds in simulated tests for soft (left) and hard (right) metallic plate data

4. FUTURE WORK

4.1 Live Fire Testing and Validation

The authors plan to further validate the limit velocity estimation method proposed by this paper. This will include augmenting the current data set by testing simple geometric penetrators against a wider range of targets, to include metals and ceramics, as well as any other materials deemed appropriate. Additionally, the authors plan on testing legacy military ammunition against various personal armour targets, to determine if the proposed methodology is robust enough to handle other complex dynamical interactions.

4.2 Test Algorithm Development

The development of the sister test algorithm to the modelling technique described in this paper is still in its early stages, but a few key differences from currently used adaptive test methods are noteworthy. In

https://doi.org/10.52202/080042-0037

3POD and other traditional binary test methods, testing requires the observation of both perforation and no-perforation results to "home in" on the velocity region of interest. If during testing only perforations or no-perforations are observed, nothing of real use can be done with this data. Due to the nature of the modelling technique described in this paper, any no-perforation is essentially a sub-optimal data point, where no continuous residual data can be gleaned. This means that the test algorithm in development will focus on adaptively approaching the point at which residual velocity is estimated to reach zero without going over, starting at higher striking velocities and moving to lower striking velocities. Optimally efficient placement of test points to formulate the model will likely be along high leverage inflection points on the logistic regression curve. The details with regard to the desired spacing, starting point, and number of samples are still in development.

5. CONCLUSION

In this paper, the authors derived and formalised a new methodology for estimating limit velocities by analysing residual velocity data. They also demonstrated via test data and numerical simulation that the inclusion of this additional continuous data significantly improves both precision and efficiency with regard to the estimation of limit velocities with respect to the metallic targets analysed.

The derivations in this paper were based on material-agnostic principles; mainly, Newton's Law, smoothness and monotonicity of the velocity retardation function, zero retardation force at termination of transient, and existence of a Taylor series expansion about the limit velocity. Thus, while the test data analysed in this paper demonstrated applicability to simple metallic targets, the authors posit that the methodology is sufficiently generic to apply to a wide range of penetrators and targets, to include personal armour targets.

Future work was discussed that will seek to continue to validate and refine the methodology, as well as investigate adaptive testing methods to further increase testing efficiency.

References

[1] C. Drake and D. Ray, Risk Management of a High Pressured Test Cartridge using Sensitivity Testing, 2016; IEEE

[2] H.J. Langlie, A Reliability Test Method for "One-Shot" Items, Aeronutronic Division of Ford Motor Company, 1962; Publication No. U-1792.

[3] C.F.J. Wu and Y. Tian, Three-phase optimal design of sensitivity experiments, Journal of Statistical Planning and Inference, 2013; Volume 149; 1-15.

[4] STAT COE-Report-08-2017

[5] P. Billingsley, Probability and Measure (John Wiley Sons, New York, 1995); pp. 362

[6] M.H. DeGroot and M. J. Schervish, Probability and Statistics (Pearson Education, Inc., New York, 2012); pp. 312-314,418-441,839-849.

https://doi.org/10.52202/080042-0037

Ballistic Assessments of the ARL Reusable Temperature Insensitive Clay (ARTIC) as a Ballistic Backing Material

J. Gardner[1], J. Cora Cruz[1], R. Mrozek[2], J. Hopping[3], E. Beaudoin[3], L. Lombardo[1]

[1]*U.S. Army Aberdeen Test Center, B400, 6943 Colleran Rd, APG, MD, 21005, USA*
[2]*DEVCOM Army Research Laboratory, 6300 Rodman Rd, APG, MD 21005, USA*
[3]*Program Executive Office – Soldier, U.S. Army, 9102 Iry Road, Fort Belvoir, VA 22060, USA*

Abstract. Backing materials play an important role in evaluating the performance of body armour materials. In 1977 Russel N. Prather et al. investigated the use of Roma Plastilina No.1 (RP1) artistic modeling clay as a surrogate for human tissue to define the penetration and deformation characteristics of soft body armour as worn. In that study, Prather nominally compared backface signature, the behind armour deformation of the clay, against a 20% ballistic gelatin formula which previous researchers correlated with the penetration resistance of human tissue. As noted by Prather then and as is still true today, clay has some notable drawbacks for simulating human tissue response, but it provides advantages over gelatin for cost-effective high-volume testing. First, is the relative preservation of the backface signature which allows for fast, direct measurement of behind armour deformation. Additional practical advantages of clay include ease of use in its handling and storage, ready commercial availability, and lower cost. Overall, Prather's studies effectively demonstrated the suitability of clay as a backface material, and the subsequent use of RP1 led to the creation of new body armour test methods which were broadly adopted by government, academic, and industrial institutions for the research, development, and testing of body armour materials and systems. Although RP1 proved the best choice at the time and has served an integral role in the testing and evaluation of body armour over nearly a half century of use, it is not the ideal ballistic clay material. Thus, the United States Army has found reason to reevaluate the continued use of RP1. The primary impetus for this reevaluation is the result of periodic changes in RP1's material constituents (due to market availability, artistic improvements, or both). Some of these changes have influenced the ballistic response of RP1. One clear indication of the changes in ballistic response over time is the noteworthy fact that RP1 must be heated to near 38°C (100°F) to achieve similar impact responses observed at ambient room temperatures nearly a half century ago. To address the issues related to the change in RP1's properties, the United States Army Research Laboratory (ARL) developed a substitute backing material called the ARL Reusable Temperature Insensitive Clay (ARTIC). Following ARL's development of ARTIC, the United States Army Aberdeen Test Center (ATC) conducted ballistic assessments of ARTIC's performance, and this paper presents the results of ATC's testing. ATC focused this ballistic evaluation of ARTIC on: (1) body armour commodities including soft armour, hard armour, and helmets; and (2) characterizing the effects of mechanical work, aging, temperature, and changes in material formulation. Where practicable, ATC completed side-by-side comparisons between ARTIC and RP1 during testing. This paper finds that in general ARTIC behaves very similarly to RP1 with some noted advantages and a few noted differences that do not likely interfere with or preclude the use of ARTIC as a snap in replacement for RP1. ATC is also supporting the future implementation of ARTIC through the development of new test protocols and material handling procedures.

1. INTRODUCTION

The United States Army, in conjunction with many partners within and outside of the body armour community, conducted a review of the clay backing material used for the ballistic testing of body armour. This review produced a recommendation for the development of a material replacement for Roma Plastilina No.1 (RP1) ballistic clay. In following up on the Army's recommendation, Army collaboration partners sought a replacement material for RP1. After review of several candidate materials, the United States Army Research Laboratory's (ARL) ARL Reusable Temperature Insensitive Clay (ARTIC) was selected as the most promising candidate for further development. Following the selection of ARTIC, these same partners began a thorough evaluation of ARTIC as a replacement for RP1 with the United States Army Aberdeen Test Center (ATC) performing most of the ballistic assessments and the primary development of material handling and test procedures. This paper discusses portions of this testing done by ATC and provides an overview of the results obtained. Through these discussions and data overview, this paper examines ARTIC's ballistic performance in testing and the suitability of ARTIC's use as a direct replacement for RP1.

RP1 is part of a commercial line of artistic modeling clays. Although not an intentional use by product design, the ballistic testing community found that RP1 was a highly practical medium to test body armour, dating to 1977 with the publication of Russell N. Prather et al.'s study "Backface Signatures of Soft Body Armors and the Associated Trauma Effects [1]." Prather's work focused mostly on the

https://doi.org/10.52202/080042-0038

identification of a material for use in production scale testing of soft body armour. Since Prather's report, the ballistic testing community has employed RP1 as a backing material for resistance to penetration and ballistic limit testing of all the major body armour commodities i.e., soft and hard amour constructions for torso and extremity protection as well as helmet systems for head protection.

Despite its suitability as an initial choice and prevalence of use by the ballistic testing community since 1977, the use of RP1 as a primary backface material for the ballistic testing of body armour has had a significantly negative effect on reproducibility in testing. One of the flaws of RP1 stems from RP1's intended use – as a modeling clay. RP1 has "about 10 constituents … include[ing] pigments or colorants, antioxidants, and other minor materials as well as an intentional blending of multiple sources of, for example, the microcrystalline wax to dampen out lot-to-lot variations from individual suppliers" [1]. Formulation changes in RP1 due to commercial constraints (e.g., constituent availability or price) or due to purposeful product improvement (e.g., in response to artistic modeler feedback) have downstream effects on the consistency and repeatability of ballistic testing using RP1 as the backface material. These changes are unannounced, can occur frequently or infrequently, are noticed by a change in clay "feel" or statistical trending data, or are not noticed at all. Out of all possible concerns, the primary concerns associated with the continued use of RP1, and that form the justification for this work, relate to how changes made in RP1's constituents have led to an increase in material hardness. RP1 must now be heated to near 38°C (100°F) to match its historical ballistic performance. This creates numerous logistical and technical challenges in testing that could be removed using a material replacement with a set, controlled formulation, and a calibrated response optimized at room temperature conditions.

Finding a replacement for RP1 that meets the required criteria is no trivial task. Finding a true surrogate for human tissue response or suitability for injury evaluation is even harder. Testing materials intended for the evaluation of protective systems for the human body must achieve an optimum balance of suitability in mechanical response, repeatability, range flexibility, and affordability despite complex, difficult, and counterproductive material performance interrelationships in the desired zone of testing. During his 1977 study, Prather also examined a secondary correlation with behind armor blunt trauma (BHBT) by comparing RP1 to 20% ballistic gelatin formula [2]; however, he did not and no one since has completed a comprehensive study on the current use of ballistic clays for assessment of human injury [3]. Body armour testing using RP1 (or other clays) as the backface material thus relies primarily on comparative metrics, historical comparisons, and operational benchmarks to evaluate performance. Like Prather's own efforts, this study assesses ARTIC as a ballistic backing material using the following approach but without attempting to address the medical questions associated with the use of ballistic clay in the testing of body armour, such as whether ballistic clay replicates an anthropomorphic response.

2. APPROACH

Ballistic performance assessments of ARTIC were made by taking backface deformation (BFD) measurements using a laser arm scanner during testing of the three major body armour commodity types including soft armour, hard armour, and head protection (helmets). Testing was also conducted to evaluate ARTIC for changes in ballistic response based on mechanical work, aging, temperature effects, and changes in material formulation. When practicable, ballistic results for ARTIC were compared with RP1 (heated to near 38°C (100°F)) by means of side-by-side i.e., concurrent testing. This study focuses on making material performance assessments of the clay under relevant ballistic strain conditions and not on the performance of the body armour test items themselves. Each body armour commodity type provides a unique assessment of ballistic clay performance based on the BFD shape it produces and the resultant strain conditions in the clay. By seeking out unique test cases and strain conditions, the tester can explore the entire breadth of the representative mechanical response tradespace without the need to make or use associations between test article response and performance specifications. Therefore, ATC did not make, use, or note associations between test article response and performance specifications during this study. This approach has several advantages including increased flexibility in test sample selection and allowance of the use of non-standard projectile types, velocities, and obliquities. This approach also enables sharing of the results of this study with the broader body armour community.

During all testing for this study, only test personnel who possess extensive work experience with RP1 were employed. No such extensive experience yet exists for the handling of ARTIC. Research personnel possessing comparatively limited (but still the highest level attained by anyone anywhere) experience working with ARTIC provided the test personnel with recommendations on the handling of ARTIC for preparation and repair of clay blocks, molds, and headforms. The research personnel's recommendations pertained primarily to handling procedures such as spreading and cutting and the researcher's experience with the effects of mechanical work on the overall workability of ARTIC. In most cases no further guidance was provided to the testers unless explicitly requested. This limitation was directed purposefully to conduct the ARTIC evaluation in a test environment that mirrors the RP1 operating conditions as closely as possible and to minimize bias in the feedback solicited from the testers. The values for "n" of each test design in tables 1-6 represent the number of shots conducted in each test.

2.1 Soft Armour

Four subtests were designed to evaluate clay performance by resistance to penetration testing (RTP) of soft armour and to characterize ballistic responses of the clay types due to periodic mechanical work (and rest), aging, the effect of changes in temperature and material formulation. For this, soft armor ballistic panels, consisting of stacked uniformly sized plies of soft armor fabric enclosed in a fabric sleeve were mounted to the face of a clay block using strapping. A unique five-shot pattern (Figure 1) was developed and adopted specifically for the soft armour testing in this study to provide the data consistency and reproducibility necessary for direct comparisons between data sets. The design of the five-shot pattern considers the boundary conditions provided by the clay block, shot obliquities, expected BFD volumes, and resultant shot order effects to maximize the data collected from the available test articles and clay blocks. Shot locations on the ballistic panel correspond with the shot locations on the clay block, and test personnel executed all shots by location in numerical order. The distance between shot locations is larger on the clay block than on the ballistic panel. Therefore, for the shot locations to align, test personnel moved the ballistic panel and remounted it to the clay block between shots. Except for the aging study (see below) which purposefully aged the clay for one year, this study used only ARTIC and RP1 that was less than one year old, and the typical material used for this study was within three to nine months of the manufacturing date.

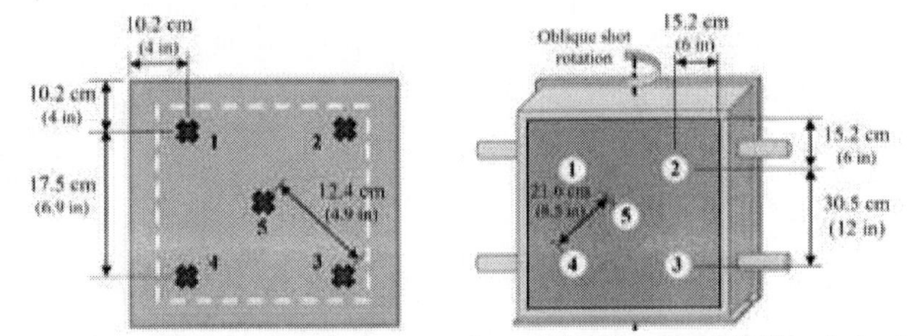

Figure 1. Five-shot pattern shot locations on a soft armor ballistic panel (left) and clay block (right).

The first soft armor subtest compared BFD results over a one-year period of material aging (1-Year Aging subtest) to evaluate the effect of shelf-life on material response. This subtest conducted shots at two distinct obliquity angles using three dedicated, initially untested blocks of ARTIC. The 1-Year Aging subtest interrogated each block of material three times per obliquity at approximately 30-day intervals between each test series. Only soft armor ballistic panels from the same design and manufacturing lot were used during this subtest and all tests were conducted using a single set of test conditions i.e., projectile type, velocity window, and shot obliquities. An evaluation of ARTIC's performance between months two to twelve was then made versus data from month 1 (test series from the first month) as a baseline comparison to assess ARTIC's ballistic aging response over a 1-year period. To compare ARTIC's ballistic aging response versus RP1, a baseline for RP1 was generated by selecting BFD data that had been collected intermittently during the 1-year lifetime of select RP1 clay blocks from subtests using a matching shot pattern, ballistic panel type, and shot conditions. Table 1 summarizes the design of the 1-Year aging subtest.

Table 1. Ballistic test design for the 1-Year Aging subtest.

Subtest Name	Time between tests (days)	Obliquity 1 n (Clay Type)	Obliquity 2 n (Clay Type)
1-Year Aging	Approximately 30	530 (ARTIC) 185 (RP1)	530 (ARTIC) 187 (RP1)

The second soft armor subtest (Temperature Effects subtest) evaluated the effect of temperatures below and above ambient (from 13.9 °C to 29.9 °C) on BFD response. ATC assessed that the selected temperature range of the evaluation is relevant to the body armour test community, although some of the

https://doi.org/10.52202/080042-0038

range evaluated (both below and above) is outside of typical ambient test requirements. This subtest used two blocks of initially untested ARTIC for each temperature condition and each block was tested twice per shot obliquity. Soft armor ballistic panel type and test conditions matched those used in the 1-Year Aging subtest, including shot obliquities. An evaluation of ARTIC's performance at temperatures below and above ambient was then made versus BFD data from the month 1 (first month tested at 21.1 °C) test series within the 1-Year Aging subtest as a baseline for ARTIC's ambient temperature performance. Table 2 summarizes the Temperature Effects subtest design.

Table 2. Ballistic test design for the Temperature Effects subtest.

Subtest Name	Temperature (°C)	Obliquity 1 n	Obliquity 2 n
Temperature Effects	Cold (13.9 °C)	20	20
	Hot (29.9 °C)	19	20

The third soft armor subtest (Formulation Changes subtest) evaluated formulation changes to the fumed silica ingredient in ARTIC for constituent percentages below and above the standard formulation. ARL produced batches of ARTIC with these two formulation changes for ATC to investigate the effects of potential alterations in ARTIC's material recipe on BFD response. This subtest used one block of initially untested ARTIC for each formulation type and tested each block nine times per shot obliquity. As in the Temperature Effects subtest, soft armor ballistic panel type and test conditions matched those used in the 1-Year Aging subtest, including shot obliquities. A performance evaluation of the two variations in ARTIC's formulation was then made versus BFD data from the month 1 (first month) test series within the 1-Year Aging subtest as a baseline comparison for the performance of ARTIC's standard formulation. Table 3 summarizes the Formulation Changes subtest design.

Table 3. Ballistic test design for the Formulation Changes subtest.

Subtest Name	Material Formulation	Obliquity 1 n	Obliquity 2 n
Formulation Changes	Soft (lower % fumed silica)	44	45
	Hard (higher % fumed silica)	45	45

The fourth soft armor subtest (Mechanical Work and Rest Time subtest) evaluated the effects of mechanical work and rest times on BFD response: (1) by varying the procedures used for block repair between hand kneading and pneumatic air hammer methods and (2) by varying block resting times between 1 hour, 1.5 hours, 3 hours, and 24 hours. This subtest used a single shot obliquity and required eight clay blocks. The clay blocks were selected and labeled to identify their use for each preparation method and associated rest time in the subtest. Two blocks (those used for the 24-hour condition) contained initially untested ARTIC, and six blocks (those used for the 1-, 1.5-, and 3-hour conditions) contained material that had been used lightly for earlier non-ballistic testing. Testers allowed all blocks to rest for a minimum of two weeks after being filled, or after previous use, prior to testing. This subtest used a different design of soft armor ballistic panels with a similar construction to those used in the aging, temperature, and formulation studies. The same projectile type was used as the previous subtests but with a slightly modified velocity window. ARTIC's performance for the various rest times and work methods was evaluated by comparing against a pooled BFD data set from the first test series of the 1-, 1.5-, and 3-hour conditions for both work methods representing ARTIC's performance in a rested condition. Table 4 summarizes the design of the Mechanical Work and Rest Time subtest.

Table 4. Ballistic test design for the Mechanical Work and Rest Time subtest.

Subtest Name	Rest Time (hr.)	Hand kneaded n	Air hammer n
Mechanical Work and Rest Time	1	23	25
	1.5	75	78
	3	24	23
	24	5	5

2.2 Hard Armour

Two subtests were conducted to assess clay performance for testing of hard armour plates. Deformation witnessed in hard armour can exceed those of soft armour and helmets due to the increased energy of rifle-fired projectiles. Therefore, the first subtest was designed to assess a range of strain and deformation conditions in the clay during hard armour RTP testing by targeting three BFD depths of approximately 30mm, 40mm, and 50mm. Multiple projectile types and velocities were used to achieve the various BFD target depths. The second subtest simulated the conditions of a hard armour V_{50} Ballistic Limit (BL) test. All hard armor test articles used were of a non-standard design. All impacts were placed in the crown region of the plates, the location on the front of the plate where the peak curvature of a curved hard armor plate is located and produced highly repeatable test results in both RTP and V_{50} BL testing. These subtests used three blocks of initially untested ARTIC, and RP1 testing was conducted concurrently with ARTIC using matching test designs. Table 5 summarizes the hard armor subtest designs.

Table 5. Hard armour ballistic subtest designs.

Subtest Name	Shot Location	Target BFD (mm)	n
RTP	Crown	30	15
		40	15
		50	15
V_{50} BL	Crown	-	36

2.3 Helmets

Two helmet RTP subtests were conducted to assess clay performance during helmet testing using the National Institute of Justice (NIJ) headform. Although there are several headforms used for helmet testing, the body armour test community uses the NIJ headform perhaps more widely than any other. The NIJ headform is constructed of solid aluminum and has channels of equal width machined symmetrically along the sagittal (front to back), and coronal (side to side) planes. When filled with ballistic clay, the NIJ headform can be used to test helmets from any of five principle shot orientations along these planes, i.e., the front, back, crown, left side, and right side. These subtests used previously untested ARTIC to fill and repair clay in the headforms. ATC also devised and used a hydraulic press molding method to ensure repeatability of the clay filling procedure. The first subtest (RTP 1) evaluated ARTIC for the crown and back shot locations using test conditions designed to produce minimally constrained BFDs, i.e., with as little contact between the deformation in the helmet and the channel posts of the NIJ headform as possible. The second subtest (RTP 2) evaluated clay BFD under comparatively higher strain and deformation conditions for the back shot location only. These subtests used two helmet types to generate the required BFD conditions and RP1 testing was conducted concurrently with ARTIC during both subtests using matching test designs. Table 6 summarizes the helmet test designs.

Table 6. Helmet ballistic subtest designs.

Subtest Name	Shot Pattern	n
RTP 1	1st Shot Crown 2nd Shot Back	22
RTP 2	Back	32

3. RESULTS

3.1 1-Year Aging (Soft Armour) Subtest

Figure 2 provides monthly ARTIC results of the 1-Year Aging subtest with a plot of average BFDs by shot location and obliquity combination (brightly colored "moving" lines) for each month (two through twelve) versus the ARTIC Month 1 BFD baseline for the same shot location and obliquity combination (light grey straight lines). This plot shows the relative (to scale) differences between the ARTIC subsequent monthly averages and the ARTIC first month BFD average values, and the listed standard deviations by shot location and obliquity combination provide a sense of the total magnitude of the observed variance for each shot series (see column noted as "σ" to the right of the plot). The plot marks with a small, open, red circle where the average ARTIC BFD by shot location and obliquity combination in that month is statistically significantly different (α=0.05) than the average ARTIC Month 1 BFD baseline for the same condition combination. The plot also marks months with a large, open, blue circle

https://doi.org/10.52202/080042-0038

where the average ARTIC BFD by condition combination is statistically significantly different than the average RP1 BFD baseline for that same condition combination. Note that RP1 was not tested concurrently with ARTIC during this subtest. Although not depicted in Figure 2, ATC also observed statistically significant differences between the ARTIC Month 1 baseline and RP1 baseline for obliquity 1 at shot locations 1 and 4, and for obliquity 2 at shot location 5. The total (pooled) average BFD differences between ARTIC and RP1 in this dataset for obliquities 1 and 2 is 0.8mm and 0.1mm, respectively and is not statistically significant.

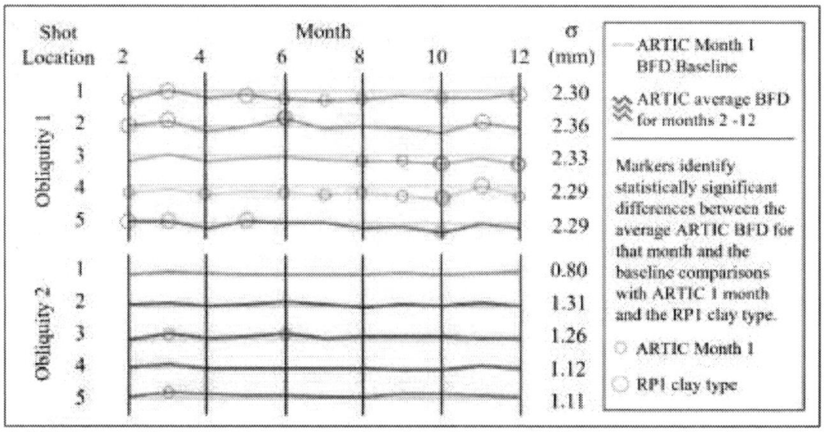

Figure 2. ARTIC monthly BFD averages by shot location and obliquity combination with significant difference notations versus the ARTIC month 1 baseline for the same shot location and obliquity combination. Statistical significance threshold for notation is α=0.05.

3.2 Temperature Effects (Soft Armour) Subtest

Figure 3 shows the results of the Temperature Effects subtest. For both obliquities, differences in the average ARTIC BFD for the cold (13.9 °C) and hot (29.9 °C) temperature conditions versus the baseline at ambient (21.1 °C) temperature are not statistically significant. For obliquity 1 cold and hot temperature conditions, the difference is 0 and 1.7mm, respectively. For obliquity 2 cold and hot temperature conditions, the difference is 0.2mm and 2.0mm, respectively.

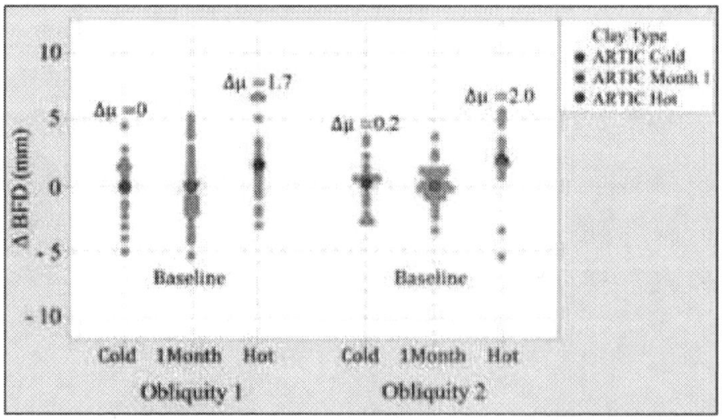

Figure 3. Sample average differences in ARTIC BFD for cold and hot temperature conditions at two shot obliquities versus the ARTIC Month 1 BFD baseline from the 1-Year Aging subtest.

3.3 Formulation Changes (Soft Armour) Subtest

Figure 4 shows results of the Formulation Changes subtest which evaluates the effects of changes in ARTIC's material formulation. Differences in the average BFD for the hard (higher percentage fumed silica) and soft (lower percentage fumed silica) formulations versus the baseline ARTIC standard formulation are not statistically significant. For obliquity 1, the average BFD differences for the hard and soft formulations versus the baseline are 0.5mm and 0.9mm respectively. For obliquity 2, the average BFD differences for the hard and soft formulations from the baseline are 0.5mm and 2.1mm respectively.

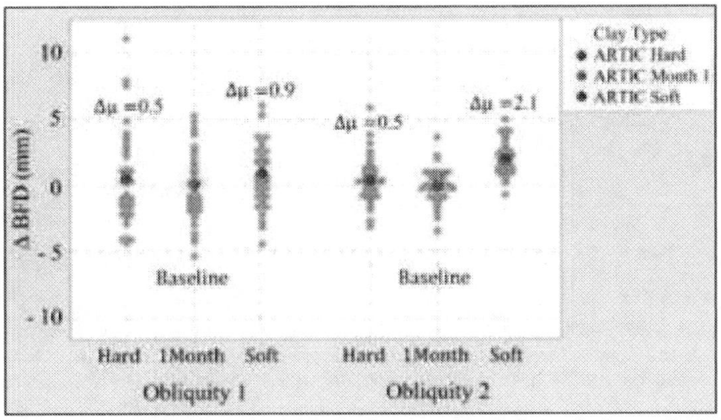

Figure 4. Sample average differences in BFD for hard and soft formulations of ARTIC at two shot obliquities versus the ARTIC Month 1 BFD baseline from the 1-Year Aging subtest.

3.4 Mechanical Work and Rest Time (Soft Armour) Subtest

Figure 5 shows the results of the Mechanical Work and Rest Time subtest. The figure plots sample averages for each rest time and work method versus a baseline of pooled data including the first test series from the 1-, 1.5-, and 3-hour conditions for both work methods. None of the observed differences in the average ARTIC BFD for each rest time interval and work method are statistically significant. However, additional testing is being planned to increase the confidence of the statistical model.

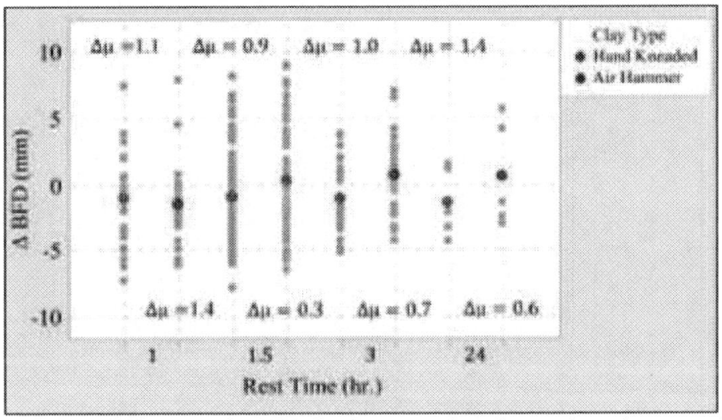

Figure 5. Sample average differences in BFD for four rest time intervals and two work methods versus the baseline of pooled data including the first test series from the 1-, 1.5-, and 3-hour conditions.

3.5 RTP and V_{50} BL (Hard Armour) Subtests

Figure 6 shows representative photos and laser surface scans from the hard armour RTP subtest in which testers evaluated ARTIC and RP1 concurrently. Color in the surface scan images indicates changes in crater depth. Testers observed highly comparable responses between ARTIC and RP1 clay types with respect to the BFD depth, shape, and overall behavior of the materials. The differences in average BFD between clay types for the target BFD depths of 30mm, 40mm, and 50mm are 1.0mm, 0.8mm, and 0.5mm, respectively. None of these differences are statistically significant. Testers noted differences between clay types in the extent of cracking on the clay surfaces after removal of test items. Table 7 shows the average BFD and standard deviations for each BFD depth and clay type from this subtest.

Table 7. BFD results from the hard armour RTP test.

Target BFD (mm)	ARTIC BFD (mm)		RP1 BFD (mm)	
	μ	σ	μ	σ
30	29.6	2.0	30.6	2.1
40	41.7	2.4	42.5	1.9
50	51.7	1.9	51.2	2.0

https://doi.org/10.52202/080042-0038

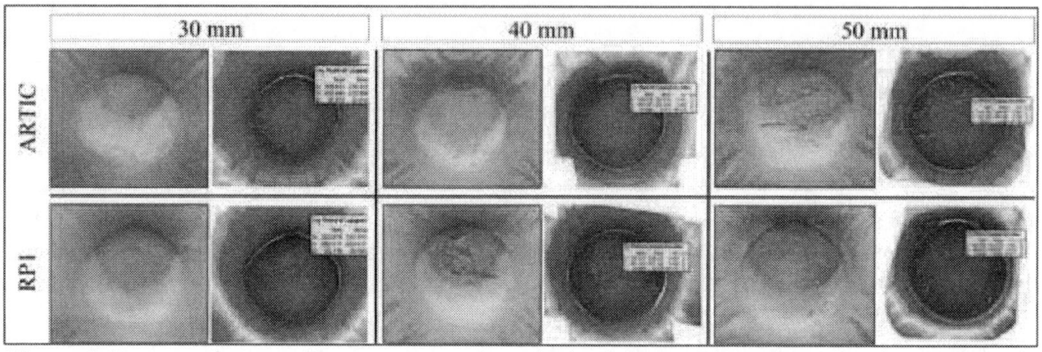

Figure 6. Post-test photos and laser arm scans from the hard armour RTP subtest showing BFD depths of approximately 30mm (left), 40mm (center), and 50mm (right) in ARTIC (top) and RP1 (bottom).

Hard armour V_{50} BL testing for ARTIC and RP1 produced a difference in V_{50} BL point estimates of 3 m/s that is not statistically significant. Figure 7 shows statistically modeled probability of penetration curves that ATC derived from this data. In addition to the test on the V50, a statistical test on the overall performance was conducted. Even though there is some difference in the tails observed in figure 7, the test indicates that the difference in overall performance is not statistically significant; however, additional study is needed prior to drawing firm conclusions.

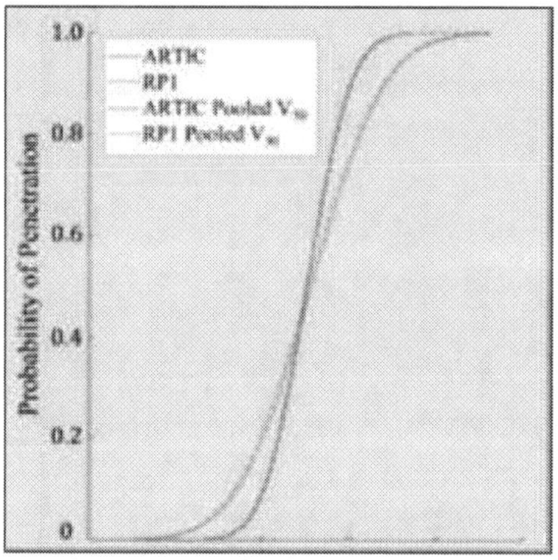

Figure 7. Probability of penetration versus projectile velocity from the hard armour V_{50} BL subtest.

3.6 RTP 1 and RTP 2 (Helmet) Subtests

Figure 8 shows representative photos from the helmet RTP 1 subtest in which ARTIC and RP1 were tested concurrently. In this subtest, testers observed highly comparable responses between ARTIC and RP1 clay types with respect to the BFD depth, shape, and overall behavior of the materials for the crown and back shot locations. The difference in average BFD between clay types for the crown and back shot locations is 0.6mm and 0.2mm respectively with standard deviations of 0.8mm and 0.32mm, as shown in Table 8.

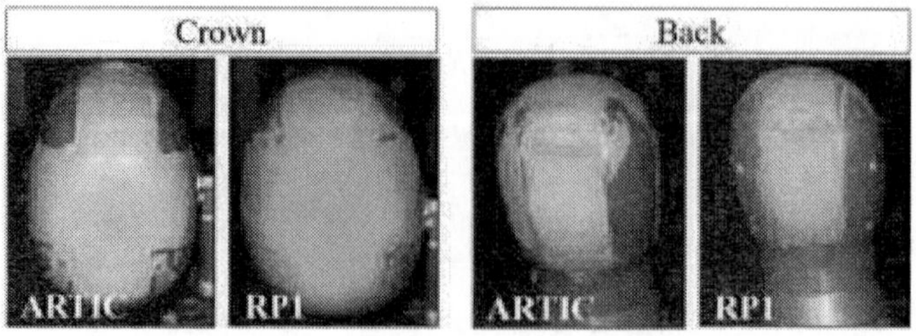

Figure 8. Post-test photos from the helmet RTP 1 subtest showing BFDs in ARTIC and RP1 for the crown (left) and back (right) shot locations.

Table 8. Difference in BFD between ARTIC and RP1 from the helmet RTP 1 subtest.

Shot location	Δ BFD Difference (mm)	
	μ	σ
Crown	0.6	0.08
Back	0.2	0.32

Testers witnessed similar results in the helmet RTP 2 subtest for comparatively greater BFD depths, as shown in Figure 9. This subtest generated significant flow in both clay types because of high compressive and shear strain conditions. Test personnel observed a highly comparable response throughout testing between ARTIC and RP1 regarding the overall behavior of the materials for the back shot location. Figure 10 shows a plot of the full data set for ARTIC and RP1 and depicts the 1.3mm difference found in the average BFDs with a difference in standard deviations of 0.09mm.

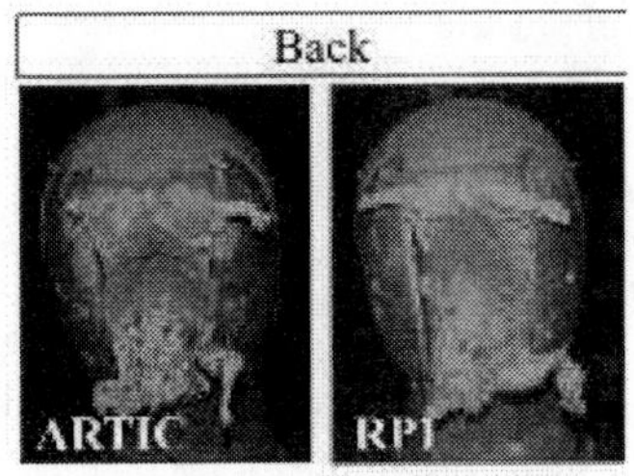

Figure 9. Post-test photos from the helmet RTP 2 subtest showing BFDs in ARTIC and RP1 for the back shot location.

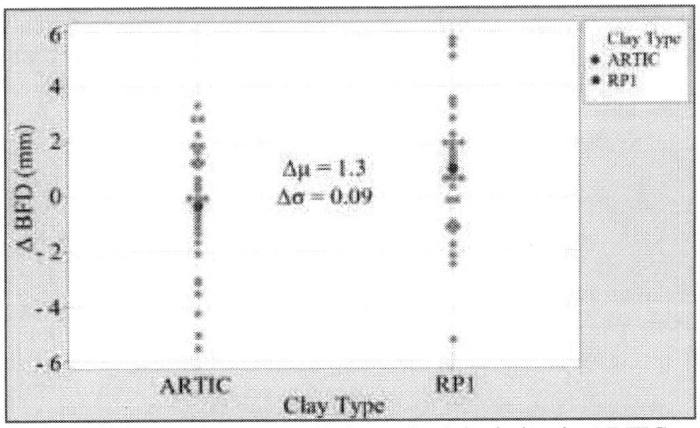

Figure 10. Differences in BFD sample average and standard deviation in ARTIC and RP1 for the back shot location from the helmet RTP 2 subtest.

4. DISCUSSION

This study uses ballistic performance assessments to evaluate the suitability of ARTIC as a material replacement for RP1 ballistic clay in the testing of body armour. To support this suitability assessment, ATC collected backface deformation (BFD) measurements using a laser scanning arm to compare clay performance during controlled conditions with soft armour, hard armour, and helmet testing. Testers also conducted four additional subtests to assess the effects of aging, temperature conditioning, changes in material formulation, mechanical work, and resting time. From a testing perspective, key considerations in the development of a material replacement for RP1 are repeatability, reproducibility, and a robust characterization of the differences in ballistic performance. Therefore, during the testing phase of ARTIC's development, the primary goals are to evaluate material performance for relevant ballistic strain conditions, reproduce those results, and characterize differences in ballistic performance with RP1.

Previously during the material development phase, ARL focused on the development of a robust manufacturing process and matching the material performance of ARTIC to RP1 (heated to near 38°C (100°F)) as closely as possible. ARTIC's performance is not an identical match for RP1 under every test condition examined; however, the results of this multi-year work demonstrate that these objectives have

https://doi.org/10.52202/080042-0038

been achieved to a high degree. This study shows that by a significant margin most (although not all) of the differences observed through ballistic comparisons of ARTIC and RP1 are not statistically significant. On the other hand, shot order, obliquity, and location do affect the ballistic response of both ARTIC and RP1, usually similarly. There are some shot conditions (for example some obliquities) where the similarities between ARTIC and RP1 begin to deviate to a statistically significant level. The most marked example of such a deviation that this study shows is in the 1-Year Aging soft armour subtest where a statistically significantly different material response between the monthly ARTIC results and the RP1 baseline BFD for 14 out of 55 total possible pairwise comparisons for the obliquity 1 portion of the subtest was found. This is well above the 2.75 total expected by chance alone when using a significance threshold of 0.05 and is strongly indicative of a true underlying difference in that test condition. The probability of penetration model created during hard armour V_{50} BL testing also predicts that for certain test designs, projectile velocities above and below the V_{50} BL of the test article may produce different results for ARTIC and RP1, but this finding is very tenuous given the relatively large size of the uncertainly intervals in the tail regions of the ballistic limit curves compared to the strength of the effect this study demonstrated. Lastly, ATC has completed a very limited study that this paper does not discuss in detail that examined the two-year shelf-life of ARTIC. That newer, more limited study does show that there are statistical differences in performance between year one and two – differences that are worth further exploration and study.

In total, the findings of this study support the use of ARTIC as a snap-in replacement for RP1; albeit ARTIC is not nor ever will be an exact match to the highly variable, formulation changing RP1. Despite these positive findings, further testing and changes to handling procedures for ARTIC are necessary prior to full-scale implementation as the replacement for RP1. ATC and others are conducting additional studies and handing procedure trials currently, and this study's authors anticipate that the additional changes recommended by these ongoing studies and trials will be minor. Ongoing and future work in testing includes a large, somewhat comprehensive, ARL led interlaboratory study and continued ATC conducted ballistic testing to increase confidence in existing data. One other element of potential fruitful additional study is in focusing on the surface cracking differences observed by this study between ARTIC and RP1. ATC and the body armour testing community may benefit from verification of similar measurement accuracies for scans taken with these cracks using proven measurement techniques or methods to ensure accuracy and repeatability across the full range of cracking observed on RP1 and ARTIC. Lastly, ATC is also evaluating potential modifications and alternatives to the current drop test calibration method due to observed tactile differences in the current calibration methods across the two clay types.

The final stages of ARTIC's development are commercialization and a phased implementation into full-scale testing. ATC is continuing the development of test procedures for that purpose and for the eventual verification of a commercially produced ARTIC product. ATC in conjunction with U.S. Army material developer partners is planning for concurrent production testing with RP1, during full-scale implementation, for the purpose of collecting additional data in the operational environment and to increase confidence by testing over an extended and more intense period of use. This additional data will add greatly to the body of knowledge presented here and prove very useful to the broader body armor testing community.

Acknowledgements

The authors would like to thank the members of the body armour community who have and continue to support this work, including our many key international partners. The countless contributions of the community, its individuals, and others who have been called upon to contribute their knowledge and expertise to improving the testing of body armour have added unquantifiable value to the academic and operational aspects of this work. We thank all the members of the U.S. and Allied armed forces, and our colleagues in the U.S. Army and other services of the U.S. federal government for their tireless effort in support of the warfighter.

References

[1] National Research Council. Testing of Body Armor Materials for Use by the U.S. Army Phase II: Letter Report. Washington, DC: The National Academies Press, 2010.

[2] R.L. Prather, C.L Swann, C.E. Hawkins, Backface Signatures of Soft Body Armors and the Associated Trauma Effects, U.S. Army, DTIC, 1977.

[3] Committee on Testing of Body Armor for Use by the U.S. Army, in "Testing of Body Armor Materials-Phase III: Report", The National Academies Press, 2012.

Beyond V_{50}: A More Comprehensive and Efficient Methodology for Assessing Armour Performance

S. Magnan[1], G. Pageau[2] and A. Bouamoul[3]

[1]Biokinetics and Associates Limited, 2470 Don Reid Drive, Ottawa, ON, Canada, magnan@biokinetics.com
[2]Calian Ltd., 2459 Route de la Bravoure, Quebec, Qc, Canada
[3]DRDC Valcartier, 2459 Route de la Bravoure, Quebec, Qc, Canada

Abstract. Assessing personal armour protection involves repeated destructive testing to infer ballistic resistance; for example, ballistic limit (V_{50}) evaluations determine the velocity-dependent perforation risk. Many common ballistic standards rely on firing procedures and analysis methods developed before the widespread integration of computer systems in ballistics testing (e.g., the up/down firing method, arithmetic V_{50}). These aspects of ballistic testing, including data acquisition and analysis can now be modernized and automated using commercially available ballistics measurement hardware and software. The new methodology benefits from modern real-time computation capabilities and reduces the cost of ballistics testing by requiring fewer shots to obtain more information, especially for hard armour systems having limited multi-hit capacity. A common objective of ballistics testing is comparing two or more armour systems to determine which is better suited for a given application. The present study proposes a framework to compare armour systems with similar ballistic limits (V_{50}) by leveraging data collected during standard testing to describe the undermatched (V_{05}) and overmatched (up two times the V_{50}, but for practical reasons, it is often limited to $1.5 \cdot V_{50}$) velocity regimes. The analysis introduces two novel data presentation tools, Yawgit, which proposes a mixed velocity-yaw perforation probability, and Ballistic Triple Plots, which fully describe the ballistic resistance characteristic in the three velocity regimes. The discussion also includes implementing and interpreting confidence intervals to differentiate the performance of two armour systems. V_{50} Assist™, a commercially available ballistic testing software that guides users through all aspects of testing described in common ballistic standards, was used in the present study for data collection, firing procedure, and analysis.

1. CURRENT STATE OF BALLISTIC TESTING STANDARDS ANALYSES

Common ballistic test standards for personal body armour systems outline the methodologies, threats, parameters, and analyses required for certification. These standards typically assume that the ballistic limits (V_{50}) metric fully describes the ballistic resistance of armour. Armour systems with a higher V_{50} do tend to provide more protection than systems with a higher V_{50}, but this conclusion only extends to the velocities near the ballistic limit. Common test standards do not attempt to quantify the relative performance of perforating and non-perforating events across the full range of feasible threat velocities. The ballistic limit is not a flawed metric but should be augmented with information obtained across a broader range of velocities. This data is critical for manufacturers to improve products, for researchers trying to understand what affects armour performance, and for acquisition officers specifying performance criteria.

The firing methods specified in standards repeatedly test armour over a range of velocities that elicit both non-perforations (partial penetrations, PPs) and perforations (complete penetrations, CPs). With sufficient tests producing velocity-outcome data, it is possible to perform statistical analyses to ascertain the V_{50} ballistic limit. Standard procedures include traditional up-and-down methods in AEP 2920 [1], NIJ 0101.06 [2], and MIL-STD-662F [3], and the Modified Langlie from [4] although more involved methods, such as three-phase optimal design (3PoD) [5] have been proposed but are not referenced yet in ballistic standards.

Efforts to assess the ballistic resistance properties are complicated by data censoring, which describes the incomplete information gained during testing. Ideally, each shot would produce a single value – the exact outcome transition velocity (V_{OT}) at which a non-perforation (PP) becomes a perforation (CP). Instead, the velocity-outcome information of every shot can only be used to reach one of the two following conclusions: the outcome was a PP; therefore, the velocity did not reach the PP/CP transition, or the outcome was a CP; therefore, the velocity was above the PP/CP transition. Unfortunately, due to variability in the armour and projectile manufacturing/composition (i.e., defects) and test conditions (i.e., projectile yaw), V_{OT} is not identical for every shot. The V_{50} metric approximates the average V_{OT} value.

Trivial metrics such as the average of the *k* highest PP velocities and *k* lowest CP velocities, seen in MIL-STD-662F [3] and Section H.1 of AEP 2920 [1], provide very little information and require

https://doi.org/10.52202/080042-0039

several caveats regarding the overlapping and grouping of results. To reduce bias when using this method, the maximum velocity spread may be constrained [1]. Consequently, advanced statistics such as logistic regression are required to approximate the distribution of the outcome transition velocity (i.e., $V_{OT} \sim N(\mu,\sigma)$). A "link function" such as Logit, Probit, or complementary Log-Log (Gompit) relates strike velocity to outcome probability [6]. Despite weighing velocity-outcome data when solving logistic regression parameters, common testing standards do not sufficiently consider the nuanced differences between two PPs (or two CPs) at different velocities. The definition of an outcome transition velocity may become less clear in specific applications, such as hard armour systems with a known shatter-gap or blunting-gap effect. Using different witness materials to assess perforation (i.e., Army vs Navy vs Protection) also influences the transition velocity. Nonetheless, the fundamental concept of censored outcome transition velocity is transferable. It illustrates the need for a more descriptive assessment of the differences between shots with the same outcome and distinct velocities. Alternate methods to logistic regression that reduce the dependency on normality assumptions and simultaneously describe the physical behaviour and statistics have also been proposed [7]. Logit is the recommended method for NIJ 0101.06 [2], with minimal information provided in the standard regarding the method of interpretation [2]. AEP 2920 (Section H.3) [1] requires a Probit analysis to compute both the mean and standard deviation of the perforation probability distribution; it also mentions leveraging the logistic regression best-fit parameters to approximate the limit velocities (V_{01} and V_{99}) and describes a method of computing confidence intervals on the two regression parameters (V_{50} and σ). The concept of confidence intervals need not be limited to just the V_{50} value but can be plotted over the full range of test velocities and can be used to assess the confidence level at any perforation probability (i.e., V_{05}) [6], [8]. However, no commonly referenced ballistic test standards for personal armour provide a framework for the computation or interpretation of the full confidence interval width at a value other than the V_{50}.

Given the high cost of ballistics testing due in part to the destructive testing of many samples and limited information gained with every shot, full use of all available data should be prioritized. Not all ballistic tests are the same – there are fundamental differences in the testing of soft armour systems compared to ballistic-resistant hard armour plates, from the mounting fixture to the type of backing/recovery media (e.g., clay, foam packs, air). Standard analyses use the velocity-outcome data with little consideration for the many other aspects of ballistic testing that can provide insight into the armour performance. For example, there is little or no consideration for the undermatched (V_{05}) or overmatched (up to $1.5 \cdot V_{50}$), where the results are predominantly PPs or CPs, respectively, but other properties that are not typically measured can vary significantly within each range. Overmatch is more often relevant to fragment-simulating projectiles (i.e., blast simulants) than bullets with velocities constrained by casing size. It is feasible to perform these additional measurements on every shot at minimal cost to quantify the ballistic performance across the three velocity regimes (undermatched, limit, and overmatched) without requiring additional testing. For example, to evaluate the undermatched velocity regime, when testing on a clay block, the backface deformation for PPs may be measured for NIJ 0101.06, or the number of plies perforated in a multi-layer soft armour can be counted. The overmatched performance can be evaluated by measuring the residual velocity (V_r) of armour in an air-backed fixture [9] or the depth of penetration into a recovery media such as multi-layered foam packs. V_r-V_s overmatch data are especially important for fragmentation-resistant armour since used in vulnerability/lethality codes for conducting casualty reduction analysis. Ballistic standards also consider projectile yaw, but only to the extent of assessing shot fairness by limiting to 5° in MIL-STD-662F [3] and NIJ 0101.06 [2], and 3° or 5° in AEP 2920 [1]. To be fully compliant with the testing requirements of these standards, yaw must be measured for every shot; therefore, the data is being recorded but is not used to its full potential.

Notably, the test standards all aim to provide a repeatable framework for assessing ballistic resistance of personal body armour. In this context, the results are taken to indicate a pass or fail at the specified certification level. However, this approach does not provide sufficient context for a researcher or purchaser to understand the difference in performance between two sample types tested according to the same methodology. Therefore, there is currently no guidance on the quantitative comparison of the two armour systems. Single-purpose tools (spreadsheets or code) designed to perform partial analyses (DRDC Ballistic Limit Calculator (BLC) [10]) or guide the user through firing procedures (GoNoGo [5]) have been developed, but no fully integrated software exists. V_{50} Assist™ (Biokinetics and Associates Ltd., Ottawa, Canada), which was used to produce all results in this paper, is a commercial-off-the-shelf (COTS) ballistics software package that walks the user through firing procedures, data collection, and analyses as described in common ballistic standards and literature.

2. METHODOLOGY

For the present study, two multi-layered soft armour systems were tested according to procedures outlined in AEP 2920. Testing was performed on a soft armour clamping fixture with force and tensions described in the DRDC-V frag vest method [9]. Witness paper positioned approximately 150 mm down range of the samples was used to assess the outcome of each shot (i.e., protection criterion) as a CP or a PP. The standard test methodologies were modified slightly with additional data collection not typically performed in ballistic testing to increase the value and knowledge gained for every shot. For each shot, the strike velocity, projectile yaw, and residual velocity (for CPs only) were measured, and the number of perforated plies was counted. Both materials were tested using the 1.1g (17 gr.) chisel nose cylinder Fragment Simulating Projectile (FSP) defined in [1].

2.1 Sample Preparation

Two materials were included in the present study. The focus of the present study is on the comparative analysis and not the relative performance of two materials with similar ballistic limits. Material A had an assembled sample areal density of 3 kg/m^2). Preliminary testing was conducted on the Material B packs to determine the ply count required to have a comparable V_{50} to Material A. Material B had an assembled sample areal density of 2.6 kg/m^2. Samples were assembled in 400mm x 400mm layers for use in the DND clamping fixture [9]. Each sample was clamped to the required 2-30 N before testing. The samples were partially stitched along the upper corners and in a U-shape along the lower perimeter to facilitate the capture of the projectiles that did not fully perforate the samples. The stitching pattern and nine-shot firing pattern are shown in Figure 33.

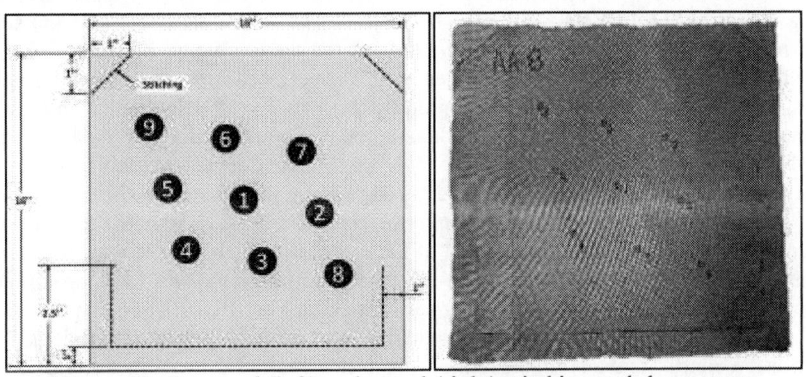

Figure 52: Proposed (left) and actual (right) stitching and shot pattern.

2.2 Firing Procedure

A total of 48 shots were conducted on each of the two materials. The shots were split into three series. First, 16 shots were performed using the AEP 2920 [1] up/down firing method with an initial velocity estimate of 520 m/s. The following series of 16 shots were conducted with a reinitialized up/down procedure where the initial velocity was the Probit V_{50} of the 16 shots from series 1. The initial velocity for the third series was determined using the combined 32-shot dataset (16 from each of series 1 and 2).

2.3 Measurement Devices

Data collection was performed using commercially available measurement devices. Projectile velocity was assessed using the SpeedTube^{TM} (Biokinetics and Associates Ltd., Ottawa, Canada), a ballistic chronograph with two pairs of light gates to redundantly measure the velocity of each shot at approximately 2.5 m before impact. The strike velocity was computed in the SpeedTube^{TM} (Biokinetics and Associates Ltd., Ottawa, Canada) software using the average drag coefficient method described in Eq II in Annex K of AEP 2920 [1]. Yaw was measured approximately 250 mm up-range of the sample using the YawBox^{TM}, which uses a single camera and mirrors to obtain two orthogonal views of the projectile. Both the YawBox^{TM} and SpeedTube^{TM} are shown in Figure 53. A Doppler radar (Infinition Inc., Trois-Rivières, Canada) was positioned to measure the residual velocity of projectiles that fully perforate the armour system. The Doppler radar was positioned at a 30° angle from the firing trajectory; therefore, a correction factor was applied to determine the residual velocity along the initial trajectory.

https://doi.org/10.52202/080042-0039

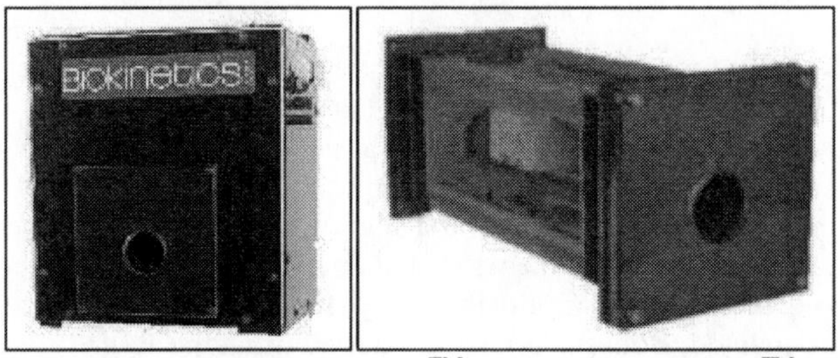

Figure 53: Pictures of the YawBox™ (left) and SpeedTube™ (right).

2.3.1 Measurement Uncertainty

All measurement devices in this study were analysed to quantify the expanded uncertainties according to the principles of the guide to the expression of uncertainty in measurement (GUM) and ISO 17025. The expanded uncertainty, reported with a coverage factor of 2 and a normally distributed coverage level of approximately 95% is ±0.12% for the SpeedTube™, ±0.46° for FSPs with the YawBox™, and ±1.1% for 17-grain FSPs with the Infinition doppler radar in its present configuration.

2.4 Data Collection

Data collection and analyses were performed using the V_{50} Assist™ COTS software, an all-in-one package to walk a user through standard ballistic testing. The initial parameters of the selected firing procedure were input into the software, which provided the next velocity. For every shot, the powder load used by the technician, the strike velocity reported by the SpeedTube™ (Biokinetics and Associates Ltd., Ottawa, Canada), the projectile yaw measured by the YawBox™ (Biokinetics and Associates Ltd., Ottawa, Canada), and, if applicable, the residual velocity from the doppler radar system, outcome (PP/CP), and the time of the test, were input into V_{50} Assist. Preceding test velocities and outcomes were used to recommend the next velocity and load required to reach that velocity. The end of each test series is indicated by the software, and individual completion criteria (e.g., CI width less than 4% of V_{50} in [1]) are updated after every shot. Following the end of the test series, each sample was dissected to identify how many layers had been perforated for each PP. Any modification of test data prompted immediate recompilation of all analyses, including all parameters specified in MIL-STD-662F [3], NIJ 0101.06 [2], and AEP 2920 [1]. Charge calibration, chronological shot velocities, logistic regressions (Logit, Probit, Gompit, Scobit, Weibull, Yawgit), unperforated ply ratio, energy absorption ratio, and several other analysis types are plotted to illustrate the benefits of recording more data during ballistic testing.

3. ANALYSIS METHODOLOGY

The present study demonstrates a methodology that maximizes the information gained during ballistic testing. This is achieved using all available data points, including performing additional measurements to adequately quantify the ballistic resistance in the three velocity regimes (under-matched, V_{50}, and over-matched) without requiring additional shots or armour samples.

3.1 Ballistic Resistance Triple Plot

The ballistic resistance triple plots are a visual representation of the armour performance in the three velocity regimes overlayed on the same axis. Different curves may be selected for the three velocity regimes depending on the armour type and test configuration. To illustrate the potential of the triple plots, the following three curves were selected and plotted as protection curves (i.e., probability of 100% at 0 m/s): unperforated ply ratio (UPR) for under-matched velocities, logistic protection probability for ballistic limit velocities, and energy absorption ratio (EAR) for over-matched velocities. The UPR is defined as the fraction of layers that were not perforated during the test. For example, if a fragment perforated 10 plies in a 40-ply sample, the UPR is 0.75. The UPR quantifies the remaining protective margin and helps assess the safety margin (i.e., how close was the sample to failure?). As is common practice in ballistics testing, any non-perforating fragments would remain in the sample for subsequent shots and dissection is only performed after all shots are completed. The data is trivial to acquire following typical testing but is generally discarded. The effects of trapped projectiles on future shots are unknown; however, shot spacing is typically selected to ensure a suitable shot-to-shot spacing. The UPR definition presented here is a special case of the residual areal density ratio (RADR), where the test samples are composed of a single material. The more general RADR, which is the sum of the areal densities of all unperforated layers normalized by the areal density of the sample, can be used if materials with different areal densities are present in the sample. A logit-inspired continuous fit was then used to determine the expected UPR across the full range of velocities.

$$P_{surv-UPR} = 1 - \left(1 + e^{-(\beta_0 + \beta_1 UPR)}\right)^{-1} \quad (1)$$

The protection probability can use any logistic regression link function and is the complement of the perforation probability. The logit link function formulation was used in this study.

$$P_{surv-logit} = 1 - \left(1 + e^{-(\beta_0 + \beta_1 V_s)}\right)^{-1} \quad (2)$$

The over-matched component is based on the residual velocity model of Lambert-Jonas described in [11] based on an energy approach with three parameters describing the magnitude, asymptotic slope, and limit velocity. The Lambert-Jonas equation was previously shown to be adequate for computing the armour effective velocity [12] characterizing the overmatch regime. The energy absorption ratio, described in [13], is the difference between the incident projectile kinetic energy and the projectile's residual kinetic energy after passing through the armour. The EAR, which is the absorbed energy (incident-residual) normalized by the incident energy, assumes the energy used to deform the projectile is negligible and may not be valid for all projectiles. The EAR version presented here is derived by inserting the Lambert-Jonas residual velocity regression in the EAR definition. After the computation of the EAR over the full range of velocities, the Armour Performance Rating (APR) can be computed as the average EAR in a velocity range [13]. Here, it is computed between V_{50} and $1.5 \cdot V_{50}$.

$$EAR = \frac{V_s^2 - \left(\alpha\left(V_s^\delta - V_l^\delta\right)^{\frac{1}{\delta}}\right)^2}{V_s^2}, \quad V_s > V_l \quad (3)$$

3.2 Yawgit

Many common ballistic testing standards call for projectile yaw to be measured shortly before impact to judge the fairness of the test. For example, if the yaw exceeds 3° or 5°, the test should be repeated according to [1] and [2]. Unfortunately, yaw cards, which are still commonly used in ballistics test facilities, provide cruder measurements than digital systems [14] and are likely less repeatable due to the potential for different measurements by different technicians. With COTS digital systems, yaw can be assessed with significantly higher certainty by eliminating user variability. Standards requiring yaw measurements as a go/no-go screening tool acknowledge that the yaw angle affects the outcome but assume that if it is close to a direct impact, the effect is small and thus negligible. To maximize the value of each test, the effects of projectile yaw can be quantified to provide meaningful information regarding the ballistic resistance, particularly when the value is already measured for fairness screening.

In ballistic testing, the Logit perforation probability depends on an expression containing a linear combination of a constant and the strike velocity (V) as the argument. Other logistic regressions (e.g., Probit, Gompit) are similarly constructed. Expanding the Logit argument to include a contribution from precise digital yaw angle measurements is now possible. The proposed modification can be applied to any logistic link function (e.g., Logit).

$$P_{yawgit} = \left(1 + e^{-(\beta_0 + \beta_1 V + \beta_2 \theta)}\right)^{-1} \quad (4)$$

The new perforation probability (Yawgit) adds a linear combination of the yaw (θ) to the original formulation. Significant insight into the validity of the small-yaw screening assumption can be gained from this logistic regression. For example, the magnitude and sign of β_2 may help determine if small yaw angles are more or less likely to perforate the armour. An analysis of the sensitivity to outliers was not performed in this study. For data representation of Yawgit, it is recommended to solve for the coefficients using maximum likelihood, then plotting the perforation probability across a range of velocities for several fixed yaw angles (isolines) at 0°, 1°, 2°, 3°, 4°, and 5° (isolines). The isolines will have the same

https://doi.org/10.52202/080042-0039

slope but will be translated along the velocity axis. Similarly, the isolines can be defined at fixed velocities while varying the yaw (i.e., V_{50}, V_{50}+50m/s)

3.3 Confidence intervals

A comparison of the perforation probability versus velocity curves of two materials cannot be performed without confidence intervals (CIs). AEP 2920 [1], which provides the most thorough description of confidence intervals in common ballistic standards, only provides a method of assessing the confidence intervals for the Probit fit coefficients (V_{50} and σ). Proper confidence intervals on probability for the full velocity range in logistic regression are constructed using Wald's test. These confidence intervals can be interpreted around the central portion of the curve (e.g., V_{25}-V_{75}). Unfortunately, they may become difficult to interpret in small series with few shots due to divergence at the upper and lower tails. The probability confidence intervals determined using Wald's test (vertical) can be transformed into confidence intervals on velocity (horizontal), which have much better convergence characteristics [15]. The interpretation of these confidence intervals is different and depends on the formulation of the research question [15]. For example, if the focus is on the range of velocities that contain the V_{50}, the horizontal formulation is suitable. Otherwise, the vertical formulation, which is used in AEP 2920 [1], must be used if the focus is on the range of perforation probabilities at a given velocity.

The selection of the coverage level of the confidence intervals is critical to the interpretation of the results. When considering a single test series, a 95% confidence level is suitable and reflects the confidence level required by AEP 2920 [1] on logistic coefficients. If two logistic regressions are being compared, the confidence interval width must be adjusted. The null hypothesis being posed is effectively the following: what is the probability of observing non-overlapping V_{50} confidence intervals for independent series if V_{50A}=V_{50B}? The answer is $p<0.006$ if using 95% confidence intervals. Instead, statistical significance (i.e., $p<0.05$) is reached when the confidence level is reduced to 83.4% [16]. Note, 83.4% confidence intervals are narrower than 95% confidence intervals. Both types of confidence intervals are implemented in V_{50} Assist™ where the confidence level can be customized to any value.

4. RESULTS AND DISCUSSION

The present analysis first computed the ballistic limit and associated parameters as specified in common personal ballistic armour standards (MIL-STD-662F [3], NIJ 0101.06 [2], and AEP 2920 [1]) using V_{50} Assist™ (Table 15). Series 1 of the Material B testing had a large zone of mixed results (ZMR) and thus failed to quantify a valid arithmetic V_{50} according to the H1 method of AEP 2920. The metrics computed according to common ballistics standards are primarily focused only on the V_{50} value. However, NIJ 0101.06 [2] discusses how to compute the velocity corresponding to any perforation probability, and AEP 2920 [1] shows equations for V_{05} and V_{95}. The benefit of many of these approaches is their simplicity and ease with which the performance can be measured. The alternative is that every test facility and research lab must apply its own statistical analyses to achieve maximum benefit in ballistics testing. Therefore, relying on fully validated and verified COTS analysis packages to walk a user through ballistics testing would provide a standardized series of traceable tools for any test facility.

Table 22. Summary of representative ballistic results as per common ballistic standards

	Material A Results (m/s)				Material B Results (m/s)			
Series	1	2	3	1-3	4	5	6	4-6
MIL-STD-662F								
6 Shot V_{50}	496.2	501.4	495.0	497.8	517.3	512.2	504.2	510.3
NIJ 0101.06								
Logit V_{50}	496.9	501.1	494.4	497.0	512.8	511.8	503.8	509.6
Logit Std. Dev.	7.3	12.2	5.6	9.0	30.7	17	23.5	23.8
AEP 2920 §H.1								
Arithmetic V_{50}	493.1	498.0	497.5	493.1	*	512.5	522.2	*
AEP 2920 §H.3								
Probit V_{50}	497.4	500.7	494.1	497.0	513.1	510.9	503.6	509.6
Probit Std. Dev.	12.5	20.7	9.3	15.7	51.6	28.7	38.5	40.0
V_{50} Lower 95% CI	484.8	485.7	486.7	489.9	479.7	490.5	478.1	494.2
V_{50} Upper 95% CI	508.9	515.2	501.5	504.0	545.6	531.2	529.1	524.9

*Convergence criteria not met

4.1 Ballistic Resistance Triple Plot

The fundamental objective of a ballistic resistance triple plot is to simultaneously present protection outcomes in the three velocity regimes (undermatched, limit, and overmatched). In the present study, the three representative curves selected to represent the velocity regimes were: the Unperforated Ply Ratio (UPR), protection probability logit, and Energy Absorption Ratio (EAR). The average EAR over the range of V_{50} to $1.5 \cdot V_{50}$ is shown as well. This can be particularly beneficial in cases where it is impossible to differentiate the performance of two armour systems using computations outlined in ballistic standards (Table 15). The triple plot with fit coefficients (Figure 54), constructed using the combined 48-shot dataset for each material, illustrates the relative ballistic performance of Material A and Material B across the three velocity regimes further supporting the assertion that Material B may have outperformed the Material A samples in this series of tests.

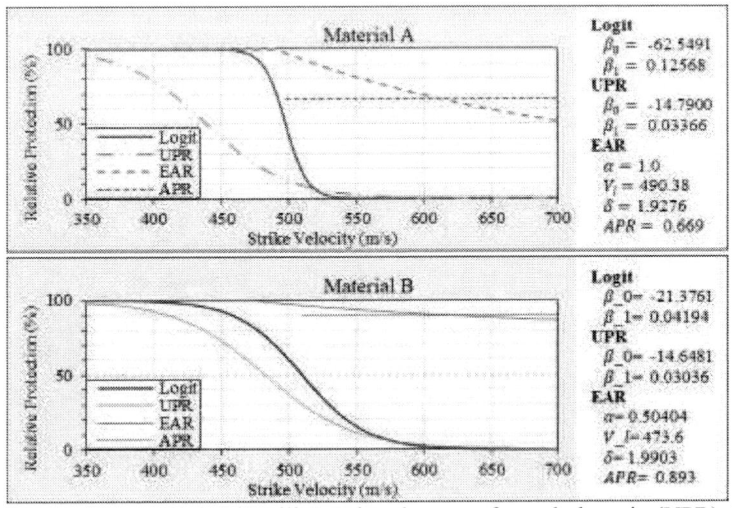

Figure 54: Ballistic resistance triple plots illustrating the unperforated ply ratio (UPR), logit protection probability, energy absorption ratio (EAR), and armour protection rating (APR) for two materials.

First, by examining the logistic regression, the slope for Material A is steeper than for Material B, which indicates a more consistent performance for perforation probability (i.e., a smaller ZMR). The undermatched performance shows similar slopes (note: the number of plies for each material was different, but the ratio of unperforated plies provides a suitable basis for comparison). Typically, the V_{05}-perforation level (i.e., V_{95} for protection) would be used to differentiate the undermatched performance of two armour systems with a similar $V_{50;}$ if additional data were not considered here, the Material A sample might have been considered superior. Another potentially interesting parameter when considering the undermatched velocity regime is the UPR at the V_{50} ballistic limit (approximately 10% for Material A and 35% for Material B). The significance of this result is unknown at this time; however, it may also be indicative of the ZMR width and consistency of results. The overmatch results interestingly converged on a similar limit velocity (from the Lambert-Jonas equation). The slope parameter indicates that Material B samples absorbed more energy than Material A. The EAR is averaged across the range of V_{50}-$1.5 \cdot V_{50}$ to produce the APR. The steeper drop-off in the EAR for the Material A samples is reflected in the lower APR. The triple plot was constructed using data obtained in three series of 16 shots on each material. Significantly more information was gained by processing the results to query the performance in the three velocity regimes than if only the logistic regression of the perforation (or protection) probability was queried. If the test velocities did not have a suitable spread, a small number of additional shots could be conducted at undermatched and overmatched velocities to improve the confidence in the corresponding curves in the triple plot. Depending on the test, equivalent plots could be formulated to include backface deformation or recovery media depth, for example, if clay or foam packs were used. The focus of this paper is the comparative methodology when two materials have similar ballistic limits. Differences in areal density between Material A and Material B were therefore not considered (i.e., no normalisation was applied to the results).

https://doi.org/10.52202/080042-0039

4.2 Yawgit

The Yawgit curves (Figure 55) is a method of quantifying the perforation resistance of an armour system as a function of both velocity and yaw. Including isolines allows for a more intuitive interpretation of two-input regression (perforation probability as a function of yaw and velocity) than a heat map. The two types of isolines correspond to inputting a constant yaw or a constant velocity into the regression line. The interpretation of the isolines depends on the regression coefficients, which describe the slope, scale, and direction of isoline spacing. For example, Material A showed a very small positive correlation between perforation probability and yaw angle as shown in Figure 55 (a), where the isolines are very close together, and in Figure 55 (b), where the slopes are negative and shallow. Material B showed a negative relationship between perforation probability and yaw in Figure 55 (c), where the isolines for higher yaw angles are translated towards lower velocities, and in Figure 55 (d), where the slopes are positive.

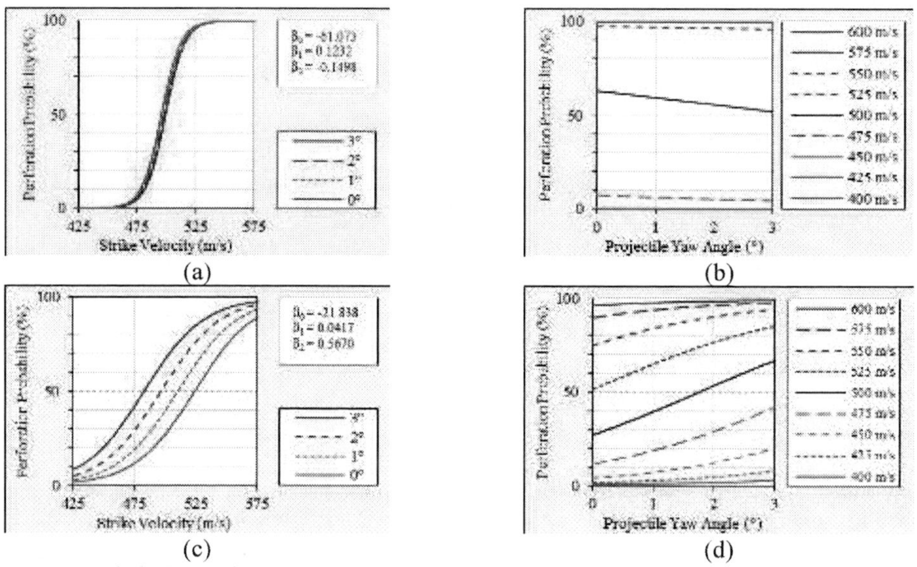

Figure 55: Logistic (Yawgit) regression for perforation probability plotted with (a) Material A yaw isolines and (b) velocity isolines, and Material B (c) yaw isolines and (d) velocity isolines.

The interpretation of Yawgit isolines should normally be accompanied by a statistical analysis to assess the significance of the regression coefficients. The interpretation may be further limited by outliers. It may be prudent to verify the cross-correlation between velocity and yaw before performing a Yawgit analysis. When used carefully, the Yawgit analysis demonstrates that additional measurements performed during ballistic testing can better characterize armour performance across test conditions and velocities while providing an additional basis for comparing systems with similar V_{50} values.

4.3 Confidence Intervals

The use of confidence intervals is essential to fully understand the perforation probability of an armour sample and the expected variability of the results. In general, the width of the confidence intervals is expected to decrease as more shots are performed. The width of the ZMR will likely also affect the confidence intervals as a shallower slope in the logistic regression indicates a more variable outcome transition velocity. The standard up-and-down firing sequence, which focuses the shots on the V_{50} may not provide the true ZMR and the corresponding CI. Better-suited approaches such as 3PoD ensure that the actual extent/width of the ZMR is sufficiently explored, leading to more reliable CIs. The formulation of confidence intervals and the coverage level significantly affect the interpretation of the expected variability in results. When considering only a single test series, a 95% confidence level should be used to bound the expected results; however, the difference between ballistic limits may still achieve statistical significance if the 95% confidence intervals overlap. Given a null hypothesis $V_{50A}=V_{50B}$, the probability of the confidence intervals not overlapping reaches statistical significance (p-value <0.05) when the confidence intervals are narrowed to 83.4% [16]. Therefore, the probability of observing two non-overlapping 83.4% confidence intervals if the ballistic limit is the same is less than 5% (i.e., 95% confidence level). This approach provides an intuitive and visual methodology for assessing the <5%

https://doi.org/10.52202/080042-0039

probability of two materials having the same ballistic limit if the pointwise +confidence intervals do not overlap. Figure 56 shows a case where the confidence level was reduced from 95% to 83.4%. This change resulted in a significantly different interpretation of the data beyond the 60% perforation probability. Therefore, inferences about the perforation probability above the V_{60} are only possible after reducing the coverage level, thereby eliminating any overlap.

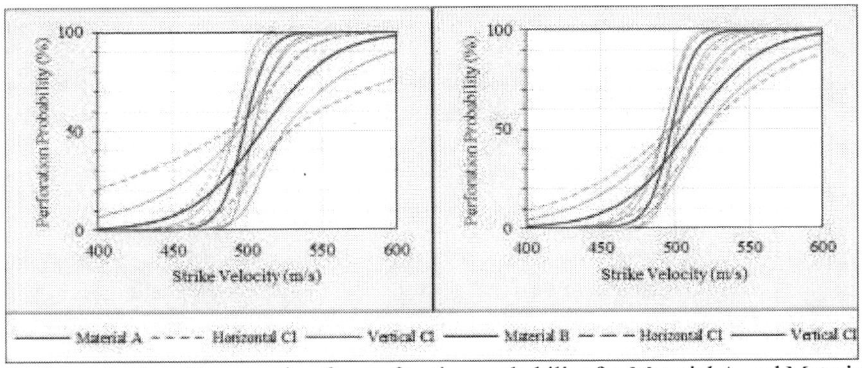

Figure 56: Logistic (Logit) regression for perforation probability for Material A and Material B test data plotted with 95% confidence intervals (left), and 83.4% confidence intervals (right).

Traditional confidence intervals computed using Wald's test typically show a much larger range of expected values towards the extremes (upper confidence bound for V_{01} and lower confidence bound for V_{99}) than at the ballistic limit (either bound for V_{50}). Transforming the confidence intervals from their original vertical formulation to the horizontal variant is a method that generally improves the convergence at the extremes of the curve. In all cases shown in Figure 56, the vertical formulation of the confidence intervals resulted in tighter confidence bands.

5. CONCLUSION AND RECOMMENDATIONS

The significant resources required to quantify the ballistic resistance of armour systems are the driving factor limiting the number of tests conducted. Existing ballistic test standards place the importance almost exclusively on velocity-perforation outcome data while focusing heavily on the V_{50} ballistic limit. There is limited emphasis on exploring other key metrics that describe the undermatched or overmatched ballistic resistance. The methodology described in this study provides the tools to quantify and interpret data that could be trivially acquired during testing with minimal additional effort yet further understanding of threat mitigation behaviour across a wide range of velocities and test conditions, which could significantly increase the value of every test.

This work introduced Yawgit, a bivariate logistic regression to quantify the perforation probability as a function of velocity and projectile yaw, leveraging yaw measurements that are already performed as a go/no-go screening tool. The proper use and interpretation of two types of confidence intervals on perforation curves (Logit) were also discussed, with a special focus on the coverage level being 95% for a single test series or 83.4% if two series are being compared with 95% confidence.

Ballistic resistance triple plots were created to describe armour performance across the three velocity ranges, providing a suitable basis for comparing two armour types. The ballistic resistance triple plots are flexible to different armour and threat types so long as the three velocity regimes are adequately represented. The tests described in this paper included an analysis of the unperforated ply ratio (UPR) for non-perforations of soft armour systems which quantifies the safety margin before failure. The energy absorption ratio (EAR) and the Armour Performance Rating (APR) were derived from the residual velocity to quantify the margin by which the armour was defeated. Examining metrics beyond the V_{50} is critical to comprehensively assessing the armour performance across all feasible velocities. These metrics, and others, can be used by manufacturers to design better armour systems or help scientists understand projectile/armour interactions and eventually could be integrated into test standards to ensure operators have the best possible protection.

The absence of computationally non-trivial metrics may be indicative that reliable, verified tools are not widely available to perform these analyses. The simplicity of concepts such as up/down firing procedures and arithmetic V_{50} does not justify their continued use when more advanced and descriptive alternatives exist. Implementing advanced ballistics analyses and statistics in test standards is long overdue. Many test facilities use various spreadsheets to track different aspects of ballistic testing,

https://doi.org/10.52202/080042-0039

including firing procedures and analyses. V_{50} Assist™ is a fully validated and traceable COTS software package that walks technicians through ballistic testing, from firing procedures to analyses described in common ballistic standards. It was developed to provide ballistic test facilities (commercial and experimental) with a set of tools that provide consistent analyses independent of the statistical or analytical background of the technicians or researchers.

Disclosure Statement
The first author is a paid employee of Biokinetics and Associates Ltd., the developer of V50 Assist™, SpeedTube™ and YawBox™, for the duration of the present study. The methodology proposed in this study can be applied using software and hardware from other manufacturers.

References

[1] North Atlantic Treaty Organization, "NATO STANDARD AEP-2920 Procedures for the Evaluations and Classification of Personal Armour. Bullet and Fragmentation Threats. Edition A Version 1," NATO Standardization Office, 2015.

[2] National Institute of Justice, "Ballistic Resistance of Body Armor NIJ Standard-0101.06," U.S. Department of Justice, Washington DC, 2008.

[3] Department of Defense (DoD), "V50 Ballistic Test for Armor (Test Method Standard MIL-STD-662F)," Department of Defense, 1997.

[4] C. C. Joseph and M. L. C. Linda, "LangMod Users Manual," Survivability/Lethality Analysis Directorate, Army Research Laboratory, Aberdeen Proving Ground, 2011.

[5] P. A. Roediger, "Gonogo: An R Implementation of Test Methods to Perform, Analyze and Simulate Sensitivity Experiments," arXiv, Ithaca, NY, 2020.

[6] D. Mauchant, K. D. Rice, M. A. Riley, D. Leber, D. Samarov and A. L. Forster, "Analysis of Three Different Regression Models to Estimate the Ballistic Performance of New and Environmentally Conditioned Body Armour," National Institue of Standards and Technology (NIST), US Department of Commerce, 2011.

[7] B. Tahenti, F. Coghe, R. Nashri, M. Pirlot and B. Lauwens, "The Brownian-Based Approach Applied to a Limited Sample Sie For Ballistic Resistance Evaluation: Two Experimental Databases," in *32nd International Symposium on Ballistics*, Reno, Nevada, USA, 2022.

[8] M. A. Riley, R. D. Kirk and F. L. Amanda, "Assessment of Uncertainty in Ballistic Response Estimates Obtained from Ballistic Limit Testing," in *Personal Armour Systems Symposium*, Nuremberg, 2012.

[9] Canadian National Defence, "Technical Puchase Description Fragmentation Protective Vests for the Canadian Forces," Department of National Defence, 2004.

[10] D. Bourget and M. Bolduc, "Calculation software for ballistic limit and Vproof tests from ballistic tests results (BLC)," Defence Research and Development Canada, 2020.

[11] J. P. Lambert and G. H. Jonas, "Towards Standardization in Terminal Ballistics Testing: Velocity Representation," USA Ballistic Research Laboratories (BRL), Aberdeen Proving Ground, Maryland, 1976.

[12] D. Bourget and G. Pageau, "The Effective Ballistic Resistance Concept, a New Approach for Assessing the Average Energy Absorption Capability of Armour Materials," in *Proceedings of the 18th International Symposium on Ballistics*, San Antonio, Texas, USA, 1999.

[13] G. Pageau and S. Ouellet, "Development of soil ejecta surrogate projectiles for laboratory testing of lightweight protective materials," in *Proceedings of the Personal Armour Systems Symposium 2020*, Copenhagen, Denmark, 2020.

[14] R. Decker, M. Duca and S. Spickert-Fulton, "Measurement of Bullet Impact Conditions Using Automated In-Flight Photography System," in *20th International Symposium on Ballistics*, Long Beach, CA, USA, 2017.

[15] T. L. McMurry and G. S. Poplin, "Statistical Considerations in the Development of Injury Risk Functions," *Traffic Injury Prevention,* pp. 618-626, 2015.

[16] M. J. Knol, W. R. Pestman and G. E. Diederick, "The (mis)use of overlap of confidence intervals to assess effect modification," *European Journal of Epidemiology,* pp. 253-254, 2011.

Data Filtering for the Analysis of Biological Tests for Behind Armor Blunt Trauma Studies

A. Shah[1], J. McMahon[2], J. Op't Eynde[3], R. Salzar[2], B. Johnson[4], and J. McEntire[4]

[1]*Medical College of Wisconsin, 8701 Watertown Plank Road, Milwaukee, WI, USA*
alokshah@mcw.edu
[2]*University of Virginia, 1827 University Avenue, Charlottesville, VA 22903*
[3]*Duke University, 2138 Campus Drive, Durham, NC 27708*
[4]*U.S. Army Aeromedical Research Laboratory, 6901 Farrel Rd, Fort Rucker, AL*

Abstract. The use of body armor has been shown to reduce the risk of penetrating injuries among military personnel. While the body armor reduces the risk of penetrating injuries, the energy from the impact can still be transferred through the armor, causing backface deformation, resulting in injuries to the underlying tissue. The resistance of the ballistic armor is tested based on the maximum allowable backface deformation limit of 44 millimeter(mm). The standard was developed over four decades ago for soft body armor from a limited set of data from goat experiments. Although the standard was developed for soft body armor, the standard was adopted for hard armor. The 44mm standard is independent of threat level, threat type, and impact location. The development of thoracoabdominal region specific injury criteria will aid in assessing and designing body armor and improve BABT safety against current and future threats. Surrogate models can be used to develop these region injury risk curves for BABT application. Cadaver swine and human models have been used in automotive studies for over 60 years for developing injury criteria. Although, automotive risk curves are not appropriate for BABT application due to lower velocity loading conditions and greater impacting mass, the hybrid approach of utilizing swine and human cadavers to develop scaling relationships can be used to develop region specific risk curves for BABT. These tests will require data collection on biological specimens at high rates. Although there are established standards for processing these data in the automotive environment, there are no established standards or consensus for processing data from BABT experiments. This study analyzed indentor accelerometer data collected from swine and human cadavers and developed a protocol for data processing to determine load applied to thoracoabdominal body regions for BABT applications.

1. INTRODUCTION

The use of body armor has become increasingly prevalent in military and law enforcement contexts, with the aim of protecting individuals from ballistic threats [1]. However, the use of body armor can also result in a phenomenon known as behind armor blunt trauma (BABT), which occurs when an individual is struck by a projectile that is unable to penetrate the armor but still causes injury. BABT can result in a range of injuries, including fractures, contusions, and internal organ damage [2]. Despite the potential severity of BABT injuries, there is a lack of comprehensive data on the incidence and nature of these injuries in peer-reviewed literature, like the automotive field. This lack of data makes it difficult to develop effective measures to mitigate the risk of BABT and improve body armor design.

The National Institute of Justice (NIJ) Standard-0101.06 is an armor test standard commonly used in the industry as a benchmark for evaluating the ability of body armor to protect against BABT [3]. Tests are conducted by placing the body armor over Roma Plastilina No. 1 (RP1) clay backing and targeted based on the prescribed conditions in the standard. Assuming there is no penetration of the armor, the armor is solely evaluated based on the maximum allowable backface deformation limit of 44 mm. It is important to note that the 44 mm limit was developed specifically for soft body armor based on a limited set of experiments on live goats [4]. The original researchers of the study underscored its limitations and emphasized the need to conduct additional tests that included animal experimentation and simulants to improve the 44 mm clay standard. The legacy standard developed over four decades ago remains the current BABT standard independent of armor type, threat level, or impact location on thoracoabdominal region covered by armor.

Biomechanical responses of the human tissue, anatomy, and loading conditions from back-face deformation play a significant role in resulting injury and the injury mechanism. In humans, the different mechanical properties of musculoskeletal structures in the thoracoabdominal region (tissues, organs, and ribcage), may lead to different responses in injury, injury tolerance and mechanisms [5]–[8]. It is important to delineate these region-based tolerances considering a single threshold of 44 mm limit may not be appropriate in determining injury risk severity. Additional testing is needed to develop risk curves for different thoracoabdominal regions. An hybrid model of cadaver and live swine tests in conjunction

https://doi.org/10.52202/080042-0040

with human cadavers have been used to develop injury risk criteria for automotive safety for decades [9]. Although these risk criteria exist, they are not applicable for BABT injures as the impact velocities are significantly greater and mass of projectile is lower compared to velocities and masses used for automotive testing. Additional testing is needed to develop region specific injury risk curves for BABT by utilizing cadaveric swine, live swine, and human cadavers, similar to automotive safety testing, to update the legacy 44 mm standard.

These experiments will require the use of sensors such as accelerometers to capture the loading event at high rates. Existing standards for processing injury biomechanics data signals are based on automotive loading cases associated with lower velocities than those seen in military environments. The objective of this study was to process data from behind armor blunt trauma (BABT) experiments on post-mortem human subjects (PMHS) and swine cadavers at varied frequencies to standardize the filtering system for use in the development of BABT injury criteria.

2. METHODS

All experiments were conducted in compliance with the Institutional Animal Care and Use Committees at all three academic institutions of the authors: Duke University, University of Virginia, Medical College of Wisconsin, and Zablocki Veterans' Administration Medical Center. Prior approval was obtained from the Animal Care and Use Review Office (ACURO) of the U. S. Department of Defense.

2.1 Impactor Design

High-speed flash x-ray images were used to determine the depth and diameter of backface deformation in hard body armor (UHMWPE) from rifle rounds (7.62-mm NATO ball round) [10]. An impactor approximately 100 mm in diameter with a dome depth of 25mm was designed based on measurements from flash x-rays (Figure 72). A triaxial accelerometer (Endevco) was mounted within the impactor using a hardwired (DTS Slice Nano) or 'onboard' (DTS Slice Pro) data acquisition system and sampled at over 100 kilohertz (kHz). (Figure 73)

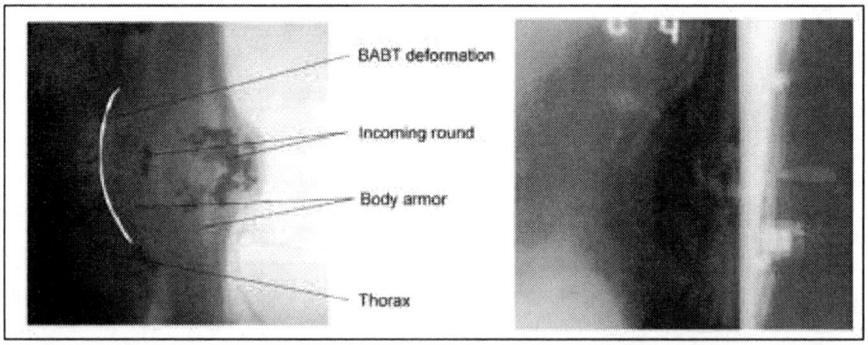

Figure 72: Backface deformation profile from rifle round in hard body armor [10]

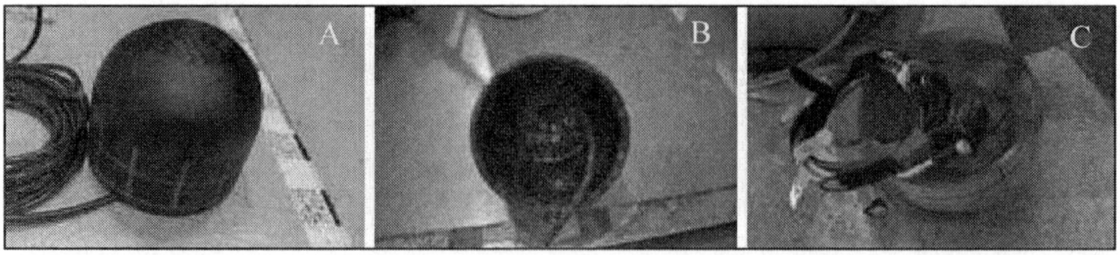

Figure 73: (A-B) Wired indentor with triaxial accelerometer. (C) Wireless indentor with triaxial accelerometer with on-bord data acquisition system

2.2 Experimental Setup

A gas-driven launching system was used to propel the indentor to the target at varying velocities. The indentor was loaded into an open-ended tube and pressurized gas was released behind the indentor propelling it towards the indented target at the exit of the tube. Human cadavers and swine cadavers were strapped to a custom harness and hoisted up using a winch such that they were positioned upright in front of the tube (Figure 3).

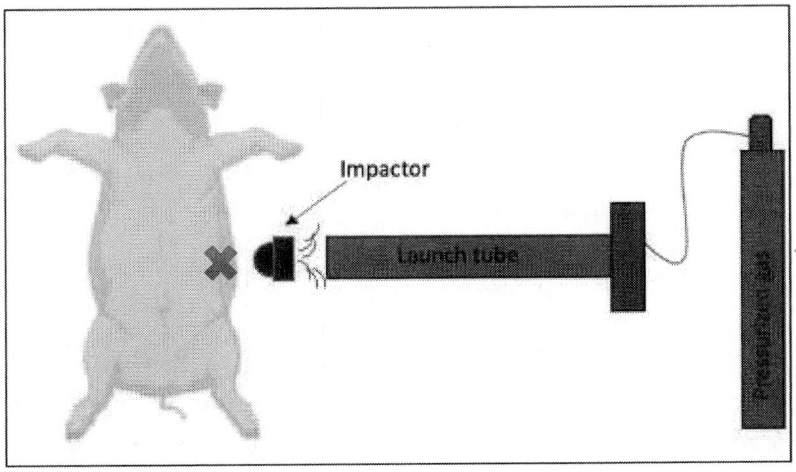

Figure 74: Experimental setup with swine position at the exit of launch tube

This impactor was used to impact multiple thoracic regions: heart, lungs, kidney, liver, sternum, and spine. The specimen position was adjusted based on the targeted location. Impacts simulated 7.62x51 NATO bullets at velocities of 311-1067 (meters/second) m/s.

2.3 Data Processing

Data from the indentor were processed with a four-pole Butterworth filter at 1, 2, 4, and 10 kHz. Power spectrum density functions for all signals and impacts were analyzed using log and natural ordinate scales. Similar experiments were conducted at the three institutions of the authors of this study with swine cadavers and human cadavers to add to the feasibility of using the simulated indentor on different surrogates and at different thoracoabdominal regions. Although other institutions used similar experimental setups, some variations existed in launching systems, acquisition systems, accelerometers, and indentors within institutions. Data from all institutions were also processed and analyzed for the present study.

3. RESULTS

Over 100 acceleration signals from the indentor impacts to targeted locations in swine and human cadavers were analyzed. Signals were examined to determine the applied load to the surrogates, focusing on the impacting event window. Small subset of impact tests were removed from the dataset due to low quality signals, dead channels, or broken cables.

Raw and filtered signal, and Power Spectral Density (PSD) of indentor accelerometer from a single swine (Section 3.1) and human (Section 3.2) cadaver tests are shown as an example of signal examination process undertaken to identify a low-pass filter appropriate for processing biomechanical signals in BABT experiments. PSD results are presented in form of two different plots, linear and log scales, for sake of clarity for the reader.

https://doi.org/10.52202/080042-0040

3.1 Swine cadaver results from institution 1

Impactor acceleration data from left lung impact to a swine cadaver are presented in **Error! Reference source not found.** and Figure 76.

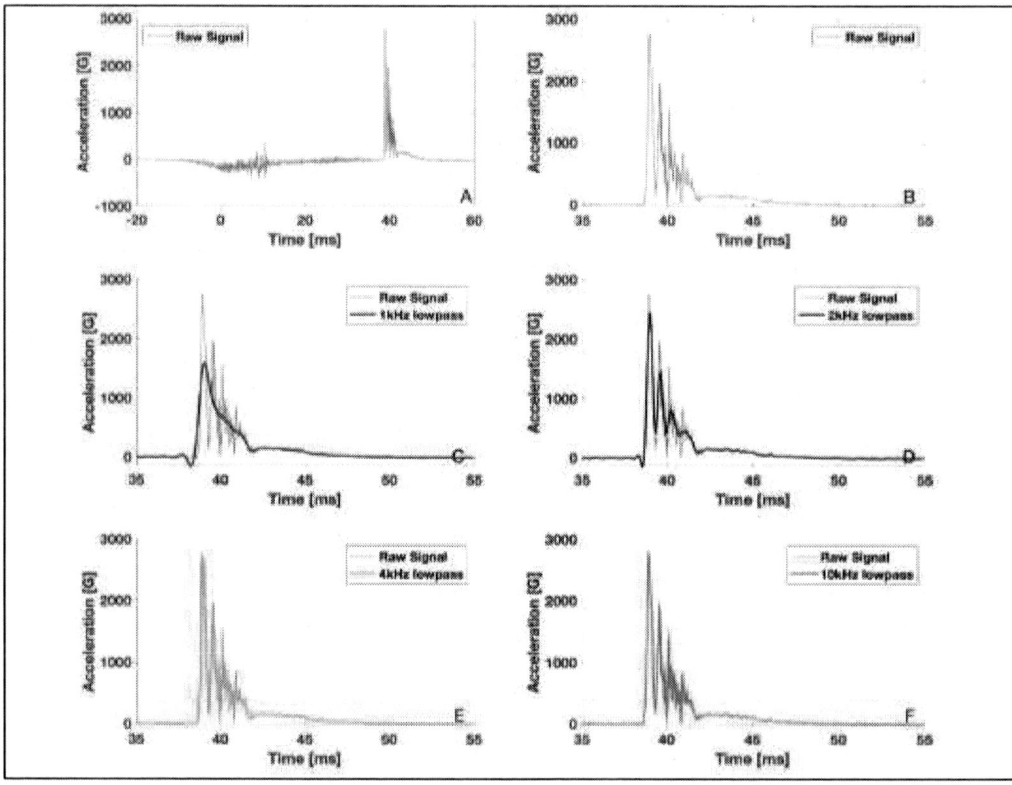

Figure 75: Raw signal (A) of indentor loading event (B) from impact to left lung on swine cadaver is filtered at 1 kHz (C), 2 kHz (D), 4 kHz (E) and 10 kHz (F)

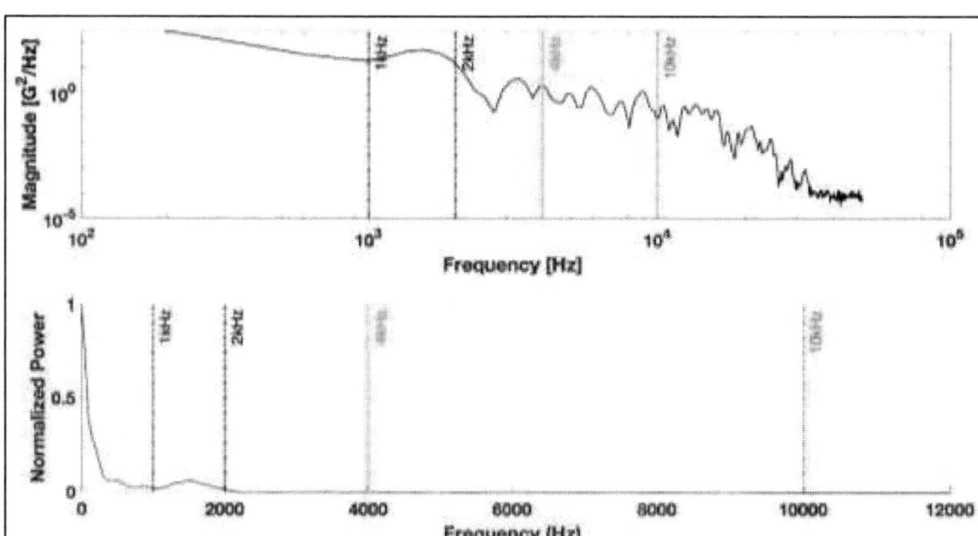

Figure 76: Power spectral density plots in log (top) and linear (bottom) scale for indentor acceleration signal from impact to left lung on swine cadaver

3.2 Human cadaver results from institution #1

Impactor acceleration data from spine impact to a human cadaver are presented in Figure 77 and Figure 78.

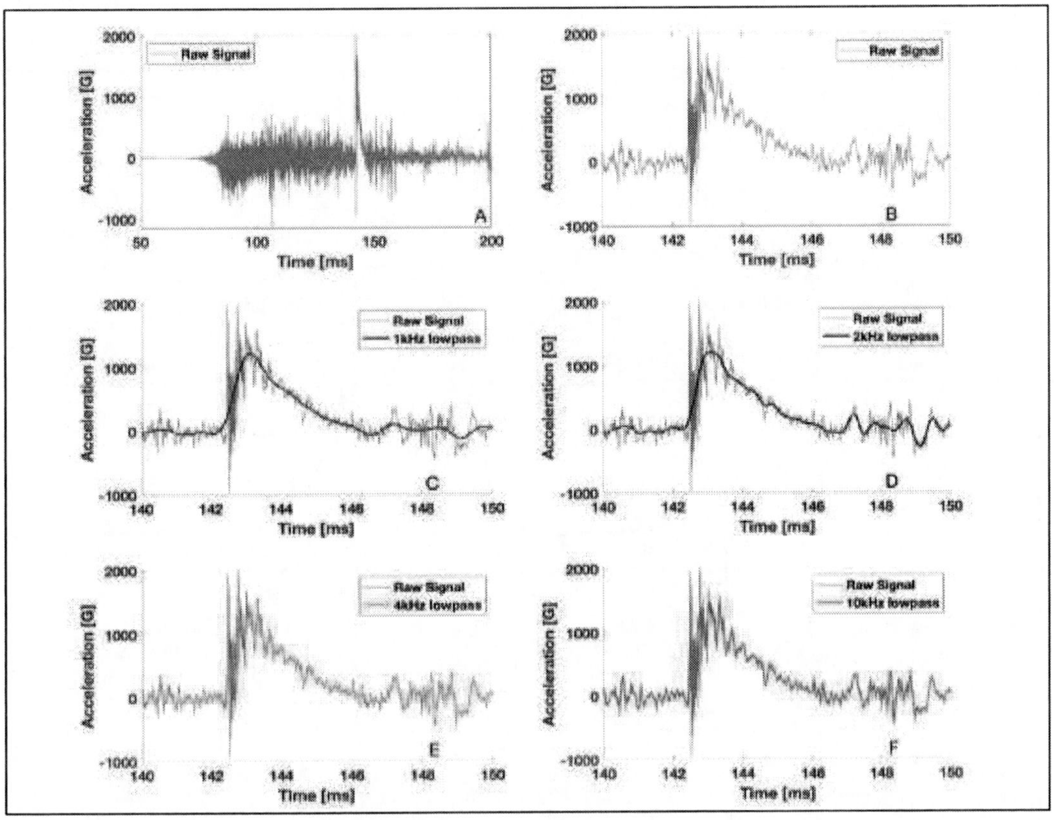

Figure 77: Raw signal (A) of indentor loading event (B) from impact to spine on PMHS is filtered at 1 kHz (C), 2 kHz (D), 4 kHz (E) and 10 kHz (F)

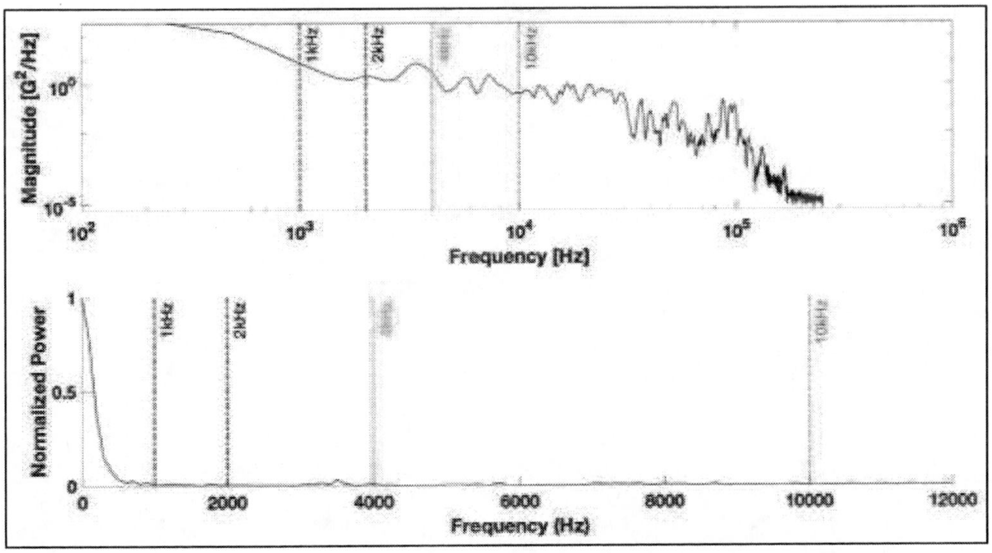

Figure 78: Power spectral density plots in log (top) and linear (bottom) scale for indentor acceleration signal from impact to spine on PMHS

https://doi.org/10.52202/080042-0040

3.3 Results from other Institutions

Similar analysis was conducted on impactor data from other two institutions of authors of this study. Impactor acceleration data from liver impact to a human cadaver are presented in Figure 79 and Figure 80 from institution #2 and data from left lung impact on swine cadaver are presented in Figure 81Figure 82 from institution #3.

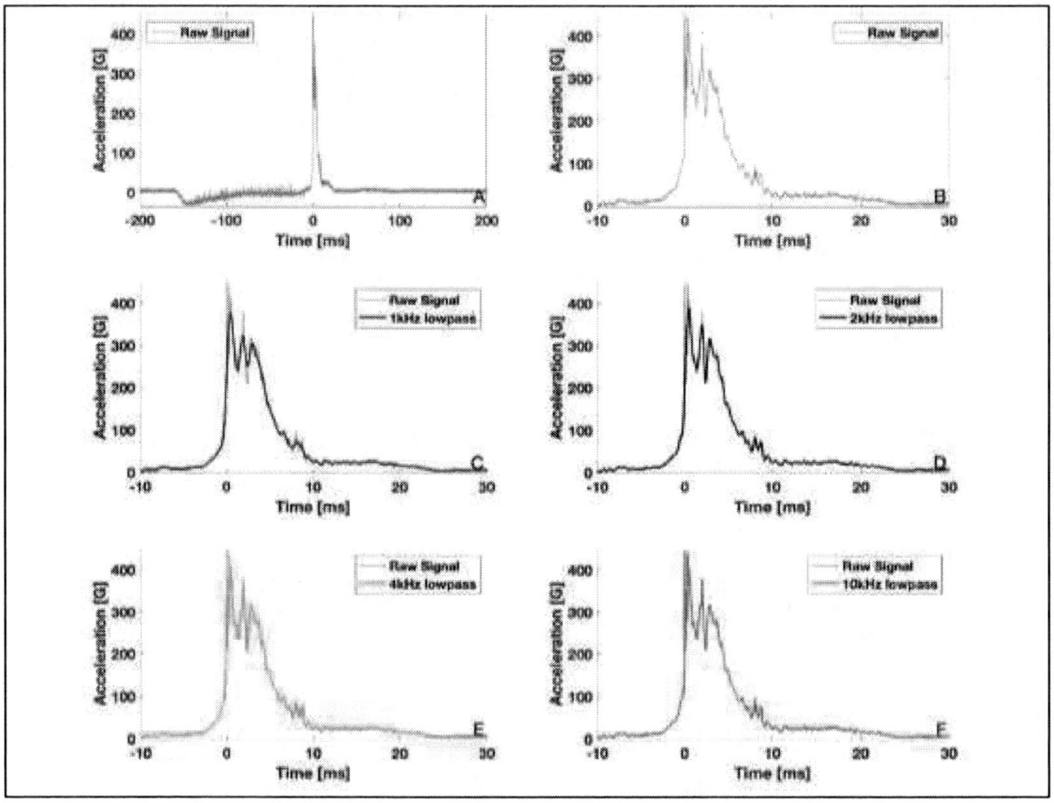

Figure 79: Institution # 2 raw signal (A) of indentor loading event (B) from impact to liver on PMHS is filtered at 1 kHz (C), 2 kHz (D), 4 kHz (E) and 10 kHz (F)

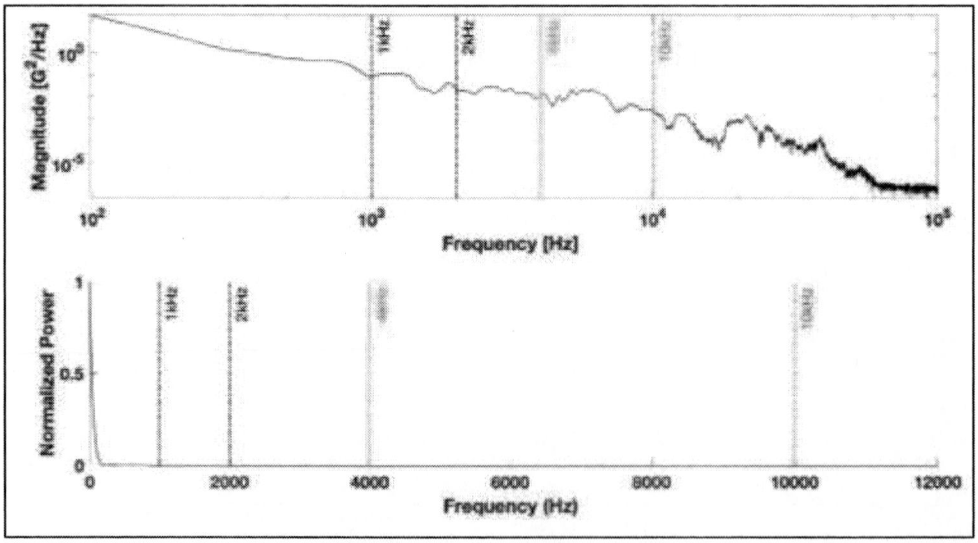

Figure 80: Power spectral density plots in log (top) and linear (bottom) scale for indentor acceleration signal from institution #2 from impact to liver on PMHS

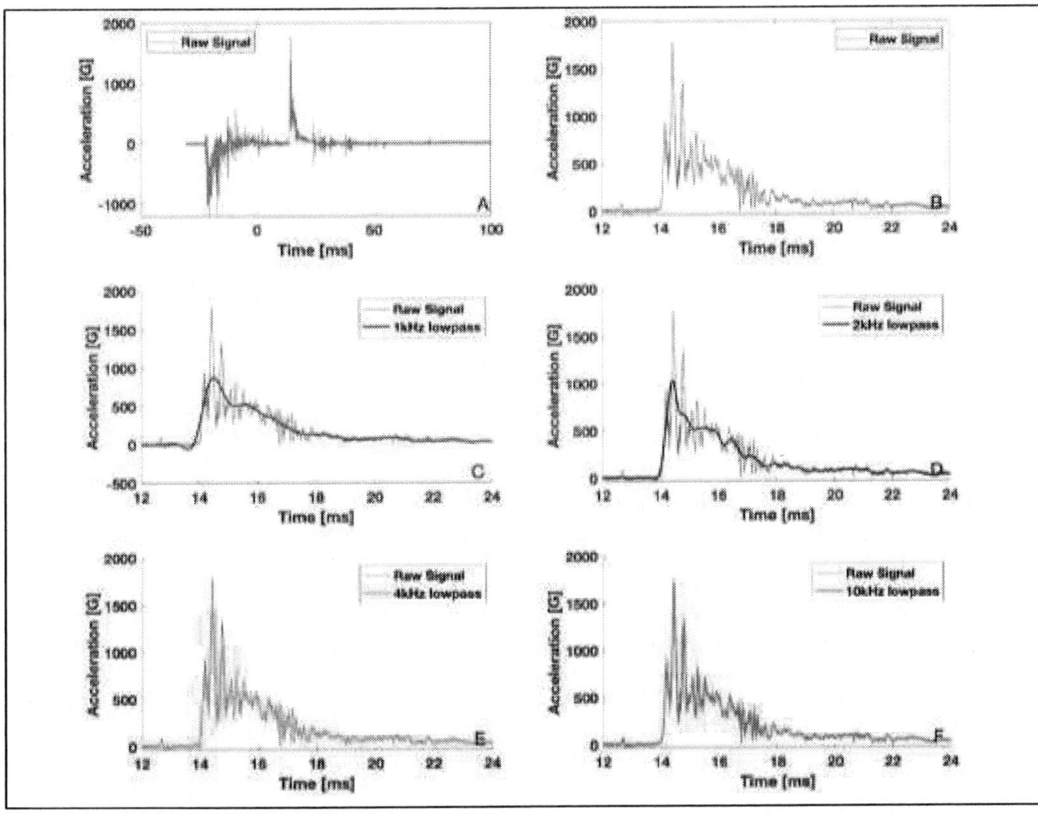

Figure 81: Institution # 3 raw signal (A) of indentor loading event (B) from impact to left lung on swine cadaver is filtered at 1 kHz (C), 2 kHz (D), 4 kHz (E) and 10 kHz (F)

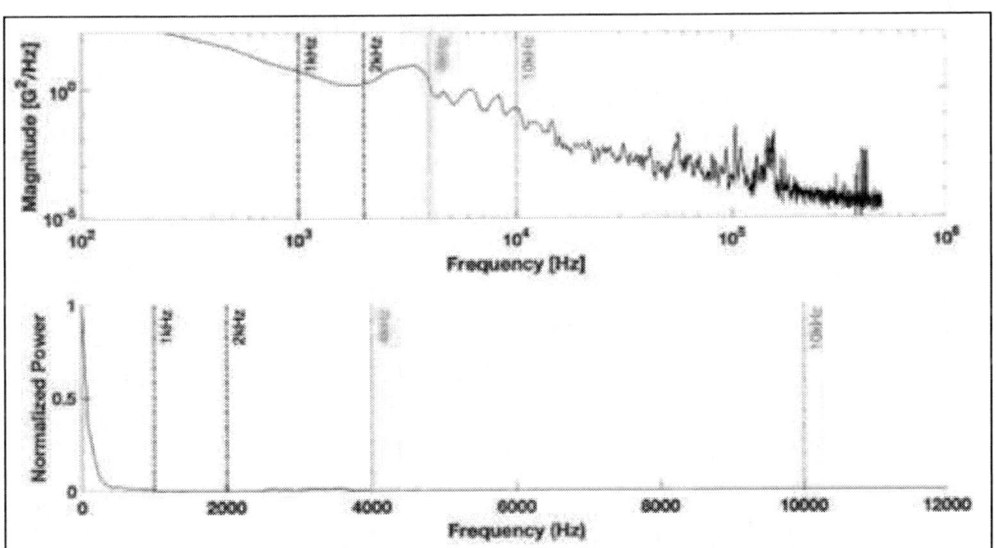

Figure 82: Power spectral density plots in log (top) and linear (bottom) scale for indentor acceleration signal from institution #3 from impact to left lung on swine cadaver

For each body region from human and swine cadavers, temporal analysis of raw and processed signals were processed at each frequency. For all body regions, in addition to the actual loading profile, raw data briefly included Gaussian noise. Filtering the signal at 10 kHz significantly reduced the noise with a concomitant peak amplitude decrease while the signal remained oscillatory under varying loading conditions. Filtering at 4kHz further decreased the amplitude and reduced the oscillatory nature for a

https://doi.org/10.52202/080042-0040

subset of data. This phenomenon was primarily observed for impacts at lower velocity (<40 m/s) impacts to lungs. While oscillations were minimal at 2kHz, the 1kHz filter produced the smoothest single wave-type pulse without any oscillatory pattern.

4. DISCUSSION

Data acquisition and processing of signals is an important component of data analysis in any dynamic loading experiment. Researchers in the automotive field including academia and industry have dealt with this critical issue for years. The Society of Automotive Engineers (SAE) assembled groups to decide the best approaches and developed standards (SAE-J211) [11]. Impacts tests conducted with sled equipment that delivers dynamic loading to biological surrogates (human cadavers in particular), pendulum that delivers localized impacts (animals and human cadavers and physical models such as the Hybrid III manikin), drop tests that applies loading similar to sled equipment (whole body and isolated component human cadavers and manikin), and other loading methods continue to use the SAE standard as the accepted procedure for signal processing and filtering. For example, sled accelerations are filtered using channel filter class CFC 60, acceleration on the head and thorax on both biological surrogates and physical models are filtered at CFC 1000, and pendulum impactor accelerations are filtered at CFC 1000 [9], [12]–[14]. Similar widely accepted international procedure for BABT impacts does not exist to the best knowledge of the authors of this study [15]. Some impact loading studies with human cadavers and manikins have used the same automotive filtering techniques. As an example, a brief review of BABT-related papers presented at the recent PASS conferences confirms the lack of consensus, and in addition, unlike automotive studies, not all BABT studies have reported filtering methods [16]–[20] . While both automotive and BABT studies apply to the load dynamically via impact to different regions of the human body, the loading magnitudes are greater in the BABT scenario. Consequently, a method to process signals for BABT applications is needed. The present study was designed with this intent, and as a first step, used experiments with a biological model as the overall aim of the project was to develop thoraco-abdominal injury criteria for different regions of live and cadaver-based human surrogates. Gathered accelerometer data from the indentor that applied the impact load to the swine and cadavers was used in the filtering analysis.

The results from the present study show that the 10 kHz filtered signals remove the Gaussian noise only to a certain extent, while decreasing the filter frequency had a larger effect; however, as expected, the peak amplitudes were lower with increasing filter frequencies. Filtering at 1 kHz produced the smoothest curve in all cases, and from all tests at the three institutions. But the peaks reduced considerably. It should be noted that acceleration signals depend on the region of impact: skeletal regions tend to produce sharper and higher rise time profiles that add to the noise while impacts to softer regions (e.g., unprotected liver) tend to spread the pulse with lower peak acceleration and less Gaussian noise. Flesh thickness is also a modulating factor. A common finding from these varied tests at three institutions is that beyond 2 kHz, the power spectral density plots show little to no signal for BABT loading event and hence, this filter can be considered as a true representation of impact responses with live and cadaver biological surrogates, and at the 2 kHz will also not considerably compromise the peak amplitudes of the acceleration signal. This frequency is also reasonable considering the fact that this is twice the filter rate used in automotive studies. A similar signal processing analysis with BABT impact has not been published to the best knowledge of the authors. While preliminary, the authors are analyzing more signals from additional tests to the three surrogates to reinforce these findings.

The advantage of filtering the collected signals that are at higher sampling rates (automotive pulses are longer and sampling frequencies of 20 kHz are acceptable, in general) is to enable post processing for secondary variables can be made with confidence. For example, it is customary to remove inertial effects of the indentor by attaching an accelerometer to the indentor to calculate the actual forces to the biological or physical model. Likewise, once the acceleration signals are filtered, parameters such as the velocity of the indentor that applied the loading to the surrogate, and compression imparted to the surrogate can be calculated by integrating the filtered accelerometer signal. As stated earlier, in automotive studies have used filtered accelerometer signals to determine deflection on human surrogates and developed injury criteria [21]. As-collected raw signals from the impact test cannot be used for this purpose. Once velocity and compression/deflection are obtained, other measures such as viscous criteria ($[VC]_{max}$, V_{max} C_{max}) and momentum transferred to the surrogate, can be obtained [22], [23]. All these measures are potential candidates for BABT injury criteria. Optical methods can be used in lieu of the indentor accelerometer-derived measures; however, they have issues such as visibility to gather data and difficulty in obtaining off-axis components, and frame rate is also an issue as greater impact velocities require greater frame rates that may involve camera resolution/pixel constraints. In addition, camera placements to capture unobstructed images throughout the experiment can be challenging, especially with biological surrogates. Live animal experiments add to this complexity as often other physiological monitoring equipment also require space in the laboratory with clinician involvement. These factors affect the accuracy of optical measurements. As sensors signals can be captured at very high frequencies, they are suitable to BABT applications, and the approach used in the present paper would serve as a first step in the process of developing a well-defined and accepted filtering algorithm.

Acknowledgments

This research was supported by the U.S. Army Medical Research and Development Command contract W81XWH-21-9-0015 and the Department of Veterans Affairs Medical Research.

References

[1] J. S. Gondusky and M. P. Reiter, "Protecting military convoys in Iraq: An examination of battle injuries sustained by a mechanized battalion during operation Iraqi Freedom II," *Mil. Med.*, vol. 170, no. 6, pp. 546–549, 2005, doi: 10.7205/MILMED.170.6.546.

[2] E. Liden, R. Berlin, B. Janzon, B. Schantz, and T. Seeman, "Some observations relating to behind-body armour blunt trauma effects caused by ballistic impact," *J. Trauma - Inj. Infect. Crit. Care*, vol. 28, no. 1, pp. S145–S148, 1988, doi: 10.1097/00005373-198801001-00029.

[3] M. B. Mukasey, J. L. Sedgwick, and D. W. Hagy, "Ballistic Resistance of Body Armor," *NIJ Stand. - 0101-16*, pp. 1–89, 2008.

[4] E. Hanlon and P. Gillich, "Origin of the 44-mm behind-armor blunt trauma standard," *Mil. Med.*, vol. 177, no. 3, pp. 333–339, 2012, doi: 10.7205/MILMED-D-11-00303.

[5] B. Sanborn, C. A. Gunnarsson, M. Foster, and T. Weerasooriya, "Quantitative Visualization of Human Cortical Bone Mechanical Response: Studies on the Anisotropic Compressive Response and Fracture Behavior as a Function of Loading Rate," *Exp. Mech.*, vol. 56, no. 1, pp. 81–95, 2016, doi: 10.1007/s11340-015-0060-y.

[6] E. F. Morgan, G. U. Unnikrisnan, and A. I. Hussein, "Bone Mechanical Properties in Healthy and Diseased States," *Annu. Rev. Biomed. Eng.*, vol. 20, pp. 119–143, Jun. 2018, doi: 10.1146/annurev-bioeng-062117-121139.

[7] N. Yoganandan, F. A. Pintar, J. R. Humm, G. W. Stadter, W. H. Curry, and K. J. Brasel, "Comparison of AIS 1990 update 98 versus AIS 2005 for describing PMHS injuries in lateral and oblique sled tests," *Ann. Adv. Automot. Med. Assoc. Adv. Automot. Med. Annu. Sci. Conf.*, vol. 57, pp. 197–208, 2013.

[8] W. Zhang, A. Capilnasiu, and D. Nordsletten, "Comparative Analysis of Nonlinear Viscoelastic Models Across Common Biomechanical Experiments," *J. Elast.*, vol. 145, no. 1, pp. 117–152, Aug. 2021, doi: 10.1007/s10659-021-09827-7.

[9] P. Prasad, "Injury crietria and motor vehicle regulations," in *Accidental Injury: Biomechanics and Prevention*, N. Yoganandan, A. M. Nahum, and J. W. Melvin, Eds., Third Edition.New York: Springer, 2015, pp. 793–809.

[10] C. R. Bass *et al.*, "Injury risk in behind armor blunt thoracic trauma," *Int. J. Occup. Saf. Ergon.*, vol. 12, no. 4, pp. 429–442, 2006, doi: 10.1080/10803548.2006.11076702.

[11] SAE, "Society of Automotive Engineers," *Instrum. Impact Test-Part 1-Electonic Instrum.- SAE J2111*, 2003.

[12] FMVSS-208, "Upgrade: Code of Federal Regulations: 571.208, Washington, DC, US Government Printing Office," Oct. 2001.

[13] FMVSS-214, "Update: Code of Federal Regulations: 571.214, Washington, DC, US Government Printing Office," Oct. 2008.

[14] N. Yoganandan, A. Nahum, and J. W. Melvin, *Accidental Injury: Biomechanics and Prevention*. New York: Springer-Verlag, 2015.

[15] M. Franklyn and P. Lee, *Military Injury Biomechanics: The Cause and Prevention of Impact Injuries*. CRC Press, 2017.

[16] A. Bracq *et al.*, "Numerical recreation of Police Field Cases on a human body FE model: first insights into BABT," *PASS*, 2020.

[17] S. Ouellet, G. Pageau, and G. James, "A new biofidelic backing for the evaluation of the ballistic performance of soft armour and lightweight protective fabrics," *PASS*, 2020.

[18] K. A. Rafaels, M. E. Lizins, K. L. Loftis, and C. A. Bir, "The Relationship between the Shape of Backface Deformation and Behind-Armour Blunt Trauma," *PASS*, 2020.

[19] Cyril Robb, Nestor Nsiampa, and Alexandre Papy, "Sensitivity studies of the bttr surrogate and comparison Between nlw and babt applications," *PASS*, 2018.

 https://doi.org/10.52202/080042-0040

[20] M. Vignos *et al.*, “Comparison of Pressure Attenuation Performance of Bomb Suit Designs during Free-Field Blasts using an Advanced Human Surrogate,” *PASS*, 2020.
[21] D. C. Viano and I. V. Lau, “A viscous tolerance criterion for soft tissue injury assessment,” *J Biomech*, vol. 21, no. 5, pp. 387–99, 1988, doi: 10.1016/0021-9290(88)90145-5.
[22] S. Kuppa, R. H. Eppinger, F. McKoy, T. Nguyen, F. A. Pintar, and N. Yoganandan, “Development of Side Impact Thoracic Injury Criteria and Their Application to the Modified ES-2 Dummy with Rib Extensions (ES-2re),” *Stapp Car Crash J*, vol. 47, pp. 189–210, Oct. 2003.
[23] J. M. Cavanaugh and N. Yoganandan, “Thoorax Injury Biomechanics,” in *Accidental Injury: Biomechanics and Prevention*, N. Yoganandan, A. M. Nahum, and J. W. Melvin, Eds., Third.New York: Springer, 2015, pp. 331–372.

Development of a physical human thorax surrogate dedicated to blunt ballistic impacts

M. Chaufer[1], R.Delille[2], B. Bourel[2], F.Lauro[2, 3], O.Mauzac[4], S.Roth[1]

[1]*Interdisciplinary Laboratory Carnot of Bourgogne, site UTBM, UMR 6303, CNRS / Université de Technologie de Belfort-Montbéliard, 4 Rue Edouard Branly 90010 Belfort France, martin.chaufer@utbm.fr*
[2]*Univ. Polytechnique Hauts-de-France, CNRS, UMR 8201–LAMIH–Laboratoire d'Automatique de Mécanique et d'Informatique Industrielles et Humaines, F-59313 Valenciennes, France*
[3]*Insa Hauts-de-France, F-59313 Valenciennes*
[4]*French Ministry of Interior, CREL/SAILMI, Place Beauvau, Paris, France*

Abstract. In the field of biomechanics, conducting experimental setups on cadavers can be challenging due to ethical constraints. To overcome this issue, physical or numerical surrogates can be used. In the case of blunt ballistics, numerical surrogates have gained significant interest as they aid in designing ballistic protections. However, the characterisation and modelling of body armour remains difficult which is why physical surrogates are sometimes preferred. In this study, the authors propose the development of a physical human thorax surrogate, named SurHUByx, dedicated to injury risk prediction in blunt ballistic impacts, such as those from Less Lethal Projectiles or bullets hitting armour. The surrogate is based on the geometry of the existing numerical model HUByx, which was validated against numerous impacts. In order to build the physical surrogate, a simplified finite element model called SurHUByx FEM was used. Replication of experimental reference cases conducted on Post Mortem Human Subjects validated this simplified FE model. Once validated, SurHUByx FEM was used as a basis to build SurHUByx, its physical twin. Once built, the physical surrogate was compared with the well-known test of Bir et al., the surrogate global behaviour of the surrogate was validated. The physical surrogate was then equipped with sensors inside internal organs and on ribs to capture local data during impact cases with Less Lethal projectiles and firearms and armour. Injury risk prediction curves were constructed based on the data obtained from the sensors and these curves can now help in injury prediction.

1. INTRODUCTION

Optimisation has become a crucial aspect of developing new technologies, especially in the field of safety and protection. The optimisation of protective devices like body armour can enhance the performance of police officers and soldiers while reducing their weight. In addition, the development of new less-lethal kinetic energy weapons requires a thorough understanding of the human thorax behaviour under blunt impact to ensure both non-lethality and sufficient stopping power. However, conducting experiments on humans or cadavers is difficult due to strict ethical guidelines. As a result, researchers have developed surrogates to mimic the human thorax behaviour behind armour or a Less Lethal Kinetic Energy (LLKE) weapon. There are two types of surrogates: numerical and physical ones. Recently, biofidelic Finite Element models such as HUByx (Hermaphrodite Universal Body YX) [1, 2] [1, 2, 3], SHTIM (Surrogate Human Thorax for Impact Model) [4], and WALT (Waterloo Thorax Model) [5] have gained popularity when studying blunt ballistic impacts. However, the difficulty in developing these models lies in the characterisation and numerical modelling of body armour. Therefore, physical surrogates are sometimes favoured to evaluate the effectiveness of armour. However, so far, only clay has been approved by the NIJ standard [6]. Other materials like 10% or 20% ballistic gelatin, Permagel, ballistic soap, Roma Plastilina No. 1 clay, or styrene-ethylene-butylene-styrene (SEBS) based-gel have also been used, but these surrogates are in the form of a cubic block [7]. Anthropomorphic human surrogates such as Ausman (Bass C. , et al., 2006), SSO (Skin-Skeleton-Organs) [9], MHS (Modular Human Surrogate) [10], HSTM (Human Surrogate Torso Model) [11], and BTTR (Blunt Trauma Torso Rig) (Bolduc & Anctil, Improved Test Methods for Better Protection, a BABT Protocol Proposal fo STANAG 2920, 2010) have been developed, but only a few physical surrogates are consistent with ballistic biomechanical corridors. Moreover, physical surrogates which do not include internal organs can only provide global data.

To create a more detailed human torso model, Roberts et al. proposed a reverse engineering method using a biofidelic numerical model called HTFEM (Human Torso Finite Element Model) to develop a biofidelic physical surrogate of the human thorax HSTM (Human Surrogate Torso Model)

 https://doi.org/10.52202/080042-0041

[13]. This method was interesting and enabled them to build numerical and physical twins but neither the initial FE model biofidelity nor the physical surrogate was checked. To overcome this limitation, the authors proposed a reverse engineering method using a biofidelic numerical surrogate as a reference (HUByx) to build a biofidelic physical surrogate of the human thorax called SurHUByx (Surrogate HUByx).

This study proposed the creation of a physical surrogate using a reverse engineering method. Once created and validated, this physical surrogate was used to replicate field impact cases to plot injury risk prediction curves.

First the SurHUByx FEM model was developed by simplifying the HUByx model and combining the cortical and trabecular parts of bones and cartilage into a single entity [14]. The spine was modelled as a single part, and only the essential organs such as lungs, heart, liver, and spleen were included. The consistency of its behaviour in terms of force and deflection over time within the established corridors was evaluated using Bir et al. experiments as a reference (Bir, Viano, & King, 2004). The anthropometry of SurHUByx FEM was consistent with a 50th percentile, indicating its potential as a basis for building a physical surrogate. Secondly, the study described the construction of the physical surrogate SurHUByx. Comparisons of its behaviour to Bir et al. experiments and corridors were conducted to assess its biofidelity and validate it for protection assessments. Finally, sensors were added to record more local data inside internal organs and on ribs. Cadaver and field impact cases were replicated with the physical surrogate and injury risk prediction curves were built for lungs, heart, and ribs based on the replications.

2. MATERIAL AND METHODS

The HUByx FEM was used as a reference to create a simplified biofidelic numerical surrogate (SurHUByx FEM) which will be used to create its physical twin named SurHUByx. Subsequently, SurHUByx FEM was used to create SurHUByx. SurHUByx was then compared to Bir et al. corridors. Once validated, it was used to construct injury risk functions using locally recorded data. The entire process is outlined in Figure 1 and described in detail in the relevant sections.

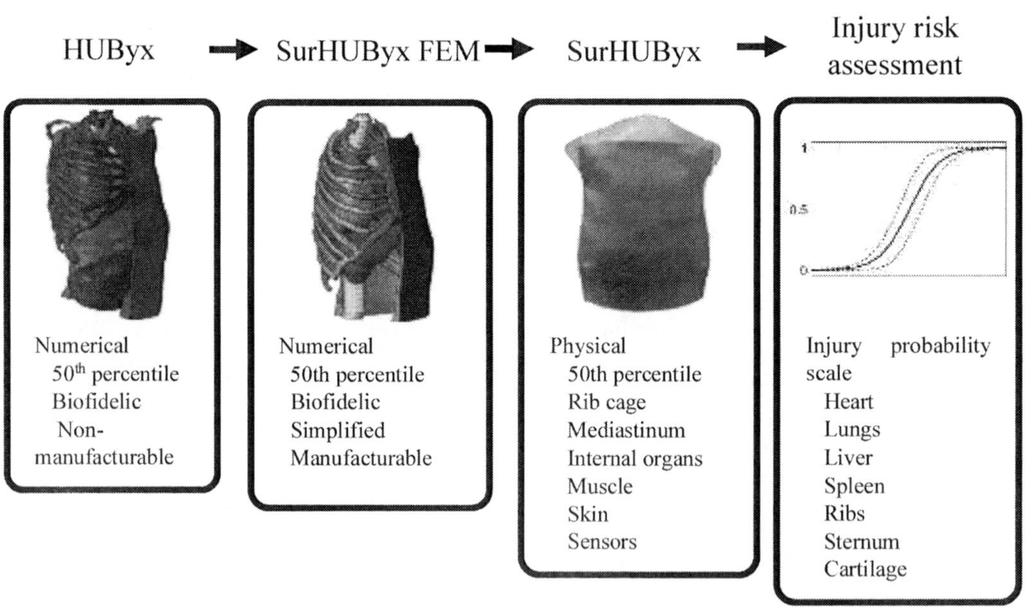

Figure 111. Reverse engineering procedure: from finite element model to its physical twin

2.1 SurHUByx FEM creation and validation process

2.1.1 Creation and simplifications

The HUByx model was used as a basis to develop SurHUByx, but due to its complexity, a simplified FE model was necessary. This simplified model was based on the removal of undesired components and geometrical simplifications. Concerning internal organs, only heart, lungs, liver and spleen were kept.

All the vertebrae were assembled as a continuous part representing the spine. In addition to these parts, SurHUByx was made of skin, muscle mediastinum and rib cage. Fat was made from muscle which resulted in a softer muscle.

Concerning the rib cage in the HUByx model the bones and cartilages were modelled using cortical and trabecular parts, which were unified in SurHUByx through an equivalence in terms of bending stiffness. This merged the cortical and trabecular parts of bones and cartilage and computed their equivalent properties.

To ensure that the physical surrogate could be built using readily available materials on the market, the material laws used in the simplified FE model needed to accurately represent the behaviour of those materials that can be feasibly manufactured. To that aim, a reverse engineering method was used. First, readily available material was mechanically tested. Then, its response was analysed and its properties computed. If this tested material had similar properties to the initial material implemented in the FE model, the properties of this manufacturable material were used to replace the initial ones in the simplified FEM. If the properties of the tested manufacturable material did not match with the initial properties, harder or softer materials were tested.

By using this method to find a surrogate for bones, the authors found that the ideal material can match either the desired Young modulus or the strain to failure. In order to avoid early bone fractures, the authors decided to use a polyurethane resin for the surrogate bones material, since it had the relevant strain to failure. In order to have consistent structure using this material, an equivalence in bending stiffness (EI), with E the Young modulus and I the moment of inertia, was conducted. The unknown was the diameter of the equivalent structure. This equivalence was performed for each rib, varying the increase of cross section along the ribcage. This material behaviour was modelled using an elasto-plastic tabulated law.

This increase in cross section resulted in a reduction of intercostal space, but previous impact case replications showed that this space was necessary to accurately represent the human thorax behaviour in case of intercostal bullet [16]. Therefore, the height of the surrogate was increased to maintain the same intercostal space. The change in ribs cross section also impacted the cartilage cross sections and material properties. Ultimately, an elastomeric resin was found to be a suitable substitute for cartilage, which was implemented in the code using an elastic law.

The Hybrid III crash test dummy vinyl skin was identified as a suitable material to simulate human skin [17]. For the internal organs, muscle, and mediastinum, a gel made of Styrene-Ethylene-Butylene-Styrene (SEBS) material was used in different concentrations. This gel has various advantages, including mechanical consistency and transparency [18, 19]. SEBS based-gel used for the internal organs were previously characterised for their hyper viscoelastic behaviour by Bracq et al. [20]. To simplify the implementation of SEBS based-gel used for muscle and mediastinum, they were modelled using an elastic law.

After its creation, the anthropometry of the SurHUByx FEM was compared to that of a 50th percentile human. The comparison revealed that the SurHUByx FEM anthropometry is similar to that of a 50th percentile male human, making it possible to validate the global behaviour of SurHUByx FEM using Bir et al. impacts.

2.1.2 Validation process

The authors aimed to compare the behaviour of their model with established biomechanical corridors by replicating the impacts performed by Bir et al. (Bir, Viano, & King, 2004). In Bir et al. study, thirteen Post Mortem Human Subjects (PMHS) were impacted with various projectiles at different speeds over the mid sternum. The impacts were categorised into three conditions: Case A (140g projectile at 20 m/s), Case B (140g projectile at 40m/s), and Case C (30g projectile at 60m/s). These tests helped establish biomechanical corridors. The authors numerically replicated these impact cases by applying an initial velocity to the impactor that struck the SurHUByx FEM in a similar manner to the experimental tests. By recreating these impact cases, the authors compared the force time, displacement time curves and VCmax (maximal viscous criterion) values between their numerical model and the biomechanical corridors.

2.2 SurHUByx construction and validation process

2.2.1 Fabrication process

The surfaces of the SurHUByx FEM mesh were used to construct the SurHUByx geometry, which was then imported into a computer-aided design (CAD) software. All components and junctions between the

https://doi.org/10.52202/080042-0041

parts were modelled using the CAD model. The junctions between the ribs/cartilage and cartilage/sternum were held together by a mortise and tenon system and glue. The spine was perforated to allow the insertion of the ribs, which were secured with small axes to allow for natural breathing movement. The mediastinum was designed with shaped holes to accommodate the organs and divided into two parts to facilitate insertion. The intercostal muscles were embedded in the surrogate muscle and mediastinum, while the skin was tightly fitted around the muscle. CATIA V5 software was used to create the CAD model for the surrogate.

After the CAD model was finalised, the CAD modelling of moulds began. Various moulding processes were used. Silicon moulds were used to create the bone parts, while moulds for the cartilage were directly printed in 3D using Polylactic Acid (PLA). High Temperature Polyamide reinforced with carbon fiber (PAHT CF15) was used to 3D print moulds for the muscle, mediastinum and internal organs as this material could withstand high temperatures during the mould casting process. Moulds for bones, costal cartilages, spleen and one part of the mediastinum are presented in Figure 112.

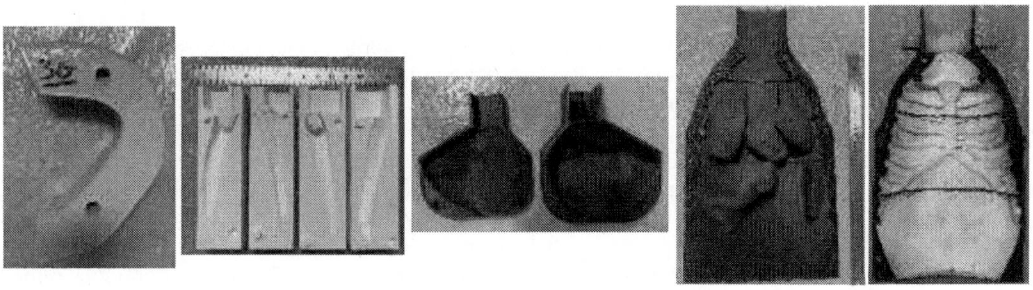

(a) (b) (c) (d)

Figure 112. Silicon moulds for bones (a), PLA moulds for costal cartilages (b), PAHT CF15 moulds for spleen and mediastinum (c and d)

All the parts were then moulded and assembled together. Figure 113 depicts the different stages of the SurHUByx FEM (a), CAD model (b), assembled surrogate (c), surrogate without skin (d), and surrogate without muscle (e). Once the surrogate was assembled, it was submitted to Bir et al. impacts to determine the thoracic wall displacement and compare it to biomechanical corridors, similar to the SurHUByx FEM.

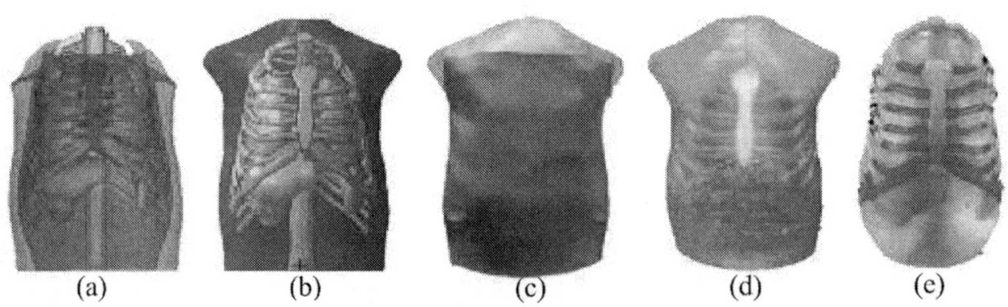

(a) (b) (c) (d) (e)

Figure 113. SurHUByx FEM (a), SurHUByx CAD model (b), SurHUByx (c), SurHUByx without skin (d) and SurHUByx without muscle (e)

2.2.2 Impact cases replication for global behaviour validation

In order to assess the physical surrogate similarity to human biomechanics, the Bir et al. tests were replicated by the authors (Bir, Viano, & King, 2004). They ensured similar conditions, such as launching projectiles with pneumatic launchers at a specific speed, and positioned the SurHUByx on an inclined surface to ensure direct anterior impact level with the 8th thoracic vertebrae. The skin was removed to accurately adjust the impact location (Figure 114a), and a distance of 50 cm was maintained between the launcher and surrogate (Figure 114b). For cases A and B, a 140 g projectile (including sabot projectile and rings) that was 100 mm long and 36.5 mm in diameter was used. The projectile for case C was 30 g (including tracking rod), 28.5 mm long (without the tracking rod) and 36.5 mm in diameter. The projectiles were made of Rubber Baton L5A7 (Pains Wessex Schermuly (UK)) (Figure 114c). Guide rings were utilised to control the speed of projectiles, and a lateral camera at 22000 fps recorded images to track the projectile displacement and record data. Once the contact between the projectile and the dummy ended, the tracking was stopped. The comparison between the cadavers, SurHUByx, and SurHUByx FEM responses included analysing force-time, deflection-time curves, and VCmax values

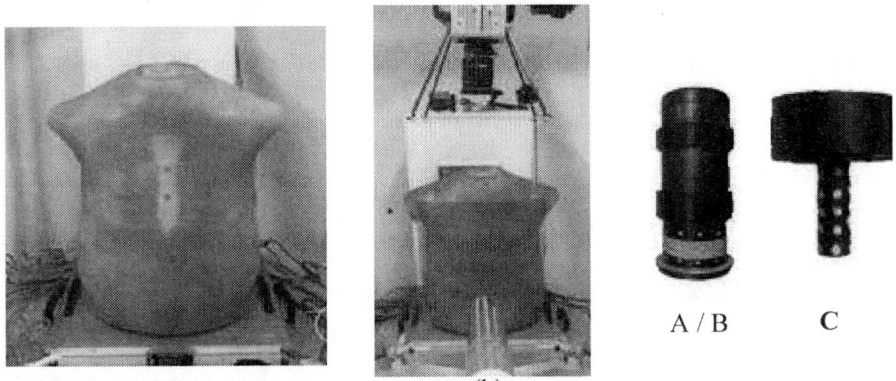

(a) (b)

over the three impact conditions.

Figure 114. Point of impact (a), experimental setup (b), projectiles (c)

2.2.3 Inclusion of sensors and acquisition channels

After validating the overall performance of the surrogate, the authors aimed to obtain more detailed data by strategically placing sensors inside organs and on ribs.

For internal organs, Interlink Electronics® FSR sensors were used and positioned to record frontal events data. These sensors were placed in the center of the lungs, heart, liver, and spleen and were powered by a 5V current generator. The data were captured at 1 MHz by a YOKOGAWA DL750 (1).

For ribs, strain gauges were placed on critical areas of the rib bones to allow for up to 3% strain, consistent with the material properties used to build the bone surrogate. Ribs 1 to 8 were instrumented with 1 to 3 strain gauges, for a total of 30 gauges placed on the surrogate ribs. The strain gauges were mounted with quarter Wheatstone bridges and data were recorded by another YOKOGAWA DL750 (2) at 1 MHz. Due to the limited number of ports available on the YOKOGAWA, only 15 gauges could be connected at a time. As a result, only the 15 closest strain gauges to the impact location were connected. Figure 115 illustrates the placement of the sensors and a schematic representation of the acquisition

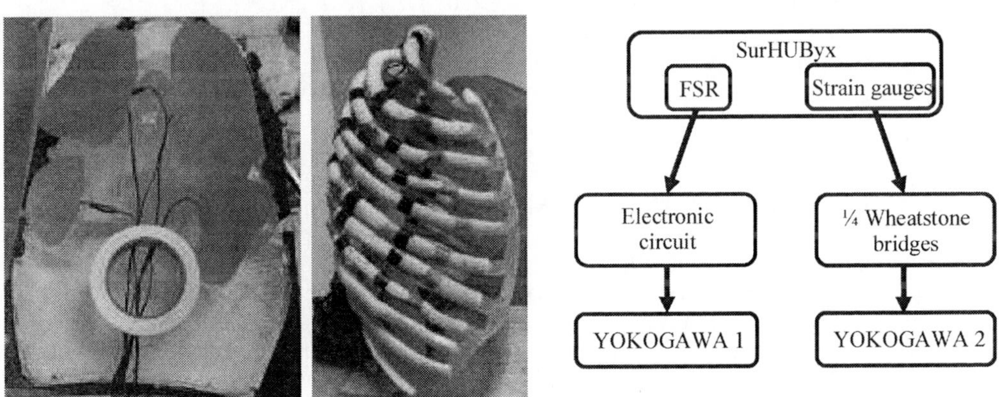

(a) (b) (c)

system.

Figure 115. FSR Sensors embedded in organs (a), strain gauges (b) and acquisition system (c)

2.2.4 Impact cases replication for injury prediction

To correlate the captured data with injury assessment, the authors conducted experiments using LLKE weapons with five impact cases. Three of these cases were taken from Bir et al. study on cadavers (Bir, Viano, & King, 2004), while the remaining two were extracted from case reports by Kobayashi and Mellen [16], and Wahl et al. [21]. Eight cases involving firearms and armour were also extracted from case reports established by Riffault [22, 23]. The ammunition used in these cases ranged from 9mm to

https://doi.org/10.52202/080042-0041

18.5mm, at velocity range from 245 m/s to 410m/s and various soft armours using Kevlar® were used as protection.

To replicate the impact cases from Bir et al., similar conditions were used as described in the study, such as shooting range and impact location. For the other two impacts with Less Lethal Weapons, the projectiles were launched using their respective weapons, Flash-Ball® and Brugger & Thomet®.

For the replication of firearm cases, a universal ballistic breech and barrels of different lengths and diameters were used to fire various projectiles at the desired speeds. Before each impact, a calibration shot was performed to ensure that the projectile was launched at the desired speed, and the bore sight was checked. The projectile speeds were measured 2 m before the impact using a HPI Doppler radar. The replication cases were validated when both the impact location and the desired speed matched with the case report. The projectiles and armours used were equivalent to the ones described in Riffault reports.

Table 32 provides an overview of the replicated cases and their corresponding AIS scores. The report of the experimental study did not specify the organ on which the AIS score was established. Impacts were replicated from the softer to the harder ones, and a total check of the surrogate was conducted between each impact to ensure its physical integrity. If any damage was detected on the SurHUByx, the necessary repairs were carried out to enable the experimental tests to continue.

Table 32 Case report details

Case	Projectile	Impact velocity [m/s]	Body armour	AIS
Bir A	37 mm – 140 g	20	-	0
Bir B	37 mm – 140 g	40	-	2
Bir C	37 mm – 30 g	60	-	0
Kobayashi	eXact iMpact	95	-	3
Wahl	Flash Ball	110	-	3
260-2	9 mm	380	16 layers	3
261	9 mm	380	16 layers	2
263	Brenneke	360	20 layers	4
264-2	Brenneke	385	2 * 16 layers	3
279	9 mm	380	10 layers	1
283	Brenneke	410	20 layers	4
287	9 mm	370	20 layers	1
289	9 mm	370	20 layers	3

3. RESULTS

3.1 Global behaviour

The thoracic displacement of SurHUByx FEM and SurHUByx were compared to biomechanical corridors for the three impact conditions, and the results are illustrated in Figure 116 and Figure 117. The parameter VCmax was also calculated, and the results were compared to cadaveric experiments as illustrated in Figure 118. SurHUByx FEM produced displacement time curves and force time curves that were consistent with Bir et al. corridors for all three impact cases. VCmax values for SurHUByx FEM were also consistent with experimental range values obtained by Bir et al. Similarities of results obtained between SurHUByx and HUByx validated the whole simplification procedure.

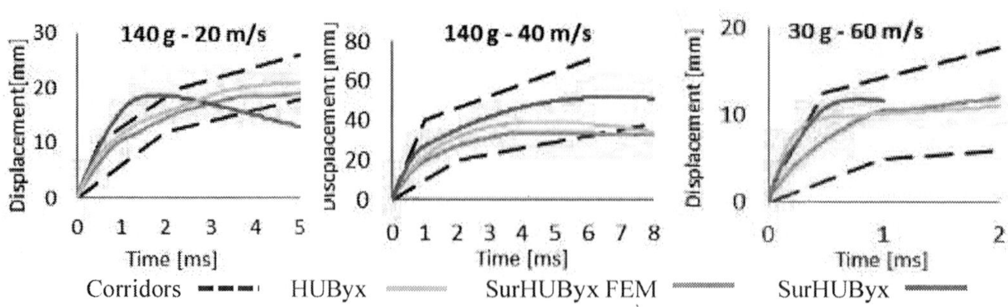

Figure 116. Displacement/time curves for the three impact cases (A, B and C)

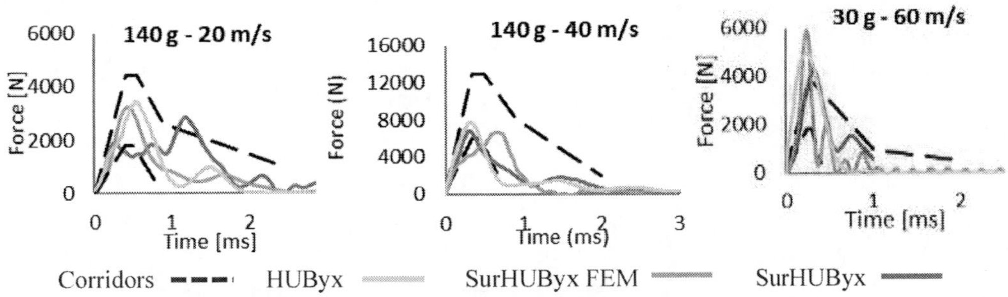

Figure 117. Force/time curves for the three impact cases (A, B and C)

The tracking method used in the experimental study produced results that were consistent with corridors for all three impact cases, with SurHUByx generally in the upper part of the displacement/time corridors for cases A and C, and in the middle of the corridor for case B. SurHUByx was in the lower part of the force/time corridors for cases A and B and in the upper part for case C. VCmax values for SurHUByx were also consistent with experimental range values reported by Bir et al. (Figure 118 left). Sternal fracture was observed on SurHUByx for case B only (Figure 118 middle and left), which is consistent with observations on cadavers. These results validated the SurHUByx behaviour in terms of global response.

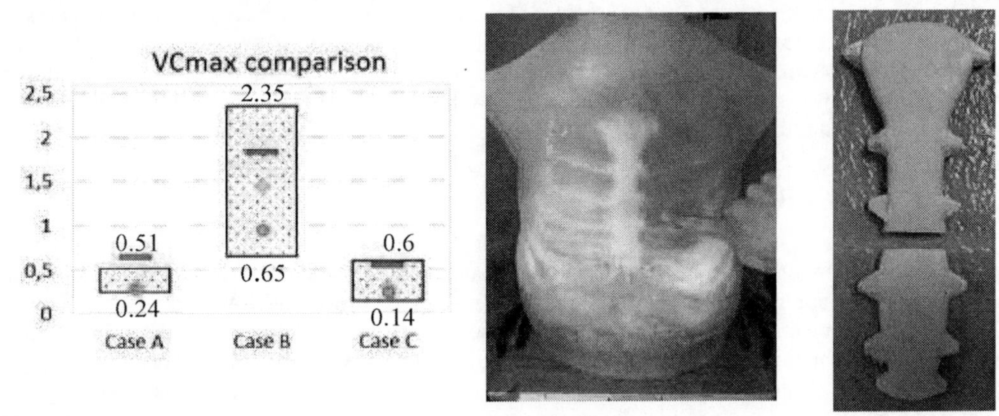

Figure 118. VCmax comparisons between cadaveric experiments, HUByx, SurHUByx FEM and SurHUByx (left) and fracture pattern over the sternum for case B: with muscle (middle), sternum only (right)

3.2 Injury prediction

Once the overall behaviour of SurHUByx was validated, the local behaviour was assessed by collecting data from 13 impact cases. It is worth noting that SurHUByx ribs did not break in any of the experiments, but if the human subject suffered from sternal or cartilage rupture, SurHUByx also exhibited sternal fracture or cartilage rupture. The strain gauge data showed that when no injury was present, small amplitude curves were observed, as depicted in Figure 9 (a). In contrast, when a fracture occurred, the

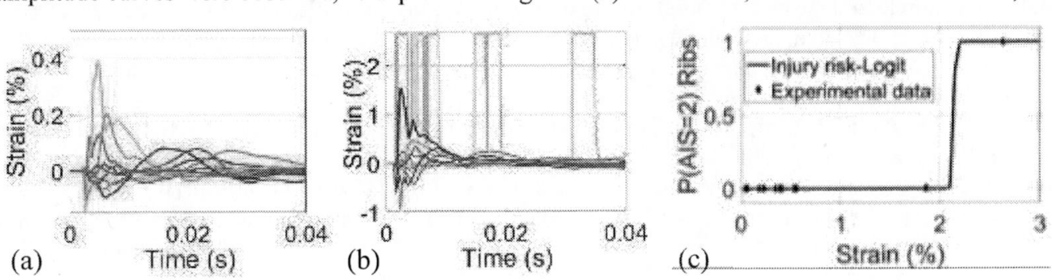

https://doi.org/10.52202/080042-0041

Figure 119. Typical curves obtained from strain gauges: no injury (a), injury (b), Injury risk function AIS=2 ribs (c)

curves showed higher amplitudes, and at least one strain gauge saturated at 2.63%, as illustrated in Figure 9 (b). The maximum value of all the connected strain gauges was used to build a logistic regression. This logistic regression was built using the LOGIT model and represents the probability of an AIS score of 2 on ribs Figure 9 (c). The prediction bounds were not defined with the actual data.

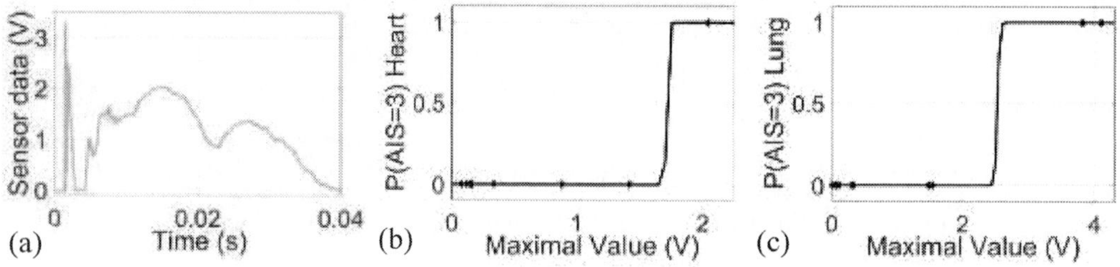

Figure 120. Typical curves obtained from FSR sensors (a), Injury risk function AIS=3 for heart (b) and lung (c)

The FSR sensors showed similar curves for all the internal organs, as seen in Figure 120 (a). These curves had two phases, with the first phase a quick and intense peak, and the second phase longer but with less amplitude. Both phases provided consistent information. For both lungs and heart, the maximum value of the sensors in the second phase was used to build the logistic regression. Injury prediction curves for an AIS score of 3 on the heart and lungs were developed and shown in Figure 120 (b and c). The prediction bounds were not defined with the actual data. Data were also recorded for the spleen and liver, but since no injury was reported, no injury curve could be plotted. There was no indication of sensor saturation in any of these organs.

4. DISCUSSION

Numerical models are often used in mechanics to replicate physical phenomena through simulations [24, 25]. The traditional approach involves creating a model to predict the behaviour and then validating it with physical experiments. This study proposes a reverse engineering method: the creation of a biofidelic numerical model, which is used to select manufacturable materials with the desired behaviour, to build a physical surrogate. A similar approach was used by Roberts et al. to develop HSTM and HTFEM [13], but these models have not been validated against animal or cadaveric data. That is why this study uses the HUByx model as a reference, which represents a 50th percentile human thorax, and which was validated regarding various impact cases [1, 3]. In order to address the wide range of variations in human morphology and response to loading, the study used biomechanical corridors as a validation method, which is a common practice in the biomechanical field. To validate a numerical model its response is generally required to fall within the experimental corridors.

The validation approach suggested in this research relies solely on cadaver tests. An alternative method to assess the performance of surrogates is through live animal experiments. These two approaches are considered complementary: while PMHS provide the closest resemblance in terms of morphology (Bir, Viano, & King, 2004), pigs are better at replicating pathophysiological responses [26] but validation using live animal models introduces additional complexity and relying solely on single biomechanical injury metrics may not provide a comprehensive solution. The physiology of live animal models is a highly complex system with numerous interrelationships and dependencies, and relying solely on a simple and easily measured metric may not capture the full range of outcomes accurately. Previous research has compared the results of ballistic impact experiments using PMHS and pigs [27]. Recent study compared the behaviour of PMHS and both living and dead pigs in a ballistic setting (Bourget D. , 2020).

In this study, the simplified finite element model of the human thorax, SurHUByx FEM, was compared to experimental data obtained on PMHS to ensure that its behaviour was consistent with established biomechanical corridors (Bir, Viano, & King, 2004). The validated SurHUByx FEM was then used to create its physical twin, SurHUByx, which was also validated against the same experimental corridors. The results showed that both the numerical and physical surrogates had a consistent mechanical response to the experimental data in terms of force time, displacement time curves and VCmax values. However, because each human behaviour is different, biomechanical corridors were used to evaluate the

https://doi.org/10.52202/080042-0041

models and no conclusion can be drawn regarding which model has a closer dynamic response to the human body. Nevertheless both of these FE models can be enhanced in terms of biofidelity.

After validation, SurHUByx was used to replicate impact cases and create injury probability risk functions based on data from local sensors embedded in the physical surrogate. The replication cases showed that SurHUByx could record local data which could be linked to a probability of injury. The creation of the injury probability curves relay on statistics as it is recommended by McMurry et al. [29]. However, it is important to use the probability curves with caution because only a few impact cases were replicated. The non-definition of the 95% prediction bounds of the injury risk curves confirms this limitation. To build accurate probability injury risk functions, a large amount of experimental data is needed, as shown in a previous study [30].

To improve the probability injury risk function developed in this study, further research is needed to find and replicate impact cases. Once enough cases are replicated, these curves could be used to assess protection using local information without the need for living animals or cadavers. Currently, sensors embedded in the surrogate can be used to compare different body armour systems using local data. Future research could focus on developing a way to measure the VC response without affecting the surrogate behaviour. In addition, creating twin surrogates dedicated to blunt ballistic impacts for various anthropometry, as in the crashworthiness field, could also be pursued.

5. CONCLUSION

A simplified version of the HUByx model, a biofidelic finite element model of the human thorax, was created by simplifying its structure to form SurHUByx FEM. The reverse engineering method was used to find manufacturable materials available in the industry with consistent properties to the initial ones, and their corresponding material laws were implemented in the code. The SurHUByx FEM behaviour was validated by numerically replicating cadaveric impact cases. The geometry and materials of SurHUByx FEM were used to create SurHUByx, its physical twin, which was then compared to the Bir et al. biomechanical corridors. The results showed good agreement with cadaveric experiments in terms of sternal force, displacement, and VCmax values, thereby validating SurHUByx. Local sensors were included in the surrogate internal organs and ribs, and impact cases involving less lethal weapons and firearms with soft armours were replicated. Injury reports and recorded data were used to construct injury risk functions for the heart, lungs, and ribs. While this study validated the method used to build SurHUByx and proved its ability to predict injuries, additional impact cases need to be recreated to enhance the accuracy of the probability injury risk functions.

Acknowledgments

This study was conducted as part of a PhD thesis research project supported by the French "Direction Générale de l'Armement" (DGA). The authors also acknowledge the financial and material support of the French Ministry of Interior.

References

[1] S. Roth, F. Torres, P. Feuerstein and K. Thoral-Pierre., «Anthropometric Dependence of the Response of a Thorax FE Model under High Speed Loading: Validation and Real World Accident Replication» Computer Methods and Programs in Biomedicine, vol. 110, n° 12, pp. 160-170, 2013.

[2] M. Chaufer, R. Delille, B. Bourel, C. Marechal, F. Lauro, O. Mauzac and S. Roth., «Enhancement of a Human Thoracic FE Model for Blunt Impact and Related Trauma» Computer Methods in Biomechanics and Biomedical Engineering, vol. 24:sup1, pp. 310-311, 2021.

[3] A. Bracq, R. Delille, B. Bourel, C. Maréchal, G. Haugou, F. Lauro, S. Roth and O. Mauzac., «Numerical Recreation of Field Cases on a Biofidelic Human FE Model Involving Deformable Less-Lethal Projectiles» Human Factors and Mechanical Engineering for Defense and Safety, 2019.

[4] N. Nsiampa, «Development of a Thorax Finite Element Model for Thoracic Injury Assessment» 8 th LS-DYNA User Conference, 2011.

[5] D. S. Cronin, M. C. Bustamante, J. Barker, D. Singh, K. A. Rafaels and C. Bir, «Assessment of Thorax Finite Element Model Response for Behind Armor Blunt Trauma Impact Loading Using an Epidemiological Database» Journal of Biomechanical Engineering, vol. 143, n° 13, 2021.

[6] National Institute of Justice, «Ballistic Resistance of Body Armor NIJ Standard-0101.06» 2008.

https://doi.org/10.52202/080042-0041

[7] J. Read, R. Hazael and R. Critchley, «Soft Tissue Simulants for Survivability Assessment - A Sustainability Focussed Review» Applied sciences, vol. 12, n° 14954, p. 14, 2022.
[8] C. R. Bass, R. S. Salzar, S. R. Lucas, M. Davis, L. Donnellan, B. Folk, E. Sanderson and S. Waclawik, «Injury Risk in Behind Armor Blunt Thoracic Trauma» International Journal of Occupational Safety and Ergonomics, vol. 12, n° 14, 2006.
[9] Y. Wenmin, X. Yao, Y. Wang, Y. Jin and W. Wei, «Experimental Study of the Mechanical Response of a Physical Human Surrogate Thoracic Model Impacted by a Rubber Ball» Journal of Physics, vol. Conf. Ser. 1507, n° 1102032, 2020.
[10] K. Sedberry and S. Foley, «Modular Human Surrogate for Non-Lethal Weapons (NLW) Testing» DSIAC Journal, vol. 6, n° 11, 2019.
[11] P. Biermann, E. Ward, R. Cain, B. Carkhuff, A. Merkle and J. Roberts, «Development of a physical Human Surrogate Torso Model (HSTM) for ballistic impact and blast» Journal of Advanced Materials, vol. 38, n° 11, pp. 3-12, 2006.
[12] M. Bolduc and B. Anctil, «Improved Test Methods for Better Protection, a BABT Protocol Proposal fo STANAG 2920» chez Proc Personal Armour System Symp, 2010.
[13] J. C. Roberts, A. C. Merkle, P. Biermann, E. E. Ward, B. Carkhuff, R. Cain and J. V. O'Connor, «Computational and experimental models of the human torso for non-penetrating ballistic impact» Journal of Biomechanics, n° 140, pp. 125-136, 2007.
[14] M. Chaufer, R. Delille, B. Bourel, C. Marechal, F. Lauro, O. Mauzac and S. Roth, «Development of a simplified human thoracic FE model for blunt impact and related trauma» chez 27th Congress of the European Society of Biomechanics, Porto, 2022.
[15] C. Bir, D. Viano and A. King, «Development of Biomechanical Response Corridors of the Thorax to Blunt Ballistic Impacts» Journal of Biomechanics, vol. 37, pp. 73-79, 2004.
[16] M. Kobayashi and P. F. Mellen, «Rubber Bullet Injury Case Report With Autopsy Observation and Literature Review» Forensic Med Pathol, vol. 30, n° 13, pp. 262-267, 2009.
[17] W. Wood Garrett, M. B. Panzer, C. R. Bass and B. S. Myers, «Viscoelastic Properties of Hybrid III Head Skin» SAE International Journal of Materials and Manufacturing, vol. 3, n° 11, pp. 186-193, 2010.
[18] R. A. Mrozek, B. Leighliter, C. S. Gold, I. R. Beringer, J. H. Yu, M. R. VanLandingham, P. Moy, M. H. Foster and J. L. Lenhart., «The Relationship between Mechanical Properties and Ballistic Penetration Depth in a Viscoelastic Gel» Journal of the Mechanical Behavior of Biomedical Materials, vol. 44, pp. 109-120, 2015.
[19] O. Mauzac, C. Paquier, E. Debord, F. Barbillon, P. Mabire and J. Jacquet, «A substitute of gelatin for the measurement of dynamic back face deformation» chez Personal armour systems symposium, Quebec, Canada, 2010.
[20] A. Bracq, G. Haugou, B. Bourel, C. Maréchal, F. Lauro, S. Roth and O. Mauzac., «On the Modeling of a Visco-Hyperelastic Polymer Gel under Blunt Ballistic Impacts» International Journal of Impact Engineering, n° 1118, pp. 78-90, 2018.
[21] P. Wahl, N. Schreyer and B. Yersin, «Injury Pattern of the Flash-Ball, A less-lethal weapon used for law enforcement: report of two cases and review of the literature» The Journal of Emergency Medicine, vol. 31, n° 13, pp. 325-330, 2006.
[22] E. Riffault, «Etude des traumatismes dus à l'impact de balles» Ministère de l'intérieur, Direction générale de la police nationale, Direction de la logistique de la police, Paris, France, 1983.
[23] «Lésions par balles avec et sans dispositif de protection» Contrat I.R.O. - D.R.E.T. 80715, 1984.
[24] M. Giglio, A. Manes and A. Gilioli, «Investigations on sandwich core properties through an experimental–numerical approach» Composites: Part B, vol. 43, n° 12, pp. 361-374, 2012.
[25] T. Kanehira, H. Mutsuda, S. Draycott, N. Taniguchi, T. Nakashima, Y. Doi and D. Ingram, «Numerical re-creation of multi-directional waves in a circular basin using a particle based method» Ocean Engineering, vol. 209, n° 1107446, 2020.
[26] N. Prat, F. Rongieras, E. Voiglio, P. Magnan, C. Destombe, E. Debord, F. Barbillon, T. Fusai and J. Sarron, «Intrathoracic pressure impulse predicts pulmonary contusion volume in ballistic blunt thoracic trauma» The Journal of Trauma, n° 169, pp. 749-755, 2010.
[27] N. Prat, F. Rongieras, H. d. Freminville, P. Magnan, E. Debord, T. Fusai, C. Destombe, J.-C. Sarron and E. J. Voiglio, «Comparison of thoracic wall behavior in large animals and human cadavers submitted to an identical ballistic blunt thoracic trauma» Forensic Science International, n° 1222, pp. 179-185, 2012.
[28] D. Bourget, «Scaling of animal and PMHS thoracic BABT data to live human data» *Proceedings of the Personal Armour Systems Symposium 2020,* p. 10, 2020.
[29] T. L. McMurry et G. S. Poplin, «Statistical Considerations in the Development of Injury Risk Functions» *Traffic Injury Prevention,* vol. 0, pp. 1-9, 2015.
[30] N. Dau, J. Cavaneugh and C. Bir, «Evaluation of Injury Criteria for the Prediction of Commotio Cordis from Lacrosse Ball Impacts» Stapp Car Crash Journal, vol. 55, n° 110, pp. 251-279, 2011.

Edge Performance of Ballistic Helmets

P.M. Fenne and P.L. Gotts

[1]Physical Protection Group, Metropolitan Police Service, 60 Albany Street, London NW1 4EE
paul.m.fenne@met.police.uk
[2]Phil Gotts Consulting Ltd, 23 Thorney Road, Capel St Mary, Ipswich, Suffolk IP9 2HL, UK

Abstract. Within Police Forces there is now a trend to specify high-cut ballistic helmets for reasons of ergonomics, mass and compatibility. This of course reduces the protective area of coverage compared to a standard-cut helmet. With reduced area of coverage, it is more critical than ever that as much of the area of the helmet as possible provides ballistic resistance. This means that it is critical to assess the performance of the helmet close to the edge of the helmet, and ultimately determine how close to the edge a shot can be defeated. There appears to be little history of edge testing of helmets, and even though an edge test was incorporated into VPAM HVN 2009, there seems to be no background evidence as to why the particular limit was chosen. To this end the Metropolitan Police Service (MPS) Physical Protection Group (PPG), and Phil Gotts Consulting Ltd (PGC), have conducted a programme to support the writing of a test method for edge testing of helmets in general, and high-cut helmets in particular. The work has been to research the effects of ballistic impacts at progressively reduced distances from the edge. The test results were analysed to understand the impact mechanisms involved. Behaviour of filaments and the effects of varying adhesive forces within the matrix were identified. Generic samples of both para-aramid and ultra-high molecular weight polyethylene (UHMWPE) were used, as they are known to behave differently. The research identifies the exact point of impact along with any bullet deflection as a result of firing close to the edge and provides both permanent and transient back-face deformation information. The impact event was recorded using high-speed video. The content of the work is relevant for compiling a test methodology and has potential with respect to the design of helmets to improve ballistic protection along the edge of helmets.

1. INTRODUCTION

Within Police Forces there is now a trend to specify high-cut ballistic helmets for reasons of ergonomics, mass and compatibility. This of course reduces the protective area of coverage compared to a standard-cut helmet. With reduced area of coverage, it is more critical than ever that as much of the area of the helmet as possible provides ballistic resistance. This means that it is critical to assess the performance of the helmet close to the edge of the helmet, and ultimately determine how close to the edge a shot can be defeated. If a ballistic helmet cannot defeat a low velocity bullet relatively close to the edge, then its effective area of coverage is reduced, burdening the wearer with added weight and discomfort for no added protection.

Figure 1 shows the different geometries of generic standard, mid and high-cut helmet designs, showing the reduction in area of coverage with the high-cut helmets.

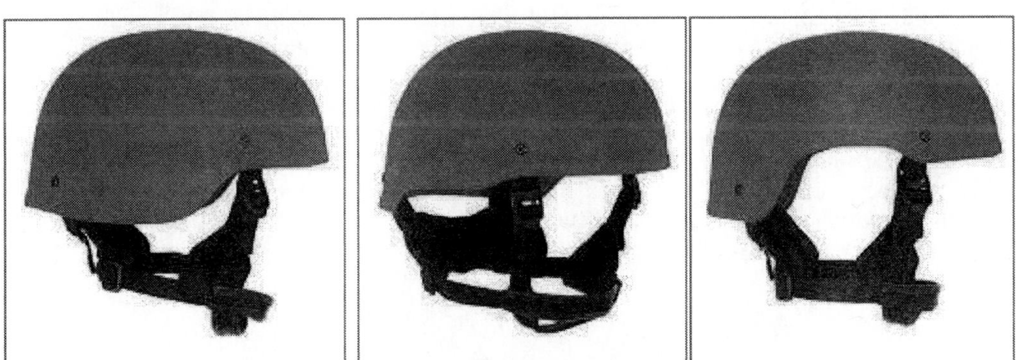

Figure 1. Standard-Cut (left), Mid-Cut (centre) and High-Cut (right)

In order to assess the ballistic edge performance, ballistic testing needs to be conducted, which is easier to do if there is a suitable test method to use. Currently only one test standard includes edge shots on ballistic helmets, which is VPAM HVN 2009 [1]. This study aims to develop a test method and not to assess the performance of specific ballistic helmets.

2. VPAM HVN 2009 (2017)

https://doi.org/10.52202/080042-0042

There appears to be little history of edge testing of helmets. A literature search conducted in early 2022 identified only a single test standard which incorporated an edge test – VPAM HVN 2009.

The VPAM (Vereinigung der Prüfstellen für angriffshemmende Materialien und Konstruktionen) organisation is an association of German-speaking test organisations, who have produced their own series of test standards and methods. VPAM HVN 2009 is a standard which specifies the test methods used for helmets (H), visors (V) and neck-guards (N). Focussing upon the helmet (H) part only, there are separate ballistic perforation-resistance and behind helmet blunt trauma parts of the testing.

As there are no threat levels included within HVN 2009, it calls upon another VPAM standard, APR 2006 [2], which in theory allows any of 14 different levels to be chosen. A number of these levels are completely irrelevant for helmets, and even pushing the boundaries of current technology, only levels 2 to 7 are likely to be specified.

The shot pattern specified for the helmet consists of 5 impacts around the helmet, 80 mm from an edge or another shot. The VPAM then states that a further shot is to be placed at 20 + 5 mm (i.e. between 20 and 25 mm from the edge) from an edge. Questioning of the current VPAM committee has not been able to acquire any evidence of why the edge shot, at that particular distance from the edge, was included for the first time, in the 2009 version of the standard. It was not part of the 2005 version, which many of the VPAM test laboratories seem to still be using.

During the development of the MPS edge test method, a single helmet specified to VPAM HVN 2009 was tested. The results in Table 1 show that the design of that particular helmet does easily meet the 20 + 5 mm edge test requirement.

Table 1. Ballistic Test Results for VPAM HVN 2009 Helmet

Shot No	*Impact Point*	*Velocity (m/s)*	*Outcome (N/P)*	*Comments*
1	Crown (front)	404	N	
2	Crown (rear)	405	N	
3	25 mm from rear edge	404	N	
4	20 mm from rear edge	415	N	
5	10 mm from front edge	415	P	jacket impact on witness
6	15 mm from front edge	413	N	
7	15 mm from side edge	408	N	
8	15 mm from side edge	415	N	

3. DEVELOPMENT OF TEST METHOD

To this end the Metropolitan Police Service (MPS) Physical Protection Group (PPG), and Phil Gotts Consulting Ltd (PGC), have conducted a programme to support the writing of a test method for edge testing of helmets in general, and high-cut helmets in particular.

The initial series of trials used the helmet mounted in its usual orientation, as shown in figure 2. It soon became obvious that this was not ideal, as it was difficult to determine whether it was a perforation (P) or non-perforation (N) after each shot, and there was a general feeling that the detail of what had actually happened during the event was something of an unknown. There was also the tendency for bullets which were a non-perforation to drop out of the helmet, either at the time of the impact, or during subsequent shots, or handling of the helmet.

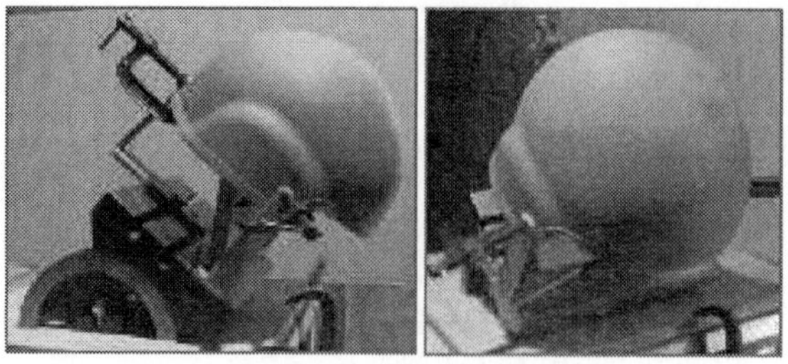

Figure 2. Helmet mounted to Test Rig in Typical Orientation

The obvious solution for the following round of testing was to invert the helmet. In fact, at the end of this particular trial this was attempted with the current rig and one of the previously tested helmets. This inversion added several advantages:

- Any non-perforation result bullets are retained in the helmet shell
- It is simple to see what the edge of the helmet looks like without removing it each time
- It allows for further instrumentation to be included easily.

The inversion of the helmet has allowed further development of the test method to include the use of high-speed cameras, which has greatly increased the ability to understand what is happening during the bullet / helmet interaction. This allows the trajectory of the bullet to be observed, and aids with the determination of whether a shot should be considered as a perforation or non-perforation. This also allows an estimate to be made of the temporary deformation of the helmet shell, as well as the permanent deformation.

The increased ability to visualise the impact event has also led to other issues which need to be considered. For example, when a bullet impacts an edge, what should be considered as a perforation or non-perforation. Previously with helmet ballistic testing, this has been determined by a witness placed within the helmet, but this may provide a false impression, due to the potentially erratic trajectories of some bullets post-perforation, as well as the possibility of bullet break up with parts taking different trajectories.

4. EDGE TESTING OF HIGH-CUT BALLISTIC HELMETS

A series of three trials were conducted using low-cost high-cut ballistic helmets procured from China. These were manufactured from either para-aramid or ultra-high molecular weight polyethylene (UHMWPE) and specified to meet NIJ-0101.04 Level IIIa. The designs were identical for both materials, with the para-aramid shells being approximately 5 % heavier than the UHMWPE ones, whereas the UHMWPE helmet shells are approximately 25 % thicker than the para-aramid ones. Surprisingly, perhaps, the para-aramid helmets were also the more expensive items. For the trials the para-aramid helmets are identified as ‘PA’ followed by a number relating to the trial series part, while the UHMWPE helmets are identified with ‘PE’ followed by the appropriate number.

All trials were conducted with the same ammunition – DAG 9 x 19 mm DM11A1B2 FMJ (full metal jacket), fired as full-charge, out-of-the-box ammunition, at 405 ± 15 m/s from an appropriate proof barrel. The bullet’s impact velocity was measured using optical sky-screens. The outcome of each shot was deemed to be either a perforation (P) or a non-perforation (N). The trial configuration was changed slightly across the three trials, but the changes were minimal and only affected the ease of mounting the helmet shells on the rig. The general configuration is as shown in Figure 3.

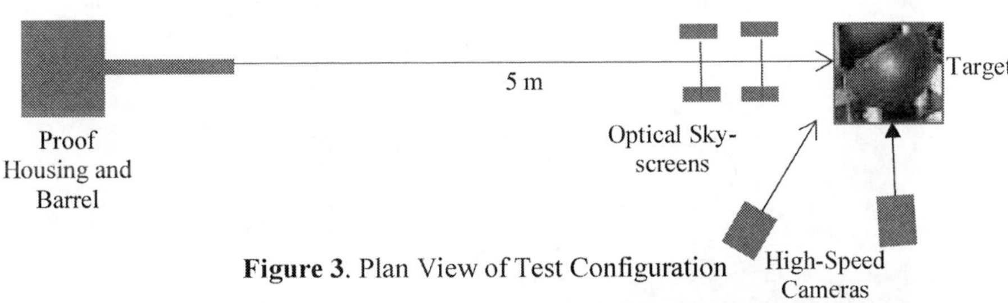

Figure 3. Plan View of Test Configuration

Figure 4 shows the helmet mounted on the rig in its inverted position. The 45 ° mirror allows one of the high-speed cameras to look down into the helmet, while the other high-speed camera shows the view of the bullet impacting the helmet and hence confirms the orientation of the bullet impact to the surface. This high-speed video view also shows the impact point should further damage make this difficult to confirm. The strawboard mounted behind is designed to capture any of the bullets which are deflected beyond and out of the helmet. The footage from the high-speed camera is used to calculate both the peak temporary and the peak permanent deformations in each case.

Figure 4. Plan View of Test Configuration

https://doi.org/10.52202/080042-0042

4.1 Part 1: 25 mm from edge of helmet shell

The first part of the trial was to conduct the edge shot testing at 25 mm from the edge of the helmets at four positions around the helmet. These were at the front, rear and both sides. For Part 1 of the trial each helmet also had a 5th shot placed at the crown, to confirm that the helmet behaved as expected well away from the edges.

Table 2. Ballistic Helmet Test Results for Part 1 (25 mm from edge)

Helmet ID	*Shot No*	*Impact Point (Distance from edge, position)*	*Bullet Impact Velocity (m/s)*	*Outcome (N/P)*	*Peak Temporary Deformation (mm)*	*Peak Permanent Deformation (mm)*
PA1	1	25 mm, front	406	N	91	8
	2	25 mm, rear	419	N	79	16
	3	25 mm, RH side	416	N	79	12
	4	25 mm, LH side	407	N	90	14
	5	Crown	405	N	25	8
PE1	1	25 mm, front	408	*P*	96	82
	2	25 mm, rear	415	N*	84	70
	3	25 mm, RH side	405	N	74	37
	4	25 mm, LH side	417	N	96	55
	5	Crown	413	N	65	35

N* - bullet broken up, with at least one part of the jacket leaving the helmet edge

The results above show that at 25 mm from the edge both helmets provided a good resistance to perforation. The 1st impact on helmet shell PE1 slightly missed the aim point and was at approximately 22 mm from the edge. Therefore, it is assumed that both helmet materials will provide predominantly non-perforations at 25 mm from the edge. One important point to note at this stage, is that although the temporary deformation of both helmets was of a similar magnitude, the para-aramid helmet recovered to a lower permanent deformation.

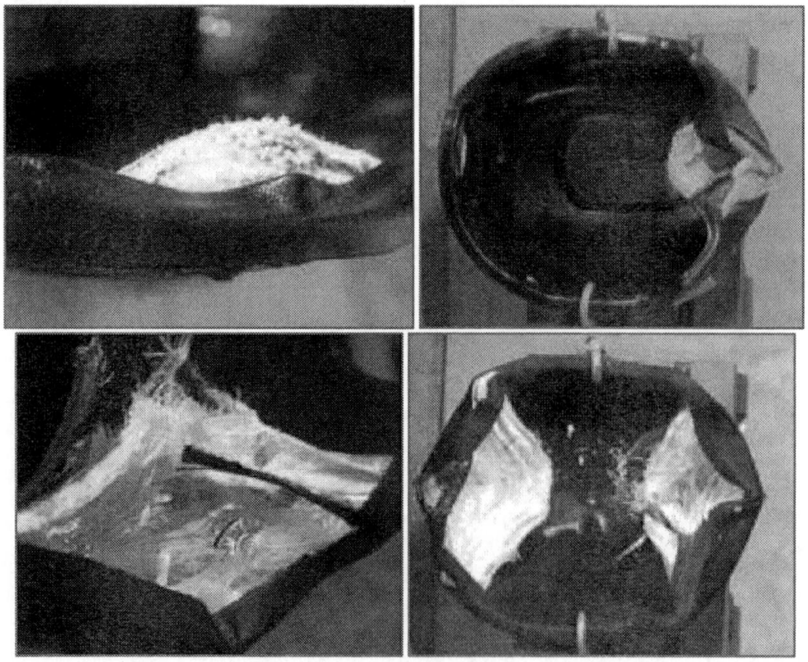

Figure 5. Shot 2 on para-aramid helmet PA1 (top) and Shot 2 UHMWPE helmet PE1 (bottom): permanent deformation (left); high-speed still of temporary deformation (right).

Figure 5 clearly shows the difference between the permanent and temporary deformation for the two helmet shell materials.

In order to add to the understanding of the perforation / non-perforation effects upon the bullets, the post-test helmet shells were X-rayed, as shown in figure 6. The figure also shows the greater delamination, withing the larger permanent deformation for the UHMWPE helmet. The X-rays give a clear indication of the differences in the deformation of the bullets between the para-aramid and UHMWPE helmet shells and show that one of the bullets in the UHMWPE shell was broken up.

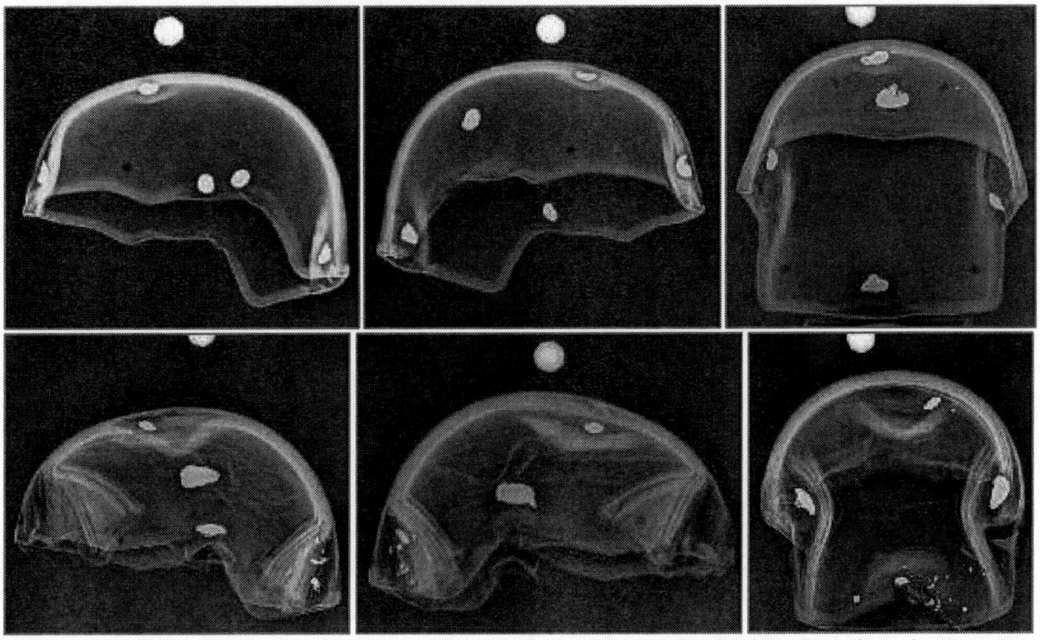

Figure 6. X-Rays of Para-Aramid Helmet PA1 (upper) and UHMWPE Helmet PE1 (lower) – Post-Test

4.2 Part 2: 15 mm from edge

Part 2 of the trial was a repeat of Part 1, but with the edge shots aimed at 15 mm from the edge. In this case, the majority of the shots produced perforations, with just one non-perforation on each helmet.

Table 3. Ballistic Helmet Test Results for Part 1 (15 mm from edge)

Helmet ID	*Shot No*	*Impact Point (Distance from edge, position)*	*Bullet Impact Velocity (m/s)*	*Outcome (N/P)*	*Peak Temporary Deformation (mm)*	*Peak Permanent Deformation (mm)*
PA2	1	Crown	416	N	21	5.2
	2	15 mm, front	405	P	72	n/a
	3	15 mm, rear	405	P	78	28
	4	15 mm, RH side	406	P	86	40
	5	15 mm, LH side	405	N	86	28
PE2	1	Crown	409	N	29	13
	2	15 mm, front	407	P	67	n/a
	3	15 mm, rear	404	N	118	109
	4	15 mm, RH side	413	P	65	42
	5	15 mm, LH side	403	P	76	75

Again these results show that there is little recovery of the shell material from temporary to permanent deformation with the UHMWPE shell, whereas there is significant recovery with the para-aramid one.

https://doi.org/10.52202/080042-0042

Figures 7 and 8 show example shots from the para-aramid and UHMWPE helmets, respectively. The red circle on the high-speed still image for the para-aramid helmet (figure 7) shows the perforating bullet in flight across the helmet, as it drags fibres from the shell with it.

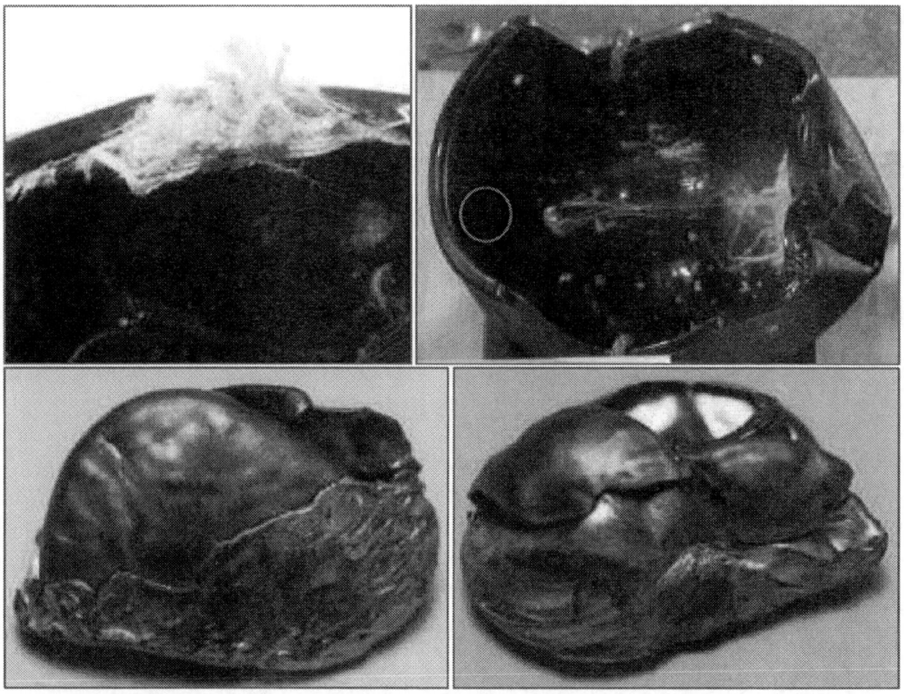

Figure 7. Shot 2 on para-aramid helmet PA2: permanent deformation (top left); high-speed still of temporary deformation (top right); recovered bullet (bottom)

Figure 8. Shot 2 on UHMWPE helmet PE2: permanent deformation (top left); high-speed still of temporary deformation (top right); recovered bullet (bottom)

In order to add to the understanding of the perforation / non-perforation effects upon the bullets,, the post-test helmet shells were X-rayed, as shown in figure 9. These X-rays also show the much greater delamination of the UHMWPE shell than the para-aramid ones, as well as a greater tendency for break-up of the bullets with the UHMWPE shell.

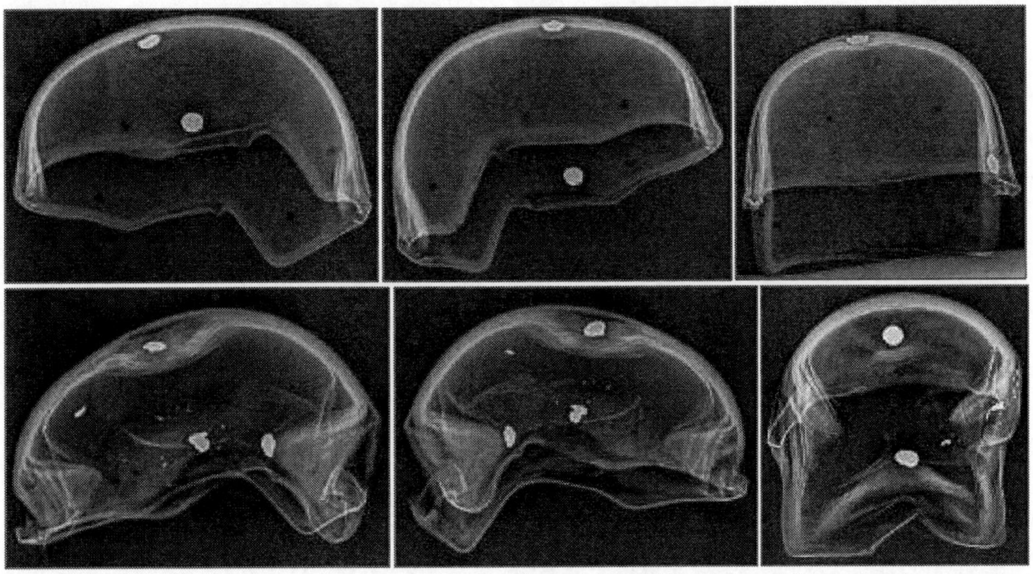

Figure 9. X-Rays of Para-Aramid Helmet PA2 (top) and UHMWPE Helmet PE2 (bottom) – Post-Test

4.3 Part 3: 20 mm from edge

With Part 1 at 25 mm from the edge producing predominantly non-perforations and Part 2 at 15 mm from the edge producing predominantly perforations, and bearing in mind that the VPAM HVN 2009 specified an edge shot at 20 mm, Part 3 of the trial conducted testing at 20 mm from the edge.

Table 4. Ballistic Helmet Test Results for Part 1 (20 mm from edge)

Helmet ID	*Shot No*	*Impact Point (Distance from edge, position)*	*Bullet Impact Velocity (m/s)*	*Outcome (N/P)*	*Peak Temporary Deformation (mm)*	*Peak Permanent Deformation (mm)*
PA3	1	20 mm, front	396	N	99	27
	2	20 mm, rear	392	N	92	52
	3	20 mm, RH side	391	P	80	25
	4	20 mm, LH side	401	P	90	28
PE3	1	20 mm, front	394	P	94	94
	2	20 mm, rear	395	P	92	85
	3	20 mm, RH side	392	P*	84	84
	4	20 mm, LH side	407	P*	79	79

P* - bullet broken up, with at least one part of the lead and the jacket exiting the helmet edge

At 20 mm from the edge there is a difference observed between the para-aramid and the UHMWPE helmet shells. For the para-aramid helmet, there has been an equal division between non-perforations and perforations. This outcome could be expected based upon the results for both 25 and 15 mm. For the UHMWPE helmet shell, all the impacts are considered to be perforations, although it is worth noting that two of the impacts broke up the bullet significantly and it was only part of the bullet that perforated. Figures 10 and 11 show equivalent shots 3 on each helmet material. Both of these shot 3s are considered to be perforations. Again, the para-aramid helmet has recovered from temporary to permanent deformation, whereas the UHMWPE helmet has shown no recovery from the temporary deformation.

https://doi.org/10.52202/080042-0042

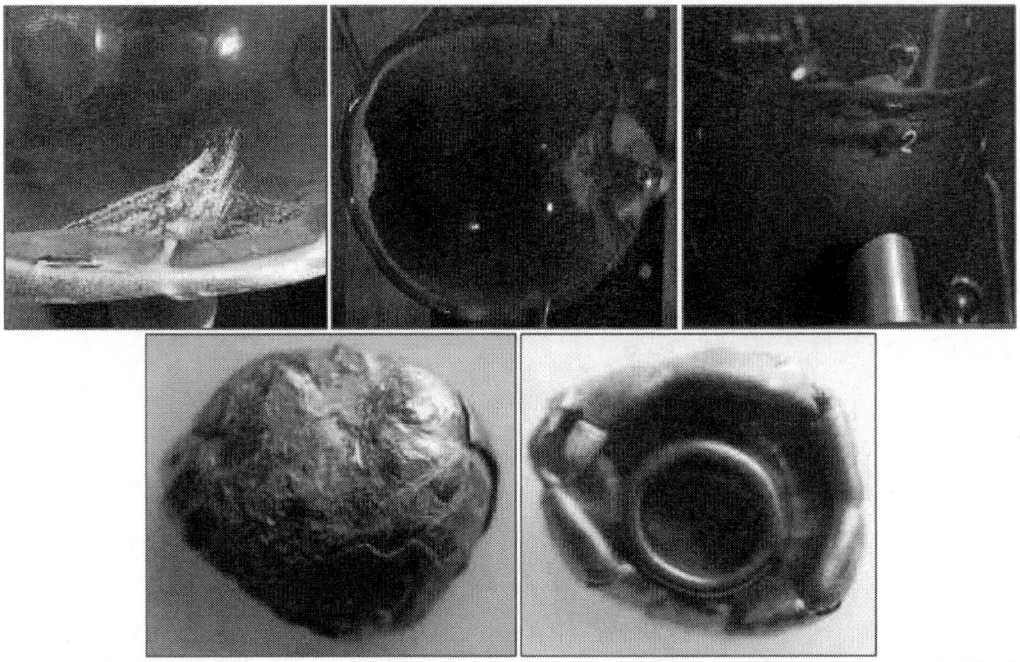

Figure 10. Shot 2 on para-aramid helmet PA3: permanent deformation (top left); high-speed still of temporary deformation (top right); recovered bullet (bottom)

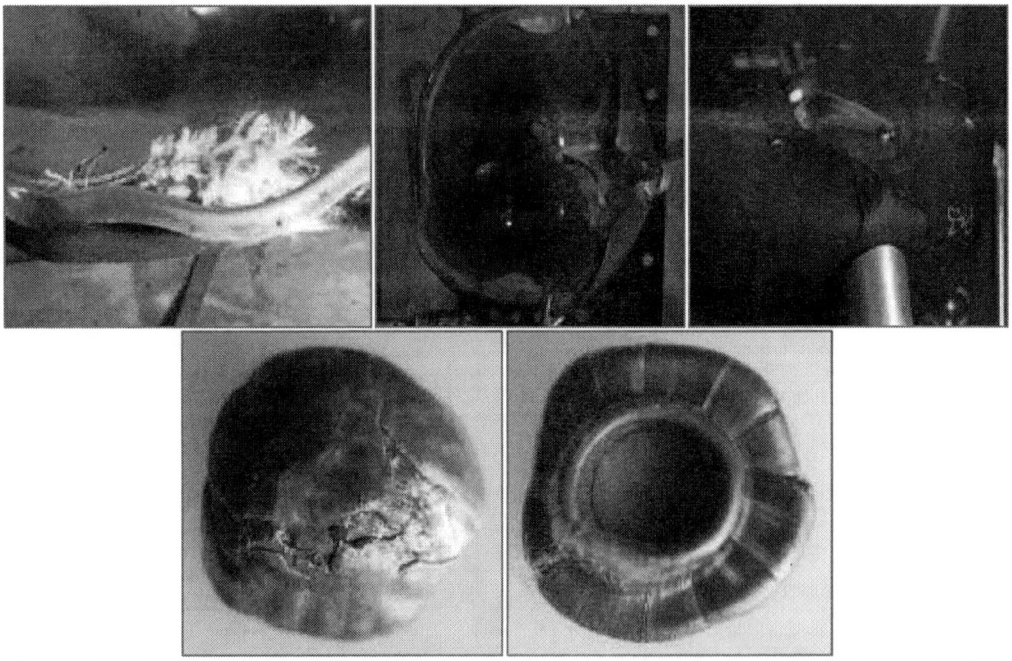

Figure 11. Shot 3 on para-aramid helmet PA3: permanent deformation (top left); high-speed still of temporary deformation (top right); recovered bullet (bottom)

Figures 12 and 13 show equivalent shots 2 on the two helmet materials. The para-aramid helmet shot 2 provided a non-perforation and exhibited significant recovery from temporary to permanent deformation. The UHMWPE helmet suffered a perforation and showed very little recovery from temporary to permanent deformation.

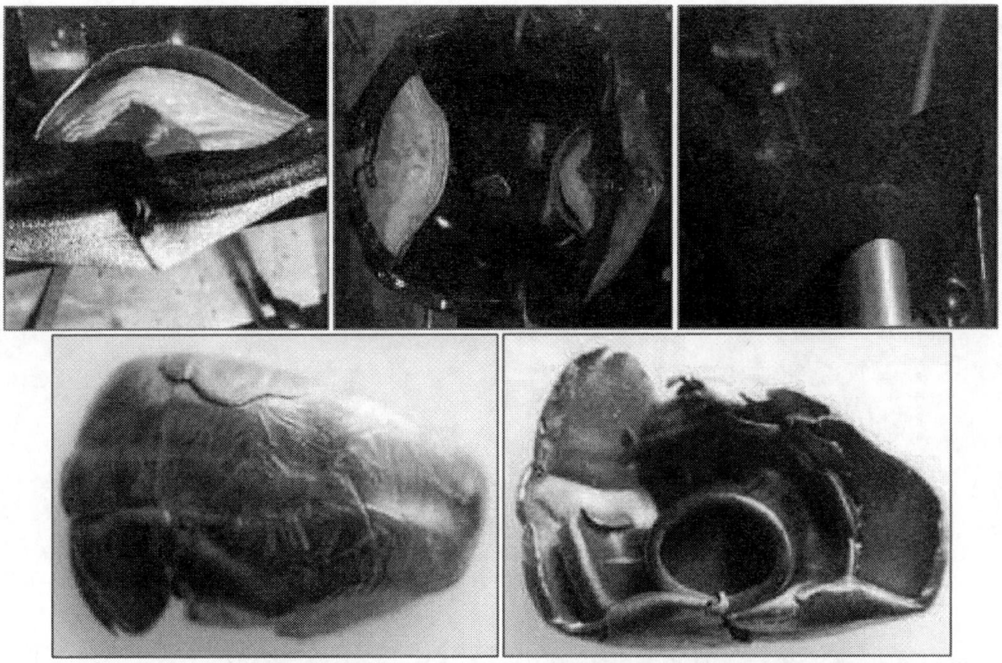

Figure 12. Shot 2 on UHMWPE helmet PE3: permanent deformation (top left); high-speed still of temporary deformation (top right); recovered bullet (bottom)

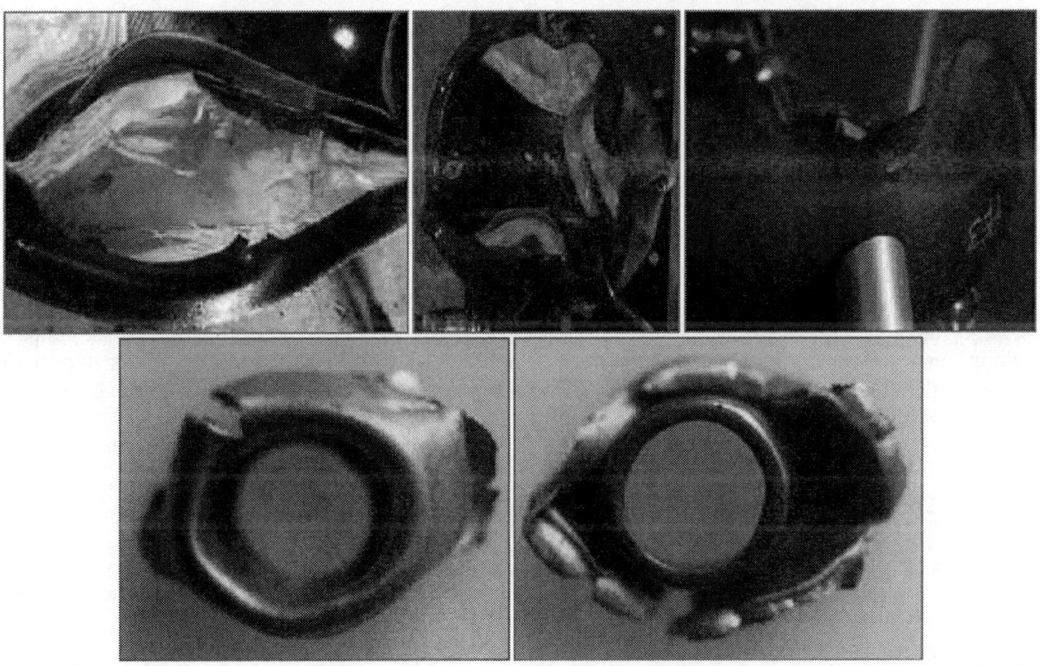

Figure 13. Shot 3 on UHMWPE helmet PE3: permanent deformation (top left); high-speed still of temporary deformation (top right); recovered bullet (bottom)

Figure 14 shows the X-rays for both para-aramid and UHMWPE helmet shells. The view of the para-aramid helmet indicates that all four edge shots were perforations. However, the first two shots left the helmet, not by perforating, or pushing aside the material, but by being pushed back and upwards by the recovering deformation. For the UHMWPE helmet, shots 3 and 4 (at the sides) show retained components of the broken-up bullet, but this was confirmed a perforation, as most of the jacket, in each case, perforated the shell material.

https://doi.org/10.52202/080042-0042

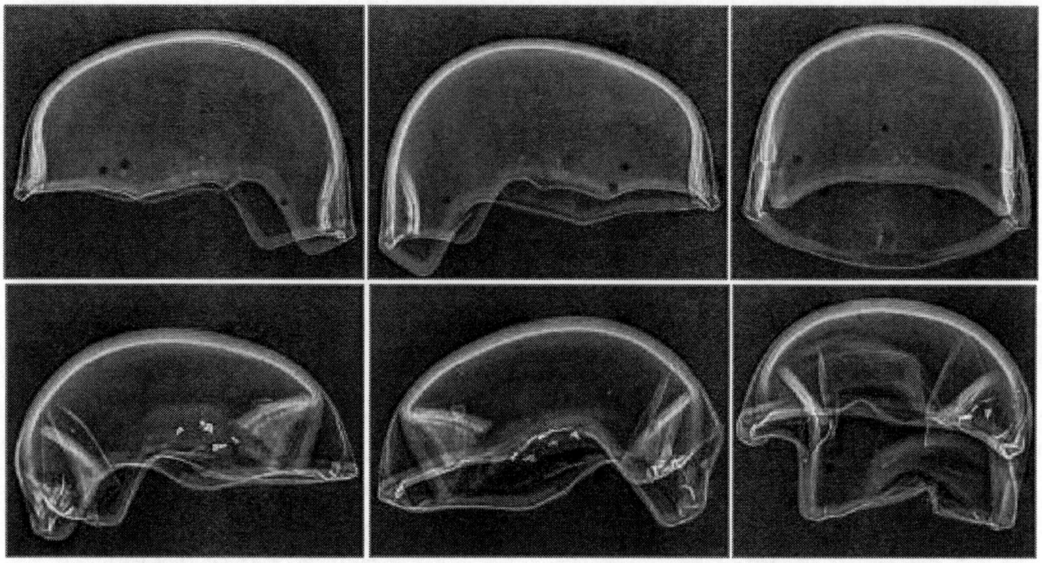

Figure 14. X-Rays of Para-Aramid Helmet PA3 (top) and UHMWPE Helmet PE3 (bottom): Post-Test

4.4 Discussion of High-Cut Ballistic Helmet Edge Test Results

One of the issues identified with the edge testing was the definition of what constitutes a perforation or a non-perforation. This issue was identified when the high-speed video was incorporated into the test method. With an impact at a greater distance from the edge it is obvious what constitutes a perforation or non-perforation, as there is either a hole in the material, or there is not. With edge impacts this is not so simple. In most cases the bullet continues its trajectory by pushing the material near the edge out of the way, without perforating it. Therefore, for edge performance a different definition of perforation and non-perforation is required. A perforation may be defined as a bullet continuing in a forward trajectory, and its lethality may need to be assessed using a witness material. These definitions have been used throughout these series of trials.

Bullet deformation in these pressed composite shapes tested (as opposed to soft body armour) demonstrated a wider variety of bullet deformation, with a number of irregular shapes. Some resulted in severe fragmenting of the bullet (most notable in the UHMWPE shells).

As the main objective of this study is to develop a suitable test method for the edge performance of helmets using inexpensive helmets as a development tool, (whilst making observations during the process), detailed discussion of the helmet performance at the different distances from the edge is not relevant to this paper.

4.5 Way Forward for Test Method

The test method developed with the inverted helmet and the two high-speed cameras has been shown to be viable and easy to conduct. Therefore, it is proposed as an on-going edge test within the MPS. In fact, it has already been used within a procurement tender for ballistic helmets.

Returning to the edge shot requirement within VPAM HVN 2009, it is felt that the 20 + 5 mm for an edge test is a target that should be achievable. However, rather than a single edge shot, it is proposed that one example of each helmet submitted is tested as here with edge shots to the front, rear and both sides, as the different locations may vary in performance.

References

[1] VPAM. 'HVN-2009 - Test Guideline - Bullet-Resistant Helmet with Visor and Neck Guard.' 2017.

[2] VPAM. 'VPAM APR 2006 - General basis for ballistic material, construction and product testing - Requirements, test levels and test procedures' Edition 2009

Increasing Confidence in the Performance of Ballistic-resistant Shields

B. Bertsch[1], J. Horlick[2], N. Roberts[3], C. Robinson[2], S. Cohen[4]

[1] *Oregon Ballistic Laboratories, 2873 22nd Street SE, Salem OR 97302 (bbertsch@oregonbl.com)*
[2] *National Institute of Standards and Technology (NIST), 100 Bureau Drive, Gaithersburg MD 20899, USA*
[3] *Unified Police Department of Greater Salt Lake, 3365 South 900 West, Salt Lake City UT 84119*
[4] *U.S. Drug Enforcement Administration, Redstone Arsenal, Huntsville AL 35808*

Abstract. ASTM International has published test methods and a specification for ballistic-resistant shields used by law enforcement officers to protect against handgun or rifle rounds. These shields are complex technologies that consist of the main shield body ballistic-resistant materials, a transparent ballistic-resistant viewport, fasteners, joints and seams, edging, appliques (intended to increase localized protection), and more features. Each of these must be tested to verify that the complete shield protects against bullets and also can withstand the conditions of use and storage. The ASTM International standards are intended to verify that ballistic shields meet this purpose. Until now, the standard typically used for assessment of protection afforded by ballistic shields was the National Institute of Justice (NIJ) Standard 0108.01, Ballistic Resistance of Materials, published in 1985. The NIJ standard was developed for materials used to fabricate protective products, not for complete products like shields, and requires limited testing of either a soft armour shoot pack or a hard armour coupon. Additionally, the test threats required in the NIJ standard are not current threats facing U.S. law enforcement; therefore, manufacturers have mixed the requirements of NIJ Standard 0108.01 with test threats from NIJ Standard 0101.06, Ballistic Resistance of Body Armor, leading to confusion, and possibly deception, in the marketplace. To alleviate these issues and develop shield-specific standards, a diverse team of more than forty stakeholders worked together, including shield manufacturers; suppliers; federal, state, and local law enforcement end users; ballistic testing and certification experts; researchers; federal ballistic protection experts; and standards professionals. The new ASTM standards are a tremendous step forward because they are specifically designed for assessing the performance of entire ballistic shields of multiple sizes. The test method specifies detailed testing procedures to assess all aspects of a ballistic shield and requires a minimum number of shots on each area: the shield body, the viewport, fasteners, weak points, and appliques. The specification details pre-conditioning and testing requirements, ballistic performance levels with associated up-to-date test threats, and performance requirements that shields must meet. These two standards form the basis for a new ASTM Verification Program, which includes independent, third-party verification by the Safety Equipment Institute, online listing of verified products, authorization to put the ASTM-verified mark on products, and annual testing requirements to assess continued compliance. This paper will describe the new standards, the new ASTM Verification Program, and the benefits to the U.S. law enforcement community.

1. INTRODUCTION

A ballistic shield is a type of protective equipment intended to protect the user against bullets and provide greater coverage than personal body armour that typically only covers the torso. These shields are complex technologies that consist of the main shield body ballistic-resistant materials, a transparent ballistic-resistant viewport, fasteners, interfaces, joints and seams, edging, appliques (intended to increase localized protection), and more features. See Figure 1 for an example of a shield and its protective components and features. To assess and verify that the complete shield stops bullets and can withstand the conditions of use and storage, each of the shield's protective components and features must be tested.

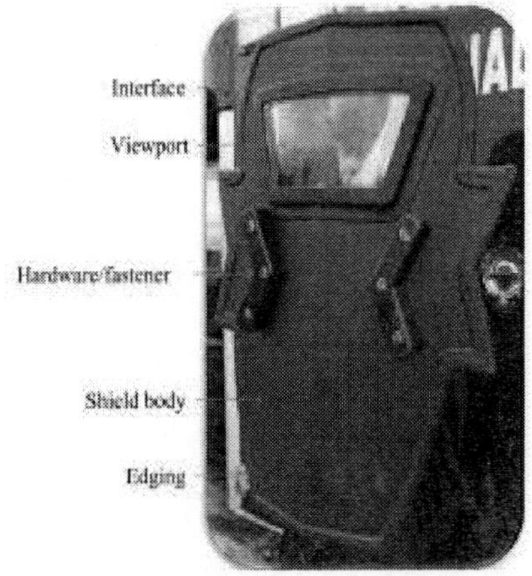

Figure 121. Examples of Protective Components and Features
Source: ASTM International

Ballistic shields for U.S. law enforcement officers have historically been assessed using the National Institute of Justice (NIJ) Standard 0108.01, *Ballistic Resistance of Materials* [1], published in 1985. That NIJ standard was developed for materials used to fabricate protective

https://doi.org/10.52202/080042-0043

products, not for complete products like shields. The NIJ standard requires only one test item conditioned at ambient temperature and humidity, at least 30.5 cm by 30.5 cm (12 inch by 12 inch) in size, per ballistic test threat, with a maximum of five shots per test item. Additionally, the test threats required in the NIJ standard are not representative of current threats facing U.S. law enforcement; therefore, manufacturers have mixed the requirements of NIJ Standard 0108.01 [1] with test threats of NIJ Standard 0101.06, *Ballistic Resistance of Body Armor* [2], leading to confusion in the marketplace and inconsistency of testing across products.

To alleviate these issues, a diverse team of stakeholders worked together to develop two ASTM International shield-specific standards and an ASTM Verification Program.

2. RECOGNIZING THE NEED

The need for a ballistic shield standard was recognized by working with the National Association of State Procurement Officials (NASPO) ValuePoint Program, a U.S. national public cooperative purchasing program. One of its public safety contracts is for personal body armour and ballistic-resistant products.[20] When the current ValuePoint contract was written, a requirement was included that ballistic-resistant shields be tested in accordance with NIJ Standard 0108.01 [1]. As previously stated, that standard is intended only for materials used to make finished products and is not appropriate nor sufficient for a shield. However, it was the only performance standard available at the time the contract was put in place. This highlighted the need for a new shield-specific standard.

Requiring that shields meet a standard then highlighted the need for a conformity assessment process for determining whether a shield met the standard. Since there was no established conformity assessment program for ballistic shields, the only option was to have an independent technical expert review test reports submitted by manufacturers and provide a recommendation to the ValuePoint contract lead. Reviewing test reports revealed several key points: (1) the NIJ standard was either loosely applied or a blend of multiple standards was used; (2) there was inconsistency of testing across products, and (3) some test reports were for something besides a shield.

This experience prompted discussions with industry, technical experts, researchers, and end users about initiating an effort to write a new shield-specific performance standard and establish a conformity assessment process.

3. IDENTIFYING AND ADDRESSING OFFICER NEEDS AND REQUIREMENTS

The first step in developing the performance standard was to identify the types of shields that are used by U.S. law enforcement officers and to understand their operational needs and requirements. The officers on this project team represented federal law enforcement agencies, local police departments, and sheriff's offices from across the U.S. The team agreed up front to match the new NIJ handgun and rifle threats specified for use with the next version of the NIJ body armour standard (NIJ Standard 0101.07, *Ballistic Resistance of Body Armor*, not yet published). The team additionally chose to add a supplemental shotgun threat as an add-on to either a handgun-rated or rifle-rated shield. The test threats will be discussed in Section 5.

Many configurations of ballistic-resistant shields are available in the marketplace, including hand-held or hand-carried shields, person-portable shields with wheels, fixed or mobile barriers, flexible shields that drape over a surface (e.g., ballistic blanket), and "accordion" shields. Officers typically use hand-held or hand-carried shields and person-portable shields, so the scope of the standards effort was limited to these two configurations. See Figure 2 for examples of shield types.

Officers were asked about issues experienced during use of shields. While no one on the team was aware of a shield failing to stop a bullet during use, ballistic protection was their highest priority. Other concerns were raised about damage caused by use or storage conditions, including delaminating of viewports and edges when stored in vehicles and cracked edges caused by dropping the shield on an edge.

Protective components and features of the shields that needed to be ballistically tested were identified (as shown in Figure 1), and the most appropriate test and performance requirements for each one was discussed at length.

---

[20] Details on the NASPO ValuePoint Contract may be found here:
https://www.naspovaluepoint.org/portfolio/body-armor-and-ballistic-resistant-products/

(a) Hand-held, rigid
Source: United Shield

(b) hand-held, folding
Source: Multi-Threat Shield

(c) person-portable, wheeled
Source: Safariland

Figure 2. Examples of Shields Typically Used by U.S. Law Enforcement Officers

4. IMPROVING TEST METHODS FOR BALLISTIC SHIELDS

A preliminary set of test methods for ballistic shields was published in 2018 as ASTM E3141/E3141M, *Standard Test Method for Ballistic Resistant Shields for Law Enforcement* [3], but these test methods were not developed with limited stakeholder input and were not put into practice. The technical experts on the team evaluated and improved each test method as described below.

Testing of the shield body was improved by adding a second 3-shot cluster with shots at 30 degrees angle of incidence (to complement the initial 3-shot cluster at 0 degrees) because the angled shots better assess the performance of newer ballistic materials. The cluster shot spacing for each type of test threat (i.e., handgun, rifle, shotgun) was specified in detail. No changes were made to the shield body edge shots from the 2018 version.

Details for testing fasteners and perceived weak points were added, with clear specifications for shot placement, angle of incidence, and number required.

The required shots on the viewports were discussed in greater length because the viewport is the most vulnerable component of the shield, especially to certain rifle projectiles, and a failure would likely result in an officer's face being impacted. Figure 3(a) shows an example of a shield viewport shot with handgun projectiles, and Figure 3(b) shows a shield viewport shot with a single 7.62x51mm M80 NATO (M80 ball) rifle projectile[21]. For the handgun shield example, multiple shots were stopped by the viewport; for the rifle shield example, the viewport would not have been able to stop a second shot due to the capabilities of current technology.

The 2018 test method had specified shots for the viewport corner, edge, and center with no mention of the interface between the shield body and viewport. The requirements for the shots in the corner,

(a) Handgun-rated shield with multiple shots

(b) Rifle-rated shield with one M80 shot

Figure 3. Examples of Viewport Ballistic Testing

[21] Figures 3(a) and 3(b) were provided by the U.S. Drug Enforcement Administration

https://doi.org/10.52202/080042-0043

edge, and center were specified in greater detail in the 2022 version, and interface shots were added at 0-degree and 45-degree angles of incidence with specificity regarding direction of each shot for each type of interface (i.e., protruding, overlapping, protruding and overlapping, flush, and recessed).

A summary of required shots, excerpted from ASTM E3141/E3141M [3], is shown in Figure 4. It can be seen that the shots required for a handgun or a rifle shield are the same, except for the number of viewport shots for a rifle shield is reduced, recognizing the limitations of current viewport technology.

The team discussed the increasingly common use of in-conjunction with (ICW) armour appliques that may be applied to a shield to enhance ballistic protection in a localized area. Test requirements were added to shoot the applique, at the claimed protection level, as well as any exposed hardware attaching the applique to the shield.

Shield Component	Shot Description	Total Number of Shots Required for Handgun Shields	Total Number of Shots Required for Rifle Shields
Viewport	Shot in center, 0°	2 shots	1 shot
Viewport	Shot in corner, 0°	2 shots per unique corner	0 shots
Viewport	0° shot on unique edge	1 shot per unique edge	1 shot per unique edge
Viewport-Body Interface	45° angled shot at interface	1 shot at interface	1 shot at interface
Viewport-Body Interface	0° shot at interface	1 shot at interface	1 shot at interface
Body	3-shot cluster shot at 0°	2 clusters per unique construction type	2 clusters per unique construction type
Body	3-shot cluster shot at 30°, with all shots in same direction	2 clusters per unique construction type	2 clusters per unique construction type
Body	Edge shot	4 shots	4 shots
Fasteners	Fastener head shot, 0°	2 shots per unique fastener	2 shots per unique fastener
Fasteners	Fastener proximity shot, 0°	2 shots per unique fastener	2 shots per unique fastener
Fasteners	Fastener shank shot, 45°	2 shots per unique fastener	2 shots per unique fastener
Weak Points	Shot on/near perceived weak points, 0°	1 shot per unique weak point	1 shot per unique weak point

Figure 4. Summary of Required Shots from ASTM E3141/E3141M

5. DEVELOPING THE BALLISTIC SHIELD SPECIFICATION

In addition to ballistic testing requirements, the team agreed to the need to develop a specification to identify the ballistic test threats and define performance requirements and testing details supplemental to the test methods of ASTM E3141/E3141M.

As previously mentioned, the team chose to match the protection categories and associated test threats as defined by NIJ, and the list, excerpted from ASTM E3347/E3347M, *Specification for Ballistic-Resistant Shields Used by Law Enforcement Officer* [4], is as shown in Figure 5.

Each protective component and feature of the shield was considered, and the team determined which test procedures should be applied and what the performance requirements should be. For all ballistic testing, the performance requirement is no complete penetration.

To assess a shield's ability to withstand storage and use conditions that could degrade ballistic performance, the team discussed potential conditioning procedures that could be included as a pre-cursor to ballistic testing. Thermal shock, submersion in water, extreme temperature exposure, and dropping on an edge were chosen as the most relevant procedures. In order to reduce the number of shield samples required for testing, the team elected to perform the conditioning as a sequence of procedures: controlled ambient, thermal shock, controlled ambient, submersion, and extreme temperature (cold for one sequence, hot for another).

ASTM Shield Ballistic Protection Level	Test Threat	Test Threat Reference Velocity
ASTM-Shield-HG2	T1: Handgun, 9 mm Luger FMJ RN 124 grains	448 ± 9.1 m/s [1470 ± 30 ft/s]
	T2: Handgun, .44 MAG JHP 240 grains	436 ± 9.1 m/s [1430 ± 30 ft/s]
ASTM-Shield-RF1	T3: Rifle, 7.62 × 51 mm M80 Ball NATO FMJ Steel Jacket, 147 +0/-3 grains (U.S. military supply or rounds meeting NATO specifications)	847 ± 9.1 m/s [2780 ± 30 ft/s]
	T4: Rifle, 7.62 × 39 mm, MSC Ball Ammunition Type 56 from Factory 31, 123 grains	732 ± 9.1 m/s [2400 ± 30 ft/s]
	T5: Rifle, 5.56 mm M193, 56 +0/-2 grains (U.S. military supply or rounds meeting NATO specifications)	990 ± 9.1 m/s [3250 ± 30 ft/s]
ASTM-Shield-RF2	T3, T4, T5 (See above)	T3, T4, T5 (See above)
	T6: Rifle, 5.56 mm M855 BT, 61.8 ± 1.5 grains (U.S. military supply or rounds meeting NATO specifications)	950 ± 9.1 m/s [3115 ± 30 ft/s]
ASTM-Shield-RF3	T7: Armor-piercing rifle, 30.06 M2 AP, 165.7 +0/-7 grains (U.S. military supply or rounds meeting NATO specifications)	878 ± 9.1 m/s [2880 ± 30 ft/s]
ASTM-Shield-SG	T8: Shotgun, 12 Gauge, 1 oz. 2¾ in. lead slug	427 ± 9.1 m/s [1400 ± 30 ft/s]

Figure 5. ASTM Shield Ballistic Protection Levels and Associated Test Threats from ASTM E3347/E3347M

To ensure actual shields are tested, the specification requires that at least one of the smallest and one of the largest sizes of a shield model available on the market be tested. This requirement also addresses any performance changes that may be caused by a change in size of the product.

By including requirements for ballistic performance, pre-conditioning procedures, and detailed shots to all parts of the shield, ASTM E3347/E3347M, *Specification for Ballistic-Resistant Shields Used by Law Enforcement Officers*, [4] is a robust standard appropriate for assessing the durability and protective capabilities of a complete ballistic-resistant shield.

6. ASTM VERIFICATION PROGRAM

Having a robust standard specification, based on standard test methods, is an excellent starting point but is not sufficient for ensuring that shields are tested and perform as expected; a conformity assessment process is necessary.

The team considered independent third-party certification because it gives the highest confidence in product performance. Certification is a decision and statement by a third-party authoritative body, based on review of test reports and documentation, that a product is compliant with a standard specification based on the following:

(1) pre-market product testing and evaluation,
(2) periodic product testing to a limited set of critical tests (e.g., post-market testing),
(3) manufacturing facility inspections, and
(4) supplier management system audits.

The certifying body has responsibility for reviewing test reports and making a determination of the product's compliance with requirements at the time of initial testing and for performing periodic testing to determine whether the products continue to meet requirements. Certified products are listed by the certifying body in a publicly accessible listing and are authorized to be labeled with the certification body's mark. Because certification offers the highest level of confidence in a product's performance, it is the preferred conformity assessment process. However, participating in a certification program is very expensive for a supplier, with up-front costs for initial certification and recurring costs for maintaining certification. This means there must be a strong driver, such as regulations or grant funding requirements, mandating or motivating a supplier to submit products for certification.

A less-costly conformity assessment option is verification, which is a decision and statement by a third-party authoritative body that a product is compliant with a product standard based on the following:

(1) pre-market product testing and evaluation and
(2) periodic product testing to a limited set of critical tests (e.g., post-market testing).

The authoritative body has responsibility for reviewing test reports and making a determination of the product's compliance with the standard. Verified products are listed in a publicly accessible listing and authorized to be labeled with the verification body's mark.

Verification of responder products to recognized standards is a fairly new concept. It offers a slightly lower level of confidence in a product's performance than certification, but the cost is significantly less, making it more likely that a supplier will be willing to submit products for verification.

The team agreed that verification is an acceptable choice for ballistic-resistant shields, and an ASTM Verification Program was established.

As with certification, there must be a driver mandating or motivating a supplier to submit products for verification. There are currently no federal regulations or grant programs that require verification, but the NASPO ValuePoint program intends to require that ballistic-resistant shields be ASTM Verified to be included in their contract. Additionally, manufacturers involved in the project team agreed to participate in the ASTM Verification Program.

7. CONCLUSIONS

The new ASTM Verification Program and related ASTM standards for ballistic shields work together to address the needs and requirements identified by U.S. law enforcement officers, balancing current technology limitations, testing costs, and necessary confidence in the performance of ballistic shields. As technology improves, ballistic threats change, and operational scenarios evolve, the ballistic shield standards and ASTM Verification Program will be updated to ensure officers continue to have the protection they need.

https://doi.org/10.52202/080042-0043

Acknowledgments

This work was funded by the National Institute of Justice (NIJ). Any opinions, findings and conclusions, or recommendations expressed in this material are those of the author(s) and do not necessarily reflect the views of NIJ. The authors acknowledge and appreciate the contributions of the ASTM International task group members who invested their time, experience, and knowledge in the development of the ASTM standards and verification program.

References

[1] NIJ Standard 0108.01, *Ballistic Resistance of Materials*, National Institute of Justice, Washington DC, 1985.

[2] NIJ Standard 0101.06, *Ballistic Resistance of Body Armor,* National Institute of Justice, Washington DC, 2008.

[3] ASTM E3141/E3141M, *Standard Test Method for Ballistic Resistant Shields for Law Enforcement,* ASTM International, West Conshohocken PA, 2023, https://www.astm.org/e3141_e3141m-23.html.

[4] ASTM E3347/E3347M, *Specification for Ballistic-Resistant Shields Used by Law Enforcement Officers*, ASTM International, West Conshohocken PA, 2022, https://www.astm.org/e3347_e3347m-22.html.

Rifle Burst Fire Testing - Probability of Number of Impacts on Hard Armour Panel

Hemant Chouhan[1], Kartikeya Kartikeya[1], Khushi Ram[1], Aisha Ahmed[1], Makhan Singh[1], **Naresh Bhatnagar[1]**

[1]*Indian Institute of Technology Delhi, Hauz Khas, New Delhi, India - 110016*
nareshb@@mech.iitd.ac.in
9811326491

Abstract. This study aimed to evaluate the feasibility of firing 6-8 bullets on a 1000 sq. cm bullet-resistant panel from a distance of 10m using a three-bullet burst fire mode of a modern rifle. The results showed that firing three bullets in burst mode within 25 mm was not feasible, but firing three bullet bursts in 25-100mm was possible. Furthermore, the study compared different rifles and found that the Avtomat Kalashnikova-47 (AK-47) was more suitable than other rifles for burst fire mode, with a higher energy density noted for bursts using the AK-47 than Indian Small Arms System (INSAS) rifle. In the case of SIG Sauer 716, its larger cartridge size and resulting higher recoil make it more difficult to aim accurately. The study concluded that highly trained military personnel could accurately fire 6 bullets on a 1000 sq. cm hard armour panel (HAP) in accordance with National Institute of Justice (NIJ) 0101.06 or Bureau of Indian Standards (BIS)17051:2018 standards from a distance of 10m. However, the study highlighted the importance of considering the recoil of each rifle and its potential impact on the soldier's shoulder tissues. The study's findings suggest that there may be difficulties in achieving a high level of accuracy when firing 6-8 bullets in burst mode to hit a 1000 sq. cm area, which raises concerns about the feasibility of using this firing method. Overall, the study provided valuable insights into the feasibility and challenges associated with the body armour design and burst fire mode of modern rifles for military personnel.

1. INTRODUCTION

Limited information is available on various firing modes available in different rifles and their applications. The firing mode of a rifle determines the number of bullets that can be fired with each pull of the trigger and can have a significant impact on the accuracy and effectiveness of the weapon. There are several different firing modes available in rifles, including [1],

(i) Single shot: A single-shot rifle allows for precise aiming and can be useful for hunting or target shooting where accuracy is paramount.
(ii) Bolt action: A bolt action rifle is a type of single-shot rifle that requires the shooter to manually cycle the bolt to eject the spent cartridge and load a new round. This firing mode is reliable and accurate and is commonly used in precision shooting competitions.
(iii) Semi-automatic: A semi-automatic rifle can fire multiple rounds with a single trigger pull, which can be useful in situations where a rapid fire is necessary, such as in combat or self-defense. However, the rate of fire may make it less accurate compared to a single shot or bolt action rifle.
(iv) Full automatic: A fully automatic rifle can continuously fire rounds as long as the trigger is held down. This firing mode is primarily used in military and law enforcement contexts.

Overall, the importance of a specific firing mode in a rifle depends on the intended use of the firearm. For example, a hunter may prioritize accuracy and choose a single-shot or bolt action rifle, while a soldier may prioritize rapid fire and choose a semi-automatic or fully automatic rifle. Owing to the excessive consumption of bullets in fully automatic mode, a limited number of bullets are fired in automatic mode that serves better and in general the same is termed as burst fire.

Burst fire is a shooting technique that involves firing a predetermined number of rounds in quick succession. Three-bullet burst fire is a specific type of burst fire that involves firing three rounds with a single pull of the trigger. This technique is commonly used with rifles and is a popular choice for both military and law enforcement personnel. The primary factors in favour of 3-bullet burst fire using a rifle are,

(i) Precision and Control: One of the primary advantages of using three-bullet burst fire with a rifle is that it allows for a high degree of precision and control. By firing only three rounds at a time, the shooter can maintain accuracy while minimizing recoil and operating the weapon.

 https://doi.org/10.52202/080042-0044

(ii) Efficient Use of Ammunition: Another advantage of using three-bullet burst fire is that it allows for efficient use of ammunition. Rather than expending a large number of rounds in a short amount of time, the shooter can conserve ammunition by firing in controlled bursts.

(iii) Training and Technique: Using a three-bullet burst fire effectively requires a high degree of training and technique. It needs the shooter to be able to quickly and accurately fire three rounds in succession, while also maintaining control of the weapon. As such, it is a technique that is typically reserved for experienced shooters who have received specialized training.

The advantages of using 3-bullet burst fire mode over a single bullet and fully automatic fire are established yet the feasibility of all the bullets hitting every time within the target area needs further investigation. The 3-bullet burst fire also has some limitations. For instance, it can be less effective in engaging multiple targets or engaging targets at longer ranges, as it may not provide enough rounds to take down a target. Moreover, the limited number of rounds in each burst can also minimize the need for reloading, which can be an advantage in situations where time is of the essence. The recoil of a rifle in burst mode is yet another area of major concern. Firing a rifle in burst mode creates a higher recoil than firing a single shot and this can cause damage to the shoulder tissues of the shooter. The impact of the recoil on the soldier's shoulder can vary depending on the rifle and the size of the bullet used. It is important to consider the recoil and its impact on the shooter when selecting a rifle for burst fire mode [2]. While it has its advantages and limitations, it is important to consider the context and specific requirements of each engagement when evaluating the effectiveness of burst fire.

Body Protective Jackets (BPJs) are typically designed with an area ranging from 735 to 1000 sq cm in accordance with standards set by organizations like the National Institute of Justice (NIJ) or the Bureau of Indian Standards (BIS) [3][4]. Generally, a BPJ encounters a single shot at a time. However, in burst fire scenarios where multiple bullets are fired within a short span of less than a second, there is a possibility that three bullets may hit the BPJ. This raises concerns about the effectiveness of the current BPJ design in adequately protecting against burst fire incidents. Consequently, there may be a need for improvements in BPJ design to enhance their ability to withstand and mitigate the impact of multiple bullets hitting within such a short timeframe.

Despite its widespread use, there is a lack of understanding of the mechanics of 3-bullet burst fire, and its potential advantages and limitations compared to other firing techniques. Also, in case the person is hit with one or more bullets with or without a suitable bullet-resistant jacket fired from a given rifle will suffer how much trauma/injury needs to be investigated. As understanding the level of trauma and injury caused by different bullet impacts is crucial for assessing the effectiveness of protective gear and developing improved strategies for soldier protection. Expanding research in this area, will enhance the understanding of the physiological effects caused by bullet impacts and ultimately enhance the survivability and well-being of individuals in combat situations.

Accordingly, the first objective of the present study is to establish the feasibility of 3-bullets hitting the target when fired using different rifle types with their respective ammunition. For the study purpose, the weapons used were SMC (Sub Machine Gun), INSAS (Indian Small Arms System), AK-47 and a newly inducted weapon by the Indian forces SIG Sauer 716 (Table 1). The ammunition used by each of the weapons were 9 x 19mm FMJ/Pb, 5.56 x 45mm FMJ/(SI+Pb), 7.62 x 39 mm HSC (Hard Steel Core) and 7.62 x 51mm FMJ/Pb (M80 NATO Ball), all the bullets were supplied by two of the primary Ordnance Factories of India (Table 2). After the burst impact, the area covered by 3 bullets is vital that will depend on several factors, including the caliber, velocity, and type of the bullets, as well as the distance between the bullets and the target. Also, the factors of stability, accuracy, health condition and ability to bear rifle recoil of personnel using the weapon will significantly affect the accuracy of burst fire. On the basis of the area under the impact, impact energy density was estimated, which will help in better designing the bullet-resistant panels. Also, the possibility of organ or muscle damage can be enumerated on the basis of impact energy density.

2. METHOD

Generally, in shooting ranges accuracy of bullet shots is vital. Accordingly, the majority of soldiers are trained to precisely fire a single round. However, when it comes to automatic firing or 3-bullet bursts, a slight reduction in accuracy is inevitable. Accordingly, to make all the soldiers comfortable with the experimentation scheme, a minimum of 10 sets of 3-bullet fires were conducted as a practice set. INSAS and AK-47 were the main experimentation weapon and a limited number of tests were also conducted using SMC and SIG Sauer 716. The total number of 3-bullet burst fire experiments conducted using each bullet and weapon type is presented in Table 1. As the recoil produced by a 9x19mm FMJ/Pb bullet when fired using SMC is presumably the lowest compared to any other firearm, the accuracy was highest and accordingly a lesser number of experiments were conducted for this case. The primary study focus was on rifles and accordingly, 30 sets of experiments were conducted with INSAS and AK-47 rifles. Owing to the highest recoil of SIG Sauer leading to significant variation in the hit locations, the experimentation was limited to 10 sets.

Table 1. Number of 3-bullet burst experimentations conducted using each weapon type

S. No.	Bullet type	Weapon	Number of 3-bullet burst fire tests conducted
1.	9 x 19mm FMJ/Pb	SMC	5
2.	5.56 x 45mm FMJ/(SI+Pb)	INSAS	30
3.	7.62 x 39mm HSC	AK-47	30
4.	7.62 x 51mm M80 Ball	SIG Sauer	10

The experimental scheme depicted in Figure 1 involves firing a bullet at a target from a distance of 10 meters. To measure the velocity of a bullet, a ballistic chronograph is placed 5 meters before the target. A ballistic chronograph is a device that measures the velocity of a bullet by detecting the interruption of two light beams as the bullet passes through them. By measuring the time, it takes for the bullet to pass through the light beams, the chronograph can calculate the bullet's velocity and is used for single shot validation purposes only.

In addition to the chronograph, a high-speed camera is placed suitably to capture the ballistic event for further analysis. The camera is set up to record the impact of the bullets on the target and to provide a detailed view of the bullet's trajectory into the target and impact dynamics. The frame rate of the high-speed camera was set at 10,000 fps to capture the ballistic event. The high-speed imaging data can be used to analyze the performance of the bullet, the rifle and bullet resistant panel, as well as to identify any potential issues or areas for improvement. Overall, this experimental setup is designed to provide detailed information about the performance of the rifle and bullets, which can be used to improve accuracy and optimize the design of the personal body armour, apart from serving as feedback to the person using the rifle.

Figure 1. Set-up used to conduct the 3-bullet burst fire (a) complete set-up with a high-speed camera, (b) Soldier firing in burst mode, and (c) ballistic chronograph

The shooters' level of experience with the various weapons can indeed play a determining role in the intershot distance. In the case of the AK-47, the first three targets of the 3-bullet burst were fired by a soldier with 10 years of experience, while the remaining 27 targets were fired by a soldier with 28 years

https://doi.org/10.52202/080042-0044

of experience. As for the INSAS rifle, the first nine targets were fired by a soldier with 28 years of experience, and the remaining 21 targets were fired by a soldier with 29 years of experience.

Table 2 presents the specifications of different bullet types used and the energies associated with the impact of each bullet type. This fact is noteworthy that not all bullets will attain equal speeds and there will always be a minor variation in the bullet mass within the permitted range. Accordingly, the kinetic energy (KE) was calculated on the basis of an average of experimental results.

Table 2. Bullet types and specifications used for the study

S. No.	Bullet	Velocity (m/s)	Mass (g)	K.E. (J)
1.	9 x 19mm FMJ/Pb	430 ± 15	7.5± 0.05	740
2.	5.56 x 45mm FMJ/(SI+Pb)	890 ± 15	4.2± 0.05	1500
3.	7.62 x 39 mm FMJ/HSC	700 ± 15	8± 0.1	2016
4.	7.62 x 51 mm FMJ/Pb	840 ± 15	9.5 ± 0.1	3350

3. RESULTS AND DISCUSSION

3-bullet burst fire testing was done using four different types of weapons with the respective bullet types. The primary focus was on the results of INSAS and AK-47. In order to examine the influence of soldiers' posture on firing accuracy, two distinct firing positions were employed during the experimentation process(i) prone position giving complete stability to the human body and a sandbag to support the rifle barrel, and (ii) standing in a trench with a foot resting against a wall and rifle barrel on a sandbag as shown Figure 2. Initial 12 experiments were conducted in the prone position and the remaining 18 experiments were conducted with soldiers in the trench. Apart from the main study focusing on INSAS and AK-47 rifles, two other studies were conducted using SMC and SIG Sauer 716.

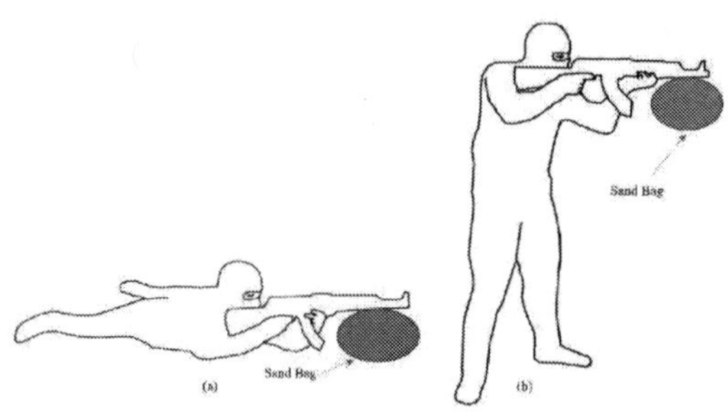

Figure 2. Firing positions used for 3 bullets burst fire experimentation (a) prone position, and (b) standing position

3.1 Sub-Machine Gun

A small study was conducted on firing 3 bullets using a Sub Machine Gun (SMC) with 9x19mm bullets in five experiments. For this experiment set, actual soft armour panels (SAP) were used instead of paper targets.

The results showed that the intershot distance with SMC was always less than 100mm and in the range of 20-100mm. However, due to the limited data on deformable SAPs, no significant conclusions could be drawn. It was observed that owing to the lowest bullet energy and lowest recoil, all the SMC bullets could comfortably hit within the target area. However, the study proves that SMC with 9x19mm bullets is a reliable option for shooting tasks where accuracy is paramount within a short distance.

3.2 INSAS Rifle

Figure 3 presents the experimental data of intershot distances of a 3-bullet burst fire when 5.56x45mm bullets were fired using an INSAS rifle. Out of 30 burst fire rounds, 29 times all the bullets successfully hit within the (paper) target area of 350 x 300 mm^2. Experiment number 25 was a rogue one wherein the third bullet couldn't hit within the target area. Apart from that another interesting fact observed was that in three of the experiments, one after the other bullet hit was in a straight-line path. Figure 3(a-c), presents the intershot distances between the 1-2, 2-3 and 3-1 bullets on a paper target. It's noteworthy that intershot distance between shot numbers 1-2 could never attain distances less than 25mm. Also, the majority of intershot distance for 1-2 bullets were in the range of 25-100mm. For four specific cases, the intershot distance was recorded as higher than 100mm. Figure 3(b) depicts that less than 25mm intershot distance happens but it's a relatively less common phenomenon. Also, intershot distance of higher than 100mm is relatively less common. By the time the third bullet comes into action, the shoulder of the human operating the rifle had already experienced two recoil jerks. Accordingly, the third bullet was expected to hit far from the first bullet location and the same was also experimentally confirmed in Figure 3(c). Also, the number of intershot distances well above 100mm increased significantly. On comparing Figure 1(a) and 1(c), it was observed that intershot distances for shot 1-2 were closer to 25mm mark, whereas intershot distances for shot 3-1 were close to 100 mm mark for the majority of cases. Figure 3(d) presents combined results of intershot distances and it can be claimed on the basis of the figure that the given security personnel were trained enough to fire the majority of 5.56x45mm bullets using INSAS rifle in the range of 25-100mm.

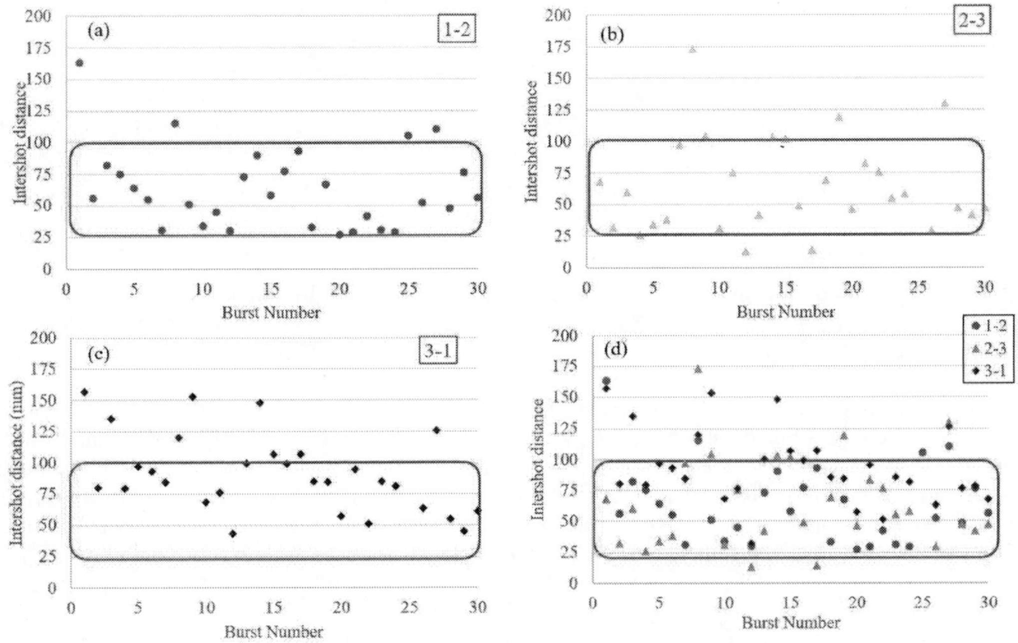

Figure 3. Intershot distances for 3-bullet burst round fired from a distance of 10m for INSAS Rifle (a) distance between shot 1-2, (b) distance between shot 2-3, (c) distance between shot 3-1, and (d) combined result of 3-bullet burst.

3.3 AK-47 Rifle

Figure 4 presents intershot distance data of 7.62x39mm bullets when fired in 3-bullet burst mode using an AK-47 rifle. Out of 30 burst fire rounds, 27 times all the bullets successfully hit within the (paper) target area of 350 x 300 mm^2. For experiment numbers 12, 19 and 28 the third bullet couldn't hit the target area. Apart from that it was observed that in five of the experiments, one after the other bullet hit was in a straight-line path. The average spread of the AK-47 rifle was measured to be 62.4 mm, 48.5 mm, and 88.7 mm between shot number 1-2, 2-3, and 3-1, respectively. On the other hand, for the INSAS rifle, the average spread was recorded as 62.2 mm, 64.2 mm, and 92.5 mm between shot number 1-2, 2-3, and 3-1, respectively. A rare instance of intershot distance for shot numbers 1-2 was noted as less than 25mm, which was not the case with INSAS rifle. Interestingly, the number of intershot distances on higher side (above 100mm) was also higher for AK-47. Intershot distance data for shot numbers 2-3 as depicted in Figure 3(b) was an interesting finding, as the majority of data points were within the range of 25-100mm. Similar to INSAS, the intershot distances for between shot 3-1 raised from 25-50mm to close to 75-100mm (Figure 4(c)). Overall, in this case also the majority of intershot distances were in the range of 25-100mm (Figure 4(d)).

https://doi.org/10.52202/080042-0044

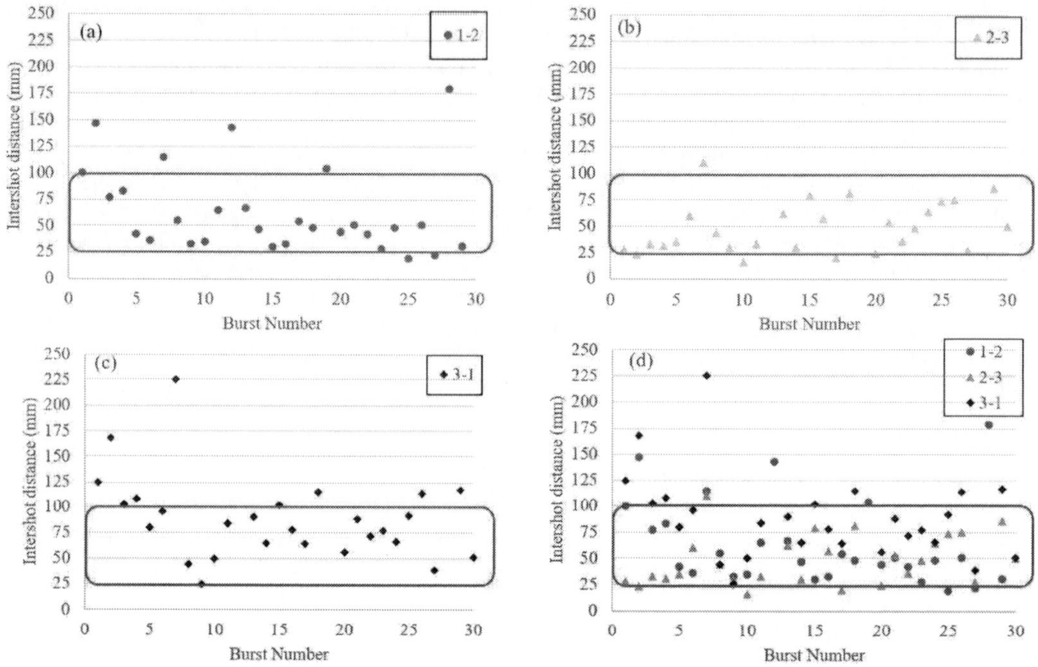

Figure 4. Intershot distances for 3-bullet burst round for AK-47 Rifle

3.4 Energy density

In real life scenario, perforation of bullet resistant panel will lead to a bullet hitting a person, however, even if 3-bullets of a burst fire are stopped by a suitable bullet-resistant panel, then the total impact energy of 3 bullets hit within a small time of ~0.3 seconds has to be borne by the localized human body area. The exact site of this local area will vary for each case and which vital human body organ is at the back of the panel can never be predicted. To evaluate the impact energy transmitted to the human/target area, the intershot distance data was used to create suitable triangles wherever feasible. The area of triangles for 3 bullet burst fire for INSAS and AK-47 are presented in Figure 5 (a, c). Knowing the average impact kinetic energy (KE) per bullet shot, the energy density for both rifles for all the experiments was established and is presented in Figure 5 (b, d). The higher intershot distances along with the lower total KE in the case of INSAS rifle led to relatively lower total energy imparted to the unit area in Joules per square mm. For AK-47, a relatively lesser impact area with higher bullet KE resulted in higher energy densities. The proposed energy density function can serve the purpose of evaluating the possibility of damage applicable to human muscles, vital organs and bones. Table 3 presents the average area of triangles and energy densities generated due to 3 bullet burst fires. As expected, the average energy to be borne by a target is higher for 3 bullets fired from an AK-47 rifle. This indicates that a person even if saved due to the presence of a suitable bullet-resistant panel, will experience a higher impact and the possibility of organ/bone damage due to behind armour blunt trauma (BABT) will be higher in case of a hit by burst fire of three bullets by a AK-47 rifle.

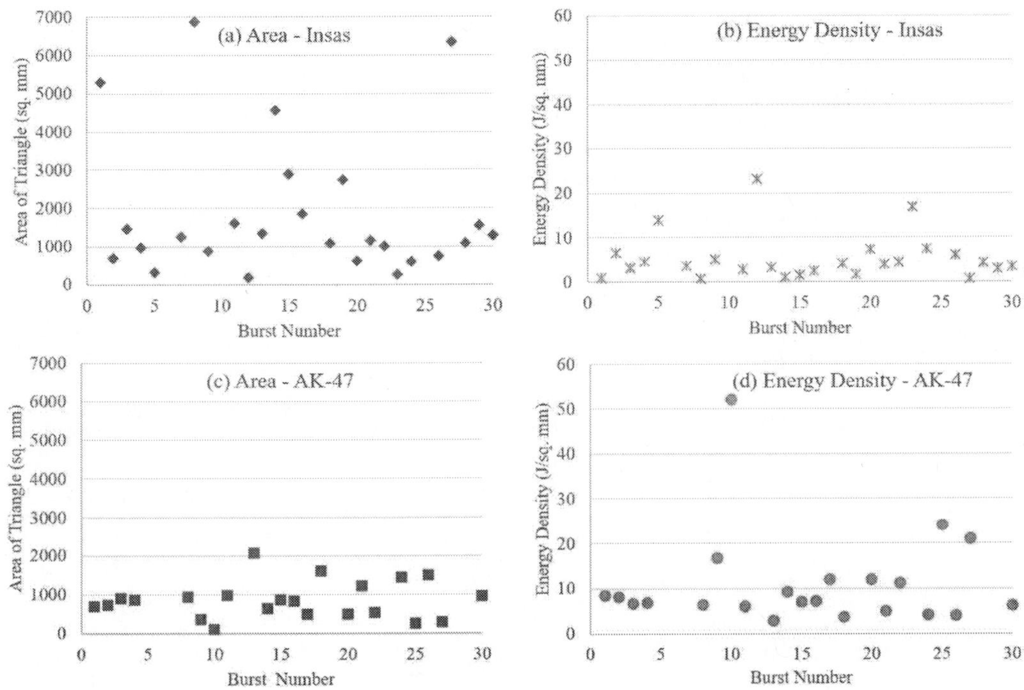

Figure 5. Comparison of triangle area and energy density for 3-bullet burst fire using INSAS and AK-47 rifle

Table 3. Average triangle areas and energy densities [5][6]

Rifle type	Average Area of Triangle (sq. mm)	Average Energy Density (J/sq. mm)	Recoil
INSAS	1943	4.5	4.43
AK-47	856	11	9.35

3.5 Post-Impact Analysis

Figure 6 shows the experimental results of firing 3 bullets in burst mode using INSAS and AK-47 rifles. In Figure 6(a), it is shown that when firing with an INSAS rifle, the bullets can hit the target in a triangle shape. The leg lengths of the triangle varied in 26 experiments. Figure 6(b) shows that it is possible for all three bullets to hit in a straight line also, but this only occurred in 10% of the total cases. Figure 6(c) depicts the rare phenomenon where the third bullet did not hit the target area.

Figures 6(d-f) show similar experimental findings for burst firing 3 bullets using an AK-47 rifle. The results indicate that, compared to the INSAS rifle, the AK-47 rifle had more variation in the results. In 5 out of 30 experiments, the bullets hit the target in a straight line, and in 10% of the cases, the third bullet did not hit the target area.

https://doi.org/10.52202/080042-0044

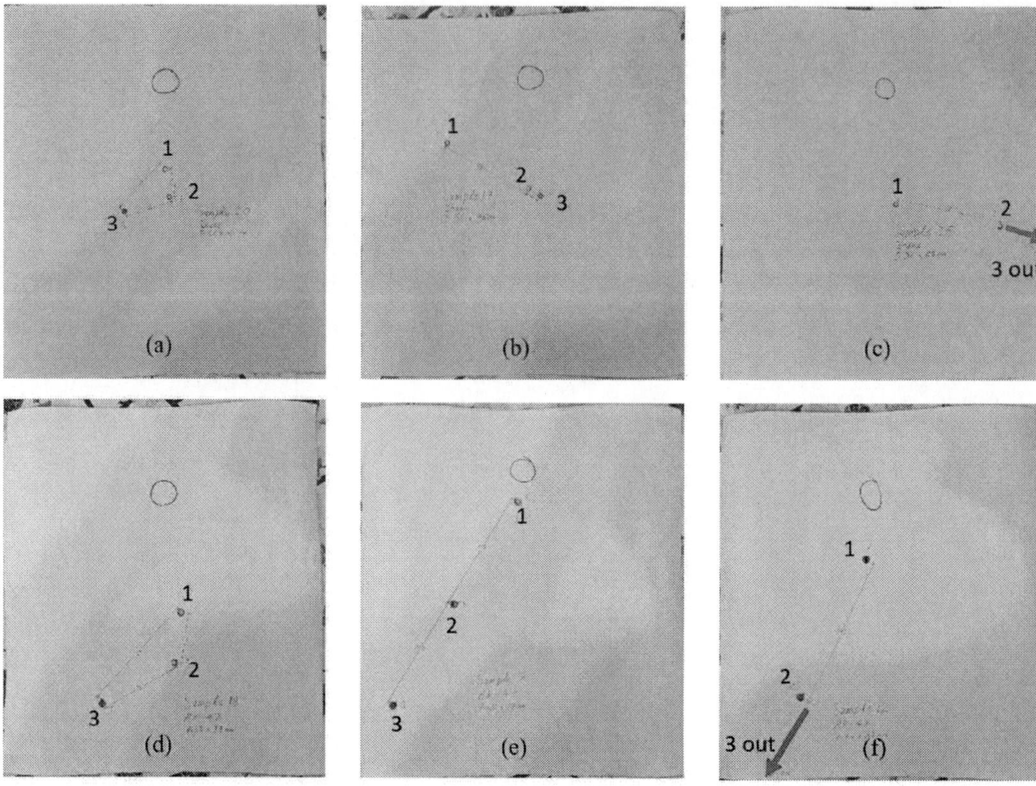

Figure 6. Sample INSAS rifle fired targets (a) 3-bullet burst inside the target forming triangle, (b) 3-bullet burst in-line, (c) first, second bullet inside and third bullet outside the target area, and sample AK-47 fired targets (d) 3-bullet burst inside the target forming triangle, (e) 3-bullet burst in-line, (f) first, second bullet inside and third bullet outside the target area.

3.6 SIG Sauer 716

A small study was conducted on firing 3 bullets using a SIG Sauer 716 with 7.62x51mm bullets in ten experiments. For this experiment set hard armour panel (HAP) was used along with the paper targets. The SIG Sauer 716 is a relatively new weapon recently introduced into the Indian army. In the experimental range of bullet types, 7.62x51mm is the largest-sized bullet. A total of ten 3-bullet burst fire tests were conducted, five on hard armour panels (HAP) and five on paper target. The results showed that, generally, only two out of the three bullets hit within the target area of 1000 sq. cm. Out of 30 burst fire rounds, 21 bullets successfully hit within the target area of 350 x 300 mm^2. Only 5 pair of bullets had intershot distances less than 100mm, while the rest of the 16 bullets had intershot distances in the range of 100-275mm.

Figure 7 shows the damage caused to the soldier's shoulder tissues due to the recoil of the SIG Sauer 716 firing 7.62x51mm bullets. Though 10ms as reported in the literature [2] is a small time yet within this small time when peak recoil of a rifle occurs, it creates difficulty in maintaining the aim of a soldier. The reddening of the shoulder portion that supports the rifle is visible in the figure, indicating the impact of the recoil on the soldier's shoulder. This, in turn, can result in a decrease in the accuracy of hits after a few rounds of bursts. It's worth noting that while both the INSAS and AK-47 have enough recoil to affect the accuracy of a soldier's aim, they don't seem to leave a notable mark on the shoulder. Therefore, it is important to take measures to reduce the impact of recoil on soldiers shoulder to maintain the accuracy and effectiveness of their shots.

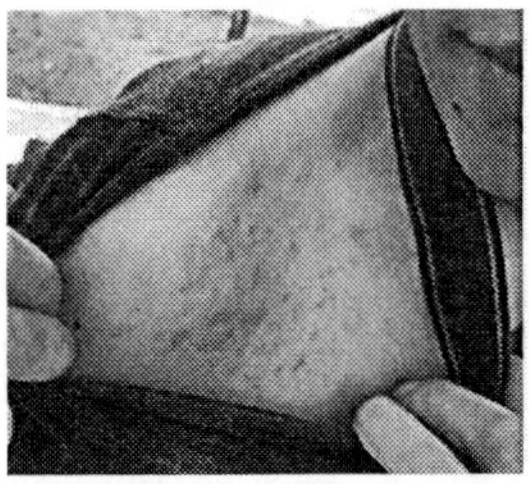

Figure 7. Effect of SIG Sauer 716 rifle recoil on the soldier's shoulder after firing over 50 bullets of 7.62 x 51mm M80 NATO Ball

Figure 8 depict the aftermath of firing two sets of three bullet bursts of 7.62x51mm using SIG Sauer 716 on a hybrid Hard Armour Panel (HAP), displaying the resulting damage on the front, rear, and side of the panel, as well as how the panel covers vital body organs. Figure 8 (c) clearly illustrates the back-face deformation caused by the impact, which has the potential to cause severe damage to any vital body organ it comes into contact with. It's important to note that the back-face deformation shown in this case is the result of an edge shot, where the distance from the side edge was less than 51mm. However, in real-world scenarios, it may not be feasible to maintain such precise dimensions during a firearm attack, which the standards like NIJ 0101.06 or BIS 17051:2018 talks about and leaves an open question of personal body armour safety guarantee. Under the given conditions, the need for 3-6 Armour Piercing Incendiary (API) bullets to hit a restricted 1000 sq. cm area is also a matter of concern, given the challenges highlighted in the study regarding the accuracy of firing bursts of bullets in this way.

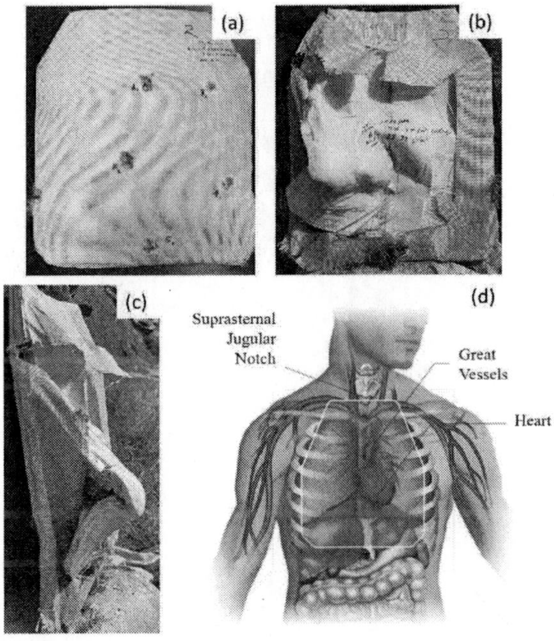

Figure 8. Images of Hard Armour Panel (HAP) after bearing two sets of 3 bullet burst (a) front, (b) rear, (c) sideview and (d) vital body organs covered by a HAP

4. CONCLUSIONS

It is important to note that firing a firearm accurately in a small area, such as 1000 sq. cm, can be challenging, especially in a burst mode. The accuracy of firing a firearm depends on various factors such as the skill level of the shooter, the type of firearm and the distance from the target.

AK-47 rifle is a more powerful weapon than the INSAS rifle. The AK-47 rifle fires a larger cartridge than the INSAS rifle, which results in more energy being transferred to the target. This increased energy can result in more damage to the target and surrounding areas leading to extensive BABT.

https://doi.org/10.52202/080042-0044

It is practically not feasible to achieve an intershot distance of 25mm in burst mode of a rifle even when the firing distance is just 10m. This distance is too small to be achieved even by skilled marksmen. In fact, the experimental data presented in this study shows that the intershot distances are much larger.

The SIG Sauer 716 rifle firing 7.62x51mm bullets showed the maximum variation in shot pattern results. On the other hand, the INSAS and AK-47 rifle shows a consistent result, with intershot distances ranging from 25mm to 200mm and a higher percentage of bullets hitting within the target area in 3-bullet burst mode. This indicates that the INSAS and AK-47 may be more efficient weapon for burst fire, at least within the experimental range of rifles tested in this study. AK-47 with HSC bullets thus becomes a more lethal weapon system against any body armour.

According to the results of the experiments, it can be inferred that firing in burst mode using INSAS, AK-47, and SIG Sauger 716 rifles leads to a certain percentage of bullets not hitting within the 1000 sq. cm target area. Specifically, the percentages of missed shots were found to be 3%, 10%, and 30% for the three rifles, respectively. It's important to note that the results may vary based on factors such as the skill and training of the shooter, environmental conditions and the specific configuration of the rifle used. Therefore, it's important to conduct further studies and more experiments to validate these findings and determine the most effective body armour systems and their testing strategies for burst fire in real-world scenarios.

One open question still remains and that is the use of sniper shots 7.62x54R (AP) using a Dragunov, which only has single shot capacity, then why there is a need of qualifying 6 shots on a 1000 sq. cm armour plate to qualify as per BIS 17051:2018 or other ballistic standards to be inducted in Indian Army and other security forces globally.

References

[1] Ilie Florin. Moşteanu Dănuţ-Eugeniu. Georgescu Bogdan-Mihai., International conference Knowledge-Based Organization, 2021; 27(3); 40-47.

[2] Matthew J. Hall. Int. Jr. of Impact Engineering, 2008; 35; 540-548.

[3] M. B. Mukasey, J. L. Sedgwick, and D. W. Hagy, "Ballistic Resistance of Body Armor," in *NIJ Standard - 0101-16*, Washington, DC, 2008, pp. 1–89.

[4] Bureau of Indian Standards, "Textiles — Bullet Resistant Jackets — Performance Requirements," *IS 17051 2018*, vol. 16, no. December 2018, 2018.

[5] Vinod., The Big Fat Indian INSAS, 2017; Weblink: http://surl.li/fmqqd

[6] Josh Montgomery., The Recoil of AK-47, 2023; Weblink: http://surl.li/fmqqi

Police versus Military Ammunition – Design, Wound Ballistics and Standards

PL Gotts

Phil Gotts Consulting Ltd, 23 Thorney Rd, Capel St Mary, Ipswich, Suffolk, IP9 2HL, UK
phil@philgotts.com

Abstract. The designs of ammunition for police use and military use are often different, as the requirements for the interaction with the target are different. Police ammunition needs to stop an individual target, without the possibility of shoot-through which could cause collateral injuries or fatalities. Military ammunition is mainly designed to cause casualties rather than fatalities, and hence there is no requirement to stop the bullet in a single target. For these reasons, military ammunition is predominantly full metal jacket (FMJ) ball ammunition, whereas much police ammunition is now of the expanding type, which, in accordance with international law, is not legal for military use. The two ammunition design types may be compared by studying the wound ballistic effects in gelatine, examining such parameters as depth of penetration, and volume of the temporary cavity. Expanding ammunition also exhibits a tendency to break up within tissue and this can also be demonstrated within gelatine. Historically, personal armour test standards, whether for police or military, have used FMJ ball ammunition, particularly for the high velocity rifle levels. A recent exception for this is the UK Home Office 2017 Body Armour standard, which uses expanding ammunition as options in most levels, including those for high velocity rifle bullets. What are the implications for personal armour of these different ammunition types? Can it always be assumed that FMJ ball ammunition will be a more severe threat to personal armour than expanding ammunition? Although the wound ballistics parameters measured do not impinge on the design of the armour, they are the results of designing the projectile differently to meet a specific operational requirement. The need for different wound ballistics characteristics leads to a necessary design of the bullet. This means that the geometry of the nose of the bullet impacting armour is therefore different between the two designs. For hard armour this design has little effect upon the terminal ballistics, but for soft armour there is more scope for differences to be observed. It is for this reason that both HO CAST and VPAM have included expanding hollow point ammunition within their standards as well as FMJ ammunition. A final consideration with police ammunition is that relating to whether it should be specified to defeat, or be stopped by, personal armour systems, including those worn by the police officers themselves. Military and Police ammunition are designed differently to meet different requirements.

1. INTRODUCTION

The designs of ammunition for police use and military use are often different, as the requirements for the interaction with the target are different. Police ammunition needs to stop an individual target, without the possibility of shoot-through which could cause collateral injuries or fatalities. Military ammunition is mainly designed to cause casualties rather than fatalities, and hence there is no requirement to stop the bullet in a single target.

For these reasons, military ammunition is predominantly full metal jacket ball ammunition, whereas much police ammunition is now of the expanding type, which, in accordance with international law [1], is not legal for military / warfare use. The two ammunition design types may be compared by studying the wound ballistic effects in gelatine, which is a method used for such purposes for many decades [2]. The gelatine may be used for assessing such parameters as depth of penetration [3], and volume of the temporary cavity. Expanding ammunition also exhibits a tendency to break up within tissue and this can also be demonstrated within gelatine, by considering the retained mass of the recovered bullet [4].

Historically, personal armour test standards, whether for police or military, have used FMJ ball ammunition, particularly for the high velocity rifle levels. A recent exception for this is the UK Home Office 2017 Body Armour standard [5], which uses expanding ammunition as options in most levels, including those for high velocity rifle bullets. Expanding ammunition is also included for some calibres in the VPAM (Vereinigung der Prüfstellen für angriffshemmende Materialien und Konstruktionen) AND-SoM [6] supplement entitled Ammunition Types for Special Tests.

A final consideration with police ammunition is that relating to whether it should be specified to defeat, or be stopped by, personal armour systems, including those worn by the police officers themselves.

https://doi.org/10.52202/080042-0045

2. AMMUNITION DESIGNS AND OPERATION

Small arms ammunition (SAA) is designed to impart kinetic energy into the target from a significant distance away.

There are many different ammunition types designed for use in different scenarios, by different users, with different requirements. Ammunition is designed to:

- Kill,
- Incapacitate,
- Injure,
- Suppress and
- Deter

The requirements are not the same for each of these. Military ammunition is predominantly designed to injure and suppress. Police ammunition is designed to incapacitate and deter. The design of the ammunition to meet these different requirements is also not the same. For suppression and as a deterrent the ammunition has no real design requirements, except perhaps noise is useful for suppression. For injuring, it is probably useful for the ammunition to have full-metal jacket (FMJ) bullets, whereas for incapacitation, where it is desirable to dump the kinetic energy into the target rapidly, expanding bullets of the soft point or hollow point design are preferable.

For a bullet to operate as an expanding bullet, it needs to expand upon, or very soon after, impact. This means that upon impact the target needs to exert an appropriate pressure upon the tip of the bullet to promote the required expansion. The exerted pressure is related to the velocity of the impact. Therefore, for the ammunition to function as per its design, there will be a velocity range over which it is expected to impact the target. For this reason, testing with ballistic gelatine is conducted with ammunition fired from specified distances, and the impact velocity measured.

3. COMPARISON OF POLICE AND MILITARY AMMUNITION

On the whole, military ammunition is designed to inflict casualties rather than fatalities. On the battlefield a casualty is a greater logistical burden than a fatality, and hence more desirable for the mission. Therefore, a bullet which passes through the target completely is fully acceptable. It should however dump sufficient energy into the target to produce a significant injury.

The aim of police ammunition is to immediately incapacitate the target, without endangering the life of innocent bystanders. This incapacitation is achieved by dumping as much of the bullet's energy as possible into the desired target. The preservation of the life of innocent bystanders is achieved by retaining the bullet within the target.

For the purposes of this paper, it is assumed that military ammunition is of the full metal jacket type, and that police ammunition is of the expanding type.

For military ammunition the muzzle velocity is advantageous to be as high as possible whilst complying with any pressure restrictions of the weapon. A high muzzle velocity is advantageous for accuracy and lethal range requirements, as the kinetic energy is related to the velocity squared.

One way of highlighting the differences between the two ammunition types is to compare them through ballistic gelatine-based experiments. The following comparison was achieved from different series of tests, conducted using identical protocols, which have now been brought together and referenced for comparison purposes.

4. COMPARISON BY WOUND BALLISTICS

Typically, the wound ballistic effects of a projectile are assessed by studying the effects as it passes through a block of ballistic gelatine. Unfortunately, there is often also a difference between how this assessment is conducted for police and military ammunition. Military ammunition tends to be tested using 20 % gelatine at room temperature, while Police ammunition tends to be assessed using 10 % gelatine at 4 °C. For the following comparison, all the firings are conducted with the same specification of gelatine, being 10 % at 4 °C. Two ammunition calibres are used as case studies, being 9 x 19 mm and 5.56 x 45 mm.

The wound ballistic assessments include three parameters:

- Depth of penetration into the gelatine
- Estimated maximum volume of the temporary cavity
- Retained mass of the recovered bullet

These parameters are particularly important for police ammunition, which have strict requirements for each of these parameters. For each of these parameters, examples of both full metal jacket (FMJ) and jacketed hollow-point (JHP) ammunition are compared.

4.1 Depth of Penetration

For police ammunition, the depth of penetration achieved is critical, and must be within specified limits, being between 250 mm and 500 mm [3]. If the depth of penetration is below 250 mm it is deemed to be insufficient to cause the required incapacitation of the target. If the depth of penetration is greater than 500 mm it is considered likely to cause a shoot-through, thus endangering the lives of innocent bystanders. There is also a requirement to dump all the bullet's energy within the target, which does not occur in the event of a shoot-through.

4.1.1 9 x 19 mm Ammunition

The following photographs show the gelatine block with the permanent cavity. For the military 9 mm FMJ the bullet passed completely through the 560 mm long block. It was captured in a follow-on block, giving a total depth of penetration of 715 mm. In contrast, the 9 mm hollow-point police ammunition stopped at 362 mm in the original block meaning that it passed the depth of penetration requirement for police ammunition.

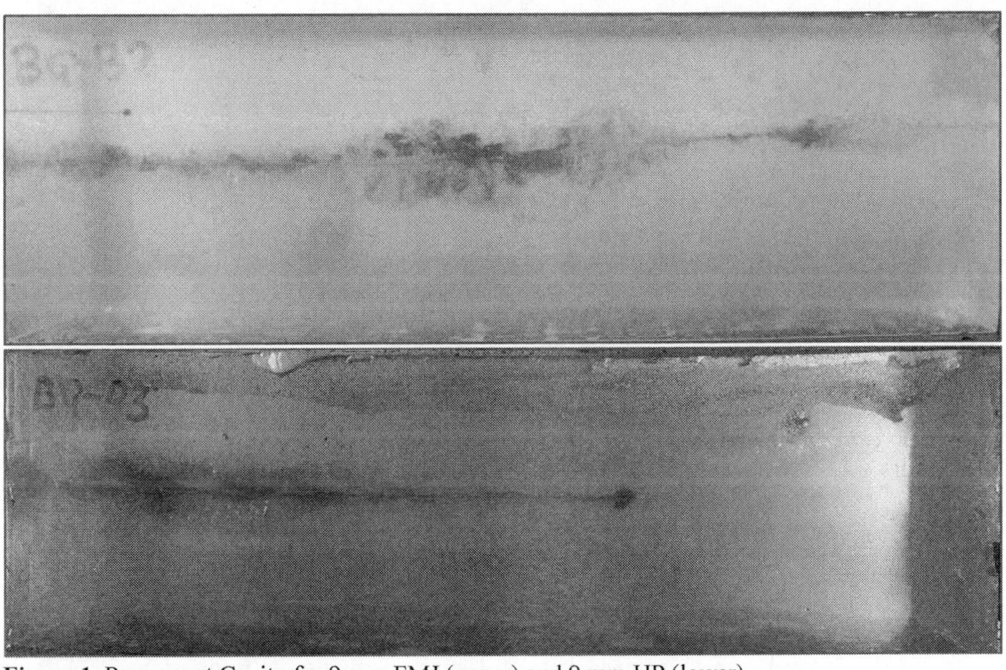

Figure 1. Permanent Cavity for 9 mm FMJ (upper) and 9 mm HP (lower)

https://doi.org/10.52202/080042-0045

4.1.2 5.56 x 45 mm Ammunition

The following photographs show the gelatine block with the permanent cavity. For the military 5.56 mm FMJ the bullet passed completely through the 560 mm long block. It was captured in a follow-on block, giving a total depth of penetration of 565 mm. In contrast, the 5.56 mm hollow-point police ammunition stopped at 433 mm in the original block meaning that it passed the depth of penetration requirement for police ammunition.

Figure 2. Permanent Cavity for 5.56 mm FMJ (upper) and 5.56 mm HP (lower)

4.2 Estimated Maximum Volume of Temporary Cavity

The maximum volume of the temporary cavity is an indication of the quantity of energy imparted to the gelatine. The assessment of the maximum volume of the temporary cavity in this case is conducted using the still image of high-speed footage, and using bespoke software to estimate the volume, by dividing the cavity into a series of truncated cones. This does, of course, assume that the cavity is symmetrical around the bullet axis. It should be noted that for police ammunition the temporary cavity is usually a single cavity, whereas with military ammunition there may be multiple temporary cavities in the length of the gelatine block.

4.2.1 9 x 19 mm Ammunition

The following photographs show the gelatine block with the temporary cavity as a still image taken from the high-speed video. For the military 9 mm FMJ the estimated maximum temporary cavity volume was 1,236 cm^3. This volume was that combined from all temporary cavities up to the stopping point. In contrast, the estimated maximum temporary cavity volume for the 9 mm hollow-point police ammunition was 1,398 cm^3, obtained from a single cavity.

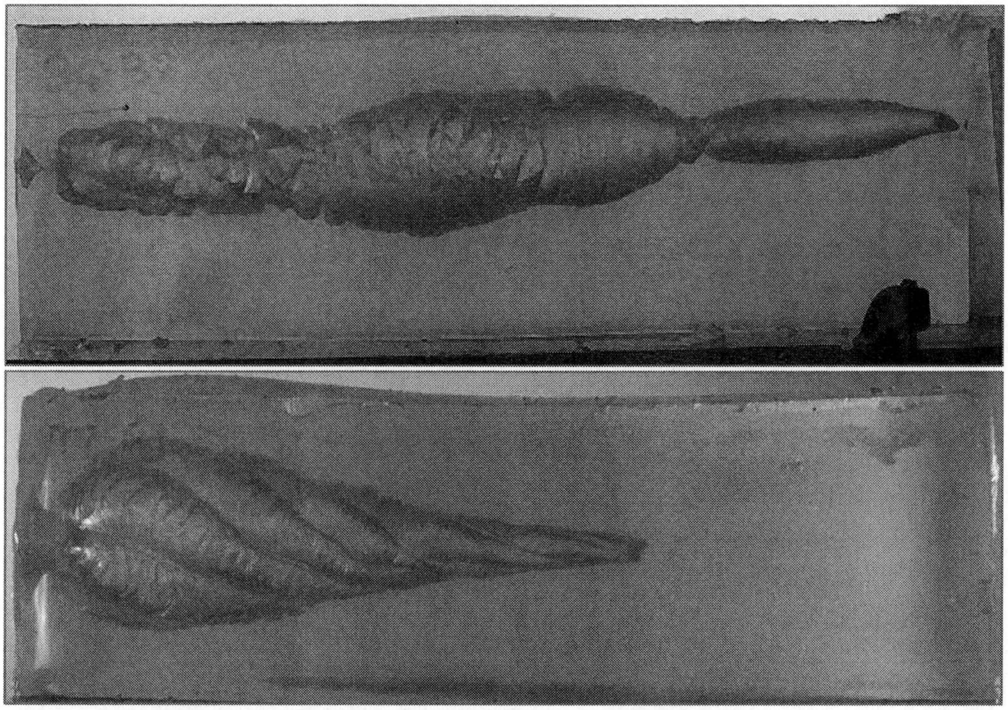

Figure 3. Temporary Cavity for 9 mm FMJ (upper) and 9 mm HP (lower)

4.2.2 5.56 x 45 mm Ammunition

The following photographs show the gelatine block with the temporary cavity as a still image taken from the high-speed video. For the military 5.56 mm FMJ the estimated maximum temporary cavity volume was 3,290 cm^3. This volume was that combined from all temporary cavities up to the stopping point. In contrast, the estimated maximum temporary cavity volume for the 5.56 mm hollow-point police ammunition was 2,592 cm^3, obtained from a single cavity.

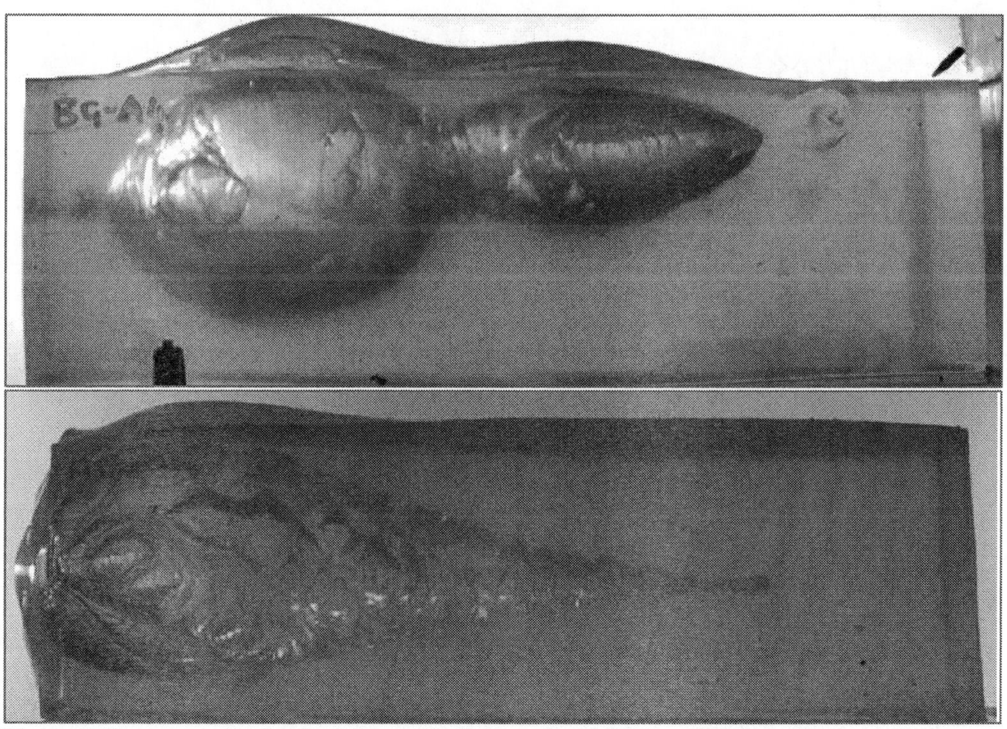

Figure 4. Temporary Cavity for 5.56 mm FMJ (upper) and 5.56 mm HP (lower)

https://doi.org/10.52202/080042-0045

4.3 Retained Mass of Recovered Bullet

If the bullet is retained within the gelatine, it may be complete, or it may have fragmented into a number of parts. For police ammunition it is desirable that the bullet in the target remains complete. Should it have broken up, the mass of the largest part is considered to be the retained mass. This is then reported as a percentage of the original pre-fired bullet mass. For police ammunition there are minimum percentage retained mass values that must be achieved.

4.3.1 9 x 19 mm Ammunition

Figure 5 below shows a comparison of the recovered 9 mm FMJ and a 9 mm JHP. The 9 mm FMJ looks very similar to the pre-fired bullet, with the obvious addition of the rifling striations. The 9 mm JHP bears little resemblance to the fired bullet. It should be noted that different designs of 9 mm JHP look quite different after recovery, whereas most 9 mm FMJ will look similar after recovery.

Figure 5. Pre-Fired (left) and Recovered 9 mm FMJ (upper) and 9 mm HP (lower)

4.3.2 5.56 x 45 mm Ammunition

Figure 6 below shows a comparison of the recovered 5.56 mm FMJ and a 5.56 mm JHP. The 5.56 mm FMJ looks very similar to the pre-fired bullet, with the obvious addition of the rifling striations. The flattening was caused by impact with the floor after leaving the gelatine block. The 5.56 mm JHP bears little resemblance to the fired bullet. It should be noted that different designs of 5.56 mm JHP look quite different after recovery, whereas most 5.56 mm FMJ will look similar after recovery.

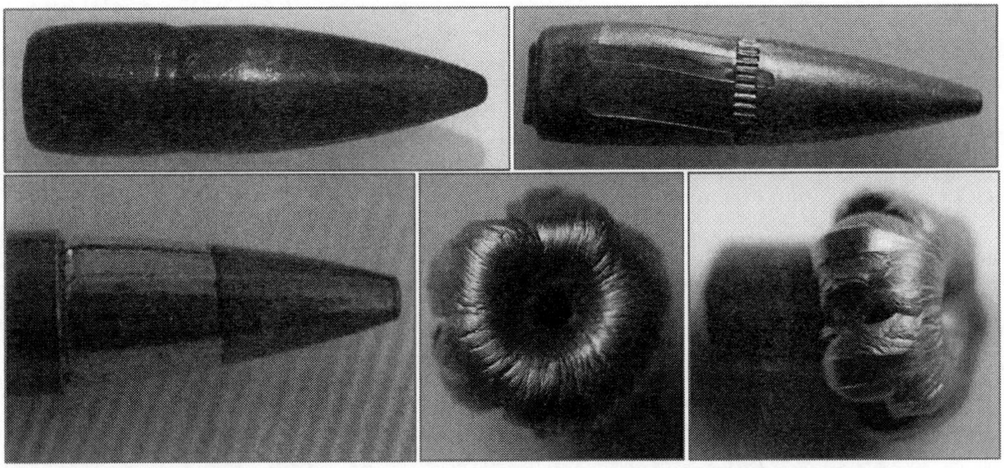

Figure 6. Pre-Fired (left) and Recovered 5.56 mm FMJ (upper) and 5.56 mm HP (lower)

4.4 Summary of Wound Ballistics Results

Table 1 below shows the summary of the results obtained from the comparison of the wound ballistics for the 2 calibres and the 2 types of ammunition for each calibre. The values quoted are the average of 5 results each.

Ammunition	Impact Velocity (m/s)	Depth of Penetration (mm)	Estimated Maximum Temporary Cavity Volume (cm^3)	Percentage Retained Mass (%)
9 mm FMJ	367	715	1,236	100.0
9 mm HP	332	362	1,398	100.0
5.56 mm FMJ	766	565	3,290	95.2
5.56 mm HP	794	433	2,592	99.5

Table 1. Wound Ballistics Results Summary

The table shows that with respect to depth of penetration, both FMJ ammunition types over-penetrate the gelatine block and significantly over-penetrate the 500 mm upper threshold value. The estimated maximum temporary cavity volume for the 9 mm HP is slightly higher than the 9 mm FMJ, and is achieved at a much lower impact velocity. The estimated maximum temporary cavity volume for the 5.56 mm HP is actually lower than the 5.56 mm FMJ, but the FMJ volume is the combined volume of a double cavity which starts further from the entrance to the block. With respect to the percentage retained mass of the recovered bullet, the only type which has lost mass is the 5.56 mm FMJ.

https://doi.org/10.52202/080042-0045

5. STANDARDS

Body armour standards have included both FMJ and expanding ammunition types for many years, but it is rare that they include direct equivalents of both types in the same level. One standard that does include direct FMJ and expanding bullet equivalents is the same level is the HO CAST Body Armour Standard of 2017. For this reason, it is worth explaining a little bit more detail of the level aspects of this standard.

5.1 HO CAST 2017 Body Armour Standard [1]

The UK Home Office published a body armour test standard in 2017, in which they included both FMJ and JHP versions of 9 mm ammunition:

- HO1 - 9 mm DM11A1B2 (MEN) at 365 ± 10 m/s
  - 9 mm Federal Premium JHP P9HST1 at 365 ± 10 m/s
- HO2 - 9 mm DM11A1B2 (MEN) at 430 ± 10 m/s
  - 9 mm Federal Premium JHP P9HST1 at 430 ± 10 m/s

These levels include 9 mm ammunition of both FMJ and JHP designs. Both projectile types are of the same mass and projected at the same two velocities. It may be expected that the most aggressive of these two ammunition types would be the FMJ, as the JHP is designed to expand on impact with the target.

Level HO3 includes two types of 7.62 mm calibre ammunition, which are both of the FMJ type. These are the 7.62 x 51 mm NATO ball (9.3 g, test velocity 830 ± 15 m/s, 3.20 kJ) and a 7.62 x 39 mm PS ball surrogate (7.9 g, test velocity 705 ± 15 m/s, 1.96 kJ).

Level HO4 includes two further 7.62 mm (.308) calibre ammunition types, which are basically two supplier designations for the same thing. They are listed as the Sako .308 480A Powerhead and the Barnes .308 TSX BT. The Home Office includes them as they are heavier bullets than the NATO or Soviet 7.62 mm military bullets of level HO3 at 10.7g (165 grains). However, they are also both hunting ammunition designed for big game, including, deer, moose, bears and big cats, and are of the solid copper expanding type. At 820 m/s test velocity they produce almost 3.6 kJ kinetic energy, but again the kinetic energy density reduces rapidly upon impact due to the expansion.

6. REQUIREMENT TO DEFEAT BODY ARMOUR, OR NOT?

When designing either ammunition or body armour for either the military or the police user there is a question raised regarding the interaction of the user's ammunition and the user's body armour. Should the user's ammunition defeat the user's armour, or should the user's armour defeat the user's ammunition? This question is considered by the user communities, and the answer is influenced by a number of stakeholders.

The user would like their ammunition to defeat the body armour of their adversary, but unless their own body armour is of a much higher performance than that of their adversary, this will mean that it will defeat their own armour as well. This, therefore, increases the risk of fratricide scenarios. In the military scenario it is probably a reasonable assumption that the enemy will be wearing body armour, whereas in the police example, this is much less likely. Those responsible for the user's health and safety, would prefer their armour would defeat their own ammunition to reduce this risk of fratricide. This however means that their ammunition may have less chance of defeating an adversary's armour.

From the author's experience with both the military and police environments, the preferable decision is for the user's body armour to be capable of defeating the user's ammunition. In the military scenario this includes the ammunition fired by other NATO allies. Therefore, most military body armour requirements specify the defeat of the user's own ammunition, and most police ammunition requirements specify that it does not defeat the police armour. This is, however, a requirement, which needs to be assessed during the specification stage of both the ammunition and the armour.

7. SUMMARY

Military and Police ammunition are designed differently to meet different requirements. Military ammunition is designed to produce casualties and to supress hostile forces. Police ammunition is designed to rapidly incapacitate, usually a single target, whilst managing the risk of injury to innocent bystanders. For this reason, military ammunition is of the FMJ design, whereas much police ammunition is now of the expanding design.

The differences in behaviour of FMJ and expanding ammunition in tissue can be demonstrated using shots into ballistic gelatine, where parameters such as depth of penetration, maximum volume of temporary cavity and recovered retained bullet mass, can be obtained. Although the wound ballistics parameters measured do not impinge on the design of the armour they are the results of the different designs of projectile, to meet a specific operational requirement. The need for different wound ballistics characteristics leads to a necessary different design of the bullet. This means that the geometry of the nose of the bullet impacting armour is therefore different between the two designs. For hard armour this design has little effect upon the terminal ballistics, but for soft armour there is more scope for differences to be observed. It is for this reason that both HO CAST and VPAM have included expanding hollow point ammunition within their standards as well as FMJ ammunition.

The question as to whether the tested ammunition should, or should not, defeat the users own armour is one of very different views. Usually, the preferable decision is for the user's body armour to be capable of defeating the user's ammunition. This is, however, a requirement, which needs to be assessed during the specification stage of both the ammunition and the armour.

References

[1] 1899 Hague Declaration concerning Expanding Bullets

[2] Kneubuehl BP (2011) Simulants. In: Kneubuehl B (ed) Wound ballistics: basics and applications. Springer, Berlin, Heidelberg, pp 136–143

[3] FBI; 'FBI Penetration Testing'; revised 04/05/11

[4] Fackler ML, Surinchak JS, Malinowski JA, Bowen RE (1984) Bullet fragmentation: a major cause of tissue disruption. J Trauma 24(1):35–39

[5] Payne, T; O'Rourke, S; Malbon, C; 'Body Armour Standard (2017)'; CAST Publication No. 012/17, July 2017

[6] VPAM AND-SoM; 'Ammunition Type for Special Tests – Supplement to APR 2006', VPAM; 15 March 2021

https://doi.org/10.52202/080042-0045

An experimental investigation into the threat posed by arrows to body armour

M. Seidl[1], K. Lehmann[2], and S. Grobert[3]

[1] *French-German Research Institute of Saint-Louis (ISL), 5 Rue du Général Cassagnou, 68300 Saint-Louis, Frankreich (marina.seidl@isl.eu)*
[2]*State Police Schleswig-Holstein, Mühlenweg 166, 24116 Kiel, Germany*
[3] *Bundeswehr Hospital Berlin, Scharnhorststraße 13, 10115 Berlin, Germany*

Abstract. Compound bows and crossbows are effective hunting tools and, hence, lethal weapons. Occasionally, light-protected security personnel, such as law enforcement officers and security guards, face attackers armed with archery weapons. While research has focused on the potential damage caused by bow attacks on the human body, limited work has been done to investigate the level of protection offered by body armour against such threats. This paper examines the performance of various types of body armour, including helmets and vests, against the impact of crossbow bolts. The study utilised a commercially available 300-pound (136 kg) draw weight crossbow and bolts with two different categories of penetration tips. Impact velocity was measured using a light barrier, and the body armour was fixed onto a 20% ballistic gelatine block serving as a backing. The bolts reached a velocity of 146 +/- 1 m/s after travelling a distance of 10 m, showing a very good repeatability during the tests. All tested body armour samples were found to be perforated, with the depth of penetration significantly influenced by the tip design. The aim of the study was to investigate the penetration depth and behaviour of bolts with two different penetration tips on body protection in the technically strongest case.

1. INTRODUCTION

Archery, dating back to 10,000 years, held significant importance in the armed forces. Bowmen, or archers, were considered superior due to the expertise required to handle such weapons. The invention of the crossbow in ancient China marked a significant shift in projectile weaponry. Its simplicity, affordability, and ease of use made it accessible to a large number of soldiers, including those with limited experience. Then, with the advent of firearms, archery gradually became obsolete in military applications during the 16th century; however, it experienced a revival in the late 18th century through the establishment of societies and competitions. Today, archery remains a popular sport, with crossbows widely used in shooting sports and hunting [1]. The growing popularity of archery [2] is evidenced by numerous private publications on social media showcasing arrow perforation on soft ballistic body protection.

2. PRELIMANARY RESULTS

The preliminary tests were conducted to validate the findings presented in open sources regarding the perforating power of arrows, particularly on soft ballistic protection. These tests employed a 70-pound (32 kg) compound bow on soft body protection.

The ballistic body armours were made of woven aramid layers and were launched perpendicular to the strike face. The preliminary tests conducted on the soft ballistic vests confirmed two findings. First, the type of arrow head significantly impacts the penetration depth. This is because hunting arrowheads with sharp edges exhibited lower penetration than solid penetration ones. The penetration tips, or sometimes referred to points, were used for all later impact configurations.

Second, the support provided to the soft body armour during impact is crucial. In these tests, a Styrofoam backing was used. Although Styrofoam is not representative of the human body, as it is too stiff, its influence on the arrowhead's penetration capability was evident, when observed in relation to no backing (**Figure 1**. a, b) [3].

https://doi.org/10.52202/080042-0046

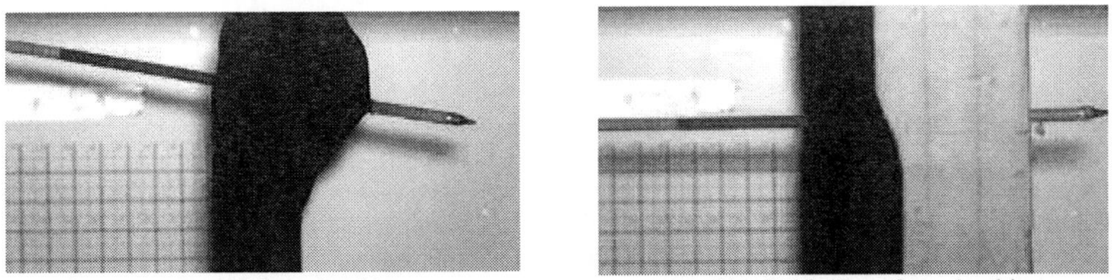

a) Without backing b) With Styrofoam backing

Figure 1. Arrow with penetration tip perforates soft body armour

Upon impact, ballistic protection without a backing attachment on the body armour behaves differently from the human body [4]. Additionally, some energy dissipates into the pendulum motion of the body armour. Therefore, to represent a realistic and elastic background material, ballistic gelatine [5] was chosen for this study.

An unfavourable load case from an engineering perspective was chosen and a Ravin R500 compound crossbow, a commercially available crossbow, was selected for the experiments.

The bolts, referred to as arrows for the crossbow, were equipped with points similar to those that demonstrated the strongest penetration performance in the compound bow. Considering the significant influence of the backing on arrow perforation, particularly with soft ballistic body protection, ballistic gelatine was utilised.

2. METHOD AND MATERIALS

2.1 Crossbow

The tests were conducted using the Ravin R500 compound crossbow (**Figure 2**). The crossbow was operated with a manual cocking system and a magnifying glass (**Table 1**).

Figure 2. Ravin R500 compound crossbow

Table 1. Technical data of the crossbow

Component	Unit	Value
Draw weight	kg (lbs)	136 (300)
Mass (without scope)	kg	3.8
Length	mm (in)	710 (30)
Width	mm (in)	190 (7.5)

2.2 Bolts and tips

Ravin bolts were used. **Figure 3** depicts one of the used carbon bolts with a training point.

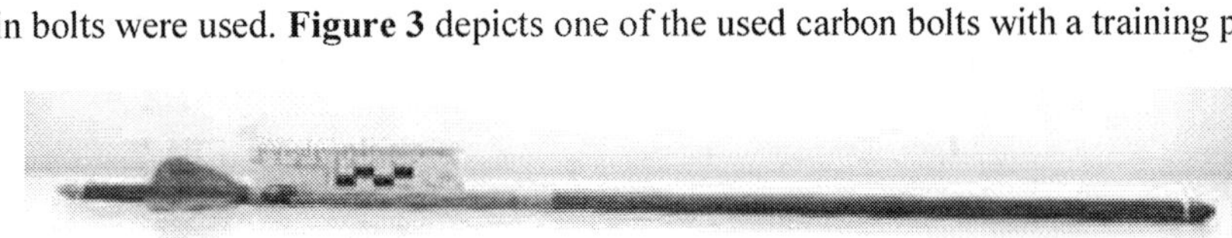

Figure 3. Bolt with training point

https://doi.org/10.52202/080042-0046

Figure 4 shows the bolt tips used. The topmost point of the first bolt represents the training point. Its ogival shape is designed to prevent excessive penetration into practice targets and facilitate easy removal. The middle bolt in the picture has a silver-coloured steel penetration tip. It features a groove after the conical tip and a gradual increase in cross-section to potentially enhance penetration into the target, referred to as 'penetration I' (PEN I). The last bolt represents another penetration tip with a longer first part and curved transitions leading to the largest cross-section, known as 'penetration II' (PEN II). These points are screwed onto the carbon bolt using threaded brass inserts.

Including the mass of tip, the bolt had a mass of 26 g with both the penetration tip and the training tip, and 27 g with the stainless steel penetration tip (**Table 2**).

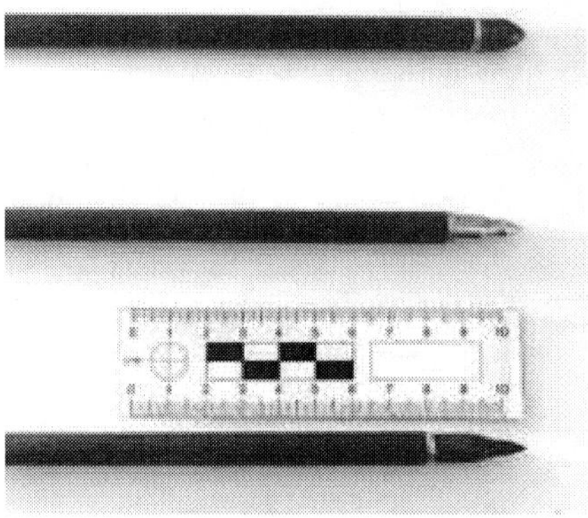

Figure 4. Bolts with training point (top), penetration tip I (PEN I) (middle), and penetration tip II (PEN II) (bottom)

Table 2. Technical data of the bolts

	Unit	Value
Mass with training point	g (oz)	27 (1.0)
Mass with PEN I	g (oz)	26 (0.9)
Mass with PEN II	g (oz)	28 (1.0)
Length	m (in)	0.51 (20)
Diameter	m (in)	0.09 (3.5)

2.3 Velocity profile

A Doppler radar was used to record the initial velocity profile of the bolt with the penetration point. Four measurements were taken, and the bolt's velocities were recorded at 1 m, 5 m, 10 m, and 50 m (**Figure 5**). Each measurement was repeated three times and the velocity results varied by ±1 m/s. The velocity data in the blue curve are partially interpolated between 0 m and 50 m and extrapolated above 50 m. At 10 m, a velocity of 146 m/s was measured, which aligns with the light barrier measurement (**Figure 6**). After 50 m, the bolt had lost 10% of its initial velocity and was travelling at 136 m/s.

The resulting energy of the arrow is illustrated in the red curve. The bolt's initial energy was 300 J, approximately half the kinetic energy of a 9 x 19 mm FMJ Luger projectile with 415 m/s velocity.

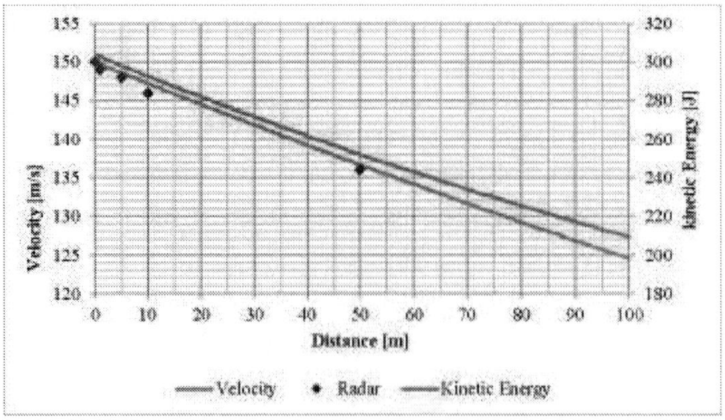

Figure 5. External ballistics of bolt (STANAG 2920 without error correction)

https://doi.org/10.52202/080042-0046

2.4 Target configurations

Eight different protective structures were considered for testing and varied in their protection class. **Table 3** provides an overview of the target configurations used. The protective structures used as the target configurations are classified according to the German technical classification [6]. The classification specifies different protection classes (German: Schutzklassen [SK]) ranging from 1, with protection against soft (lead core) 9 mm full metal jacket (FMJ), to 4, with protection against hard (steel) core rifle ammunition.

Protection class SK2 is certified for 9 mm hard-core bullet protection. In addition to the soft ballistic vest, this class requires an aramid in conjunction with (ICW) plate. A soft composite layer is also added to reduce behind armour blunt trauma (BABT).

Stab protection (German: Stichschutz [ST]) is achieved by using a metal foil or chain mail with ultra-high-molecular-weight polyethylene (UHMWPE) armour [7].

Table 3. Target configurations

Ref.	Configuration (from attack to body side)	Protection class*
1	ICW Plate + SK1 soft body armour + BABT reducing element	TR SK2 (German)
2	Standalone ICW aramid plate	none
3	Soft aramid layers + metal foil stab protection	TR SK1 ST (German)
4	Soft UHMWPE layers + metal chainmail stab protection	TR SK1 ST (German)

**Material parameters are under disclosure*

The ballistic body armour was placed on a 0.40 x 0.30 x 0.20 m block of 20% ballistic gelatine, which is internationally recognised as a representative soft tissue simulant [5]. The gelatine block was positioned upright and supported by wood (**Figure 7**).

2.5 Experimental setup

Figure 6 shows the schematic test setup. The shooting distance of 10 m was chosen based on international test standards for body armour, such as VPAM (Vereinigung der Prüfstellen für angriffshemmende Materialien und Konstruktionen) [8]. The crossbow (1) was treated as a rifle, and efforts were made to align the arrows with the target setup in an approximately perpendicular trajectory. The study did not consider the actual arrow angle incidence (yaw).

The velocities of the crossbow bolts were measured using a light barrier positioned 1.5 m ($v_{1.5}$) in front of the target (2). A coloured high-speed camera (3) captured the bolt impact on the protective body armour and the gelatine block (4).

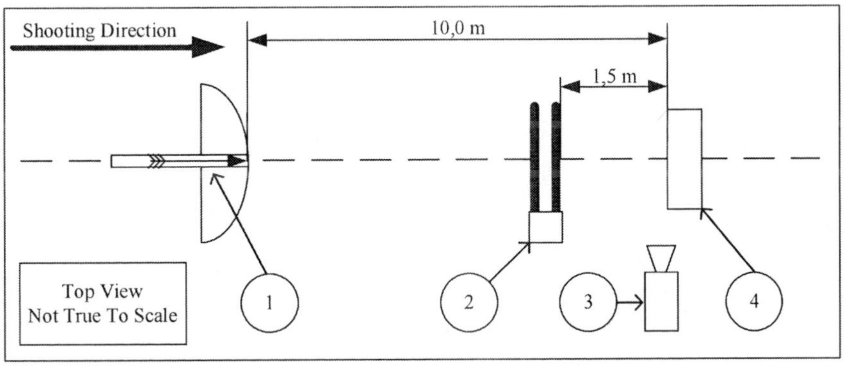

Figure 6 Schematic representation of experimental setup

3. RESULTS

Table 4 starts with the highest certified ballistic protection and is lowered step by step, as was described in **Table 3**. Different bolt tips were used for the different target configurations (**Figure 4**). The velocity is given by its reproducibility as an average value, as well as the resulting kinetic energy E (**Figure 5**). The bolt's impulse was derived.

The target perforation was considered when the bolt point was visible at the back face of the target, while penetration in the gelatine was determined by observing surface damage.

https://doi.org/10.52202/080042-0046

Table 4. Bolt penetration results

Test ID	Target config.	Tip	Velocity v [m/s]	Energy E [J]	Impulse I [Ns]	Target perforation	Gelatine penetration
A	3	PEN II				Yes	Yes
B	4	PEN II				Yes	Yes
C	1	PEN II				No[#]	No
D	1	PEN II	$147^{\pm 2}$	$292^{\pm 8}$	$3{,}97^{\pm 0{,}068}$	Yes	No
E	1	PEN I				Yes	No
F	Helmet	PEN II				Yes	Yes
G	Helmet	PEN II				Yes	Yes
H	Helmet	PEN II				Yes	Perforation

[#] *Perforation of ICW plate only*

3.1 Test A: Soft ballistic body armour with metal foil

The bolt with the PEN II penetration tip impacted the soft ballistic body armour, which consisted of aramid layers and a metal foil for stab protection (**Table 3**, Target configuration 3).

Figure 7 shows the high-speed camera images depicting the impact on the ballistic gelatine at time t and t + 0.8 ms. Both the body armour and the 0.2 m thick gelatine were perforated, with the bolt being stopped by the wooden backing (**Figure 7** b). The bolt created a circular hole in the body armour.

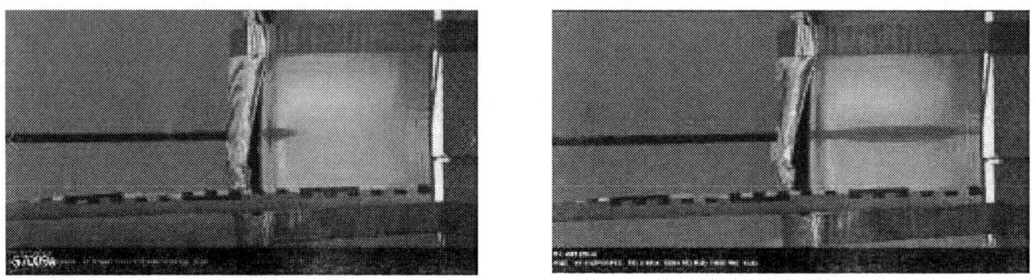

a) Impact at t b) Impact at t+0.8 ms

Figure 7. Bolt PEN II perforates soft ballistic body armour with stab protection foil

3.2 Test B: Soft ballistic body armour with chain mail

The bolt with the PEN II penetration tip was launched at 146 m/s onto the soft ballistic body armour (**Table 3**, Target configuration 4), which was supported by chain mail for stab protection. **Figure 8** shows the images of the perforation in the body armour at t = 0 and 0.7 ms. The full length of 0.2 m gelatine was perforated, and the bolt was stopped by the wooden backing (**Figure 8** b).

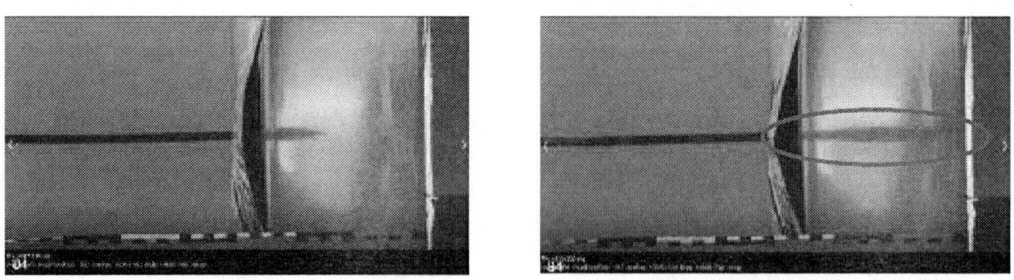

a) Impact at t = 0 ms b) Impact at t = 0.7 ms

Figure 8. Bolt PEN II perforates soft ballistic body armour with chain mail stab protection

Figure 9 shows the chain mail damage, with the failure of a single link. The bolt impact bent the link, leaving an unprotected area large enough for the bolt to penetrate the soft ballistic protection [8].

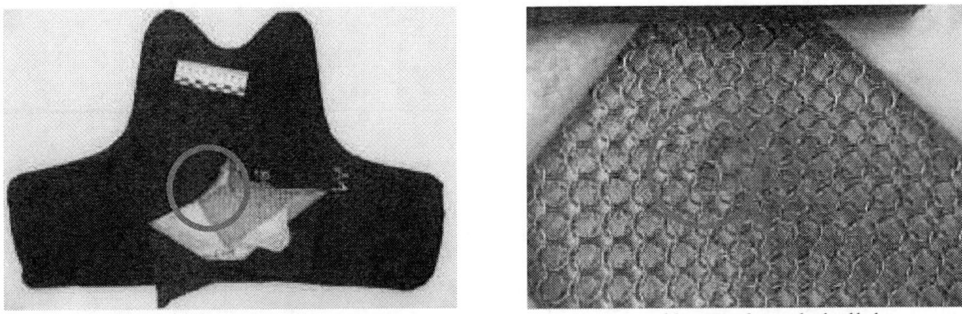

a) Perforated body armour b) Broken chain link

Figure 9. Damage of the chainmail upon perforation of a bolt with PEN II at 146 m/s

3.3 Tests C and D: Hard ballistic armour (SK2)

Test C (**Table 3**, Target configuration 1) examined the influence of yaw angle on the bolt's perforation capability. The bolt with the penetration tip PEN II impacted with yaw relative to the flight axis (**Figure 10** a). It perforated the aramid ICW plate but not the soft ballistic body protection behind it (**Figure 10** b). Due to oblique shear stresses on the longitudinal axis, the arrow broke directly at the protective plate.

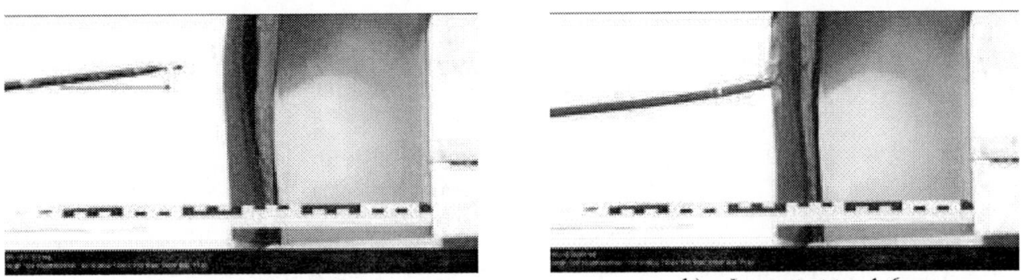

a) Impact at t = 0 ms b) Impact at t = 1.6 ms

Figure 10. Bolt PEN II penetrates SK2 body armour at impact with yaw

In Test D (**Table 3**, Target configuration 1), high-speed imaging showed a perpendicular impact without bolt nutation (**Figure 11** a), resulting in perforation of the ballistic body protection and complete penetration of the gelatine block until the bolt was stopped by the wooden backing (**Figure 11** b).

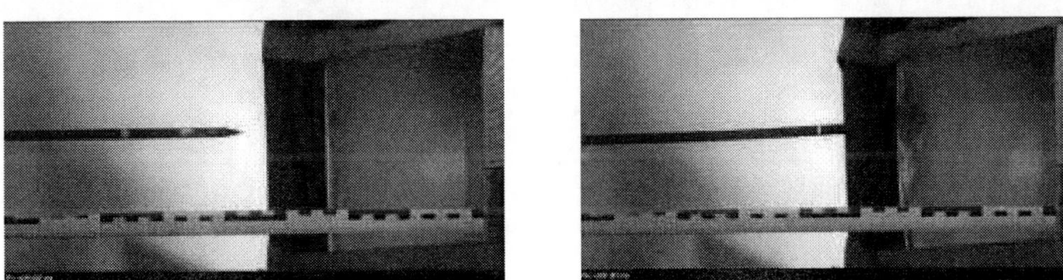

a) Impact at t = 0 ms b) Impact at t = 4.5 ms

Figure 11. Bolt PEN II perforates SK2 body armour at perpendicular impact

On the strike face of the protective plate, relatively small material bulges caused a permanent deflection (**Figure 12** a, b). No pronounced delamination was observed in the protective plate.

In the soft ballistic protection package, a clear material displacement was observed around the area where the bolt perforated the soft armour (**Figure 12** c). The contact surfaces of the bolt were hardened and raised (**Figure 12** d).

https://doi.org/10.52202/080042-0046

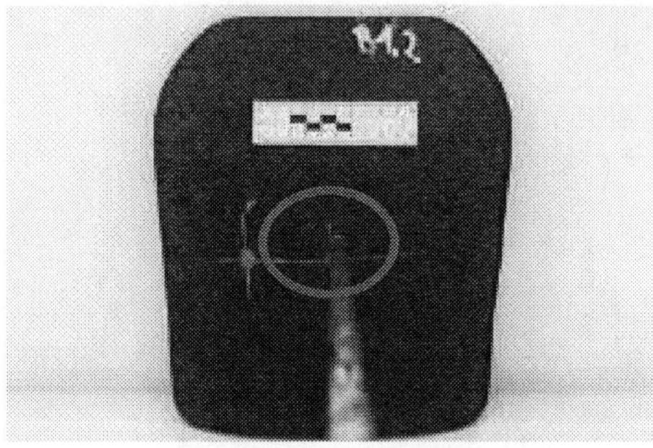

a) Front view, strike face

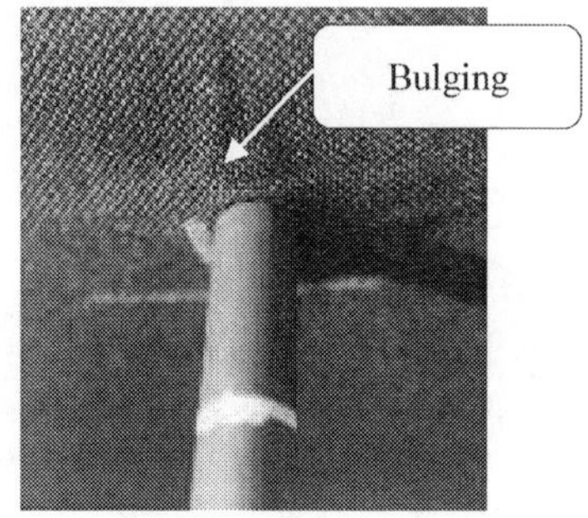

b) Material bulging

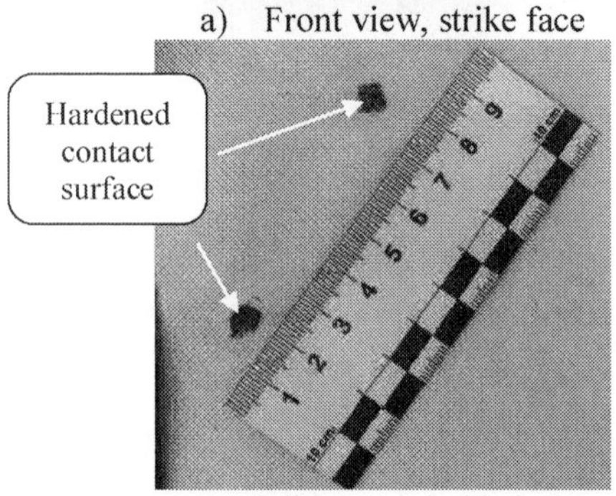

c) Soft armour behind plate

d) Contact surfaces of plate with bolt

Figure 12. Damage signature on hard ballistic body armour SK2

3.4 Test E: Exploration of the influence of the bolt penetration tip

The steel penetration tip (PEN I) was launched towards the target configuration (**Table 3,** Target configuration 1) at a constant velocity of 146 m/s from a 10 m distance. The bolt followed a straight flight path and perforated the target but not the gelatine block (**Figure 13** a). PEN I exhibited 40% less depth of penetration than PEN II under the same conditions (**Figure 13** b).

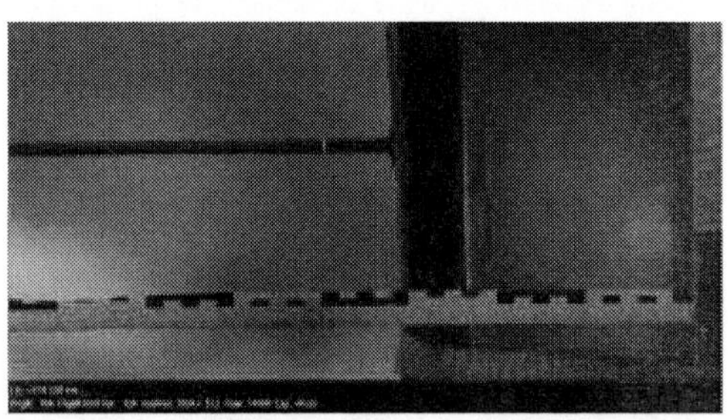

a) High speed image of impact

b) Penetrated plate

Figure 13. Bolt PEN I perforates ICW plate at 149 m/s

3.6 Tests F, G, and H: Helmets

Three types of combat helmets were tested. The bolts with PEN II were launched at $v_{1.5}$ = 146 m/s from a distance of 10 m.

A head simulant, consisting of a spherical skull substitute filled with 20% gelatine, was fixed inside the helmet using helmet straps (**Figure 14** b). The bolt perforated the helmet and the skull simulant [9, 10].

In Test F (**Table 4**), the helmet has the classification of VPAM 3 (9 mm FMJ, E < 25 J) lead bullet [9] and was impacted parietally by the bolt with the velocity $v_{1.5}$. The bolt perforated the helmet shell and interior and penetrated the head simulant (**Figure 14** a).

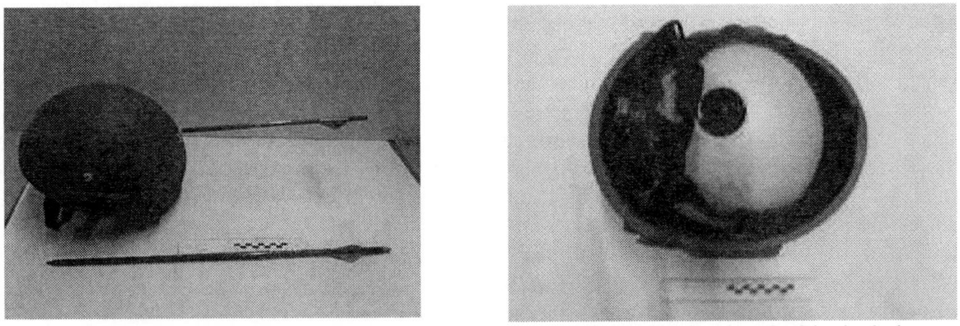

a) Parietal bolt impact b) Head simulant inside the helmet

Figure 14. Bolt PEN II perforates parietally a VPAM 3 aramid combat helmet

In test G (**Table 4**), a standard aramid shell was impacted laterally (**Figure 15** a). The lateral position was the most straightforward to test on this helmet model. Further tests would be required to determine the influence on the damage to the head simulant between the impact positions on the helmets and models. The bolt perforated the helmet and penetrated the head simulant. The damage to the head simulant for this test is represents the results of tests F and H, respectively (**Figure 15** b).

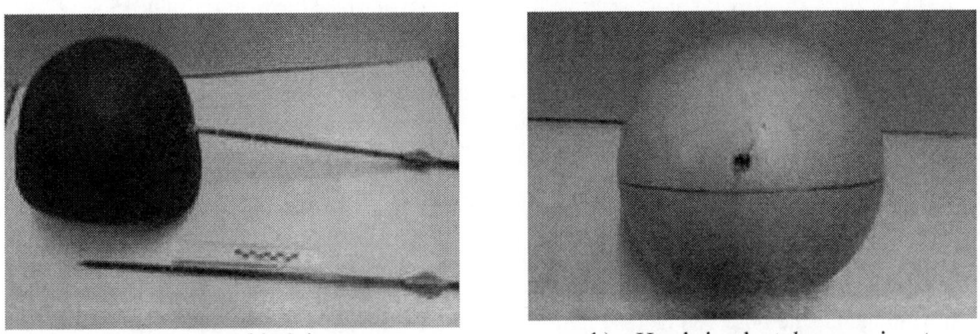

a) Lateral bolt impact b) Head simulant damage signature

Figure 15. Bolt PEN II perforates laterally a standard aramid combat helmet

Test H (**Table 4**) involved a light combat helmet made of UHMWPE. The helmet was perforated at the parietal region (**Figure 16** a). The bolt perforated the helmet and the head simulant. The head simulant had an entry hole damage signature as described in the prior test G (**Figure 15** b) and petaling failure on the exit. The helmet showed delamination on the inner layers, which was not observed in the previous test configurations and clean cut hole on the strike face (**Figure 16** b).

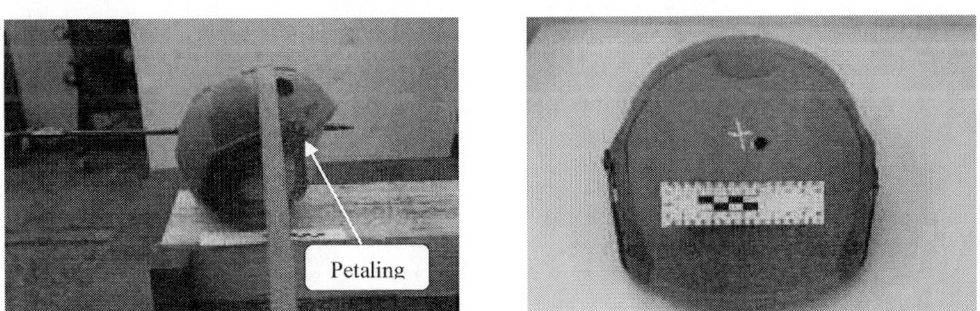

a) Parietal bolt impact b) Helmet damage signature

Figure 16. Bolt PEN II perforates parietally an UHMWPE helmet and the head simulant

4. DISCUSSION

The Ravin R500 compound crossbow was chosen because it is commercially available and comes with basic equipment. The mechanical crank had a loading time of approximately 30 seconds. While there are gadgets that speed up the loading process, they were not included in this study. The crossbow

https://doi.org/10.52202/080042-0046

was placed on a mounting block and launched by hand, so precision and accuracy were not emphasised in this study. The bolts were launched perpendicular to the target surface.

The velocity $v_{1.5}$ of the bolts was determined for each shot with light barrier placed 1.5 m before the strike face and had an error band of $v_{1.5} \pm 2$ m/s. In a free flight test, the bolt was tracked using radar. After 10 m, the bolt had a velocity of 146 m/s, verifying the measured $v_{1.5}$ form the light barrier results. It was observed that the velocity dropped by 10% after 50 m. Due to the high length-to-diameter (L/D) ratio, arrows or bolts are less affected by drag, allowing them to maintain their velocity over longer distances. However, this high L/D ratio can impact flight stability. It was demonstrated that the inclination of the bolt influences its penetration capability upon striking the target. It is noteworthy that the strong angle of incidence occurred only in one test.

The bolts were equipped with two different penetration tips, either the PEN I, or the PEN II. These specially shaped tips have a two-fold gradual increase in the overall diameter of the arrow. Due to the sharp tip, which maintains its shape and does not deform upon entry, there is a consistent and high cross-sectional load on the arrow throughout penetration. After perforating the material, the arrow can slide off the first level and further enhance its penetration ability on the second level. This occurs under a lower cross-sectional load but under a constant, powerful thrust of the heavy arrow. As a result, arrows can bypass the protective properties of vests and helmets, similar to hard-core bullets from higher protection class handguns. PEN II demonstrated the highest capacity to penetrate the body armour.

The soft body armour backing consisted of a 20% gelatine block. The gelatine block size was estimated to represent the size of a medium sized male torso. The torso shape was not considered, as the gelatine block was primarily for most accurate representation of a backing material. Therefore, additional clothing and skin that would have an influence on the penetration, was not considered.

The first two tests (Test A and B) were performed using soft ballistic body armour with gelatine backing. One piece of soft ballistic body armour was equipped with a metal foil for stab protection, while the second one had chain mail. In this experimental setup, the influence of stab protection, whether it be metal foil or chain mail, on the penetration of the bolt was not observed. Next, the hard ballistic body armour was tested (Test C and D). Despite completely perforating the hard ballistic plate, the bolt was stopped at a shooting distance of 10 m. Based on these results, it was expected that bolts would penetrate a helmet.

Consequently, the depth of penetration into helmets was investigated. The bolts perforated the helmets and penetrated the skull simulant. In the case of the UHMWPE helmet, the bolt also perforated the skull simulant. The observed crack lengths in the skull simulant were approximately comparable to those caused by 9 x 19 mm FMJ bullets in gelatine. This indicates a risk of injury beyond the penetration zone, reaching into the surrounding tissue. It is noteworthy that this temporary cavity occurred even after the protective plates in the gelatine block were penetrated, suggesting a high effectiveness of the bolts.

Although bolts are slower and have roughly half the kinetic energy of a 9 x 19 mm FMJ bullet, their rigid tip allows them to penetrate the protective body armour. Their end-ballistic properties differ from those of bullets, and the energy dissipating mechanisms of the target material fall somewhere between traditional penetration (stabbing) and bullet ballistics.

5. CONCLUSION

From a terminal ballistic perspective, the high repeatability and relatively easy handling make crossbow an interesting threat to test body armour. It does not require pyrotechnic considerations, and different bolt tips exhibit distinct failure patterns, which provide valuable insights into material modelling.

The velocity range of 100–150 m/s is not typically tested and gives an insight on a underrepresented load case, helping for better understanding on the material behaviour and therefore, to better body protection design in the future.

The experimental set-up was chosen to investigate penetration and perforation of hard and soft ballistic armour. However, the chosen crossbow had too much energy and questions for instance of how much a chainmail protects the wearer could not be answered within this set-up. A thorough investigation on material behaviour and damage requires a lighter crossbow with around 100 - 200 pound draw weight would be sufficient within the experimental set-up described in this study.

Radar measurement showed that a bolt loses 10% of its initial velocity during flight over a distance of 50 m and based on the results, implying that soft ballistic armour and helmets are perforated within this range.

Finally, it can be stated, that the crossbow represents an impact case which could be taken into consideration for future experiments.

6. RESEARCH RECOMMENDATIONS

Figure 17 illustrates the depth of penetration of the bolt launched with the penetration point. Two blocks of 20% gelatine, totalling 0.8 m in length, were used. The bolt is visible at the back face of the second block.

In future investigations, the resulting injury patterns, particularly the temporary processes during bolt penetration, will be examined, along with the observed crack lines [12].

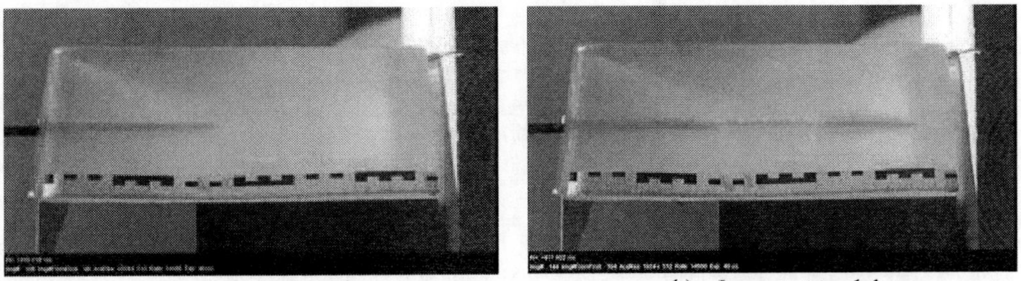

a) Impact at t = 0 ms b) Impact at t = 1.1 ms

Figure 17. Depth of penetration test in gelatine

Acknowledgments

The Authors would like to thank the Police Berlin and the German Federal Police for their support.

References

[1] T. R. Whitman, The History of Archery, US: CreateSpace Independent Publishing Platform, 2017.

[2] A. Eriksson,WOR B. Georén and Ö. M., "Work-Place Homicide by Bow and Arrow," *J Forensic Sci,* vol. 45, no. 4, pp. 911-916, 2000.

[3] T.-T. N. Nguyen, H. Tsukada, J. Breeze and S. D. Masouros, "The Critical Role of a Backing Material in Assessing the Performance of Soft Ballistic Protection," *Hum Factors Mech Eng Def Saf,* vol. 6, no. 13, pp. 1-11, 2022.

[4] N. MacPhee, A. Savage, N. Noton, E. Beattie, L. Milne and J. Fraser, "A comparison of penetration and damage caused by different types of arrowheads on loose and tight fit clothing," *Sci. Justice,* vol. 58, no. 2, pp. 109-120, 2018.

[5] B. P. Kneubuehl , R. M. Coupland , M. A. Rothschild and M. J. Thali, Wound ballistics basics and applications., Berlin:: Springer, 2011.

[6] DHPol/PTI, Ballistische Schutzwesten, Technische Richtlinie (TR), Muenster, GER: Polizeitechnisches Institut (PTI) , 2008.

[7] VPAM-KDIW, Stab and Impact Resistance, www.vpam.eu: Association of test laboratories for bullet resistant materials and constructions, 2011.

[8] VPAM-APR, General Basics of Ballistic Material,Design and Product Tests, www.vpam.eu: Association of Test Laboratories for Attack Resistant Materials and Constructions, 2021.

[9] VPAM-HVN, Durchschusshemmender Helm mit Visier, www.vpam.eu, 2009.

[10] M. J. Thali, B. P. Kneubuehl and R. Dirnhofer, "A ''Skin–Skull–Brain Model'' for the Biomechanical Reconstruction of Blunt Forces to the Human Head," *Forensic Sci.,* vol. 125, no. 1, pp. 195-200, 2002.

[11] A. M. Ruchonnet, Y. H. Tang and E. F. Kranioti, "Cranial Blunt Force Trauma in," *Forensic Sci.,* vol. 301, no. 1, pp. 350-357, 2019.

[12] J. C. Downs, C. A. Nichols, D. Scala-Barnett and B. D. Lifschultz, "Handling and interpretation of crossbow injuries.," *J Forensic Sci. ,* vol. 39, no. 2, pp. 428-445, 1994.

https://doi.org/10.52202/080042-0046

Development of a New Type of Laser Protection Glasses for Aviation Crews: Results of Combining Absorptive and Reflective Filters

M. Kalter[1], P. Hank[1], D. Pertsch[1], R. Mönikes[1] and F. M. Jakobs[2]

[1]ESG Elektroniksystem- und Logistik-GmbH, Livry-Gargan-Strasse 6, Fuerstenfeldbruck, Germany
marcel.kalter@esg.de
[2]Zentrum für Luft- und Raumfahrtmedizin der Luftwaffe, Fachabteilung I und II, Strasse der Luftwaffe 322, Köln/Fuerstenfeldbruck, Germany

Abstract. Dazzling attacks on aircrafts with lasers are a serious problem regarding aviation security. Although a broad variety of laser protection devices are commercially available for usage in research laboratories and industrial settings, the requirements for police and military missions are highly different. These requirements are not addressed in current standards. Therefore, a unique type of laser protection glasses for the usage in aircrafts was developed in close cooperation with the operational user. The concept is premised on an absorptive dyed polymer base body to provide protection against three discrete laser lines in the visible spectrum in combination with an interference coating to filter off laser radiation in spectral bands in the near infrared region. The absorbing dyes were balanced to maximize colour fidelity and visible light transmission and to ensure that information and warning lights of the cockpit as well as the helmet displays can be seen correctly by the pilot. The coating was optimized to have no discernible impact in the visible spectrum. The performance was evaluated in laboratory, ground and flight tests and was approved for flight by the Bundeswehr Federal Office of Equipment, Information Technology an In-Service Support (BAAINBw) as well as by the German AirForce Center of Aerospace Medicine. The presented results demonstrate the feasibility of designing laser protection glasses with simultaneous high-level protection in five spectral regions in the UV/VIS/NIR range without adverse effect on the user's flight safety with respect to colour vision.

1. INTRODUCTION

There is little doubt that the way in which future military conflicts may be decided will be completely different from the way recent wars have been waged. Apart from changed strategical concepts and political doctrines, this will primarily be the result of newly developed technologies such as drones, hypersonic missiles, cyber campaigns, and directed energy weapons including lasers. The overall tendency is clearly diverging from uncontrolled mass destruction and kinetic weaponry towards an integrated concept of faster, more precise, longer range, and interconnected 'smart' weapon systems. [1] In conjunction with improved sensors, autonomous robotics, and artificial intelligence, these systems will dominate the battlefields of the future.

Future combat lasers will be operating at very high energies up to several hundreds of kilowatts in the near-infrared range of the electromagnetic spectrum which is invisible to the human eye. [2] Although operationally not yet available, a number of demonstrators have been developed and successfully tested by different nations including the United States [3–5], the UK, [6] and Germany. [7] Besides these high energy lasers (HEL) that will be deployed within the next decade, a multitude of low energy laser (LEL) applications have already been available for a long time, including laser jammers, laser range finders (LRF), laser radars, and laser target designators. [8] Other devices such as beam riders are used to precisely direct laser guided weapons (LGWs) against military targets. The first LRF was based on Maiman's ruby laser and introduced in 1965, just four years after successful demonstration of the first operational laser by the inventor himself. [9–11]

Visible lasers are utilized as laser pointers, target designators mounted on ballistic weapons, and military laser dazzlers. Notably, the most trivial of these applications, the common laser pointer, has captured the highest level of attention in the past. Once considered a childish prank, misuse of laser pointers in the public for illumination of man and materials has reached epidemic dimensions which is particularly true for aircraft illumination. As of 2022, a cumulative rate of 100,000 events worldwide since 2004 has been exceeded, with increasing incidences despite of reduced air traffic due to the COVID-19 lockdowns. [12] Military aircraft are affected on home bases as well as on duty abroad, as has been confirmed by numerous official reports and press releases. [13–16] The continuously increasing emissive power of devices, the worldwide availability, and the diversification of wavelengths make it difficult to officially control this contemporary phenomenon by effective public health intervention.

https://doi.org/10.52202/080042-0047

Military laser dazzlers, on the other hand, are the logical derivatives of handheld dazzling devices, optimized for strategical purposes and operating at another order of magnitude. Based on the same diode technologies, they have been used, for instance, in the Falkland War by the British and at Operation Desert Storm by the US military. [17,18]

As laser technologies emerge, engaged soldiers, police officers, and particularly aviators are calling for eye protective devices. Unfortunately, however, these safety technologies did not keep pace with the diversification and proliferation of lasers in the modern world. Most laser protection glasses (LPG) available are blocking single wavelengths of the electromagnetic spectrum, which is sufficient under laboratory conditions, but raises two major problems: The first problem is that unlike laser protection in scientific or industrial environments, the wavelength possibly impacting the human eye is not known under military conditions. Even when green lasers are considered the most frequently used emitters, as is confirmed by nearly all civil and military aviation authorities, there will always remain a certain possibility that other wavelengths are used for illumination. [19] Of particular concern in this context are blue lasers that according to the respective FAA statistics have continuously been increasing from 2 – 3 % to nearly 10 % during the last decade. [20] The traditional red laser pointer, for unknown reasons, nowadays appears to be of lower public interest. The second problem is that filtering a single wavelength or sequence out of the visible light spectrum will considerably change individual colour perception resulting in an LPG-induced deuter- or protanomalous colour shift. This problem mainly affects pilots, for instance through reduced perception of external airfield lightings such as PAPI[18] systems, or internal colour displays, both requiring normal trichromatic colour discrimination.

Since most commercial laser pointers are operating either in the red, green, or blue wavelength range, a trichromatic approach would be useful which can be technically achieved by application of dye filters absorbing the wavelengths desired. Under these conditions, however, another problem will occur: the more pigment is present in the polycarbonate LPG layers, the lower the visible light transmission (VLT) will be, resulting in a functional degradation of visual acuity and contrast sensitivity. The effect is further enhanced by increasing optical density (OD) which defines the LPG protection level and is inversely log-related to transmission. [21] Hence, a cut-off decision between OD and VLT has to be made. Most LPGs are available in optical densities between 1 and 3 meaning that the incoming light is attenuated by a factor of 10 (OD1), 100 (OD2), or 1000 (OD3).

Such attenuation factors are sufficient for visible glare reduction, for the purpose of blocking invisible laser irradiation, however, much higher attenuation factors are needed. Considering the enormous infrared emission of a 100 kW laser weapon, at least OD6 is needed, even if only the part of radiation that is reflected by an encountered target is taken into account. A promising approach to fulfilling this requirement could be the implementation of a dichroic or interference filter consisting of multiple layers of dielectric materials with different refractive indices that reflects one or more spectral bands and transmits others, while maintaining a nearly zero coefficient of absorption for the wavelengths needed for visual field perception. [22] The conceptual result of such a device would be a physical composite of laser glare protection (LGP) and laser eye protection (LEP).

Here, we describe the technological concept of a combined approach of visible and invisible light attenuation including the extent of visual degradation induced by realization of this concept. The first completed visAIRion LPGs were already delivered in 2022.

2. TECHNICAL ASPECTS

From a technical point of view, the main challenge was the development of a filter technology tasked with the blocking of about 99 % of the statistically ascertained laser attacks on aircraft by attenuating the three most common visible wavelengths and blocking the three main invisible wavelengths in the UV-A, UV-B and NIR spectral region. As a result, the filter technology is based on a combination of a dyed absorptive polymer coated with a reflective interference filter. The basics of the according technologies are briefly resumed in the following. Please note that detailed information on the filter construction and performance data are subjects to confidentiality.

2.1 Dye Filters

The working principle of dye filters is primarily based on the attenuation of light by absorption in the filter material itself while the reflection amount is basically limited by Fresnel reflection at interfaces. The spectral transmission factor $\tau(\lambda)$ of a filter with an absorption coefficient $\chi(\lambda)$, refraction index $n(\lambda)$,

[18] Precision Approach Path Indicator

https://doi.org/10.52202/080042-0047

thickness d, and dye concentration c is described in good approximation by the product of the reflection factor $P(\lambda)$ and the internal transmission $\tau_i(\lambda)$ following Lambert-Beer's law:

$$\tau(\lambda) = \underbrace{\frac{2 \cdot n(\lambda)}{n^2(\lambda)+1}}_{P(\lambda)} \cdot \underbrace{10^{-\chi(\lambda) \cdot c \cdot d}}_{\tau_i(\lambda)} \quad (1)$$

Contrary to coloured glass filters where heavy metal or rare earth ions are utilized, the absorption of the presented filter system in the UV/VIS region is implemented by organic dyes. The corresponding absorption mechanism is based on wavelength specific excitation of valence π-electron systems. [23] In modern design of synthetic dyes the affinity to the substrate as well as the energy gap between ground and excited states can be individually adjusted by direct molecular modeling of the π-electron system. For example, increasing the degree of conjugation as well as the introduction of auxochromes like hydroxyl, amino or aldehyde functional groups (electron donors) induces a shift of the absorption maximum to longer wavelength, whereas anti-auxochromes like carbonyl, nitro or carboxyl groups (electron acceptors) cause an inverse effect. [24–26]

Furthermore, a large quantity of existing dyes with either more or less selective absorption features is already available. Dye filters are an attractive option for their simplicity in manufacturing from a technical point of view but typically suffer from an adverse effect on the VLT due to relatively broad absorptions with unfavorable transition steepness when used for the blocking of discrete laser lines in the visible spectral range. [27]

2.2 Interference Filters

The operating principle of interference coatings is based on wavelength selective constructive and destructive superposition of coherent light by reflection and transmission at interfaces of dielectric layers. Considering the simplest case of one layer with plane parallel surfaces and thickness d (Figure 1), an incident beam of coherent light I with an incident angle θ is partially reflected and refracted at the surface according to the Fresnel-formula and Fermat's principle, respectively. [28]

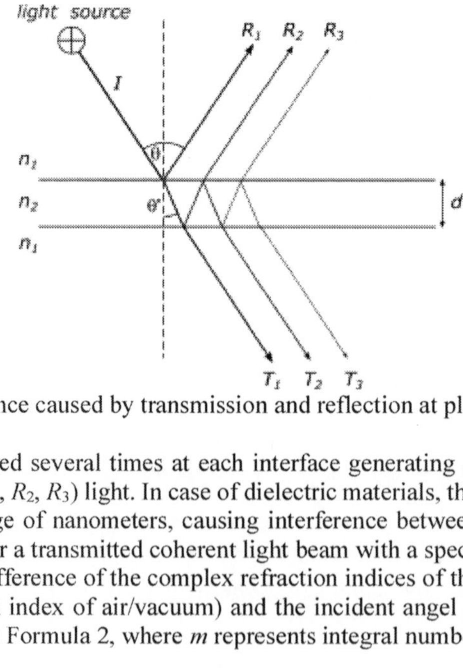

Figure 1. Interference caused by transmission and reflection at plane parallel surfaces.

This process is repeated several times at each interface generating parallel beams of transmitted (T_1, T_2, T_3) and reflected (R_1, R_2, R_3) light. In case of dielectric materials, the absorption can be neglected and d is usually in the range of nanometers, causing interference between adjacent rays of light. The nature of the interference for a transmitted coherent light beam with a specific wavelength λ depends on the layer thickness d, the difference of the complex refraction indices of the two materials (n_1 is usually considered as the refraction index of air/vacuum) and the incident angel θ. Destructive interference is induced by accordance with Formula 2, where m represents integral numbers:

$$2 \cdot d\sqrt{n_2^2 - n_1^2 \cdot sin^2(\theta)} = \left(m + \frac{1}{2}\right) \cdot \lambda \quad (2)$$

This principle can be extended to filters blocking whole spectral bands by multi-layer systems of alternating dielectric materials with high and low refractive indices and defined thicknesses calculated by software employing matrices methods. Thus, extremely effective filters can be constructed featuring transmission values close to 100 % in the passband region with simultaneous values of OD6+ in the blocking region accompanied by an outstanding transition steepness. Furthermore, the damage threshold regarding laser radiation is extremely high compared to dye filters because the intrinsic minimal absorption leads to a minor conversion of the incident light to thermal energy. Unfortunately, the dependence of the interfering wavelength on varying travelling distances of light inside the dielectric material due to different incident angles θ (see Figure 1) limits the applicability to eyewear in the visible spectral range because of angular dependent colour distortions. [29]

Interference coatings are usually applied by physical vapor deposition (PVD) methods like ion beam sputtering or thermal/plasma enhanced chemical vapor deposition (CVD) methods and are among other aspects limited to the coating complexity, substrate geometries and the minimal layer-thickness regarding the error propagation in sequences with up to several hundreds of layers. [30]

2.3 Filter Design and Spectral Properties

The filter concept of the presented LPG is based on a combination of both filter technologies. Blocking of spectral bands in the UV-A (315 - 400 nm) and UV-B region (280 - 315 nm), as well as attenuation of three major laser lines in the visible spectral range is realized by a specially developed dye-combination. The crucial advantage regarding the choice of absorption filters in the VIS range is the low sensitivity on the incident angle of light compared to interference coatings ensuring the prevention of colour distortions. Conversely, a relatively low VLT of typically 32 % was accepted leading to a visual perception comparable to that of moderate sunglasses. Proceeding from this prototype, the protection level can be tuned by changing the concentration of the dye-mixture (formula 1). Moreover, the proportions of the individual dyes were deliberately balanced in order to achieve an optimum in colour fidelity. Typical values of attenuation coefficients relative to the global VLT for red, yellow, green and blue standardized signal lights are 0.86; 1.22; 0.82 and 0.85, respectively. [31] Additionally, compatibility with night vision goggles availing low-light amplification by green or white emission of phosphorus compounds is considered by transmission values of >32 % at the emission maximum frequencies.

Dyes are directly incorporated into the blank by mixing with the polycarbonate precursor prior to injection molding. The result and functionality are shown in Figure 2:

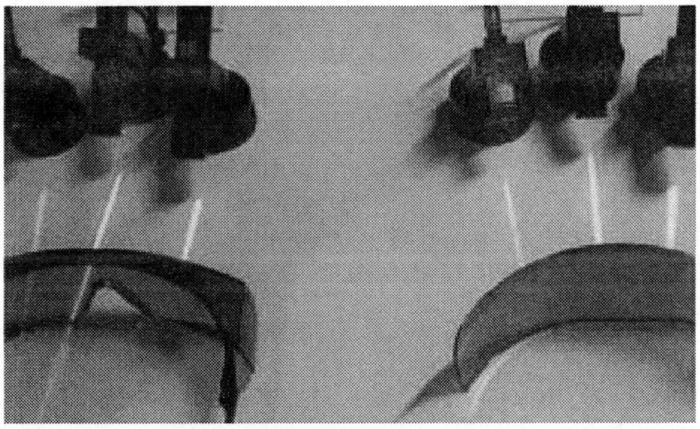

Figure 2. Experimental evidence of successful RGB blocking by the blank (right, absorption filter) as compared to conventional laser protection glasses for usage in a laboratory environment (left).

Blocking in the NIR spectral region, where incident angle dependent colour distortions play a minor role on the visual cognition of the protected person was achieved by an interference coating of 120 layers with an overall thickness of approximately 14.5 μm applied by a magnetron sputtering PVD-process. In this way high protection levels in the NIR region of OD6+ could be realized.

For the application of PVD coatings some crucial factors have to be considered in terms of the overall stability of the filter against thermal or mechanical stress. [32] Potentially, low adhesion of the PVD coating to the substrate or internal strain may cause delamination. [33] This is usually accomplished by choosing substrates and dielectric materials with similar mechanical properties like hardness or thermal expansion coefficients. [34] Since polycarbonate and the dielectric materials used for filter construction exhibit significantly different properties, a hard coat based on polysiloxanes was applied to the blank by dip coating as an adhesion promoter prior to the PVD process. Additionally, a special anti-fog coating is applied on the inner side of the lenses to prevent mist formation in environments with high atmospheric humidity.

https://doi.org/10.52202/080042-0047

2.4 Spectacle Design

Usually, commercially available laser protection glasses can be purchased in a single configuration, or a filter system is offered with different spectacle frames and lens geometries. In the current case a different approach was chosen because of the complexity of the coating process and its dependency on the lens geometry. Therefore, one type of spectacles was developed offering numerous options for the customization like different frame and nose bridge sizes, as well as length, shape, and material of the temple stems. Altogether, 96 modifications are available to cover almost all personal requirements and facial contours. A prototype of visAIRion LPG is depicted in Figure 3. Shape and design consider occupational safety requirements such as aviation related operational aspects. The field of view was enlarged as compared to usual commercial design, and the curvature was manufactured to prevent lateral intrusion of laser irradiation underneath of protection. Low weight, temperature resistance and shatter protection are guaranteed by the polycarbonate composites of the base body.

Figure 3. Prototype of the visAIRion laser protection glasses.

Special attention was dedicated to the fit. The temple joint areas are positioned inward in order to narrow the frame ensuring compatibility with visor systems as being used in the German Air Force. The temples were designed as flat as possible to avoid inconvenient pressure in the contact area when wearing them in combination with ear protectors of the integrated helmet system. Moreover, the temples are much shorter compared to conventional LPGs and the typical side curvature was removed to simplify sliding them underneath the headphones or the inner lining of the helmet. Nose bridges are attached and exchangeable per clip mechanism and can be optionally equipped with correction lenses without modifying the basic protection device.

3. MEDICAL TEST RESULTS

The development of the final prototype of the visAIRion LPG presented in this paper took a total of five years to meet requirements like compatibility with the helmet system and to figure out the ideal compromise between visible light transmission (VLT) and the protection level. For the latter task, a set of 6 prototypes with incrementally graded transmission properties was developed in collaboration with various eyewear manufacturers and independent research institutes and was provided to the AirForce Center of Aerospace Medicine for visual testing and fine tuning. The final cut-off settled in a 30 - 40 % VLT range. Experimental increase of transmission resulted in a decrease of optical density to values below the targeted minimum. Conversely, experimental increase of optical density introduced a disturbing image obscuration, which, as a result of reduced contrast vision, was rated operationally unsatisfactory by test pilots, especially under twilight conditions. The finally achieved VLT has to be regarded the maximum of glare protection currently feasible with the technology described in this paper.

https://doi.org/10.52202/080042-0047

3.1 Visual Acuity

Central visual acuity was tested in non-spectacle or contact lens wearing persons only using standard Landolt's C-projection[19] at a distance of 5 m. Tests were performed under mesopic ambient light conditions with room lights turned off. There were $n = 20$ male persons tested, with a mean age of 31,3 years and an age range between 18 and 39 years. All participants were tested with and without the laser protection device.

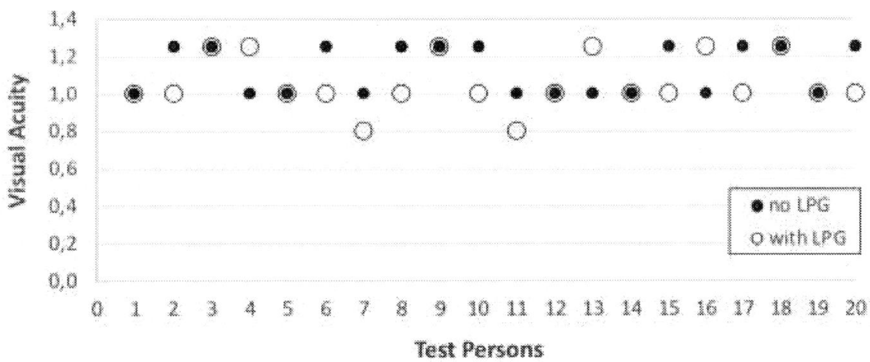

Figure 4. Visual acuity with and without laser protection glasses (right eyes only)

There was no significant difference between both groups using Mann-Whitney U-Test (1.125 ± 0.128 vs. 1.055 ± 0.144; $p = 0.781$). In two cases, a visual acuity impairment was noted while LPG wearing (VA = 0,8 or 20/25), whereas all remaining cases coincidentally scattered between 1,0 (20/20) and 1,25 (25/20).

3.2 Contrast Sensitivity

For measuring contrast sensitivity (CS), Pelly-Robson Charts were used under photopic (lights on) and mesopic (lights off) conditions. Under mesopic conditions, all objects within the test room were fairly visible including details such as door handles or name badges. All volunteers ($n = 8$) were measured binocularly with and without LPG at a distance of 2 m. The test measures optotype visibility using a single large letter size optotype with continuously decreasing grayscales across groups of letters. To determine contrast sensitivity, the letter-by-letter scoring system as indicated on the referring scoring sheets was used, whereby each letter correctly identified was scored as 0.05 log units. A Pelli-Robson score of 2.0 indicates normal contrast sensitivity of 100 %, whereas scores less than 2.0 indicates reduced contrast sensitivity. Scores of less than 1.5 are considered visual impairment. Importantly, there were no light boxes used in this test as we were interested in the differentiation of photopic and mesopic visual environments.

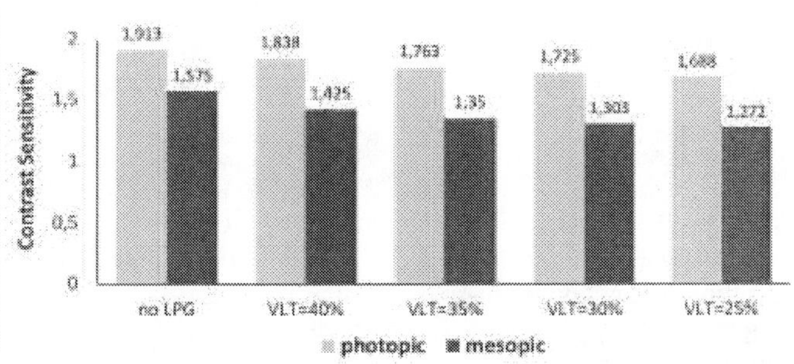

Figure 5: Contrast sensitivity with and without laser protection glasses (mean values from binocular testing in 8 test subjects)

[19] The Landolt C is an optotype defined as a ring with a gap at varying positions (left, right, bottom, top and the 45° positions in between). The size of this optotype is reduced until the tested persons makes a specified rate of errors.

https://doi.org/10.52202/080042-0047

Under photopic conditions, use of LPG did not impair vision even when VLT was reduced to 25 %; all scores were evaluated with a mean CS > 1.5. Under mesopic conditions, all scores except the one without wearing any LPG resulted in values < 1.5 indicating significantly reduced contrast sensitivity. Backward calculation revealed an LPG-induced reduction of contrast sensitivity by 11.8 % under photopic, and 19.2 % under mesopic conditions.

3.3 Colour Perception

Colour vision was approached by anomaloscopy. The test procedure has been described first by Nagel in 1917 and relies on subjective comparison of two separated hemicycles. In the upper part a mixed colour of green and red (548 and 666 nm) is offered, while in the lower part an orange to yellow colour (589 nm) is set. The colouring of the upper part can be varied by the examiner and matched by the tested person through operating a red-green screw. [35] The purpose is to determine the anomalous quotient (AQ) i.e., the subjectively required ratio of green and red colour shares. Calculations were performed using the Rayleigh equation [36], where P corresponds to the study participant and N to the (normal-sighted) examiner:

$$AQ = \frac{(73-P)/P}{(73-N)/N} \quad (3)$$

This corresponds to the relationship:

$$\frac{Green(P)\cdot Red(N)}{Red(P)\cdot Green(N)} \quad (4)$$

where 73 is the scale value of the mixed colour for green-free presentation, and 0 that for red-free presentation. A colour normal subject will perceive the two half-fields as equal if the mean standard equation 40/15 (i.e., mixture = 40 and brightness = 15) is set. Since colour discrimination is subject to a certain degree of variance (range of adjustment), ratios between 0.65 and 1.32 are considered normal. A deuteranomalous subject will match with too much green ($P < 40$; $AQ > 1.32$), a protanomalous one with too much red ($P > 40$; $AQ < 0.65$). Tests were performed in 10 volunteers.

Results indicate that monochromatic as well as dichromatic filters had a stronger effect on colour perception than trichromatic filters. Commercially available LPGs tested for comparison (Figure 6) introduced a significant impairment of colour discrimination, which apparently resulted less from filter types used than from specific wavelengths selectively or cumulatively blocked. With simultaneous blue and green blocking, colour perception shifted into the deuteranomal range (average adjustment range 1.08 - 1.46), whereas under red and green blockade, a shift into the protanomal range (average adjustment range 0.39 - 0.48) resulted. This corresponds to a deutan shift between + 31.6 and + 39.1 % (A) and a protan shift between -52.4 and -54.7 % (B), respectively. In contrast, with triple RGB blocking (C), colour perception remained approximately normal (average adjustment range 0.77 - 0.94) with relative shifts between -10.2 and -12.4%. On the behalf of these results, subjective colour perception of a pilot in duty can be simulated as shown in Figure 7.

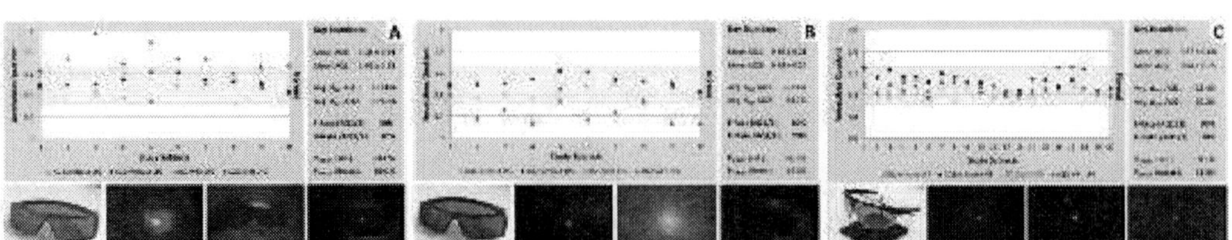

Figure 6. Results of colour vision testing. The diagrams show the anomalous quotients (red dots) of the respective glasses imaged underneath, as compared to the individual reference values without glasses (black dots). The transillumination images display the filter capacities of test glasses in a dark room. The residual dot remaining at effective attenuation levels is intentional in order to enable the pilot to realize and to report the respective laser event. **Mean**: Mean plus/minus standard deviation. **Avg. Δ_{Ref} AQ1/2**: Mean deviation from reference value (= AQ without LSB). **F-Rate (AQ1/2)**: Medical fitness rate considering AQ1 and AQ2. **U-Rate (AQ1/2)**: Medical unfitness rate, inverse function of T-rate. **PERROR (sat.)**: Error probability 15-Hue saturated.. **PERROR (desat)**: Error probability 15-Hue desaturated. Test results of monochromatic filters have been omitted in this depiction.

In order to evaluate the best cut-off between VLT and OD, 6 VLT variants of the developed prototype were tested for their colour fidelity and image quality. The tests were performed blinded, i.e., the respective transmission of the provided test glasses was not known at time of testing. Examinations were performed on 20 male study subjects, with each subject evaluating all glasses according to the same proceeding and order. Participants were asked to sort the glasses based on their quality of vision in order from 1 to 6. For evaluation, an inversely proportional score from 6 to 1 was assigned. From this, a total ranking score was calculated for each of the 6 glasses. After unblinding and assignment of transmission levels, an almost perfect correlation between both parameters was found ($R^2 = 0.98$), i.e., the higher the light transmission, the better the subjective visual impression was rated by the subjects (Figure 8). There was remarkably low variance in this result.

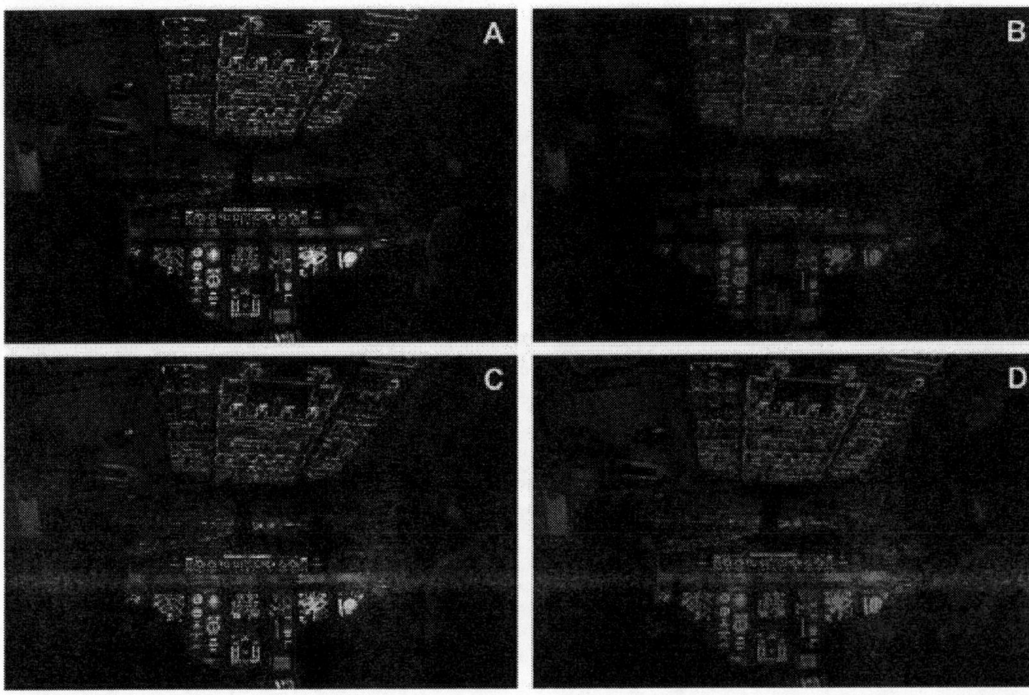

Figure 7. Simulation of colour perception with different LPG filters: (A) View at night without LPG. (B) Selective blocking of green results in over-representation of red and violet hues (deuteranomalia); reliable differentiation of cockpit displays under these conditions is not possible. (C) Selective blocking of red results in over-representation of green and blue hues (protanomalia); reliable differentiation of cockpit displays under these conditions is not possible. (D) Simultaneous blocking of red, green, and blue under equilibrated conditions results in a reduction of contrast vision and colour saturation while maintaining a normal colour perception; reliable differentiation of cockpit displays under these conditions is possible.

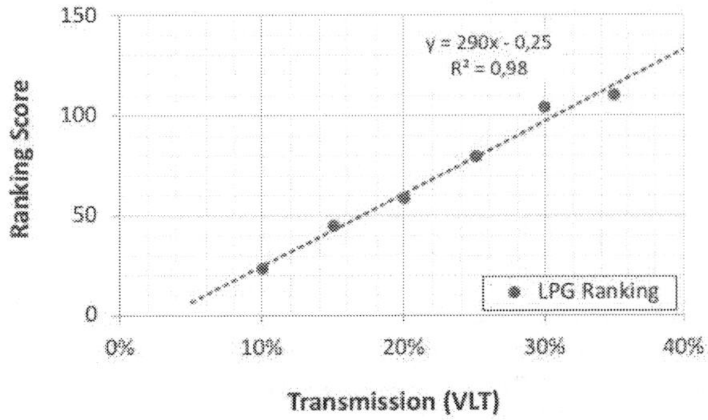

Figure 8. Subjective ranking of quality of vision at different transmission levels.

4. Discussion

https://doi.org/10.52202/080042-0047

The combination of improved sensors and automated controls with advanced laser technologies will produce a generation of weapon systems that will change the appearance and strategical concepts of future battlefields. These systems will primarily be available to advanced military nations with some of them in reach of smaller state and non-state actors including asymmetric forces and terrorists. Hence, it can be predicted that safety precautions and counter measures will be major concerns in this instance.

The main hazard of laser exposure is ocular injury which makes the topic a highly sensitive issue with considerable psychological impact. Although the anti-personal use of high-energy lasers has been banned by protocol IV of the UN Declaration of Helsinki, the remote possibility of collateral eye damage by scattered or reflected laser radiation remains. Given the destructive power of a 100 kW laser in operation, the 1/millionth part of the emitted energy reaching a soldier's eye might be sufficient to cause irreversible eye injury. In contrast to the continuous radiation emitted by HELs, low energy lasers rely on pulsed emissions which makes it difficult to predict remote interaction with biological tissues. Historical experiences suggest that most cases of ocular injury in the past have been accidental due to inattentive, careless, or untrained handling of the respective devices. Taken together, the ocular hazard from future invisible lasers will be primarily coincidental and accidental, if deliberate anti-personal use is excluded.

Meanwhile, there is consent that in case of visible lasers the primary hazard is visual impairment rather than retinal injury. From a physical perspective, focused retinal injury at larger distances is nearly impossible due to atmospheric disturbances, beam divergence, target movements, and scatter effects at transmissive surfaces such as windows and visors. Although it might appear injudicious to deny the remote effects of a device that was actually designed for distant operation, the missing evidence of ocular injury in pilots illuminated in flight appears to confirm that medical incapacitation is not a primary concern of visible laser exposure, at least when evoked by traditional laser pointers. However, the remaining effects of glare and disruption may cause psychophysical impairment possibly resulting in a reduction of human performance. With respect to aviation, the main concern is the pilot's ability to maintain control during take-off and landing maneuvers when operational skills are critical. [37] According to the Federal Aviation Regulations, aircrew interactions in these phases of flight are restricted on operationally relevant communication in order to maximize attention and to minimize the potential of human error. [38] This safety principle is called the sterile cockpit environment which is agreed to be violated in any case of accidental or deliberate laser illumination, thus interfering with flight safety even if no physical injury might be expected.

Concerning the unique characteristics of lasers and their interaction with the human eye, design and construction of advanced laser-protective eyewear is a demanding task that requires multiple considerations in terms of safety and efficacy on the one, and cut-off decisions on technological feasibility on the other hand. [39] Examples include visibility of the outer world under daylight as well as twilight or night conditions, implementation of optional refractive corrections, and the customization of spectacle fit. Military considerations include anti-fog coatings, scratch resistance for operational use in desert environments, technical compatibility with integrated helmet/visor systems, and visual compatibility with helmet mounted displays, night vision goggles (NVG), and optical combiner technologies.

The prototype presented in this paper was developed by the ESG Elektroniksystem- und Logistik-GmbH, Fuerstenfeldbruck, in close collaboration with the German Air Force (GAF) Centre of Aerospace Medicine in Cologne and Fuerstenfeldbruck. Operational testing was performed by combat mission pilots of the GAF test facilities in Manching and Fritzlar, Germany. The device as pictured in Figure 3 belongs to a new generation of LPG combining laser glare protection (LGP) with laser eye protection (LEP). Glare protection is provided by visible light attenuation at three distinct wavelengths (RGB), while eye protection is provided by additional high-range attenuation of three invisible spectral bands (UV-A, UV-B, NIR). To our knowledge, this concept has not been described or realized to date for use in aviation or related tasks. Further information of attenuation factors and wavelength specifications are classified restricted according to governmental regulations.

The three key features to be considered from the ophthalmological point of view are visual acuity (VA), contrast sensitivity (CS), and colour perception. Unlike tinted sunglasses, wearing of the LPG did not introduce a significant reduction of visual acuity. In two test persons, a VA of 0,8 (20/25) resulted in one eye which, if confirmed, would not meet the VA criteria required for active aircrew. Contrast sensitivity showed a decrease ranging from 11.8 to 19.2 % depending on VLT and ambient light conditions which would reflect a very reasonable, hence acceptable grade of visual degradation. The question is whether the testing standards of routine VA and CS measurements under photopic conditions may be applied to a device that will be primarily used under mesopic and scotopic conditions. Both, VA, and CS measurements, rely on spatial resolution which is expected to considerably increase when optotypes are exposed on background-illuminated plates or light boxes as foreseen for test routines using Landolt C or Pelly-Robson Charts.

As expected, the most challenging issues to be handled were the cut-off decisions regarding colour perception. Effective attenuation of dichromatic filters induced iatrogenic colour deficiencies in every constellation tested, to an extent that was not acceptable for demanding visual tasks such as flying an aircraft. The colour shifts recorded by conventional anomaloscopy produced deuteranomalous and protanomalous results in up to 100% of cases (Figure 6) dependent on wavelengths blocked and optical densities applied. The best results were received in greyish-toned glasses indicating a role of colour equilibration with regard to simultaneous attenuation of red, green, and blue colour spaces. After multiple refinements of wavelength peaks and ranges based on this RGB equilibration principle we finally elaborated a profile that is compatible with high-end visual tasks including aircraft-related operations.

From the military point of view, not all operational interferences and limitations under diverging ambient light conditions have been tested at this point. Standard optical requirements including related regulations such as DIN EN 207, [40] DIN EN 167, [41] DIN EN 168, [42] etc. however, are met. In addition, the built-in dye and dichroic filters do not restrict the application of military image intensifiers such as night vision goggles (NVG) or forward looking infrared (FLIR) used in aviation-related as well as ground-based operations. Future developments will include optimization for different types of aircraft and special forces requirements, including colour saturation, contrast vision, and enhanced protection levels for invisible lasers. At the end, the remaining question will always be how much safety is desired, and which extent of visual degradation is acceptable for this purpose.

In conclusion, we have shown that it is possible to design laser safety eyewear that provides simultaneous protection in three visible and three invisible wavelength ranges without affecting colour vision to an extent that would be incompatible with actual flight safety requirements. The finalized product is currently being delivered after completion of operational tests and final approval by the German Ministry of Defense.

Acknowledgments

We would like to thank the test persons and combat mission pilots of the GAF for their effort and their assistance during development of the presented LPG.

References

[1] The Future of the Battlefield, in Global Trends – Emerging and Evolving Dynamics, NIC-2021-02493, US National Intelligence Council, 2021.

[2] Zohuri B., Laser Weapons, in Directed Energy Weapons: Physics of High Energy Lasers (HEL), (Springer International Publishing, Cham, 2016); pp. 47–77.

[3] High Energy Laser Tactical Vehicle Demonstrator (HEL TVD), U.S. Army Space and Missile Defense Command, https://www.smdc.army.mil/Portals/38/Documents/Publications/Fact_Sheets/Archived_Fact_Sheets/HEL_TVD.pdf (accessed on Mar 4, 2023).

[4] O'Rourke R., Navy Shipboard Lasers: Background and Issues for Congress, Congressional Research Service (CRS), US Department of Defense, 2022.

[5] Raytheon Delivers First Laser Counter-UAS System to USAF, Military Aerospace Press Release (2019). https://www.militaryaerospace.com/commercial-aerospace/article/14230740/laser-counter-uas-uav-drone-military. (accessed on Mar 4, 2023).

[6] Scott R., UK Dragonfire Laser Weapon Demonstrator Completes First High-Power Firings, Janes Press Release (2022). https://www.janes.com/defence-news/news-detail/uk-dragonfire-laser-weapon-demonstrator-completes-first-high-power-firings. (accessed on Mar 4, 2023).

[7] German Armed Forces Conduct First Operational Tests of High-Energy Laser Weapon against Drones, Rheinmetall Press Release (2022). https://www.rheinmetall.com/media/editor_media/rheinmetallag/press/pressearchiv_2022/2022_10/2022-10-27_Rheinmetall_MBDA_HEL_FGS_Sachsen_en.pdf (accessed Mar 6, 2023).

[8] Titterton D. H., Military Laser Technology and Systems (Artech House Publishers, Norwood (UK), 2015).

[9] Barr J. R. M, Challenges for New Laser Sources in the Defense Industry, in Advances in Lasers and Applications (CRC, Boca Raton, 1999); pp. 253–270.

[10] Maiman T. H., Nature, 1960; 187; 493-494.

https://doi.org/10.52202/080042-0047

[11] Maiman T. H, Ruby Laser Systems, U.S. Patent Number 3,353,115 (1961).
[12] Jakobs F. M., Hering D.,Hoffmann M. A., Waizenegger C. and Bressem L., Why Does the Incidence of Laser Attacks on Aircraft Increase Despite Reduced Numbers of Flights during the COVID-19 Pandemic?, Annual Scientific Meeting of the Aerospace Medical Association (ASMA), Peppermill Resort Hotel Reno (NV, USA), 2022.
[13] Roblin S., China Blinded a Coast Guard Crew with a Military-Grade Laser, Popular Mechanics Press Release (2023). https://www.popularmechanics.com/military/weapons/a42863850/china-uses-military-grade-laser-blind-philippine-coast-guard/ (accessed Mar 11, 2023).
[14] Seidel J., Chinese Fishing Boats Are Blinding US Military Pilots with Lasers, News Corp Australia Network Press Release (2023). https://www.news.com.au/technology/innovation/military/chinese-fishing-boats-are-blinding-us-military-pilots-with-lasers/news-story/0e4ca5befb6c21c3d7928dd8c00392a4 (accessed Mar 11, 2023).
[15] US Accuses China of Pointing Lasers at its Pilots from Djibouti Base, BBC Press Release (2018). https://www.bbc.com/news/world-asia-china-43999502 (accessed Mar 11, 2023).
[16] Laser Incidents from 2014-2021 Including Events That Involve UK Operations Overseas, Civil Aviation Authority (CAA) UK, https://www.caa.co.uk/data-and-analysis/safety-and-security (accessed Mar 11, 2023).
[17] UK Deployed Falklands "dazzle" Laser, Documents Show, BBC News Press Release (2013). https://www.bbc.com/news/uk-politics-23518592 (accessed Mar 11, 2023).
[18] Morrison D. C., Lasers & Optronics, 1991; 10; 43-45.
[19] Piloten Fordern Laser-Verbot, Aero. de Luftfahrtnachrichten Press Release (2016). https://www.aero.de/news-23338/Piloten-fordern-Laser-Verbot.html (accessed on Mar 11, 2023).
[20] United States Federal Aviation Administration (FAA): Laser Incidents, https://www.faa.gov/about/initiatives/lasers/laws (accessed on Mar 4, 2023).
[21] Guidance on Mitigation Strategies Against Laser Illumination Effects. Aerospace Recommended Practice, No. ARP6378, SAE International, 2018.
[22] Dobrowolski J. A., Optical Properties of Films and Coatings, in Handbook of Optics (3rd Edition) Vol IV: Optical Properties of Materials, Nonlinear Optics, Quantum Optics (McGraw Hill LLC, New York, 2009); pp. 32-113.
[23] Bafana A., Devi S.S. and Chakrabarti T., Environ. Rev., 2011; 19; 350-371.
[24] Freeman H.S., Front. Biosci. (Landmark Ed), 2013; 18(1); 145-164.
[25] Laurent A.D., Adamo C. and Jacquemin D., Phys. Chem. Chem. Phys., 2014; 16; 14334-14356.
[26] Bricks J.L., Kachkovskii A.D., Slominskii Y.L., Gerasov A.O. and Popov S.V., Dyes Pigments, 2015; 121; 238-255.
[27] Mang-Löffler M., Naumann H. and Schröder G., Absorptionsfilter, in Handbuch Bauelemente der Optik (8th Edition, Carl Hanser Verlag, München, 2020); pp. 251–256.
[28] Hering E. and Martin R., Interferenz an Dielektrischen Schichten, in Optik für Ingenieure und Naturwissenschaftler - Grundlagen und Anwendungen (Fachbuchverlag Leipzig, Carl Hanser Verlag, Leipzig, 2017); pp. 251–264.
[29] Mang-Löffler M., Naumann H., and Schröder G., Interferenzfilter, in Handbuch Bauelemente der Optik, (8th Edition, Carl Hanser Verlag, München, 2020); pp. 256–261.
[30] Ehlers H. and Ristau D., Production Strategies for High-Precision Optical Coatings, in Optical Thin Films and Coatings (2nd Edition, Woodhead Publishing, Cambridge, 2018); pp. 103–140.
[31] DIN e.V., DIN EN 172:2002, Sonnenschutzfilter für Den Betrieblichen Gebrauch (Beuth Verlag, Berlin, 2002).
[32] Kern W. and Vossen J.L., Thin Film Processes II (Elsevier Science, Amsterdam, 2012).
[33] Martinu L. and Poitras D., J. Vac. Sci, Technol. A, 2000; 18; 2619-2645.
[34] Martin P. M., Handbook of Deposition Technologies for Films and Coatings: Science, Applications and Technology (Elsevier Science, Amsterdam, 2009).
[35] Nagel W., Two Devices for Ophthalmological Colour Testing, Z Augenheilkd, 1907; 17; 201-222.
[36] G. Crawford, Rayleigh Equations on an Anomaloscope, Br. J. Physiol. Opt., 1951; 8; 225.
[37] Nakagawara V.B. and Montgomery R.W., Aviat Space Environ Med., 2000; 71(10); 1060-1062.
[38] Standard Operating Procedures and Pilot Monitoring Duties for Flight Deck Crewmembers, AC No. 120-71, US Department of Transportation, 2017.
[39] Aircrew Laser Eye Protection (ALEP), No. AFI11-301 Vol. 4, US Air Force, 2008.

[40] DIN e.V., DIN EN 207:2017-05, Persönlicher Augenschutz – Filter Und Augenschutzgeräte Gegen Laserstrahlung (Laserschutzbrillen); Deutsche Fassung EN 207:2017.
[41] DIN e.V., DIN EN 167:2001, Persönlicher Augenschutz – Optische Prüfverfahren; Deutsche Fassung EN 167:2001.
[42] DIN e.V., DIN EN 168:2001, Persönlicher Augenschutz – Nichtoptische Prüfverfahren; Deutsche Fassung EN 168:2001.

 https://doi.org/10.52202/080042-0047

AUTHOR INDEX

Royal Military Academy (Belgium)
Department of Weapon Systems
Renaissance Avenue 30
1000 Brussels
Belgium

ISBN 979-8-3313-1852-9
9 798331 318529

CURRAN ASSOCIATES INC.
proceedings.com

6th Workshop on Statistical Machine Translation 2011

(WMT 2011)

Held at EMNLP 2011

Edinburgh, Scotland, UK
30-31 July 2011

ISBN: 978-1-61839-255-8